INTRODUCTION TO LINEAR ALGEBRA

HARCOURT BRACE JOVANOVICH COLLEGE OUTLINE SERIES

INTRODUCTION TO LINEAR ALGEBRA

Pauline Lowman
Joseph Stokes

Department of Mathematics
Western Kentucky University
Bowling Green, Kentucky

Books for Professionals
Harcourt Brace Jovanovich, Publishers
San Diego New York London

Printed in the United States of America

Library of Congress Cataloging-in-Publication Data

Lowman, Pauline, 1929-
Introduction to linear algebra / Pauline Lowman, Joseph Stokes.
p. cm. — (Harcourt Brace Jovanovich college outline series)
(Books for professionals)
Includes index.
ISBN 0-15-601527-7 : $12.95
1. Algebras, Linear. I. Stokes, Joseph, 1934- . II. Title.
III. Series. IV. Series: Books for professionals.
QA184.L69 1991
512'.5—dc20

91-8671
CIP

ISBN 0-15-601527-7

First Edition

A B C D E

PREFACE

Do not *read* this Outline—*use* it. You can't learn linear algebra simply by reading it: you have to *do* it. Solving specific, practical problems is the best way to master—and to demonstrate your mastery of—the definitions, laws, and principles upon which the mathematical techniques and geometrical interpretations of linear algebra are based. Outside the classroom, you need three tools to do linear algebra: a pencil, paper, and a calculator. Add a fourth tool, this Outline, and you're all set. *Remember*, linear algebra is not a spectator sport—you must be a participant.

This HBJ College Outline has been designed as a tool to sharpen your problem-solving skills in linear algebra. Each chapter covers a unit of material whose fundamental principles are broken down in outline form for easy reference. The outline text is heavily interspersed with worked-out examples, so you can see immediately how each new concept is applied in problem form. Except for Chapters 5 and 19, each of the chapters contains a Summary and a Raise Your Grades section, which (taken together) give you an opportunity to review the primary principles of a topic and the problem-solving techniques implicit in those principles.

Most important, this Outline gives you plenty of problems to practice on. Work through each Solved Problem, and check yourself against the step-by-step solution provided. Test your mastery of the material in each chapter by doing the Supplementary Problems. (In the Supplementary Problems, you're given answers only—the details of the solution are up to you.) Finally, you can review all of the topics covered in the Outline by working the problems in the five periodic Exams and the Final Exam. (Use the Exam problems to diagnose your own strengths and weaknesses.)

Having the tools is one thing; knowing how to use them is another. The study of linear algebra is so essential to many branches of science and business that a knowledge of this subject is invaluable to anyone pursuing a career in engineering, physics, computer graphics, production planning, in fact almost every technical field. Also, a knowledge of linear algebra is necessary for the study of advanced calculus, differential equations, graph theory, and operations research—almost every phase of advanced mathematics.

This Outline is intended to provide you with a comprehensive review of the concepts of elementary linear algebra. It can be used as a supplement to your linear algebra textbook—or, if the purpose of a course is primarily practical, it can serve as a textbook. Little attempt has been made to show derivations of theorems and formulas used in this Outline. Applications have been used throughout the chapters, with Chapter 5 devoted entirely to various applications. Chapter 18 is devoted to applications to geometric transformations—very useful in computer graphics; and Chapter 19 contains computer programs (in BASIC) for matrix algebra. You will find in Chapter 3 the Sylvester eliminant method for evaluating a determinant. In Chapter 5, a new technique for row echelon reduction is introduced—the proof is in Appendix C. In Chapter 13, a section on using a partitioned matrix to change bases is introduced—this is a short-cut device. The four Appendices give you a handy reference of the Greek alphabet, basic trigonometry, the row echelon reduction proof used in Chapter 5, and models of quadric surfaces.

We wish to thank the HBJ editor, Emily B. Thompson, for her advice, encouragement, and suggestions. We also thank Larry McPherson of the North Carolina DOT for providing engineering problems.

During these many months, our families have shown their patient support. We dedicate this book to them.

Bowling Green, Kentucky

PAULINE LOWMAN
JOSEPH F. STOKES

CONTENTS

CHAPTER 1 Systems of Linear Equations **1**

1-1: Linear Equations 1
1-2: Systems of Linear Equations 2
1-3: Elimination Method of Solving Linear Equations 3
1-4: Matrices 6
1-5: Matrix Addition and Scalar Multiplication 7
1-6: Matrix Multiplication 9
1-7: Special Matrices 12
Solved Problems *16*

CHAPTER 2 Matrix Solutions of Linear Systems **28**

2-1: Matrix Representation of a Linear System 28
2-2: Row Equivalent Matrices 30
2-3: Gaussian and Gauss–Jordan Elimination 33
2-4: Inverse of a Matrix 36
2-5: Properties of Inverse Matrices 38
2-6: Homogeneous Systems of Linear Equations 43
2-7: Invertible Matrix Method 45
2-8: Equivalent Statements 46
Solved Problems *48*

CHAPTER 3 Determinants **61**

3-1: Permutations 61
3-2: The Determinant Function 62
3-3: Properties of Determinants 64
3-4: Evaluating Determinants by Row Reduction 68
3-5: Evaluating Determinants by Cofactor Expansion 70
3-6: Evaluating Determinants by the Sylvester Eliminant 72
Solved Problems *76*

CHAPTER 4 Additional Topics of Determinants and Matrices **87**

4-1: Cramer's Rule 87
4-2: Adjoint of a Matrix 89
4-3: Elementary Matrices 92
4-4: Properties of Elementary Matrices 94
4-5: Equivalent Matrices 97
4-6: Partitioned Matrices 99
Solved Problems *102*

EXAM 1 — 113

CHAPTER 5 **Applications of Matrices and Determinants** 116
5-1: Linear Systems 116
5-2: Matrices 119
5-3: Determinants 123
5-4: Additional Uses of Determinants 129

CHAPTER 6 **Vector Spaces** 134
6-1: Vector Spaces 134
6-2: Subspaces 138
6-3: Some Properties of Vector Spaces 142
6-4: Linear Independence 143
6-5: Some Properties of Linear Independence and Dependence 147
Solved Problems *149*

CHAPTER 7 **Representing Vector Spaces by Subsets** 159
7-1: Spanning Sets 159
7-2: Bases and Dimension 162
7-3: Properties of Bases 164
Solved Problems *169*

CHAPTER 8 **Row and Column Spaces of a Matrix** 181
8-1: Rank of a Matrix 181
8-2: Properties of Row and Column Spaces 182
8-3: Equivalent Statements 188
Solved Problems *191*

EXAM 2 — 200

CHAPTER 9 **Inner Products** 203
9-1: Inner Product 203
9-2: Length of a Vector 205
9-3: Distance between Vectors 206
9-4: Cauchy–Schwarz Inequality 206
9-5: Triangle Inequality 207
9-6: Unit Vectors 207
Solved Problems *208*

CHAPTER 10 **Geometry of Vectors** 214
10-1: Geometric Representation of Vectors in R_2 and R_3 214
10-2: Geometric Interpretations of Operations on Vectors in R_2 and R_3 214
10-3: Angle between Two Vectors 215
10-4: Vector Equation of a Plane 216
10-5: Lines in R_3 217
10-6: Cross Product 219
Solved Problems *223*

CHAPTER 11 **Orthogonality and Gram–Schmidt** 232
11-1: Orthonormal Basis 232

	11-2: Orthogonal Projection	235
	11-3: Gram–Schmidt Process	236
	Solved Problems	*240*
EXAM 3		**249**
CHAPTER 12	**Linear Transformations**	**251**
	12-1: Linear Transformation	251
	12-2: Properties of Linear Transformations	253
	12-3: Isomorphism	254
	12-4: Properties of Isomorphic Vector Spaces	255
	12-5: Kernel and Range	256
	Solved Problems	*260*
CHAPTER 13	**Matrices of Linear Transformations**	**269**
	13-1: Matrix Transformation	269
	13-2: Coordinates	270
	13-3: Transition Matrices	271
	13-4: Matrix of a Linear Transformation	275
	13-5: Change of Bases for Matrix Transformation	277
	13-6: Partitioned-Matrix Method for Change of Bases	281
	Solved Problems	*284*
CHAPTER 14	**Similarity of Matrices**	**298**
	14-1: Similarity	298
	14-2: Properties of Similarity	299
	14-3: Similarity of Matrices of a Linear Transformation	300
	14-4: Rank and Nullity of a Linear Transformation	304
	Solved Problems	*309*
EXAM 4		**323**
CHAPTER 15	**Eigenvalues and Eigenvectors**	**326**
	15-1: The Matrix Eigenproblem	326
	15-2: The Linear Transformation Eigenproblem	329
	15-3: Properties of Eigenvalues and Eigenvectors	333
	Solved Problems	*340*
CHAPTER 16	**Diagonalization**	**357**
	16-1: The Meaning of Diagonalization	357
	16-2: Diagonalization of Matrices	358
	16-3: Diagonalization of Symmetric Matrices	363
	Solved Problems	*370*

CHAPTER 17 Applications to Quadratic Forms **384**
17-1: Quadratic Forms 384
17-2: Principal Axes Property for Quadratic Forms 385
17-3: Conics 386
17-4: Quadric Surfaces 390
17-5: Definiteness of Quadratic Forms 391
Solved Problems *395*

CHAPTER 18 Applications to Geometric Transformations **403**
18-1: Transformations in Euclidean Space R_2 403
18-2: Properties of R_2 Transformations 410
18-3: Transformations in Euclidean Space R_3 413
18-4: Properties of R_3 Transformations 418
18-5: Isometries 421
Solved Problems *425*

EXAM 5 **438**

CHAPTER 19 Computer Applications **441**
19-1: Matrix Statements 441
19-2: Selected Programs 442

FINAL EXAM **451**

APPENDIX A: Greek Alphabet **457**
APPENDIX B: Trigonometry **457**
APPENDIX C: Row Echelon Form—The Lowman Method **458**
APPENDIX D: Models of Quadratic Forms **460**

INDEX **461**

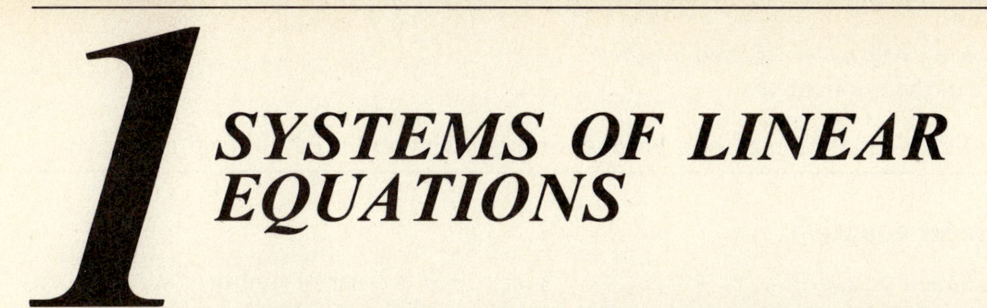

1 SYSTEMS OF LINEAR EQUATIONS

THIS CHAPTER IS ABOUT

- ☑ **Linear Equations**
- ☑ **Systems of Linear Equations**
- ☑ **Elimination Method of Solving Linear Equations**
- ☑ **Matrices**
- ☑ **Matrix Addition and Scalar Multiplication**
- ☑ **Matrix Multiplication**
- ☑ **Special Matrices**

1-1. Linear Equations

A. Definition of a linear equation

An equation in n variables $x_1, x_2, \ldots, x_n$ is called a **linear equation** if it can be written in the form

$$a_1x_1 + a_2x_2 + \cdots + a_nx_n = b$$

where $a_1, a_2, \ldots, a_n$ and b are constants.

In a linear equation, all of the variables must be first-degree, and none of the variables may be products or quotients of each other.

If an equation is not linear, it is called a **nonlinear equation**.

note: A variable is just what its name implies—its value varies with respect to the relation it's in. It's common practice to denote variables as $x_1, x_2, \ldots, x_n$, where the subscripts indicate different variables. Or, we can use x, y, and z to denote three variables.

A constant is also just what its name implies—its value stays the same, regardless of the expression it's in. Thus, all real numbers are constants including such irrational numbers as "pi" ($\pi = 3.14156\ldots$).

EXAMPLE 1-1: Given that x, y, and z are variables, classify the following equations as linear or nonlinear. Explain your answer.

(a) $x + y - z = 3$ **(b)** $x^2 + y - z = 3$ **(c)** $x + |y| = 2$

(d) $x + \sqrt{y^2} = 6$ **(e)** $4x - 5y + 6z = 0$ **(f)** $xy + x + y = 2$

(g) $\pi x + y - z = \pi^2$ **(h)** $x + y/4 - z = 5$ **(i)** $x + y/z + 2 = 3$

Solution

(a) Linear, with constants $a_1 = 1$, $a_2 = 1$, $a_3 = -1$, and $b = 3$.
(b) Nonlinear because of x^2.
(c) Nonlinear because of $|y|$, which equals y when $y \geq 0$ and $-y$ when $y < 0$.
(d) Nonlinear because of $\sqrt{y^2}$, which equals y when $y \geq 0$ and $-y$ when $y < 0$.
(e) Linear, with constants $a_1 = 4$, $a_2 = -5$, $a_3 = 6$, and $b = 0$.
(f) Nonlinear because of the variable product xy.
(g) Linear, with $a_1 = \pi$ (which is a constant), $a_2 = 1$, $a_3 = -1$, and $b = \pi^2$.

(h) Linear, with $a_1 = 1$, $a_2 = \frac{1}{4}$, $a_3 = -1$, and $b = 5$.
(i) Nonlinear because of the quotient y/z.

note: A constant in a linear equation may be a product or a quotient and need not be first-degree.

B. Solution of a linear equation

A solution of the linear equation $a_1x_1 + a_2x_2 + \cdots + a_nx_n = b$ is a set of numbers $s_1, s_2, \ldots, s_n$ such that if x_1 is replaced by s_1, x_2 by s_2, and so on, the equation becomes a true statement. The set of numbers $s_1, s_2, \ldots, s_n$ is also called an **ordered *n*-tuple**; and the set of all solutions of the equation is called the **solution set**.

EXAMPLE 1-2: Find the solution set of each of the following linear equations:

(a) $2x = 3$ **(b)** $2x + y = 3$ **(c)** $4x - 5y + 6z = 0$

Solution

(a) To find the solution of $2x = 3$, you simply solve the equation for x, getting $x = 3/2$. The solution set contains a single element.
(b) To find the solution set of $2x + y = 3$, you can assign an arbitrary number to x and solve for y or assign an arbitrary number to y and solve for x. Let $x = c$; then $y = 3 - 2c$. Therefore, the solution set is all ordered pairs of the form $(c, 3 - 2c)$, where c is arbitrary. If you let $y = d$, then $x = (3 - d)/2$ and the ordered pairs are of the form $(3 - d)/2$. The set of all such ordered pairs is the same in each case.
(c) Let $x = c$ and $y = d$. Then $z = (5d - 4c)/6$. Therefore, the solution set is all ordered triples of the form $[c, d, (5d - 4c)/6]$, where c and d are arbitrary. Observe that the number of arbitrary values is one less than the number of variables in the equation.

1-2. Systems of Linear Equations

A. Definition of a system of linear equations

A finite set of two or more linear equations in n variables $x_1, x_2, \ldots, x_n$ is called a **system of linear equations**. A solution of the linear system is an ordered n-tuple of numbers $s_1, s_2, \ldots, s_n$ such that if x_1 is replaced by s_1, x_2 by s_2, and so on, each equation becomes a true statement. The set of all solutions of the system is called the solution set. A system that has one or more solutions is called **consistent**. If it has no solution, it is called **inconsistent**.

B. Geometric interpretation of two linear equations in two unknowns

Consider the following system:

$$a_1x + b_1y = c_1 \qquad \ell_1$$

$$a_2x + b_2y = c_2 \qquad \ell_2$$

where a_1, b_1, c_1, a_2, b_2, and c_2 are real numbers and the coefficients of x and y are not both zero in either equation. In the xy plane the graph of each of these equations is a line. Since a solution of the system is a pair of numbers (x, y) that satisfies each of the two equations, a solution corresponds to a common point of the two lines. There are three possibilities:

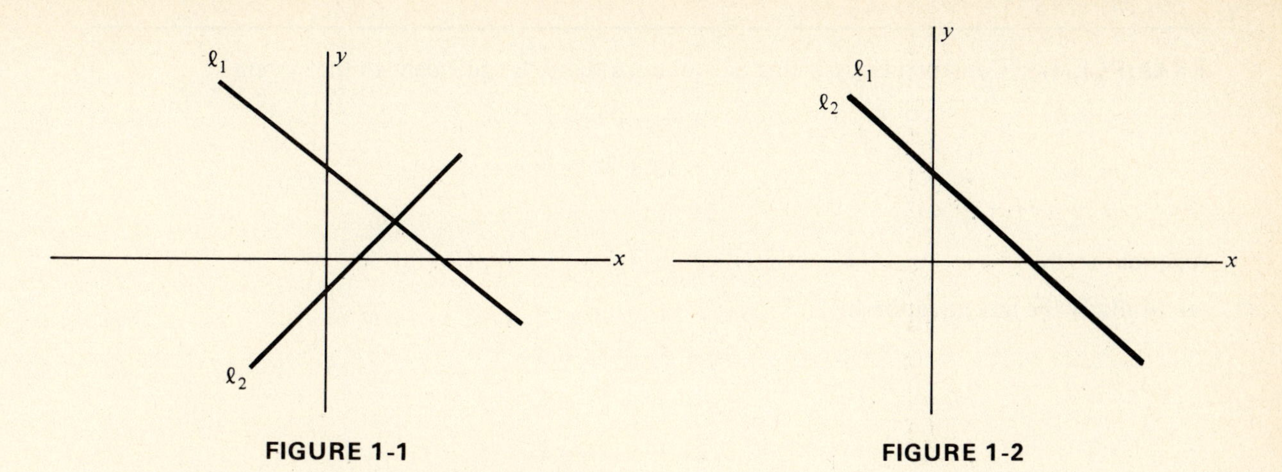

FIGURE 1-1

FIGURE 1-2

(a) ℓ_1 and ℓ_2 intersect at a single point (Figure 1-1). In this case there is exactly one solution, and the system is consistent.
(b) ℓ_1 and ℓ_2 are coincident lines—that is, the same line (Figure 1-2). In this case there are an infinite number of common points and, therefore, an infinite number of solutions. This is a consistent system that is often called a **dependent** system.
(c) ℓ_1 and ℓ_2 are parallel lines (Figure 1-3). In this case there are no common points and, therefore, no solution. This system is inconsistent.

Only a system of two linear equations in two unknowns was considered here, but every system of m linear equations in n unknowns has **exactly one solution, an infinite number of solutions,** or **no solution**.

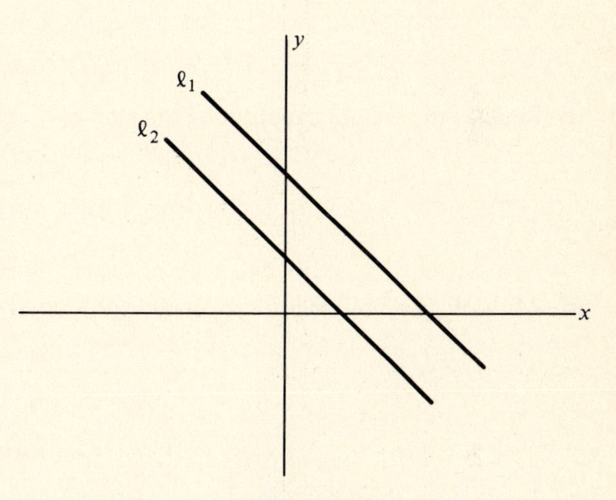

FIGURE 1-3

1-3. Elimination Method of Solving Linear Equations

A. Definition of equivalent systems

Two systems of linear equations are equivalent if they have exactly the same solution set. Equivalent systems are obtained if

(a) an equation of the system is multiplied by a nonzero constant;
(b) two equations in the system are interchanged; or
(c) one equation of the system is replaced by the sum of it and a constant multiple of another equation of the system.

When any sequence of these manipulations is used to obtain a new, simpler system of linear equations, the process is called the **elimination method**.

EXAMPLE 1-3: Construct five systems of equations that are equivalent to the system

$$x + y + z = 2$$
$$x + 3y + 2z = 1$$
$$2x + y - z = 2$$

Solution

(a) Multiply the first equation by 3:

$$3x + 3y + 3z = 6$$
$$x + 3y + 2z = 1$$
$$2x + y - z = 2$$

(b) Interchange the second and third equations:

$$x + y + z = 2$$
$$2x + y - z = 2$$
$$x + 3y + 2z = 1$$

(c) Replace the second equation by the sum of it and twice the first equation:

$$x + y + z = 2$$
$$3x + 5y + 4z = 5$$
$$2x + y - z = 2$$

(d) Multiply the first equation by 2 and the second equation by -1:

$$2x + 2y + 2z = 4$$
$$-x - 3y - 2z = -1$$
$$2x + y - z = 2$$

(e) Replace the third equation by the sum of it and -2 times the first equation:

$$x + y + z = 2$$
$$x + 3y + 2z = 1$$
$$-y - 3z = -2$$

EXAMPLE 1-4: Use the elimination method to solve the system

$$x + y + z = 2$$
$$x + 3y + 2z = 1$$
$$y + 3z = 2$$

Solution: Replace the second equation by the sum of it and -1 times the first equation:

$$x + y + z = 2$$
$$2y + z = -1$$
$$y + 3z = 2$$

Interchange the second and third equations:

$$x + y + z = 2$$
$$y + 3z = 2$$
$$2y + z = -1$$

Multiply the second equation by -2 and add it to the third equation:

$$\begin{aligned} x + y + z &= 2 \\ y + 3z &= 2 \\ -5z &= -5 \end{aligned}$$

Multiply the third equation by $-1/5$:

$$\begin{aligned} x + y + z &= 2 \\ y + 3z &= 2 \\ z &= 1 \end{aligned}$$

Substitute $z = 1$ into the first two equations to obtain

$$\begin{aligned} x + y + 1 &= 2 \\ y + 3 &= 2 \end{aligned}$$

or

$$\begin{aligned} x + y &= 1 \\ y &= -1 \\ \therefore \quad x - 1 &= 1 \\ x &= 2 \end{aligned}$$

Therefore, the solution of the system is $(2, -1, 1)$.

B. Determination of consistency or inconsistency by the elimination method

Use of the elimination method results in exactly one of the following three situations.

(a) You obtain a unique solution of the system, and hence the system is consistent.
(b) You obtain an equivalent system in which one of the equations results in a false statement, meaning that there is no solution and the system is inconsistent.
(c) You obtain an equivalent system having fewer equations than unknowns, but no false statements occur. This means that the system is consistent but has an infinite number of solutions.

EXAMPLE 1-5: Determine if each of the following systems is consistent or inconsistent. If consistent, find the solution.

(a) $x + y = 3$, $3x - y = 5$ **(b)** $x + y = 3$, $x + y = 4$ **(c)** $x + y = 3$, $2x + 2y = 6$

Solution

(a) Adding the two equations, you get $4x = 8$. Hence, $x = 2$. By substituting $x = 2$ into the first equation, you get $y = 1$. The same result occurs if you substitute $x = 2$ into the second equation. Hence, the system is consistent with unique solution $(2, 1)$. The two lines intersect in a point.

(b) Replacing the second equation by the sum of it and -1 times the first equation, you get

$$\begin{aligned} x + y &= 3 \\ 0 &= 1 \end{aligned}$$

This results in the false statement $0 = 1$. Therefore, the system is inconsistent and there is no solution. The two lines are parallel.

(c) Replacing the second equation by the sum of it and -2 times the first equation, you get

$$\begin{aligned} x + y &= 3 \\ 0 &= 0 \end{aligned}$$

You have only one equation in two variables with no false statement. Therefore, the system is consistent with an infinite number of solutions. Assign the arbitrary number t to y, and solve $x + t = 3$ for x to obtain $x = 3 - t$. The solution set is all ordered pairs $(3 - t, t)$ where t is arbitrary. The two lines coincide.

1-4. Matrices

If you use matrices, you can remove much of the "excess baggage" that appears in solving systems of linear equations by the elimination method.

A. Definition of a matrix

A **matrix** is a rectangular array of numbers denoted by

$$\begin{bmatrix} a_{11} & a_{12} & \cdots & a_{1n} \\ a_{21} & a_{22} & \cdots & a_{2n} \\ \vdots & \vdots & & \vdots \\ a_{m1} & a_{m2} & \cdots & a_{mn} \end{bmatrix}$$

A number in the matrix is called an **entry** or an **element** of the matrix. The horizontal lines of elements are called **rows**, and the vertical lines of elements are called **columns**. A pair of subscripts is used to indicate the location of an element. For example, the element a_{ij} (read "a sub $i\ j$") is in the ith row and jth column and the matrix is often written as $[a_{ij}]$. Observe the order row–column. For a 3×4 (read "3 by 4") matrix the rows, columns, and subscripts for elements are as follows:

$$\begin{array}{r} \\ \text{row 1} \\ \text{row 2} \\ \text{row 3} \end{array} \begin{array}{c} \text{col. 1} \quad \text{col. 2} \quad \text{col. 3} \quad \text{col. 4} \\ \begin{bmatrix} a_{11} & a_{12} & a_{13} & a_{14} \\ a_{21} & a_{22} & a_{23} & a_{24} \\ a_{31} & a_{32} & a_{33} & a_{34} \end{bmatrix} \end{array}$$

A matrix is classified by its size. If it has m rows and n columns it is called an $m \times n$ matrix. If it has n rows and n columns it is called a **square** matrix of order n, and the elements $a_{11}, a_{22}, \ldots, a_{nn}$ are said to be the **main diagonal** of the matrix. If a matrix has only one row (column) and two or more columns (rows), it is called a **vector**.

EXAMPLE 1-6: Classify the following matrices by size, and if it is a square matrix, give its order:

(a) $A = \begin{bmatrix} 1 & 0 & 2 \\ -2 & 1 & 3 \end{bmatrix}$ **(b)** $B = \begin{bmatrix} 3 & -2 \\ 4 & -1 \end{bmatrix}$ **(c)** $C = \begin{bmatrix} 2 \\ -1 \\ 3 \end{bmatrix}$

(d) $D = \begin{bmatrix} 1 & -2 & 3 \\ 0 & 3 & 1 \\ 2 & -1 & 2 \end{bmatrix}$ **(e)** $E = [2 \ \ -1 \ \ 3]$ **(f)** $F = [-4]$

Solution

(a) A is a 2×3 matrix since it contains two rows and three columns.
(b) B is a 2×2 square matrix. Its order is 2.
(c) C is a 3×1 matrix since it contains three rows and one column. This is also called a **column matrix** or a vector.
(d) D is a 3×3 square matrix. Its order is 3.
(e) E is a 1×3 matrix. This is also called a **row matrix** or vector.
(f) F is a 1×1 square matrix. Its order is 1. This is a **single-element matrix**.

B. Definition of equality of two matrices

Two matrices A and B are equal ($A = B$) if and only if they have the same size and the corresponding elements are identical.

EXAMPLE 1-7: If $A = \begin{bmatrix} 1 & 3 \\ -4 & 5 \end{bmatrix}$ and $B = \begin{bmatrix} 1 & 3 \\ x & y \end{bmatrix}$, find x and y such that $A = B$.

Solution: Since the corresponding elements must be identical, $x = -4$ and $y = 5$.

1-5. Matrix Addition and Scalar Multiplication

A. Definition of matrix addition

If matrices A and B are the same size, the sum is denoted by $A + B$ and is the matrix obtained by adding corresponding elements in the two matrices.

EXAMPLE 1-8: If

$$A = \begin{bmatrix} 1 & 0 & 4 \\ 2 & 3 & -1 \\ 5 & -1 & 3 \end{bmatrix} \qquad B = \begin{bmatrix} 3 & 2 & 1 \\ 3 & 5 & 7 \\ 0 & 1 & 2 \end{bmatrix} \qquad C = \begin{bmatrix} 2 & 3 & 1 & 5 \\ 0 & 4 & -1 & 0 \\ 5 & 3 & 2 & 1 \end{bmatrix}$$

find:

(a) $A + B$ **(b)** $A + C$

Solution

(a)
$$A + B = \begin{bmatrix} 1+3 & 0+2 & 4+1 \\ 2+3 & 3+5 & -1+7 \\ 5+0 & -1+1 & 3+2 \end{bmatrix} = \begin{bmatrix} 4 & 2 & 5 \\ 5 & 8 & 6 \\ 5 & 0 & 5 \end{bmatrix}$$

(b) $A + C$ is undefined because the two matrices are not the same size.

B. Definition of scalar multiplication

If an $m \times n$ matrix A is multiplied by a real number k (called a **scalar**), then each element of A is multiplied by k.

EXAMPLE 1-9: Perform the following scalar multiplications:

(a) $3\begin{bmatrix} 1 & 0 & 4 \\ 2 & 3 & -1 \end{bmatrix}$ **(b)** $-2\begin{bmatrix} 3 & -2 & 1 \\ -3 & 5 & -7 \\ 0 & 1 & 2 \end{bmatrix}$.

Solution

(a)
$$3\begin{bmatrix} 1 & 0 & 4 \\ 2 & 3 & -1 \end{bmatrix} = \begin{bmatrix} 3 & 0 & 12 \\ 6 & 9 & -3 \end{bmatrix}$$

(b)
$$-2\begin{bmatrix} 3 & -2 & 1 \\ -3 & 5 & -7 \\ 0 & 1 & 2 \end{bmatrix} = \begin{bmatrix} -6 & 4 & -2 \\ 6 & -10 & 14 \\ 0 & -2 & -4 \end{bmatrix}$$

C. Properties of matrix addition and scalar multiplication

Let A, B, and C be any $m \times n$ matrices and a and b any real numbers; then the following are true statements.

(a) $A + B = B + A$
(b) $(A + B) + C = A + (B + C)$
(c) $a(A + B) = aA + aB$
(d) $(a + b)A = aA + bA$
(e) $a(bA) = (ab)A$
(f) If ${}_m0_n$ is the $m \times n$ matrix with all entries zero, then $A + 0 = 0 + A$.
(g) There exists a unique $m \times n$ matrix B' such that $A + B' = B' + A = {}_m0_n$. The matrix B' is $(-1)A$ and is called the **negative** of A.

EXAMPLE 1-10: Verify each of the properties in Section C for

$$A = \begin{bmatrix} 1 & 3 \\ -4 & 5 \end{bmatrix}, \qquad B = \begin{bmatrix} 2 & 4 \\ 3 & -1 \end{bmatrix}, \qquad C = \begin{bmatrix} 3 & 0 \\ 2 & 2 \end{bmatrix},$$

$$a = 2, \qquad b = 3.$$

Solution

(a)

$$A + B = \begin{bmatrix} 1 & 3 \\ -4 & 5 \end{bmatrix} + \begin{bmatrix} 2 & 4 \\ 3 & -1 \end{bmatrix} = \begin{bmatrix} 3 & 7 \\ -1 & 4 \end{bmatrix}$$

$$B + A = \begin{bmatrix} 2 & 4 \\ 3 & -1 \end{bmatrix} + \begin{bmatrix} 1 & 3 \\ -4 & 5 \end{bmatrix} = \begin{bmatrix} 3 & 7 \\ -1 & 4 \end{bmatrix}$$

$$\therefore \quad A + B = B + A$$

(b)

$$(A + B) + C = \left(\begin{bmatrix} 1 & 3 \\ -4 & 5 \end{bmatrix} + \begin{bmatrix} 2 & 4 \\ 3 & -1 \end{bmatrix}\right) + \begin{bmatrix} 3 & 0 \\ 2 & 2 \end{bmatrix}$$

$$= \begin{bmatrix} 3 & 7 \\ -1 & 4 \end{bmatrix} + \begin{bmatrix} 3 & 0 \\ 2 & 2 \end{bmatrix} = \begin{bmatrix} 6 & 7 \\ 1 & 6 \end{bmatrix}$$

$$A + (B + C) = \begin{bmatrix} 1 & 3 \\ -4 & 5 \end{bmatrix} + \left(\begin{bmatrix} 2 & 4 \\ 3 & -1 \end{bmatrix} + \begin{bmatrix} 3 & 0 \\ 2 & 2 \end{bmatrix}\right)$$

$$= \begin{bmatrix} 1 & 3 \\ -4 & 5 \end{bmatrix} + \begin{bmatrix} 5 & 4 \\ 5 & 1 \end{bmatrix} = \begin{bmatrix} 6 & 7 \\ 1 & 6 \end{bmatrix}$$

$$\therefore \quad (A + B) + C = A + (B + C)$$

(c)

$$a(A + B) = 2\left(\begin{bmatrix} 1 & 3 \\ -4 & 5 \end{bmatrix} + \begin{bmatrix} 2 & 4 \\ 3 & -1 \end{bmatrix}\right) = 2\begin{bmatrix} 3 & 7 \\ -1 & 4 \end{bmatrix} = \begin{bmatrix} 6 & 14 \\ -2 & 8 \end{bmatrix}$$

$$aA + aB = 2\begin{bmatrix} 1 & 3 \\ -4 & 5 \end{bmatrix} + 2\begin{bmatrix} 2 & 4 \\ 3 & -1 \end{bmatrix} = \begin{bmatrix} 2 & 6 \\ -8 & 10 \end{bmatrix} + \begin{bmatrix} 4 & 8 \\ 6 & -2 \end{bmatrix} = \begin{bmatrix} 6 & 14 \\ -2 & 8 \end{bmatrix}$$

$$\therefore \quad a(A + B) = aA + aB$$

(d) $$(a+b)A = (2+3)\begin{bmatrix} 1 & 3 \\ -4 & 5 \end{bmatrix} = 5\begin{bmatrix} 1 & 3 \\ -4 & 5 \end{bmatrix} = \begin{bmatrix} 5 & 15 \\ -20 & 25 \end{bmatrix}$$

$$aA + bA = 2\begin{bmatrix} 1 & 3 \\ -4 & 5 \end{bmatrix} + 3\begin{bmatrix} 1 & 3 \\ -4 & 5 \end{bmatrix} = \begin{bmatrix} 2 & 6 \\ -8 & 10 \end{bmatrix} + \begin{bmatrix} 3 & 9 \\ -12 & 15 \end{bmatrix} = \begin{bmatrix} 5 & 15 \\ -20 & 25 \end{bmatrix}$$

$$\therefore \quad (a+b)A = aA + bA$$

(e) $$a(bA) = 2\left(3\begin{bmatrix} 1 & 3 \\ -4 & 5 \end{bmatrix}\right) = 2\begin{bmatrix} 3 & 9 \\ -12 & 15 \end{bmatrix} = \begin{bmatrix} 6 & 18 \\ -24 & 30 \end{bmatrix}$$

$$(ab)A = (2)(3)\begin{bmatrix} 1 & 3 \\ -4 & 5 \end{bmatrix} = 6\begin{bmatrix} 1 & 3 \\ -4 & 5 \end{bmatrix} = \begin{bmatrix} 6 & 18 \\ -24 & 30 \end{bmatrix}$$

$$\therefore \quad a(bA) = (ab)A$$

(f) $$A + {}_2 0_2 = \begin{bmatrix} 1 & 3 \\ -4 & 5 \end{bmatrix} + \begin{bmatrix} 0 & 0 \\ 0 & 0 \end{bmatrix} = \begin{bmatrix} 1 & 3 \\ -4 & 5 \end{bmatrix} = A$$

$${}_2 0_2 + A = \begin{bmatrix} 0 & 0 \\ 0 & 0 \end{bmatrix} + \begin{bmatrix} 1 & 3 \\ -4 & 5 \end{bmatrix} = \begin{bmatrix} 1 & 3 \\ -4 & 5 \end{bmatrix} = A$$

$$\therefore \quad A + {}_2 0_2 = {}_2 0_2 + A$$

(g) $$B' = (-1)A = (-1)\begin{bmatrix} 1 & 3 \\ -4 & 5 \end{bmatrix} = \begin{bmatrix} -1 & -3 \\ 4 & -5 \end{bmatrix}$$

$$A + B' = \begin{bmatrix} 1 & 3 \\ -4 & 5 \end{bmatrix} + \begin{bmatrix} -1 & -3 \\ 4 & -5 \end{bmatrix} = \begin{bmatrix} 0 & 0 \\ 0 & 0 \end{bmatrix} = {}_2 0_2$$

$$B' + A = \begin{bmatrix} -1 & -3 \\ 4 & -5 \end{bmatrix} + \begin{bmatrix} 1 & 3 \\ -4 & 5 \end{bmatrix} = \begin{bmatrix} 0 & 0 \\ 0 & 0 \end{bmatrix} = {}_2 0_2$$

Therefore, for A there exists B' such that

$$A + B' = B' + A = {}_2 0_2$$

note: None of these is a proof. They simply show that, for the three matrices and two real numbers given, the properties are true.

1-6. Matrix Multiplication

A. Definition of matrix multiplication

If $A = [a_{ij}]$ is an $m \times n$ matrix and $B = [b_{ij}]$ is an $n \times p$ matrix, then the product AB is the matrix $C = [c_{ij}]$, where

$$c_{ij} = a_{i1}b_{1j} + a_{i2}b_{2j} + a_{i3}b_{3j} + \cdots + a_{in}b_{nj}$$
$$= \sum_{k=1}^{n} a_{ik}b_{kj}$$

This means that the element in the ith row and jth column of the product AB is obtained by multiplying the elements in the ith row of A by the corresponding elements of the jth column of B and adding up these products.

To determine whether the product AB is defined and to find the size of AB, you will find the following procedure helpful:

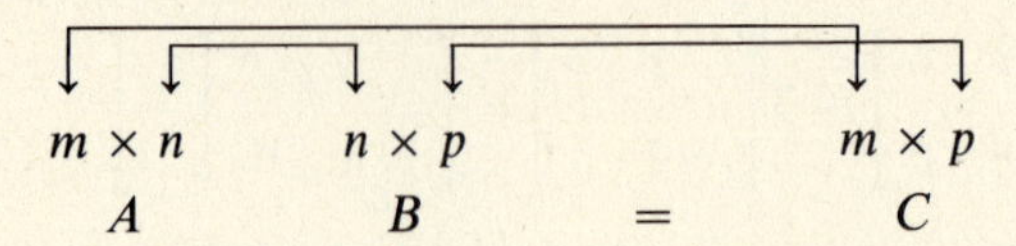

The product is defined only if the inside dimensions are equal. If the product is defined, the number of rows of C is the same as the number of rows of A, and the number of columns of C is the numbers of columns of B. As indicated in the above device, the size of C is $m \times p$.

EXAMPLE 1-11: Let A be 2×3, B be 3×5, and C be 5×4 matrices. Determine whether the following products are defined and, if so, find the size of the products

(a) AB, **(b)** BA, **(c)** AC, **(d)** $(AB)C$

Solution

(a) $\underset{2\times 3}{A}\ \underset{3\times 5}{B}$ It is defined and the product is 2×5.

(b) $\underset{3\times 5}{B}\ \underset{2\times 3}{A}$, $5 \neq 2$ Therefore, the product is not defined.

(c) $\underset{2\times 3}{A}\ \underset{5\times 4}{C}$, $3 \neq 5$ Therefore, the product is not defined.

(d) $(\underset{2\times 3}{A}\ \underset{3\times 5}{B})\ \underset{5\times 4}{C} = (\underset{2\times 5}{A\ B})\ \underset{5\times 4}{C}$ It is defined, and the product is 2×4.

EXAMPLE 1-12: If $A = \begin{bmatrix} 2 & 3 & 1 \\ 5 & 0 & 6 \end{bmatrix}$ and $B = \begin{bmatrix} 1 & 0 & 4 & 2 \\ 2 & 3 & -1 & 3 \\ -1 & 2 & 0 & -2 \end{bmatrix}$, find AB.

Solution

$$AB = \begin{bmatrix} 2 & 3 & 1 \\ 5 & 0 & 6 \end{bmatrix} \begin{bmatrix} 1 & 0 & 4 & 2 \\ 2 & 3 & -1 & 3 \\ -1 & 2 & 0 & -2 \end{bmatrix}$$

$$= \begin{bmatrix} 2(1) + 3(2) + 1(-1) & 2(0) + 3(3) + 1(2) & 2(4) + 3(-1) + 1(0) & 2(2) + 3(3) + 1(-2) \\ 5(1) + 0(2) + 6(-1) & 5(0) + 0(3) + 6(2) & 5(4) + 0(-1) + 6(0) & 5(2) + 0(3) + 6(-2) \end{bmatrix}$$

$$= \begin{bmatrix} 2 + 6 - 1 & 0 + 9 + 2 & 8 - 3 + 0 & 4 + 9 - 2 \\ 5 + 0 - 6 & 0 + 0 + 12 & 20 + 0 + 0 & 10 + 0 - 12 \end{bmatrix}$$

$$= \begin{bmatrix} 7 & 11 & 5 & 11 \\ -1 & 12 & 20 & -2 \end{bmatrix}$$

Observe that the element in row 1, column 2 of AB, $11 = 0 + 9 + 2$, is the sum of the products of the elements in row 1 of A by the corresponding elements in column 2 of B; $2(0) + 3(3) + 1(2)$.

EXAMPLE 1-13: If $A = \begin{bmatrix} 3 & -2 \\ 5 & 1 \end{bmatrix}$ and $B = \begin{bmatrix} 2 & -1 \\ 1 & 3 \end{bmatrix}$, find AB and BA.

Solution

$$AB = \begin{bmatrix} 3 & -2 \\ 5 & 1 \end{bmatrix} \begin{bmatrix} 2 & -1 \\ 1 & 3 \end{bmatrix} = \begin{bmatrix} 3(2) + (-2)(1) & 3(-1) + (-2)(3) \\ 5(2) + 1(1) & 5(-1) + 1(3) \end{bmatrix}$$

$$= \begin{bmatrix} 6 - 2 & -3 - 6 \\ 10 + 1 & -5 + 3 \end{bmatrix} = \begin{bmatrix} 4 & -9 \\ 11 & -2 \end{bmatrix}$$

$$BA = \begin{bmatrix} 2 & -1 \\ 1 & 3 \end{bmatrix}\begin{bmatrix} 3 & -2 \\ 5 & 1 \end{bmatrix} = \begin{bmatrix} 2(3) + (-1)(5) & 2(-2) + (-1)(1) \\ 1(3) + 3(5) & 1(-12) + 3(1) \end{bmatrix}$$

$$= \begin{bmatrix} 6 - 5 & -4 - 1 \\ 3 + 15 & -2 + 3 \end{bmatrix} = \begin{bmatrix} 1 & -5 \\ 18 & 1 \end{bmatrix}$$

We must conclude that the multiplication of matrices is not commutative.

B. Properties of matrix multiplication

Let the sizes of matrices A, B, and C be such that the indicated operations are defined and let a be a real number.

(a) $A(BC) = (AB)C$
(b) $A(B + C) = AB + AC$
(c) $(A + B)C = AC + BC$
(d) $a(AB) = (aA)B = A(aB)$

EXAMPLE 1-14: Verify each of these properties for

$$A = \begin{bmatrix} 1 & 3 \\ -4 & 5 \end{bmatrix}, \qquad B = \begin{bmatrix} 2 & 4 \\ 3 & -1 \end{bmatrix}, \qquad C = \begin{bmatrix} 3 & 0 \\ 2 & 2 \end{bmatrix},$$

and $a = 2$.

Solution

(a)

$$BC = \begin{bmatrix} 2 & 4 \\ 3 & -1 \end{bmatrix}\begin{bmatrix} 3 & 0 \\ 2 & 2 \end{bmatrix} = \begin{bmatrix} 14 & 8 \\ 7 & -2 \end{bmatrix}$$

$$\therefore \quad A(BC) = \begin{bmatrix} 1 & 3 \\ -4 & 5 \end{bmatrix}\begin{bmatrix} 14 & 8 \\ 7 & -2 \end{bmatrix} = \begin{bmatrix} 35 & 2 \\ -21 & -42 \end{bmatrix}$$

$$AB = \begin{bmatrix} 1 & 3 \\ -4 & 5 \end{bmatrix}\begin{bmatrix} 2 & 4 \\ 3 & -1 \end{bmatrix} = \begin{bmatrix} 11 & 1 \\ 7 & -21 \end{bmatrix}$$

$$\therefore \quad AB(C) = \begin{bmatrix} 11 & 1 \\ 7 & -21 \end{bmatrix}\begin{bmatrix} 3 & 0 \\ 2 & 2 \end{bmatrix} = \begin{bmatrix} 35 & 2 \\ -21 & -42 \end{bmatrix}$$

$$\therefore \quad A(BC) = (AB)C$$

(b)

$$B + C = \begin{bmatrix} 2 & 4 \\ 3 & -1 \end{bmatrix} + \begin{bmatrix} 3 & 0 \\ 2 & 2 \end{bmatrix} = \begin{bmatrix} 5 & 4 \\ 5 & 1 \end{bmatrix}$$

$$\therefore \quad A(B + C) = \begin{bmatrix} 1 & 3 \\ -4 & 5 \end{bmatrix}\begin{bmatrix} 5 & 4 \\ 5 & 1 \end{bmatrix} = \begin{bmatrix} 20 & 7 \\ 5 & -11 \end{bmatrix}$$

$$AB + AC = \begin{bmatrix} 1 & 3 \\ -4 & 5 \end{bmatrix}\begin{bmatrix} 2 & 4 \\ 3 & -1 \end{bmatrix} + \begin{bmatrix} 1 & 3 \\ -4 & 5 \end{bmatrix}\begin{bmatrix} 3 & 0 \\ 2 & 2 \end{bmatrix}$$

$$= \begin{bmatrix} 11 & 1 \\ 7 & -21 \end{bmatrix} + \begin{bmatrix} 9 & 6 \\ -2 & 10 \end{bmatrix} = \begin{bmatrix} 20 & 7 \\ 5 & -11 \end{bmatrix}$$

$$\therefore \quad A(B + C) = AB + AC$$

(c)

$$A + B = \begin{bmatrix} 1 & 3 \\ -4 & 5 \end{bmatrix} + \begin{bmatrix} 2 & 4 \\ 3 & -1 \end{bmatrix} = \begin{bmatrix} 3 & 7 \\ -1 & 4 \end{bmatrix}$$

$$(A + B)C = \begin{bmatrix} 3 & 7 \\ -1 & 4 \end{bmatrix}\begin{bmatrix} 3 & 0 \\ 2 & 2 \end{bmatrix} = \begin{bmatrix} 23 & 14 \\ 5 & 8 \end{bmatrix}$$

$$AC = \begin{bmatrix} 1 & 3 \\ -4 & 5 \end{bmatrix}\begin{bmatrix} 3 & 0 \\ 2 & 2 \end{bmatrix} = \begin{bmatrix} 9 & 6 \\ -2 & 10 \end{bmatrix}$$

$$BC = \begin{bmatrix} 2 & 4 \\ 3 & -1 \end{bmatrix}\begin{bmatrix} 3 & 0 \\ 2 & 2 \end{bmatrix} = \begin{bmatrix} 14 & 8 \\ 7 & -2 \end{bmatrix}$$

$$\therefore \quad AC + BC = \begin{bmatrix} 9 & 6 \\ -2 & 10 \end{bmatrix} + \begin{bmatrix} 14 & 8 \\ 7 & -2 \end{bmatrix} = \begin{bmatrix} 23 & 14 \\ 5 & 8 \end{bmatrix}$$

$$\therefore \quad (A + B)C = AC + BC$$

(d)

$$AB = \begin{bmatrix} 1 & 3 \\ -4 & 5 \end{bmatrix}\begin{bmatrix} 2 & 4 \\ 3 & -1 \end{bmatrix} = \begin{bmatrix} 11 & 1 \\ 7 & -21 \end{bmatrix}$$

$$\therefore \quad 2(AB) = 2\begin{bmatrix} 11 & 1 \\ 7 & -21 \end{bmatrix} = \begin{bmatrix} 22 & 2 \\ 14 & -42 \end{bmatrix}$$

$$2A = 2\begin{bmatrix} 1 & 3 \\ -4 & 5 \end{bmatrix} = \begin{bmatrix} 2 & 6 \\ -8 & 10 \end{bmatrix}$$

$$\therefore \quad (2A)B = \begin{bmatrix} 2 & 6 \\ -8 & 10 \end{bmatrix}\begin{bmatrix} 2 & 4 \\ 3 & -1 \end{bmatrix} = \begin{bmatrix} 22 & 2 \\ 14 & -42 \end{bmatrix}$$

$$2B = 2\begin{bmatrix} 2 & 4 \\ 3 & -1 \end{bmatrix} = \begin{bmatrix} 4 & 8 \\ 6 & -2 \end{bmatrix}$$

$$\therefore \quad A(2B) = \begin{bmatrix} 1 & 3 \\ -4 & 5 \end{bmatrix}\begin{bmatrix} 4 & 8 \\ 6 & -2 \end{bmatrix} = \begin{bmatrix} 22 & 2 \\ 14 & -42 \end{bmatrix}$$

$$\therefore \quad 2(AB) = (2A)B = A(2B)$$

1-7. Special Matrices

Even though some of these special matrices are mentioned elsewhere in this book, we are including them here for the completeness of the topic and review. The following definitions are stated.

A **zero matrix** is a matrix with all zero entries. If it is $m \times n$ and you find it necessary to emphasize its size, denote it by ${}_m0_n$. If the size of the zero matrix is clear, you may denote it by 0.

The **transpose** of a matrix A is the matrix A^T obtained by interchanging the rows and columns of A.

A **square matrix** is one that has the same number of rows as columns. Its size is called **order** and is the number of rows (columns).

The **main diagonal** of a square matrix is the diagonal from the upper left to the lower right. It is the diagonal with entries $a_{11}, a_{22}, a_{33}, \ldots, a_{nn}$.

A **diagonal matrix** is a square matrix in which all entries off the main diagonal are zeros.

A **scalar matrix** is a diagonal matrix whose diagonal entries are equal.

The $n \times n$ **identity matrix** is a scalar matrix in which all diagonal entries are ones. It is denoted by I_n, except if its order is clear; then you may use I.

A square matrix is **tridiagonal** if all nonzero entries are on the main diagonal or on a diagonal immediately adjacent to the main diagonal. (The matrix may not contain any nonzero elements.)

A square matrix is **upper triangular** if all entries below the main diagonal are zeros. It is **lower triangular** if all entries above the main diagonal are zeros. If it is either upper or lower triangular, it is called **triangular**. Sometimes upper triangular is also called **echelon**.

A square matrix A is **symmetric** if $A^T = A$.

A square matrix A is **skew-symmetric** if $A^T = -A$.

EXAMPLE 1-15: Find the transpose of

(a) $A = \begin{bmatrix} 3 & 5 & -6 \\ 4 & 2 & 0 \end{bmatrix}$ **(b)** $B = [1 \quad -2 \quad 3]$ **(c)** $C = \begin{bmatrix} 2 \\ 1 \\ -1 \end{bmatrix}$.

Solution

(a) $A^T = \begin{bmatrix} 3 & 4 \\ 5 & 2 \\ -6 & 0 \end{bmatrix}$ **(b)** $B^T = \begin{bmatrix} 1 \\ -2 \\ 3 \end{bmatrix}$ **(c)** $C^T = [2 \quad 1 \quad -1]$.

EXAMPLE 1-16: Indicate which of the following is a diagonal, tridiagonal, scalar, identity, upper triangular, or lower triangular matrix. (More than one term may apply to a given matrix.)

(a) $\begin{bmatrix} 1 & 0 & 0 \\ 0 & 1 & 0 \\ 0 & 0 & 1 \end{bmatrix}$ **(b)** $\begin{bmatrix} 1 & 0 & 0 \\ 0 & 2 & 0 \\ 0 & 0 & 3 \end{bmatrix}$ **(c)** $\begin{bmatrix} 3 & 0 & 0 \\ 0 & 3 & 0 \\ 0 & 0 & 3 \end{bmatrix}$ **(d)** $\begin{bmatrix} 1 & 2 & 3 \\ 0 & 1 & 2 \\ 0 & 0 & 1 \end{bmatrix}$

(e) $\begin{bmatrix} 1 & 2 & 0 \\ 5 & 3 & 3 \\ 0 & 6 & -4 \end{bmatrix}$ **(f)** $\begin{bmatrix} 3 & 0 & 0 \\ -2 & 1 & 0 \\ 5 & 4 & 2 \end{bmatrix}$ **(g)** $\begin{bmatrix} 1 & -1 & 0 & 0 \\ 2 & 2 & -2 & 0 \\ 0 & -1 & 3 & 1 \\ 0 & 0 & -2 & 4 \end{bmatrix}$ **(h)** $\begin{bmatrix} 0 & 0 & 0 \\ 0 & 0 & 0 \\ 0 & 0 & 0 \end{bmatrix}$

Solution

(a) It is diagonal, tridiagonal, scalar, identity, upper triangular, and lower triangular. This is called the 3×3 identity matrix (I_3) unless one of the other characteristics is being emphasized.
(b) It is diagonal, tridiagonal, upper triangular, and lower triangular. Unless one of the other characteristics is being emphasized, this is called a diagonal matrix.
(c) It is diagonal, tridiagonal, scalar, upper triangular, and lower triangular—or simply diagonal, unless another characteristic is emphasized.
(d) It is upper triangular.
(e) It is tridiagonal.
(f) It is lower triangular.
(g) It is tridiagonal.
(h) It is diagonal, tridiagonal, scalar, upper triangular, and lower triangular.

EXAMPLE 1-17: Determine whether each of the following matrices is symmetric or nonsymmetric.

(a) $A = \begin{bmatrix} 2 & 3 & 5 \\ 3 & 1 & -2 \\ 5 & -2 & 3 \end{bmatrix}$ **(b)** $B = \begin{bmatrix} 4 & 1 & 0 & 2 \\ 1 & 3 & -2 & 4 \\ 0 & -2 & 6 & 7 \\ 2 & 4 & 7 & -3 \end{bmatrix}$ **(c)** $C = \begin{bmatrix} 1 & 0 & 0 \\ 0 & 1 & 0 \\ 0 & 0 & 1 \end{bmatrix}$

(d) $D = \begin{bmatrix} 1 & 2 & 3 \\ 0 & 3 & 5 \\ 4 & -1 & 2 \end{bmatrix}$

Solution

(a) $A^T = \begin{bmatrix} 2 & 3 & 5 \\ 3 & 1 & -2 \\ 5 & -2 & 3 \end{bmatrix}$ Since $A^T = A$, A is symmetric.

(b) $B^T = \begin{bmatrix} 4 & 1 & 0 & 2 \\ 1 & 3 & -2 & 4 \\ 0 & -2 & 6 & 7 \\ 2 & 4 & 7 & -3 \end{bmatrix}$ Since $B^T = B$, B is symmetric.

(c) $C^T = \begin{bmatrix} 1 & 0 & 0 \\ 0 & 1 & 0 \\ 0 & 0 & 1 \end{bmatrix}$ Since $C^T = C$, C is symmetric.

(d) $D^T = \begin{bmatrix} 1 & 0 & 4 \\ 2 & 3 & -1 \\ 3 & 5 & 2 \end{bmatrix}$ Since $D^T \neq D$, D is not symmetric.

EXAMPLE 1-18: Determine whether each of the following matrices is skew-symmetric.

(a) $A = \begin{bmatrix} 0 & -5 & 4 \\ 5 & 0 & 2 \\ -4 & -2 & 0 \end{bmatrix}$ **(b)** $B = \begin{bmatrix} 0 & 0 & 0 \\ 0 & 0 & 0 \\ 0 & 0 & 0 \end{bmatrix}$ **(c)** $C = \begin{bmatrix} 1 & -5 & 4 \\ 5 & 1 & 2 \\ -4 & -2 & 1 \end{bmatrix}$

Solution

(a) $A^T = \begin{bmatrix} 0 & 5 & -4 \\ -5 & 0 & -2 \\ 4 & 2 & 0 \end{bmatrix}$ Since $A^T = -A$, A is skew-symmetric.

(b) $B^T = \begin{bmatrix} 0 & 0 & 0 \\ 0 & 0 & 0 \\ 0 & 0 & 0 \end{bmatrix}$ Since $B^T = -B$, B is skew-symmetric.

(c) $C^T = \begin{bmatrix} 1 & 5 & -4 \\ -5 & 1 & -2 \\ 4 & 2 & 1 \end{bmatrix}$ Since $C^T \neq C$, C is not skew-symmetric.

EXAMPLE 1-19: Express the following matrix as the sum of a symmetric matrix and a skew-symmetric matrix:

$$A = \begin{bmatrix} 1 & -2 & 3 \\ 0 & 3 & 1 \\ 2 & -1 & 2 \end{bmatrix}$$

Solution: We need $A = S + K$, where S is symmetric and K is skew-symmetric.

$A^T = (S + K)^T = S^T + K^T$ by properties of transpose. Since S is symmetric, $S^T = S$, and since K is skew-symmetric, $K^T = -K$. Thus $A^T = S - K$, and we have the system of equations:

$$A^T = S - K$$

$$A = S + K$$

Solving this system for S and K gives

$$S = 1/2(A + A^T) \quad \text{and} \quad K = 1/2(A - A^T)$$

$$A^T = \begin{bmatrix} 1 & 0 & 2 \\ -2 & 3 & -1 \\ 3 & 1 & 2 \end{bmatrix}$$

$$S = \frac{1}{2}\left(\begin{bmatrix} 1 & -2 & 3 \\ 0 & 3 & 1 \\ 2 & -1 & 2 \end{bmatrix} + \begin{bmatrix} 1 & 0 & 2 \\ -2 & 3 & -1 \\ 3 & 1 & 2 \end{bmatrix}\right) = \frac{1}{2}\begin{bmatrix} 2 & -2 & 5 \\ -2 & 6 & 0 \\ 5 & 0 & 4 \end{bmatrix} = \begin{bmatrix} 1 & -1 & 5/2 \\ -1 & 3 & 0 \\ 5/2 & 0 & 2 \end{bmatrix}$$

$$K = \frac{1}{2}\left(\begin{bmatrix} 1 & -2 & 3 \\ 0 & 3 & 1 \\ 2 & -1 & 2 \end{bmatrix} - \begin{bmatrix} 1 & 0 & 2 \\ -2 & 3 & -1 \\ 3 & 1 & 2 \end{bmatrix}\right) = \frac{1}{2}\begin{bmatrix} 0 & -2 & 1 \\ 2 & 0 & 2 \\ -1 & -2 & 0 \end{bmatrix} = \begin{bmatrix} 0 & -1 & 1/2 \\ 1 & 0 & 1 \\ -1/2 & -1 & 0 \end{bmatrix}$$

$$S + K = \begin{bmatrix} 1 & -1 & 5/2 \\ -1 & 3 & 0 \\ 5/2 & 0 & 2 \end{bmatrix} + \begin{bmatrix} 0 & -1 & 1/2 \\ 1 & 0 & 1 \\ -1/2 & -1 & 0 \end{bmatrix} = \begin{bmatrix} 1 & -2 & 3 \\ 0 & 3 & 1 \\ 2 & -1 & 2 \end{bmatrix} = A$$

SUMMARY

1. An equation in n variables $x_1, x_2, \ldots, x_n$ is linear if it can be expressed in the form $a_1x_1 + a_2x_2 + \cdots + a_nx_n = b$, where $a_1, a_2, \ldots, a_n$ and b are constants. A solution of the linear equation is an ordered n-tuple of numbers $s_1, s_2, \ldots, s_n$ such that when x_1 is replaced by s_1, x_2 by s_2, and so on, the equation becomes a true statement.
2. A set of two or more linear equations is called a system. To be a solution of the system, an ordered n-tuple must make each equation a true statement.
3. A system of m linear equations in n unknowns has no solution, exactly one solution, or an infinite number of solutions.
4. Two systems of linear equations are equivalent if they have exactly the same solution set. Equivalent systems are obtained if
   - an equation of the system is multiplied by a nonzero constant,
   - two equations in the system are interchanged, or
   - one equation of the system is replaced by the sum of it and a constant multiple of another equation of the system.
5. An $m \times n$ matrix A is a rectangular array of numbers denoted by

$$A = [a_{ij}] = \begin{bmatrix} a_{11} & a_{12} & \cdots & a_{1n} \\ a_{21} & a_{22} & \cdots & a_{2n} \\ \vdots & & \cdots & \vdots \\ a_{m1} & a_{m2} & \cdots & a_{mn} \end{bmatrix}$$

6. A matrix is of size $m \times n$ if it has m rows and n columns.
7. Matrices are equal if their corresponding elements are identical.
8. Matrices of the same size may be added or subtracted.
9. A matrix may be multiplied by a real number.
10. If A, B, and C are $m \times n$ matrices and a and b are real numbers, then
    - $A + B = B + A$
    - $(A + B) + C = A + (B + C)$
    - $a(A + B) = aA + aB$

- $(a + b)A = aA + bA$
- $a(bA) = (ab)A$
- $A + {}_m0_n = {}_m0_n + A = A$
- $A + (-A) = {}_m0_n$

11. If the number of columns of matrix A is n and the number of rows of B is n, then the product AB is matrix C, where $c_{ij} = a_{i1}b_{1j} + a_{i2}b_{2j} + \cdots + a_{in}b_{nj}$.

12. If the sizes of matrices A, B, and C are such that the indicated operations are defined and a is a real number, then

- $A(BC) = (AB)C$
- $A(B + C) = AB + AC$
- $(A + B)C = AC + BC$
- $a(AB) = (aA)B = A(aB)$
- $AI = IA = A$

13. A zero matrix is one with all zero entries.

14. The transpose of a matrix A is the matrix A^T obtained by interchanging the rows and columns of A.

15. A square matrix has the same number of rows and columns. This common number is called the order of the matrix.

16. The main diagonal of a square matrix extends from upper left to lower right.

17. A diagonal matrix is a square matrix in which all entries off the main diagonal are zeros.

18. A scalar matrix is a diagonal matrix whose diagonal entries are equal.

19. A square matrix is upper triangular (or echelon) if all entries below the main diagonal are zeros. If all entries above the main diagonal are zeros, it is called lower triangular.

20. A square matrix A is symmetric if $A^T = A$.

21. A square matrix A is skew-symmetric if $A^T = -A$.

RAISE YOUR GRADES

Can you ...?

☑ determine whether an equation is linear and, if so, find its solution set

☑ determine the consistency or inconsistency of a system of linear equations and, if consistent, find the solution set by the elimination method

☑ define a matrix and use subscripts to denote its elements

☑ find the size of a matrix and, if square, its order

☑ multiply a matrix by a real number and determine if the two matrices are equal

☑ add, subtract, and multiply matrices of the proper size

☑ identify whether a given matrix is a zero, identity, diagonal, scalar, tridiagonal, upper triangular, lower triangular, symmetric, or skew-symmetric

☑ find the transpose of a matrix

SOLVED PROBLEMS

PROBLEM 1-1 Determine which of the following equations are linear: **(a)** $2x - 3y = 4$; **(b)** $x - 3y + z^2 = 5$; **(c)** $2|x| - y = 4$; **(d)** $2\sqrt{x^2} - y = 5$; **(e)** $2x + 3y - 5z = 0$; **(f)** $x + 2x^2y + y = 4$.

Solution

(a) This is linear with $a_1 = 2$, $a_2 = -3$, and $b = 4$.
(b) This is nonlinear because of z^2.
(c) This is nonlinear because of $|x|$, which equals x if $x \geq 0$ and $-x$ if $x < 0$.
(d) This is nonlinear because of $\sqrt{x^2}$, which equals x if $x \geq 0$ and $-x$ if $x < 0$.
(e) This is linear with $a_1 = 2$, $a_2 = 3$, $a_3 = -5$, and $b = 0$.
(f) This is nonlinear because of the product x^2y.

PROBLEM 1-2 Find the solution set of each of the following linear equations: **(a)** $2x - 3y = 4$; **(b)** $2x + 3y - 5z = 0$; **(c)** $\pi x + 3y = 2$.

Solution

(a) To find the solution set of $2x - 3y = 4$, we can assign an arbitrary number to x and solve for y. Let $x = k$, then $y = (2k - 4)/3$. The solution set consists of all ordered pairs $[k, (2k - 4)/3]$ where k is arbitrary. If we let $y = m$, then $x = (3m + 4)/2$ and the solution set consists of all pairs of form $[(3m + 4)/2, m]$.
(b) Let $x = k$ and $y = m$, then $z = (2k + 3m)/5$. The solution set consists of all ordered triples of form $[k, m, (2k + 3m)/5]$, where k and m are arbitrary.
(c) Let $x = k$, then $y = (-\pi k + 2)/3$. The solution set consists of all ordered pairs $[k, (-\pi k + 2)/3]$, where k is arbitrary. (π is a constant.)

PROBLEM 1-3 Construct five systems of equations that are equivalent to

$$\begin{aligned} x - 2y + 3z &= 4 \\ 2x + y + z &= 3 \\ 3x - y + 2z &= 1 \end{aligned}$$

Solution

(a) Multiply the first equation by 2:

$$\begin{aligned} 2x - 4y + 6z &= 8 \\ 2x + y + z &= 3 \\ 3x - y + 2z &= 1 \end{aligned}$$

(b) Interchange the first and second equations:

$$\begin{aligned} 2x + y + z &= 3 \\ x - 2y - 3z &= 4 \\ 3x - y + 2z &= 1 \end{aligned}$$

(c) Replace the second equation by the sum of it and -2 times the first equation:

$$\begin{aligned} x - 2y + 3z &= 4 \\ 5y - 5z &= -5 \\ 3x - y + 2z &= 1 \end{aligned}$$

(d) Multiply the first equation by 3 and the third equation by -1:

$$\begin{aligned} 3x - 6y + 9z &= 12 \\ 2x + y + z &= 3 \\ -3x + y - 2z &= -1 \end{aligned}$$

(e) Replace the third equation by the sum of it and -3 times the first equation:

$$x - 2y + 3z = 4$$
$$2x + y + z = 3$$
$$5y - 7z = -11$$

PROBLEM 1-4 Solve by the elimination method:

$$x - 2y + 3z = 4$$
$$2x + y + z = 3$$
$$5y - 7z = -11$$

Solution Replace the second equation by the sum of it and -2 times the first equation:

$$x - 2y + 3z = 4$$
$$5y - 5z = -5$$
$$5y - 7z = -11$$

Interchange the second and third equations:

$$x - 2y + 3z = 4$$
$$5y - 7z = -11$$
$$5y - 5z = -5$$

Multiply the second equation by -1 and add to the third equation:

$$x - 2y + 3z = 4$$
$$5y - 7z = -11$$
$$2z = 6$$

Multiply the third equation by 1/2:

$$x - 2y + 3z = 4$$
$$5y - 7z = -11$$
$$z = 3$$

Substitute $z = 3$ into the first two equations to obtain

$$x - 2y + 9 = 4$$
$$5y - 21 = -11$$

or

$$x - 2y = -5$$
$$5y = 10$$

or

$$x - 2y = -5$$
$$y = 2$$
$$\therefore \quad x - 4 = -5$$
$$x = -1$$

The solution set of the system is $(-1, 2, 3)$.

PROBLEM 1-5 Determine whether each of the following systems is consistent or inconsistent. If consistent, find the solution. Graph each system: **(a)** $x - y = 2, 2x + y = 7$; **(b)** $x + 2y = 5, 2x + 4y = 7$; **(c)** $3x - 2y = 4, 6x - 4y = 8$.

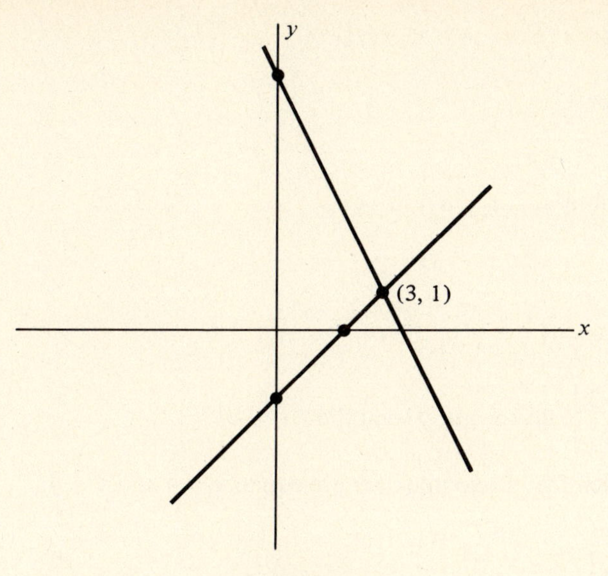

FIGURE 1-4

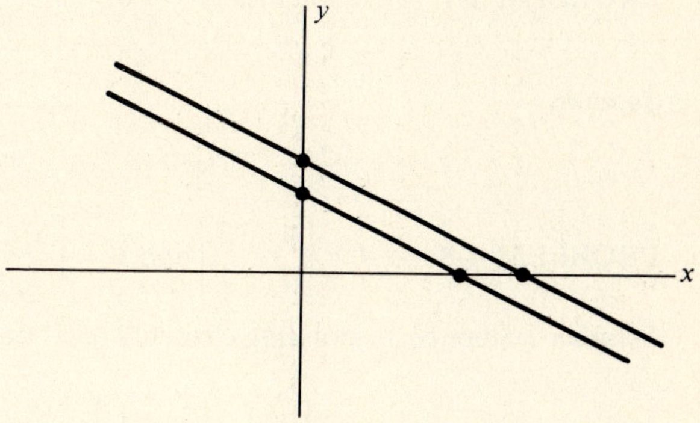

FIGURE 1-5

Solution

(a) Adding the first equation to the second equation, you get $3x = 9$. Hence $x = 3$. Thus by substituting $x = 3$ into the first equation (or the second) you get $3 - y = 2$ or $y = 1$. The system is **consistent** and the lines intersect at (3, 1) (Figure 1-4).

(b) Subtracting 2 times the first equation from the second equation, you get

$$x + 2y = 5$$
$$0 = -3$$

This gives the false statement that $0 = -3$. Therefore, the system is inconsistent, and there are no solutions. The lines are parallel (Figure 1-5).

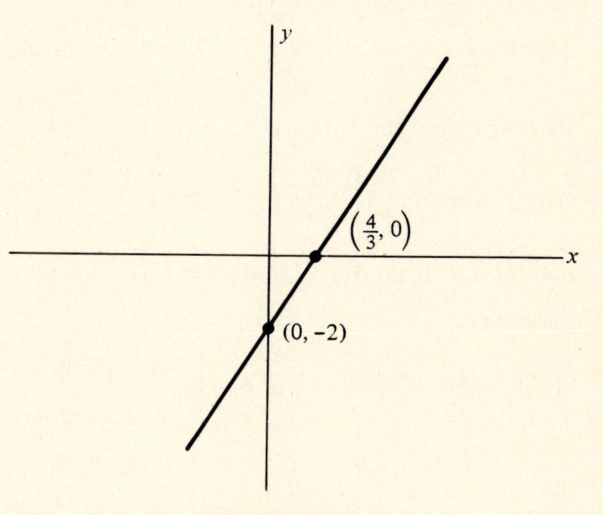

FIGURE 1-6

(c) Replacing the second equation by the sum of it and -2 times the first equation, you get

$$3x - 2y = 4$$
$$0 = 0$$

You have only one equation with two variables but no false statement. The system is consistent, with an infinite number of solutions. Let $y = k$; then $3x - 2k = 4$ and $x = (2k + 4)/3$. The solution set is all ordered pairs $[(2k + 4)/3, k]$, where k is arbitrary. The lines **coincide** (Figure 1-6).

PROBLEM 1-6 Classify the following matrices by size. If it is a square matrix, give its order: **(a)** $A = [3]$; **(b)** $B = \begin{bmatrix} 3 & 0 \\ 4 & -1 \end{bmatrix}$; **(c)** $C = \begin{bmatrix} 5 & 6 & 3 & -2 \\ 2 & 1 & 0 & 5 \\ 3 & 4 & 5 & 0 \end{bmatrix}$; **(d)** $D = [3 \quad 5 \quad 6]$; **(e)** $E = \begin{bmatrix} 4 \\ 2 \\ 1 \end{bmatrix}$;
(f) $F = \begin{bmatrix} 1 & -2 & 3 \\ 0 & 3 & 1 \\ 2 & -1 & 2 \end{bmatrix}$.

Solution

(a) A is a 1×1 matrix. It is a square matrix of order 1.
(b) B is a 2×2 matrix. It is a square matrix of order 2.
(c) C is a 3×4 matrix.

(d) D is a 1×3 matrix. It is also a vector.
(e) E is a 3×1 matrix. It is a vector.
(f) F is a 3×3 matrix. It is a square matrix of order 3.

PROBLEM 1-7 Given matrix $A = \begin{bmatrix} 1 & -2 & 3 \\ 0 & 3 & 1 \\ 2 & -1 & 2 \end{bmatrix}$ find $a_{12}, a_{22}, a_{32}, a_{21}, a_{33}$.

Solution

$$a_{12} = -2 \qquad a_{22} = 3 \qquad a_{32} = -1 \qquad a_{21} = 0 \qquad a_{33} = 2$$

PROBLEM 1-8 If $A = \begin{bmatrix} 4 & -3 \\ 2 & 1 \end{bmatrix}$ and $B = \begin{bmatrix} x & -3 \\ 2 & y \end{bmatrix}$, find x and y such that $A = B$.

Solution Since corresponding elements must be identical if two matrices are equal, $x = 4$ and $y = 1$.

PROBLEM 1-9 If $A = \begin{bmatrix} 1 & -2 & -1 \\ 3 & -1 & 4 \\ -2 & -3 & 2 \end{bmatrix}$ and $B = \begin{bmatrix} 1 & x & -1 \\ y & -1 & 4 \\ z & -3 & w \end{bmatrix}$, find x, y, z, and w such that $A = B$.

Solution $x = -2$, $y = 3$, $z = -2$, and $w = 2$, since corresponding elements must be identical.

PROBLEM 1-10 Given the matrices

$$A = \begin{bmatrix} 1 & 0 & 2 \\ -2 & 1 & 3 \end{bmatrix} \qquad B = \begin{bmatrix} 3 & 2 & 1 \\ 4 & 1 & -1 \end{bmatrix} \qquad C = \begin{bmatrix} 1 & 0 & 0 \\ 0 & 1 & 1 \end{bmatrix} \qquad D = \begin{bmatrix} 2 & -3 \\ 4 & 5 \end{bmatrix}$$

find **(a)** $A + B$, **(b)** $A + C$, and **(c)** $B + D$.

Solution

(a)
$$A + B = \begin{bmatrix} 1+3 & 0+2 & 2+1 \\ -2+4 & 1+1 & 3+(-1) \end{bmatrix} = \begin{bmatrix} 4 & 2 & 3 \\ 2 & 2 & 2 \end{bmatrix}$$

(b)
$$A + C = \begin{bmatrix} 1+1 & 0+0 & 2+0 \\ -2+0 & 1+1 & 3+1 \end{bmatrix} = \begin{bmatrix} 2 & 0 & 2 \\ -2 & 2 & 4 \end{bmatrix}$$

(c) $B + D$ is undefined because B and D have different sizes.

PROBLEM 1-11 Given $A = \begin{bmatrix} 1 & -2 & -1 \\ 3 & -1 & 4 \\ -2 & -3 & 2 \end{bmatrix}$ and $B = \begin{bmatrix} 2 & -4 \\ 6 & 2 \end{bmatrix}$, find

(a) $2\begin{bmatrix} 1 & -2 & -1 \\ 3 & -1 & 4 \\ -2 & -3 & 2 \end{bmatrix}$ **(b)** $\frac{1}{2}\begin{bmatrix} 2 & -4 \\ 6 & 2 \end{bmatrix}$

Solution

(a)
$$2\begin{bmatrix} 1 & -2 & -1 \\ 3 & -1 & 4 \\ -2 & -3 & 2 \end{bmatrix} = \begin{bmatrix} 2(1) & 2(-2) & 2(-1) \\ 2(3) & 2(-1) & 2(4) \\ 2(-2) & 2(-3) & 2(2) \end{bmatrix} = \begin{bmatrix} 2 & -4 & -2 \\ 6 & -2 & 8 \\ -4 & -6 & 4 \end{bmatrix}$$

(b)

$$\frac{1}{2}\begin{bmatrix} 2 & -4 \\ 6 & 2 \end{bmatrix} = \begin{bmatrix} (\frac{1}{2})(2) & (\frac{1}{2})(-4) \\ (\frac{1}{2})(6) & (\frac{1}{2})(2) \end{bmatrix} = \begin{bmatrix} 1 & -2 \\ 3 & 1 \end{bmatrix}$$

PROBLEM 1-12 Given $A = \begin{bmatrix} 2 & -3 \\ 3 & 5 \end{bmatrix}$, $B = \begin{bmatrix} 4 & 0 \\ -1 & 2 \end{bmatrix}$, $C = \begin{bmatrix} 1 & -2 \\ -4 & 3 \end{bmatrix}$, $a = 3$, and $b = -2$, verify the following: **(a)** $A + B = B + A$. **(b)** $(A + B) + C = A + (B + C)$. **(c)** $a(A + B) = aA + aB$. **(d)** $(a + b)A = aA + bA$. **(e)** $a(bA) = (ab)A$.

Solution

(a)

$$A + B = \begin{bmatrix} 2 & -3 \\ 3 & 5 \end{bmatrix} + \begin{bmatrix} 4 & 0 \\ -1 & 2 \end{bmatrix} = \begin{bmatrix} 6 & -3 \\ 2 & 7 \end{bmatrix}$$

$$B + A = \begin{bmatrix} 4 & 0 \\ -1 & 2 \end{bmatrix} + \begin{bmatrix} 2 & -3 \\ 3 & 5 \end{bmatrix} = \begin{bmatrix} 6 & -3 \\ 2 & 7 \end{bmatrix}$$

$$\therefore \quad A + B = B + A$$

(b)

$$(A + B) + C = A + (B + C)$$

$$(A + B) + C = \left(\begin{bmatrix} 2 & -3 \\ 3 & 5 \end{bmatrix} + \begin{bmatrix} 4 & 0 \\ -1 & 2 \end{bmatrix}\right) + \begin{bmatrix} 1 & -2 \\ -4 & 3 \end{bmatrix} = \begin{bmatrix} 6 & -3 \\ 2 & 7 \end{bmatrix} + \begin{bmatrix} 1 & -2 \\ -4 & 3 \end{bmatrix} = \begin{bmatrix} 7 & -5 \\ -2 & 10 \end{bmatrix}$$

$$A + (B + C) = \begin{bmatrix} 2 & -3 \\ 3 & 5 \end{bmatrix} + \left(\begin{bmatrix} 4 & 0 \\ -1 & 2 \end{bmatrix} + \begin{bmatrix} 1 & -2 \\ -4 & 3 \end{bmatrix}\right) = \begin{bmatrix} 2 & -3 \\ 3 & 5 \end{bmatrix} + \begin{bmatrix} 5 & -2 \\ -5 & 5 \end{bmatrix} = \begin{bmatrix} 7 & -5 \\ -2 & 10 \end{bmatrix}$$

$$\therefore \quad (A + B) + C = A + (B + C)$$

(c)

$$a(A + B) = aA + aB$$

$$a(A + B) = 3\left(\begin{bmatrix} 2 & -3 \\ 3 & 5 \end{bmatrix} + \begin{bmatrix} 4 & 0 \\ -1 & 2 \end{bmatrix}\right) = 3\begin{bmatrix} 6 & -3 \\ 2 & 7 \end{bmatrix} = \begin{bmatrix} 18 & -9 \\ 6 & 21 \end{bmatrix}$$

$$aA + aB = 3\begin{bmatrix} 2 & -3 \\ 3 & 5 \end{bmatrix} + 3\begin{bmatrix} 4 & 0 \\ -1 & 2 \end{bmatrix} = \begin{bmatrix} 6 & -9 \\ 9 & 15 \end{bmatrix} + \begin{bmatrix} 12 & 0 \\ -3 & 6 \end{bmatrix} = \begin{bmatrix} 18 & -9 \\ 6 & 21 \end{bmatrix}$$

$$\therefore \quad a(A + B) = aA + aB$$

(d)

$$(a + b)A = aA + bA$$

$$(a + b)A = (3 - 2)\begin{bmatrix} 2 & -3 \\ 3 & 5 \end{bmatrix} = 1\begin{bmatrix} 2 & -3 \\ 3 & 5 \end{bmatrix} = \begin{bmatrix} 2 & -3 \\ 3 & 5 \end{bmatrix}$$

$$aA + bA = 3\begin{bmatrix} 2 & -3 \\ 3 & 5 \end{bmatrix} + (-2)\begin{bmatrix} 2 & -3 \\ 3 & 5 \end{bmatrix} = \begin{bmatrix} 6 & -9 \\ 9 & 15 \end{bmatrix} + \begin{bmatrix} -4 & 6 \\ -6 & -10 \end{bmatrix} = \begin{bmatrix} 2 & -3 \\ 3 & 5 \end{bmatrix}$$

$$\therefore \quad (a + b)A = aA + bA$$

(e)

$$a(bA) = (ab)A$$

$$a(bA) = 3\left\{(-2)\begin{bmatrix} 2 & -3 \\ 3 & 5 \end{bmatrix}\right\} = 3\begin{bmatrix} -4 & 6 \\ -6 & -10 \end{bmatrix} = \begin{bmatrix} -12 & 18 \\ -18 & -30 \end{bmatrix}$$

$$(ab)A = [(3)(-2)]\begin{bmatrix} 2 & -3 \\ 3 & 5 \end{bmatrix} = -6\begin{bmatrix} 2 & -3 \\ 3 & 5 \end{bmatrix} = \begin{bmatrix} -12 & 18 \\ -18 & -30 \end{bmatrix}$$

$$\therefore \quad a(bA) = (ab)A$$

PROBLEM 1-13 If A is a 3×4, B a 4×5, and C a 5×3 matrix, find the size of each product where it is defined: **(a)** AB, **(b)** BA, **(c)** CA, **(d)** BC, and **(e)** $(AB)C$.

Solution

(a) $\underset{3\times 4}{A}\ \underset{4\times 5}{B}$ Product is defined and is 3×5.

(b) $\underset{4\times 5}{B}\ \underset{3\times 4}{A}$ $5 \neq 3$. Product is not defined.

(c) $\underset{5\times 3}{C}\ \underset{3\times 4}{A}$ Product is defined and is 5×4.

(d) $\underset{4\times 5}{B}\ \underset{5\times 3}{C}$ Product is defined and is 4×3.

(e) $(\underset{3\times 4}{A}\ \underset{4\times 5}{B})\ \underset{5\times 3}{C}$ Product is defined and is 3×3.

PROBLEM 1-14 If $A = \begin{bmatrix} 1 & 0 & 2 \\ -2 & 1 & 3 \end{bmatrix}$ and $B = \begin{bmatrix} 3 & 2 \\ 4 & 1 \\ -1 & -3 \end{bmatrix}$, find AB and BA.

Solution

$$AB = \begin{bmatrix} 1 & 0 & 2 \\ -2 & 1 & 3 \end{bmatrix}\begin{bmatrix} 3 & 2 \\ 4 & 1 \\ -1 & -3 \end{bmatrix} = \begin{bmatrix} 1(3) + 0(4) + 2(-1) & 1(2) + 0(1) + 2(-3) \\ -2(3) + 1(4) + 3(-1) & -2(2) + 1(1) + 3(-3) \end{bmatrix} = \begin{bmatrix} 1 & -4 \\ -5 & -12 \end{bmatrix}$$

$$BA = \begin{bmatrix} 3 & 2 \\ 4 & 1 \\ -1 & -3 \end{bmatrix}\begin{bmatrix} 1 & 0 & 2 \\ -2 & 1 & 3 \end{bmatrix} = \begin{bmatrix} 3(1) + 2(-2) & 3(0) + 2(1) & 3(2) + 2(3) \\ 4(1) + 1(-2) & 4(0) + 1(1) & 4(2) + 1(3) \\ -1(1) + (-3)(-2) & -1(0) + (-3)(1) & -1(2) + (-3)(3) \end{bmatrix}$$

$$= \begin{bmatrix} -1 & 2 & 12 \\ 2 & 1 & 11 \\ 5 & -3 & -11 \end{bmatrix}$$

PROBLEM 1-15 If $A = \begin{bmatrix} 5 & 1 \\ -2 & 4 \end{bmatrix}$ and $B = \begin{bmatrix} 3 & -4 \\ -2 & 1 \end{bmatrix}$, find AB and BA.

Solution

$$AB = \begin{bmatrix} 5 & 1 \\ -2 & 4 \end{bmatrix}\begin{bmatrix} 3 & -4 \\ -2 & 1 \end{bmatrix} = \begin{bmatrix} 5(3) + 1(-2) & 5(-4) + 1(1) \\ -2(3) + 4(-2) & -2(-4) + 4(1) \end{bmatrix} = \begin{bmatrix} 13 & -19 \\ -14 & 12 \end{bmatrix}$$

$$BA = \begin{bmatrix} 3 & -4 \\ -2 & 1 \end{bmatrix}\begin{bmatrix} 5 & 1 \\ -2 & 4 \end{bmatrix} = \begin{bmatrix} 3(5) + (-4)(-2) & 3(1) + (-4)(4) \\ -2(5) + 1(-2) & -2(1) + 1(4) \end{bmatrix} = \begin{bmatrix} 23 & -13 \\ -12 & 2 \end{bmatrix}$$

This verifies that multiplication of matrices is not commutative.

PROBLEM 1-16 Given $A = \begin{bmatrix} 3 & -1 \\ -2 & 4 \end{bmatrix}$, $B = \begin{bmatrix} 2 & 3 \\ 1 & 2 \end{bmatrix}$, $C = \begin{bmatrix} 0 & 2 \\ 3 & 2 \end{bmatrix}$, and $a = 3$, verify the following: **(a)** $A(BC) = (AB)C$. **(b)** $A(B + C) = AB + AC$. **(c)** $(A + B)C = AC + BC$. **(d)** $a(AB) = (aA)B = A(aB)$.

Solution

(a)

$$A(BC) = \begin{bmatrix} 3 & -1 \\ -2 & 4 \end{bmatrix}\left(\begin{bmatrix} 2 & 3 \\ 1 & 2 \end{bmatrix}\begin{bmatrix} 0 & 2 \\ 3 & 2 \end{bmatrix}\right) = \begin{bmatrix} 3 & -1 \\ -2 & 4 \end{bmatrix}\begin{bmatrix} 9 & 10 \\ 6 & 6 \end{bmatrix}$$

$$= \begin{bmatrix} 21 & 24 \\ 6 & 4 \end{bmatrix}$$

$$(AB)C = \left(\begin{bmatrix} 3 & -1 \\ -2 & 4 \end{bmatrix}\begin{bmatrix} 2 & 3 \\ 1 & 2 \end{bmatrix}\right)\begin{bmatrix} 0 & 2 \\ 3 & 2 \end{bmatrix} = \begin{bmatrix} 5 & 7 \\ 0 & 2 \end{bmatrix}\begin{bmatrix} 0 & 2 \\ 3 & 2 \end{bmatrix}$$

$$= \begin{bmatrix} 21 & 24 \\ 6 & 4 \end{bmatrix} \quad \therefore \quad A(BC) = (AB)C$$

(b)

$$A(B + C) = \begin{bmatrix} 3 & -1 \\ -2 & 4 \end{bmatrix}\left(\begin{bmatrix} 2 & 3 \\ 1 & 2 \end{bmatrix} + \begin{bmatrix} 0 & 2 \\ 3 & 2 \end{bmatrix}\right) = \begin{bmatrix} 3 & -1 \\ -2 & 4 \end{bmatrix}\begin{bmatrix} 2 & 5 \\ 4 & 4 \end{bmatrix}$$

$$= \begin{bmatrix} 2 & 11 \\ 12 & 6 \end{bmatrix}$$

$$AB + AC = \begin{bmatrix} 3 & -1 \\ -2 & 4 \end{bmatrix}\begin{bmatrix} 2 & 3 \\ 1 & 2 \end{bmatrix} + \begin{bmatrix} 3 & -1 \\ -2 & 4 \end{bmatrix}\begin{bmatrix} 0 & 2 \\ 3 & 2 \end{bmatrix} = \begin{bmatrix} 5 & 7 \\ 0 & 2 \end{bmatrix} + \begin{bmatrix} -3 & 4 \\ 12 & 4 \end{bmatrix}$$

$$= \begin{bmatrix} 2 & 11 \\ 12 & 6 \end{bmatrix} \quad \therefore \quad A(B + C) = AB + AC$$

(c)

$$(A + B)C = \left(\begin{bmatrix} 3 & -1 \\ -2 & 4 \end{bmatrix} + \begin{bmatrix} 2 & 3 \\ 1 & 2 \end{bmatrix}\right)\begin{bmatrix} 0 & 2 \\ 3 & 2 \end{bmatrix} = \begin{bmatrix} 5 & 2 \\ -1 & 6 \end{bmatrix}\begin{bmatrix} 0 & 2 \\ 3 & 2 \end{bmatrix} = \begin{bmatrix} 6 & 14 \\ 18 & 10 \end{bmatrix}$$

$$AC + BC = \begin{bmatrix} 3 & -1 \\ -2 & 4 \end{bmatrix}\begin{bmatrix} 0 & 2 \\ 3 & 2 \end{bmatrix} + \begin{bmatrix} 2 & 3 \\ 1 & 2 \end{bmatrix}\begin{bmatrix} 0 & 2 \\ 3 & 2 \end{bmatrix} = \begin{bmatrix} -3 & 4 \\ 12 & 4 \end{bmatrix} + \begin{bmatrix} 9 & 10 \\ 6 & 6 \end{bmatrix} = \begin{bmatrix} 6 & 14 \\ 18 & 10 \end{bmatrix}$$

$$\therefore \quad (A + B)C = AC + BC$$

(d)

$$a(AB) = 3\left(\begin{bmatrix} 3 & -1 \\ -2 & 4 \end{bmatrix}\begin{bmatrix} 2 & 3 \\ 1 & 2 \end{bmatrix}\right) = 3\begin{bmatrix} 5 & 7 \\ 0 & 2 \end{bmatrix} = \begin{bmatrix} 15 & 21 \\ 0 & 6 \end{bmatrix}$$

$$(aA)B = \left(3\begin{bmatrix} 3 & -1 \\ -2 & 4 \end{bmatrix}\right)\begin{bmatrix} 2 & 3 \\ 1 & 2 \end{bmatrix} = \begin{bmatrix} 9 & -3 \\ -6 & 12 \end{bmatrix}\begin{bmatrix} 2 & 3 \\ 1 & 2 \end{bmatrix} = \begin{bmatrix} 15 & 21 \\ 0 & 6 \end{bmatrix}$$

$$A(aB) = \begin{bmatrix} 3 & -1 \\ -2 & 4 \end{bmatrix}\left(3\begin{bmatrix} 2 & 3 \\ 1 & 2 \end{bmatrix}\right) = \begin{bmatrix} 3 & -1 \\ -2 & 4 \end{bmatrix}\begin{bmatrix} 6 & 9 \\ 3 & 6 \end{bmatrix} = \begin{bmatrix} 15 & 21 \\ 0 & 6 \end{bmatrix}$$

$$\therefore \quad a(AB) = (aA)B = A(aB)$$

PROBLEM 1-17 Given $A = \begin{bmatrix} -1 & 2 & 1 \\ 0 & 3 & -2 \\ 1 & 4 & -3 \end{bmatrix}$ and $B = \begin{bmatrix} 2 & -3 \\ -1 & 2 \\ 5 & 2 \end{bmatrix}$, find **(a)** A^T, **(b)** B^T, and **(c)** $(A^T)^T$.

Solution

(a) $A^T = \begin{bmatrix} -1 & 0 & 1 \\ 2 & 3 & 4 \\ 1 & -2 & -3 \end{bmatrix}$ **(b)** $B^T = \begin{bmatrix} 2 & -1 & 5 \\ -3 & 2 & 2 \end{bmatrix}$

(c) $(A^T)^T = \begin{bmatrix} -1 & 2 & 1 \\ 0 & 3 & -2 \\ 1 & 4 & -3 \end{bmatrix}$

This shows $(A^T)^T = A$.

PROBLEM 1-18 Given $A = \begin{bmatrix} 3 & -2 \\ -1 & 4 \end{bmatrix}$, $B = \begin{bmatrix} 2 & 3 \\ 1 & 2 \end{bmatrix}$, and $a = 2$, verify each of the following: **(a)** $(A^T)^T = A$. **(b)** $(A + B)^T = A^T + B^T$. **(c)** $(AB)^T = B^TA^T$. **(d)** $(aA)^T = aA^T$.

Solution

(a)
$$(A^T)^T = \begin{bmatrix} 3 & -1 \\ -2 & 4 \end{bmatrix}^T = \begin{bmatrix} 3 & -2 \\ -1 & 4 \end{bmatrix} = A$$
$$\therefore \quad (A^T)^T = A$$

(b)
$$(A + B)^T = \left(\begin{bmatrix} 3 & -2 \\ -1 & 4 \end{bmatrix} + \begin{bmatrix} 2 & 3 \\ 1 & 2 \end{bmatrix}\right)^T = \begin{bmatrix} 5 & 1 \\ 0 & 6 \end{bmatrix}^T = \begin{bmatrix} 5 & 0 \\ 1 & 6 \end{bmatrix}$$
$$A^T + B^T = \begin{bmatrix} 3 & -2 \\ -1 & 4 \end{bmatrix}^T + \begin{bmatrix} 2 & 3 \\ 1 & 2 \end{bmatrix}^T = \begin{bmatrix} 3 & -1 \\ -2 & 4 \end{bmatrix} + \begin{bmatrix} 2 & 1 \\ 3 & 2 \end{bmatrix} = \begin{bmatrix} 5 & 0 \\ 1 & 6 \end{bmatrix}$$
$$\therefore \quad (A + B)^T = A^T + B^T$$

(c)
$$(AB)^T = \left(\begin{bmatrix} 3 & -2 \\ -1 & 4 \end{bmatrix}\begin{bmatrix} 2 & 3 \\ 1 & 2 \end{bmatrix}\right)^T = \begin{bmatrix} 4 & 5 \\ 2 & 5 \end{bmatrix}^T = \begin{bmatrix} 4 & 2 \\ 5 & 5 \end{bmatrix}$$
$$B^T + A^T = \begin{bmatrix} 2 & 3 \\ 1 & 2 \end{bmatrix}^T \begin{bmatrix} 3 & -2 \\ -1 & 4 \end{bmatrix}^T = \begin{bmatrix} 2 & 1 \\ 3 & 2 \end{bmatrix}\begin{bmatrix} 3 & -1 \\ -2 & 4 \end{bmatrix} = \begin{bmatrix} 4 & 2 \\ 5 & 5 \end{bmatrix}$$
$$\therefore \quad (AB)^T = B^T A^T$$

(d)
$$(aA)^T = \left(2\begin{bmatrix} 3 & -2 \\ -1 & 4 \end{bmatrix}\right)^T = \begin{bmatrix} 6 & -4 \\ -2 & 8 \end{bmatrix}^T = \begin{bmatrix} 6 & -2 \\ -4 & 8 \end{bmatrix}$$
$$aA^T = 2\begin{bmatrix} 3 & -2 \\ -1 & 4 \end{bmatrix}^T = 2\begin{bmatrix} 3 & -1 \\ -2 & 4 \end{bmatrix} = \begin{bmatrix} 6 & -2 \\ -4 & 8 \end{bmatrix}$$
$$\therefore \quad (aA)^T = aA^T$$

PROBLEM 1-19 Determine which of the following matrices is symmetric:

(a) $A = \begin{bmatrix} 2 & 3 \\ 3 & 1 \end{bmatrix}$ **(b)** $B = \begin{bmatrix} 2 & -3 & 1 \\ 3 & -4 & 1 \\ -1 & 1 & 6 \end{bmatrix}$ **(c)** $C = \begin{bmatrix} 5 & 1 & 2 \\ 1 & 4 & -3 \\ 2 & -3 & 1 \end{bmatrix}$

(d) $D = \begin{bmatrix} 4 & -2 & 0 & 1 \\ -2 & 1 & 3 & 0 \\ 0 & 3 & -2 & 5 \\ 1 & 0 & 5 & 2 \end{bmatrix}$

Solution

(a) $A^T = \begin{bmatrix} 2 & 3 \\ 3 & 1 \end{bmatrix}$ Since $A^T = A$, A is symmetric.

(b) $B^T = \begin{bmatrix} 2 & 3 & -1 \\ -3 & -4 & 1 \\ 1 & 1 & 6 \end{bmatrix}$ $B^T \neq B$. B is nonsymmetric.

(c) $C^T = \begin{bmatrix} 5 & 1 & 2 \\ 1 & 4 & -3 \\ 2 & -3 & 1 \end{bmatrix}$ $C^T = C$. C is symmetric.

(d)
$$D^T = \begin{bmatrix} 4 & -2 & 0 & 1 \\ -2 & 1 & 3 & 0 \\ 0 & 3 & -2 & 5 \\ 1 & 0 & 5 & 2 \end{bmatrix} \qquad D^T = D. \ D \text{ is symmetric.}$$

PROBLEM 1-20 Determine which of the following matrices is skew-symmetric: **(a)** $A = \begin{bmatrix} 0 & -1 & 2 \\ 1 & 0 & -4 \\ -2 & 4 & 0 \end{bmatrix}$, **(b)** $B = \begin{bmatrix} 0 & 0 \\ 0 & 0 \end{bmatrix}$, and **(c)** $C = \begin{bmatrix} 0 & 0 & 1 \\ 0 & 0 & 0 \\ -1 & 0 & 1 \end{bmatrix}$.

Solution

(a)
$$A^T = \begin{bmatrix} 0 & 1 & -2 \\ -1 & 0 & 4 \\ 2 & -4 & 0 \end{bmatrix} \qquad A^T = -A. \ A \text{ is skew-symmetric.}$$

(b)
$$B^T = \begin{bmatrix} 0 & 0 \\ 0 & 0 \end{bmatrix} \qquad B^T = -B. \ B \text{ is skew-symmetric.}$$

(c)
$$C^T = \begin{bmatrix} 0 & 0 & -1 \\ 0 & 0 & 0 \\ 1 & 0 & 1 \end{bmatrix} \qquad C^T \neq -C. \ C \text{ is not skew-symmetric.}$$

PROBLEM 1-21 Given $A = \begin{bmatrix} 1 & 3 & -2 \\ 4 & -2 & 1 \end{bmatrix}$, find AA^T and A^TA.

Solution

$$AA^T = \begin{bmatrix} 1 & 3 & -2 \\ 4 & -2 & 1 \end{bmatrix} \begin{bmatrix} 1 & 4 \\ 3 & -2 \\ -2 & 1 \end{bmatrix} = \begin{bmatrix} 14 & -4 \\ -4 & 21 \end{bmatrix}$$

$$A^TA = \begin{bmatrix} 1 & 4 \\ 3 & -2 \\ -2 & 1 \end{bmatrix} \begin{bmatrix} 1 & 3 & -2 \\ 4 & -2 & 1 \end{bmatrix} = \begin{bmatrix} 17 & -5 & 2 \\ -5 & 13 & -8 \\ 2 & -8 & 5 \end{bmatrix}$$

Observe that both AA^T and A^TA are symmetric but not equal.

PROBLEM 1-22 Solve for x and y.

$$\begin{bmatrix} 2 & -1 \\ 1 & 3 \end{bmatrix} \begin{bmatrix} x \\ y \end{bmatrix} = \begin{bmatrix} 5 \\ 7 \end{bmatrix}$$

Solution

$$\begin{bmatrix} 2x - y \\ x + y \end{bmatrix} = \begin{bmatrix} 5 \\ 7 \end{bmatrix}$$

This is true only if

$$2x - y = 5$$
$$x + y = 7$$

Adding the two equations gives $3x = 12$ or $x = 4$. Hence, $2(4) - y = 5$ or $y = 3$.

Supplementary Problems

PROBLEM 1-23 Find the solution set of each of the following linear equations: **(a)** $2x - y = 6$ and **(b)** $x + 2y - 2z = 4$.

PROBLEM 1-24 Solve by the elimination process:

(a) $a_{11}x + a_{12}y = b_1$
$a_{21}x + a_{22}y = b_2$

(b) $x + y - 2z = -3$
$2x - y - z = 0$
$x + 2y + 3z = 13$

PROBLEM 1-25 Determine whether each of the following systems is consistent or inconsistent. If consistent, find the solution.

(a) $3x - 2y = 5$
$3x + y = -4$

(b) $1/2x + 1/3y = 2$
$3x + 2y = 4$

(c) $3x - 2y = 4$
$6x - 4y = 8$

Use the following matrices in Problems 1-26 through 1-31:

$$A = \begin{bmatrix} 1 & -2 & 3 \\ 0 & 3 & 4 \end{bmatrix} \quad B = \begin{bmatrix} -2 & 1 \\ 4 & -3 \end{bmatrix} \quad C = \begin{bmatrix} 3 \\ -2 \\ 4 \end{bmatrix} \quad D = \begin{bmatrix} 3 & -2 & -1 \\ -2 & -2 & 3 \\ -1 & 3 & 2 \end{bmatrix}$$

$$E = [3 \quad -2 \quad 4] \quad F = [-3] \quad G = \begin{bmatrix} 3 & 0 \\ 0 & -4 \end{bmatrix} \quad H = \begin{bmatrix} -1 & 2 \\ 2 & -1 \\ -4 & 3 \end{bmatrix}$$

$$J = \begin{bmatrix} 1 & 0 & 0 \\ 0 & -2 & 0 \\ 0 & 0 & 5 \end{bmatrix} \quad K = \begin{bmatrix} 1 & -2 & 3 & -1 \\ 2 & 4 & -1 & 5 \\ 0 & 2 & 1 & 3 \end{bmatrix}$$

PROBLEM 1-26 Give the size of each of the following and, if the matrix is square, give its order: **(a)** A, **(b)** B, **(c)** D, **(d)** E, **(e)** F, **(f)** H, **(g)** J, **(h)** K, **(i)** AK.

PROBLEM 1-27 Find the following, if possible: **(a)** $B + G$, **(b)** AC, **(c)** AD, **(d)** $BG + GB$, **(e)** $2D - 3J$, **(f)** BB, **(g)** HJ.

PROBLEM 1-28 Find the following, if possible: **(a)** $(BA)D$, **(b)** $A(HB)$ and $(AH)B$, **(c)** $CF + 2E^T$, **(d)** $(DJ)C$, **(e)** $C(DJ)$, **(f)** $H(B + G)$.

PROBLEM 1-29 Find the following, if possible: **(a)** A^T, **(b)** D^T, **(c)** F^T, **(d)** G^T, **(e)** J^T, **(f)** $B + B^T$.

PROBLEM 1-30 Compute the following, if possible: **(a)** $(D + J)^T$ and $D^T + J^T$, **(b)** $(3B + 2G)^T$, **(c)** $(FE)^T + C$, **(d)** $(AD)^T$ and D^TA^T, **(e)** $H^T + A$, **(f)** H^TA and AH^T, **(g)** $(AK)^T$.

PROBLEM 1-31 Which of the given matrices are symmetric?

Answers to Supplementary Problems

1-23 **(a)** $(k, 2k - 6)$ **(b)** $[k, m, (k + 2m - 4)/2]$

1-24 **(a)** $x = (a_{22}b_1 - a_{12}b_2)/(a_{11}a_{22} - a_{12}a_{21})$, $y = (a_{11}b_2 - a_{21}b_1)/(a_{11}a_{22} - a_{12}a_{21})$, where $a_{11}a_{22} - a_{12}a_{21} \neq 0$
(b) $x = 2, y = 1, z = 3$

1-25 **(a)** Consistent. $x = -\frac{1}{3}, y = -3$ or $(-\frac{1}{3}, 3)$
(b) Inconsistent: parallel lines
(c) Consistent. Infinite number of solutions of form $[k, (3k - 4)/2]$.

1-26 **(a)** 2×3
(b) 2×2, order 2
(c) 3×3, order 3
(d) 1×3
(e) 1×1, order 1
(f) 3×2
(g) 3×3, order 3
(h) 3×4
(i) 2×4

1-27 **(a)** $\begin{bmatrix} 1 & 1 \\ 4 & -7 \end{bmatrix}$
(b) $\begin{bmatrix} 19 \\ 10 \end{bmatrix}$
(c) $\begin{bmatrix} 4 & 11 & -1 \\ -10 & 6 & 17 \end{bmatrix}$
(d) $\begin{bmatrix} -12 & -1 \\ -4 & 24 \end{bmatrix}$
(e) $\begin{bmatrix} 3 & -4 & -2 \\ -4 & 2 & 6 \\ -2 & 6 & -11 \end{bmatrix}$
(f) $\begin{bmatrix} 8 & -5 \\ -20 & 13 \end{bmatrix}$
(g) Impossible

1-28 **(a)** $\begin{bmatrix} -18 & -16 & 19 \\ 46 & 26 & -55 \end{bmatrix}$
(b) $\begin{bmatrix} 86 & -56 \\ 56 & -37 \end{bmatrix}$
(c) $\begin{bmatrix} -3 \\ 2 \\ -4 \end{bmatrix}$
(d) $\begin{bmatrix} -19 \\ 46 \\ 49 \end{bmatrix}$
(e) Impossible
(f) $\begin{bmatrix} 7 & -15 \\ -2 & 9 \\ 8 & -25 \end{bmatrix}$

1-29 **(a)** $\begin{bmatrix} 1 & 0 \\ -2 & 3 \\ 3 & 4 \end{bmatrix}$
(b) $\begin{bmatrix} 3 & -2 & -1 \\ -2 & -2 & 3 \\ -1 & 3 & 2 \end{bmatrix}$
(c) [-3]
(d) $\begin{bmatrix} 3 & 0 \\ 0 & -4 \end{bmatrix}$
(e) $\begin{bmatrix} 1 & 0 & 0 \\ 0 & -2 & 0 \\ 0 & 0 & 5 \end{bmatrix}$
(f) $\begin{bmatrix} -4 & 5 \\ 5 & -6 \end{bmatrix}$

1-30 **(a)** $\begin{bmatrix} 4 & -2 & -1 \\ -2 & -4 & 3 \\ -1 & 3 & 7 \end{bmatrix}$ and $\begin{bmatrix} 4 & -2 & -1 \\ -2 & -4 & 3 \\ -1 & 3 & 7 \end{bmatrix}$
(b) $\begin{bmatrix} 0 & 12 \\ 3 & -17 \end{bmatrix}$
(c) $\begin{bmatrix} -6 \\ 4 \\ -8 \end{bmatrix}$
(d) $\begin{bmatrix} 4 & -10 \\ 11 & 6 \\ -1 & 17 \end{bmatrix}$ and $\begin{bmatrix} 4 & -10 \\ 11 & 6 \\ -1 & 17 \end{bmatrix}$
(e) $\begin{bmatrix} 0 & 0 & -1 \\ 2 & 2 & 7 \end{bmatrix}$
(f) Impossible and impossible
(g) $\begin{bmatrix} -3 & 6 \\ -4 & 20 \\ 8 & 1 \\ -2 & 27 \end{bmatrix}$

1-31 *D, F, G, J*

2 MATRIX SOLUTIONS OF LINEAR SYSTEMS

THIS CHAPTER IS ABOUT

- ☑ Matrix Representation of a Linear System
- ☑ Row Equivalent Matrices
- ☑ Gaussian and Gauss–Jordan Elimination
- ☑ Inverse of a Matrix
- ☑ Properties of Inverse Matrices
- ☑ Homogeneous Systems of Linear Equations
- ☑ Invertible Matrix Method
- ☑ Equivalent Statements

In Chapter 1 we defined a matrix, developed the operations on matrices, and verified some properties of matrices. We are now in a position to use matrices in solving systems of linear equations.

2-1. Matrix Representation of a Linear System

Nomenclature

Consider the following system of m linear equations in n unknowns where $x_1, x_2, \ldots, x_n$ are the unknowns and the a's and b's denote constants:

$$\begin{array}{ccccccccc} a_{11}x_1 & + & a_{12}x_2 & + & \cdots & + & a_{1n}x_n & = & b_1 \\ a_{21}x_1 & + & a_{22}x_2 & + & \cdots & + & a_{2n}x_n & = & b_2 \\ \vdots & & \vdots & & \cdots & & \vdots & & \vdots \\ a_{m1}x_1 & + & a_{m2}x_2 & + & \cdots & + & a_{mn}x_n & = & b_m \end{array}$$

The matrix

$$A = \begin{bmatrix} a_{11} & a_{12} & \cdots & a_{1n} \\ a_{21} & a_{22} & \cdots & a_{2n} \\ \vdots & & \cdots & \vdots \\ a_{m1} & a_{m2} & \cdots & a_{mn} \end{bmatrix}$$

whose entries are the coefficients of the x's is called the **coefficient matrix**; the matrix

$$X = \begin{bmatrix} x_1 \\ x_2 \\ \vdots \\ x_m \end{bmatrix}$$

is the **column matrix of variables**; the matrix

$$B = \begin{bmatrix} b_1 \\ b_2 \\ \vdots \\ b_m \end{bmatrix}$$

is the **column matrix of constants**; and the matrix

$$[A|B] = \begin{bmatrix} a_{11} & a_{12} & \cdots & a_{1n} & b_1 \\ a_{21} & a_{22} & \cdots & a_{2n} & b_2 \\ \vdots & & \cdots & \cdots & \\ a_{m1} & a_{m2} & \cdots & a_{mn} & b_m \end{bmatrix}$$

is called the **augmented matrix**. Sometimes a solid or dotted line separates the coefficient matrix from the column matrix. $AX = B$ is called the **matrix form** of the system.

note: Problem 1-22 is stated in matrix form.

EXAMPLE 2-1: Given the linear system

$$\begin{aligned} x + 3y - z &= 5 \\ 2x - y + z &= 2 \\ x + 2y + 2z &= 3 \end{aligned}$$

Find the following:

(a) the coefficient matrix
(b) the augmented matrix
(c) the matrix form of the system.

Solution

(a) $\begin{bmatrix} 1 & 3 & -1 \\ 2 & -1 & 1 \\ 1 & 2 & 2 \end{bmatrix}$ **(b)** $\begin{bmatrix} 1 & 3 & -1 & 5 \\ 2 & -1 & 1 & 2 \\ 1 & 2 & 2 & 3 \end{bmatrix}$ **(c)** $\begin{bmatrix} 1 & 3 & -1 \\ 2 & -1 & 1 \\ 1 & 2 & 2 \end{bmatrix} \begin{bmatrix} x \\ y \\ z \end{bmatrix} = \begin{bmatrix} 5 \\ 2 \\ 3 \end{bmatrix}$

EXAMPLE 2-2: Find a linear system whose augmented matrix is:

(a) $\begin{bmatrix} 2 & 3 & 4 \\ 4 & -1 & 2 \end{bmatrix}$ **(b)** $\begin{bmatrix} 1 & 2 & 3 & 4 & -3 \\ 4 & -2 & 1 & 0 & 4 \end{bmatrix}$ **(c)** $\begin{bmatrix} 1 & 0 & 0 & 0 & 5 \\ 0 & 1 & 0 & 0 & 2 \\ 0 & 0 & 1 & 0 & 3 \\ 0 & 0 & 0 & 1 & -4 \end{bmatrix}$

Solution

(a) $\begin{aligned} 2x + 3y &= 4 \\ 4x - y &= 2 \end{aligned}$

(b) $\begin{aligned} x_1 + 2x_2 + 3x_3 + 4x_4 &= -3 \\ 4x_1 - 2x_2 + x_3 &= 4 \end{aligned}$

(c) $\begin{aligned} x_1 &= 5 \\ x_2 &= 2 \\ x_3 &= 3 \\ x_4 &= -4 \end{aligned}$

EXAMPLE 2-3: Given the linear system

$$x - 3z = 4$$
$$-2y + z = 3$$
$$2x - 3y = 5$$

find

(a) the coefficient matrix
(b) the augmented matrix.

Solution: Rewrite the system as

$$x + 0y - 3z = 4$$
$$0x - 2y + z = 3$$
$$2x - 3y + 0z = 5$$

Then

(a) $\begin{bmatrix} 1 & 0 & -3 \\ 0 & -2 & 1 \\ 2 & -3 & 0 \end{bmatrix}$ (b) $\begin{bmatrix} 1 & 0 & -3 & 4 \\ 0 & -2 & 3 & 1 \\ 2 & -3 & 0 & 5 \end{bmatrix}$

2-2. Row Equivalent Matrices

A. Definition of elementary row operation

An elementary row operation on a matrix A is any one of the following.

(a) Multiply any row of A by a nonzero constant.
(b) Interchange two rows of A.
(c) Add a constant multiple of one row to another row.

We will use the following notation for these operations

Row operation	Notation
Multiply row i by k	$R_i \to kR_i$
Interchange rows i and j	$R_i \leftrightarrow R_j$
Add k times row j to row i	$R_i \to R_i + kR_j$

You may think of $R_i \to kR_i$ as meaning, "replace row i by k times row i," and $R_i \to R_i + kR_j$ as meaning, "replace row i by row i plus k times row j."

B. Definition of row equivalent matrices

An $m \times n$ matrix A is **row equivalent** to an $m \times n$ matrix B if B can be obtained from A by applying a finite sequence of elementary row operations to A.

If A is the augmented matrix of a system of linear equations, then any matrix B that is row equivalent to A is the augmented matrix of an equivalent system of linear equations. This means that the two systems have exactly the same solution set.

EXAMPLE 2-4: Use elementary row operations to change the augmented matrix of the following system of equations to five different equivalent matrices, each of which will be the augmented matrix of an equivalent system of equations. (Compare this with Example 1-3.)

$$x + y + z = 2$$
$$x + 3y + 2z = 1$$
$$2x + y - z = 2$$

Solution

augmented matrix — **system of equations**

$$A = \begin{bmatrix} 1 & 1 & 1 & 2 \\ 1 & 3 & 2 & 1 \\ 2 & 1 & -1 & 2 \end{bmatrix} \qquad \text{given system}$$

$$A \xrightarrow{R_1 \to 3R_1} \begin{bmatrix} 3 & 3 & 3 & 6 \\ 1 & 3 & 2 & 1 \\ 2 & 1 & -1 & 2 \end{bmatrix} \qquad \begin{aligned} 3x + 3y + 3z &= 6 \\ x + 3y + 2z &= 1 \\ 2x + y - z &= 2 \end{aligned}$$

$$A \xrightarrow[R_2 \leftrightarrow R_3]{} \begin{bmatrix} 1 & 1 & 1 & 2 \\ 2 & 1 & -1 & 2 \\ 1 & 3 & 2 & 1 \end{bmatrix} \qquad \begin{aligned} x + y + z &= 2 \\ 2x + y - z &= 2 \\ x + 3y + 2z &= 1 \end{aligned}$$

$$A \xrightarrow{R_2 \to R_2 + 2R_1} \begin{bmatrix} 1 & 1 & 1 & 2 \\ 3 & 5 & 4 & 5 \\ 2 & 1 & -1 & 2 \end{bmatrix} = B \qquad \begin{aligned} x + y + z &= 2 \\ 3x + 5y + 4z &= 5 \\ 2x + y - z &= 2 \end{aligned}$$

We do not have to begin with A each time. For the next two equivalent matrices, let us begin with the last one obtained above, denoted by B:

$$B \xrightarrow[R_3 \to R_3 - 2R_1]{R_2 \to R_2 - 3R_1} \begin{bmatrix} 1 & 1 & 1 & 2 \\ 0 & 2 & 1 & -1 \\ 0 & -1 & -3 & -2 \end{bmatrix} \qquad \begin{aligned} x + y + z &= 2 \\ 2y + z &= -1 \\ -y - 3z &= -2 \end{aligned}$$

$$\xrightarrow{R_2 \to R_2 + 2R_3} \begin{bmatrix} 1 & 1 & 1 & 2 \\ 0 & 0 & -5 & -5 \\ 0 & -1 & -3 & -2 \end{bmatrix} \qquad \begin{aligned} x + y + z &= 2 \\ -5z &= -5 \\ -y - 3z &= -2 \end{aligned}$$

C. Definition of row echelon form of a matrix

A matrix A is in **row echelon** form if

(a) all rows consisting only of zeros, if any, are at the bottom of the matrix;
(b) from the left the first nonzero entry, if any, in each row is 1 (this 1 is also called the **leading entry** of its row); and
(c) the leading entry of each row after row 1 occurs to the right of the leading entry of the previous row.

If, in addition to these three conditions, any column containing a leading entry of some row has all zeros above the leading entry, then we say the matrix is in **reduced row echelon** form. Every $m \times n$ matrix A is row equivalent to a matrix in reduced row echelon form.

EXAMPLE 2-5: Which of the following matrices are in row echelon form?

(a) $\begin{bmatrix} 1 & 3 & 2 \\ 0 & 1 & 5 \\ 0 & 0 & 1 \end{bmatrix}$ **(b)** $\begin{bmatrix} 1 & 3 & 2 \\ 0 & 0 & 0 \\ 0 & 0 & 1 \end{bmatrix}$ **(c)** $\begin{bmatrix} 0 & 0 & 0 & 0 \\ 0 & 0 & 0 & 0 \\ 0 & 0 & 0 & 0 \end{bmatrix}$ **(d)** $\begin{bmatrix} 1 & 5 & -3 \\ 0 & 1 & 2 \\ 0 & 0 & 0 \end{bmatrix}$

(e) $\begin{bmatrix} 1 & 4 & 7 \\ 0 & 1 & 6 \\ 0 & 1 & 8 \end{bmatrix}$ **(f)** $\begin{bmatrix} 1 & 4 & 7 & -3 \\ 0 & 1 & 2 & 3 \end{bmatrix}$ **(g)** $\begin{bmatrix} 1 & -3 & 4 \\ 0 & 2 & 5 \\ 0 & 0 & 1 \end{bmatrix}$

Solution

(a) There are no rows of all zeros. The first nonzero number in each row is 1. All numbers below the 1 in the first column are zeros, and all numbers below the 1 in the second column are zeros. Therefore, it is row echelon.
(b) The second row has all zeros and is not at the bottom of the matrix. It is not row echelon.
(c) All rows consist only of zeros, and they are at the bottom of the matrix. It is row echelon.
(d) It is row echelon.
(e) It is not row echelon because the a_{32} entry is 1 and not 0.
(f) It is row echelon.
(g) It is not row echelon because a_{22} entry is 2 and not 1.

EXAMPLE 2-6: Which of the following matrices are in reduced row echelon form?

(a) $\begin{bmatrix} 1 & 0 & 0 \\ 0 & 1 & 0 \\ 0 & 0 & 1 \end{bmatrix}$ **(b)** $\begin{bmatrix} 1 & 0 & 0 & -4 \\ 0 & 1 & 0 & 2 \\ 0 & 0 & 0 & 6 \end{bmatrix}$ **(c)** $\begin{bmatrix} 1 & 5 & 0 & 3 & 6 \\ 0 & 0 & 1 & 1 & 2 \\ 0 & 0 & 0 & 0 & 0 \end{bmatrix}$

(d) $\begin{bmatrix} 1 & 3 & 0 & 1 \\ 0 & 1 & 4 & 3 \\ 0 & 0 & 1 & 2 \end{bmatrix}$ **(e)** $\begin{bmatrix} 1 & 0 & 0 & 1 \\ 0 & 1 & 0 & -3 \\ 0 & 0 & 4 & 6 \end{bmatrix}$ **(f)** $\begin{bmatrix} 1 & 0 & 0 & 1 \\ 0 & 1 & 0 & -2 \\ 0 & 0 & 1 & 3 \end{bmatrix}$

Solution

(a) It is reduced row echelon.
(b) It is not reduced row echelon because the first nonzero entry in the third row is 6 and not 1. Also, the elements above the 6 are 2 and -4, not zeros.
(c) It is reduced row echelon.
(d) It is not reduced row echelon because of the 3 in the second column and the 4 in the third column.
(e) It is not reduced row echelon because the leading entry in the third row is 4 rather than 1.
(f) It is reduced row echelon.

EXAMPLE 2-7: Use elementary row operations to change each of the following matrices to an equivalent matrix in reduced row echelon form.

(a) $\begin{bmatrix} 1 & 3 \\ 2 & 8 \end{bmatrix}$ **(b)** $\begin{bmatrix} 5 & 2 & -4 \\ 3 & 4 & 1 \end{bmatrix}$ **(c)** $\begin{bmatrix} 2 & 3 & 1 \\ 1 & 2 & 3 \\ 4 & 1 & 6 \end{bmatrix}$ **(d)** $\begin{bmatrix} 1 & -2 & 4 & 0 \\ 3 & 2 & 1 & -3 \\ -1 & 4 & 3 & 9 \end{bmatrix}$

Solution

(a)
$$\begin{bmatrix}1&3\\2&8\end{bmatrix}\xrightarrow{R_2\to R_2-2R_1}\begin{bmatrix}1&3\\0&2\end{bmatrix}\xrightarrow{R_2\to 1/2R_2}\begin{bmatrix}1&3\\0&1\end{bmatrix}\xrightarrow{R_1\to R_1-3R_2}\begin{bmatrix}1&0\\0&1\end{bmatrix}$$

(b)
$$\begin{bmatrix}5&2&-4\\3&4&1\end{bmatrix}\xrightarrow{R_1\to R_1-2R_2}\begin{bmatrix}-1&-6&-6\\3&4&1\end{bmatrix}\xrightarrow{R_1\to -R_1}\begin{bmatrix}1&6&6\\3&4&1\end{bmatrix}$$
$$\xrightarrow{R_2\to R_2-3R_1}\begin{bmatrix}1&6&6\\0&-14&-17\end{bmatrix}\xrightarrow{R_2\to -1/14R_2}\begin{bmatrix}1&6&6\\0&1&\frac{17}{14}\end{bmatrix}\xrightarrow{R_1\to R_1-6R_2}\begin{bmatrix}1&0&-\frac{9}{7}\\0&1&\frac{17}{14}\end{bmatrix}$$

(c)
$$\begin{bmatrix}2&3&1\\1&2&3\\4&1&6\end{bmatrix}\xrightarrow{R_1\leftrightarrow R_2}\begin{bmatrix}1&2&3\\2&3&1\\4&1&6\end{bmatrix}\xrightarrow[R_3\to R_3-4R_1]{R_2\to R_2-2R_1}\begin{bmatrix}1&2&3\\0&-1&-5\\0&-7&-6\end{bmatrix}$$
$$\xrightarrow{R_2\to -R_2}\begin{bmatrix}1&2&3\\0&1&5\\0&-7&-6\end{bmatrix}\xrightarrow{R_3\to R_3+7R_2}\begin{bmatrix}1&2&3\\0&1&5\\0&0&29\end{bmatrix}\xrightarrow{R_3\to 1/29R_3}\begin{bmatrix}1&2&3\\0&1&5\\0&0&1\end{bmatrix}$$
$$\xrightarrow[R_2\to R_2-5R_3]{R_1\to R_1-3R_3}\begin{bmatrix}1&2&0\\0&1&0\\0&0&1\end{bmatrix}\xrightarrow{R_1\to R_1-2R_2}\begin{bmatrix}1&0&0\\0&1&0\\0&0&1\end{bmatrix}$$

(d)
$$\begin{bmatrix}1&-2&4&0\\3&2&1&-3\\-1&4&3&9\end{bmatrix}\xrightarrow[R_3\to R_3+R_1]{R_2\to R_2-3R_1}\begin{bmatrix}1&-2&4&0\\0&8&-11&-3\\0&2&7&9\end{bmatrix}\xrightarrow{R_2\to R_2-4R_3}\begin{bmatrix}1&-2&4&0\\0&0&-39&-39\\0&2&7&9\end{bmatrix}$$
$$\xrightarrow{R_2\to -1/39R_2}\begin{bmatrix}1&-2&4&0\\0&0&1&1\\0&2&7&9\end{bmatrix}\xrightarrow{R_2\leftrightarrow R_3}\begin{bmatrix}1&-2&4&0\\0&2&7&9\\0&0&1&1\end{bmatrix}$$
$$\xrightarrow[R_2\to R_2-7R_3]{R_1\to R_1-4R_3}\begin{bmatrix}1&-2&0&-4\\0&2&0&2\\0&0&1&1\end{bmatrix}\xrightarrow{R_1\to R_1+R_2}\begin{bmatrix}1&0&0&-2\\0&2&0&2\\0&0&1&1\end{bmatrix}$$
$$\xrightarrow{R_2\to 1/2R_2}\begin{bmatrix}1&0&0&-2\\0&1&0&1\\0&0&1&1\end{bmatrix}$$

2-3. Gaussian and Gauss–Jordan Elimination

The process of solving a system of linear equations by reducing the augmented matrix to an equivalent matrix in row echelon form is called **Gaussian elimination.** If the augmented matrix is changed to reduced row echelon form, the method is called **Gauss–Jordan elimination.**

EXAMPLE 2-8: Use Gaussian elimination to solve the system:

$$x + y + z = 2$$
$$x + 3y + 2z = 1$$
$$y + 3z = 2$$

(You may wish to refer back to Example 1-4 and notice that this method uses only the augmented matrix of each of the systems developed there.)

Solution

$$\begin{bmatrix} 1 & 1 & 1 & 2 \\ 1 & 3 & 2 & 1 \\ 0 & 1 & 3 & 2 \end{bmatrix} \xrightarrow{R_2 \to R_2 - R_1} \begin{bmatrix} 1 & 1 & 1 & 2 \\ 0 & 2 & 1 & -1 \\ 0 & 1 & 3 & 2 \end{bmatrix} \xrightarrow{R_2 \leftrightarrow R_3} \begin{bmatrix} 1 & 1 & 1 & 2 \\ 0 & 1 & 3 & 2 \\ 0 & 2 & 1 & -1 \end{bmatrix}$$

$$\xrightarrow{R_3 \to R_3 - 2R_2} \begin{bmatrix} 1 & 1 & 1 & 2 \\ 0 & 1 & 3 & 2 \\ 0 & 0 & -5 & -5 \end{bmatrix} \xrightarrow{R \to -1/5 R_3} \begin{bmatrix} 1 & 1 & 1 & 2 \\ 0 & 1 & 3 & 2 \\ 0 & 0 & 1 & 1 \end{bmatrix}$$

$$\therefore \quad \begin{aligned} x + y + z &= 2 \\ y + 3z &= 2 \\ z &= 1 \end{aligned}$$

Substituting $z = 1$ into the first two equations, we obtain

$$x + y = 1 \qquad \text{and} \qquad y = -1$$

Therefore, $x = 2$, $y = -1$, and $z = 1$, which may be written as

$$[2, -1, 1] \qquad \text{or} \qquad \begin{bmatrix} 2 \\ -1 \\ 1 \end{bmatrix}$$

EXAMPLE 2-9: Use Gauss–Jordan elimination to solve the system given in Example 2-8.

Solution: The procedure is identical to that in Example 2-8 until we get to the final matrix. We begin with that matrix:

$$\begin{bmatrix} 1 & 1 & 1 & 2 \\ 0 & 1 & 3 & 2 \\ 0 & 0 & 1 & 1 \end{bmatrix} \xrightarrow[R_2 \to R_2 - 3R_3]{R_1 \to R_1 - R_3} \begin{bmatrix} 1 & 1 & 0 & 1 \\ 0 & 1 & 0 & -1 \\ 0 & 0 & 1 & 1 \end{bmatrix} \xrightarrow{R_1 \to R_1 - R_2} \begin{bmatrix} 1 & 0 & 0 & 2 \\ 0 & 1 & 0 & -1 \\ 0 & 0 & 1 & 1 \end{bmatrix}$$

The last column in this matrix gives us the answers $x = 2$, $y = -1$, and $z = 1$.

EXAMPLE 2-10: Use Gauss–Jordan elimination to solve each of the following linear systems:

(a) $x - y + 5z = -6$ **(b)** $x + 3y - 6z = 7$

$3x + 3y - z = 10$ $2x - y + 2z = 0$

$x + 3y + 2z = 5$ $3x + 2y - 4z = 7$

Solution

(a)
$$\begin{bmatrix} 1 & -1 & 5 & -6 \\ 3 & 3 & -1 & 10 \\ 1 & 3 & 2 & 5 \end{bmatrix} \xrightarrow[R_3 \to R_3 - R_1]{R_2 \to R_2 - 3R_1} \begin{bmatrix} 1 & -1 & 5 & -6 \\ 0 & 6 & -16 & 28 \\ 0 & 4 & -3 & 11 \end{bmatrix}$$

$$\xrightarrow{R_2 \to 1/2 R_2} \begin{bmatrix} 1 & -1 & 5 & -6 \\ 0 & 3 & -8 & 14 \\ 0 & 4 & -3 & 11 \end{bmatrix} \xrightarrow{R_2 \to R_2 - R_3} \begin{bmatrix} 1 & -1 & 5 & -6 \\ 0 & -1 & -5 & 3 \\ 0 & 4 & -3 & 11 \end{bmatrix}$$

$$\xrightarrow[R_3 \to R_3+4R_2]{} \begin{bmatrix} 1 & -1 & 5 & -6 \\ 0 & -1 & -5 & 3 \\ 0 & 0 & -23 & 23 \end{bmatrix} \xrightarrow[R_3 \to -1/23R_3]{R_2 \to -R_2} \begin{bmatrix} 1 & -1 & 5 & -6 \\ 0 & 1 & 5 & -3 \\ 0 & 0 & 1 & -1 \end{bmatrix}$$

$$\xrightarrow[R_2 \to R_2-5R_3]{R_1 \to R_1-5R_3} \begin{bmatrix} 1 & -1 & 0 & -1 \\ 0 & 1 & 0 & 2 \\ 0 & 0 & 1 & -1 \end{bmatrix} \xrightarrow{R_1 \to R_1+R_2} \begin{bmatrix} 1 & 0 & 0 & 1 \\ 0 & 1 & 0 & 2 \\ 0 & 0 & 1 & -1 \end{bmatrix}$$

$$\therefore \quad x = 1, \quad y = 2, \quad z = -1$$

(b)

$$\begin{bmatrix} 1 & 3 & -6 & 7 \\ 2 & -1 & 2 & 0 \\ 3 & 2 & -4 & 7 \end{bmatrix} \xrightarrow[R_3 \to R_3-3R_1]{R_2 \to R_2-2R_1} \begin{bmatrix} 1 & 3 & -6 & 7 \\ 0 & -7 & 14 & -14 \\ 0 & -7 & 14 & -14 \end{bmatrix} \xrightarrow[R_3 \to R_3-R_2]{} \begin{bmatrix} 1 & 3 & -6 & 7 \\ 0 & -7 & 14 & -14 \\ 0 & 0 & 0 & 0 \end{bmatrix}$$

$$\xrightarrow{R_2 \to -1/7R_2} \begin{bmatrix} 1 & 3 & -6 & 7 \\ 0 & 1 & -2 & 2 \\ 0 & 0 & 0 & 0 \end{bmatrix} \xrightarrow{R_1 \to R_1-3R_2} \begin{bmatrix} 1 & 0 & 0 & 1 \\ 0 & 1 & -2 & 2 \\ 0 & 0 & 0 & 0 \end{bmatrix}$$

The linear system associated with this augmented matrix is

$$x = 1$$

$$y - 2z = 2$$

If we let z be any arbitrary number t, then $x = 1$, $y = 2t + 2$, and $z = t$. The solution set is all ordered triples $(1, 2t + 2, t)$, where t is arbitrary.

EXAMPLE 2-11 (Geometry-Technology): A rectangular coordinate system has been superimposed over a frontal view of a piece of machinery that contains a circular pulley. Find, in general form, the equation of the circle formed by the pulley if it is observed that the circle passes through the three noncollinear points $(1, 2)$, $(-2, 1)$, and $(-3, -1)$ of the coordinate system.

Solution: The general form of the equation of a circle is $x^2 + y^2 + Ax + By + C = 0$. Substituting $(1, 2)$, we get $A + 2B + C = -5$. Likewise, $(-2, 1)$ gives $-2A + B + C = -5$ and $(-3, -1)$ results in $-3A - B + C = -10$.

We will use the Gauss–Jordan elimination to solve this system.

$$\begin{bmatrix} 1 & 2 & 1 & -5 \\ -2 & 1 & 1 & -5 \\ -3 & -1 & 1 & -10 \end{bmatrix} \xrightarrow[R_3 \to R_3+3R_1]{R_2 \to R_2+2R_1} \begin{bmatrix} 1 & 2 & 1 & -5 \\ 0 & 5 & 3 & -15 \\ 0 & 5 & 4 & -25 \end{bmatrix}$$

$$\xrightarrow[R_3 \to R_3-R_2]{} \begin{bmatrix} 1 & 2 & 1 & -5 \\ 0 & 5 & 3 & -15 \\ 0 & 0 & 1 & -10 \end{bmatrix} \xrightarrow[R_2 \to R_2-3R_3]{R_1 \to R_1-R_3} \begin{bmatrix} 1 & 2 & 0 & 5 \\ 0 & 5 & 0 & 15 \\ 0 & 0 & 1 & -10 \end{bmatrix}$$

$$\xrightarrow{R_2 \to 1/5R_2} \begin{bmatrix} 1 & 2 & 0 & 5 \\ 0 & 1 & 0 & 3 \\ 0 & 0 & 1 & -10 \end{bmatrix} \xrightarrow{R_1 \to R_1-2R_2} \begin{bmatrix} 1 & 0 & 0 & -1 \\ 0 & 1 & 0 & 3 \\ 0 & 0 & 1 & -10 \end{bmatrix}$$

Therefore, $A = -1$, $B = 3$, $C = -10$, and the equation is

$$x^2 + y^2 - x + 3y - 10 = 0$$

2-4. Inverse of a Matrix

Definition of an inverse matrix

An $n \times n$ matrix A is called **nonsingular** or **invertible** if there is an $n \times n$ matrix B such that $AB = BA = I_n$. The matrix B is denoted by A^{-1} and is called the **inverse** of A. Observe that A and B are inverses of each other. If no such matrix B exists, then A is said to be **singular** or **noninvertible**.

note: To find A^{-1}, change by elementary row operations $[A|I]$ to $[I|B]$. $B = A^{-1}$.

EXAMPLE 2-12: Show that $\begin{bmatrix} \frac{1}{3} & \frac{2}{3} \\ \frac{2}{3} & \frac{7}{3} \end{bmatrix}$ and $\begin{bmatrix} 7 & -2 \\ -2 & 1 \end{bmatrix}$ are inverses.

Solution

$$\begin{bmatrix} 7 & -2 \\ -2 & 1 \end{bmatrix}\begin{bmatrix} \frac{1}{3} & \frac{2}{3} \\ \frac{2}{3} & \frac{7}{3} \end{bmatrix} = \begin{bmatrix} \frac{3}{3} & 0 \\ 0 & \frac{3}{3} \end{bmatrix} = \begin{bmatrix} 1 & 0 \\ 0 & 1 \end{bmatrix}$$

$$\begin{bmatrix} \frac{1}{3} & \frac{2}{3} \\ \frac{2}{3} & \frac{7}{3} \end{bmatrix}\begin{bmatrix} 7 & -2 \\ -2 & 1 \end{bmatrix} = \begin{bmatrix} 1 & 0 \\ 0 & 1 \end{bmatrix}$$

$$\therefore \quad \begin{bmatrix} \frac{1}{3} & \frac{2}{3} \\ \frac{2}{3} & \frac{7}{3} \end{bmatrix} \text{ is the inverse of } \begin{bmatrix} 7 & -2 \\ -2 & 1 \end{bmatrix} \text{ and}$$

$$\begin{bmatrix} 7 & -2 \\ -2 & 1 \end{bmatrix} \text{ is the inverse of } \begin{bmatrix} \frac{1}{3} & \frac{2}{3} \\ \frac{2}{3} & \frac{7}{3} \end{bmatrix}$$

EXAMPLE 2-13: Show that $\begin{bmatrix} 2 & 3 \\ 4 & 6 \end{bmatrix}$ has no inverse.

Solution: Suppose there is an inverse $\begin{bmatrix} a & b \\ c & d \end{bmatrix}$. Then $\begin{bmatrix} 2 & 3 \\ 4 & 6 \end{bmatrix}\begin{bmatrix} a & b \\ c & d \end{bmatrix} = \begin{bmatrix} 1 & 0 \\ 0 & 1 \end{bmatrix}$. Multiplying the two matrices on the left side, we get

$$\begin{bmatrix} 2a + 3c & 2b + 3d \\ 4a + 6c & 4b + 6d \end{bmatrix} = \begin{bmatrix} 1 & 0 \\ 0 & 1 \end{bmatrix}$$

Since the equality of two matrices implies that the corresponding elements are equal, we get

$$2a + 3c = 1$$

$$2b + 3d = 0$$

$$4a + 6c = 0$$

$$4b + 6d = 1$$

The first equation implies that $2a + 3c = 1$, and the third equation implies that $2a + 3c = 0$; these are contradictory. Likewise, the second equation implies that $2b + 3d = 0$ and the fourth equation implies that $2b + 3d = 1/2$; these are also contradictory. Therefore, there is no solution and thus no inverse for $\begin{bmatrix} 2 & 3 \\ 4 & 6 \end{bmatrix}$.

EXAMPLE 2-14: Find the inverse of $A = \begin{bmatrix} 1 & 2 \\ 0 & 1 \end{bmatrix}$

Solution

$$[A|I_2] = \left[\begin{array}{cc|cc} 1 & 2 & 1 & 0 \\ 0 & 1 & 0 & 1 \end{array}\right] \xrightarrow{R_1 \to R_1 - 2R_2} \left[\begin{array}{cc|cc} 1 & 0 & 1 & -2 \\ 0 & 1 & 0 & 1 \end{array}\right]$$

$$\therefore \quad A^{-1} = \begin{bmatrix} 1 & -2 \\ 0 & 1 \end{bmatrix}$$

EXAMPLE 2-15: Find the inverse of

$$A = \begin{bmatrix} 2 & 1 & 1 \\ 1 & 2 & 3 \\ -1 & 0 & 2 \end{bmatrix}$$

Solution

$$\left[\begin{array}{ccc|ccc} 2 & 1 & 1 & 1 & 0 & 0 \\ 1 & 2 & 3 & 0 & 1 & 0 \\ -1 & 0 & 2 & 0 & 0 & 1 \end{array}\right] \xrightarrow{R_1 \leftrightarrow R_3} \left[\begin{array}{ccc|ccc} -1 & 0 & 2 & 0 & 0 & 1 \\ 1 & 2 & 3 & 0 & 1 & 0 \\ 2 & 1 & 1 & 1 & 0 & 0 \end{array}\right] \xrightarrow[R_3 \to R_3 + 2R_1]{R_2 \to R_2 + R_1} \left[\begin{array}{ccc|ccc} -1 & 0 & 2 & 0 & 0 & 1 \\ 0 & 2 & 5 & 0 & 1 & 1 \\ 0 & 1 & 5 & 1 & 0 & 2 \end{array}\right]$$

$$\xrightarrow[R_2 \to R_2 - R_3]{R_1 \to -R_1} \left[\begin{array}{ccc|ccc} 1 & 0 & -2 & 0 & 0 & -1 \\ 0 & 1 & 0 & -1 & 1 & -1 \\ 0 & 1 & 5 & 1 & 0 & 2 \end{array}\right] \xrightarrow[R_3 \to R_3 - R_2]{} \left[\begin{array}{ccc|ccc} 1 & 0 & -2 & 0 & 0 & -1 \\ 0 & 1 & 0 & -1 & 1 & -1 \\ 0 & 0 & 5 & 2 & -1 & 3 \end{array}\right]$$

$$\xrightarrow[R_3 \to 1/5 R_3]{} \left[\begin{array}{ccc|ccc} 1 & 0 & -2 & 0 & 0 & -1 \\ 0 & 1 & 0 & -1 & 1 & -1 \\ 0 & 0 & 1 & \frac{2}{5} & -\frac{1}{5} & \frac{3}{5} \end{array}\right] \xrightarrow{R_1 \to R_1 + 2R_3} \left[\begin{array}{ccc|ccc} 1 & 0 & 0 & \frac{4}{5} & -\frac{2}{5} & \frac{1}{5} \\ 0 & 1 & 0 & -1 & 1 & -1 \\ 0 & 0 & 1 & \frac{2}{5} & -\frac{1}{5} & \frac{3}{5} \end{array}\right]$$

$$\therefore \quad A^{-1} = \begin{bmatrix} \frac{4}{5} & -\frac{2}{5} & \frac{1}{5} \\ -1 & 1 & -1 \\ \frac{2}{5} & -\frac{1}{5} & \frac{3}{5} \end{bmatrix}$$

EXAMPLE 2-16: Show that the matrix

$$A = \begin{bmatrix} 1 & 1 & 1 \\ -3 & 2 & 3 \\ -7 & 3 & 5 \end{bmatrix}$$

has no inverse.

Solution

$$\left[\begin{array}{ccc|ccc} 1 & 1 & 1 & 1 & 0 & 0 \\ -3 & 2 & 3 & 0 & 1 & 0 \\ -7 & 3 & 5 & 0 & 0 & 1 \end{array}\right] \xrightarrow[R_3 \to R_3 + 7R_1]{R_2 \to R_2 + 3R_1} \left[\begin{array}{ccc|ccc} 1 & 1 & 1 & 1 & 0 & 0 \\ 0 & 5 & 6 & 3 & 1 & 0 \\ 0 & 10 & 12 & 7 & 0 & 1 \end{array}\right] \xrightarrow[R_3 \to R_3 - 2R_2]{} \left[\begin{array}{ccc|ccc} 1 & 1 & 1 & 1 & 0 & 0 \\ 0 & 5 & 6 & 3 & 1 & 0 \\ 0 & 0 & 0 & 1 & -2 & 1 \end{array}\right]$$

We stop at this stage because there is a row of zeros to the left of the solid line. This row of zeros tells us that A has no inverse.

2-5. Properties of Inverse Matrices

Property A Although, in general, matrix multiplication is not commutative (that is, $AB \neq BA$) if $B = A^{-1}$, then $AA^{-1} = A^{-1}A$. (This means you do not have to check both $AB = I$ and $BA = I$.)

EXAMPLE 2-17: Verify that $\begin{bmatrix} \cos\theta & \sin\theta \\ -\sin\theta & \cos\theta \end{bmatrix}$ and $\begin{bmatrix} \cos\theta & -\sin\theta \\ \sin\theta & \cos\theta \end{bmatrix}$ are inverses for all real values of θ.

Solution

$$\begin{bmatrix} \cos\theta & \sin\theta \\ -\sin\theta & \cos\theta \end{bmatrix}\begin{bmatrix} \cos\theta & -\sin\theta \\ \sin\theta & \cos\theta \end{bmatrix} = \begin{bmatrix} \cos^2\theta + \sin^2\theta & -\cos\theta\sin\theta + \sin\theta\cos\theta \\ -\sin\theta\cos\theta + \cos\theta\sin\theta & \sin^2\theta + \cos^2\theta \end{bmatrix}$$

$$= \begin{bmatrix} 1 & 0 \\ 0 & 1 \end{bmatrix}$$

since $\cos^2\theta + \sin^2\theta = 1$.

Property B If a matrix is invertible, it has only one inverse.

Property C If A and B are invertible matrices of the same size, then AB is invertible and $(AB)^{-1} = B^{-1}A^{-1}$. More generally, the product of any number of invertible matrices is invertible, and the inverse is the product of the inverses in reverse order.

EXAMPLE 2-18

(a) Verify: if $A = \begin{bmatrix} 3 & 1 & 0 \\ 1 & -1 & 2 \\ 1 & 1 & 1 \end{bmatrix}$, then $A^{-1} = \begin{bmatrix} \frac{3}{8} & \frac{1}{8} & -\frac{2}{8} \\ -\frac{1}{8} & -\frac{3}{8} & \frac{6}{8} \\ -\frac{2}{8} & \frac{2}{8} & \frac{4}{8} \end{bmatrix}$

(b) Verify: if $B = \begin{bmatrix} 2 & 1 & -1 \\ 1 & 3 & 2 \\ -1 & 2 & 1 \end{bmatrix}$, then $B^{-1} = \begin{bmatrix} \frac{1}{10} & \frac{3}{10} & -\frac{5}{10} \\ \frac{3}{10} & -\frac{1}{10} & \frac{5}{10} \\ -\frac{5}{10} & \frac{5}{10} & -\frac{5}{10} \end{bmatrix}$

(c) Find the product AB.

(d) Find the product $B^{-1}A^{-1}$.

(e) Find $(AB)^{-1}$ and show that it is $B^{-1}A^{-1}$.

Solution

(a) $$\begin{bmatrix} 3 & 1 & 0 \\ 1 & -1 & 2 \\ 1 & 1 & 1 \end{bmatrix}\begin{bmatrix} \frac{3}{8} & \frac{1}{8} & -\frac{2}{8} \\ -\frac{1}{8} & -\frac{3}{8} & \frac{6}{8} \\ -\frac{2}{8} & \frac{2}{8} & \frac{4}{8} \end{bmatrix} = \begin{bmatrix} 1 & 0 & 0 \\ 0 & 1 & 0 \\ 0 & 0 & 1 \end{bmatrix} \quad \therefore \text{ they are inverses}$$

(b) $$\begin{bmatrix} 2 & 1 & -1 \\ 1 & 3 & 2 \\ -1 & 2 & 1 \end{bmatrix}\begin{bmatrix} \frac{1}{10} & \frac{3}{10} & -\frac{5}{10} \\ \frac{3}{10} & -\frac{1}{10} & \frac{5}{10} \\ -\frac{5}{10} & \frac{5}{10} & -\frac{5}{10} \end{bmatrix} = \begin{bmatrix} 1 & 0 & 0 \\ 0 & 1 & 0 \\ 0 & 0 & 1 \end{bmatrix} \quad \therefore \text{ they are inverses}$$

(c)

$$AB = \begin{bmatrix} 3 & 1 & 0 \\ 1 & -1 & 2 \\ 1 & 1 & 1 \end{bmatrix} \begin{bmatrix} 2 & 1 & -1 \\ 1 & 3 & 2 \\ -1 & 2 & 1 \end{bmatrix} = \begin{bmatrix} 7 & 6 & -1 \\ -1 & 2 & -1 \\ 2 & 6 & 2 \end{bmatrix}$$

(d)

$$B^{-1}A^{-1} = \begin{bmatrix} \frac{1}{10} & \frac{3}{10} & -\frac{5}{10} \\ \frac{3}{10} & -\frac{1}{10} & \frac{5}{10} \\ -\frac{5}{10} & \frac{5}{10} & -\frac{5}{10} \end{bmatrix} \begin{bmatrix} \frac{3}{8} & \frac{1}{8} & -\frac{2}{8} \\ -\frac{1}{8} & -\frac{3}{8} & \frac{6}{8} \\ -\frac{2}{8} & \frac{2}{8} & \frac{4}{8} \end{bmatrix} = \begin{bmatrix} \frac{10}{80} & -\frac{18}{80} & -\frac{4}{80} \\ 0 & \frac{16}{80} & \frac{8}{80} \\ -\frac{10}{80} & -\frac{30}{80} & \frac{20}{80} \end{bmatrix}$$

$$= \begin{bmatrix} \frac{5}{40} & -\frac{9}{40} & -\frac{2}{40} \\ 0 & \frac{8}{40} & \frac{4}{40} \\ -\frac{5}{40} & -\frac{15}{40} & \frac{10}{40} \end{bmatrix}$$

(e)

$$\left[\begin{array}{ccc|ccc} 7 & 6 & -1 & 1 & 0 & 0 \\ -1 & 2 & -1 & 0 & 1 & 0 \\ 2 & 6 & 2 & 0 & 0 & 1 \end{array}\right] \xrightarrow{R_2 \to -R_2} \left[\begin{array}{ccc|ccc} 7 & 6 & -1 & 1 & 0 & 0 \\ 1 & -2 & 1 & 0 & -1 & 0 \\ 2 & 6 & 2 & 0 & 0 & 1 \end{array}\right]$$

$$\xrightarrow{R_1 \leftrightarrow R_2} \left[\begin{array}{ccc|ccc} 1 & -2 & 1 & 0 & -1 & 0 \\ 7 & 6 & -1 & 1 & 0 & 0 \\ 2 & 6 & 2 & 0 & 0 & 1 \end{array}\right] \xrightarrow[R_3 \to R_3 - 2R_1]{R_2 \to R_2 - 7R_1} \left[\begin{array}{ccc|ccc} 1 & -2 & 1 & 0 & -1 & 0 \\ 0 & 20 & -8 & 1 & 7 & 0 \\ 0 & 10 & 0 & 0 & 2 & 1 \end{array}\right]$$

$$\xrightarrow{R_2 \to R_2 - 2R_3} \left[\begin{array}{ccc|ccc} 1 & -2 & 1 & 0 & -1 & 0 \\ 0 & 0 & -8 & 1 & 3 & -2 \\ 0 & 10 & 0 & 0 & 2 & 1 \end{array}\right] \xrightarrow[R_3 \leftrightarrow R_2]{} \left[\begin{array}{ccc|ccc} 1 & -2 & 1 & 0 & -1 & 0 \\ 0 & 10 & 0 & 0 & 2 & 1 \\ 0 & 0 & -8 & 1 & 3 & -2 \end{array}\right]$$

$$\xrightarrow[R_3 \to -1/8R_3]{R_2 \to 1/10R_2} \left[\begin{array}{ccc|ccc} 1 & -2 & 1 & 0 & -1 & 0 \\ 0 & 1 & 0 & 0 & \frac{2}{10} & \frac{1}{10} \\ 0 & 0 & 1 & -\frac{1}{8} & -\frac{3}{8} & \frac{2}{8} \end{array}\right] \xrightarrow{R_1 \to R_1 - R_3} \left[\begin{array}{ccc|ccc} 1 & -2 & 0 & \frac{1}{8} & -\frac{5}{8} & -\frac{2}{8} \\ 0 & 1 & 0 & 0 & \frac{2}{10} & \frac{1}{10} \\ 0 & 0 & 1 & -\frac{1}{8} & -\frac{3}{8} & \frac{2}{8} \end{array}\right]$$

$$\xrightarrow{R_1 \to R_1 + 2R_2} \left[\begin{array}{ccc|ccc} 1 & 0 & 0 & \frac{1}{8} & -\frac{9}{40} & -\frac{2}{40} \\ 0 & 1 & 0 & 0 & \frac{2}{10} & \frac{1}{10} \\ 0 & 0 & 1 & -\frac{1}{8} & -\frac{3}{8} & \frac{2}{8} \end{array}\right]$$

$$\therefore \quad (AB)^{-1} = \begin{bmatrix} \frac{1}{8} & -\frac{9}{40} & -\frac{2}{40} \\ 0 & \frac{2}{10} & \frac{1}{10} \\ -\frac{1}{8} & -\frac{3}{8} & \frac{2}{8} \end{bmatrix} = \begin{bmatrix} \frac{5}{40} & -\frac{9}{40} & -\frac{2}{40} \\ 0 & \frac{8}{40} & \frac{4}{40} \\ -\frac{5}{40} & -\frac{15}{40} & \frac{10}{40} \end{bmatrix} = B^{-1}A^{-1}$$

EXAMPLE 2-19: If A, B, C, D, E, and F are invertible matrices of the same order, what is the inverse of the product $ABCDEF$?

Solution

$$(ABCDEF)^{-1} = F^{-1}E^{-1}D^{-1}C^{-1}B^{-1}A^{-1}$$

Property D If A and B are row equivalent matrices, then B is invertible if and only if A is invertible. In general, the inverses of A and B are not equal.

EXAMPLE 2-20: Verify that $A = \begin{bmatrix} 1 & 2 & 3 \\ 0 & 1 & 5 \\ 0 & 0 & 1 \end{bmatrix}$ and $B = \begin{bmatrix} 1 & 2 & 0 \\ 0 & 1 & 0 \\ 0 & 0 & 1 \end{bmatrix}$ are row equivalent and find A^{-1} and B^{-1}.

Solution

$$\begin{bmatrix} 1 & 2 & 3 \\ 0 & 1 & 5 \\ 0 & 0 & 1 \end{bmatrix} \xrightarrow[R_2 \to R_2 - 5R_3]{R_1 \to R_1 - 3R_3} \begin{bmatrix} 1 & 2 & 0 \\ 0 & 1 & 0 \\ 0 & 0 & 1 \end{bmatrix}$$

$\therefore$ A and B are row equivalent

$$\left[\begin{array}{ccc|ccc} 1 & 2 & 3 & 1 & 0 & 0 \\ 0 & 1 & 5 & 0 & 1 & 0 \\ 0 & 0 & 1 & 0 & 0 & 1 \end{array}\right] \xrightarrow[R_2 \to R_2 - 5R_3]{R_1 \to R_1 - 3R_3} \left[\begin{array}{ccc|ccc} 1 & 2 & 0 & 1 & 0 & -3 \\ 0 & 1 & 0 & 0 & 1 & -5 \\ 0 & 0 & 1 & 0 & 0 & 1 \end{array}\right]$$

$$\xrightarrow{R_1 \to R_1 - 2R_2} \left[\begin{array}{ccc|ccc} 1 & 0 & 0 & 1 & -2 & 7 \\ 0 & 1 & 0 & 0 & 1 & -5 \\ 0 & 0 & 1 & 0 & 0 & 1 \end{array}\right]$$

$$\therefore \quad A^{-1} = \begin{bmatrix} 1 & -2 & 7 \\ 0 & 1 & -5 \\ 0 & 0 & 1 \end{bmatrix}$$

$$\left[\begin{array}{ccc|ccc} 1 & 2 & 0 & 1 & 0 & 0 \\ 0 & 1 & 0 & 0 & 1 & 0 \\ 0 & 0 & 1 & 0 & 0 & 1 \end{array}\right] \xrightarrow{R_1 \to R_1 - 2R_2} \left[\begin{array}{ccc|ccc} 1 & 0 & 0 & 1 & -2 & 0 \\ 0 & 1 & 0 & 0 & 1 & 0 \\ 0 & 0 & 1 & 0 & 0 & 1 \end{array}\right]$$

$$\therefore \quad B^{-1} = \begin{bmatrix} 1 & -2 & 0 \\ 0 & 1 & 0 \\ 0 & 0 & 1 \end{bmatrix}$$

Property E A square matrix A of order n is invertible if and only if it is row equivalent to I_n. A is singular, that is, noninvertible, if and only if it is row equivalent to a matrix B which has a row of all zeros.

EXAMPLE 2-21: Show that $A = \begin{bmatrix} 1 & 2 & 3 \\ 5 & -3 & 4 \\ 6 & -1 & 7 \end{bmatrix}$ is singular.

Solution

$$\begin{bmatrix} 1 & 2 & 3 \\ 5 & -3 & 4 \\ 6 & -1 & 7 \end{bmatrix} \xrightarrow[R_3 \to R_3 - 6R_1]{R_2 \to R_2 - 5R_1} \begin{bmatrix} 1 & 2 & 3 \\ 0 & -13 & -11 \\ 0 & -13 & -11 \end{bmatrix} \xrightarrow{R_3 \to R_3 - R_2} \begin{bmatrix} 1 & 2 & 3 \\ 0 & -13 & -11 \\ 0 & 0 & 0 \end{bmatrix}$$

Therefore, A is singular because it is row equivalent to a matrix having a row of all zeros.

Property F If A is an invertible matrix, then

(a) A^{-1} is invertible and $(A^{-1})^{-1} = A$.
(b) A^T is invertible and $(A^T)^{-1} = (A^{-1})^T$.
(c) A^k is invertible for $k = 1, 2, 3, \ldots$ and $(A^k)^{-1} = (A^{-1})^k$ where A^k means the product of k matrices A.
(d) For any nonzero scalar s, sA is invertible and $(sA)^{-1} = (1/s)A^{-1}$.

EXAMPLE 2-22: Let $A = \begin{bmatrix} 1 & 1 & 1 \\ 2 & 3 & 6 \\ -1 & 0 & 1 \end{bmatrix}$.

(a) Verify that $A^{-1} = \begin{bmatrix} -\frac{3}{2} & \frac{1}{2} & -\frac{3}{2} \\ 4 & -1 & 2 \\ -\frac{3}{2} & \frac{1}{2} & -\frac{1}{2} \end{bmatrix}$.

(b) What is $(A^{-1})^{-1}$?

(c) Find $(A^T)^{-1}$.

(d) Find $(A^2)^{-1}$.

(e) Find $(1/4A)^{-1}$.

Solution

(a) $$\begin{bmatrix} 1 & 1 & 1 \\ 2 & 3 & 6 \\ -1 & 0 & 1 \end{bmatrix}\begin{bmatrix} -\frac{3}{2} & \frac{1}{2} & -\frac{3}{2} \\ 4 & -1 & 2 \\ -\frac{3}{2} & \frac{1}{2} & -\frac{1}{2} \end{bmatrix} = \begin{bmatrix} 1 & 0 & 0 \\ 0 & 1 & 0 \\ 0 & 0 & 1 \end{bmatrix} \quad \therefore \quad A^{-1} = \begin{bmatrix} -\frac{3}{2} & \frac{1}{2} & -\frac{3}{2} \\ 4 & -1 & 2 \\ -\frac{3}{2} & \frac{1}{2} & -\frac{1}{2} \end{bmatrix}$$

(b) $$(A^{-1})^{-1} = A = \begin{bmatrix} 1 & 1 & 1 \\ 2 & 3 & 6 \\ -1 & 0 & 1 \end{bmatrix}$$

(c) $$(A^T)^{-1} = (A^{-1})^T = \begin{bmatrix} -\frac{3}{2} & \frac{1}{2} & -\frac{3}{2} \\ 4 & -1 & 2 \\ -\frac{3}{2} & \frac{1}{2} & -\frac{1}{2} \end{bmatrix}^T = \begin{bmatrix} -\frac{3}{2} & 4 & -\frac{3}{2} \\ \frac{1}{2} & -1 & \frac{1}{2} \\ -\frac{3}{2} & 2 & -\frac{1}{2} \end{bmatrix}$$

(d) $$(A^2)^{-1} = (A^{-1})^2 = \begin{bmatrix} -\frac{3}{2} & \frac{1}{2} & -\frac{3}{2} \\ 4 & -1 & 2 \\ -\frac{3}{2} & \frac{1}{2} & -\frac{1}{2} \end{bmatrix}\begin{bmatrix} -\frac{3}{2} & \frac{1}{2} & -\frac{3}{2} \\ 4 & -1 & 2 \\ -\frac{3}{2} & \frac{1}{2} & -\frac{1}{2} \end{bmatrix} = \begin{bmatrix} \frac{13}{2} & -2 & 4 \\ -13 & 4 & -10 \\ 5 & -\frac{3}{2} & \frac{7}{2} \end{bmatrix}$$

(e) $$(1/4A)^{-1} = 4A^{-1} = 4\begin{bmatrix} -\frac{3}{2} & \frac{1}{2} & -\frac{3}{2} \\ 4 & -1 & 2 \\ -\frac{3}{2} & \frac{1}{2} & -\frac{1}{2} \end{bmatrix} = \begin{bmatrix} -6 & 2 & -6 \\ 16 & -4 & 8 \\ -6 & 2 & -2 \end{bmatrix}$$

Property G If A and B are square matrices of the same order and the product AB is invertible, then both A and B are invertible.

EXAMPLE 2-23: Let $A = \begin{bmatrix} 1 & -1 & 2 \\ 2 & 3 & 4 \\ -1 & 2 & 1 \end{bmatrix}$ and $B = \begin{bmatrix} 1 & 1 & 2 \\ 0 & 1 & 1 \\ 0 & -1 & 1 \end{bmatrix}$.

(a) Find AB.

(b) Show that AB is nonsingular—that is, invertible.

(c) What can you conclude about A and B?

Solution

(a) $$AB = \begin{bmatrix} 1 & -1 & 2 \\ 2 & 3 & 4 \\ -1 & 2 & 1 \end{bmatrix}\begin{bmatrix} 1 & 1 & 2 \\ 0 & 1 & 1 \\ 0 & -1 & 1 \end{bmatrix} = \begin{bmatrix} 1 & -2 & 3 \\ 2 & 1 & 11 \\ -1 & 0 & 1 \end{bmatrix}$$

(b)

$$\begin{bmatrix} 1 & -2 & 3 \\ 2 & 1 & 11 \\ -1 & 0 & 1 \end{bmatrix} \xrightarrow[R_3 \to R_3 + R_1]{R_2 \to R_2 - 2R_1} \begin{bmatrix} 1 & -2 & 3 \\ 0 & 5 & 5 \\ 0 & -2 & 4 \end{bmatrix} \xrightarrow[R_3 \to 1/2 R_3]{R_2 \to 1/5 R_2} \begin{bmatrix} 1 & -2 & 3 \\ 0 & 1 & 1 \\ 0 & -1 & 2 \end{bmatrix} \xrightarrow[R_3 \to R_3 + R_2]{} \begin{bmatrix} 1 & -2 & 3 \\ 0 & 1 & 1 \\ 0 & 0 & 3 \end{bmatrix}$$

$$\xrightarrow[R_3 \to 1/3 R_3]{} \begin{bmatrix} 1 & -2 & 3 \\ 0 & 1 & 1 \\ 0 & 0 & 1 \end{bmatrix} \xrightarrow[R_2 \to R_2 - R_3]{R_1 \to R - 3R_3} \begin{bmatrix} 1 & -2 & 0 \\ 0 & 1 & 0 \\ 0 & 0 & 1 \end{bmatrix} \xrightarrow{R_1 \to R_1 + 2R_2} \begin{bmatrix} 1 & 0 & 0 \\ 0 & 1 & 0 \\ 0 & 0 & 1 \end{bmatrix} = I_3.$$

Therefore, AB is nonsingular since it is row equivalent to I_3. (Observe that this method simply tells us of the existence of an inverse, but does not give the inverse.)

(c) Since AB is invertible, then both A and B are invertible.

Property H If A is a diagonal matrix, then it is invertible if and only if each diagonal entry is nonzero. Its inverse is a diagonal matrix in which each diagonal element is the reciprocal of the corresponding element of A.

EXAMPLE 2-24: Determine whether each of the following diagonal matrices is nonsingular and, if so, find its inverse.

(a) $A = \begin{bmatrix} 4 & 0 \\ 0 & 2 \end{bmatrix}$ **(b)** $B = \begin{bmatrix} 3 & 0 & 0 \\ 0 & 0 & 0 \\ 0 & 0 & 7 \end{bmatrix}$ **(c)** $C = \begin{bmatrix} 6 & 0 & 0 & 0 \\ 0 & \frac{1}{2} & 0 & 0 \\ 0 & 0 & 3 & 0 \\ 0 & 0 & 0 & 8 \end{bmatrix}$

Solution

(a) A is nonsingular and $A^{-1} = \begin{bmatrix} \frac{1}{4} & 0 \\ 0 & \frac{1}{2} \end{bmatrix}$.

(b) B is singular, since $a_{22} = 0$.

(c) C is nonsingular and $C^{-1} = \begin{bmatrix} \frac{1}{6} & 0 & 0 & 0 \\ 0 & 2 & 0 & 0 \\ 0 & 0 & \frac{1}{3} & 0 \\ 0 & 0 & 0 & \frac{1}{8} \end{bmatrix}$

Property I If A is invertible and $AB = AC$, then $B = C$.

EXAMPLE 2-25: Use $A = \begin{bmatrix} 1 & 0 & 0 \\ 0 & 1 & 0 \\ 0 & 2 & 0 \end{bmatrix}$, $B = \begin{bmatrix} 2 & 1 & 4 \\ 3 & 2 & -1 \\ 4 & 3 & 3 \end{bmatrix}$, and $C = \begin{bmatrix} 2 & 1 & 4 \\ 3 & 2 & -1 \\ -6 & 4 & 7 \end{bmatrix}$ to show that Property I is not true when A is singular.

Solution

$$AB = \begin{bmatrix} 1 & 0 & 0 \\ 0 & 1 & 0 \\ 0 & 2 & 0 \end{bmatrix} \begin{bmatrix} 2 & 1 & 4 \\ 3 & 2 & -1 \\ 4 & 3 & 3 \end{bmatrix} = \begin{bmatrix} 2 & 1 & 4 \\ 3 & 2 & -1 \\ 6 & 4 & -2 \end{bmatrix}$$

$$AC = \begin{bmatrix} 1 & 0 & 0 \\ 0 & 1 & 0 \\ 0 & 2 & 0 \end{bmatrix} \begin{bmatrix} 2 & 1 & 4 \\ 3 & 2 & -1 \\ -6 & 4 & 7 \end{bmatrix} = \begin{bmatrix} 2 & 1 & 4 \\ 3 & 2 & -1 \\ 6 & 4 & -2 \end{bmatrix}$$

Therefore, $AB = AC$, but $B \neq C$.

The trouble arises because A is singular, which is evident from

$$A = \begin{bmatrix} 1 & 0 & 0 \\ 0 & 1 & 0 \\ 0 & 2 & 0 \end{bmatrix} \xrightarrow[R_3 \to R_3 - 2R_2]{} \begin{bmatrix} 1 & 0 & 0 \\ 0 & 1 & 0 \\ 0 & 0 & 0 \end{bmatrix}$$

2-6. Homogeneous Systems of Linear Equations

A. Definition of a homogeneous system

The system

$$\begin{aligned} a_{11}x_1 + a_{12}x_2 + \cdots + a_{1n}x_n &= b_1 \\ a_{21}x_1 + a_{22}x_2 + \cdots + a_{2n}x_n &= b_2 \\ \vdots \qquad \cdots \qquad \cdots \quad &\;\; \vdots \\ a_{m1}x_1 + a_{m2}x_2 + \cdots + a_{mn}x_n &= b_m \end{aligned}$$

of m linear equations in n unknowns is **homogeneous** if $b_1 = b_2 = \cdots = b_n = 0$. The system is often written in matrix form $AX = 0$ where A is the coefficient matrix.

Every homogeneous system has the **trivial solution** $x_1 = x_2 = x_3 = \cdots = x_n = 0$. A solution in which not all of the variables are zero is called a **nontrivial solution**.

B. Properties of homogeneous systems

(1) Every homogeneous linear system has either the trivial solution only or infinitely many solutions.
(2) Every homogeneous linear system with fewer equations than unknowns has infinitely many solutions.
(3) The homogeneous system $AX = 0$ having the same number of unknowns as equations has only the trivial solution if and only if A is invertible.
(4) The homogeneous system $AX = 0$ having the same number of unknowns as equations has a nontrivial solution and hence an infinite number of solutions if and only if A is singular.

EXAMPLE 2-26: Solve each of the following homogeneous systems of linear equations.

(a) $$\begin{aligned} x + y + z &= 0 \\ x - y + 3z &= 0 \end{aligned}$$

(b) $$\begin{aligned} x + y + z &= 0 \\ x - y + 3z &= 0 \\ x \qquad + 2z &= 0 \\ y - z &= 0 \end{aligned}$$

(c) $$\begin{aligned} x + y + z &= 0 \\ x - y + 2z &= 0 \\ 3x + y \qquad &= 0 \end{aligned}$$

(d) $$\begin{aligned} x_1 + x_2 + x_3 + x_4 &= 0 \\ x_1 - x_2 + x_3 - x_4 &= 0 \\ x_1 \qquad + x_3 + x_4 &= 0 \\ 3x_1 \qquad + 3x_3 + x_4 &= 0 \end{aligned}$$

Solution

(a) Since there are three unknowns but only two equations, by Property **(2)** there exists an infinite number of solutions. We start with the augmented matrix.

$$\begin{bmatrix} 1 & 1 & 1 & 0 \\ 1 & -1 & 3 & 0 \end{bmatrix} \xrightarrow{R_2 \to R_2 - R_1} \begin{bmatrix} 1 & 1 & 1 & 0 \\ 0 & -2 & 2 & 0 \end{bmatrix} \xrightarrow{R_2 \to -1/2 R_2} \begin{bmatrix} 1 & 1 & 1 & 0 \\ 0 & 1 & -1 & 0 \end{bmatrix}$$

$$\xrightarrow{R_1 \to R_1 - R_2} \begin{bmatrix} 1 & 0 & 2 & 0 \\ 0 & 1 & -1 & 0 \end{bmatrix}$$

$$\therefore \quad x + 2z = 0 \quad \text{or} \quad x = -2z$$

$$y - z = 0 \qquad\qquad y = z$$

Let $z = t$ where t is an arbitrary real number. Therefore, any ordered triple of form $(-2t, t, t)$ is a solution of the system.

(b) All we know for certain from the properties in Section B is that the system has at least the trivial solution. We begin again with the augmented matrix.

$$\begin{bmatrix} 1 & 1 & 1 & 0 \\ 1 & -1 & 3 & 0 \\ 1 & 0 & 2 & 0 \\ 0 & 1 & -1 & 0 \end{bmatrix} \xrightarrow[R_3 \to R_3 - R_1]{R_2 \to R_2 - R_1} \begin{bmatrix} 1 & 1 & 1 & 0 \\ 0 & -2 & 2 & 0 \\ 0 & -1 & 1 & 0 \\ 0 & 1 & -1 & 0 \end{bmatrix} \xrightarrow[R_3 \to R_3 + R_4]{R_2 \to R_2 - R_3} \begin{bmatrix} 1 & 1 & 1 & 0 \\ 0 & 0 & 0 & 0 \\ 0 & 0 & 0 & 0 \\ 0 & 1 & -1 & 0 \end{bmatrix}$$

$$\xrightarrow{R_2 \leftrightarrow R_4} \begin{bmatrix} 1 & 1 & 1 & 0 \\ 0 & 1 & -1 & 0 \\ 0 & 0 & 0 & 0 \\ 0 & 0 & 0 & 0 \end{bmatrix} \xrightarrow{R_1 \to R_1 - R_2} \begin{bmatrix} 1 & 0 & 2 & 0 \\ 0 & 1 & -1 & 0 \\ 0 & 0 & 0 & 0 \\ 0 & 0 & 0 & 0 \end{bmatrix}$$

$$\therefore \quad x + 2z = 0$$

$$y - z = 0$$

Let $z = t$ where t is an arbitrary real number. Therefore, any ordered triple of form $(-2t, t, t)$ is a solution of the system.

(c) Let A be the coefficient matrix.

$$A = \begin{bmatrix} 1 & 1 & 1 \\ 1 & -1 & 2 \\ 3 & 1 & 0 \end{bmatrix} \xrightarrow[R_3 \to R_3 - 3R_1]{R_2 \to R_2 - R_1} \begin{bmatrix} 1 & 1 & 1 \\ 0 & -2 & 1 \\ 0 & -2 & -1 \end{bmatrix} \xrightarrow[R_3 \to R_3 - R_2]{} \begin{bmatrix} 1 & 1 & 1 \\ 0 & -2 & 1 \\ 0 & 0 & -2 \end{bmatrix}$$

$$\xrightarrow[R_3 \to -1/2 R_3]{} \begin{bmatrix} 1 & 1 & 1 \\ 0 & -2 & 1 \\ 0 & 0 & 1 \end{bmatrix} \xrightarrow[R_2 \to R_2 - R_3]{R_1 \to R_1 - R_3} \begin{bmatrix} 1 & 1 & 0 \\ 0 & -2 & 0 \\ 0 & 0 & 1 \end{bmatrix} \xrightarrow[R_2 \to -1/2 R_2]{R_1 \to R_1 + 1/2 R_2} \begin{bmatrix} 1 & 0 & 0 \\ 0 & 1 & 0 \\ 0 & 0 & 1 \end{bmatrix} = I_3$$

Since A is row equivalent to I_3, then A is invertible. Therefore, by Property **(3)** the system has only the trivial solution $(0, 0, 0)$.

(d) Start with the augmented matrix.

$$\begin{bmatrix} 1 & 1 & 1 & 1 & 0 \\ 1 & -1 & 1 & -1 & 0 \\ 1 & 0 & 1 & 1 & 0 \\ 3 & 0 & 3 & 1 & 0 \end{bmatrix} \xrightarrow[\substack{R_3 \to R_3 - R_1 \\ R_4 \to R_4 - 3R_1}]{R_2 \to R_2 - R_1} \begin{bmatrix} 1 & 1 & 1 & 1 & 0 \\ 0 & -2 & 0 & -2 & 0 \\ 0 & -1 & 0 & 0 & 0 \\ 0 & -3 & 0 & -2 & 0 \end{bmatrix}$$

$$\xrightarrow{R_2 \leftrightarrow -R_3} \begin{bmatrix} 1 & 1 & 1 & 1 & 0 \\ 0 & 1 & 0 & 0 & 0 \\ 0 & -2 & 0 & -2 & 0 \\ 0 & -3 & 0 & -2 & 0 \end{bmatrix} \xrightarrow[R_4 \to R_4 + 3R_2]{R_3 \to R_3 + 2R_2} \begin{bmatrix} 1 & 1 & 1 & 1 & 0 \\ 0 & 1 & 0 & 0 & 0 \\ 0 & 0 & 0 & -2 & 0 \\ 0 & 0 & 0 & -2 & 0 \end{bmatrix}$$

$$\xrightarrow[R_4 \to R_4 - R_3]{R_3 \to -1/2R_3} \begin{bmatrix} 1 & 1 & 1 & 1 & 0 \\ 0 & 1 & 0 & 0 & 0 \\ 0 & 0 & 0 & 1 & 0 \\ 0 & 0 & 0 & 0 & 0 \end{bmatrix} \xrightarrow{R_1 \to R_1 - R_2 - R_3} \begin{bmatrix} 1 & 0 & 1 & 0 & 0 \\ 0 & 1 & 0 & 0 & 0 \\ 0 & 0 & 0 & 1 & 0 \\ 0 & 0 & 0 & 0 & 0 \end{bmatrix}$$

$$\therefore \quad x_1 + x_3 = 0$$
$$x_2 = 0$$
$$x_4 = 0$$

Since the coefficient matrix

$$A = \begin{bmatrix} 1 & 1 & 1 & 1 \\ 1 & -1 & 1 & -1 \\ 1 & 0 & 1 & 1 \\ 3 & 0 & 3 & 1 \end{bmatrix}$$

is row equivalent to

$$\begin{bmatrix} 1 & 0 & 1 & 0 \\ 0 & 1 & 0 & 0 \\ 0 & 0 & 0 & 1 \\ 0 & 0 & 0 & 0 \end{bmatrix}$$

A is noninvertible or singular and thus the system has an infinite number of solutions.

Let $x_3 = t$, where t is any arbitrary real number. Therefore, any ordered 4-tuple of form $(-t, 0, t, 0)$ is a solution of the system.

2-7. Invertible Matrix Method

Consider the matrix form $AX = B$ of a linear system of n equations in n unknowns. If A is invertible, then $A^{-1}(AX) = A^{-1}B$ and the solution is $X = A^{-1}B$. If $B = 0$, the system is homogeneous and has been discussed in the previous section. We shall assume $B \neq 0$.

EXAMPLE 2-27: Solve each of the following systems of linear equations by using the invertible matrix method.

(a)
$$\begin{aligned} 2x + y + z &= 3 \\ x + 2y + 3z &= 5 \\ -x \qquad + 2z &= -4 \end{aligned}$$

(b)
$$\begin{aligned} 2x + y - z &= -30 \\ x + 3y + 2z &= 40 \\ -x + 2y + z &= 10 \end{aligned}$$

Solution

(a) $\begin{bmatrix} 2 & 1 & 1 \\ 1 & 2 & 3 \\ -1 & 0 & 2 \end{bmatrix} \begin{bmatrix} x \\ y \\ z \end{bmatrix} = \begin{bmatrix} 3 \\ 5 \\ -4 \end{bmatrix}$. In Example 2-15 we found

$$\begin{bmatrix} 2 & 1 & 1 \\ 1 & 2 & 3 \\ -1 & 0 & 2 \end{bmatrix}^{-1} = \begin{bmatrix} \frac{4}{5} & -\frac{2}{5} & \frac{1}{5} \\ -1 & 1 & -1 \\ \frac{2}{5} & -\frac{1}{5} & \frac{3}{5} \end{bmatrix}$$

$$\therefore \quad \begin{bmatrix} x \\ y \\ z \end{bmatrix} = \begin{bmatrix} \frac{4}{5} & -\frac{2}{5} & \frac{1}{5} \\ -1 & 1 & -1 \\ \frac{2}{5} & -\frac{1}{5} & \frac{3}{5} \end{bmatrix} \begin{bmatrix} 3 \\ 5 \\ -4 \end{bmatrix} = \begin{bmatrix} -\frac{2}{5} \\ 6 \\ -\frac{11}{5} \end{bmatrix}$$

$\therefore \quad x = -2/5, \quad y = 6, \quad z = -11/5$ is the solution.

(b) $\begin{bmatrix} 2 & 1 & -1 \\ 1 & 3 & 2 \\ -1 & 2 & 1 \end{bmatrix} \begin{bmatrix} x \\ y \\ z \end{bmatrix} = \begin{bmatrix} -30 \\ 40 \\ 10 \end{bmatrix}$. In Example 2-18 we were given

$$\begin{bmatrix} 2 & 1 & -1 \\ 1 & 3 & 2 \\ -1 & 2 & 1 \end{bmatrix}^{-1} = \begin{bmatrix} \frac{1}{10} & \frac{3}{10} & -\frac{5}{10} \\ \frac{3}{10} & -\frac{1}{10} & \frac{5}{10} \\ -\frac{5}{10} & \frac{5}{10} & -\frac{5}{10} \end{bmatrix}$$

$$\therefore \quad \begin{bmatrix} x \\ y \\ z \end{bmatrix} = \begin{bmatrix} \frac{1}{10} & \frac{3}{10} & -\frac{5}{10} \\ \frac{3}{10} & -\frac{1}{10} & \frac{5}{10} \\ -\frac{5}{10} & \frac{5}{10} & -\frac{5}{10} \end{bmatrix} \begin{bmatrix} -30 \\ 40 \\ 10 \end{bmatrix} = \begin{bmatrix} 4 \\ -8 \\ 30 \end{bmatrix}$$

$\therefore \quad x = 4, \quad y = -8, \quad z = 30$

2-8. Equivalent Statements

The following statements are equivalent for an $n \times n$ matrix; that is, they are all true or they are all false.

(1) A is invertible.
(2) A is row equivalent to I_n.
(3) The homogeneous system $AX = 0$ has only the trivial solution.
(4) The linear system $AX = B$ has a unique solution for every $n \times 1$ matrix B.

SUMMARY

1. A system of m linear equations in n unknowns can be written as

$$\begin{bmatrix} a_{11} & a_{12} & \cdots & a_{1n} \\ a_{21} & a_{22} & \cdots & a_{2n} \\ \vdots & & \cdots & \vdots \\ a_{m1} & a_{m2} & \cdots & a_{mn} \end{bmatrix} \begin{bmatrix} x_1 \\ x_2 \\ \vdots \\ x_n \end{bmatrix} = \begin{bmatrix} b_1 \\ b_2 \\ \vdots \\ b_n \end{bmatrix}$$

   or symbolically as $AX = B$ in matrix form.
2. The coefficient matrix is A and $[A|B]$ is the augmented matrix.
3. Elementary row operations performed on a matrix produces row equivalent matrices. Two linear systems with row equivalent augmented matrices have the exact same solution set.
4. A matrix A is in row echelon form if
   - all rows consisting of all zeros, if any, are at the bottom of the matrix;
   - from the left, the first nonzero entry, if any, is 1 and is called the leading entry of its row; and
   - the leading entry of each row after row one occurs to the right of the leading entry of the previous row.

5. A matrix A is reduced row echelon if it is row echelon and any column containing a leading entry has zeros elsewhere.
6. Gaussian and Gauss–Jordan elimination are applied by reducing the augmented matrix of a linear system to row echelon or reduced row echelon, respectively.
7. An $n \times n$ matrix A is nonsingular or invertible if there exists an $n \times n$ matrix B called the inverse of A and denoted by A^{-1} such that $AA^{-1} = A^{-1}A = I_n$.
8. A matrix having no inverse is called singular or noninvertible.
9. To find the inverse of A, perform elementary row operations on $[A|I_n]$ to change to $[I_n|B]$. The resulting matrix B to the right of the solid line is A^{-1}.
10. A square matrix is singular if and only if it is row equivalent to a matrix having a row of all zeros.
11. If A and B are invertible and of the same order, then AB is invertible and $(AB)^{-1} = B^{-1}A^{-1}$. The converse statement is also true.
12. If A and B are row equivalent matrices, then B is invertible if and only if A is invertible.
13. If A is an invertible matrix, then
    - its inverse is unique;
    - A^{-1} is invertible and $(A^{-1})^{-1} = A$;
    - A^T is invertible and $(A^T)^{-1} = (A^{-1})^T$;
    - for any positive integer k, A^k is invertible and $(A^k)^{-1} = (A^{-1})^k$; and
    - for any nonzero scalar s, sA is invertible and $(sA)^{-1} = 1/s\ A^{-1}$.
14. The inverse of a diagonal matrix A having nonzero diagonal elements is a diagonal matrix in which each diagonal element is the reciprocal of the corresponding element of A.
15. If A is invertible and $AB = AC$, then $B = C$.
16. Every homogeneous linear system has either the trivial solution only or infinitely many solutions.
17. Every homogeneous linear system with fewer equations than unknowns has infinitely many solutions.
18. The homogeneous system $AX = 0$ having the same number of unknowns as equations has only the trivial solution if and only if A is invertible.
19. The homogeneous system $AX = 0$ having the same number of unknowns as equations has a nontrivial solution and hence an infinite number of solutions if and only if A is singular.
20. If A is invertible, then the system of linear equations $AX = B$ has solution $X = A^{-1}B$.
21. The following statements are equivalent.
    - A is invertible.
    - A is row equivalent to I_n.
    - The homogeneous system $AX = 0$ has only the trivial solution.
    - The linear system $AX = B$ has a unique solution for every $n \times 1$ matrix B.

RAISE YOUR GRADES

Can you ...?

☑ use elementary row operations to find equivalent matrices
☑ find the row and reduced row echelon forms of a matrix
☑ use Gaussian and Gauss–Jordan elimination to solve a linear system
☑ determine whether matrix A is invertible and, if so, find its inverse
☑ find the inverse of a product of invertible matrices
☑ find $(A^{-1})^{-1}$, $(A^T)^{-1}$, $(A^k)^{-1}$, $(sA)^{-1}$ when A is invertible, k is a positive integer and s is a nonzero scalar
☑ determine whether a homogeneous system has only the trivial solution or infinitely many solutions and then find the solution(s)
☑ solve the system $AX = B$ with A invertible

SOLVED PROBLEMS

PROBLEM 2-1 For the following linear system find **(a)** the coefficient matrix, **(b)** the augmented matrix, and **(c)** write the matrix form of the system:

$$\begin{aligned} x - 2y + 3z &= 4 \\ 2x + y + z &= 3 \\ 3x - y + 2z &= 1 \end{aligned}$$

Solution

(a) $\begin{bmatrix} 1 & -2 & 3 \\ 2 & 1 & 1 \\ 3 & -1 & 2 \end{bmatrix}$ **(b)** $\begin{bmatrix} 1 & -2 & 3 & 4 \\ 2 & 1 & 1 & 3 \\ 3 & -1 & 2 & 1 \end{bmatrix}$ **(c)** $\begin{bmatrix} 1 & -2 & 3 \\ 2 & 1 & 1 \\ 3 & -1 & 2 \end{bmatrix}\begin{bmatrix} x \\ y \\ z \end{bmatrix} = \begin{bmatrix} 4 \\ 3 \\ 1 \end{bmatrix}$

PROBLEM 2-2 Find a linear system whose augmented matrix is

(a) $\begin{bmatrix} 3 & 1 & 4 \\ 2 & 1 & 2 \end{bmatrix}$ **(b)** $\begin{bmatrix} 1 & -2 & 1 & -1 & -4 \\ 3 & 0 & 2 & 1 & 2 \end{bmatrix}$ **(c)** $\begin{bmatrix} 1 & 0 & 0 & 2 \\ 0 & 1 & 0 & -3 \\ 0 & 0 & 1 & 1 \end{bmatrix}$

Solution

(a) $\begin{aligned} 3x + y &= 4 \\ 2x + y &= 2 \end{aligned}$ **(b)** $\begin{aligned} x_1 - 2x_2 + x_3 - x_4 &= -4 \\ 3x_1 \qquad + 2x_3 + x_4 &= 2 \end{aligned}$ **(c)** $\begin{aligned} x &= 2 \\ y &= -3 \\ z &= 1 \end{aligned}$

PROBLEM 2-3 Given the linear system

$$\begin{aligned} x + 2y + 5z &= -5 \\ x + w &= 6 \\ 3x - y + z &= -4 \\ 2x + 3z + 2w &= 3 \end{aligned}$$

find **(a)** the coefficient matrix, and **(b)** the augmented matrix.

Solution Rewrite the system as

$$\begin{aligned} x + 2y + 5z + 0w &= -5 \\ x + 0y + 0z + w &= 6 \\ 3x - y + z + 0w &= -4 \\ 2x + 0y + 3z + 2w &= 3 \end{aligned}$$

Then

(a) $\begin{bmatrix} 1 & 2 & 5 & 0 \\ 1 & 0 & 0 & 1 \\ 3 & -1 & 1 & 0 \\ 2 & 0 & 3 & 2 \end{bmatrix}$ **(b)** $\begin{bmatrix} 1 & 2 & 5 & 0 & -5 \\ 1 & 0 & 0 & 1 & 6 \\ 3 & -1 & 1 & 0 & -4 \\ 2 & 0 & 3 & 2 & 3 \end{bmatrix}$

PROBLEM 2-4 Use elementary row operations to change the augmented matrix of the given system of equations to five different matrices:

$$x + y + 2z = 4$$
$$2x - y - z = 2$$
$$x + 2y + 3z = 6$$

Solution

$$A = \begin{bmatrix} 1 & 1 & 2 & 4 \\ 2 & -1 & -1 & 2 \\ 1 & 2 & 3 & 6 \end{bmatrix} \qquad \text{given system}$$

$$A \xrightarrow[R_3 \to 2R_3]{} \begin{bmatrix} 1 & 1 & 2 & 4 \\ 2 & -1 & -1 & 2 \\ 2 & 4 & 6 & 12 \end{bmatrix} \qquad \begin{aligned} x + y + 2z &= 4 \\ 2x - y - z &= 2 \\ 2x + 4y + 6z &= 12 \end{aligned}$$

$$A \xrightarrow{R_2 \leftrightarrow R_3} \begin{bmatrix} 1 & 1 & 2 & 4 \\ 1 & 2 & 3 & 6 \\ 2 & -1 & -1 & 2 \end{bmatrix} \qquad \begin{aligned} x + y + 2z &= 4 \\ x + 2y + 3z &= 6 \\ 2x - y - z &= 2 \end{aligned}$$

$$A \xrightarrow{R_2 \to R_2 - R_1} \begin{bmatrix} 1 & 1 & 2 & 4 \\ 1 & -2 & -3 & -2 \\ 1 & 2 & 3 & 6 \end{bmatrix} \qquad \begin{aligned} x + y + 2z &= 4 \\ x - 2y - 3z &= -2 \\ x + 2y + 3z &= 6 \end{aligned}$$

$$A \xrightarrow[R_3 \to R_3 - R_1]{R_2 \to R_2 - 2R_1} \begin{bmatrix} 1 & 1 & 2 & 4 \\ 0 & -3 & -5 & -6 \\ 0 & 1 & 1 & 2 \end{bmatrix} \qquad \begin{aligned} x + y + 2z &= 4 \\ -3y - 5z &= -6 \\ y + z &= 2 \end{aligned}$$

We have found four matrices equivalent to A by beginning with A each time. We may begin with any equivalent matrix. Choosing the last one obtained,

$$\begin{bmatrix} 1 & 1 & 2 & 4 \\ 0 & -3 & -5 & -6 \\ 0 & 1 & 1 & 2 \end{bmatrix} \xrightarrow[R_3 \to R_3 - R_1]{} \begin{bmatrix} 1 & 1 & 2 & 4 \\ 0 & -3 & -5 & -6 \\ -1 & 0 & 1 & -2 \end{bmatrix} \qquad \begin{aligned} x + y + 2z &= 4 \\ -3y - 5z &= -6 \\ x - z &= -2 \end{aligned}$$

PROBLEM 2-5 Which of the following matrices are in row echelon form?

(a) $\begin{bmatrix} 1 & 0 & 3 \\ 0 & 1 & 2 \\ 0 & 0 & 1 \end{bmatrix}$ **(b)** $\begin{bmatrix} 1 & 0 & 0 \\ 0 & 0 & 0 \\ 0 & 1 & 0 \end{bmatrix}$ **(c)** $\begin{bmatrix} 2 & 1 & 0 \\ 0 & 2 & 1 \\ 0 & 0 & 2 \end{bmatrix}$ **(d)** $\begin{bmatrix} 1 & 2 & 3 \\ 0 & 1 & 4 \end{bmatrix}$

(e) $\begin{bmatrix} 0 & 0 & 1 \\ 0 & 1 & 0 \\ 1 & 0 & 0 \end{bmatrix}$ **(f)** $\begin{bmatrix} 1 & 2 & 3 & 4 \\ 0 & 1 & -1 & 2 \\ 0 & 0 & 1 & 3 \end{bmatrix}$

Solution

(a) There are no rows of all zeros. The first nonzero number in each row is 1 and all numbers below the main diagonal are zeros. This is row echelon.

(b) The row of zeros is not at the bottom of the matrix. It is not row echelon.

(c) The first nonzero number in each row is 2. This prevents the matrix from being row echelon.

(d) This is row echelon.
(e) This is not row echelon because the 1 in the second row occurs to the left of the column containing the 1 in the first row. Likewise, the 1 in the third row occurs to the left of the second column.
(f) This is row echelon.

PROBLEM 2-6 Which of the following matrices are in reduced row echelon form?

(a) $\begin{bmatrix} 1 & 0 & 0 & 3 \\ 0 & 1 & 0 & 2 \\ 0 & 0 & 1 & 1 \end{bmatrix}$ **(b)** $\begin{bmatrix} 1 & -3 & 0 & 4 \\ 0 & 0 & 2 & 1 \\ 0 & 0 & 0 & 1 \end{bmatrix}$ **(c)** $\begin{bmatrix} 1 & -3 & 0 & -1 \\ 0 & 0 & 1 & 2 \\ 0 & 0 & 0 & 0 \end{bmatrix}$

(d) $\begin{bmatrix} 1 & 2 & 0 & 0 & 1 \\ 0 & 0 & 1 & 0 & -2 \\ 0 & 0 & 0 & 1 & 3 \end{bmatrix}$ **(e)** $\begin{bmatrix} 1 & 0 & 0 & 1 \\ 0 & 0 & 1 & 0 \\ 0 & 0 & 0 & 1 \end{bmatrix}$

Solution

(a) This is reduced row echelon.
(b) The leading entry in the second row is a 2 and the numbers above the leading 1 in the third row are not zeros. This is not reduced row echelon.
(c) This is reduced row echelon.
(d) This is reduced row echelon.
(e) Column four has a 1 above the leading 1 in the fourth row. This is not reduced row echelon.

PROBLEM 2-7 Use elementary row operations to change each of the given matrices into reduced row echelon form.

(a) $\begin{bmatrix} 0 & 0 & 1 \\ 0 & 1 & 0 \\ 1 & 0 & 0 \end{bmatrix}$ **(b)** $\begin{bmatrix} 2 & 1 & 0 \\ 0 & 2 & 1 \\ 0 & 0 & 2 \end{bmatrix}$ **(c)** $\begin{bmatrix} 1 & 1 & 2 & 4 \\ 0 & -3 & -5 & -6 \\ 0 & 1 & 1 & 2 \end{bmatrix}$ **(d)** $\begin{bmatrix} 1 & -2 & 3 & 4 \\ 2 & 1 & 1 & 3 \\ 3 & -1 & 2 & 1 \end{bmatrix}$

Solution

(a)
$$\begin{bmatrix} 0 & 0 & 1 \\ 0 & 1 & 0 \\ 1 & 0 & 0 \end{bmatrix} \xrightarrow{R_1 \leftrightarrow R_3} \begin{bmatrix} 1 & 0 & 0 \\ 0 & 1 & 0 \\ 0 & 0 & 1 \end{bmatrix}$$

(b)
$$\begin{bmatrix} 2 & 1 & 0 \\ 0 & 2 & 1 \\ 0 & 0 & 2 \end{bmatrix} \xrightarrow[R_3 \to 1/2R_3]{} \begin{bmatrix} 2 & 1 & 0 \\ 0 & 2 & 1 \\ 0 & 0 & 1 \end{bmatrix} \xrightarrow{R_2 \to R_2 - R_3} \begin{bmatrix} 2 & 1 & 0 \\ 0 & 2 & 0 \\ 0 & 0 & 1 \end{bmatrix}$$
$$\xrightarrow{R_2 \to 1/2R_2} \begin{bmatrix} 2 & 1 & 0 \\ 0 & 1 & 0 \\ 0 & 0 & 1 \end{bmatrix} \xrightarrow{R_1 \to R_1 - R_2} \begin{bmatrix} 2 & 0 & 0 \\ 0 & 1 & 0 \\ 0 & 0 & 1 \end{bmatrix} \xrightarrow{R_1 \to 1/2R_1} \begin{bmatrix} 1 & 0 & 0 \\ 0 & 1 & 0 \\ 0 & 0 & 1 \end{bmatrix}$$

(c)
$$\begin{bmatrix} 1 & 1 & 2 & 4 \\ 0 & -3 & -5 & -6 \\ 0 & 1 & 1 & 2 \end{bmatrix} \xrightarrow{R_2 \leftrightarrow R_3} \begin{bmatrix} 1 & 1 & 2 & 4 \\ 0 & 1 & 1 & 2 \\ 0 & -3 & -5 & -6 \end{bmatrix} \xrightarrow[R_3 \to R_3 + 3R_2]{} \begin{bmatrix} 1 & 1 & 2 & 4 \\ 0 & 1 & 1 & 2 \\ 0 & 0 & -2 & 0 \end{bmatrix}$$
$$\xrightarrow[R_3 \to -1/2R_3]{R_1 \to R_1 - R_2} \begin{bmatrix} 1 & 0 & 1 & 2 \\ 0 & 1 & 1 & 2 \\ 0 & 0 & 1 & 0 \end{bmatrix} \xrightarrow[R_2 \to R_2 - R_3]{R_1 \to R_1 - R_3} \begin{bmatrix} 1 & 0 & 0 & 2 \\ 0 & 1 & 0 & 2 \\ 0 & 0 & 1 & 0 \end{bmatrix}$$

(d)

$$\begin{bmatrix} 1 & -2 & 3 & 4 \\ 2 & 1 & 1 & 3 \\ 3 & -1 & 2 & 1 \end{bmatrix} \xrightarrow[R_3 \to R_3 - 3R_1]{R_2 \to R_2 - 2R_1} \begin{bmatrix} 1 & -2 & 3 & 4 \\ 0 & 5 & -5 & -5 \\ 0 & 5 & -7 & -11 \end{bmatrix} \xrightarrow[R_3 \to R_3 - R_2]{} \begin{bmatrix} 1 & -2 & 3 & 4 \\ 0 & 5 & -5 & -5 \\ 0 & 0 & -2 & -6 \end{bmatrix}$$

$$\xrightarrow[R_3 \to -1/2R_3]{R_2 \to 1/5R_2} \begin{bmatrix} 1 & -2 & 3 & 4 \\ 0 & 1 & -1 & -1 \\ 0 & 0 & 1 & 3 \end{bmatrix} \xrightarrow[R_2 \to R_2 + R_3]{R_1 \to R_1 + 2R_2} \begin{bmatrix} 1 & 0 & 1 & 2 \\ 0 & 1 & 0 & 2 \\ 0 & 0 & 1 & 3 \end{bmatrix}$$

$$\xrightarrow{R_1 \to R_1 - R_3} \begin{bmatrix} 1 & 0 & 0 & -1 \\ 0 & 1 & 0 & 2 \\ 0 & 0 & 1 & 3 \end{bmatrix}$$

PROBLEM 2-8 Use Gaussian elimination to solve the system:

$$x - 2y + 3z = 4$$
$$2x + y + z = 3$$
$$3x - y + 2z = 1$$

Solution We reduce the augmented matrix to row echelon form.

$$\begin{bmatrix} 1 & -2 & 3 & 4 \\ 2 & 1 & 1 & 3 \\ 3 & -1 & 2 & 1 \end{bmatrix} \xrightarrow[R_3 \to R_3 - 3R_1]{R_2 \to R_2 - 2R_1} \begin{bmatrix} 1 & -2 & 3 & 4 \\ 0 & 5 & -5 & -5 \\ 0 & 5 & -7 & -11 \end{bmatrix}$$

From Problem 2-7 (**d**),

$$\begin{bmatrix} 1 & -2 & 3 & 4 \\ 0 & 5 & -5 & -5 \\ 0 & 5 & -7 & -11 \end{bmatrix} \to \begin{bmatrix} 1 & -2 & 3 & 4 \\ 0 & 1 & -1 & -1 \\ 0 & 0 & 1 & 3 \end{bmatrix}$$

$$\therefore \quad x - 2y + 3z = 4$$
$$y - z = -1$$
$$z = 3$$

Substituting $z = 3$ in the first two equations we get

$$x - 2y = -5$$
$$y = 2$$
$$\therefore \quad x = -1, \quad y = 2, \quad \text{and} \quad z = 3.$$

PROBLEM 2-9 Use Gauss–Jordan elimination to solve the system given in Problem 2-8.

Solution In Problem 2-7 (**d**) we reduced the augmented matrix to reduced row echelon form, which is $\begin{bmatrix} 1 & 0 & 0 & -1 \\ 0 & 1 & 0 & 2 \\ 0 & 0 & 1 & 3 \end{bmatrix}$. This represents the system

$$x = -1$$
$$y = 2$$
$$z = 3$$

PROBLEM 2-10 Use Gauss–Jordan elimination to solve the following linear system:

$$x + 2y - z = 5$$
$$2x + y + z = -5$$

Solution

$$\begin{bmatrix} 1 & 2 & -1 & 5 \\ 2 & 1 & 1 & -5 \end{bmatrix} \xrightarrow{R_2 \to R_2 - 2R_1} \begin{bmatrix} 1 & 2 & -1 & 5 \\ 0 & -3 & 3 & -15 \end{bmatrix} \xrightarrow{R_2 \to -1/3R_2} \begin{bmatrix} 1 & 2 & -1 & 5 \\ 0 & 1 & -1 & 5 \end{bmatrix}$$

$$\xrightarrow{R_1 \to R_1 - 2R_2} \begin{bmatrix} 1 & 0 & 1 & -5 \\ 0 & 1 & -1 & 5 \end{bmatrix}$$

This represents the system

$$x + z = -5$$
$$y - z = 5$$

Let $z = t$, any arbitrary real number, then $x = -t - 5$ and $y = t + 5$. The solution set is all ordered triples of form $(-t - 5, t + 5, t)$, where t is arbitrary.

PROBLEM 2-11 (Production Planning—Nutrition) A nursing home dietition is to prepare a meal consisting of three basic foods A, B, and C. The diet is to include 25 units of potassium, 22 units of zinc, and 18 units of thiamine. The number of units of potassium in one ounce of food A is 3, of food B is 3, and of food C is 2. The number of units of zinc in one ounce of food A is 3, of food B is 2, and of food C is 3. The number of units of thiamine in one ounce of food A is 2, of food B is 1, and of food C is 4. How many ounces of each food must be used to meet the diet requirements?

Solution First we will set up a table to show the information

	Units per ounce			
	Food *A*	Food *B*	Food *C*	Diet requirements
Potassium	3	3	2	25
Zinc	3	2	3	22
Thiamine	2	1	4	18

Let x be the number ounces of A needed, y the number ounces of B needed, and z the number ounces of C needed. The problem can now be written in the following matrix form:

$$\begin{bmatrix} 3 & 3 & 2 \\ 3 & 2 & 3 \\ 2 & 1 & 4 \end{bmatrix} \begin{bmatrix} x \\ y \\ z \end{bmatrix} = \begin{bmatrix} 25 \\ 22 \\ 18 \end{bmatrix} \quad \text{or} \quad \begin{aligned} 3x + 3y + 2z &= 25 \\ 3x + 2y + 3z &= 22 \\ 2x + y + 4z &= 18 \end{aligned}$$

We will solve this by Gauss–Jordan elimination.

$$\begin{bmatrix} 3 & 3 & 2 & 25 \\ 3 & 2 & 3 & 22 \\ 2 & 1 & 4 & 18 \end{bmatrix} \xrightarrow{R_1 \to R_1 - R_3} \begin{bmatrix} 1 & 2 & -2 & 7 \\ 3 & 2 & 3 & 22 \\ 2 & 1 & 4 & 18 \end{bmatrix} \xrightarrow[R_3 \to R_3 - 2R_1]{R_2 \to R_2 - 3R_1} \begin{bmatrix} 1 & 2 & -2 & 7 \\ 0 & -4 & 9 & 1 \\ 0 & -3 & 8 & 4 \end{bmatrix}$$

$$\xrightarrow{R_2 \to R_2 - R_3} \begin{bmatrix} 1 & 2 & -2 & 7 \\ 0 & -1 & 1 & -3 \\ 0 & -3 & 8 & 4 \end{bmatrix} \xrightarrow[R_3 \to R_3 - 3R_2]{R_2 \to -1R_2} \begin{bmatrix} 1 & 2 & -2 & 7 \\ 0 & 1 & -1 & 3 \\ 0 & 0 & 5 & 13 \end{bmatrix}$$

$$\xrightarrow[R_3 \to 1/5R_3]{R_1 \to R_1 - 2R_2} \begin{bmatrix} 1 & 0 & 0 & 1 \\ 0 & 1 & -1 & 3 \\ 0 & 0 & 1 & \frac{13}{5} \end{bmatrix} \xrightarrow{R_2 \to R_2 + R_3} \begin{bmatrix} 1 & 0 & 0 & 1 \\ 0 & 1 & 0 & \frac{28}{5} \\ 0 & 0 & 1 & \frac{13}{5} \end{bmatrix}$$

So $x = 1$, $y = 28/5$, and $z = 13/5$ or $x = 1$, $y = 5.6$, and $z = 2.6$. Therefore,

- 1 ounce of food A,
- 5.6 ounces of food B, and
- 2.6 ounces of food C

are needed to prepare the meal.

PROBLEM 2-12 Find the inverse of each of the following, if it exists. If it does not exist, explain why.

(a) $\begin{bmatrix} 1 & -2 \\ 2 & 4 \end{bmatrix}$ **(b)** $\begin{bmatrix} 1 & -1 & 2 \\ 2 & -1 & 1 \\ -3 & 2 & 3 \end{bmatrix}$ **(c)** $\begin{bmatrix} 1 & 2 & -1 \\ 1 & 1 & -2 \\ 1 & 2 & 0 \end{bmatrix}$

(d) $\begin{bmatrix} 2 & 4 \\ 1 & 2 \end{bmatrix}$ **(e)** $\begin{bmatrix} 1 & -1 & 0 & 2 \\ 2 & 2 & 0 & 1 \\ -3 & -1 & -5 & 1 \\ 1 & -1 & 0 & 0 \end{bmatrix}$ **(f)** $\begin{bmatrix} \frac{1}{2} & \frac{1}{4} \\ -\frac{1}{4} & \frac{1}{8} \end{bmatrix}$

Solution

(a)

$$\left[\begin{array}{cc|cc} 1 & -2 & 1 & 0 \\ 2 & 4 & 0 & 1 \end{array}\right] \xrightarrow{R_2 \to R_2 - 2R_1} \left[\begin{array}{cc|cc} 1 & -2 & 1 & 0 \\ 0 & 8 & -2 & 1 \end{array}\right] \xrightarrow{R_2 \to 1/8 R_2} \left[\begin{array}{cc|cc} 1 & -2 & 1 & 0 \\ 0 & 1 & -\frac{1}{4} & \frac{1}{8} \end{array}\right]$$

$$\xrightarrow{R_1 \to R_1 + 2R_2} \left[\begin{array}{cc|cc} 1 & 0 & \frac{1}{2} & \frac{1}{4} \\ 0 & 1 & -\frac{1}{4} & \frac{1}{8} \end{array}\right]$$

$$\therefore \text{ its inverse is } \begin{bmatrix} \frac{1}{2} & \frac{1}{4} \\ -\frac{1}{4} & \frac{1}{8} \end{bmatrix}$$

(b)

$$\left[\begin{array}{ccc|ccc} 1 & -1 & 2 & 1 & 0 & 0 \\ 2 & -1 & 1 & 0 & 1 & 0 \\ -3 & 2 & 3 & 0 & 0 & 1 \end{array}\right] \xrightarrow[R_3 \to R_3 + 3R_1]{R_2 \to R_2 - 2R_1} \left[\begin{array}{ccc|ccc} 1 & -1 & 2 & 1 & 0 & 0 \\ 0 & 1 & -3 & -2 & 1 & 0 \\ 0 & -1 & 9 & 3 & 0 & 1 \end{array}\right]$$

$$\xrightarrow[R_3 \to R_3 + R_2]{R_1 \to R_1 + R_2} \left[\begin{array}{ccc|ccc} 1 & 0 & -1 & -1 & 1 & 0 \\ 0 & 1 & -3 & -2 & 1 & 0 \\ 0 & 0 & 6 & 1 & 1 & 1 \end{array}\right]$$

$$\xrightarrow[R_3 \to 1/6 R_3]{} \left[\begin{array}{ccc|ccc} 1 & 0 & -1 & -1 & 1 & 0 \\ 0 & 1 & -3 & -2 & 1 & 0 \\ 0 & 0 & 1 & \frac{1}{6} & \frac{1}{6} & \frac{1}{6} \end{array}\right]$$

$$\xrightarrow[R_2 \to R_2 + 3R_3]{R_1 \to R_1 + R_3} \left[\begin{array}{ccc|ccc} 1 & 0 & 0 & -\frac{5}{6} & \frac{7}{6} & \frac{1}{6} \\ 0 & 1 & 0 & -\frac{9}{6} & \frac{9}{6} & \frac{3}{6} \\ 0 & 0 & 1 & \frac{1}{6} & \frac{1}{6} & \frac{1}{6} \end{array}\right]$$

$$\therefore \text{ its inverse is } \begin{bmatrix} -\frac{5}{6} & \frac{7}{6} & \frac{1}{6} \\ -\frac{9}{6} & \frac{9}{6} & \frac{3}{6} \\ \frac{1}{6} & \frac{1}{6} & \frac{1}{6} \end{bmatrix}$$

(c)

$$\left[\begin{array}{ccc|ccc} 1 & 2 & -1 & 1 & 0 & 0 \\ 1 & 1 & -2 & 0 & 1 & 0 \\ 1 & 2 & 0 & 0 & 0 & 1 \end{array}\right] \xrightarrow[R_3 \to R_3 - R_1]{R_2 \to R_2 - R_1} \left[\begin{array}{ccc|ccc} 1 & 2 & -1 & 1 & 0 & 0 \\ 0 & -1 & -1 & -1 & 1 & 0 \\ 0 & 0 & 1 & -1 & 0 & 1 \end{array}\right]$$

$$\xrightarrow{R_1\to R_1+2R_2}\left[\begin{array}{ccc|ccc}1&0&-3&-1&2&0\\0&-1&-1&-1&1&0\\0&0&1&-1&0&1\end{array}\right]$$

$$\xrightarrow[R_2\to -R_2]{}\left[\begin{array}{ccc|ccc}1&0&-3&-1&2&0\\0&1&1&1&-1&0\\0&0&1&-1&0&1\end{array}\right]$$

$$\xrightarrow[R_2\to R_2-R_3]{R_1\to R_1+3R_3}\left[\begin{array}{ccc|ccc}1&0&0&-4&2&3\\0&1&0&2&-1&-1\\0&0&1&-1&0&1\end{array}\right]$$

$$\therefore \quad \text{its inverse is } \begin{bmatrix}-4&2&3\\2&-1&-1\\-1&0&1\end{bmatrix}$$

(d) $$\left[\begin{array}{cc|cc}2&4&1&0\\1&2&0&1\end{array}\right]\xrightarrow{R_1\leftrightarrow R_2}\left[\begin{array}{cc|cc}1&2&0&1\\2&4&1&0\end{array}\right]\xrightarrow{R_2\to R_2-2R_1}\left[\begin{array}{cc|cc}1&2&0&1\\0&0&1&-1\end{array}\right]$$

This cannot be changed so that $\begin{bmatrix}1&0\\0&1\end{bmatrix}$ occurs on the left side of solid line.

$\therefore$ The matrix has no inverse and is singular.

(e) $$\left[\begin{array}{cccc|cccc}1&-1&0&2&1&0&0&0\\2&2&0&1&0&1&0&0\\-3&-1&-5&1&0&0&1&0\\1&-1&0&0&0&0&0&1\end{array}\right]\xrightarrow[\substack{R_3\to R_3+3R_1\\ R_4\to R_4-R_1}]{R_2\to R_2-2R_1}\left[\begin{array}{cccc|cccc}1&-1&0&2&1&0&0&0\\0&4&0&-3&-2&1&0&0\\0&-4&-5&7&3&0&1&0\\0&0&0&-2&-1&0&0&1\end{array}\right]$$

$$\xrightarrow{R_3\to R_3+R_2}\left[\begin{array}{cccc|cccc}1&-1&0&2&1&0&0&0\\0&4&0&-3&-2&1&0&0\\0&0&-5&4&1&1&1&0\\0&0&0&-2&-1&0&0&1\end{array}\right]\xrightarrow[R_3\to R_3+2R_4]{\substack{R_1\to R_1+R_4\\ R_2\to R_2-2R_4}}\left[\begin{array}{cccc|cccc}1&-1&0&0&0&0&0&1\\0&4&0&1&0&1&0&-2\\0&0&-5&0&-1&1&1&2\\0&0&0&-2&-1&0&0&1\end{array}\right]$$

$$\xrightarrow[\substack{R_3\to -1/5R_3\\ R_4\to -1/2R_4}]{R_2\to 1/4R_2}\left[\begin{array}{cccc|cccc}1&-1&0&0&0&0&0&1\\0&1&0&\frac{1}{4}&0&\frac{1}{4}&0&-\frac{2}{4}\\0&0&1&0&\frac{1}{5}&-\frac{1}{5}&-\frac{1}{5}&-\frac{2}{5}\\0&0&0&1&\frac{1}{2}&0&0&-\frac{1}{2}\end{array}\right]\xrightarrow{R_2\to R_2-1/4R_4}\left[\begin{array}{cccc|cccc}1&-1&0&0&0&0&0&1\\0&1&0&0&-\frac{1}{8}&\frac{1}{4}&0&-\frac{3}{8}\\0&0&1&0&\frac{1}{5}&-\frac{1}{5}&-\frac{1}{5}&-\frac{2}{5}\\0&0&0&1&\frac{1}{2}&0&0&-\frac{1}{2}\end{array}\right]$$

$$\xrightarrow{R_1\to R_1+R_2}\left[\begin{array}{cccc|cccc}1&0&0&0&-\frac{1}{8}&\frac{1}{4}&0&\frac{5}{8}\\0&1&0&0&-\frac{1}{8}&\frac{1}{4}&0&-\frac{3}{8}\\0&0&1&0&\frac{1}{5}&-\frac{1}{5}&-\frac{1}{5}&-\frac{2}{5}\\0&0&0&1&\frac{1}{2}&0&0&-\frac{1}{2}\end{array}\right]$$

$$\therefore \quad \text{its inverse is } \begin{bmatrix}-\frac{1}{8}&\frac{1}{4}&0&\frac{5}{8}\\-\frac{1}{8}&\frac{1}{4}&0&-\frac{3}{8}\\\frac{1}{5}&-\frac{1}{5}&-\frac{1}{5}&-\frac{2}{5}\\\frac{1}{2}&0&0&-\frac{1}{2}\end{bmatrix}$$

(f)

$$\left[\begin{array}{cc|cc} \frac{1}{2} & \frac{1}{4} & 1 & 0 \\ -\frac{1}{4} & \frac{1}{8} & 0 & 1 \end{array}\right] \xrightarrow[R_2 \to 8R_2]{R_1 \to 4R_1} \left[\begin{array}{cc|cc} 2 & 1 & 4 & 0 \\ -2 & 1 & 0 & 8 \end{array}\right] \xrightarrow[R_2 \to R_2 + R_1]{} \left[\begin{array}{cc|cc} 2 & 1 & 4 & 0 \\ 0 & 2 & 4 & 8 \end{array}\right]$$

$$\xrightarrow[R_2 \to 1/2R_2]{} \left[\begin{array}{cc|cc} 2 & 1 & 4 & 0 \\ 0 & 1 & 2 & 4 \end{array}\right] \xrightarrow{R_1 \to R_1 - R_2} \left[\begin{array}{cc|cc} 2 & 0 & 2 & -4 \\ 0 & 1 & 2 & 4 \end{array}\right]$$

$$\xrightarrow{R_1 \to 1/2R_1} \left[\begin{array}{cc|cc} 1 & 0 & 1 & -2 \\ 0 & 1 & 2 & 4 \end{array}\right]$$

$$\therefore \quad \text{its inverse is } \begin{bmatrix} 1 & -2 \\ 2 & 4 \end{bmatrix}$$

This should be no surprise if you check the **(a)** part of this problem.

PROBLEM 2-13 Given the matrix $A = \begin{bmatrix} 1 & 2 & 0 \\ -1 & 3 & 0 \\ 2 & -1 & a \end{bmatrix}$, find all values a for which A^{-1} exists.

Solution

$$\left[\begin{array}{ccc|ccc} 1 & 2 & 0 & 1 & 0 & 0 \\ -1 & 3 & 0 & 0 & 1 & 0 \\ 2 & -1 & a & 0 & 0 & 1 \end{array}\right] \xrightarrow[R_3 \to R_3 - 2R_1]{R_2 \to R_2 + R_1} \left[\begin{array}{ccc|ccc} 1 & 2 & 0 & 1 & 0 & 0 \\ 0 & 5 & 0 & 1 & 1 & 0 \\ 0 & -5 & a & -2 & 0 & 1 \end{array}\right]$$

$$\xrightarrow[R_3 \to R_3 + R_2]{} \left[\begin{array}{ccc|ccc} 1 & 2 & 0 & 1 & 0 & 0 \\ 0 & 5 & 0 & 1 & 1 & 0 \\ 0 & 0 & a & -1 & 1 & 1 \end{array}\right]$$

$$\xrightarrow{R_2 \to 1/5R_2} \left[\begin{array}{ccc|ccc} 1 & 2 & 0 & 1 & 0 & 0 \\ 0 & 1 & 0 & \frac{1}{5} & \frac{1}{5} & 0 \\ 0 & 0 & a & -1 & 1 & 1 \end{array}\right]$$

$$\xrightarrow[R_3 \to 1/aR_3]{R_1 \to R_1 - 2R_2} \left[\begin{array}{ccc|ccc} 1 & 0 & 0 & \frac{3}{5} & -\frac{2}{5} & 0 \\ 0 & 1 & 0 & \frac{1}{5} & \frac{1}{5} & 0 \\ 0 & 0 & 1 & -\frac{1}{a} & \frac{1}{a} & \frac{1}{a} \end{array}\right]$$

A^{-1} exists for any value of a that is not zero.

$$\therefore \quad A^{-1} = \begin{bmatrix} \frac{3}{5} & -\frac{2}{5} & 0 \\ \frac{1}{5} & \frac{1}{5} & 0 \\ -\frac{1}{a} & \frac{1}{a} & \frac{1}{a} \end{bmatrix}$$

PROBLEM 2-14 For the given matrix A, find A^{-1} and then show that $A^{-1}A = I$:

$$A = \begin{bmatrix} 2 & -3 \\ 4 & 3 \end{bmatrix}$$

Solution

$$\left[\begin{array}{cc|cc} 2 & -3 & 1 & 0 \\ 4 & 3 & 0 & 1 \end{array}\right] \xrightarrow{R_2 \to R_2 - 2R_1} \left[\begin{array}{cc|cc} 2 & -3 & 1 & 0 \\ 0 & 9 & -2 & 1 \end{array}\right] \xrightarrow[R_2 \to 1/9R_2]{R_1 \to 1/2R_1} \left[\begin{array}{cc|cc} 1 & -\frac{3}{2} & \frac{1}{2} & 0 \\ 0 & 1 & -\frac{2}{9} & \frac{1}{9} \end{array}\right]$$

$$\xrightarrow{R_1 \to R_1 + 3/2R_2} \left[\begin{array}{cc|cc} 1 & 0 & \frac{1}{6} & \frac{1}{6} \\ 0 & 1 & -\frac{2}{9} & \frac{1}{9} \end{array}\right]$$

$$\therefore \quad A^{-1} = \begin{bmatrix} \frac{1}{6} & \frac{1}{6} \\ -\frac{2}{9} & \frac{1}{9} \end{bmatrix}$$

$$\therefore \quad A^{-1}A = \begin{bmatrix} \frac{1}{6} & \frac{1}{6} \\ -\frac{2}{9} & \frac{1}{9} \end{bmatrix}\begin{bmatrix} 2 & -3 \\ 4 & 3 \end{bmatrix} = \begin{bmatrix} 1 & 0 \\ 0 & 1 \end{bmatrix}$$

PROBLEM 2-15 Given the two matrices $A = \begin{bmatrix} 1 & -3 \\ 2 & 4 \end{bmatrix}$ and $B = \begin{bmatrix} 2 & 5 \\ 1 & 3 \end{bmatrix}$, find **(a)** A^{-1} and B^{-1}, **(b)** AB and $(AB)^{-1}$, **(c)** $B^{-1}A^{-1}$ and $B^{-1}A^{-1} = (AB)^{-1}$, and **(d)** $(B^T)^{-1}$ and $(B^{-1})^T$.

Solution

(a)

$$\left[\begin{array}{cc|cc} 1 & -3 & 1 & 0 \\ 2 & 4 & 0 & 1 \end{array}\right] \xrightarrow{R_2 \to R_2 - 2R_1} \left[\begin{array}{cc|cc} 1 & -3 & 1 & 0 \\ 0 & 10 & -2 & 1 \end{array}\right]$$

$$\xrightarrow{R_2 \to 1/10 R_2} \left[\begin{array}{cc|cc} 1 & -3 & 1 & 0 \\ 0 & 1 & -\frac{1}{5} & \frac{1}{10} \end{array}\right] \xrightarrow{R_1 \to R_1 + 3R_2} \left[\begin{array}{cc|cc} 1 & 0 & \frac{2}{5} & \frac{3}{10} \\ 0 & 1 & -\frac{1}{5} & \frac{1}{10} \end{array}\right]$$

$$\therefore \quad A^{-1} = \begin{bmatrix} \frac{2}{5} & \frac{3}{10} \\ -\frac{1}{5} & \frac{1}{10} \end{bmatrix}$$

$$\left[\begin{array}{cc|cc} 2 & 5 & 1 & 0 \\ 1 & 3 & 0 & 1 \end{array}\right] \xrightarrow{R_1 \to R_1 - R_2} \left[\begin{array}{cc|cc} 1 & 2 & 1 & -1 \\ 1 & 3 & 0 & 1 \end{array}\right] \xrightarrow{R_2 \to R_2 - R_1} \left[\begin{array}{cc|cc} 1 & 2 & 1 & -1 \\ 0 & 1 & -1 & 2 \end{array}\right]$$

$$\xrightarrow{R_1 \to R_1 - 2R_2} \left[\begin{array}{cc|cc} 1 & 0 & 3 & -5 \\ 0 & 1 & -1 & 2 \end{array}\right]$$

$$\therefore \quad B^{-1} = \begin{bmatrix} 3 & -5 \\ -1 & 2 \end{bmatrix}$$

(b)

$$AB = \begin{bmatrix} 1 & -3 \\ 2 & 4 \end{bmatrix}\begin{bmatrix} 2 & 5 \\ 1 & 3 \end{bmatrix} = \begin{bmatrix} -1 & -4 \\ 8 & 22 \end{bmatrix}$$

$$(AB)^{-1}: \left[\begin{array}{cc|cc} -1 & -4 & 1 & 0 \\ 8 & 22 & 0 & 1 \end{array}\right] \xrightarrow{R_2 \to R_2 + 8R_1} \left[\begin{array}{cc|cc} -1 & -4 & 1 & 0 \\ 0 & -10 & 8 & 1 \end{array}\right]$$

$$\xrightarrow[R_2 \to -1/10 R_2]{R_1 \to -1R_1} \left[\begin{array}{cc|cc} 1 & 4 & -1 & 0 \\ 0 & 1 & -\frac{4}{5} & -\frac{1}{10} \end{array}\right] \xrightarrow{R_1 \to R_1 - 4R_2} \left[\begin{array}{cc|cc} 1 & 0 & \frac{11}{5} & \frac{2}{5} \\ 0 & 1 & -\frac{4}{5} & -\frac{1}{10} \end{array}\right]$$

$$\therefore \quad (AB)^{-1} = \begin{bmatrix} \frac{11}{5} & \frac{2}{5} \\ -\frac{4}{5} & -\frac{1}{10} \end{bmatrix}$$

(c)

$$B^{-1}A^{-1} = \begin{bmatrix} 3 & -5 \\ -1 & 2 \end{bmatrix}\begin{bmatrix} \frac{2}{5} & \frac{3}{10} \\ -\frac{1}{5} & \frac{1}{10} \end{bmatrix} = \begin{bmatrix} \frac{11}{5} & \frac{4}{10} \\ -\frac{4}{5} & -\frac{1}{10} \end{bmatrix} = \begin{bmatrix} \frac{11}{5} & \frac{2}{5} \\ -\frac{4}{5} & -\frac{1}{10} \end{bmatrix}$$

$$\therefore \quad B^{-1}A^{-1} = (AB)^{-1}$$

(d)

$$B^T = \begin{bmatrix} 2 & 5 \\ 1 & 3 \end{bmatrix}^T = \begin{bmatrix} 2 & 1 \\ 5 & 3 \end{bmatrix}$$

$$(B^T)^{-1}: \left[\begin{array}{cc|cc} 2 & 1 & 1 & 0 \\ 5 & 3 & 0 & 1 \end{array}\right] \xrightarrow{R_2 \to R_2 - 2R_1} \left[\begin{array}{cc|cc} 2 & 1 & 1 & 0 \\ 1 & 1 & -2 & 1 \end{array}\right] \xrightarrow{R_1 \leftrightarrow R_2} \left[\begin{array}{cc|cc} 1 & 1 & -2 & 1 \\ 2 & 1 & 1 & 0 \end{array}\right]$$

$$\xrightarrow{R_2 \to R_2 - 2R_1} \left[\begin{array}{cc|cc} 1 & 1 & -2 & 1 \\ 0 & -1 & 5 & -2 \end{array}\right] \xrightarrow{R_1 \to R_1 + R_2} \left[\begin{array}{cc|cc} 1 & 0 & 3 & -1 \\ 0 & -1 & 5 & -2 \end{array}\right]$$

$$\xrightarrow{R_2 \to -1R_2} \left[\begin{array}{cc|cc} 1 & 0 & 3 & -1 \\ 0 & 1 & -5 & 2 \end{array}\right]$$

$$\therefore \quad (B^T)^{-1} = \begin{bmatrix} 3 & -1 \\ -5 & 2 \end{bmatrix}$$

$$(B^{-1})^T = \begin{bmatrix} 3 & -5 \\ -1 & 2 \end{bmatrix}^T = \begin{bmatrix} 3 & -1 \\ -5 & 2 \end{bmatrix}$$

$$\therefore \quad (B^T)^{-1} = (B^{-1})^T$$

PROBLEM 2-16 Given $A = \begin{bmatrix} \frac{1}{2} & \frac{1}{4} \\ -\frac{1}{4} & \frac{1}{8} \end{bmatrix}$, find $(\frac{1}{8}A)^{-1}$.

Solution $(\frac{1}{8}A)^{-1} = 8A^{-1}$. From Problem 2-12 (**a**), $A^{-1} = \begin{bmatrix} 1 & -2 \\ 2 & 4 \end{bmatrix}$. Therefore, $8A^{-1} = 8\begin{bmatrix} 1 & -2 \\ 2 & 4 \end{bmatrix} = \begin{bmatrix} 8 & -16 \\ 16 & 32 \end{bmatrix}$.

PROBLEM 2-17 Show that $B = \begin{bmatrix} \frac{1}{5} & \frac{2}{5} & 0 \\ -\frac{1}{5} & \frac{3}{5} & 0 \\ \frac{2}{5} & -\frac{1}{5} & 1 \end{bmatrix}$ and $C = \begin{bmatrix} 3 & -2 & 0 \\ 1 & 1 & 0 \\ -1 & 1 & 1 \end{bmatrix}$ are invertible without finding their inverses.

Solution If $BC = I$, B and C are invertible:

$$\begin{bmatrix} \frac{1}{5} & \frac{2}{5} & 0 \\ -\frac{1}{5} & \frac{3}{5} & 0 \\ \frac{2}{5} & -\frac{1}{5} & 1 \end{bmatrix} \begin{bmatrix} 3 & -2 & 0 \\ 1 & 1 & 0 \\ -1 & 1 & 1 \end{bmatrix} = \begin{bmatrix} 1 & 0 & 0 \\ 0 & 1 & 0 \\ 0 & 0 & 1 \end{bmatrix}$$

$$\therefore \quad B \text{ and } C \text{ are invertible.}$$

PROBLEM 2-18 Solve the following homogeneous systems of linear equations:

(**a**) $x - 2y + z = 0$
$x + y - 2z = 0$

(**b**) $x + y + z = 0$
$2x - y = 0$
$x - 2y - z = 0$

Solution

(**a**)

$$\begin{bmatrix} 1 & -2 & 1 & 0 \\ 1 & 1 & -2 & 0 \end{bmatrix} \xrightarrow{R_2 \to R_2 - R_1} \begin{bmatrix} 1 & -2 & 1 & 0 \\ 0 & 3 & -3 & 0 \end{bmatrix} \xrightarrow{R_2 \to 1/3R_2} \begin{bmatrix} 1 & -2 & 1 & 0 \\ 0 & 1 & -1 & 0 \end{bmatrix}$$

$$\xrightarrow{R_1 \to R_1 + 2R_2} \begin{bmatrix} 1 & 0 & -1 & 0 \\ 0 & 1 & -1 & 0 \end{bmatrix}$$

$$\therefore \quad \begin{matrix} x - z = 0 \\ y - z = 0 \end{matrix} \quad \text{or} \quad \begin{matrix} x = z \\ y = z \end{matrix}$$

Let $z = t$, where t is any arbitrary real number. Therefore, any ordered triple of form (t, t, t) is a solution of the system.

(b) Begin with the coefficient matrix A:

$$A = \begin{bmatrix} 1 & 1 & 1 \\ 2 & -1 & 0 \\ 1 & 2 & -1 \end{bmatrix} \xrightarrow[R_3 \to R_3 - R_1]{R_2 \to R_2 - 2R_1} \begin{bmatrix} 1 & 1 & 1 \\ 0 & -3 & -2 \\ 0 & 1 & -2 \end{bmatrix} \xrightarrow{R_2 \leftrightarrow R_3} \begin{bmatrix} 1 & 1 & 1 \\ 0 & 1 & -2 \\ 0 & -3 & -2 \end{bmatrix}$$

$$\xrightarrow[R_3 \to R_3 + 3R_2]{R_1 \to R_1 - R_2} \begin{bmatrix} 1 & 0 & 3 \\ 0 & 1 & -2 \\ 0 & 0 & -8 \end{bmatrix} \xrightarrow[R_3 \to -1/8 R_3]{} \begin{bmatrix} 1 & 0 & 3 \\ 0 & 1 & -2 \\ 0 & 0 & 1 \end{bmatrix} \xrightarrow[R_2 \to R_2 + 2R_3]{R_1 \to R_1 - 3R_3} \begin{bmatrix} 1 & 0 & 0 \\ 0 & 1 & 0 \\ 0 & 0 & 1 \end{bmatrix}$$

Since A is row equivalent to I_3, A is invertible and the only solution is the trivial solution $(0, 0, 0)$.

PROBLEM 2-19 Solve each of the following systems by using the invertible matrix method.

(a)
$$\begin{aligned} x - y + z &= 3 \\ 2x + y - 2z &= 4 \\ x + y - z &= -5 \end{aligned}$$

(b)
$$\begin{bmatrix} 1 & -2 & -1 \\ 3 & -1 & 4 \\ -2 & -3 & 2 \end{bmatrix} \begin{bmatrix} x \\ y \\ z \end{bmatrix} = \begin{bmatrix} 2 \\ -1 \\ 3 \end{bmatrix}$$

Solution

(a)
$$\begin{bmatrix} 1 & -1 & 1 \\ 2 & 1 & -2 \\ 1 & 1 & -1 \end{bmatrix} \begin{bmatrix} x \\ y \\ z \end{bmatrix} = \begin{bmatrix} 3 \\ 4 \\ -5 \end{bmatrix} \qquad \left[\begin{array}{ccc|ccc} 1 & -1 & 1 & 1 & 0 & 0 \\ 2 & 1 & -2 & 0 & 1 & 0 \\ 1 & 1 & -1 & 0 & 0 & 1 \end{array}\right]$$

$$\xrightarrow[R_3 \to R_3 - R_1]{R_2 \to R_2 - 2R_1} \left[\begin{array}{ccc|ccc} 1 & -1 & 1 & 1 & 0 & 0 \\ 0 & 3 & -4 & -2 & 1 & 0 \\ 0 & 2 & -2 & -1 & 0 & 1 \end{array}\right] \xrightarrow{R_2 \to R_2 - R_3} \left[\begin{array}{ccc|ccc} 1 & -1 & 1 & 1 & 0 & 0 \\ 0 & 1 & -2 & -1 & 1 & -1 \\ 0 & 2 & -2 & -1 & 0 & 1 \end{array}\right]$$

$$\xrightarrow[R_3 \to R_3 - 2R_2]{} \left[\begin{array}{ccc|ccc} 1 & -1 & 1 & 1 & 0 & 0 \\ 0 & 1 & -2 & -1 & 1 & -1 \\ 0 & 0 & 2 & 1 & -2 & 3 \end{array}\right] \xrightarrow[R_2 \to R_2 + R_3]{R_1 \to R_1 + R_2} \left[\begin{array}{ccc|ccc} 1 & 0 & -1 & 0 & 1 & -1 \\ 0 & 1 & 0 & 0 & -1 & 2 \\ 0 & 0 & 2 & 1 & -2 & 3 \end{array}\right]$$

$$\xrightarrow[R_3 \to 1/2 R_3]{} \left[\begin{array}{ccc|ccc} 1 & 0 & -1 & 0 & 1 & -1 \\ 0 & 1 & 0 & 0 & -1 & 2 \\ 0 & 0 & 1 & \frac{1}{2} & -1 & \frac{3}{2} \end{array}\right] \xrightarrow{R_1 \to R_1 + R_3} \left[\begin{array}{ccc|ccc} 1 & 0 & 0 & \frac{1}{2} & 0 & \frac{1}{2} \\ 0 & 1 & 0 & 0 & -1 & 2 \\ 0 & 0 & 1 & \frac{1}{2} & -1 & \frac{3}{2} \end{array}\right]$$

$$\begin{bmatrix} \frac{1}{2} & 0 & \frac{1}{2} \\ 0 & -1 & 2 \\ \frac{1}{2} & -1 & \frac{3}{2} \end{bmatrix} \begin{bmatrix} 3 \\ 4 \\ -5 \end{bmatrix} = \begin{bmatrix} -\frac{2}{2} \\ -14 \\ -10 \end{bmatrix} = \begin{bmatrix} -1 \\ -14 \\ -10 \end{bmatrix}$$

$$\therefore \quad x = -1, y = -14, z = -10$$

(b)
$$\left[\begin{array}{ccc|ccc} 1 & -2 & -1 & 1 & 0 & 0 \\ 3 & -1 & 4 & 0 & 1 & 0 \\ -2 & -3 & 2 & 0 & 0 & 1 \end{array}\right] \xrightarrow[R_3 \to R_3 + 2R_1]{R_2 \to R_2 - 3R_1} \left[\begin{array}{ccc|ccc} 1 & -2 & -1 & 1 & 0 & 0 \\ 0 & 5 & 7 & -3 & 1 & 0 \\ 0 & -7 & 0 & 2 & 0 & 1 \end{array}\right]$$

$$\xrightarrow[R_3 \to -1/7 R_3]{} \left[\begin{array}{ccc|ccc} 1 & -2 & -1 & 1 & 0 & 0 \\ 0 & 5 & 7 & -3 & 1 & 0 \\ 0 & 1 & 0 & -\frac{2}{7} & 0 & -\frac{1}{7} \end{array}\right] \xrightarrow{R_2 \leftrightarrow R_3} \left[\begin{array}{ccc|ccc} 1 & -2 & -1 & 1 & 0 & 0 \\ 0 & 1 & 0 & -\frac{2}{7} & 0 & -\frac{1}{7} \\ 0 & 5 & 7 & -3 & 1 & 0 \end{array}\right]$$

$$\xrightarrow[R_3 \to R_3 - 5R_2]{R_1 \to R_1 + 2R_2} \left[\begin{array}{ccc|ccc} 1 & 0 & -1 & \frac{3}{7} & 0 & -\frac{2}{7} \\ 0 & 1 & 0 & -\frac{2}{7} & 0 & -\frac{1}{7} \\ 0 & 0 & 7 & -\frac{11}{7} & 1 & \frac{5}{7} \end{array}\right] \xrightarrow[R_3 \to 1/7 R_3]{} \left[\begin{array}{ccc|ccc} 1 & 0 & -1 & \frac{3}{7} & 0 & -\frac{2}{7} \\ 0 & 1 & 0 & -\frac{2}{7} & 0 & -\frac{1}{7} \\ 0 & 0 & 1 & -\frac{11}{49} & \frac{1}{7} & \frac{5}{49} \end{array}\right]$$

$$\xrightarrow{R_1 \to R_1 + R_3} \left[\begin{array}{ccc|ccc} 1 & 0 & 0 & \frac{10}{49} & \frac{1}{7} & -\frac{9}{49} \\ 0 & 1 & 0 & -\frac{2}{7} & 0 & -\frac{1}{7} \\ 0 & 0 & 1 & -\frac{11}{49} & \frac{1}{7} & \frac{5}{49} \end{array}\right]$$

$$\begin{bmatrix} \frac{10}{49} & \frac{1}{7} & -\frac{9}{49} \\ -\frac{2}{7} & 0 & -\frac{1}{7} \\ -\frac{11}{49} & \frac{1}{7} & \frac{5}{49} \end{bmatrix} \begin{bmatrix} 2 \\ -1 \\ 3 \end{bmatrix} = \begin{bmatrix} -\frac{14}{49} \\ -\frac{7}{7} \\ -\frac{14}{49} \end{bmatrix} = \begin{bmatrix} -\frac{2}{7} \\ -1 \\ -\frac{2}{7} \end{bmatrix}$$

$$\therefore \quad x = -2/7, y = -1, z = -2/7$$

PROBLEM 2-20 Find the conditions on a so that the linear system has a unique solution.

$$x + y + 2z = 0$$
$$2x - 3y - z = 1$$
$$3x - 2y + 3z = a$$

Solution The coefficient matrix $A = \begin{bmatrix} 1 & 1 & 2 \\ 2 & -3 & -1 \\ 3 & -2 & 3 \end{bmatrix}$ has inverse

$$\begin{bmatrix} \frac{11}{10} & \frac{7}{10} & -\frac{5}{10} \\ \frac{9}{10} & \frac{3}{10} & -\frac{5}{10} \\ -\frac{5}{10} & -\frac{5}{10} & \frac{5}{10} \end{bmatrix} = A^{-1}$$

$$\therefore \quad \begin{bmatrix} \frac{11}{10} & \frac{7}{10} & -\frac{5}{10} \\ \frac{9}{10} & \frac{3}{10} & -\frac{5}{10} \\ -\frac{5}{10} & -\frac{5}{10} & \frac{5}{10} \end{bmatrix} \begin{bmatrix} 0 \\ 1 \\ a \end{bmatrix} = \begin{bmatrix} \frac{7}{10} - \frac{5}{10}a \\ \frac{3}{10} - \frac{5}{10}a \\ -\frac{5}{10} + \frac{5}{10}a \end{bmatrix} = \begin{bmatrix} x \\ y \\ z \end{bmatrix}$$

$$\therefore \quad x = 7/10 - (5/10)a, y = 3/10 - (5/10)a, z = -5/10 + (5/10)a$$

There are no restrictions on a, so a can take on any value and the system would have a unique solution.

Supplementary Problems

PROBLEM 2-21 Find a linear system whose augmented matrix is:

(a) $\begin{bmatrix} 1 & -4 & 3 \\ 2 & 3 & -2 \end{bmatrix}$ **(b)** $\begin{bmatrix} 3 & -1 & -3 & -1 \\ 2 & 1 & 1 & -5 \\ 2 & -3 & 2 & -11 \end{bmatrix}$

PROBLEM 2-22 Use elementary row operations to change each of the following matrices into reduced row echelon form.

(a) $\begin{bmatrix} 2 & -3 & 4 \\ 1 & 2 & -1 \end{bmatrix}$ **(b)** $\begin{bmatrix} 1 & -2 & 3 & 2 \\ 2 & 1 & 1 & -1 \\ 3 & -1 & 2 & 3 \end{bmatrix}$

PROBLEM 2-23 Find the inverses of the following matrices:

(a) $\begin{bmatrix} \frac{1}{2} & -\frac{1}{3} \\ 3 & 2 \end{bmatrix}$ **(b)** $\begin{bmatrix} 1 & 1 & 2 \\ 2 & -1 & -1 \\ 1 & 2 & 3 \end{bmatrix}$ **(c)** $\begin{bmatrix} 2 & -3 \\ -4 & 6 \end{bmatrix}$

PROBLEM 2-24 Given $A = \begin{bmatrix} 1 & -2 \\ 3 & 4 \end{bmatrix}$ and $B = \begin{bmatrix} 1 & 3 \\ -2 & 5 \end{bmatrix}$, find **(a)** A^{-1}, **(b)** B^{-1}, **(c)** $(AB)^{-1}$, **(d)** $(1/4B)^{-1}$, and **(e)** $(A^2)^{-1}$.

PROBLEM 2-25 Solve the following homogeneous systems of linear equations:

(a)
$$\begin{aligned} 2x - y + z &= 0 \\ x \qquad - 2z &= 0 \end{aligned}$$

(b)
$$\begin{aligned} x - 2y + 3z &= 0 \\ 2x - y + z &= 0 \\ x + y - 2z &= 0 \end{aligned}$$

PROBLEM 2-26 Solve the following systems by using any matrix method:

(a)
$$\begin{aligned} 3x - y &= 7 \\ 2x + y &= 5 \end{aligned}$$

(b)
$$\begin{aligned} x - 2y + z &= 4 \\ 3x \qquad + 2z &= 2 \\ y - z &= -1 \end{aligned}$$

PROBLEM 2-27 Solve the linear system by any method.

$$x + 2y - z = 5$$

$$2x + y + z = -5$$

PROBLEM 2-28 Does the homogeneous system having $\begin{bmatrix} 1 & -1 & 2 \\ 1 & 2 & 0 \\ 2 & -1 & 3 \end{bmatrix}$ as coefficient matrix have a nontrivial solution? Give a reason for your answer.

Answers to Supplementary Problems

2-21 **(a)** $x - 4y = 3$, $2x + 3y = -2$
(b) $3x - y - 3z = -1$, $2x + y + z = -5$, $2x - 3y + 2z = -11$

2-22 **(a)** $\begin{bmatrix} 1 & 0 & \frac{5}{7} \\ 0 & 1 & -\frac{6}{7} \end{bmatrix}$ **(b)** $\begin{bmatrix} 1 & 0 & 0 & 1 \\ 0 & 1 & 0 & -2 \\ 0 & 0 & 1 & -1 \end{bmatrix}$

2-23 **(a)** $\begin{bmatrix} 1 & \frac{1}{6} \\ -\frac{3}{2} & \frac{1}{4} \end{bmatrix}$ **(c)** Has no inverse

(b) $\begin{bmatrix} -\frac{1}{2} & \frac{1}{2} & \frac{1}{2} \\ -\frac{7}{2} & \frac{1}{2} & \frac{5}{2} \\ \frac{5}{2} & -\frac{1}{2} & -\frac{3}{2} \end{bmatrix}$

2-24 **(a)** $A^{-1} = \begin{bmatrix} \frac{4}{10} & \frac{2}{10} \\ -\frac{3}{10} & \frac{1}{10} \end{bmatrix}$

(b) $B^{-1} = \begin{bmatrix} \frac{5}{11} & -\frac{3}{11} \\ \frac{2}{11} & \frac{1}{11} \end{bmatrix}$

(c) $(AB)^{-1} = \begin{bmatrix} \frac{29}{110} & \frac{7}{110} \\ \frac{5}{110} & \frac{5}{110} \end{bmatrix} = B^{-1}A^{-1}$

(d) $(\frac{1}{4}B)^{-1} = 4(B^{-1}) = \begin{bmatrix} \frac{20}{11} & -\frac{12}{11} \\ \frac{8}{11} & \frac{4}{11} \end{bmatrix}$

(e) $(A^2)^{-1} = (A^{-1})^2 = \begin{bmatrix} \frac{4}{10} & \frac{2}{10} \\ -\frac{3}{10} & \frac{1}{10} \end{bmatrix}^2$

$= \begin{bmatrix} \frac{10}{100} & \frac{10}{100} \\ -\frac{15}{100} & -\frac{5}{100} \end{bmatrix} = \begin{bmatrix} \frac{1}{10} & \frac{1}{10} \\ -\frac{3}{20} & -\frac{1}{20} \end{bmatrix}$

2-25 **(a)** $(2t, 5t, t)$, t an arbitrary number
(b) $(\frac{1}{3}t, \frac{5}{3}t, t)$, t an arbitrary number

2-26 **(a)** $x = \frac{12}{5}$, $y = \frac{1}{5}$, or $(\frac{12}{5}, \frac{1}{5})$
(b) $x = 6/5$, $y = -9/5$, $z = -4/5$, or $(6/5, -9/5, -4/5)$

2-27 $x = -t$, $y = t$, $z = t - 5$, or triples of form $(-t, t, t - 5)$, t an arbitrary number

2-28 No. The matrix is row equivalent to I_3, so the matrix is invertible, and the system has only the trivial solution $(0, 0, 0)$.

3 DETERMINANTS

THIS CHAPTER IS ABOUT

- ☑ **Permutations**
- ☑ **The Determinant Function**
- ☑ **Properties of Determinants**
- ☑ **Evaluating Determinants by Row Reduction**
- ☑ **Evaluating Determinants by Cofactor Expansion**
- ☑ **Evaluating Determinants by the Sylvester Eliminant**

3-1. Permutations

A permutation is simply an ordered arrangement of the elements in a set without any repetitions. Thus

- An ordered arrangement of all the integers in the set $\{1, 2, \ldots, n\}$ without any repetitions is a **permutation**. The number of permutations of n distinct integers in the set $\{1, 2, \ldots, n\}$ is $n!$, which is the product of all the integers 1 through n, inclusive.

EXAMPLE 3-1: Find the number of permutations of the set $\{1, 2, 3\}$ and list them.

Solution: There are 3 integers in the set, so there are $3! = 3(2)(1) = 6$ permutations. The permutations are different arrangements of the integers 1, 2, and 3:

123 132 213 231 312 321

When the integers in a permutation are in ascending order, as in 1 2 3 4 5, it is said to be in *natural order*. But when a larger integer precedes a smaller one, as 4 does in the permutation 1 2 4 3 5, we have an **inversion**. A permutation may have several inversions.

EXAMPLE 3-2: Find the number of inversions in the permutation 4 1 6 3 2 5.

Solution: Start with 4, the digit on the left, and count all the digits to the right that are smaller than 4. They are 1, 3, and 2. Now, move to the right one position and count the number of digits to the right of 1 that are smaller than 1. Repeat the process until you have done the same for all digits in the permutation.

Digit	Smaller digits to its right	Number
416325	1, 3, 2	3
16325	none	0
6325	3, 2, 5	3
325	2	1
25	none	0
5	none	0
	Total inversions	7

3-2. The Determinant Function

A. Definition of a determinant

A determinant is a number associated with a square ($n \times n$) matrix.

- If $A = [a_{ij}]$ is an $n \times n$ matrix, its **determinant** is a number denoted by $\det(A)$ or $|A|$, which is defined by

$$\det(A) = \sum (\pm) a_{1j_1} a_{2j_2} \cdots a_{nj_n},$$

where the summation is over all permutations of $j_1, j_2, \ldots, j_n$. When the permutation of j_i's in a term has an even $(0, 2, 4, \ldots)$ number of inversions, the term is positive; and when the permutation of j_i's has an odd number $(1, 3, 5, \ldots)$ of inversions, the term is negative.

- If A is a 1×1 matrix, then $\det(A) = |a_{11}| = a_{11}$.

B. Evaluating determinants

1. The determinants of 2 × 2 matrices

EXAMPLE 3-3: Use the definition to evaluate $\det(A)$ where

$$A = \begin{bmatrix} a_{11} & a_{12} \\ a_{21} & a_{22} \end{bmatrix}$$

Solution: Since n is 2, there are $2! = 2$ permutations. Thus, you begin by writing two terms with i's in ascending order:

$$\det(A) = a_{1_}a_{2_} \quad a_{1_}a_{2_}$$

For the second (j_i) subscripts, insert the permutations of j_i's, 12, and 21:

$$\det(A) = a_{1\underline{1}}a_{2\underline{2}} \quad a_{1\underline{2}}a_{2\underline{1}}$$

Finally, since the number of inversions in 12 is 0 and the number of inversions in 21 is 1, the "+" sign goes with the term whose permutation is 12 and the "−" sign with the term whose permutation is 21:

$$\det(A) = \underbrace{a_{1\underline{1}}a_{2\underline{2}}}_{\text{0 inversions}} - \underbrace{a_{1\underline{2}}a_{2\underline{1}}}_{\text{1 inversions}}$$

From Example 3-3 you can see that

- If A is a 2×2 matrix, then

$$\det(A) = \begin{vmatrix} a_{11} & a_{12} \\ a_{21} & a_{22} \end{vmatrix} = a_{11}a_{22} - a_{21}a_{12}$$

which is the difference of the products on the diagonals, where the diagonal going from upper left to lower right is positive and the diagonal going from lower left to upper right is negative.

EXAMPLE 3-4: Evaluate the determinants of:

(a) $A = [7]$ **(b)** $B = \begin{bmatrix} 3 & -1 \\ 7 & 4 \end{bmatrix}$ **(c)** $C = \begin{bmatrix} -3 & 1 \\ -7 & -4 \end{bmatrix}$

Solution

(a) $\det(A) = 7$ **(b)** $|B| = 3(4) - 7(-1) = 12 + 7 = 19$

(c) $|C| = -3(-4) - (-7)(1) = 12 + 7 = 19$

---

2. **The determinants of 3 × 3 matrices: Using the definition**

EXAMPLE 3-5: Use the definition to evaluate det(A) where

$$A = \begin{bmatrix} a_{11} & a_{12} & a_{13} \\ a_{21} & a_{22} & a_{23} \\ a_{31} & a_{32} & a_{33} \end{bmatrix}$$

Solution: Here n is 3 and $3! = 6$. So you begin by writing the following six terms (where the permutations of i's are in ascending order):

$$|A| = a_{1_}a_{2_}a_{3_} \qquad a_{1_}a_{2_}a_{3_} \qquad a_{1_}a_{2_}a_{3_} \qquad a_{1_}a_{2_}a_{3_} \qquad a_{1_}a_{2_}a_{3_} \qquad a_{1_}a_{2_}a_{3_}$$

Fill in the blanks in each term with one of the permutations of j_i's 123, 132, 213, 231, 312, and 321. Then choose the sign of the term to be positive if the number of inversions in the permutation of j_i's is even and negative otherwise:

inversions: 0+0+0=0 0+0+1=1 1+0+0=1 0+1+1=2 1+1+0=2 1+1+1=3

$$|A| = a_{1\underline{1}}a_{2\underline{2}}a_{3\underline{3}} - a_{1\underline{1}}a_{2\underline{3}}a_{3\underline{2}} - a_{1\underline{2}}a_{2\underline{1}}a_{3\underline{3}} + a_{1\underline{2}}a_{2\underline{3}}a_{3\underline{1}} + a_{1\underline{3}}a_{2\underline{1}}a_{3\underline{2}} - a_{1\underline{3}}a_{2\underline{2}}a_{3\underline{1}}$$

note: Each product contains a factor from each row and each column.

---

3. **The determinants of 3 × 3 matrices: Using diagonals products**

We've seen that the determinant of a 2 × 2 matrix is the difference of the products on the diagonals. We can use a variation of this method to evaluate the determinants of 3 × 3 matrices, as follows:

(1) Repeat the first and second columns of the matrix and draw the diagonals:

$$A = \left[\begin{array}{ccc|cc} a_{11} & a_{12} & a_{13} & a_{11} & a_{12} \\ a_{21} & a_{22} & a_{23} & a_{21} & a_{22} \\ a_{31} & a_{32} & a_{33} & a_{31} & a_{32} \end{array}\right]$$

(2) Form the products of the elements on the diagonals from upper left to lower right and give each product a plus sign:

$$+ a_{11}a_{22}a_{33} + a_{12}a_{23}a_{31} + a_{13}a_{21}a_{32}$$

(3) Form the products of the elements on the diagonals from lower left to upper right and give each product a negative sign:

$$- a_{31}a_{22}a_{13} - a_{32}a_{23}a_{11} - a_{33}a_{21}a_{12}$$

(4) Sum the products obtained in steps (2) and (3) to get the determinant:

$$\det(A) = \left|\begin{array}{ccc|cc} a_{11} & a_{12} & a_{13} & a_{11} & a_{12} \\ a_{21} & a_{22} & a_{23} & a_{21} & a_{22} \\ a_{31} & a_{32} & a_{33} & a_{31} & a_{32} \end{array}\right|$$

(with + signs above the first three columns and − signs below the first three columns)

$$= a_{11}a_{22}a_{33} + a_{12}a_{23}a_{31} + a_{13}a_{21}a_{32} - a_{31}a_{22}a_{13} - a_{32}a_{23}a_{11} - a_{33}a_{21}a_{12}$$

We call this method **diagonal expansion.**

EXAMPLE 3-6: Evaluate the determinant of each of the following matrices:

(a) $A = \begin{bmatrix} -4 & 3 \\ 6 & 5 \end{bmatrix}$ **(b)** $B = \begin{bmatrix} 1 & 2 & 3 \\ 3 & -1 & 2 \\ 5 & 4 & 3 \end{bmatrix}$ **(c)** $C = \begin{bmatrix} 4 & 5 & 6 \\ 2 & 0 & -1 \\ 3 & 1 & 4 \end{bmatrix}$

Solution

(a)

$$\det(A) = \begin{vmatrix} -4 & 3 \\ 6 & 5 \end{vmatrix} = -4(5) - 6(3) = -38$$

(b)

$$\det(B) = \begin{vmatrix} 1 & 2 & 3 \\ 3 & -1 & 2 \\ 5 & 4 & 3 \end{vmatrix} \begin{matrix} 1 & 2 \\ 3 & -1 \\ 5 & 4 \end{matrix}$$

$$= 1(-1)(3) + 2(2)(5) + 3(3)(4) - 5(-1)(3) - 4(2)(1) - 3(3)(2) = 42$$

(c)

$$\det(C) = \begin{vmatrix} 4 & 5 & 6 \\ 2 & 0 & -1 \\ 3 & 1 & 4 \end{vmatrix} \begin{matrix} 4 & 5 \\ 2 & 0 \\ 3 & 1 \end{matrix}$$

$$= 4(0)(4) + 5(-1)(3) + 6(2)(1) - 3(0)(6) - 1(-1)(4) - 4(2)(5) = -39$$

3-3. Properties of Determinants

Since evaluating an $n \times n$ determinant by the definition involves calculating $n!$ products, it can be very messy to evaluate, say, a 4×4 determinant with $4! = 24$ products to be summed. However, there are some properties of determinants that can aid us in finding the value of a large determinant.

Property A *If all the elements of a row (column) are zero, the value of the determinant is zero.*

EXAMPLE 3-7: Verify that the determinant of each of the following matrices is zero:

$$A = \begin{bmatrix} -1 & 2 \\ 0 & 0 \end{bmatrix} \qquad B = \begin{bmatrix} -1 & 0 \\ 2 & 0 \end{bmatrix}$$

Solution

$$\det(A) = \begin{vmatrix} -1 & 2 \\ 0 & 0 \end{vmatrix} = -1(0) - 0(2) = 0$$

$$\det(B) = \begin{vmatrix} -1 & 0 \\ 2 & 0 \end{vmatrix} = -1(0) - 2(0) = 0$$

EXAMPLE 3-8: Evaluate the determinants of

(a) $A = \begin{bmatrix} 1 & 0 & 3 \\ 2 & 0 & 1 \\ -1 & 0 & 2 \end{bmatrix}$ **(b)** $B = \begin{bmatrix} 0 & 0 & 0 & 0 \\ 5 & 7 & 9 & 3 \\ 3 & -9 & 7 & 5 \\ 2 & 4 & -6 & 8 \end{bmatrix}$

Solution

(a) You could multiply this out using the diagonals method; but since all the elements in the second column are 0, then

$$\det(A) = \begin{vmatrix} 1 & 0 & 3 \\ 2 & 0 & 1 \\ -1 & 0 & 2 \end{vmatrix} = 0$$

(b) All the elements in row 1 are 0, so

$$\det(B) = \begin{vmatrix} 0 & 0 & 0 & 0 \\ 5 & 7 & 9 & 3 \\ 3 & -9 & 7 & 5 \\ 2 & 4 & -6 & 8 \end{vmatrix} = 0$$

note: Recall from the definition of a determinant that each product in the summation contains a factor from each row and each column. So if all the elements in a row or column are zero, the determinant *must* have a value of zero.

Property B *If two rows (columns) have corresponding elements that are identical, the value of the determinant is zero.*

EXAMPLE 3-9: Verify that the determinant of each of the following matrices is zero:

(a) $A = \begin{bmatrix} 1 & 1 \\ 2 & 2 \end{bmatrix}$ **(b)** $B = \begin{bmatrix} 1 & 2 & 3 \\ 2 & -1 & 4 \\ 1 & 2 & 3 \end{bmatrix}$

Solution

(a) $\det(A) = \begin{vmatrix} 1 & 1 \\ 2 & 2 \end{vmatrix} = 1(2) - 1(2) = 0$

(b) $\det(B) = \begin{vmatrix} 1 & 2 & 3 \\ 2 & -1 & 4 \\ 1 & 2 & 3 \end{vmatrix} = 1(-1)3 + 2(4)(1) + 3(2)(2) - 1(-1)(3) - 2(4)(1) - 3(2)(2)$

$= 0$

EXAMPLE 3-10: Evaluate the determinants of

(a) $A = \begin{bmatrix} 2 & 4 \\ 2 & 4 \end{bmatrix}$ **(b)** $B = \begin{bmatrix} 1 & 2 & 1 \\ 2 & -1 & 2 \\ 3 & 4 & 3 \end{bmatrix}$ **(c)** $C = \begin{bmatrix} a & b & c & d \\ e & f & g & h \\ a & b & c & d \\ i & j & k & l \end{bmatrix}$

Solution

(a) Rows 1 and 2 are identical: $\det(A) = 0$
(b) Columns 1 and 3 are identical: $\det(B) = 0$
(c) Rows 1 and 3 are identical: $\det(C) = 0$

Property C *Multiplying all the elements of a row (column) by a constant k is equivalent to multiplying the value of the determinant by k.*

EXAMPLE 3-11: Verify that $\begin{vmatrix} ka & kb \\ c & d \end{vmatrix} = k\begin{vmatrix} a & b \\ c & d \end{vmatrix}$.

Solution: $kad - kbc = k(ad - bc) = kad - kbc$.

EXAMPLE 3-12: Given $A = \begin{bmatrix} 2 & 1 \\ 6 & 2 \end{bmatrix}$ and $B = \begin{bmatrix} 2 & 2 \\ 6 & 4 \end{bmatrix}$, find $|B|$ without evaluating it directly.

Solution: $|A| = \begin{vmatrix} 2 & 1 \\ 6 & 2 \end{vmatrix} = 2(2) - 6(1) = 4 - 6 = -2$. Since the second column of B is twice the second column of A, then $|B| = 2|A| = 2(-2) = -4$.

EXAMPLE 3-13: If $\det(A) = 7$ and a matrix B is obtained from A by multiplying each element in the third row of A by 6, find $\det(B)$.

Solution: $\det(B) = 6\det(A) = 6(7) = 42$.

Property D *The value of a determinant is unchanged if a constant multiple of any row (column) is added to another row (column).*

EXAMPLE 3-14: Show that the following are true:

(a) $\begin{vmatrix} 2 & 3 \\ -1 & 4 \end{vmatrix} = \begin{vmatrix} 2 & 3 \\ 3 & 10 \end{vmatrix}$ where 2 times row one has been added to the second row

(b) $\begin{vmatrix} 2 & 1 & -3 \\ -4 & -1 & 2 \\ 3 & 2 & 4 \end{vmatrix} = \begin{vmatrix} -4 & 1 & -3 \\ 0 & -1 & 2 \\ 11 & 2 & 4 \end{vmatrix}$ where 2 times the third column has been added to the first column

Solution

(a)

$$\begin{vmatrix} 2 & 3 \\ -1 & 4 \end{vmatrix} = 2(4) - (-1)(3) = 11$$

$$\begin{vmatrix} 2 & 3 \\ 3 & 10 \end{vmatrix} = 2(10) - 3(3) = 11$$

(b)

$$\begin{vmatrix} 2 & 1 & -3 \\ -4 & -1 & 2 \\ 3 & 2 & 4 \end{vmatrix} \begin{matrix} 2 & 1 \\ -4 & -1 \\ 3 & 2 \end{matrix} = -8 + 6 + 24 - 9 - 8 + 16 = 21$$

$$\begin{vmatrix} -4 & 1 & -3 \\ 0 & -1 & 2 \\ 11 & 2 & 4 \end{vmatrix} \begin{matrix} -4 & 1 \\ 0 & -1 \\ 11 & 2 \end{matrix} = 16 + 22 + 0 - 33 + 16 + 0 = 21$$

Property E *If two rows (columns) of a determinant are interchanged, only the sign of the determinant is changed.*

EXAMPLE 3-15: Verify the following:

(a)

$$\begin{vmatrix} 2 & -3 \\ -1 & 4 \end{vmatrix} = -\begin{vmatrix} -1 & 4 \\ 2 & -3 \end{vmatrix}$$

(b)

$$\begin{vmatrix} -3 & 2 & 1 \\ 2 & 1 & 3 \\ -4 & -2 & 2 \end{vmatrix} = -\begin{vmatrix} 2 & -3 & 1 \\ 1 & 2 & 3 \\ -2 & -4 & 2 \end{vmatrix}$$

Solution

(a)

$$\begin{vmatrix} 2 & -3 \\ -1 & 4 \end{vmatrix} = 2(4) - (-1)(-3) = 8 - 3 = +5$$

$$\begin{vmatrix} -1 & 4 \\ 2 & -3 \end{vmatrix} = (-1)(-3) - (2)(4) = 3 - 8 = -5$$

(b)

$$\left|\begin{array}{ccc|cc} -3 & 2 & 1 & -3 & 2 \\ 2 & 1 & 3 & 2 & 1 \\ -4 & -2 & 2 & -4 & -2 \end{array}\right. = -6 - 24 - 4 + 4 - 18 - 8 = -56$$

$$\left|\begin{array}{ccc|cc} 2 & -3 & 1 & 2 & -3 \\ 1 & 2 & 3 & 1 & 2 \\ -2 & -4 & 2 & -2 & -4 \end{array}\right. = 8 + 18 - 4 + 4 + 24 + 6 = +56$$

Property F *The value of a determinant is unchanged if all corresponding rows and columns are interchanged.*

note: The interchange of the rows and columns in matrix A results in its **transpose,** denoted by A^T. This, property can be written as $\det(A^T) = \det(A)$.

EXAMPLE 3-16: Verify the following:

(a) $\begin{vmatrix} 2 & -3 \\ -1 & 4 \end{vmatrix} = \begin{vmatrix} 2 & -1 \\ -3 & 4 \end{vmatrix}$ **(b)** $\begin{vmatrix} 2 & 1 & 1 \\ 1 & 2 & 3 \\ -1 & 0 & 2 \end{vmatrix} = \begin{vmatrix} 2 & 1 & -1 \\ 1 & 2 & 0 \\ 1 & 3 & 2 \end{vmatrix}$

Solution

(a)

$$\begin{vmatrix} 2 & -3 \\ -1 & 4 \end{vmatrix} = 2(4) - (-1)(-3) = 8 - 3 = 5$$

$$\begin{vmatrix} 2 & -1 \\ -3 & 4 \end{vmatrix} = 2(4) - (-3)(-1) = 8 - 3 = 5$$

(b)

$$\left|\begin{array}{ccc|cc} 2 & 1 & 1 & 2 & 1 \\ 1 & 2 & 3 & 1 & 2 \\ -1 & 0 & 2 & -1 & 0 \end{array}\right. = 8 - 3 + 0 + 2 + 0 - 2 = 5$$

$$\left|\begin{array}{ccc|cc} 2 & 1 & -1 & 2 & 1 \\ 1 & 2 & 0 & 1 & 2 \\ 1 & 3 & 2 & 1 & 3 \end{array}\right. = 8 + 0 - 3 + 2 + 0 - 2 = 5$$

EXAMPLE 3-17: Use properties of determinants and the given value of $|A|$ to evaluate **(a)** $|B|$ and **(b)** $|C|$:

$$|A| = \begin{vmatrix} 5 & -6 & 3 \\ 4 & 2 & -1 \\ 3 & -5 & 4 \end{vmatrix} = 51$$

(a) $|B| = \begin{vmatrix} 5 & -6 & 3 \\ 3 & -5 & 4 \\ 4 & 2 & -1 \end{vmatrix}$ (b) $|C| = \begin{vmatrix} 8 & 4 & -2 \\ 5 & -6 & 3 \\ 3 & -5 & 4 \end{vmatrix}$

Solution

(a) Rows 2 and 3 of $|A|$ are interchanged to get $|B|$:

$$|B = -|A| = -51$$

(b) Row 2 of $|A|$ is multiplied by 2, and then rows 1 and 2 are interchanged to get $|C|$:

$$|C| = 2(-|A|) = 2(-51) = -102$$

EXAMPLE 3-18: What property or properties would you use to verify that each of the determinants is zero?

(a) $|A| = \begin{vmatrix} 1 & 2 & 3 & 4 \\ -2 & 1 & 4 & 2 \\ 1 & 2 & 3 & 4 \\ 2 & 5 & 0 & 3 \end{vmatrix}$ (b) $|B| = \begin{vmatrix} -3 & 2 & 3 & 1 \\ 2 & 3 & -2 & 1 \\ 5 & 4 & -5 & 2 \\ 1 & 1 & -1 & 3 \end{vmatrix}$

Solution

(a) Rows 1 and 3 are identical: $|A| = 0$.
(b) Multiply column 3 by -1. Then,

$$|B| = -\begin{vmatrix} -3 & 2 & -3 & 1 \\ 2 & 3 & 2 & 1 \\ 5 & 4 & 5 & 2 \\ 1 & 1 & 1 & 3 \end{vmatrix}$$

Since columns 1 and 3 now are identical, $|B| = -(0) = 0$.

3-4. Evaluating Determinants by Row Reduction

A. Definition of triangular matrix

The main diagonal of a square matrix is the diagonal $a_{11}, a_{22}, \ldots, a_{nn}$ from the upper left to the lower right. A matrix in which all of the elements above or below the main diagonal are 0's is said to be in **triangular form.** An **upper triangular** matrix is a square matrix having only zeros *below* its main diagonal. A **lower triangular** matrix is a square matrix having only zeros *above* its main diagonal.

note: Upper triangular form is also called **echelon form.**

EXAMPLE 3-19: Identify whether each of the following is upper triangular, lower triangular, both, or neither:

(a) $\begin{bmatrix} 2 & 5 & 6 \\ 0 & 1 & 4 \\ 0 & 0 & 3 \end{bmatrix}$ (b) $\begin{bmatrix} 2 & 0 & 0 \\ 0 & 1 & 0 \\ 0 & 0 & 3 \end{bmatrix}$ (c) $\begin{bmatrix} 2 & 5 & 0 \\ 4 & 1 & 0 \\ 0 & 0 & 3 \end{bmatrix}$ (d) $\begin{bmatrix} 3 & 0 & 0 \\ 4 & 1 & 0 \\ 6 & 5 & 2 \end{bmatrix}$

Solution

(a) Upper triangular: The elements below the main diagonal are all 0's.
(b) Both: The elements above and below the main diagonal are all 0's.

(c) Neither: There are elements on both sides of the main diagonal that are not 0's.
(d) Lower triangular: The elements above the main diagonal are all 0's

B. Property of a triangular matrix

- *The determinant of a triangular matrix is equal to the product of the numbers in its main diagonal.*

EXAMPLE 3-20: Evaluate the determinants of

(a) $A = \begin{bmatrix} -3 & 8 & 5 & 4 \\ 0 & 4 & 1 & -2 \\ 0 & 0 & 2 & 5 \\ 0 & 0 & 0 & 3 \end{bmatrix}$ **(b)** $B = \begin{bmatrix} -3 & 0 & 0 & 0 \\ 8 & 4 & 0 & 0 \\ 5 & 1 & 2 & 0 \\ 4 & -2 & 5 & 3 \end{bmatrix}$

Solution

(a) Since the matrix is upper triangular, its determinant is the product of the numbers on the main diagonal, -3, 4, 2, and 3. Hence

$$|A| = (-3)(4)(2)(3) = -72.$$

(b) Since the matrix is lower triangular, its determinant is the product of the numbers -3, 4, 2, and 3. Hence $|B| = -72$.

note: Here, $\det(B) = \det(A^T)$, so it comes as no surprise that $|A| = |B|$.

C. Row reduction

We saw in Section 3-3 the effects of performing certain types of elementary operations on determinants of square matrices. In particular,

If matrix B is obtained from matrix A by

(1) multiplying a row of A by k, then $|B| = k|A|$ (observe that this is the same as factoring k from the row, so that $|A| = (1/k)|B|$);
(2) interchanging two rows of A, then $|B| = -|A|$;
(3) adding a multiple of a row of A to another row of A, then $|B| = |A|$.

When these three operations are performed on rows only, they are called **elementary row operations**. By using these elementary row operations, you can change a square matrix of any order (size) into a triangular matrix. Its determinant can then be evaluated by multiplying the numbers on the main diagonal.

note: These operations are called elementary *row* operations, but because $\det(A) = \det(A^T)$, the words "row" and "column" are interchangeable.

recall: In the notation for row operations, we use "$\leftrightarrow$" for "interchange" and "$\rightarrow$" for "replace," e.g.,

(1) $R_1 \leftrightarrow R_2$ means "interchange rows one and two."
(2) $R_2 \rightarrow 4R_2$ means "replace row two by 4 time row two."
(3) $R_1 \rightarrow 3R_2 + R_1$ means "replace row one by 3 times row two added to row one."

EXAMPLE 3-21: Evaluate the determinant of each of the following matrices by changing to a triangular form:

(a) $\begin{bmatrix} 1 & 2 & 3 \\ 0 & 1 & 5 \\ 1 & 2 & 1 \end{bmatrix}$ **(b)** $\begin{bmatrix} 0 & 1 & 2 & 3 \\ 3 & 4 & 1 & -2 \\ 1 & 3 & 2 & 1 \\ -2 & 0 & 1 & 3 \end{bmatrix}$

Solution

(a) $$\begin{vmatrix} 1 & 2 & 3 \\ 0 & 1 & 5 \\ 1 & 2 & 1 \end{vmatrix} \underset{R_3 \to (-1)R_1 + R_3}{=} \begin{vmatrix} 1 & 2 & 3 \\ 0 & 1 & 5 \\ 0 & 0 & -2 \end{vmatrix} = 1(1)(-2) = -2$$

(b) $$\begin{vmatrix} 0 & 1 & 2 & 3 \\ 3 & 4 & 1 & -2 \\ 1 & 3 & 2 & 1 \\ -2 & 0 & 1 & 3 \end{vmatrix} \underset{R_1 \leftrightarrow R_3}{=} - \begin{vmatrix} 1 & 3 & 2 & 1 \\ 3 & 4 & 1 & -2 \\ 0 & 1 & 2 & 3 \\ -2 & 0 & 1 & 3 \end{vmatrix} \underset{\substack{R_2 \to -3R_1 + R_2 \\ R_4 \to 2R_1 + R_4}}{=} - \begin{vmatrix} 1 & 3 & 2 & 1 \\ 0 & -5 & -5 & -5 \\ 0 & 1 & 2 & 3 \\ 0 & 6 & 5 & 5 \end{vmatrix}$$

$$\underset{R_2 \to (-1/5)R_2}{=} -5 \begin{vmatrix} 1 & 3 & 2 & 1 \\ 0 & 1 & 1 & 1 \\ 0 & 1 & 2 & 3 \\ 0 & 6 & 5 & 5 \end{vmatrix} \underset{\substack{R_3 \to -R_2 + R_3 \\ R_4 \to -6R_2 + R_4}}{=} 5 \begin{vmatrix} 1 & 3 & 2 & 1 \\ 0 & 1 & 1 & 1 \\ 0 & 0 & 1 & 2 \\ 0 & 0 & -1 & -1 \end{vmatrix}$$

$$\underset{R_4 \to R_3 + R_4}{=} 5 \begin{vmatrix} 1 & 3 & 2 & 1 \\ 0 & 1 & 1 & 1 \\ 0 & 0 & 1 & 2 \\ 0 & 0 & 0 & 1 \end{vmatrix} = 5(1)(1)(1)(1) = 5$$

3-5. Evaluating Determinants by Cofactor Expansion

A. Definition of a minor

- Let $A = [a_{ij}]$ be an $n \times n$ matrix. If the ith row and jth column of A are deleted, the determinant of the remaining $(n-1) \times (n-1)$ matrix is called the **minor** M_{ij} of entry a_{ij}.

EXAMPLE 3-22: Form and evaluate the minors **(a)** M_{23} and **(b)** M_{31} of the matrix

$$A = \begin{bmatrix} 3 & -2 & 6 \\ 0 & 2 & 4 \\ 5 & 1 & 3 \end{bmatrix}$$

Solution

(a) $$M_{23} = \begin{vmatrix} 3 & -2 & \not{6} \\ \not{0} & \not{2} & \not{4} \\ 5 & 1 & \not{3} \end{vmatrix} = \begin{vmatrix} 3 & -2 \\ 5 & 1 \end{vmatrix} = 3(1) - (5)(-2) = 13$$

(b) $$M_{31} = \begin{vmatrix} \not{3} & -2 & 6 \\ \not{0} & 2 & 4 \\ \not{5} & \not{1} & \not{3} \end{vmatrix} = \begin{vmatrix} -2 & 6 \\ 2 & 4 \end{vmatrix} = -2(4) - 2(6) = -20$$

B. Definition of a cofactor

- If M_{ij} is the minor of entry a_{ij} of square matrix A, then the **cofactor** *of* a_{ij} is $C_{ij} = (-1)^{i+j} M_{ij}$.

EXAMPLE 3-23: Find the cofactors C_{23} and C_{31} of matrix A in Example 3-22.

Solution

(a) $C_{23} = (-1)^{2+3}M_{23} = -M_{23} = -13$
(b) $C_{31} = (-1)^{3+1}M_{31} = M_{31} = -20.$

C. Cofactor expansion

The determinant of a square matrix larger than 1×1 can be evaluated by multiplying the numbers in any row (column) by their corresponding cofactors and adding the resulting products. This technique is called **cofactor expansion**.

EXAMPLE 3-24: Use the elements in the third row and their corresponding cofactors to evaluate the determinant of

$$A = \begin{bmatrix} a_{11} & a_{12} & a_{13} \\ a_{21} & a_{22} & a_{23} \\ a_{31} & a_{32} & a_{33} \end{bmatrix}$$

Solution

$$\begin{aligned} \det A &= a_{31}C_{31} + a_{32}C_{32} + a_{33}C_{33} \\ &= (-1)^{3+1}a_{31}M_{31} + (-1)^{3+2}a_{32}M_{32} + (-1)^{3+3}a_{33}M_{33} \\ &= a_{31}M_{31} - a_{32}M_{32} + a_{33}M_{33} \\ &= a_{31}\begin{vmatrix} a_{12} & a_{13} \\ a_{22} & a_{23} \end{vmatrix} - a_{32}\begin{vmatrix} a_{11} & a_{13} \\ a_{21} & a_{23} \end{vmatrix} + a_{33}\begin{vmatrix} a_{11} & a_{12} \\ a_{21} & a_{22} \end{vmatrix} \\ &= a_{31}(a_{12}a_{23} - a_{22}a_{13}) - a_{32}(a_{11}a_{23} - a_{21}a_{13}) + a_{33}(a_{11}a_{22} - a_{21}a_{12}) \\ &= a_{31}a_{12}a_{23} - a_{31}a_{22}a_{13} - a_{32}a_{11}a_{23} + a_{32}a_{21}a_{13} + a_{33}a_{11}a_{22} - a_{33}a_{21}a_{12} \end{aligned}$$

EXAMPLE 3-25: Use cofactor expansions to evaluate the following determinants:

(a) $\begin{vmatrix} 3 & 0 & 4 \\ 8 & -2 & -1 \\ 4 & 0 & 3 \end{vmatrix}$ **(b)** $\begin{vmatrix} 2 & 0 & 0 & 0 \\ 3 & 0 & 4 & 0 \\ 2 & 3 & 5 & 4 \\ 3 & 0 & 6 & 1 \end{vmatrix}$

Solution

(a) Since the second column has only one nonzero number, let's expand by that column (this will minimize the amount of computations needed):

$$\begin{aligned} \begin{vmatrix} 3 & 0 & 4 \\ 8 & -2 & -1 \\ 4 & 0 & 3 \end{vmatrix} &= 0C_{12} + (-2)C_{22} + 0C_{32} \\ &= 0 + (-2)(-1)^{2+2}M_{22} + 0 \\ &= (-2)(-1)^{2+2}\begin{vmatrix} 3 & 4 \\ 4 & 3 \end{vmatrix} \\ &= -2[3(3) - 4(4)] = 14 \end{aligned}$$

Notice that you don't have to find the cofactors C_{12} *and* C_{32}, since each is multiplied by 0.

(b) To minimize the work, we may expand either by the first now or by the second column. Since we used a column in part (a), let's choose the first row here:

$$\begin{vmatrix} 2 & 0 & 0 & 0 \\ 3 & 0 & 4 & 0 \\ 2 & 3 & 5 & 4 \\ 3 & 0 & 6 & 1 \end{vmatrix} = 2C_{11} = 2(-1)^{1+1}M_{11} = 2(-1)^{1+1}\begin{vmatrix} 0 & 4 & 0 \\ 3 & 5 & 4 \\ 0 & 6 & 1 \end{vmatrix}$$

Repeat the process and choose row 1 again:

$$2\begin{vmatrix} 0 & 4 & 0 \\ 3 & 5 & 4 \\ 0 & 6 & 1 \end{vmatrix} = 2(-1)^{1+1}(4)(-1)^{1+2}M_{12} = 2(-1)^{1+1}(4)(-1)^{1+2}\begin{vmatrix} 3 & 4 \\ 0 & 1 \end{vmatrix} = -8[3(1) - 0(4)] = -24$$

EXAMPLE 3-26: Evaluate the determinant of the matrix

$$A = \begin{bmatrix} 2 & 1 & 3 & 2 \\ 1 & 2 & -1 & 3 \\ -1 & 0 & 2 & 2 \\ 4 & 3 & -2 & -3 \end{bmatrix}.$$

Solution: Since zero appears in the second column as well as in the third row, we can minimize our work by first applying row reduction to replace the 2 and 3 in the second column by zeros:

$$\begin{vmatrix} 2 & 1 & 3 & 2 \\ 1 & 2 & -1 & 3 \\ -1 & 0 & 2 & 2 \\ 4 & 3 & -2 & -3 \end{vmatrix} \underset{R_4 \to -3R_1 + R_4}{\overset{R_2 \to -2R_1 + R_2}{=}} \begin{vmatrix} 2 & 1 & 3 & 2 \\ -3 & 0 & -7 & -1 \\ -1 & 0 & 2 & 2 \\ -2 & 0 & -11 & -9 \end{vmatrix}$$

$$= 1(-1)^{1+2}\begin{vmatrix} -3 & -7 & -1 \\ -1 & 2 & 2 \\ -2 & -11 & -9 \end{vmatrix} \underset{R_3 \to -2R_2 + R_3}{\overset{R_1 \to -3R_2 + R_1}{=}} (-1)\begin{vmatrix} 0 & -13 & -7 \\ -1 & 2 & 2 \\ 0 & -15 & -13 \end{vmatrix}$$

$$= (-1)(-1)(-1)^{2+1}\begin{vmatrix} -13 & -7 \\ -15 & -13 \end{vmatrix} = -1[(-13)(-13) - (-15)(-7)] = -64$$

3-6. Evaluating Determinants by the Sylvester Eliminant

If we have an $n \times n$ matrix, where $n > 2$ and $a_{11} \neq 0$, we can evaluate the determinant of the matrix by a method that results in repeated reductions of the order of the determinant. This method, sometimes called **reduction by the Sylvester eliminant**, is commonly known as the **pivotal method** of expansion. The procedure is as follows:

$$\begin{vmatrix} a_{11} & a_{12} & a_{13} & \cdots & a_{1n} \\ a_{21} & a_{22} & a_{23} & \cdots & a_{2n} \\ a_{31} & a_{32} & a_{33} & \cdots & a_{3n} \\ \cdots & & \cdots & \cdots & \\ a_{n1} & a_{n2} & a_{n3} & \cdots & a_{nn} \end{vmatrix} = \frac{1}{a_{11}^{n-2}} \begin{vmatrix} \begin{vmatrix} a_{11} & a_{12} \\ a_{21} & a_{22} \end{vmatrix} & \begin{vmatrix} a_{11} & a_{13} \\ a_{21} & a_{23} \end{vmatrix} & \cdots & \begin{vmatrix} a_{11} & a_{1n} \\ a_{21} & a_{2n} \end{vmatrix} \\ \begin{vmatrix} a_{11} & a_{12} \\ a_{31} & a_{32} \end{vmatrix} & \begin{vmatrix} a_{11} & a_{13} \\ a_{31} & a_{33} \end{vmatrix} & \cdots & \begin{vmatrix} a_{11} & a_{2n} \\ a_{31} & a_{3n} \end{vmatrix} \\ \cdots & \cdots & & \cdots \\ \begin{vmatrix} a_{11} & a_{12} \\ a_{n1} & a_{n2} \end{vmatrix} & \begin{vmatrix} a_{11} & a_{13} \\ a_{n1} & a_{n3} \end{vmatrix} & \cdots & \begin{vmatrix} a_{11} & a_{1n} \\ a_{n1} & a_{nn} \end{vmatrix} \end{vmatrix}$$

The result is an $(n-1) \times (n-1)$ determinant whose elements are the values of the $(n-1)^2$ determinants of order 2. (This method is easier if you use a calculator for multiplying out large numbers that may appear.)

note: A 2×2 matrix is order 2, a 3×3 matrix is order 3, and so on. Thus, for example, the determinant of a 3×3 matrix can be reduced by the pivotal method to $(3-1)^2 = 4$ determinants of order 2.

EXAMPLE 3-27: Use the pivotal method to evaluate

$$\textbf{(a)} \begin{vmatrix} 3 & -1 & 2 \\ 2 & 3 & -1 \\ 4 & 2 & -3 \end{vmatrix} \qquad \textbf{(b)} \begin{vmatrix} 2 & -1 & 3 & 1 \\ -1 & 2 & 1 & 2 \\ 3 & 4 & -1 & -2 \\ 1 & -3 & 2 & 4 \end{vmatrix}.$$

Solution

(a) Here, the order is 3, so $n = 3$; and $a_{11} = 3$, so $1/a_{11}^{n-2} = 1/3^{3-2}$. Thus

$$\begin{vmatrix} 3 & -1 & 2 \\ 2 & 3 & -1 \\ 4 & 2 & -3 \end{vmatrix} = \frac{1}{3^{3-2}} \begin{vmatrix} \begin{vmatrix} 3 & -1 \\ 2 & 3 \end{vmatrix} & \begin{vmatrix} 3 & 2 \\ 2 & -1 \end{vmatrix} \\ \begin{vmatrix} 3 & -1 \\ 4 & 2 \end{vmatrix} & \begin{vmatrix} 3 & 2 \\ 4 & -3 \end{vmatrix} \end{vmatrix} = \frac{1}{3} \begin{vmatrix} 11 & -7 \\ 10 & -17 \end{vmatrix}$$

$$= (1/3)[11(-17) - 10(-7)] = -39$$

Check: By diagonals products, we get

$$\begin{vmatrix} 3 & -1 & 2 & 3 & -1 \\ 2 & 3 & -1 & 2 & 3 \\ 4 & 2 & -3 & 4 & 2 \end{vmatrix} = -27 + 4 + 8 - 24 + 6 - 6 = -39$$

(b) $n = 4$ and $a_{11} = 2$, so

$$\begin{vmatrix} 2 & -1 & 3 & 1 \\ -1 & 2 & 1 & 2 \\ 3 & 4 & -1 & -2 \\ 1 & -3 & 2 & 4 \end{vmatrix} = \frac{1}{2^{4-2}} \begin{vmatrix} \begin{vmatrix} 2 & -1 \\ -1 & 2 \end{vmatrix} & \begin{vmatrix} 2 & 3 \\ -1 & 1 \end{vmatrix} & \begin{vmatrix} 2 & 1 \\ -1 & 2 \end{vmatrix} \\ \begin{vmatrix} 2 & -1 \\ 3 & 4 \end{vmatrix} & \begin{vmatrix} 2 & 3 \\ 3 & -1 \end{vmatrix} & \begin{vmatrix} 2 & 1 \\ 3 & -2 \end{vmatrix} \\ \begin{vmatrix} 2 & -1 \\ 1 & -3 \end{vmatrix} & \begin{vmatrix} 2 & 3 \\ 1 & 2 \end{vmatrix} & \begin{vmatrix} 2 & 1 \\ 1 & 4 \end{vmatrix} \end{vmatrix}$$

$$= \frac{1}{4} \begin{vmatrix} 3 & 5 & 5 \\ 11 & -11 & -7 \\ -5 & 1 & 7 \end{vmatrix} = \frac{1}{4}\left(\frac{1}{3^{3-2}}\right) \begin{vmatrix} \begin{vmatrix} 3 & 5 \\ 11 & -11 \end{vmatrix} & \begin{vmatrix} 3 & 5 \\ 11 & -7 \end{vmatrix} \\ \begin{vmatrix} 3 & 5 \\ -5 & 1 \end{vmatrix} & \begin{vmatrix} 3 & 5 \\ -5 & 7 \end{vmatrix} \end{vmatrix}$$

$$= \frac{1}{4}\left(\frac{1}{3}\right) \begin{vmatrix} -88 & -76 \\ 28 & 46 \end{vmatrix} = \frac{1}{12}(-4048 + 2128) = -160$$

note: With a little practice, you can accomplish the same result in fewer steps simply by writing the values of the order-2 determinants as elements in successively smaller determinants, like this:

$$\begin{vmatrix}2&-1\\-1&2\end{vmatrix}\begin{vmatrix}2&3\\-1&1\end{vmatrix}\ \text{etc.} \qquad \begin{vmatrix}3&5\\11&-11\end{vmatrix}\ \text{etc.}$$

$$\begin{vmatrix} 2&-1&3&1\\ -1&2&1&2\\ 3&4&-1&-2\\ 1&-3&2&4 \end{vmatrix} = \frac{1}{4}\begin{vmatrix} 3&5&5\\ 11&-11&-7\\ -5&1&7 \end{vmatrix} = \frac{1}{12}\begin{vmatrix} -88&-76\\ 28&46 \end{vmatrix}$$

$$= \frac{1}{12}(-4048 + 2128) = -160.$$

- If a_{11} is zero, you must perform an elementary row operation before proceeding with the pivotal expansion.

EXAMPLE 3-28: Use the pivotal expansion method to evaluate the determinant of Example 3-21 **(b)**.

Solution

$$\begin{vmatrix} 0&1&2&3\\ 3&4&1&-2\\ 1&3&2&1\\ -2&0&1&3 \end{vmatrix} \overset{R_1 \leftrightarrow R_3}{=} - \begin{vmatrix} 1&3&2&1\\ 3&4&1&-2\\ 0&1&2&3\\ -2&0&1&3 \end{vmatrix}$$

$$= -\left(\frac{1}{1^{4-2}}\right)\begin{vmatrix} \begin{vmatrix}1&3\\3&4\end{vmatrix} & \begin{vmatrix}1&2\\3&1\end{vmatrix} & \begin{vmatrix}1&1\\3&-2\end{vmatrix}\\ \begin{vmatrix}1&3\\0&1\end{vmatrix} & \begin{vmatrix}1&2\\0&2\end{vmatrix} & \begin{vmatrix}1&1\\0&3\end{vmatrix}\\ \begin{vmatrix}1&3\\-2&0\end{vmatrix} & \begin{vmatrix}1&2\\-2&1\end{vmatrix} & \begin{vmatrix}1&1\\-2&3\end{vmatrix} \end{vmatrix}$$

$$= -\begin{vmatrix} -5&-5&-5\\ 1&2&3\\ 6&5&5 \end{vmatrix} = -\left(\frac{1}{(-5)^{3-2}}\right)\begin{vmatrix} \begin{vmatrix}-5&-5\\1&2\end{vmatrix} & \begin{vmatrix}-5&-5\\1&3\end{vmatrix}\\ \begin{vmatrix}-5&-5\\6&5\end{vmatrix} & \begin{vmatrix}-5&-5\\6&5\end{vmatrix} \end{vmatrix}$$

$$= -\left(\frac{1}{-5}\right)\begin{vmatrix} -5&-10\\ 5&5 \end{vmatrix} = \frac{1}{5}(-25 + 50) = 5$$

SUMMARY

1. A permutation is an ordered arrangement of objects. For n distinct objects, the number of permutations is $n!$
2. There is an inversion in a permutation when a larger integer precedes a small integer.
3. For every square ($n \times n$) matrix A, there is a number called the determinant of A associated with the matrix. It is denoted by $\det(A)$ or $|A|$ and defined by

$$\det(A) = \sum(\pm)a_{1j_1}a_{2j_2}\cdots a_{nj_n}$$

where the summation is over all permutations of $j_1, j_2, \ldots, j_n$.

4. • The value of a 1×1 determinant $|a_{11}|$ is a_{11}.
   • The value of a 2×2 determinant may be found by taking the difference of the products on the diagonals:

$$\begin{vmatrix} a_{11} & a_{12} \\ a_{21} & a_{22} \end{vmatrix} = a_{11}a_{22} - a_{21}a_{12}$$

- The value of a 3 × 3 determinant may be found by diagonal expansion:

$$\begin{vmatrix} a_{11} & a_{12} & a_{13} \\ a_{21} & a_{22} & a_{23} \\ a_{31} & a_{32} & a_{33} \end{vmatrix} \begin{matrix} a_{11} & a_{12} \\ a_{21} & a_{22} \\ a_{31} & a_{32} \end{matrix} = \begin{matrix} a_{11}a_{22}a_{33} + a_{12}a_{23}a_{31} + a_{13}a_{21}a_{32} - a_{31}a_{22}a_{13} \\ - a_{32}a_{23}a_{11} - a_{33}a_{21}a_{12} \end{matrix}$$

5. - If all the elements of a row are zero, the value of the determinant is zero.
   - If two rows have corresponding elements that are identical, the value of the determinant is zero.
6. If matrix B is obtained from $n \times n$ matrix A by:
   - multiplying a row of A by k, then $|B| = k|A|$;
   - interchanging two rows of A, then $|B| = -|A|$;
   - adding a multiple of a row of A to another row of A, then $|B| = |A|$.
7. If all the corresponding rows and columns are interchanged, the value of the determinant is unchanged:

$$\det(A^T) = \det(A).$$

[Hence, in the statement of properties of determinants the word row(s) may always be replaced by the word column(s).

8. By elementary row operations a square matrix can be changed to triangular form where all the elements above or below the main diagonal are 0's; then its determinant evaluated as the product of the numbers on its main diagonal.
9. The minor M_{ij} of element a_{ij} in a square matrix A is the $(n - 1) \times (n - 1)$ determinant that results when row i and column j are deleted.
10. The cofactor of element a_{ij} in a square matrix is $C_{ij} = (-1)^{i+j}M_{ij}$, where M_{ij} is the minor of a_{ij}.
11. An $n \times n$ determinant can be evaluated by multiplying the numbers a_{ij} in any row (column) by their respective cofactors C_{ij} and adding the resulting products.
12. An $n \times n$ determinant can be evaluated by the Sylvester eliminant or pivotal method, which results in $(n - 1) \times (n - 1)$ determinant whose elements are the values of the $(n - 1)^2$ determinants of order 2:

$$\begin{vmatrix} a_{11} & a_{12} & a_{13} & \cdots & a_{1n} \\ a_{21} & a_{22} & a_{23} & \cdots & a_{2n} \\ a_{31} & a_{32} & a_{33} & \cdots & a_{3n} \\ & \cdots & & \cdots & \cdots \\ a_{n1} & a_{n2} & a_{n3} & \cdots & a_{nn} \end{vmatrix} = \frac{1}{a_{11}^{n-2}} \begin{vmatrix} \begin{vmatrix} a_{11} & a_{12} \\ a_{21} & a_{22} \end{vmatrix} & \begin{vmatrix} a_{11} & a_{13} \\ a_{21} & a_{23} \end{vmatrix} & \cdots & \begin{vmatrix} a_{11} & a_{1n} \\ a_{21} & a_{2n} \end{vmatrix} \\ \begin{vmatrix} a_{11} & a_{12} \\ a_{31} & a_{32} \end{vmatrix} & \begin{vmatrix} a_{11} & a_{13} \\ a_{31} & a_{33} \end{vmatrix} & \cdots & \begin{vmatrix} a_{11} & a_{1n} \\ a_{31} & a_{3n} \end{vmatrix} \\ \cdots & \cdots & & \cdots \\ \begin{vmatrix} a_{11} & a_{12} \\ a_{n1} & a_{n2} \end{vmatrix} & \begin{vmatrix} a_{11} & a_{13} \\ a_{n1} & a_{n3} \end{vmatrix} & \cdots & \begin{vmatrix} a_{11} & a_{1n} \\ a_{n1} & a_{nn} \end{vmatrix} \end{vmatrix}$$

13. Evaluation of a determinant is easier if elementary row operations are used to change some of the numbers in the determinant to zero.

RAISE YOUR GRADES

Can you ...?

- ☑ list all the permutations of a set of numbers
- ☑ find the number of inversions for the digits in a number
- ☑ use the definition of a determinant to find its value
- ☑ use the diagonals products to evaluate 3 × 3 determinants

☑ list the three elementary row operation properties of determinants
☑ list the two properties from which we can immediately conclude that the value of the determinant is zero
☑ find the value of a determinant when the rows and columns are interchanged
☑ change a square matrix to triangular form by row reduction and evaluate its determinant
☑ find a minor and a cofactor of a determinant
☑ evaluate a determinant by using cofactor expansion
☑ evaluate a determinant by the pivotal method

SOLVED PROBLEMS

PROBLEM 3-1 List all the permutations of the set $\{1, 2, 3, 4\}$.

Solution Since $4! = 4(3)(2)(1) = 24$, there are 24 permutations. (The key to listing them is to develop some systematic pattern. A good pattern is to begin with 1 and list the six permutations which begin with 1, then begin with 2, etc.)

1 2 3 4, 1 2 4 3, 1 3 2 4, 1 3 4 2, 1 4 3 2, 1 4 2 3
2 1 3 4, 2 1 4 3, 2 3 1 4, 2 3 4 1, 2 4 1 3, 2 4 3 1
3 1 2 4, 3 1 4 2, 3 2 1 4, 3 2 4 1, 3 4 1 2, 3 4 2 1
4 1 2 3, 4 1 3 2, 4 2 1 3, 4 2 3 1, 4 3 1 2, 4 3 2 1

PROBLEM 3-2 Find the number of inversions in 4 1 3 2.

Solution

Digit	Smaller digits to its right	Number
4132	1, 3, 2	3
132	none	0
32	2	1
2	none	0
	total inversions	4

PROBLEM 3-3 Evaluate the determinant of each of the following matrices.

(a) $A = [-8]$ **(b)** $B = \begin{bmatrix} 2 & 3 \\ -5 & 6 \end{bmatrix}$ **(c)** $C = \begin{bmatrix} 2 & 3 & 5 \\ 3 & -1 & 0 \\ -4 & 2 & 1 \end{bmatrix}$

Solution

(a) Since A is a 1×1 matrix $[-8]$, then $\det(A) = -8$.
(b) Since B is a 2×2 matrix, $\det(B)$ is the difference of products on diagonals:

$$\det(B) = \begin{vmatrix} 2 & 3 \\ -5 & 6 \end{vmatrix} = 2(6) - (-5)(3) = 12 + 15 = 27$$

(c) Since B is a 3×3 matrix, $\det(B)$ can be found by diagonal expansion:

$$\det(C) = \begin{vmatrix} 2 & 3 & 5 \\ 3 & -1 & 0 \\ -4 & 2 & 1 \end{vmatrix} \begin{matrix} 2 & 3 \\ 3 & -1 \\ -4 & 2 \end{matrix}$$

$$= 2(-1)(1) + 3(0)(-4) + 5(3)(2) - (-4)(-1)(5) - 2(0)(2) - (1)(3)(3)$$

$$= -2 + 0 + 30 - 20 - 0 - 9 = -1$$

PROBLEM 3-4 Find the values of x for which $\begin{vmatrix} x-2 & 1 \\ 4 & x+1 \end{vmatrix} = 0$.

Solution Evaluating the determinant by the difference of products on diagonals, we have

$$(x-2)(x+1) - 4(1) = 0$$
$$x^2 - x - 2 - 4 = 0$$
$$x^2 - x - 6 = 0$$
$$(x-3)(x+2) = 0$$
$$x - 3 = 0 \quad \text{or} \quad x + 2 = 0$$
$$x = 3 \quad \text{or} \quad x = -2$$

PROBLEM 3-5 Without expanding, find the value of each of the following determinants by using properties of determinants. State your reason in each case.

(a) $|A| = \begin{vmatrix} 2 & 0 & 6 \\ -5 & 0 & 1 \\ 6 & 0 & 4 \end{vmatrix}$ **(b)** $|B| = \begin{vmatrix} 1 & 1 & 1 \\ 0 & -2 & 3 \\ 0 & 0 & 4 \end{vmatrix}$ **(c)** $|C| = \begin{vmatrix} 1 & 2 & 3 & 4 \\ -1 & 4 & 0 & 2 \\ 1 & 2 & 3 & 4 \\ 5 & 4 & 3 & 2 \end{vmatrix}$

Solution

(a) $|A| = 0$: All the elements in the second column are zeros.
(b) $|B| = 1(-2)(4) = -8$: All the elements below the main diagonal are zeros, and hence B is upper triangular, and $|B|$ is the product of the numbers on the main diagonal.
(c) $|C| = 0$: The elements in the first and third rows are identical.

PROBLEM 3-6 Use row reduction to replace the circled element by a zero in each of the following determinants.

(a) $\begin{vmatrix} 1 & -2 & 3 \\ 2 & 4 & -1 \\ \textcircled{4} & 1 & -2 \end{vmatrix}$ **(b)** $\begin{vmatrix} 2 & -3 & 4 \\ -1 & \textcircled{2} & -1 \\ 3 & -1 & 3 \end{vmatrix}$ **(c)** $\begin{vmatrix} 1 & 2 & 3 & -4 \\ -2 & 1 & -3 & \textcircled{-1} \\ 2 & 3 & -4 & 2 \\ 3 & -1 & 4 & 2 \end{vmatrix}$

Solution

(a)
$$\begin{vmatrix} 1 & -2 & 3 \\ 2 & 4 & -1 \\ \textcircled{4} & 1 & -2 \end{vmatrix} \underset{R_3 \to -4R_1 + R_3}{=} \begin{vmatrix} 1 & -2 & 3 \\ 2 & 4 & -1 \\ 0 & 9 & -14 \end{vmatrix}$$

(b)

$$\begin{vmatrix} 2 & -3 & 4 \\ -1 & \textcircled{2} & -1 \\ 3 & -1 & 3 \end{vmatrix} \underset{R_2 \to 2R_3 + R_2}{=} \begin{vmatrix} 2 & -3 & 4 \\ 5 & 0 & 5 \\ 3 & -1 & 3 \end{vmatrix}$$

(c)

$$\begin{vmatrix} 1 & 2 & 3 & -4 \\ -2 & 1 & -3 & \textcircled{-1} \\ 2 & 3 & -4 & 2 \\ 3 & -1 & 4 & 2 \end{vmatrix} \underset{R_2 \to 2R_2}{=} \frac{1}{2} \begin{vmatrix} 1 & 2 & 3 & -4 \\ -4 & 2 & -6 & -2 \\ 2 & 3 & -4 & 2 \\ 3 & -1 & 4 & 2 \end{vmatrix}$$

note: Replacing R_2 by $2R_2$ and multiplying by 1/2 prevents the introduction of fractions within the determinant. (You *can* have fractions in a determinant, but they can be awkward.)

$$\underset{R_2 \to R_2 + R_3}{=} \frac{1}{2} \begin{vmatrix} 1 & 2 & 3 & -4 \\ -2 & 5 & -10 & 0 \\ 2 & 3 & -4 & 2 \\ 3 & -1 & 4 & 2 \end{vmatrix}$$

PROBLEM 3-7 Use row reduction to change each of the following determinants to upper triangular form and then evaluate:

(a) $\begin{vmatrix} 1 & -1 & 3 \\ 2 & 4 & 1 \\ -3 & 0 & 2 \end{vmatrix}$ **(b)** $\begin{vmatrix} 2 & 1 & 0 & 4 \\ -1 & 3 & 2 & 1 \\ 3 & 1 & -4 & -2 \\ -2 & 1 & 3 & 2 \end{vmatrix}$ **(c)** $\begin{vmatrix} 1 & 2 & -3 & 4 \\ 2 & 1 & -2 & 3 \\ -2 & 3 & -1 & -4 \\ 3 & -2 & 4 & -5 \end{vmatrix}$

Solution

(a)

$$\begin{vmatrix} 1 & -1 & 3 \\ 2 & 4 & 1 \\ -3 & 0 & 2 \end{vmatrix} \underset{\substack{R_2 \to -2R_1 + R_2 \\ R_3 \to 3R_1 + R_3}}{=} \begin{vmatrix} 1 & -1 & 3 \\ 0 & 6 & -5 \\ 0 & -3 & 11 \end{vmatrix} \underset{R_2 \leftrightarrow R_3}{=} - \begin{vmatrix} 1 & -1 & 3 \\ 0 & -3 & 11 \\ 0 & 6 & -5 \end{vmatrix}$$

$$\underset{R_3 \to -2R_2 + R_3}{=} - \begin{vmatrix} 1 & -1 & 3 \\ 0 & -3 & 11 \\ 0 & 0 & 17 \end{vmatrix} = (-1)(1)(-3)(17) = 51$$

(b)

$$\begin{vmatrix} 2 & 1 & 0 & 4 \\ -1 & 3 & 2 & 1 \\ 3 & 1 & -4 & -2 \\ -2 & 1 & 3 & 2 \end{vmatrix} \underset{R_1 \leftrightarrow R_2}{=} - \begin{vmatrix} -1 & 3 & 2 & 1 \\ 2 & 1 & 0 & 4 \\ 3 & 1 & -4 & -2 \\ -2 & 1 & 3 & 2 \end{vmatrix} \underset{\substack{R_2 \to 2R_1 + R_2 \\ R_3 \to 3R_1 + R_3 \\ R_4 \to -2R_1 + R_4}}{=} - \begin{vmatrix} -1 & 3 & 2 & 1 \\ 0 & 7 & 4 & 6 \\ 0 & 10 & 2 & 1 \\ 0 & -5 & -1 & 0 \end{vmatrix}$$

$$\underset{R_3 \to 2R_4 + R_3}{=} - \begin{vmatrix} -1 & 3 & 2 & 1 \\ 0 & 7 & 4 & 6 \\ 0 & 0 & 0 & 1 \\ 0 & -5 & -1 & 0 \end{vmatrix} \underset{R_3 \leftrightarrow R_4}{=} (-1)(-1) \begin{vmatrix} -1 & 3 & 2 & 1 \\ 0 & 7 & 4 & 6 \\ 0 & -5 & -1 & 0 \\ 0 & 0 & 0 & 1 \end{vmatrix}$$

$$\underset{R_3 \to (5/7)R_2 + R_3}{=} \begin{vmatrix} -1 & 3 & 2 & 1 \\ 0 & 7 & 4 & 6 \\ 0 & 0 & \frac{13}{7} & \frac{30}{7} \\ 0 & 0 & 0 & 1 \end{vmatrix} = (-1)(7)(13/7)(1) = -13$$

(c)
$$\begin{vmatrix} 1 & 2 & -3 & 4 \\ 2 & 1 & -2 & 3 \\ -2 & 3 & -1 & -4 \\ 3 & -2 & 4 & -5 \end{vmatrix} \underset{\substack{R_2 \to -2R_1 + R_2 \\ R_3 \to 2R_1 + R_3 \\ R_4 \to -3R_1 + R_4}}{=} \begin{vmatrix} 1 & 2 & -3 & 4 \\ 0 & -3 & 4 & -5 \\ 0 & 7 & -7 & 4 \\ 0 & -8 & 13 & -17 \end{vmatrix} \underset{\substack{R_3 \to 2R_2 + R_3 \\ R_4 \to -3R_2 + R_4}}{=} \begin{vmatrix} 1 & 2 & -3 & 4 \\ 0 & -3 & 4 & -5 \\ 0 & 1 & 1 & -6 \\ 0 & 1 & 1 & -2 \end{vmatrix}$$

$$\underset{R_3 \leftrightarrow R_2}{=} - \begin{vmatrix} 1 & 2 & -3 & 4 \\ 0 & 1 & 1 & -6 \\ 0 & -3 & 4 & -5 \\ 0 & 1 & 1 & -2 \end{vmatrix} \underset{\substack{R_3 \to 3R_2 + R_3 \\ R_4 \to -R_2 + R_4}}{=} - \begin{vmatrix} 1 & 2 & -3 & 4 \\ 0 & 1 & 1 & -6 \\ 0 & 0 & 7 & -23 \\ 0 & 0 & 0 & 4 \end{vmatrix}$$

$$= (-1)(1)(1)(7)(4) = -28$$

PROBLEM 3-8 Given the following matrix, find **(a)** the minors and **(b)** the cofactors of all its elements. **(c)** Use cofactor expansion to evaluate its determinant.

$$A = \begin{bmatrix} 1 & -1 & 3 \\ 2 & 4 & 1 \\ -3 & 0 & 2 \end{bmatrix}$$

Solution

(a) The minor M_{ij} of a_{ij} is the determinant remaining after row i and column j have been deleted:

$$M_{11} = \begin{vmatrix} 1 & -1 & 3 \\ 2 & 4 & 1 \\ -3 & 0 & 2 \end{vmatrix} = \begin{vmatrix} 4 & 1 \\ 0 & 2 \end{vmatrix} = 8 \qquad M_{12} = \begin{vmatrix} 1 & -1 & 3 \\ 2 & 4 & 1 \\ -3 & 0 & 2 \end{vmatrix} = \begin{vmatrix} 2 & 1 \\ -3 & 2 \end{vmatrix} = 7$$

$$M_{13} = \begin{vmatrix} 1 & -1 & 3 \\ 2 & 4 & 1 \\ -3 & 0 & 2 \end{vmatrix} = \begin{vmatrix} 2 & 4 \\ -3 & 0 \end{vmatrix} = 12$$

$$M_{21} = \begin{vmatrix} 1 & -1 & 3 \\ 2 & 4 & 1 \\ -3 & 0 & 2 \end{vmatrix} = \begin{vmatrix} -1 & 3 \\ 0 & 2 \end{vmatrix} = -2 \qquad M_{22} = \begin{vmatrix} 1 & -1 & 3 \\ 2 & 4 & 1 \\ -3 & 0 & 2 \end{vmatrix} = \begin{vmatrix} 1 & 3 \\ -3 & 2 \end{vmatrix} = 11$$

$$M_{23} = \begin{vmatrix} 1 & -1 & 3 \\ 2 & 4 & 1 \\ -3 & 0 & 2 \end{vmatrix} = \begin{vmatrix} 1 & -1 \\ -3 & 0 \end{vmatrix} = -3$$

$$M_{31} = \begin{vmatrix} 1 & -1 & 3 \\ 2 & 4 & 1 \\ -3 & 0 & 2 \end{vmatrix} = \begin{vmatrix} -1 & 3 \\ 4 & 1 \end{vmatrix} = -13 \qquad M_{32} = \begin{vmatrix} 1 & -1 & 3 \\ 2 & 4 & 1 \\ -3 & 0 & 2 \end{vmatrix} = \begin{vmatrix} 1 & 3 \\ 2 & 1 \end{vmatrix} = -5$$

$$M_{33} = \begin{vmatrix} 1 & -1 & 3 \\ 2 & 4 & 1 \\ -3 & 0 & 2 \end{vmatrix} = \begin{vmatrix} 1 & -1 \\ 2 & 4 \end{vmatrix} = 6$$

(b) $C_{ij} = (-1)^{i+j} M_{ij}$, so

$$C_{11} = (-1)^{1+1} M_{11} = 8 \qquad C_{12} = (-1)^{1+2} M_{12} = -7 \qquad C_{13} = (-1)^{1+3} M_{13} = 12$$

$$C_{21} = (-1)^{2+1} M_{21} = 2 \qquad C_{22} = (-1)^{2+2} M_{22} = 11 \qquad C_{23} = (-1)^{2+3} M_{23} = 3$$

$$C_{31} = (-1)^{3+1} M_{31} = -13 \qquad C_{32} = (-1)^{3+2} M_{32} = 5 \qquad C_{33} = (-1)^{3+3} M_{33} = 6$$

(c) Since element $a_{32} = 0$, it's easiest to add the products of cofactors and the elements of the second column:

$$\det(A) = a_{12}C_{12} + a_{22}C_{22} + a_{32}C_{32} = (-1)(-7) + 4(11) + 0(5) = 51$$

PROBLEM 3-9 Use row reduction to replace by zero all elements except one in a column; then use cofactor expansions to evaluate the following determinants:

$$\textbf{(a)}\ |A| = \begin{vmatrix} 2 & 1 & 0 & 4 \\ -1 & 3 & 2 & 1 \\ 3 & 1 & -4 & -2 \\ -2 & 1 & 3 & 2 \end{vmatrix} \qquad \textbf{(b)}\ |B| = \begin{vmatrix} 1 & 2 & -3 & 4 \\ 2 & 1 & -2 & 3 \\ -2 & 3 & -1 & -4 \\ 3 & -2 & 4 & -5 \end{vmatrix}.$$

note: This problem is the same as Problem 3-7 except for the requested method of solution.

Solution

(a) Begin with the third column, since it already contains a zero:

$$A = \begin{vmatrix} 2 & 1 & 0 & 4 \\ -1 & 3 & 2 & 1 \\ 3 & 1 & -4 & -2 \\ -2 & 1 & 3 & 2 \end{vmatrix} \underset{\substack{R_3 \to 2R_2 + R_3 \\ R_4 \to -R_2 + R_4}}{=} \begin{vmatrix} 2 & 1 & 0 & 4 \\ -1 & 3 & 2 & 1 \\ 1 & 7 & 0 & 0 \\ -1 & -2 & 1 & 1 \end{vmatrix}$$

$$\xrightarrow[R_2 \to -2R_4 + R_2]{} = \begin{vmatrix} 2 & 1 & 0 & 4 \\ 1 & 7 & 0 & -1 \\ 1 & 7 & 0 & 0 \\ -1 & -2 & 1 & 1 \end{vmatrix}$$

$$= 0C_{13} + 0C_{23} + 0C_{33} + (-1)^{4+3}a_{43}M_{43} = (-1)^{4+3}(1)\begin{vmatrix} 2 & 1 & 4 \\ 1 & 7 & -1 \\ 1 & 7 & 0 \end{vmatrix}$$

Now repeat the process of row reduction and cofactor expansion:

$$-\begin{vmatrix} 2 & 1 & 4 \\ 1 & 7 & -1 \\ 1 & 7 & 0 \end{vmatrix} \overset{R_1 \to 4R_2 + R_1}{=} -\begin{vmatrix} 6 & 29 & 0 \\ 1 & 7 & -1 \\ 1 & 7 & 0 \end{vmatrix}$$

$$= 0C_{13} - (-1)^{2+3}a_{23}M_{23} + 0C_{33} = -(-1)^{2+3}(-1)\begin{vmatrix} 6 & 29 \\ 1 & 7 \end{vmatrix}$$

$$= -1[6(7) - 1(29)] = -13$$

(b) Since no zeros occur, you might as well begin with the first column:

$$|B| = \begin{vmatrix} 1 & 2 & -3 & 4 \\ 2 & 1 & -2 & 3 \\ -2 & 3 & -1 & -4 \\ 3 & -2 & 4 & -5 \end{vmatrix} \underset{\substack{R_2 \to -2R_1 + R_2 \\ R_3 \to 2R_1 + R_3 \\ R_4 \to -3R_1 + R_4}}{=} \begin{vmatrix} 1 & 2 & -3 & 4 \\ 0 & -3 & 4 & -5 \\ 0 & 7 & -7 & 4 \\ 0 & -8 & 13 & -17 \end{vmatrix} = (-1)^{1+1}(1)\begin{vmatrix} -3 & 4 & -5 \\ 7 & -7 & 4 \\ -8 & 13 & -17 \end{vmatrix}$$

$$\underset{\substack{R_2 \to 2R_1 + R_2 \\ R_3 \to -3R_1 + R_3}}{=} \begin{vmatrix} -3 & 4 & -5 \\ 1 & 1 & -6 \\ 1 & 1 & -2 \end{vmatrix} \overset{\substack{R_1 \to 3R_3 + R_1 \\ R_2 \to -1R_3 + R_2}}{=} \begin{vmatrix} 0 & 7 & -11 \\ 0 & 0 & -4 \\ 1 & 1 & -2 \end{vmatrix}$$

$$= (-1)^{3+1}(1)\begin{vmatrix} 7 & -11 \\ 0 & -4 \end{vmatrix} = -28$$

PROBLEM 3-10 Find the value of the determinant of each of the following matrices by any suitable method:

(a) $A = \begin{bmatrix} 6 & 7 & 6 \\ 0 & 9 & 8 \\ 0 & 5 & 4 \end{bmatrix}$ **(b)** $B = \begin{bmatrix} 4.2 & 5.3 & 2.7 \\ -6.2 & 7.2 & 4.5 \\ 8.4 & 10.6 & 5.4 \end{bmatrix}$ **(c)** $C = \begin{bmatrix} -3 & 2 & 0 & 0 \\ 4 & 3 & 1 & 0 \\ 2 & 1 & 0 & 2 \\ 1 & 3 & 0 & 3 \end{bmatrix}$

(d) $D = \begin{bmatrix} 1 & 0 & 1 & 0 & 1 \\ 1 & 1 & 0 & 1 & 0 \\ 0 & 1 & 1 & 0 & 1 \\ 1 & 0 & 1 & 1 & 0 \\ 0 & 1 & 0 & 1 & 1 \end{bmatrix}$ **(e)** $E = \begin{bmatrix} x-2 & 3 & -1 \\ 0 & x+2 & 3 \\ 0 & 0 & x-3 \end{bmatrix}$

Solution

(a) Since the first column contains only one element that is different from zero, it is easy to use that with cofactor expansion:

$$\det(A) = \begin{vmatrix} 6 & 7 & 6 \\ 0 & 9 & 8 \\ 0 & 5 & 4 \end{vmatrix} = (-1)^{1+1}(6)\begin{vmatrix} 9 & 8 \\ 5 & 4 \end{vmatrix} = 6[9(4) - 5(8)] = -24$$

(b) Observe that the elements in the third row are twice the corresponding elements in the first row. Factoring 2 from the third row, you get

$$\det(B) = 2\begin{vmatrix} 4.2 & 5.3 & 2.7 \\ -6.2 & 7.2 & 4.5 \\ 4.2 & 5.3 & 2.7 \end{vmatrix} = 0$$

because the first and third rows are identical.

(c) Since the third column contains only one element that is different from zero, you can use that column with cofactor expansion:

$$\det(C) = (-1)^{2+3}(1)\begin{vmatrix} -3 & 2 & 0 \\ 2 & 1 & 2 \\ 1 & 3 & 3 \end{vmatrix}$$

Now use the first row with cofactor expansion:

$$-\begin{vmatrix} -3 & 2 & 0 \\ 2 & 1 & 2 \\ 1 & 3 & 3 \end{vmatrix} = (-1)\left\{(-1)^{1+1}(-3)\begin{vmatrix} 1 & 2 \\ 3 & 3 \end{vmatrix} + (-1)^{1+2}(2)\begin{vmatrix} 2 & 2 \\ 1 & 3 \end{vmatrix}\right\}$$

$$= -1[-3(3-6) - 2(6-2)] = -1$$

(d) Use row reduction to change the determinant to triangular form:

$$\det(D) = \begin{vmatrix} 1 & 0 & 1 & 0 & 1 \\ 1 & 1 & 0 & 1 & 0 \\ 0 & 1 & 1 & 0 & 1 \\ 1 & 0 & 1 & 1 & 0 \\ 0 & 1 & 0 & 1 & 1 \end{vmatrix} \underset{R_4 \to -R_1 + R_4}{\overset{R_2 \to -R_1 + R_2}{=}} \begin{vmatrix} 1 & 0 & 1 & 0 & 1 \\ 0 & 1 & -1 & 1 & -1 \\ 0 & 1 & 1 & 0 & 1 \\ 0 & 0 & 0 & 1 & -1 \\ 0 & 1 & 0 & 1 & 1 \end{vmatrix} \underset{\substack{R_3 \to -R_2 + R_3 \\ R_5 \to -R_2 + R_5}}{=} \begin{vmatrix} 1 & 0 & 1 & 0 & 1 \\ 0 & 1 & -1 & 1 & -1 \\ 0 & 0 & 2 & -1 & 2 \\ 0 & 0 & 0 & 1 & -1 \\ 0 & 0 & 1 & 0 & 2 \end{vmatrix}$$

$$\underset{R_3 \leftrightarrow R_5}{=} -\begin{vmatrix} 1 & 0 & 1 & 0 & 1 \\ 0 & 1 & -1 & 1 & -1 \\ 0 & 0 & 1 & 0 & 2 \\ 0 & 0 & 0 & 1 & -1 \\ 0 & 0 & 2 & -1 & 2 \end{vmatrix} \underset{R_5 \to -2R_3+R_5}{=} -\begin{vmatrix} 1 & 0 & 1 & 0 & 1 \\ 0 & 1 & -1 & 1 & -1 \\ 0 & 0 & 1 & 0 & 2 \\ 0 & 0 & 0 & 1 & -1 \\ 0 & 0 & 0 & -1 & -2 \end{vmatrix}$$

$$\underset{R_5 \to R_4+R_5}{=} -\begin{vmatrix} 1 & 0 & 1 & 0 & 1 \\ 0 & 1 & -1 & 1 & -1 \\ 0 & 0 & 1 & 0 & 2 \\ 0 & 0 & 0 & 1 & -1 \\ 0 & 0 & 0 & 0 & -3 \end{vmatrix}$$

$= -(1)(1)(1)(1)(-3) = 3$ because it is upper triangular

(e) This matrix is upper triangular; hence $\det(E)$ is the product of the numbers in its main diagonal:

$$\det(E) = (x - 2)(x + 2)(x - 3)$$

PROBLEM 3-11 Use the pivotal method of expanding a determinant to find

(a) $\begin{vmatrix} 1 & -2 & 3 \\ 2 & 3 & -2 \\ -1 & 4 & 1 \end{vmatrix}$ **(b)** $\begin{vmatrix} 2 & -3 & 4 \\ 4 & -5 & -2 \\ -3 & 1 & 3 \end{vmatrix}$ **(c)** $\begin{vmatrix} 0 & -1 & 2 \\ 2 & 0 & -3 \\ -2 & 1 & 0 \end{vmatrix}$

Solution Since these are all 3×3 determinants, we can use the following procedure:

$$\begin{vmatrix} a_{11} & a_{12} & a_{13} \\ a_{21} & a_{22} & a_{23} \\ a_{31} & a_{32} & a_{33} \end{vmatrix} = \frac{1}{a_{11}^{3-2}} \begin{Vmatrix} \begin{vmatrix} a_{11} & a_{12} \\ a_{21} & a_{22} \end{vmatrix} & \begin{vmatrix} a_{11} & a_{13} \\ a_{21} & a_{23} \end{vmatrix} \\ \begin{vmatrix} a_{11} & a_{12} \\ a_{31} & a_{32} \end{vmatrix} & \begin{vmatrix} a_{11} & a_{13} \\ a_{31} & a_{33} \end{vmatrix} \end{Vmatrix}$$

(a)
$$\begin{vmatrix} 1 & -2 & 3 \\ 2 & 3 & -2 \\ -1 & 4 & 1 \end{vmatrix} = \frac{1}{1^{3-2}} \begin{Vmatrix} \begin{vmatrix} 1 & -2 \\ 2 & 3 \end{vmatrix} & \begin{vmatrix} 1 & 3 \\ 2 & -2 \end{vmatrix} \\ \begin{vmatrix} 1 & -2 \\ -1 & 4 \end{vmatrix} & \begin{vmatrix} 1 & 3 \\ -1 & 1 \end{vmatrix} \end{Vmatrix} = \begin{vmatrix} 7 & -8 \\ 2 & 4 \end{vmatrix} = 28 + 16 = 44$$

(b)
$$\begin{vmatrix} 2 & -3 & 4 \\ 4 & -5 & -2 \\ -3 & 1 & 3 \end{vmatrix} = \frac{1}{2^{3-2}} \begin{Vmatrix} \begin{vmatrix} 2 & -3 \\ 4 & -5 \end{vmatrix} & \begin{vmatrix} 2 & 4 \\ 4 & -2 \end{vmatrix} \\ \begin{vmatrix} 2 & -3 \\ -3 & 1 \end{vmatrix} & \begin{vmatrix} 2 & 4 \\ -3 & 3 \end{vmatrix} \end{Vmatrix} = \frac{1}{2} \begin{vmatrix} 2 & -20 \\ -7 & 18 \end{vmatrix}$$

$$= (1/2)[2(18) - (-7)(-20)] = -52$$

(c)
$$\begin{vmatrix} 0 & -1 & 2 \\ 2 & 0 & -3 \\ -2 & 1 & 0 \end{vmatrix} \overset{R_1 \leftrightarrow R_2}{=} -\begin{vmatrix} 2 & 0 & -3 \\ 0 & -1 & 2 \\ -2 & 1 & 0 \end{vmatrix} = -\frac{1}{2^{3-2}} \begin{Vmatrix} \begin{vmatrix} 2 & 0 \\ 0 & -1 \end{vmatrix} & \begin{vmatrix} 2 & -3 \\ 0 & 2 \end{vmatrix} \\ \begin{vmatrix} 2 & 0 \\ -2 & 1 \end{vmatrix} & \begin{vmatrix} 2 & -3 \\ -2 & 0 \end{vmatrix} \end{Vmatrix}$$

$$= -\frac{1}{2} \begin{vmatrix} -2 & 4 \\ 2 & -6 \end{vmatrix} = -(1/2)(12 - 8) = -2$$

There is a zero in a_{11} position, so you can interchange the first and second rows, then use the pivotal method.

PROBLEM 3-12 Use the pivotal method of expanding a determinant to find the value of the following larger determinants:

(a) $\begin{vmatrix} 1 & 3 & 5 & 7 \\ 3 & 5 & 7 & 1 \\ 5 & 7 & 1 & 3 \\ 7 & 1 & 3 & 5 \end{vmatrix}$ **(b)** $\begin{vmatrix} 2 & 1 & 0 & 4 \\ -1 & 3 & 2 & 1 \\ 3 & 1 & -4 & -2 \\ -2 & 1 & 3 & 2 \end{vmatrix}$

Solution

(a)

$$\begin{vmatrix} 1 & 3 & 5 & 7 \\ 3 & 5 & 7 & 1 \\ 5 & 7 & 1 & 3 \\ 7 & 1 & 3 & 5 \end{vmatrix} = \frac{1}{1^{4-2}} \begin{Vmatrix} \begin{vmatrix} 1 & 3 \\ 3 & 5 \end{vmatrix} & \begin{vmatrix} 1 & 5 \\ 3 & 7 \end{vmatrix} & \begin{vmatrix} 1 & 7 \\ 3 & 1 \end{vmatrix} \\ \begin{vmatrix} 1 & 3 \\ 5 & 7 \end{vmatrix} & \begin{vmatrix} 1 & 5 \\ 5 & 1 \end{vmatrix} & \begin{vmatrix} 1 & 7 \\ 5 & 3 \end{vmatrix} \\ \begin{vmatrix} 1 & 3 \\ 7 & 1 \end{vmatrix} & \begin{vmatrix} 1 & 5 \\ 7 & 3 \end{vmatrix} & \begin{vmatrix} 1 & 7 \\ 7 & 5 \end{vmatrix} \end{Vmatrix} = \begin{vmatrix} -4 & -8 & -20 \\ -8 & -24 & -32 \\ -20 & -32 & -44 \end{vmatrix}$$

Factor -4 from R_1, -8 from R_2, and -4 from R_3

$$=(-4)(-8)(-4)\begin{vmatrix} 1 & 2 & 5 \\ 1 & 3 & 4 \\ 5 & 8 & 11 \end{vmatrix}$$

Now use the pivotal method again:

$$= -128\left(\frac{1}{1^{3-2}}\right)\begin{Vmatrix} \begin{vmatrix} 1 & 2 \\ 1 & 3 \end{vmatrix} & \begin{vmatrix} 1 & 5 \\ 1 & 4 \end{vmatrix} \\ \begin{vmatrix} 1 & 2 \\ 5 & 8 \end{vmatrix} & \begin{vmatrix} 1 & 5 \\ 5 & 11 \end{vmatrix} \end{Vmatrix} = -128\begin{vmatrix} 1 & -1 \\ -2 & -14 \end{vmatrix}$$

$$= -128[-14-2] = 2048$$

(b)

$$\begin{vmatrix} 2 & 1 & 0 & 4 \\ -1 & 3 & 2 & 1 \\ 3 & 1 & -4 & -2 \\ -2 & 1 & 3 & 2 \end{vmatrix} = \frac{1}{2^{4-2}} \begin{Vmatrix} \begin{vmatrix} 2 & 1 \\ -1 & 3 \end{vmatrix} & \begin{vmatrix} 2 & 0 \\ -1 & 2 \end{vmatrix} & \begin{vmatrix} 2 & 4 \\ -1 & 1 \end{vmatrix} \\ \begin{vmatrix} 2 & 1 \\ 3 & 1 \end{vmatrix} & \begin{vmatrix} 2 & 0 \\ 3 & -4 \end{vmatrix} & \begin{vmatrix} 2 & 4 \\ 3 & -2 \end{vmatrix} \\ \begin{vmatrix} 2 & 1 \\ -2 & 1 \end{vmatrix} & \begin{vmatrix} 2 & 0 \\ -2 & 3 \end{vmatrix} & \begin{vmatrix} 2 & 4 \\ -2 & 2 \end{vmatrix} \end{Vmatrix} = \frac{1}{2^2}\begin{vmatrix} 7 & 4 & 6 \\ -1 & -8 & -16 \\ 4 & 6 & 12 \end{vmatrix}$$

$$= \frac{1}{4}\cdot\frac{1}{7}\begin{Vmatrix} \begin{vmatrix} 7 & 4 \\ -1 & -8 \end{vmatrix} & \begin{vmatrix} 7 & 6 \\ -1 & -16 \end{vmatrix} \\ \begin{vmatrix} 7 & 4 \\ 4 & 6 \end{vmatrix} & \begin{vmatrix} 7 & 6 \\ 4 & 12 \end{vmatrix} \end{Vmatrix} = \frac{1}{28}\begin{vmatrix} -52 & -106 \\ 26 & 60 \end{vmatrix}$$

$$= (1/28)[(-52)(60) - (26)(-106)] = -13$$

PROBLEM 3-13 Verify the pivotal expansion for the 3×3 determinant

$$\det(A) = \begin{vmatrix} a_{11} & a_{12} & a_{13} \\ a_{21} & a_{22} & a_{23} \\ a_{31} & a_{32} & a_{33} \end{vmatrix}$$

Solution Multiply both the second and third rows by a_{11}:

$$\begin{vmatrix} a_{11} & a_{12} & a_{13} \\ a_{21} & a_{22} & a_{23} \\ a_{31} & a_{32} & a_{33} \end{vmatrix} \underset{\substack{R_2 \to a_{11}R_2 \\ R_3 \to a_{11}R_3}}{=} \frac{1}{(a_{11})^2} \begin{vmatrix} a_{11} & a_{12} & a_{13} \\ a_{11}a_{21} & a_{11}a_{22} & a_{11}a_{23} \\ a_{11}a_{31} & a_{11}a_{32} & a_{11}a_{33} \end{vmatrix}$$

$$\underset{\substack{R_2 \to -a_{21}R_1 + R_2 \\ R_3 \to -a_{31}R_1 + R_3}}{=} \frac{1}{(a_{11})^2} \begin{vmatrix} a_{11} & a_{12} & a_{13} \\ 0 & a_{11}a_{22} - a_{21}a_{12} & a_{11}a_{23} - a_{21}a_{13} \\ 0 & a_{11}a_{32} - a_{31}a_{12} & a_{11}a_{33} - a_{31}a_{13} \end{vmatrix}$$

Now, use cofactor expansion:

$$\det(A) = a_{11}C_{11} + a_{21}C_{21} + a_{31}C_{31} = (-1)^{1+1}a_{11}M_{11}\left[\frac{1}{(a_{11})^2}\right] + 0C_{21} + 0C_{31} = \left(\frac{1}{a_{11}}\right)M_{11}$$

$$= \left(\frac{1}{a_{11}}\right) \begin{vmatrix} a_{11}a_{22} - a_{21}a_{12} & a_{11}a_{23} - a_{21}a_{13} \\ a_{11}a_{32} - a_{31}a_{12} & a_{11}a_{33} - a_{31}a_{13} \end{vmatrix}$$

Now, since

$$\begin{vmatrix} a_{11} & a_{12} \\ a_{21} & a_{22} \end{vmatrix} = a_{11}a_{22} - a_{21}a_{12}, \qquad \begin{vmatrix} a_{11} & a_{13} \\ a_{21} & a_{23} \end{vmatrix} = a_{11}a_{23} - a_{21}a_{13}$$

and so on, you can write

$$\frac{1}{a_{11}} \begin{vmatrix} \begin{vmatrix} a_{11} & a_{12} \\ a_{21} & a_{22} \end{vmatrix} & \begin{vmatrix} a_{11} & a_{13} \\ a_{21} & a_{23} \end{vmatrix} \\ \begin{vmatrix} a_{11} & a_{12} \\ a_{31} & a_{32} \end{vmatrix} & \begin{vmatrix} a_{11} & a_{13} \\ a_{31} & a_{33} \end{vmatrix} \end{vmatrix}$$

Supplementary Problems

PROBLEM 3-14 Find the number of inversions in 8 4 2 7 6 3 5 1.

PROBLEM 3-15 Find the value of $\det(A)$ if $A = \begin{bmatrix} a & b \\ c & d \end{bmatrix}$.

PROBLEM 3-16 Evaluate the determinant $|-13|$.

PROBLEM 3-17 Evaluate the determinant $\begin{vmatrix} 2 & 5 \\ -1 & 3 \end{vmatrix}$.

PROBLEM 3-18 Use diagonals' products to find the value of $\det(A)$ if $A = \begin{bmatrix} 1 & 0 & -4 \\ 0 & -2 & 3 \\ 2 & 3 & 1 \end{bmatrix}$.

PROBLEM 3-19 Find the value of the determinant of each of the following matrices by using any suitable method:

$$A = \begin{bmatrix} 2 & 3 & 5 \\ 1 & 4 & 3 \\ 4 & 0 & 1 \end{bmatrix} \qquad B = \begin{bmatrix} 3 & 8 & -5 \\ 15 & -6 & 2 \\ 9 & 24 & -15 \end{bmatrix} \qquad C = \begin{bmatrix} 2.4 & 0.8 & 0 \\ 3.1 & -6.7 & 1 \\ 6.3 & 5.4 & 0 \end{bmatrix}$$

PROBLEM 3-20 Without expanding, find the value of each of the following determinants by using properties of determinants:

$$\text{(a)} \begin{vmatrix} 5 & 6 & 2 & 1 \\ -1 & 3 & 4 & 5 \\ 0 & 0 & 0 & 0 \\ 2 & 3 & -1 & 4 \end{vmatrix} \qquad \text{(b)} \begin{vmatrix} 5 & 6 & 2 & 1 \\ -1 & 3 & 4 & 5 \\ 5 & 6 & 2 & 1 \\ 2 & 3 & -1 & 4 \end{vmatrix} \qquad \text{(c)} \begin{vmatrix} 5 & 2 & 6 & -3 \\ 0 & 2 & -1 & 4 \\ 0 & 0 & -3 & 4 \\ 0 & 0 & 0 & 6 \end{vmatrix}$$

PROBLEM 3-21 If

$$A = \begin{bmatrix} 1 & 2 & 0 & 4 \\ 3 & 4 & 1 & 6 \\ -1 & 5 & 2 & 0 \\ 0 & 1 & 2 & -7 \end{bmatrix}$$

and $\det(A) = 25$, find $\det(A^T)$.

PROBLEM 3-22 Find the minors M_{21} and M_{33} for the matrix $A = \begin{bmatrix} 2 & 3 & 5 \\ 1 & 4 & 3 \\ 4 & 0 & 1 \end{bmatrix}$.

PROBLEM 3-23 Find the cofactors C_{21} and C_{33} for the matrix $A = \begin{bmatrix} 2 & 3 & 5 \\ 1 & 4 & 3 \\ 4 & 0 & 1 \end{bmatrix}$.

PROBLEM 3-24 Use cofactor expansion to evaluate the following determinants:

$$\text{(a)} \begin{vmatrix} 3 & 4 & 5 & 0 \\ -1 & 3 & 2 & 2 \\ 0 & 0 & 3 & 0 \\ 2 & -4 & 1 & 0 \end{vmatrix} \qquad \text{(b)} \begin{vmatrix} 1 & 7 & 3 \\ 2 & 0 & -6 \\ -1 & -3 & 5 \end{vmatrix} \qquad \text{(c)} \begin{vmatrix} 2 & -1 & 5 & 4 & 0 \\ 6 & 7 & 8 & 9 & -1 \\ 0 & 0 & 0 & 2 & 0 \\ 3 & 2 & 4 & 5 & 0 \\ 3 & 0 & 0 & -7 & 0 \end{vmatrix}$$

PROBLEM 3-25 If $\det \begin{bmatrix} 3 & -4 & 5 \\ a & b & c \\ 2 & 1 & 3 \end{bmatrix} = 7$, find

$$\text{(a)} \det \begin{bmatrix} a & b & c \\ 3 & -4 & 5 \\ 2 & 1 & 3 \end{bmatrix} \qquad \text{(b)} \det \begin{bmatrix} 3 & -4 & 5 \\ a & b & c \\ 2+3a & 1+3b & 3+3c \end{bmatrix}$$

PROBLEM 3-26 Use cofactor expansion with row one to solve the equation $\begin{vmatrix} x^2 & x & 1 \\ -3 & 2 & 3 \\ -7 & 3 & 5 \end{vmatrix} = 0$.

PROBLEM 3-27 Use row reduction to change each of the following determinants to upper triangular form and then evaluate:

$$\text{(a)} \begin{vmatrix} 1 & -3 & 2 \\ 4 & 3 & 2 \\ 2 & 0 & 1 \end{vmatrix} \qquad \text{(b)} \begin{vmatrix} 1 & -1 & 3 \\ 4 & 4 & 1 \\ 3 & 2 & 0 \end{vmatrix} \qquad \text{(c)} \begin{vmatrix} 1 & 1 & 2 & 1 \\ 2 & -1 & 1 & 1 \\ 3 & 1 & 0 & 2 \\ 1 & 4 & 1 & 4 \end{vmatrix} \qquad \text{(d)} \begin{vmatrix} -1 & 0 & 1 & 0 & 1 \\ 1 & 1 & 2 & 1 & -1 \\ 1 & 2 & 1 & -1 & 0 \\ -1 & 3 & -2 & 0 & 1 \\ 0 & 0 & 4 & 15 & 2 \end{vmatrix}$$

PROBLEM 3-28 Use the pivotal method to evaluate the determinant

$$\begin{vmatrix} 3 & -2 & 0 & 1 \\ 3 & -2 & 1 & -2 \\ 1 & 0 & -2 & -3 \\ 2 & 1 & -3 & 4 \end{vmatrix}$$

Answers to Supplementary Problems

3-14 20

3-15 $\det(A) = ad - bc$

3-16 -13

3-17 11

3-18 -27

3-19 $\det(A) = -39$, $\det(B) = 0$, $\det(C) = -7.92$

3-20 **(a)** 0 **(b)** 0 **(c)** -180

3-21 25

3-22 $M_{21} = \begin{vmatrix} 3 & 5 \\ 0 & 1 \end{vmatrix}$, $M_{33} = \begin{vmatrix} 2 & 3 \\ 1 & 4 \end{vmatrix}$

3-23 $C_{21} = (-1)^{2+1}\begin{vmatrix} 3 & 5 \\ 0 & 1 \end{vmatrix} = -\begin{vmatrix} 3 & 5 \\ 0 & 1 \end{vmatrix}$,

$C_{33} = (-1)^{3+3}\begin{vmatrix} 2 & 3 \\ 1 & 4 \end{vmatrix} = \begin{vmatrix} 2 & 3 \\ 1 & 4 \end{vmatrix}$

3-24 **(a)** 120 **(b)** -64 **(c)** -84

3-25 **(a)** -7 **(b)** 7

3-26 $x = 1$ or $x = 5$

3-27 **(a)** -9 **(b)** -17 **(c)** 28 **(d)** 216

3-28 -54

4 ADDITIONAL TOPICS OF DETERMINANTS AND MATRICES

THIS CHAPTER IS ABOUT

☑ **Cramer's Rule**
☑ **Adjoint of a Matrix**
☑ **Elementary Matrices**
☑ **Properties of Elementary Matrices**
☑ **Equivalent Matrices**
☑ **Partitioned Matrices**

4-1. Cramer's Rule

A. Matrix invertibility and determinant value

An $n \times n$ matrix A is invertible if and only if $\det(A) \neq 0$.

EXAMPLE 4-1: Determine whether $A = \begin{bmatrix} 3 & -1 & 2 \\ 2 & 4 & -1 \\ 3 & 2 & -3 \end{bmatrix}$ is invertible.

Solution

$$\begin{aligned} |A| &= 3\begin{vmatrix} 4 & -1 \\ 2 & -3 \end{vmatrix} - (-1)\begin{vmatrix} 2 & -1 \\ 3 & -3 \end{vmatrix} + 2\begin{vmatrix} 2 & 4 \\ 3 & 2 \end{vmatrix} \\ &= 3(-12 + 2) + 1(-6 + 3) + 2(4 - 12) \\ &= 3(-10) + 1(-3) + 2(-8) = -49 \end{aligned}$$

Therefore, A is invertible.

B. Statement of Cramer's rule

Consider the linear system having n equations and n unknowns, that is

$$\begin{aligned} a_{11}x_1 + a_{12}x_2 + \cdots + a_{1n}x_n &= b_1 \\ a_{21}x_1 + a_{22}x_2 + \cdots + a_{2n}x_n &= b_2 \\ \vdots \qquad\qquad \cdots \qquad\qquad &\vdots \\ a_{n1}x_1 + a_{n2}x_2 + \cdots + a_{nn}x_n &= b_n. \end{aligned}$$

If the matrix notation $AX = B$ is used and $\det(A) \neq 0$, then **Cramer's rule** gives

$$x_1 = \frac{\det(A_1)}{\det(A)}, x_2 = \frac{\det(A_2)}{\det(A)}, \ldots, x_n = \frac{\det(A_n)}{\det(A)}$$

as the unique solution of the system, where A_i is the matrix obtained by replacing the ith column of A by the column of constants B.

You might observe that Gaussian elimination and Gauss–Jordan elimination, discussed in Chapter 2, did not require the coefficient matrix to be invertible. Neither did the number of equations have to be the same as the number of unknowns. Cramer's rule and the method denoted by $X = A^{-1}B$ require the same number of equations as unknowns and A must be invertible. For a large number of equations the Gaussian method is more efficient than Cramer's rule. However, if n is small or the system involves parameters Cramer's rule is very useful.

EXAMPLE 4-2: Solve the following system by Cramer's rule:

$$6x + 5y = 3$$
$$2x - 3y = 4$$

Solution

$$x = \frac{\det(A_1)}{\det(A)} = \frac{\begin{vmatrix} 3 & 5 \\ 4 & -3 \end{vmatrix}}{\begin{vmatrix} 6 & 5 \\ 2 & -3 \end{vmatrix}} = \frac{-9-20}{-18-10} = \frac{29}{28}$$

$$y = \frac{\det(A_2)}{\det(A)} = \frac{\begin{vmatrix} 6 & 3 \\ 2 & 4 \end{vmatrix}}{\begin{vmatrix} 6 & 5 \\ 2 & -3 \end{vmatrix}} = \frac{24-6}{-28} = -\frac{18}{28}$$

EXAMPLE 4-3: Solve the following 3×3 system by Cramer's rule:

$$x + y - 2z = -3$$
$$2x - y - z = 0$$
$$x + 2y + 3z = 13$$

Solution

$$x = \frac{\begin{vmatrix} -3 & 1 & -2 \\ 0 & -1 & -1 \\ 13 & 2 & 3 \end{vmatrix}}{\begin{vmatrix} 1 & 1 & -2 \\ 2 & -1 & -1 \\ 1 & 2 & 3 \end{vmatrix}} = \frac{9 - 13 - 26 - 6}{-3 - 8 - 1 - 2 + 2 - 6} = \frac{-36}{-18} = 2$$

$$y = \frac{\begin{vmatrix} 1 & -3 & -2 \\ 2 & 0 & -1 \\ 1 & 13 & 3 \end{vmatrix}}{\begin{vmatrix} 1 & 1 & -2 \\ 2 & -1 & -1 \\ 1 & 2 & 3 \end{vmatrix}} = \frac{-52 + 3 + 13 + 18}{-18} = \frac{-18}{-18} = 1$$

$$z = \frac{\begin{vmatrix} 1 & 1 & -3 \\ 2 & -1 & 0 \\ 1 & 2 & 13 \end{vmatrix}}{\begin{vmatrix} 1 & 1 & -2 \\ 2 & -1 & -1 \\ 1 & 2 & 3 \end{vmatrix}} = \frac{-13 - 12 - 3 - 26}{-18} = \frac{-54}{-18} = 3$$

$$\therefore \quad x = 2, y = 1, z = 3$$

4-2. Adjoint of a Matrix

A. Definition of adjoint

If $A = [a_{ij}]$ is any $n \times n$ matrix and C_{ij} is the cofactor of a_{ij}, then the matrix

$$\begin{bmatrix} C_{11} & C_{21} & \cdots & C_{n1} \\ C_{12} & C_{22} & \cdots & C_{n2} \\ \vdots & & & \\ C_{1n} & C_{2n} & \cdots & C_{nn} \end{bmatrix}$$

is called the **adjoint** of A and is denoted by adj(A). Observe that this is the matrix obtained by replacing a_{ij} by C_{ij} and then taking the transpose of the resulting matrix. (See Section 3-5 for a review of cofactors.)

EXAMPLE 4-4: Find the adjoint of $A = \begin{bmatrix} 1 & 2 & 3 \\ 0 & 1 & 5 \\ 1 & 2 & 1 \end{bmatrix}$.

Solution

$$C_{11} = (-1)^{1+1}M_{11} = \begin{vmatrix} 1 & 5 \\ 2 & 1 \end{vmatrix} = 1 - 10 = -9$$

$$C_{12} = (-1)^{1+2}M_{12} = -\begin{vmatrix} 0 & 5 \\ 1 & 1 \end{vmatrix} = -(-5) = 5$$

$$C_{13} = (-1)^{1+3}M_{13} = \begin{vmatrix} 0 & 1 \\ 1 & 2 \end{vmatrix} = -1$$

$$C_{21} = (-1)^{2+1}M_{21} = -\begin{vmatrix} 2 & 3 \\ 2 & 1 \end{vmatrix} = -(2 - 6) = 4$$

$$C_{22} = (-1)^{2+2}M_{22} = \begin{vmatrix} 1 & 3 \\ 1 & 1 \end{vmatrix} = 1 - 3 = -2$$

$$C_{23} = (-1)^{2+3}M_{23} = -\begin{vmatrix} 1 & 2 \\ 1 & 2 \end{vmatrix} = 0$$

$$C_{31} = (-1)^{3+1}M_{31} = \begin{vmatrix} 2 & 3 \\ 1 & 5 \end{vmatrix} = 10 - 3 = 7$$

$$C_{32} = (-1)^{3+2}M_{32} = -\begin{vmatrix} 1 & 3 \\ 0 & 5 \end{vmatrix} = -5$$

$$C_{33} = (-1)^{3+3}M_{33} = \begin{vmatrix} 1 & 2 \\ 0 & 1 \end{vmatrix} = 1$$

$\therefore$ the adjoint of A is

$$\text{adj}(A) = \begin{bmatrix} C_{11} & C_{12} & C_{13} \\ C_{21} & C_{22} & C_{23} \\ C_{31} & C_{32} & C_{33} \end{bmatrix}^T = \begin{bmatrix} -9 & 5 & -1 \\ 4 & -2 & 0 \\ 7 & -5 & 1 \end{bmatrix}^T = \begin{bmatrix} -9 & 4 & 7 \\ 5 & -2 & -5 \\ -1 & 0 & 1 \end{bmatrix}$$

B. Using adj(A) to find A^{-1}

If A is invertible, then $A^{-1} = [1/\det(A)]\,\text{adj}(A)$.

EXAMPLE 4-5: Find the inverse of matrix A in Example 4-4.

Solution: $A = \begin{bmatrix} 1 & 2 & 3 \\ 0 & 1 & 5 \\ 1 & 2 & 1 \end{bmatrix}$. Let us find $\det(A)$ by expanding by the elements in column one multiplied by their cofactors:

$$\det(A) = 1C_{11} + 0C_{21} + 1C_{31}$$

$$= (1)(-1)^{1+1}\begin{vmatrix} 1 & 5 \\ 2 & 1 \end{vmatrix} + 0 + 1(-1)^{3+1}\begin{vmatrix} 2 & 3 \\ 1 & 5 \end{vmatrix} = 1 - 10 + 10 - 3 = -2$$

From Example 4-4,

$$\text{adj}(A) = \begin{bmatrix} -9 & 4 & 7 \\ 5 & -2 & -5 \\ -1 & 0 & 1 \end{bmatrix}$$

$$\therefore \quad A^{-1} = \frac{1}{\det(A)}\text{adj}(A) = \frac{1}{-2}\begin{bmatrix} -9 & 4 & 7 \\ 5 & -2 & -5 \\ -1 & 0 & 1 \end{bmatrix} = \begin{bmatrix} \frac{9}{2} & -2 & -\frac{7}{2} \\ -\frac{5}{2} & 1 & \frac{5}{2} \\ \frac{1}{2} & 0 & -\frac{1}{2} \end{bmatrix}$$

Checking to see if $AA^{-1} = I_3$, we get

$$\begin{bmatrix} 1 & 2 & 3 \\ 0 & 1 & 5 \\ 1 & 2 & 1 \end{bmatrix}\begin{bmatrix} \frac{9}{2} & -2 & -\frac{7}{2} \\ -\frac{5}{2} & 1 & \frac{5}{2} \\ \frac{1}{2} & 0 & -\frac{1}{2} \end{bmatrix} = \begin{bmatrix} 1 & 0 & 0 \\ 0 & 1 & 0 \\ 0 & 0 & 1 \end{bmatrix}$$

C. Other properties of adjoints

(1) If A is any matrix of order n, then $A[\text{adj}(A)] = [\text{adj}(A)]A = [\det(A)]I_n$. That is, it is a diagonal matrix of order n in which each diagonal element is $\det(A)$.

(2) The adjoint of any matrix A is invertible if and only if A is invertible. If A is invertible, then $[\text{adj}(A)]^{-1} = [1/\det(A)]A = \text{adj}(A^{-1})$.

EXAMPLE 4-6: If A is the matrix given in Example 4-4, find

(a) $A[\text{adj}(A)]$
(b) $[\det(A)]I_3$
(c) $[\text{adj}(A)]^{-1}$
(d) $[1/\det(A)]A$
(e) $\text{adj}(A^{-1})$

Solution: From Examples 4-4 and 4-5,

$$\det(A) = -2, \qquad \text{adj}(A) = \begin{bmatrix} -9 & 4 & 7 \\ 5 & -2 & -5 \\ -1 & 0 & 1 \end{bmatrix}, \qquad A^{-1} = \begin{bmatrix} \frac{9}{2} & -2 & -\frac{7}{2} \\ -\frac{5}{2} & 1 & \frac{5}{2} \\ \frac{1}{2} & 0 & -\frac{1}{2} \end{bmatrix}$$

(a)
$$A[\text{adj}(A)] = \begin{bmatrix} 1 & 2 & 3 \\ 0 & 1 & 5 \\ 1 & 2 & 1 \end{bmatrix}\begin{bmatrix} -9 & 4 & 7 \\ 5 & -2 & -5 \\ -1 & 0 & 1 \end{bmatrix} = \begin{bmatrix} -2 & 0 & 0 \\ 0 & -2 & 0 \\ 0 & 0 & -2 \end{bmatrix}$$

(b)
$$[\det(A)]I_3 = -2\begin{bmatrix} 1 & 0 & 0 \\ 0 & 1 & 0 \\ 0 & 0 & 1 \end{bmatrix} = \begin{bmatrix} -2 & 0 & 0 \\ 0 & -2 & 0 \\ 0 & 0 & -2 \end{bmatrix}$$

(c)
$$[\text{adj}(A)|I_3] = \left[\begin{array}{ccc|ccc} -9 & 4 & 7 & 1 & 0 & 0 \\ 5 & -2 & -5 & 0 & 1 & 0 \\ -1 & 0 & 1 & 0 & 0 & 1 \end{array}\right]$$

$$\xrightarrow{R_1 \leftrightarrow -R_3} \left[\begin{array}{ccc|ccc} 1 & 0 & -1 & 0 & 0 & -1 \\ 5 & -2 & -5 & 0 & 1 & 0 \\ -9 & 4 & 7 & 1 & 0 & 0 \end{array}\right] \xrightarrow[R_3 \to R_3 + 9R_1]{R_2 \to R_2 - 5R_1} \left[\begin{array}{ccc|ccc} 1 & 0 & -1 & 0 & 0 & -1 \\ 0 & -2 & 0 & 0 & 1 & 5 \\ 0 & 4 & -2 & 1 & 0 & -9 \end{array}\right]$$

$$\xrightarrow[R_3 \to R_3 + 2R_2]{} \left[\begin{array}{ccc|ccc} 1 & 0 & -1 & 0 & 0 & -1 \\ 0 & -2 & 0 & 0 & 1 & 5 \\ 0 & 0 & -2 & 1 & 2 & 1 \end{array}\right] \xrightarrow[R_3 \to -1/2R_3]{R_2 \to -1/2R_2} \left[\begin{array}{ccc|ccc} 1 & 0 & -1 & 0 & 0 & -1 \\ 0 & 1 & 0 & 0 & -\frac{1}{2} & -\frac{5}{2} \\ 0 & 0 & 1 & -\frac{1}{2} & -1 & -\frac{1}{2} \end{array}\right]$$

$$\xrightarrow{R_1 \to R_1 + R_3} \left[\begin{array}{ccc|ccc} 1 & 0 & 0 & -\frac{1}{2} & -1 & -\frac{3}{2} \\ 0 & 1 & 0 & 0 & -\frac{1}{2} & -\frac{5}{2} \\ 0 & 0 & 1 & -\frac{1}{2} & -1 & -\frac{1}{2} \end{array}\right]$$

$$\therefore \quad [\text{adj}(A)]^{-1} = \begin{bmatrix} -\frac{1}{2} & -1 & -\frac{3}{2} \\ 0 & -\frac{1}{2} & -\frac{5}{2} \\ -\frac{1}{2} & -1 & -\frac{1}{2} \end{bmatrix}$$

(d)
$$\frac{1}{\det(A)}A = -\frac{1}{2}\begin{bmatrix} 1 & 2 & 3 \\ 0 & 1 & 5 \\ 1 & 2 & 1 \end{bmatrix} = \begin{bmatrix} -\frac{1}{2} & -1 & -\frac{3}{2} \\ 0 & -\frac{1}{2} & -\frac{5}{2} \\ -\frac{1}{2} & -1 & -\frac{1}{2} \end{bmatrix}$$

(e) We begin by computing all the cofactors of A^{-1}:

$$C_{11} = (-1)^{1+1}M_{11} = \begin{vmatrix} 1 & \frac{5}{2} \\ 0 & -\frac{1}{2} \end{vmatrix} = -\frac{1}{2}$$

$$C_{12} = (-1)^{1+2}M_{12} = -\begin{bmatrix} -\frac{5}{2} & \frac{5}{2} \\ \frac{2}{1} & -\frac{1}{2} \end{bmatrix} = 0$$

$$C_{13} = (-1)^{1+3}M_{13} = \begin{bmatrix} -\frac{5}{2} & 1 \\ \frac{1}{2} & 0 \end{bmatrix} = -\frac{1}{2}$$

$$C_{21} = (-1)^{2+1}M_{21} = -\begin{bmatrix} -2 & -\frac{7}{2} \\ 0 & -\frac{1}{2} \end{bmatrix} = -1$$

$$C_{22} = (-1)^{2+2}M_{22} = \begin{bmatrix} \frac{9}{2} & -\frac{7}{2} \\ \frac{1}{2} & -\frac{1}{2} \end{bmatrix} = -\frac{1}{2}$$

$$C_{23} = (-1)^{2+3}M_{23} = -\begin{bmatrix} \frac{9}{2} & -2 \\ \frac{1}{2} & 0 \end{bmatrix} = -1$$

$$C_{31} = (-1)^{3+1}M_{31} = \begin{bmatrix} -2 & -\frac{7}{2} \\ 1 & \frac{5}{2} \end{bmatrix} = -\frac{3}{2}$$

$$C_{32} = (-1)^{3+2}M_{32} = -\begin{bmatrix} \frac{9}{2} & -\frac{7}{2} \\ -\frac{5}{2} & \frac{5}{2} \end{bmatrix} = -\frac{5}{2}$$

$$C_{33} = (-1)^{3+3}M_{33} = \begin{bmatrix} \frac{9}{2} & -2 \\ -\frac{5}{2} & 1 \end{bmatrix} = -\frac{1}{2}$$

$$\text{adj}(A^{-1}) = \begin{bmatrix} C_{11} & C_{12} & C_{13} \\ C_{21} & C_{22} & C_{23} \\ C_{31} & C_{32} & C_{33} \end{bmatrix}^T = \begin{bmatrix} -\frac{1}{2} & 0 & -\frac{1}{2} \\ -1 & -\frac{1}{2} & -1 \\ -\frac{3}{2} & -\frac{5}{2} & -\frac{1}{2} \end{bmatrix}^T$$

$$= \begin{bmatrix} -\frac{1}{2} & -1 & -\frac{3}{2} \\ 0 & -\frac{1}{2} & -\frac{5}{2} \\ -\frac{1}{2} & -1 & -\frac{1}{2} \end{bmatrix}$$

note: Observe by comparing **(c)**, **(d)**, and **(e)** that the easiest way to find $[\text{adj}(A)]^{-1}$ is by using $[1/\det(A)]A$.

4-3. Elementary Matrices

Recall that the three elementary row operations on a matrix A are to

(a) multiply any row of A by a nonzero constant,
(b) interchange any two rows of A, or
(c) add a constant multiple of one row to another row.

Corresponding to each of these is an elementary column operation. There are many cases, however, where we restrict ourselves to row operations; Gauss–Jordan elimination is such a case.

Definition of elementary matrix

The $n \times n$ matrix E is called an **elementary matrix** if it can be obtained from the identity matrix I_n by performing a single elementary row or column operation on I_n.

EXAMPLE 4-7: Which of the following matrices are elementary?

(a) $\begin{bmatrix} 3 & 0 \\ 0 & 1 \end{bmatrix}$ **(b)** $\begin{bmatrix} 0 & 0 & 1 \\ 0 & 1 & 0 \\ 1 & 0 & 0 \end{bmatrix}$ **(c)** $\begin{bmatrix} 1 & 0 & 4 \\ 0 & 1 & 0 \\ 0 & 0 & 1 \end{bmatrix}$ **(d)** $\begin{bmatrix} 3 & 0 & 0 \\ 0 & 3 & 0 \\ 0 & 0 & 1 \end{bmatrix}$

(e) $\begin{bmatrix} 1 & 0 & 0 & 0 \\ 0 & 1 & 0 & 0 \\ 0 & 0 & 1 & 0 \\ -4 & 0 & 0 & 1 \end{bmatrix}$ **(f)** $\begin{bmatrix} 0 & 0 & 4 \\ 0 & 1 & 0 \\ 1 & 0 & 1 \end{bmatrix}$

Solution: Start with the appropriate order identity and determine whether a single elementary row or column operation exists that will change the identity matrix into the given matrix.

(a)
$$\begin{bmatrix} 1 & 0 \\ 0 & 1 \end{bmatrix} \xrightarrow{R_1 \to 3R_1} \begin{bmatrix} 3 & 0 \\ 0 & 1 \end{bmatrix}$$

Therefore, the given matrix is elementary. We could have used a column operation to get

$$\begin{bmatrix} 1 & 0 \\ 0 & 1 \end{bmatrix} \xrightarrow{C_1 \to 3C_1} \begin{bmatrix} 3 & 0 \\ 0 & 1 \end{bmatrix}$$

which also shows that the matrix is elementary.

(b)
$$\begin{bmatrix} 1 & 0 & 0 \\ 0 & 1 & 0 \\ 0 & 0 & 1 \end{bmatrix} \xrightarrow{R_1 \leftrightarrow R_3} \begin{bmatrix} 0 & 0 & 1 \\ 0 & 1 & 0 \\ 1 & 0 & 0 \end{bmatrix}$$

Therefore, the given matrix is elementary. Also, we could use a column operation to get

$$\begin{bmatrix} 1 & 0 & 0 \\ 0 & 1 & 0 \\ 0 & 0 & 1 \end{bmatrix} \xrightarrow{C_1 \leftrightarrow C_3} \begin{bmatrix} 0 & 0 & 1 \\ 0 & 1 & 0 \\ 1 & 0 & 0 \end{bmatrix}$$

(c)
$$\begin{bmatrix} 1 & 0 & 0 \\ 0 & 1 & 0 \\ 0 & 0 & 1 \end{bmatrix} \xrightarrow{R_1 \to R_1 + 4R_3} \begin{bmatrix} 1 & 0 & 4 \\ 0 & 1 & 0 \\ 0 & 0 & 1 \end{bmatrix}$$

or

$$\begin{bmatrix} 1 & 0 & 0 \\ 0 & 1 & 0 \\ 0 & 0 & 1 \end{bmatrix} \xrightarrow{C_3 \to C_3 + 4C_1} \begin{bmatrix} 1 & 0 & 4 \\ 0 & 1 & 0 \\ 0 & 0 & 1 \end{bmatrix}$$

Therefore, the given matrix is elementary.

(d)
$$\begin{bmatrix} 1 & 0 & 0 \\ 0 & 1 & 0 \\ 0 & 0 & 1 \end{bmatrix} \xrightarrow[R_2 \to 3R_2]{R_1 \to 3R_1} \begin{bmatrix} 3 & 0 & 0 \\ 0 & 3 & 0 \\ 0 & 0 & 1 \end{bmatrix}$$

or

$$\begin{bmatrix} 1 & 0 & 0 \\ 0 & 1 & 0 \\ 0 & 0 & 1 \end{bmatrix} \xrightarrow[C_2 \to 3C_2]{C_1 \to 3C_1} \begin{bmatrix} 3 & 0 & 0 \\ 0 & 3 & 0 \\ 0 & 0 & 1 \end{bmatrix}$$

The given matrix is *not* elementary because *two* elementary operations are required to change I_3 into the given matrix.

(e)

$$\begin{bmatrix} 1 & 0 & 0 & 0 \\ 0 & 1 & 0 & 0 \\ 0 & 0 & 1 & 0 \\ 0 & 0 & 0 & 1 \end{bmatrix} \xrightarrow[R_4 \to R_4 - 4R_1]{} \begin{bmatrix} 1 & 0 & 0 & 0 \\ 0 & 1 & 0 & 0 \\ 0 & 0 & 1 & 0 \\ -4 & 0 & 0 & 1 \end{bmatrix}$$

Therefore it is elementary.

(f)

$$\begin{bmatrix} 1 & 0 & 0 \\ 0 & 1 & 0 \\ 0 & 0 & 1 \end{bmatrix} \xrightarrow{R_1 \leftrightarrow R_3} \begin{bmatrix} 0 & 0 & 1 \\ 0 & 1 & 0 \\ 1 & 0 & 0 \end{bmatrix} \xrightarrow{R_3 \to R_3 + R_1} \begin{bmatrix} 0 & 0 & 1 \\ 0 & 1 & 0 \\ 1 & 0 & 1 \end{bmatrix} \xrightarrow{R_1 \to 4R_1} \begin{bmatrix} 0 & 0 & 4 \\ 0 & 1 & 0 \\ 1 & 0 & 1 \end{bmatrix}$$

Therefore, it is not elementary. We could have done this:

$$\begin{bmatrix} 1 & 0 & 0 \\ 0 & 1 & 0 \\ 0 & 0 & 1 \end{bmatrix} \xrightarrow[R_3 \to 4R_3]{} \begin{bmatrix} 1 & 0 & 0 \\ 0 & 1 & 0 \\ 0 & 0 & 4 \end{bmatrix} \xrightarrow{R_1 \leftrightarrow R_3} \begin{bmatrix} 0 & 0 & 4 \\ 0 & 1 & 0 \\ 1 & 0 & 0 \end{bmatrix} \xrightarrow{R_3 \to R_3 + 1/4 R_1} \begin{bmatrix} 0 & 0 & 4 \\ 0 & 1 & 0 \\ 1 & 0 & 1 \end{bmatrix}$$

4-4. Properties of Elementary Matrices

Example 4-7 serves as an illustration to show that, for each elementary row operation performed on an identity matrix, an elementary column operation exists that will do the same job. Furthermore, it illustrates that if a sequence of elementary operations is performed on a matrix, then a different sequence may also give the same result.

Property A If an $m \times n$ matrix A is multiplied on the *left* by an $m \times m$ elementary matrix E to obtain B, that is, $EA = B$, then B can be obtained from A by performing the *same elementary row operation* on A as was performed on I_m to get E.

EXAMPLE 4-8: Let

$$A = \begin{bmatrix} 3 & -2 & 0 \\ 1 & 0 & 1 \\ 2 & -1 & 2 \\ 0 & 3 & 5 \end{bmatrix}$$

Find, in each case, the elementary matrix E that, as a premultiplier of A, that is, EA, performs the following elementary row operation on A:

(a) multiplies the third row by 7,
(b) interchanges the first and fourth rows, or
(c) adds five times the second row to the third row.

Solution: Perform the same operation on I_4 as was requested to be done on A.

(a) $$\begin{bmatrix} 1 & 0 & 0 & 0 \\ 0 & 1 & 0 & 0 \\ 0 & 0 & 1 & 0 \\ 0 & 0 & 0 & 1 \end{bmatrix} \xrightarrow{R_3 \to 7R_3} \begin{bmatrix} 1 & 0 & 0 & 0 \\ 0 & 1 & 0 & 0 \\ 0 & 0 & 7 & 0 \\ 0 & 0 & 0 & 1 \end{bmatrix} = E$$

(b) $$\begin{bmatrix} 1 & 0 & 0 & 0 \\ 0 & 1 & 0 & 0 \\ 0 & 0 & 1 & 0 \\ 0 & 0 & 0 & 1 \end{bmatrix} \xrightarrow{R_1 \leftrightarrow R_4} \begin{bmatrix} 0 & 0 & 0 & 1 \\ 0 & 1 & 0 & 0 \\ 0 & 0 & 1 & 0 \\ 1 & 0 & 0 & 0 \end{bmatrix} = E$$

(c) $$\begin{bmatrix} 1 & 0 & 0 & 0 \\ 0 & 1 & 0 & 0 \\ 0 & 0 & 1 & 0 \\ 0 & 0 & 0 & 1 \end{bmatrix} \xrightarrow{R_3 \to R_3 + 5R_2} \begin{bmatrix} 1 & 0 & 0 & 0 \\ 0 & 1 & 0 & 0 \\ 0 & 5 & 1 & 0 \\ 0 & 0 & 0 & 1 \end{bmatrix} = E$$

Property B If an $m \times n$ matrix A is multiplied on the *right* by an $n \times n$ elementary matrix E to obtain B, that is, $AE = B$, then B can be obtained from A by performing the *same elementary column operation* on A as was performed on I_n to get E.

EXAMPLE 4-9: Let

$$A = \begin{bmatrix} 3 & -2 & 0 \\ 1 & 0 & 1 \\ 2 & -1 & 2 \\ 0 & 3 & 5 \end{bmatrix}$$

Find, in each case, the elementary matrix E which, as a postmultiplier of A, performs the following elementary column operation on A:

(a) multiplies the second column by -6,
(b) interchanges the first and second columns,
(c) adds four times the second column to the third column.

Solution: Perform the same operation on I_3 as was requested to be done on A.

(a) $$\begin{bmatrix} 1 & 0 & 0 \\ 0 & 1 & 0 \\ 0 & 0 & 1 \end{bmatrix} \xrightarrow{C_2 \to -6C_2} \begin{bmatrix} 1 & 0 & 0 \\ 0 & -6 & 0 \\ 0 & 0 & 1 \end{bmatrix} = E$$

(b) $$\begin{bmatrix} 1 & 0 & 0 \\ 0 & 1 & 0 \\ 0 & 0 & 1 \end{bmatrix} \xrightarrow{C_1 \leftrightarrow C_2} \begin{bmatrix} 0 & 1 & 0 \\ 1 & 0 & 0 \\ 0 & 0 & 1 \end{bmatrix} = E$$

(c) $$\begin{bmatrix} 1 & 0 & 0 \\ 0 & 1 & 0 \\ 0 & 0 & 1 \end{bmatrix} \xrightarrow[C_3 \to C_3 + 4C_2]{} \begin{bmatrix} 1 & 0 & 0 \\ 0 & 1 & 4 \\ 0 & 0 & 1 \end{bmatrix} = E$$

Property C Each elementary matrix is invertible and its inverse is an elementary matrix. The various cases are

(a) E obtained from I by

(1) multiplying row i by $k \neq 0$
(2) interchanging rows i and j
(3) adding k times row i to row j

(b) E^{-1} obtained from I by

(1) multiplying row i by $1/k$
(2) interchanging rows i and j
(3) adding $-k$ times row i to row j.

Replace the words row(s) by column(s) to obtain the cases for columns.

EXAMPLE 4-10: Find the inverse of each of the following elementary matrices:

(a) $E_1 = \begin{bmatrix} 4 & 0 \\ 0 & 1 \end{bmatrix}$ **(b)** $E_2 = \begin{bmatrix} 1 & 0 & 0 & 0 \\ -3 & 1 & 0 & 0 \\ 0 & 0 & 1 & 0 \\ 0 & 0 & 0 & 1 \end{bmatrix}$ **(c)** $E_3 = \begin{bmatrix} 0 & 0 & 1 \\ 0 & 1 & 0 \\ 1 & 0 & 0 \end{bmatrix}$

Solution

(a) $E_1^{-1} = \begin{bmatrix} \frac{1}{4} & 0 \\ 0 & 1 \end{bmatrix}$ **(b)** $E_2^{-1} = \begin{bmatrix} 1 & 0 & 0 & 0 \\ 3 & 1 & 0 & 0 \\ 0 & 0 & 1 & 0 \\ 0 & 0 & 0 & 1 \end{bmatrix}$ **(c)** $E_3^{-1} = \begin{bmatrix} 0 & 0 & 1 \\ 0 & 1 & 0 \\ 1 & 0 & 0 \end{bmatrix}$

Property D An $n \times n$ matrix A is invertible if and only if it can be written as a product of elementary matrices.

EXAMPLE 4-11: In Example 2-15 we found that $A = \begin{bmatrix} 2 & 1 & 1 \\ 1 & 2 & 3 \\ -1 & 0 & 2 \end{bmatrix}$ is invertible with $A^{-1} = \begin{bmatrix} \frac{4}{5} & -\frac{2}{5} & \frac{1}{5} \\ -1 & 1 & -1 \\ \frac{2}{5} & -\frac{1}{5} & \frac{3}{5} \end{bmatrix}$. Express both A and A^{-1} as a product of elementary matrices.

Solution

Step 1: Use elementary row operations to reduce A to I_3.

$$\begin{bmatrix} 2 & 1 & 1 \\ 1 & 2 & 3 \\ -1 & 0 & 2 \end{bmatrix} \xrightarrow{R_1 \leftrightarrow R_3} \begin{bmatrix} -1 & 0 & 2 \\ 1 & 2 & 3 \\ 2 & 1 & 1 \end{bmatrix} \xrightarrow[R_3 \to R_3 + 2R_1]{R_2 \to R_2 + R_1} \begin{bmatrix} -1 & 0 & 2 \\ 0 & 2 & 5 \\ 0 & 1 & 5 \end{bmatrix}$$

$$\xrightarrow[R_2 \to R_2 - R_3]{R_1 \to -R_1} \begin{bmatrix} 1 & 0 & -2 \\ 0 & 1 & 0 \\ 0 & 1 & 5 \end{bmatrix} \xrightarrow{R_3 \to R_3 - R_2} \begin{bmatrix} 1 & 0 & -2 \\ 0 & 1 & 0 \\ 0 & 0 & 5 \end{bmatrix}$$

$$\xrightarrow[R_3 \to 1/5 R_3]{} \begin{bmatrix} 1 & 0 & -2 \\ 0 & 1 & 0 \\ 0 & 0 & 1 \end{bmatrix} \xrightarrow{R_1 \to R_1 + 2R_3} \begin{bmatrix} 1 & 0 & 0 \\ 0 & 1 & 0 \\ 0 & 0 & 1 \end{bmatrix}$$

Step 2: Write in sequence the row operations performed and the corresponding elementary matrix with its inverse.

$R_1 \leftrightarrow R_3$ $\quad E_1 = \begin{bmatrix} 0 & 0 & 1 \\ 0 & 1 & 0 \\ 1 & 0 & 0 \end{bmatrix} \quad E_1^{-1} = \begin{bmatrix} 0 & 0 & 1 \\ 0 & 1 & 0 \\ 1 & 0 & 0 \end{bmatrix}$

$R_2 \to R_2 + R_1$ $\quad E_2 = \begin{bmatrix} 1 & 0 & 0 \\ 1 & 1 & 0 \\ 0 & 0 & 1 \end{bmatrix} \quad E_2^{-1} = \begin{bmatrix} 1 & 0 & 0 \\ -1 & 1 & 0 \\ 0 & 0 & 1 \end{bmatrix}$

$R_3 \to R_3 + 2R_1$ $\quad E_3 = \begin{bmatrix} 1 & 0 & 0 \\ 0 & 1 & 0 \\ 2 & 0 & 1 \end{bmatrix} \quad E_3^{-1} = \begin{bmatrix} 1 & 0 & 0 \\ 0 & 1 & 0 \\ -2 & 0 & 1 \end{bmatrix}$

$R_1 \to -R_1$ $\quad E_4 = \begin{bmatrix} -1 & 0 & 0 \\ 0 & 1 & 0 \\ 0 & 0 & 1 \end{bmatrix} \quad E_4^{-1} = \begin{bmatrix} -1 & 0 & 0 \\ 0 & 1 & 0 \\ 0 & 0 & 1 \end{bmatrix}$

$R_2 \to R_2 - R_3$ $\quad E_5 = \begin{bmatrix} 1 & 0 & 0 \\ 0 & 1 & -1 \\ 0 & 0 & 1 \end{bmatrix} \quad E_5^{-1} = \begin{bmatrix} 1 & 0 & 0 \\ 0 & 1 & 1 \\ 0 & 0 & 1 \end{bmatrix}$

$R_3 \to R_3 - R_2$ $\quad E_6 = \begin{bmatrix} 1 & 0 & 0 \\ 0 & 1 & 0 \\ 0 & -1 & 1 \end{bmatrix} \quad E_6^{-1} = \begin{bmatrix} 1 & 0 & 0 \\ 0 & 1 & 0 \\ 0 & 1 & 1 \end{bmatrix}$

$R_3 \to \frac{1}{5}R_3$ $\quad E_7 = \begin{bmatrix} 1 & 0 & 0 \\ 0 & 1 & 0 \\ 0 & 0 & \frac{1}{5} \end{bmatrix} \quad E_7^{-1} = \begin{bmatrix} 1 & 0 & 0 \\ 0 & 1 & 0 \\ 0 & 0 & 5 \end{bmatrix}$

$R_1 \to R_1 + 2R_3$ $\quad E_8 = \begin{bmatrix} 1 & 0 & 2 \\ 0 & 1 & 0 \\ 0 & 0 & 1 \end{bmatrix} \quad E_8^{-1} = \begin{bmatrix} 1 & 0 & -2 \\ 0 & 1 & 0 \\ 0 & 0 & 1 \end{bmatrix}$

$$\therefore \quad A = E_1^{-1}E_2^{-1}E_3^{-1}E_4^{-1}E_5^{-1}E_6^{-1}E_7^{-1}E_8^{-1}$$

and

$$A^{-1} = E_8E_7E_6E_5E_4E_3E_2E_1$$

With a little practice, you will be able to find the answer without writing out the first eight parts of Step 2. You will want to observe that the sequences for A and A^{-1} are not unique.

4-5. Equivalent Matrices

Row equivalent matrices were covered in Chapter 2, Section 2-2. The concept can now be extended to column equivalent matrices where a matrix B can be obtained from matrix A by applying a finite sequence of elementary column operations on A.

Definition of equivalent matrices

An $m \times n$ matrix A is **equivalent** to an $m \times n$ matrix B if B can be obtained from A by applying a finite sequence of elementary row (column) operations to A; i.e., $B = PAQ$, P and Q invertible.

EXAMPLE 4-12:

(a) Show that $A = \begin{bmatrix} 1 & 5 & -3 \\ 0 & 1 & 2 \\ 1 & 6 & -1 \end{bmatrix}$ is equivalent to $B = \begin{bmatrix} 1 & 0 & 0 \\ 0 & 1 & 0 \\ 0 & 0 & 0 \end{bmatrix}$.

(b) Find invertible matrices P_1, P_2, Q_1, and Q_2 such that $B = P_1 A Q_1$ and $A = P_2 B Q_2$.

Solution

(a) Use elementary row and column operations to reduce A to B.

$$\begin{bmatrix} 1 & 5 & -3 \\ 0 & 1 & 2 \\ 1 & 6 & -1 \end{bmatrix} \xrightarrow{R_3 \to R_3 - R_1} \begin{bmatrix} 1 & 5 & -3 \\ 0 & 1 & 2 \\ 0 & 1 & 2 \end{bmatrix} \xrightarrow{R_3 \to R_3 - R_2} \begin{bmatrix} 1 & 5 & -3 \\ 0 & 1 & 2 \\ 0 & 0 & 0 \end{bmatrix} \xrightarrow{R_1 \to R_1 - 5R_2} \begin{bmatrix} 1 & 0 & -13 \\ 0 & 1 & 2 \\ 0 & 0 & 0 \end{bmatrix}$$

$$\xrightarrow{C_3 \to C_3 + 13C_1} \begin{bmatrix} 1 & 0 & 0 \\ 0 & 1 & 2 \\ 0 & 0 & 0 \end{bmatrix} \xrightarrow{C_3 \to C_3 - 2C_2} \begin{bmatrix} 1 & 0 & 0 \\ 0 & 1 & 0 \\ 0 & 0 & 0 \end{bmatrix}$$

(b) Write in sequence the row operations performed and the corresponding elementary matrix with its inverse. Do the same for the column operations.

$$R_3 \to R_3 - R_1 \qquad E_1 = \begin{bmatrix} 1 & 0 & 0 \\ 0 & 1 & 0 \\ -1 & 0 & 1 \end{bmatrix} \qquad E_1^{-1} = \begin{bmatrix} 1 & 0 & 0 \\ 0 & 1 & 0 \\ 1 & 0 & 1 \end{bmatrix}$$

$$R_3 \to R_3 - R_2 \qquad E_2 = \begin{bmatrix} 1 & 0 & 0 \\ 0 & 1 & 0 \\ 0 & -1 & 1 \end{bmatrix} \qquad E_2^{-1} = \begin{bmatrix} 1 & 0 & 0 \\ 0 & 1 & 0 \\ 0 & 1 & 1 \end{bmatrix}$$

$$R_1 \to R_1 - 5R_2 \qquad E_3 = \begin{bmatrix} 1 & -5 & 0 \\ 0 & 1 & 0 \\ 0 & 0 & 1 \end{bmatrix} \qquad E_3^{-1} = \begin{bmatrix} 1 & 5 & 0 \\ 0 & 1 & 0 \\ 0 & 0 & 1 \end{bmatrix}$$

$$C_3 \to C_3 + 13C_1 \qquad E_4 = \begin{bmatrix} 1 & 0 & 13 \\ 0 & 1 & 0 \\ 0 & 0 & 1 \end{bmatrix} \qquad E_4^{-1} = \begin{bmatrix} 1 & 0 & -13 \\ 0 & 1 & 0 \\ 0 & 0 & 1 \end{bmatrix}$$

$$C_3 \to C_3 - 2C_2 \qquad E_5 = \begin{bmatrix} 1 & 0 & 0 \\ 0 & 1 & -2 \\ 0 & 0 & 1 \end{bmatrix} \qquad E_5^{-1} = \begin{bmatrix} 1 & 0 & 0 \\ 0 & 1 & 2 \\ 0 & 0 & 1 \end{bmatrix}$$

Let

$$P_1 = E_3 E_2 E_1 = \begin{bmatrix} 1 & -5 & 0 \\ 0 & 1 & 0 \\ -1 & -1 & 1 \end{bmatrix}$$

$$Q_1 = E_4 E_5 = \begin{bmatrix} 1 & 0 & 13 \\ 0 & 1 & -2 \\ 0 & 0 & 1 \end{bmatrix}$$

Therefore, $B = P_1 A Q_1$. Multiply both sides of this on the left by P_1^{-1} and on the right by Q_1^{-1}. Hence, $P_1^{-1} B Q_1^{-1} = P_1^{-1}(P_1 A Q_1) Q_1^{-1} = A$. Let

$$P_2 = P_1^{-1} = (E_3E_2E_1)^{-1} = E_1^{-1}E_1^{-1}E_3^{-1} = \begin{bmatrix} 1 & 5 & 0 \\ 0 & 1 & 0 \\ 1 & 6 & 1 \end{bmatrix}$$

and

$$Q_2 = Q_1^{-1} = (E_4E_5)^{-1} = E_5^{-1}E_4^{-1} = \begin{bmatrix} 1 & 0 & -13 \\ 0 & 1 & 2 \\ 0 & 0 & 1 \end{bmatrix}$$

$$\therefore \quad A = P_2BQ_2$$

You should observe that P_1, P_2, Q_1, and Q_2 are all invertible since each is the product of elementary matrices.

As illustrated in part (**b**) of this example, if A is equivalent to B, then B is equivalent to A. Because of this symmetric property, we often shorten the statement to "A and B are equivalent matrices."

4-6. Partitioned Matrices

A. Definition of partitioned matrices

A matrix in which the rows or columns or both have been divided into two or more sets of rows or columns is said to be **partitioned** into submatrices. Generally, the rows are divided by either solid or dotted horizontal lines and the columns by vertical lines of the same kind.

EXAMPLE 4-13: Partition matrix A in two different ways.

$$A = \begin{bmatrix} 1 & 2 & 3 & 4 & 5 \\ 4 & -3 & 4 & 2 & 1 \\ 0 & 4 & 5 & -3 & 2 \\ -2 & 5 & 6 & 0 & 7 \end{bmatrix}$$

Solution

$$A = \left[\begin{array}{cc|ccc} 1 & 2 & 3 & 4 & 5 \\ 4 & -3 & 4 & 2 & 1 \\ \hline 0 & 4 & 5 & -3 & 2 \\ -2 & 5 & 6 & 0 & 7 \end{array}\right] = \begin{bmatrix} A_{11} & A_{12} \\ A_{21} & A_{22} \end{bmatrix}$$

where

$$A_{11} = \begin{bmatrix} 1 & 2 \\ 4 & -3 \end{bmatrix}, \quad A_{12} = \begin{bmatrix} 3 & 4 & 5 \\ 4 & 2 & 1 \end{bmatrix}, \quad A_{21} = \begin{bmatrix} 0 & 4 \\ -2 & 5 \end{bmatrix}, \quad A_{22} = \begin{bmatrix} 5 & -3 & 2 \\ 6 & 0 & 7 \end{bmatrix}$$

For a different partition, you could have

$$A = \left[\begin{array}{c|cc|cc} 1 & 2 & 3 & 4 & 5 \\ 4 & -3 & 4 & 2 & 1 \\ 0 & 4 & 5 & -3 & 2 \\ \hline -2 & 5 & 6 & 0 & 7 \end{array}\right] = \begin{bmatrix} A_{11} & A_{12} & A_{13} \\ A_{21} & A_{22} & A_{23} \end{bmatrix}$$

where

$$A_{11} = \begin{bmatrix} 1 \\ 4 \\ 0 \end{bmatrix}, \quad A_{12} = \begin{bmatrix} 2 & 3 \\ -3 & 4 \\ 4 & 5 \end{bmatrix}, \quad A_{13} = \begin{bmatrix} 4 & 5 \\ 2 & 1 \\ -3 & 2 \end{bmatrix},$$

$$A_{21} = [-2], \quad A_{22} = [5 \quad 6], \quad A_{23} = [0 \quad 7]$$

B. Addition of partitioned matrices

Two $m \times n$ partitioned matrices may be added if the corresponding submatrices are the same size.

EXAMPLE 4-14: Add the following partitioned matrices:

$$A = \left[\begin{array}{rr|rrr} 1 & 2 & 3 & 4 & 5 \\ 4 & -3 & 4 & 2 & 1 \\ \hline 0 & 4 & 5 & -3 & 2 \\ -2 & 5 & 6 & 0 & 7 \end{array}\right] = \begin{bmatrix} A_{11} & A_{12} \\ A_{21} & A_{22} \end{bmatrix}$$

$$B = \left[\begin{array}{rr|rrr} 5 & 0 & 4 & 3 & -2 \\ -3 & 4 & 5 & -2 & 0 \\ \hline 1 & 2 & 3 & -3 & -2 \\ 2 & 1 & 0 & -1 & -2 \end{array}\right] = \begin{bmatrix} B_{11} & B_{12} \\ B_{21} & B_{22} \end{bmatrix}$$

Solution

$$A + B = \begin{bmatrix} A_{11} + B_{11} & A_{12} + B_{12} \\ A_{21} + B_{21} & A_{22} + B_{22} \end{bmatrix} = \left[\begin{array}{rr|rrr} 6 & 2 & 7 & 7 & 3 \\ 1 & 1 & 9 & 0 & 1 \\ \hline 1 & 6 & 8 & -6 & 0 \\ 0 & 6 & 6 & -1 & 5 \end{array}\right]$$

C. Multiplication of partitioned matrices

Let A and B be $m \times n$ and $n \times p$ matrices, respectively, so that the product AB is defined. Matrix A may be partitioned into submatrices in any convenient manner. The columns of B may be partitioned in any desired way, but the rows of B must be partitioned in exactly the same way that the columns of A were partitioned. The product AB can then be found by the row–column multiplication process.

Multiplication of partitioned matrices is especially useful if at least one submatrix is either the zero matrix or identity matrix. It is quite useful if the sizes of matrices are such that the memory capacity of your computer is exceeded.

EXAMPLE 4-15: Find the product AB by partitioning the two matrices in conformable ways:

$$A = \begin{bmatrix} 1 & 0 & 0 & 2 & 3 \\ 0 & 1 & 0 & -1 & 4 \\ 0 & 0 & 1 & 3 & -1 \\ 2 & 3 & 4 & 0 & 1 \end{bmatrix} \qquad B = \begin{bmatrix} 1 & 2 & 3 & 0 & 0 & 0 \\ 4 & -3 & 4 & 0 & 0 & 0 \\ 0 & 4 & 5 & 0 & 0 & 0 \\ 0 & 0 & 0 & 5 & -3 & 2 \\ 0 & 0 & 0 & 6 & 0 & 7 \end{bmatrix}$$

Solution: There are many possible ways to partition A and then B, but a little observation shows us to use identity and zero matrices as follows:

$$A = \left[\begin{array}{ccc|cc} 1 & 0 & 0 & 2 & 3 \\ 0 & 1 & 0 & -1 & 4 \\ 0 & 0 & 1 & 3 & -1 \\ \hline 2 & 3 & 4 & 0 & 1 \end{array}\right] = \begin{bmatrix} A_{11} & A_{12} \\ A_{21} & A_{22} \end{bmatrix}$$

$$B = \left[\begin{array}{ccc|ccc} 1 & 2 & 3 & 0 & 0 & 0 \\ 4 & -3 & 4 & 0 & 0 & 0 \\ 0 & 4 & 5 & 0 & 0 & 0 \\ \hline 0 & 0 & 0 & 5 & -3 & 2 \\ 0 & 0 & 0 & 6 & 0 & 7 \end{array}\right] = \begin{bmatrix} B_{11} & B_{12} \\ B_{21} & B_{22} \end{bmatrix}$$

$$AB = \begin{bmatrix} A_{11} & A_{12} \\ A_{21} & A_{22} \end{bmatrix}\begin{bmatrix} B_{11} & B_{12} \\ B_{21} & B_{22} \end{bmatrix} = \begin{bmatrix} A_{11}B_{11} + A_{12}B_{21} & A_{11}B_{12} + A_{12}B_{22} \\ A_{21}B_{11} + A_{22}B_{21} & A_{21}B_{12} + A_{22}B_{22} \end{bmatrix}$$

where

$$A_{11}B_{11} + A_{12}B_{21} = \begin{bmatrix} 1 & 0 & 0 \\ 0 & 1 & 0 \\ 0 & 0 & 1 \end{bmatrix}\begin{bmatrix} 1 & 2 & 3 \\ 4 & -3 & 4 \\ 0 & 4 & 5 \end{bmatrix} + \begin{bmatrix} 2 & 3 \\ -1 & 4 \\ 3 & -1 \end{bmatrix}\begin{bmatrix} 0 & 0 & 0 \\ 0 & 0 & 0 \end{bmatrix} = \begin{bmatrix} 1 & 2 & 3 \\ 4 & -3 & 4 \\ 0 & 4 & 5 \end{bmatrix}$$

$$A_{11}B_{12} + A_{12}B_{22} = \begin{bmatrix} 1 & 0 & 0 \\ 0 & 1 & 0 \\ 0 & 0 & 1 \end{bmatrix}\begin{bmatrix} 0 & 0 & 0 \\ 0 & 0 & 0 \\ 0 & 0 & 0 \end{bmatrix} + \begin{bmatrix} 2 & 3 \\ -1 & 4 \\ 3 & -1 \end{bmatrix}\begin{bmatrix} 5 & -3 & 2 \\ 6 & 0 & 7 \end{bmatrix} = \begin{bmatrix} 28 & -6 & 25 \\ 19 & 3 & 26 \\ 9 & -9 & -1 \end{bmatrix}$$

$$A_{21}B_{11} + A_{22}B_{21} = [2 \quad 3 \quad 4]\begin{bmatrix} 1 & 2 & 3 \\ 4 & -3 & 4 \\ 0 & 4 & 5 \end{bmatrix} + [0 \quad 1]\begin{bmatrix} 0 & 0 & 0 \\ 0 & 0 & 0 \end{bmatrix} = [14 \quad 11 \quad 38]$$

$$A_{21}B_{12} + A_{22}B_{22} = [2 \quad 3 \quad 4]\begin{bmatrix} 0 & 0 & 0 \\ 0 & 0 & 0 \\ 0 & 0 & 0 \end{bmatrix} + [0 \quad 1]\begin{bmatrix} 5 & -3 & 2 \\ 6 & 0 & 7 \end{bmatrix} = [6 \quad 0 \quad 7]$$

$$\therefore \quad AB = \left[\begin{array}{ccc|ccc} 1 & 2 & 3 & 28 & -6 & 25 \\ 4 & -3 & 4 & 19 & 3 & 26 \\ 0 & 4 & 5 & 9 & -9 & -1 \\ \hline 14 & 11 & 38 & 6 & 0 & 7 \end{array}\right] = \begin{bmatrix} 1 & 2 & 3 & 28 & -6 & 25 \\ 4 & -3 & 4 & 19 & 3 & 26 \\ 0 & 4 & 5 & 9 & -9 & -1 \\ 14 & 11 & 38 & 6 & 0 & 7 \end{bmatrix}$$

SUMMARY

1. An $n \times n$ matrix A is invertible if and only if $\det(A) \neq 0$. Also, A is invertible if and only if it can be written as a product of elementary matrices.
2. Cramer's rule is a method of solving systems of linear equations having the same number of equations as unknowns and an invertible coefficient matrix. For the 3×3 case,

$$x = \frac{\det(A_1)}{\det(A)}, \qquad y = \frac{\det(A_2)}{\det(A)}, \qquad z = \frac{\det(A_3)}{\det(A)}$$

   where A is the coefficient matrix and A_1, A_2, A_3 are obtained, respectively, from A by replacing the column of coefficients of x, y, and z, respectively, by the column of constants.
3. The adjoint of a matrix A is the matrix denoted by $\text{adj}(A)$ and obtained by replacing each a_{ij} of A by the cofactor C_{ij} and then taking the transpose of the result.
4. If A is invertible, then $A^{-1} = [1/\det(A)]\,\text{adj}(A)$.
5. If A is any matrix of order n, then $A[\text{adj}(A)] = [\text{adj}(A)]A = [\det(A)]I_n$.

6. The adjoint of A is invertible if and only if A is invertible.
7. If A is invertible, then $[\text{adj}(A)]^{-1} = [1/\det(A)]A = \text{adj}(A^{-1})$.
8. A matrix E obtained by performing a single elementary row or column operation on I_n is called elementary.
9. An elementary row operation on matrix A is equivalent to multiplying A on the left by the elementary matrix E corresponding to the row operation.
10. An elementary column operation on matrix A is equivalent to multiplying A on the right by the elementary matrix E corresponding to the column operation.
11. Each elementary matrix is invertible and its inverse is an elementary matrix. If E is obtained from I by multiplying row i by $k \neq 0$, then E^{-1} is obtained from I by multiplying row i by $1/k$. If E is obtained from I by interchanging rows i and j, then E is its own inverse. If E is obtained from I by adding k times row i to row j, then E^{-1} is obtained from I by adding $-k$ times row i to row j.
12. Two $m \times n$ matrices A and B are equivalent if and only if there exist invertible matrices P and Q such that $B = PAQ$.
13. A matrix may be partitioned into submatrices.
14. Partitioned matrices may be added if corresponding submatrices are the same size.
15. Partitioned matrices may be multiplied by the row–column procedure if the number of columns of a submatrix from the left matrix is conformable with the number of rows of the appropriate submatrix from the right matrix.

RAISE YOUR GRADES

Can you ...?

☑ determine whether an $n \times n$ matrix is invertible
☑ use Cramer's rule to solve a system of linear equations
☑ find the adjoint of a matrix
☑ find the inverse of a nonsingular matrix by using the adjoint
☑ find the inverse of the adjoint of a nonsingular matrix
☑ find an elementary matrix that corresponds to each elementary row or column operation
☑ express an invertible matrix as a product of elementary matrices
☑ find the inverse of an elementary matrix
☑ find invertible matrices P and Q such that $B = PAQ$ when A and B are equivalent
☑ add and multiply appropriately partitioned matrices

SOLVED PROBLEMS

PROBLEM 4-1 Determine whether the given matrices are invertible:

(a) $A = \begin{bmatrix} 2 & -1 & 3 \\ 0 & 1 & -2 \\ 3 & -1 & 2 \end{bmatrix}$ **(b)** $B = \begin{bmatrix} 1 & 0 & -1 & 3 \\ 2 & 1 & 1 & -2 \\ -1 & 2 & 0 & 1 \\ 3 & 0 & -1 & 2 \end{bmatrix}$

Solution

(a) $$\det(A) = \begin{vmatrix} 2 & -1 & 3 \\ 0 & 1 & -2 \\ 3 & -1 & 2 \end{vmatrix} = 4 + 6 - 9 - 4 = -3 \qquad \therefore \quad A \text{ is invertible}$$

(b)
$$\det(B) = \begin{vmatrix} 1 & 0 & -1 & 3 \\ 2 & 1 & 1 & -2 \\ -1 & 2 & 0 & 1 \\ 3 & 0 & -1 & 2 \end{vmatrix} = \frac{1}{1^2}\begin{vmatrix} 1 & 3 & -8 \\ 2 & -1 & 4 \\ 0 & 2 & -7 \end{vmatrix}$$
$$= 7 - 32 - 8 + 42 = 9$$
$$\therefore \quad B \text{ is invertible}$$

PROBLEM 4-2 Solve the following systems of equations by Cramer's rule:

(a) $x + y + z = 3$
$2x + 3y - 2z = 3$
$x + 2y + z = 0$

(b) $(\lambda - 1)x + y = 1$
$x + (\lambda - 1)y = 1$

Solution

(a)
$$x = \frac{\begin{vmatrix} 3 & 1 & 1 \\ 3 & 3 & -2 \\ 0 & 2 & 1 \end{vmatrix}}{\begin{vmatrix} 1 & 1 & 1 \\ 2 & 3 & -2 \\ 1 & 2 & 1 \end{vmatrix}} = \frac{9 + 6 + 12 - 3}{3 + 4 - 2 - 3 + 4 - 2} = \frac{24}{4} = 6$$

$$y = \frac{\begin{vmatrix} 1 & 3 & 1 \\ 2 & 3 & -2 \\ 1 & 0 & 1 \end{vmatrix}}{\begin{vmatrix} 1 & 1 & 1 \\ 2 & 3 & -2 \\ 1 & 2 & 1 \end{vmatrix}} = \frac{3 - 6 - 3 - 6}{4} = \frac{-12}{4} = -3$$

$$z = \frac{\begin{vmatrix} 1 & 1 & 3 \\ 2 & 3 & 3 \\ 1 & 2 & 0 \end{vmatrix}}{4} = \frac{12 + 3 - 9 - 6}{4} = \frac{0}{4} = 0 \qquad \therefore \quad x = 6, y = -3, z = 0$$

(b)
$$x = \frac{\begin{vmatrix} 1 & 1 \\ 1 & \lambda - 1 \end{vmatrix}}{\begin{vmatrix} \lambda - 1 & 1 \\ 1 & \lambda - 1 \end{vmatrix}} = \frac{\lambda - 1 - 1}{(\lambda - 1)^2 - 1} = \frac{\lambda - 2}{\lambda^2 - 2\lambda} = \frac{\lambda - 2}{\lambda(\lambda - 2)} = \frac{1}{\lambda}$$

$$y = \frac{\begin{vmatrix} \lambda - 1 & 1 \\ 1 & 1 \end{vmatrix}}{\begin{vmatrix} \lambda - 1 & 1 \\ 1 & \lambda - 1 \end{vmatrix}} = \frac{\lambda - 1 - 1}{\lambda^2 - 2\lambda} = \frac{\lambda - 2}{\lambda(\lambda - 2)} = \frac{1}{\lambda}$$

$$\therefore \quad x = \frac{1}{\lambda}, y = \frac{1}{\lambda} \qquad \text{where } \lambda \neq 0 \quad \text{and} \quad \lambda \neq 2$$

PROBLEM 4-3 Find the adjoint of **(a)** $A = \begin{bmatrix} 1 & -2 \\ 3 & 4 \end{bmatrix}$ and **(b)** $B = \begin{bmatrix} -2 & 1 & 1 \\ 4 & 2 & 0 \\ 2 & -2 & 3 \end{bmatrix}$.

Solution

(a)

$$C_{11} = (-1)^{1+1}M_{11} = |4| = 4 \qquad C_{12} = (-1)^{1+2}M_{12} = -|3| = -3$$

$$C_{21} = (-1)^{2+1}M_{21} = -|-2| = 2 \qquad C_{22} = (-1)^{2+2}M_{22} = |1| = 1$$

$$\operatorname{adj}(A) = \begin{bmatrix} 4 & -3 \\ 2 & 1 \end{bmatrix}^T = \begin{bmatrix} 4 & 2 \\ -3 & 1 \end{bmatrix}$$

(b)

$$C_{11} = (-1)^{1+1}M_{11} = \begin{vmatrix} 2 & 0 \\ -2 & 3 \end{vmatrix} = 6$$

$$C_{12} = (-1)^{1+2}M_{12} = -\begin{vmatrix} 4 & 0 \\ 2 & 3 \end{vmatrix} = -12$$

$$C_{13} = (-1)^{1+3}M_{13} = \begin{vmatrix} 4 & 2 \\ 2 & -2 \end{vmatrix} = -12$$

$$C_{21} = (-1)^{2+1}M_{21} = -\begin{vmatrix} 1 & 1 \\ -2 & 3 \end{vmatrix} = -5$$

$$C_{22} = (-1)^{2+2}M_{22} = \begin{vmatrix} -2 & 1 \\ 2 & 3 \end{vmatrix} = -8$$

$$C_{23} = (-1)^{2+3}M_{23} = -\begin{vmatrix} -2 & 1 \\ 2 & -2 \end{vmatrix} = -2$$

$$C_{31} = (-1)^{3+1}M_{31} = \begin{vmatrix} 1 & 1 \\ 2 & 0 \end{vmatrix} = -2$$

$$C_{32} = (-1)^{3+2}M_{32} = -\begin{vmatrix} -2 & 1 \\ 4 & 0 \end{vmatrix} = 4$$

$$C_{33} = (-1)^{3+3}M_{33} = \begin{vmatrix} -2 & 1 \\ 4 & 2 \end{vmatrix} = -8$$

$$\operatorname{adj}(B) = \begin{bmatrix} 6 & -12 & -12 \\ -5 & -8 & -2 \\ -2 & 4 & -8 \end{bmatrix}^T = \begin{bmatrix} 6 & -5 & -2 \\ -12 & -8 & 4 \\ -12 & -2 & -8 \end{bmatrix}$$

PROBLEM 4-4 Find the inverses of the matrices in Problem 4-3.

Solution

(a)

$$\operatorname{adj}(A) = \begin{bmatrix} 4 & 2 \\ -3 & 1 \end{bmatrix} \qquad \det(A) = \begin{vmatrix} 1 & -2 \\ 3 & 4 \end{vmatrix} = 10$$

$$A^{-1} = \frac{1}{10}\begin{bmatrix} 4 & 2 \\ -3 & 1 \end{bmatrix} = \begin{bmatrix} \frac{4}{10} & \frac{2}{10} \\ -\frac{3}{10} & \frac{1}{10} \end{bmatrix}$$

(b)

$$\operatorname{adj}(B) = \begin{bmatrix} 6 & -5 & -2 \\ -12 & -8 & 4 \\ -12 & -2 & -8 \end{bmatrix} \qquad \det(B) = \begin{vmatrix} -2 & 1 & 1 \\ 4 & 2 & 0 \\ 2 & -2 & 3 \end{vmatrix} = -36$$

$$B^{-1} = -\frac{1}{36}\begin{bmatrix} 6 & -5 & -2 \\ -12 & -8 & 4 \\ -12 & -2 & -8 \end{bmatrix} = \begin{bmatrix} -\frac{1}{6} & \frac{5}{36} & \frac{1}{18} \\ \frac{1}{3} & \frac{2}{9} & -\frac{1}{9} \\ \frac{1}{3} & \frac{1}{18} & \frac{2}{9} \end{bmatrix}$$

PROBLEM 4-5 Using matrix B in Problem 4-3 show that $B[\mathrm{adj}(B)] = \det(B)I_3$.

Solution

$$\begin{bmatrix} -2 & 1 & 1 \\ 4 & 2 & 0 \\ 2 & -2 & 3 \end{bmatrix}\begin{bmatrix} 6 & -5 & -2 \\ -12 & -8 & 4 \\ -12 & -2 & -8 \end{bmatrix} = \begin{bmatrix} -36 & 0 & 0 \\ 0 & -36 & 0 \\ 0 & 0 & -36 \end{bmatrix} = -36\begin{bmatrix} 1 & 0 & 0 \\ 0 & 1 & 0 \\ 0 & 0 & 1 \end{bmatrix}$$

$$\therefore \quad B[\mathrm{adj}(B)] = \det(B)I_3$$

PROBLEM 4-6 Which of the following matrices are elementary matrices? **(a)** $\begin{bmatrix} 0 & -1 \\ 1 & 0 \end{bmatrix}$, **(b)** $\begin{bmatrix} 1 & 0 \\ -2 & 1 \end{bmatrix}$, **(c)** $\begin{bmatrix} 1 & -3 \\ 0 & 1 \end{bmatrix}$, **(d)** $\begin{bmatrix} 1 & -2 & 0 \\ 0 & 0 & 1 \\ 0 & 0 & 1 \end{bmatrix}$, **(e)** $\begin{bmatrix} 1 & 0 & 1 \\ 0 & 1 & 0 \\ -3 & 0 & -2 \end{bmatrix}$, **(f)** $\begin{bmatrix} -1 & 0 & 0 \\ 0 & 1 & 0 \\ 0 & 0 & 1 \end{bmatrix}$.

Solution

(a) $$\begin{bmatrix} 1 & 0 \\ 0 & 1 \end{bmatrix} \xrightarrow{C_1 \leftrightarrow C_2} \begin{bmatrix} 0 & 1 \\ 1 & 0 \end{bmatrix} \xrightarrow{C_2 \rightarrow -C_2} \begin{bmatrix} 0 & -1 \\ 1 & 0 \end{bmatrix}$$

Therefore, this is *not* elementary. It took *two* operations on I_2 to get the given matrix.

(b) $$\begin{bmatrix} 1 & 0 \\ 0 & 1 \end{bmatrix} \xrightarrow[R_2 \rightarrow R_2 - 2R_1]{} \begin{bmatrix} 1 & 0 \\ -2 & 1 \end{bmatrix} \qquad \therefore \text{ it is elementary}$$

(c) $$\begin{bmatrix} 1 & 0 \\ 0 & 1 \end{bmatrix} \xrightarrow{C_2 \rightarrow C_2 - 3C_1} \begin{bmatrix} 1 & -3 \\ 0 & 1 \end{bmatrix} \qquad \therefore \text{ it is elementary}$$

(d) $$\begin{bmatrix} 1 & 0 & 0 \\ 0 & 1 & 0 \\ 0 & 0 & 1 \end{bmatrix} \xrightarrow{C_2 \rightarrow C_2 - 2C_1} \begin{bmatrix} 1 & -2 & 0 \\ 0 & 1 & 0 \\ 0 & 0 & 1 \end{bmatrix} \qquad \therefore \text{ it is elementary}$$

(e) $$\begin{bmatrix} 1 & 0 & 0 \\ 0 & 1 & 0 \\ 0 & 0 & 1 \end{bmatrix} \xrightarrow{R_1 \rightarrow R_1 + R_3} \begin{bmatrix} 1 & 0 & 1 \\ 0 & 1 & 0 \\ 0 & 0 & 1 \end{bmatrix} \xrightarrow[R_3 \rightarrow R_3 - 3R_1]{} \begin{bmatrix} 1 & 0 & 1 \\ 0 & 1 & 0 \\ -3 & 0 & -2 \end{bmatrix}$$

$$\therefore \text{ it is not elementary}$$

(f) $$\begin{bmatrix} 1 & 0 & 0 \\ 0 & 1 & 0 \\ 0 & 0 & 1 \end{bmatrix} \xrightarrow{R_1 \leftrightarrow -R_1} \begin{bmatrix} -1 & 0 & 0 \\ 0 & 1 & 0 \\ 0 & 0 & 1 \end{bmatrix} \qquad \therefore \text{ it is elementary}$$

PROBLEM 4-7 Let

$$A = \begin{bmatrix} 1 & -2 & 1 \\ 0 & 3 & -1 \\ -2 & 1 & -2 \\ 3 & 0 & -1 \end{bmatrix}$$

Find the elementary matrix E such that EA performs the following elementary row operation on A: **(a)** multiplies the third row by 3; **(b)** interchanges the first and third rows; **(c)** adds two times the first row to the fourth row.

Solution This is easy to do by performing the same operation on I_4 as asked to perform on A.

(a)

$$\begin{bmatrix} 1 & 0 & 0 & 0 \\ 0 & 1 & 0 & 0 \\ 0 & 0 & 1 & 0 \\ 0 & 0 & 0 & 1 \end{bmatrix} \xrightarrow{R_3 \to 3R_3} \begin{bmatrix} 1 & 0 & 0 & 0 \\ 0 & 1 & 0 & 0 \\ 0 & 0 & 3 & 0 \\ 0 & 0 & 0 & 1 \end{bmatrix} = E$$

$$EA = \begin{bmatrix} 1 & 0 & 0 & 0 \\ 0 & 1 & 0 & 0 \\ 0 & 0 & 3 & 0 \\ 0 & 0 & 0 & 1 \end{bmatrix} \begin{bmatrix} 1 & -2 & 1 \\ 0 & 3 & -1 \\ -2 & 1 & -2 \\ 3 & 0 & -1 \end{bmatrix} = \begin{bmatrix} 1 & -2 & 1 \\ 0 & 3 & -1 \\ -6 & 3 & -6 \\ 3 & 0 & -1 \end{bmatrix}$$

(b)

$$\begin{bmatrix} 1 & 0 & 0 & 0 \\ 0 & 1 & 0 & 0 \\ 0 & 0 & 1 & 0 \\ 0 & 0 & 0 & 1 \end{bmatrix} \xrightarrow{R_1 \leftrightarrow R_3} \begin{bmatrix} 0 & 0 & 1 & 0 \\ 0 & 1 & 0 & 0 \\ 1 & 0 & 0 & 0 \\ 0 & 0 & 0 & 1 \end{bmatrix} = E$$

$$EA = \begin{bmatrix} 0 & 0 & 1 & 0 \\ 0 & 1 & 0 & 0 \\ 1 & 0 & 0 & 0 \\ 0 & 0 & 0 & 1 \end{bmatrix} \begin{bmatrix} 1 & -2 & 1 \\ 0 & 3 & -1 \\ -2 & 1 & -2 \\ 3 & 0 & -1 \end{bmatrix} = \begin{bmatrix} -2 & 1 & -2 \\ 0 & 3 & -1 \\ 1 & -2 & 1 \\ 3 & 0 & -1 \end{bmatrix}$$

(c)

$$\begin{bmatrix} 1 & 0 & 0 & 0 \\ 0 & 1 & 0 & 0 \\ 0 & 0 & 1 & 0 \\ 0 & 0 & 0 & 1 \end{bmatrix} \xrightarrow[R_4 \to 2R_1 + R_4]{} \begin{bmatrix} 1 & 0 & 0 & 0 \\ 0 & 1 & 0 & 0 \\ 0 & 0 & 1 & 0 \\ 2 & 0 & 0 & 1 \end{bmatrix} = E$$

$$EA = \begin{bmatrix} 1 & 0 & 0 & 0 \\ 0 & 1 & 0 & 0 \\ 0 & 0 & 1 & 0 \\ 2 & 0 & 0 & 1 \end{bmatrix} \begin{bmatrix} 1 & -2 & 1 \\ 0 & 3 & -1 \\ -2 & 1 & -2 \\ 3 & 0 & -1 \end{bmatrix} = \begin{bmatrix} 1 & -2 & 1 \\ 0 & 3 & -1 \\ -2 & 1 & -2 \\ 5 & -4 & 1 \end{bmatrix}$$

PROBLEM 4-8 In Problem 2-12 **(c)** we found that $A = \begin{bmatrix} 1 & 2 & -1 \\ 1 & 1 & -2 \\ 1 & 2 & 0 \end{bmatrix}$ has inverse $A^{-1} = \begin{bmatrix} -4 & 2 & 3 \\ 2 & -1 & -1 \\ -1 & 0 & 1 \end{bmatrix}$. Express both A and A^{-1} as a product of elementary matrices.

Solution First use elementary row operations to reduce A to I_3:

$$A = \begin{bmatrix} 1 & 2 & -1 \\ 1 & 1 & -2 \\ 1 & 2 & 0 \end{bmatrix} \xrightarrow[R_3 \to R_3 - R_1]{R_2 \to R_2 - R_1} \begin{bmatrix} 1 & 2 & -1 \\ 0 & -1 & -1 \\ 0 & 0 & 1 \end{bmatrix} \xrightarrow{R_1 \to R_1 + 2R_2} \begin{bmatrix} 1 & 0 & -3 \\ 0 & -1 & -1 \\ 0 & 0 & 1 \end{bmatrix}$$

$$\xrightarrow{R_2 \to -R_2} \begin{bmatrix} 1 & 0 & -3 \\ 0 & 1 & 1 \\ 0 & 0 & 1 \end{bmatrix} \xrightarrow[R_2 \to R_2 - R_3]{R_1 \to R_1 + 3R_3} \begin{bmatrix} 1 & 0 & 0 \\ 0 & 1 & 0 \\ 0 & 0 & 1 \end{bmatrix}$$

Next write in sequence the row operations performed and the corresponding elementary matrix with its inverse.

$$R_2 \to R_2 - R_1 \qquad E_1 = \begin{bmatrix} 1 & 0 & 0 \\ -1 & 1 & 0 \\ 0 & 0 & 1 \end{bmatrix} \qquad E_1^{-1} = \begin{bmatrix} 1 & 0 & 0 \\ 1 & 1 & 0 \\ 0 & 0 & 1 \end{bmatrix}$$

$$R_3 \to R_3 - R_1 \qquad E_2 = \begin{bmatrix} 1 & 0 & 0 \\ 0 & 1 & 0 \\ -1 & 0 & 1 \end{bmatrix} \qquad E_2^{-1} = \begin{bmatrix} 1 & 0 & 0 \\ 0 & 1 & 0 \\ 1 & 0 & 1 \end{bmatrix}$$

$$R_1 \to R_1 + 2R_1 \qquad E_3 = \begin{bmatrix} 1 & 2 & 0 \\ 0 & 1 & 0 \\ 0 & 0 & 1 \end{bmatrix} \qquad E_3^{-1} = \begin{bmatrix} 1 & -2 & 0 \\ 0 & 1 & 0 \\ 0 & 0 & 1 \end{bmatrix}$$

$$R_2 \to -R_2 \qquad E_4 = \begin{bmatrix} 1 & 0 & 0 \\ 0 & -1 & 0 \\ 0 & 0 & 1 \end{bmatrix} \qquad E_4^{-1} = \begin{bmatrix} 1 & 0 & 0 \\ 0 & -1 & 0 \\ 0 & 0 & 1 \end{bmatrix}$$

$$R_1 \to R_1 + 3R_3 \qquad E_5 = \begin{bmatrix} 1 & 0 & 3 \\ 0 & 1 & 0 \\ 0 & 0 & 1 \end{bmatrix} \qquad E_5^{-1} = \begin{bmatrix} 1 & 0 & -3 \\ 0 & 1 & 0 \\ 0 & 0 & 1 \end{bmatrix}$$

$$R_2 \to R_2 - R_3 \qquad E_6 = \begin{bmatrix} 1 & 0 & 0 \\ 0 & 1 & -1 \\ 0 & 0 & 1 \end{bmatrix} \qquad E_6^{-1} = \begin{bmatrix} 1 & 0 & 0 \\ 0 & 1 & 1 \\ 0 & 0 & 1 \end{bmatrix}$$

$$\therefore \quad A = E_1^{-1}E_2^{-1}E_3^{-1}E_4^{-1}E_5^{-1}E_6^{-1} \qquad \text{and} \qquad A^{-1} = E_6E_5E_4E_3E_2E_1$$

PROBLEM 4-9 Show that $A = \begin{bmatrix} -1 & -3 & 1 \\ 0 & 2 & 1 \\ 1 & -1 & 0 \end{bmatrix}$ is equivalent to $B = \begin{bmatrix} 1 & 0 & 0 \\ 0 & 1 & 0 \\ 0 & 0 & 1 \end{bmatrix}$.

Solution We must find invertible matrices P_1, P_2, Q_1, and Q_2 such that

$$B = P_1AQ_1 \qquad \text{and} \qquad A = P_2BQ_2$$

Use elementary operations to reduce A to B.

$$A = \begin{bmatrix} -1 & -3 & 1 \\ 0 & 2 & 1 \\ 1 & -1 & 0 \end{bmatrix} \xrightarrow{R_1 \leftrightarrow R_3} \begin{bmatrix} 1 & -1 & 0 \\ 0 & 2 & 1 \\ -1 & -3 & 1 \end{bmatrix} \xrightarrow{R_3 \to R_3 + R_1} \begin{bmatrix} 1 & -1 & 0 \\ 0 & 2 & 1 \\ 0 & -4 & 1 \end{bmatrix}$$

$$\xrightarrow{R_3 \to R_3 + 2R_2} \begin{bmatrix} 1 & -1 & 0 \\ 0 & 2 & 1 \\ 0 & 0 & 3 \end{bmatrix} \xrightarrow[R_3 \to 1/3 R_3]{} \begin{bmatrix} 1 & -1 & 0 \\ 0 & 2 & 1 \\ 0 & 0 & 1 \end{bmatrix} \xrightarrow{C_2 \to C_2 + C_1} \begin{bmatrix} 1 & 0 & 0 \\ 0 & 2 & 1 \\ 0 & 0 & 1 \end{bmatrix}$$

$$\xrightarrow{C_3 \to C_3 - 1/2 C_2} \begin{bmatrix} 1 & 0 & 0 \\ 0 & 2 & 0 \\ 0 & 0 & 1 \end{bmatrix} \xrightarrow{C_2 \to 1/2 C_2} \begin{bmatrix} 1 & 0 & 0 \\ 0 & 1 & 0 \\ 0 & 0 & 1 \end{bmatrix} = B$$

Summarizing these results, we get

$$R_1 \leftrightarrow R_3 \qquad E_1 = \begin{bmatrix} 0 & 0 & 1 \\ 0 & 1 & 0 \\ 1 & 0 & 0 \end{bmatrix} \qquad E_1^{-1} = \begin{bmatrix} 0 & 0 & 1 \\ 0 & 1 & 0 \\ 1 & 0 & 0 \end{bmatrix}$$

$$R_3 \to R_3 + R_1 \qquad E_2 = \begin{bmatrix} 1 & 0 & 0 \\ 0 & 1 & 0 \\ 1 & 0 & 1 \end{bmatrix} \qquad E_2^{-1} = \begin{bmatrix} 1 & 0 & 0 \\ 0 & 1 & 0 \\ -1 & 0 & 1 \end{bmatrix}$$

$$R_3 \to R_3 + 2R_2 \qquad E_3 = \begin{bmatrix} 1 & 0 & 0 \\ 0 & 1 & 0 \\ 0 & 2 & 1 \end{bmatrix} \qquad E_3^{-1} = \begin{bmatrix} 1 & 0 & 0 \\ 0 & 1 & 0 \\ 0 & -2 & 1 \end{bmatrix}$$

$$R_3 \to \tfrac{1}{3}R_3 \qquad E_4 = \begin{bmatrix} 1 & 0 & 0 \\ 0 & 1 & 0 \\ 0 & 0 & \frac{1}{3} \end{bmatrix} \qquad E_4^{-1} = \begin{bmatrix} 1 & 0 & 0 \\ 0 & 1 & 0 \\ 0 & 0 & 3 \end{bmatrix}$$

$$C_2 \to C_2 + C_1 \qquad E_5 = \begin{bmatrix} 1 & 1 & 0 \\ 0 & 1 & 0 \\ 0 & 0 & 1 \end{bmatrix} \qquad E_5^{-1} = \begin{bmatrix} 1 & -1 & 0 \\ 0 & 1 & 0 \\ 0 & 0 & 1 \end{bmatrix}$$

$$C_3 \to C_3 - \tfrac{1}{2}C_2 \qquad E_6 = \begin{bmatrix} 1 & 0 & 0 \\ 0 & 1 & -\frac{1}{2} \\ 0 & 0 & 1 \end{bmatrix} \qquad E_6^{-1} = \begin{bmatrix} 1 & 0 & 0 \\ 0 & 1 & \frac{1}{2} \\ 0 & 0 & 1 \end{bmatrix}$$

$$C_2 \to \tfrac{1}{2}C_2 \qquad E_7 = \begin{bmatrix} 1 & 0 & 0 \\ 0 & \frac{1}{2} & 0 \\ 0 & 0 & 1 \end{bmatrix} \qquad E_7^{-1} = \begin{bmatrix} 1 & 0 & 0 \\ 0 & 2 & 0 \\ 0 & 0 & 1 \end{bmatrix}$$

Let

$$P_1 = E_4E_3E_2E_1 = \begin{bmatrix} 0 & 0 & 1 \\ 0 & 1 & 0 \\ \frac{1}{3} & \frac{2}{3} & \frac{1}{3} \end{bmatrix}$$

$$Q_1 = E_5E_6E_7 = \begin{bmatrix} 1 & \frac{1}{2} & -\frac{1}{2} \\ 0 & \frac{1}{2} & -\frac{1}{2} \\ 0 & 0 & 1 \end{bmatrix}$$

$$\therefore \quad B = P_1AQ_1 \qquad \text{and hence} \qquad A = P_1^{-1}BQ_1^{-1}$$

Let

$$P_2 = P_1^{-1} = (E_4E_3E_2E_1)^{-1} = E_1^{-1}E_2^{-1}E_3^{-1}E_4^{-1} = \begin{bmatrix} -1 & -2 & 3 \\ 0 & 1 & 0 \\ 1 & 0 & 0 \end{bmatrix}$$

and

$$Q_2 = Q_1^{-1} = (E_5E_6E_7)^{-1} = E_7^{-1}E_6^{-1}E_5^{-1} = \begin{bmatrix} 1 & -1 & 0 \\ 0 & 2 & 1 \\ 0 & 0 & 1 \end{bmatrix}$$

$$\therefore \quad A = P_2BQ_2$$

PROBLEM 4-10 Partition $A = \begin{bmatrix} 1 & 2 & 3 & 4 \\ -1 & 0 & 1 & 2 \\ 3 & 2 & 0 & 1 \end{bmatrix}$ in two different ways.

Solution

(1)

$$A = \left[\begin{array}{rr|rr} 1 & 2 & 3 & 4 \\ -1 & 0 & 1 & 2 \\ \hline 3 & 2 & 0 & 1 \end{array}\right] = \begin{bmatrix} A_{11} & A_{12} \\ A_{21} & A_{22} \end{bmatrix}$$

where

$$A_{11} = \begin{bmatrix} 1 & 2 \\ -1 & 0 \end{bmatrix} \qquad A_{12} = \begin{bmatrix} 3 & 4 \\ 1 & 2 \end{bmatrix} \qquad A_{21} = [3 \quad 2] \qquad A_{22} = [0 \quad 1]$$

(2) Also

$$A = \left[\begin{array}{rrr|r} 1 & -2 & 3 & 4 \\ -1 & 0 & 1 & 2 \\ \hline 3 & 2 & 0 & 1 \end{array}\right] = \begin{bmatrix} A_{11} & A_{12} \\ A_{21} & A_{22} \end{bmatrix}$$

where

$$A_{11} = \begin{bmatrix} 1 & -2 & 3 \\ -1 & 0 & 1 \end{bmatrix} \qquad A_{12} = \begin{bmatrix} 4 \\ 2 \end{bmatrix} \qquad A_{21} = [3 \quad 2 \quad 0] \qquad A_{22} = [1]$$

PROBLEM 4-11 Add the following partitioned matrices:

$$A = \left[\begin{array}{rr|rr} 1 & 2 & 0 & 1 \\ -3 & 2 & 1 & 2 \\ \hline 3 & 1 & 0 & -1 \end{array}\right] \qquad B = \left[\begin{array}{rr|rr} 1 & 0 & 3 & 1 \\ 2 & -1 & 1 & 2 \\ \hline 3 & 2 & 4 & 0 \end{array}\right]$$

Solution

$$A + B = \begin{bmatrix} A_{11} + B_{11} & A_{12} + B_{12} \\ A_{21} + B_{21} & A_{22} + B_{12} \end{bmatrix} = \left[\begin{array}{rr|rr} 2 & 2 & 3 & 2 \\ -1 & 1 & 2 & 4 \\ \hline 6 & 3 & 4 & -1 \end{array}\right]$$

PROBLEM 4-12 Find the product AB by partitioning the matrices in conformable ways:

$$A = \begin{bmatrix} 1 & 0 & 0 & 2 & 1 \\ 0 & 1 & 1 & 0 & 1 \\ 2 & 1 & -1 & 1 & 0 \\ 0 & 1 & 2 & 0 & 1 \end{bmatrix} \qquad B = \begin{bmatrix} 1 & 0 & 1 & 3 \\ 1 & -2 & 1 & 2 \\ 2 & -1 & 0 & 0 \\ 0 & 1 & 0 & 0 \\ 1 & 0 & 0 & 0 \end{bmatrix}$$

Solution Look for the easy way out. Notice the presence of an identity matrix and a zero matrix:

$$A = \left[\begin{array}{rr|rrr} 1 & 0 & 0 & 2 & 1 \\ 0 & 1 & 1 & 0 & 1 \\ \hline 2 & 1 & -1 & 1 & 0 \\ 0 & 1 & 2 & 0 & 1 \end{array}\right] \qquad B = \left[\begin{array}{rr|rr} 1 & 0 & 1 & 3 \\ 1 & -2 & 1 & 2 \\ \hline 2 & -1 & 0 & 0 \\ 0 & 1 & 0 & 0 \\ 1 & 0 & 0 & 0 \end{array}\right]$$

$$AB = \begin{bmatrix} A_{11} & A_{12} \\ A_{21} & A_{22} \end{bmatrix} \begin{bmatrix} B_{11} & B_{12} \\ B_{21} & B_{22} \end{bmatrix} = \begin{bmatrix} A_{11}B_{11} + A_{12}B_{21} & A_{11}B_{12} + A_{12}B_{22} \\ A_{21}B_{11} + A_{22}B_{21} & A_{21}B_{12} + A_{22}B_{22} \end{bmatrix}$$

where

$$A_{11}B_{11} + A_{12}B_{21} = \begin{bmatrix} 1 & 0 \\ 0 & 1 \end{bmatrix}\begin{bmatrix} 1 & 0 \\ 1 & -2 \end{bmatrix} + \begin{bmatrix} 0 & 2 & 1 \\ 1 & 0 & 1 \end{bmatrix}\begin{bmatrix} 2 & -1 \\ 0 & 1 \\ 1 & 0 \end{bmatrix} = \begin{bmatrix} 1 & 0 \\ 1 & -2 \end{bmatrix} + \begin{bmatrix} 1 & 2 \\ 3 & -1 \end{bmatrix} = \begin{bmatrix} 2 & 2 \\ 4 & -3 \end{bmatrix}$$

$$A_{11}B_{12} + A_{12}B_{22} = \begin{bmatrix} 1 & 0 \\ 0 & 1 \end{bmatrix}\begin{bmatrix} 1 & 3 \\ 1 & 2 \end{bmatrix} + \begin{bmatrix} 0 & 2 & 1 \\ 1 & 0 & 1 \end{bmatrix}\begin{bmatrix} 0 & 0 \\ 0 & 0 \\ 0 & 0 \end{bmatrix} = \begin{bmatrix} 1 & 3 \\ 1 & 2 \end{bmatrix}$$

$$A_{21}B_{11} + A_{22}B_{21} = \begin{bmatrix} 2 & 1 \\ 0 & 1 \end{bmatrix}\begin{bmatrix} 1 & 0 \\ 1 & -2 \end{bmatrix} + \begin{bmatrix} -1 & 1 & 0 \\ 2 & 0 & 1 \end{bmatrix}\begin{bmatrix} 2 & -1 \\ 0 & 1 \\ 1 & 0 \end{bmatrix}$$

$$= \begin{bmatrix} 3 & -2 \\ 1 & -2 \end{bmatrix} + \begin{bmatrix} -2 & 2 \\ 5 & -2 \end{bmatrix} = \begin{bmatrix} 1 & 0 \\ 6 & -4 \end{bmatrix}$$

$$A_{21}B_{12} + A_{22}B_{22} = \begin{bmatrix} 2 & 1 \\ 0 & 1 \end{bmatrix}\begin{bmatrix} 1 & 3 \\ 1 & 2 \end{bmatrix} + \begin{bmatrix} -1 & 1 & 0 \\ 2 & 0 & 1 \end{bmatrix}\begin{bmatrix} 0 & 0 \\ 0 & 0 \\ 0 & 0 \end{bmatrix} = \begin{bmatrix} 3 & 8 \\ 1 & 2 \end{bmatrix}$$

$$\therefore \quad AB = \left[\begin{array}{cc|cc} 2 & 2 & 1 & 3 \\ 4 & -3 & 1 & 2 \\ \hline 1 & 0 & 3 & 8 \\ 6 & -4 & 1 & 2 \end{array}\right] = \begin{bmatrix} 2 & 2 & 1 & 3 \\ 4 & -3 & 1 & 2 \\ 1 & 0 & 3 & 8 \\ 6 & -4 & 1 & 2 \end{bmatrix}$$

Supplementary Problems

PROBLEM 4-13 Determine which of the following matrices are invertible without finding the inverse:

$$A = \begin{bmatrix} 2 & 1 \\ 4 & 2 \end{bmatrix} \qquad B = \begin{bmatrix} 2 & -1 & 0 \\ 1 & 3 & -2 \\ 4 & 5 & 1 \end{bmatrix} \qquad C = \begin{bmatrix} -1 & 2 & -3 & 2 \\ 0 & 1 & 2 & -1 \\ 1 & 2 & 3 & -4 \\ 4 & 0 & 1 & 2 \end{bmatrix}$$

PROBLEM 4-14 Solve the systems of linear equations using Cramer's rule:

(a)
$$\begin{aligned} x - 2y + z &= 3 \\ 2x + y - z &= 2 \\ x + y + 2z &= 0 \end{aligned}$$

(b)
$$\begin{aligned} sX + Y &= (2s - 1)/(s - 2) \\ X + sY &= 0 \end{aligned}$$

PROBLEM 4-15 Given $A = \begin{bmatrix} 2 & 3 & 4 \\ 3 & 1 & 0 \\ 0 & 2 & 1 \end{bmatrix}$, find **(a)** $\det(A)$, **(b)** $\operatorname{adj}(A)$, **(c)** $A[\operatorname{adj}(A)]$, **(d)** A^{-1}, and **(e)** show $A[\operatorname{adj}(A)] = \det(A)I_3$.

PROBLEM 4-16 Determine whether the following are elementary matrices: **(a)** $\begin{bmatrix} 1 & 3 \\ 0 & -1 \end{bmatrix}$, **(b)** $\begin{bmatrix} 1 & 0 & 0 \\ 2 & 1 & 0 \\ 0 & 0 & 1 \end{bmatrix}$, **(c)** $\begin{bmatrix} 2 & 0 & 1 \\ 0 & 1 & 0 \\ 0 & 0 & 1 \end{bmatrix}$, **(d)** $\begin{bmatrix} -2 & 0 & 0 \\ 0 & 1 & 0 \\ 0 & 0 & 1 \end{bmatrix}$

PROBLEM 4-17 Find the elementary matrix E that will perform the following elementary row operation on $A = \begin{bmatrix} -1 & 2 & 3 \\ 4 & -2 & 1 \\ 3 & 2 & -1 \end{bmatrix}$ when EA is computed: **(a)** multiplies the third row by 2; **(b)** interchanges the second and third rows; **(c)** adds three times the second row to the first row.

PROBLEM 4-18 Show that the two given matrices are equivalent and give the elementary matrices needed to transform

$$A = \begin{bmatrix} 1 & 0 & 2 \\ -2 & 1 & 0 \\ 0 & 0 & 1 \end{bmatrix} \quad \text{into} \quad B = \begin{bmatrix} 1 & 0 & -2 \\ 2 & 1 & -4 \\ 0 & 0 & 1 \end{bmatrix}$$

PROBLEM 4-19 In Problem 4-18, express A and B as products of elementary matrices.

PROBLEM 4-20 Partition the matrix

$$A = \begin{bmatrix} 2 & 1 & -3 & 1 \\ 3 & 0 & 1 & -2 \\ 1 & 2 & -1 & 3 \\ -1 & 3 & 0 & 1 \end{bmatrix}$$

into four square matrices.

PROBLEM 4-21 Given the matrices

$$A = \begin{bmatrix} 2 & 1 & -3 & 1 \\ 3 & 0 & 1 & -2 \\ 1 & 2 & -1 & 3 \end{bmatrix} \qquad B = \begin{bmatrix} 4 & -1 \\ 2 & 3 \\ 0 & 1 \\ 1 & -2 \end{bmatrix} \qquad C = \begin{bmatrix} 2 & -1 & 3 \\ 1 & 3 & 4 \end{bmatrix}$$

determine which of the following products exists: **(a)** AB, **(b)** AC, **(c)** CA, **(d)** BC.

PROBLEM 4-22 If the elements of a square matrix are probabilities and the elements of each row add up to 1, the matrix is called a **stochastic matrix**. Determine which of the following are stochastic:

(a) $\begin{bmatrix} 1 & 0 & 0 \\ \frac{1}{2} & \frac{1}{4} & \frac{1}{4} \\ \frac{1}{8} & \frac{1}{2} & \frac{3}{8} \end{bmatrix}$ **(b)** $\begin{bmatrix} \frac{1}{2} & 0 \\ -1 & 2 \end{bmatrix}$ **(c)** $\begin{bmatrix} \frac{7}{9} & 0 & \frac{2}{9} \\ \frac{1}{5} & \frac{2}{5} & \frac{2}{5} \\ 0 & \frac{1}{2} & \frac{1}{2} \end{bmatrix}$

Answers to Supplementary Problems

4-13 B and C are invertible

4-14 (a) $x = 19/14, y = -13/14, z = -3/14$
(b) $X = s(2s-1)/(s^2-1)(s-2)$,
$Y = (1-2s)/(s^2-1)(s-2)$

4-15 (a) $\det(A) = 17$

(b) $$\text{adj}(A) = \begin{bmatrix} 1 & 5 & -4 \\ -3 & 2 & 12 \\ 6 & -4 & -7 \end{bmatrix}$$

(c) $$A[\text{adj}(A)] = \begin{bmatrix} 17 & 0 & 0 \\ 0 & 17 & 0 \\ 0 & 0 & 17 \end{bmatrix}$$

(d) $$A^{-1} = 1/17\begin{bmatrix} 1 & 5 & -4 \\ -3 & 2 & 12 \\ 6 & -4 & -7 \end{bmatrix}$$

$$= \begin{bmatrix} \frac{1}{17} & \frac{5}{17} & -\frac{4}{17} \\ -\frac{3}{17} & \frac{2}{17} & \frac{12}{17} \\ \frac{6}{17} & -\frac{4}{17} & -\frac{7}{17} \end{bmatrix}$$

(e) $$\begin{bmatrix} 17 & 0 & 0 \\ 0 & 17 & 0 \\ 0 & 0 & 17 \end{bmatrix} = 17\begin{bmatrix} 1 & 0 & 0 \\ 0 & 1 & 0 \\ 0 & 0 & 1 \end{bmatrix}$$

4-16 (a) Not elementary (b) Elementary (c) Not elementary (d) Elementary

4-17 (a) $$E = \begin{bmatrix} 1 & 0 & 0 \\ 0 & 1 & 0 \\ 0 & 0 & 2 \end{bmatrix}$$ (c) $$\begin{bmatrix} 1 & 3 & 0 \\ 0 & 1 & 0 \\ 0 & 0 & 1 \end{bmatrix}$$

(b) $$E = \begin{bmatrix} 1 & 0 & 0 \\ 0 & 0 & 1 \\ 0 & 1 & 0 \end{bmatrix}$$

4-18 $$E_1 = \begin{bmatrix} 1 & 0 & 0 \\ 2 & 1 & 0 \\ 0 & 0 & 1 \end{bmatrix} \quad E_2 = \begin{bmatrix} 1 & 0 & -2 \\ 0 & 1 & 0 \\ 0 & 0 & 1 \end{bmatrix}$$

$$E_3 = \begin{bmatrix} 1 & 0 & 0 \\ 0 & 1 & -4 \\ 0 & 0 & 1 \end{bmatrix}$$

4-19 $A = E_1^{-1}E_2^{-1}E_3^{-1}$

$$= \begin{bmatrix} 1 & 0 & 0 \\ -2 & 1 & 0 \\ 0 & 0 & 1 \end{bmatrix}\begin{bmatrix} 1 & 0 & 2 \\ 0 & 1 & 0 \\ 0 & 0 & 1 \end{bmatrix}\begin{bmatrix} 1 & 0 & 0 \\ 0 & 1 & 4 \\ 0 & 0 & 1 \end{bmatrix}$$

$$B = E_3E_2E_1 = \begin{bmatrix} 1 & 0 & -2 \\ 2 & 1 & -4 \\ 0 & 0 & 1 \end{bmatrix} \quad \therefore \quad B = A^{-1}$$

4-20 $$\left[\begin{array}{cc|cc} 2 & 1 & -3 & 1 \\ 3 & 0 & 1 & -2 \\ \hline 1 & 2 & -1 & 3 \\ -1 & 3 & 0 & 1 \end{array}\right]$$

4-21 (a) AB (c) CA (d) BC

4-22 (a) and (c) are stochastic; (b) is not stochastic for several reasons: $1/2 + 0 \neq 1$ and probabilities cannot be negative or larger than 1.

EXAM 1 (Chapters 1–4)

1. If $A = \begin{bmatrix} 1 & 3 \\ -4 & 5 \end{bmatrix}$ and $B = \begin{bmatrix} 1 & 3 \\ x+y & y \end{bmatrix}$, find x and y such that $A = B$.

2. If $A = \begin{bmatrix} 3 & 1 \\ 5 & 2 \end{bmatrix}$, $B = \begin{bmatrix} 2 & -3 \\ 4 & 4 \end{bmatrix}$, and $C = \begin{bmatrix} 2 & 0 \\ 0 & 3 \end{bmatrix}$, find $A(B - C)$.

3. If $A = \begin{bmatrix} 2 & 3 & 1 \\ 5 & 0 & 6 \end{bmatrix}$ and $B = \begin{bmatrix} 1 & 0 & 4 & 2 \\ 2 & 3 & -1 & 3 \\ -1 & 2 & 0 & -2 \end{bmatrix}$, find AB.

4. Indicate whether each of the following is a diagonal, scalar, identity, upper triangular, or lower triangular matrix. (A given matrix may have several of these properties.)

(a) $\begin{bmatrix} 1 & 0 & 0 \\ 0 & 1 & 0 \\ 0 & 0 & 1 \end{bmatrix}$ **(b)** $\begin{bmatrix} 1 & 0 & 0 \\ 0 & 2 & 0 \\ 0 & 0 & 3 \end{bmatrix}$ **(c)** $\begin{bmatrix} 1 & 2 & 3 \\ 0 & 1 & 2 \\ 0 & 0 & 1 \end{bmatrix}$

5. Find A^T and determine whether A is symmetric if $A = \begin{bmatrix} 1 & 2 & 3 \\ 0 & 3 & 5 \\ 4 & -1 & 2 \end{bmatrix}$.

6. Solve the following system by the Gauss–Jordan elimination:

$$\begin{aligned} x - y + 2z &= 1 \\ x + y + z &= 2 \\ 2x - y + z &= 5 \end{aligned}$$

7. Given the homogeneous system

$$\begin{aligned} x + 2y + 3z &= 0 \\ y + 5z &= 0 \\ x + 2y + z &= 0 \end{aligned}$$

(a) Find the inverse of the coefficient matrix A.
(b) Does the given system have a nontrivial solution? Why or why not?

8. Evaluate the determinant of the matrix $A = \begin{bmatrix} 4 & -2 & 5 \\ 5 & 2 & 0 \\ 2 & 0 & 4 \end{bmatrix}$.

9. If A is an invertible matrix with $\det(A) = 5$, find

(a) $\det A^T$
(b) $\det A^{-1}$
(c) $\det(kA)$

10. Find the elementary matrix E which as a premultiplier transforms $\begin{bmatrix} -1 & 2 & 3 & -2 \\ 2 & 2 & -1 & 3 \\ 1 & 3 & 2 & 1 \end{bmatrix}$ into $\begin{bmatrix} 1 & 3 & 2 & 1 \\ 2 & 2 & -1 & 3 \\ -1 & 2 & 3 & -2 \end{bmatrix}$.

11. Express the following matrix A as a product of elementary matrices.

$$A = \begin{bmatrix} 1 & 2 & 0 \\ 0 & 1 & 0 \\ 2 & 0 & 3 \end{bmatrix}$$

12. Use Cramer's rule to solve the system

$$\begin{aligned} x + y + z &= 3 \\ x \quad - z &= 1 \\ y - z &= -4 \end{aligned}$$

Solutions to Exam 1

1. Setting corresponding entries equal, we get

$$x + y = -4 \qquad \text{and} \qquad y = 5$$

$$\therefore \quad x = -9 \text{ and } y = 5$$

2. $B - C = \begin{bmatrix} 2 & -3 \\ 4 & 4 \end{bmatrix} - \begin{bmatrix} 2 & 0 \\ 0 & 3 \end{bmatrix} = \begin{bmatrix} 0 & -3 \\ 4 & 1 \end{bmatrix}$. Therefore, $A(B - C) = \begin{bmatrix} 3 & 1 \\ 5 & 2 \end{bmatrix}\begin{bmatrix} 0 & -3 \\ 4 & 1 \end{bmatrix} = \begin{bmatrix} 4 & -8 \\ 8 & -13 \end{bmatrix}$.

3. $$AB = \begin{bmatrix} 2 & 3 & 1 \\ 5 & 0 & 6 \end{bmatrix}\begin{bmatrix} 1 & 0 & 4 & 2 \\ 2 & 3 & -1 & 3 \\ -1 & 2 & 0 & -2 \end{bmatrix} = \begin{bmatrix} 7 & 11 & 5 & 11 \\ -1 & 12 & 20 & -2 \end{bmatrix}$$

4. **(a)** Diagonal, scalar, identity, upper triangular, and lower triangular.
(b) Diagonal, upper triangular, and lower triangular.
(c) Upper triangular.

5. Interchanging the rows and columns, we get $A^T = \begin{bmatrix} 1 & 0 & 4 \\ 2 & 3 & -1 \\ 3 & 5 & 2 \end{bmatrix}$. Since $A^T \neq A$, then A is not symmetric.

6. Applying elementary row operations to the augmented matrix we get the following:

$$\begin{bmatrix} 1 & -1 & 2 & 1 \\ 1 & 1 & 1 & 2 \\ 2 & -1 & 1 & 5 \end{bmatrix} \xrightarrow[R_3 \to R_3 - 2R_1]{R_2 \to R_2 - R_1} \begin{bmatrix} 1 & -1 & 2 & 1 \\ 0 & 2 & -1 & 1 \\ 0 & 1 & -3 & 3 \end{bmatrix}$$

$$\xrightarrow{R_2 \leftrightarrow R_3} \begin{bmatrix} 1 & -1 & 2 & 1 \\ 0 & 1 & -3 & 3 \\ 0 & 2 & -1 & 1 \end{bmatrix}$$

$$\xrightarrow[R_3 \to R_3 - 2R_2]{} \begin{bmatrix} 1 & -1 & 2 & 1 \\ 0 & 1 & -3 & 3 \\ 0 & 0 & 5 & -5 \end{bmatrix}$$

$$\xrightarrow[R_3 \to 1/5 R_3]{} \begin{bmatrix} 1 & -1 & 2 & 1 \\ 0 & 1 & -3 & 3 \\ 0 & 0 & 1 & -1 \end{bmatrix}$$

$$\xrightarrow[R_2 \to R_2 + 3R_3]{R_1 \to R_1 - 2R_3} \begin{bmatrix} 1 & -1 & 0 & 3 \\ 0 & 1 & 0 & 0 \\ 0 & 0 & 1 & -1 \end{bmatrix}$$

$$\xrightarrow{R_1 \to R_1 + R_2} \begin{bmatrix} 1 & 0 & 0 & 3 \\ 0 & 1 & 0 & 0 \\ 0 & 0 & 1 & -1 \end{bmatrix}$$

$$\therefore \quad \begin{aligned} x &= 3 \\ y &= 0 \\ z &= -1 \end{aligned}$$

7. **(a)** $$\left[\begin{array}{ccc|ccc} 1 & 2 & 3 & 1 & 0 & 0 \\ 0 & 1 & 5 & 0 & 1 & 0 \\ 1 & 2 & 1 & 0 & 0 & 1 \end{array}\right]$$

$$\xrightarrow[R_3 \to R_3 - R_1]{} \left[\begin{array}{ccc|ccc} 1 & 2 & 3 & 1 & 0 & 0 \\ 0 & 1 & 5 & 0 & 1 & 0 \\ 0 & 0 & -2 & -1 & 0 & 1 \end{array}\right]$$

$$\xrightarrow[R_3 \to -1/2 R_3]{} \left[\begin{array}{ccc|ccc} 1 & 2 & 3 & 1 & 0 & 0 \\ 0 & 1 & 5 & 0 & 1 & 0 \\ 0 & 0 & 1 & \frac{1}{2} & 0 & -\frac{1}{2} \end{array}\right]$$

$$\xrightarrow[R_2 \to R_2 - 5R_3]{R_1 \to R_1 - 3R_3} \left[\begin{array}{ccc|ccc} 1 & 2 & 0 & -\frac{1}{2} & 0 & \frac{3}{2} \\ 0 & 1 & 0 & -\frac{5}{2} & 1 & \frac{5}{2} \\ 0 & 0 & 1 & \frac{1}{2} & 0 & -\frac{1}{2} \end{array}\right]$$

$$\xrightarrow{R_1 \to R_1 - 2R_2} \left[\begin{array}{ccc|ccc} 1 & 0 & 0 & \frac{9}{2} & -2 & -\frac{7}{2} \\ 0 & 1 & 0 & -\frac{5}{2} & 1 & \frac{5}{2} \\ 0 & 0 & 1 & \frac{1}{2} & 0 & -\frac{1}{2} \end{array}\right]$$

$$\therefore \quad A^{-1} = \begin{bmatrix} \frac{9}{2} & -2 & -\frac{7}{2} \\ -\frac{5}{2} & 1 & \frac{5}{2} \\ \frac{1}{2} & 0 & -\frac{1}{2} \end{bmatrix}$$

(b) No, it has only the trivial solution because the coefficient matrix is invertible.

8. Expand by the elements of the third row and their cofactors.

$$\det(A) = \begin{vmatrix} 4 & -2 & 5 \\ 5 & 2 & 0 \\ 2 & 0 & 4 \end{vmatrix} = 2\begin{vmatrix} -2 & 5 \\ 2 & 0 \end{vmatrix} + 4\begin{vmatrix} 4 & -2 \\ 5 & 2 \end{vmatrix}$$

$$= 2(-10) + 4(8 + 10) = -20 + 72 = 52.$$

9. **(a)** $\det(A^T) = \det(A) = 5$
(b) $\det(A^{-1}) = 1/\det(A) = 1/5$
(c) Since kA means each row has been multiplied by k, then $\det(kA) = k^3 \det(A) = 5k^3$.

10. Since the first and third rows have been interchanged, we interchange the corresponding rows in $\begin{bmatrix} 1 & 0 & 0 \\ 0 & 1 & 0 \\ 0 & 0 & 1 \end{bmatrix}$ to obtain $E = \begin{bmatrix} 0 & 0 & 1 \\ 0 & 1 & 0 \\ 1 & 0 & 0 \end{bmatrix}$.

11. Use elementary row operations to reduce A to I_3:

$$\begin{bmatrix} 1 & 2 & 0 \\ 0 & 1 & 0 \\ 2 & 0 & 3 \end{bmatrix} \xrightarrow{R_3 \to R_3 - 2R_1} \begin{bmatrix} 1 & 2 & 0 \\ 0 & 1 & 0 \\ 0 & -4 & 3 \end{bmatrix}$$

$$\xrightarrow{R_3 \to R_3 + 4R_2} \begin{bmatrix} 1 & 2 & 0 \\ 0 & 1 & 0 \\ 0 & 0 & 3 \end{bmatrix}$$

$$\xrightarrow[R_3 \to 1/3R_3]{R_1 \to R_1 - 2R_2} \begin{bmatrix} 1 & 0 & 0 \\ 0 & 1 & 0 \\ 0 & 0 & 1 \end{bmatrix}$$

Write in sequence each row operation performed and the corresponding elementary matrix with its inverse.

$R_3 \to R_3 - 2R_1$

$$E_1 = \begin{bmatrix} 1 & 0 & 0 \\ 0 & 1 & 0 \\ -2 & 0 & 1 \end{bmatrix} \qquad E_1^{-1} = \begin{bmatrix} 1 & 0 & 0 \\ 0 & 1 & 0 \\ 2 & 0 & 1 \end{bmatrix}$$

$R_3 \to R_3 + 4R_2$

$$E_2 = \begin{bmatrix} 1 & 0 & 0 \\ 0 & 1 & 0 \\ 0 & 4 & 1 \end{bmatrix} \qquad E_2^{-1} = \begin{bmatrix} 1 & 0 & 0 \\ 0 & 1 & 0 \\ 2 & -4 & 1 \end{bmatrix}$$

$R_1 \to R_1 - 2R_2$

$$E_3 = \begin{bmatrix} 1 & -2 & 0 \\ 0 & 1 & 0 \\ 0 & 0 & 1 \end{bmatrix} \qquad E_3^{-1} = \begin{bmatrix} 1 & 2 & 0 \\ 0 & 1 & 0 \\ 0 & 0 & 1 \end{bmatrix}$$

$R_3 \to \frac{1}{3}R_3$

$$E_4 = \begin{bmatrix} 1 & 0 & 0 \\ 0 & 1 & 0 \\ 0 & 0 & \frac{1}{3} \end{bmatrix} \qquad E_4^{-1} = \begin{bmatrix} 1 & 0 & 0 \\ 0 & 1 & 0 \\ 0 & 0 & 3 \end{bmatrix}$$

$$A = E_1^{-1}E_2^{-1}E_3^{-1}E_4^{-1}$$

$$= \begin{bmatrix} 1 & 0 & 0 \\ 0 & 1 & 0 \\ 2 & 0 & 1 \end{bmatrix}\begin{bmatrix} 1 & 0 & 0 \\ 0 & 1 & 0 \\ 0 & -4 & 1 \end{bmatrix}\begin{bmatrix} 1 & 2 & 0 \\ 0 & 1 & 0 \\ 0 & 0 & 1 \end{bmatrix}$$

$$\times \begin{bmatrix} 1 & 0 & 0 \\ 0 & 1 & 0 \\ 0 & 0 & 3 \end{bmatrix}$$

12. $$x = \frac{\begin{vmatrix} 3 & 1 & 1 \\ 1 & 0 & -1 \\ -4 & 1 & -1 \end{vmatrix}}{\begin{vmatrix} 1 & 1 & 1 \\ 1 & 0 & -1 \\ 0 & 1 & -1 \end{vmatrix}} = \frac{9}{3} = 3$$

$$y = \frac{\begin{vmatrix} 1 & 3 & 1 \\ 1 & 1 & -1 \\ 0 & -4 & -1 \end{vmatrix}}{3} = -\frac{6}{3} = -2$$

$$z = \frac{\begin{vmatrix} 1 & 1 & 3 \\ 1 & 0 & 1 \\ 0 & 1 & -4 \end{vmatrix}}{3} = \frac{6}{3} = 2$$

5 APPLICATIONS OF MATRICES AND DETERMINANTS

THIS CHAPTER IS ABOUT

☑ **Linear Systems**
☑ **Matrices**
☑ **Determinants**
☑ **Additional Uses of Determinants**

The growth of linear algebra over the past thirty years is largely due to the many applications that make use of linear algebra concepts, especially the use of matrices and matrix algebra. With the increased use of computers, we are finding an explosion of applications of linear algebra in engineering, business, graph theory, biology, physics, nutrition, production planning, sociology, computer science, and geometry—you name the field. This chapter contains problems from a number of these areas.

5-1. Linear Systems

PROBLEM 5-1 (Business) An appliance store sells three different models of dishwasher. If models A, B, and C sell for \$350, \$450, and \$550, respectively, write a linear equation to find the daily revenue (R) from the sale of the dishwashers.

Solution Let

$$x = \text{number of model } A \text{ sold daily}$$
$$y = \text{number of model } B \text{ sold daily}$$
$$z = \text{number of model } C \text{ sold daily}$$

$$\therefore \quad R = [350 \quad 450 \quad 550]\begin{bmatrix} x \\ y \\ z \end{bmatrix} \begin{matrix} \text{model } A \\ \text{model } B \\ \text{model } C \end{matrix}$$

$$\text{or} \quad R = 350x + 450y + 550z$$

PROBLEM 5-2 (Business) The sale of cartons of cola soft drinks for a week at each of the two B. G. Convenience Stores is indicated below. Find the total for the various types of soft drinks sold by the two stores during the week.

$$\begin{matrix} & \textbf{with caffeine} & & \textbf{caffeine-free} & & \\ & \text{regular} & \text{diet} & \text{regular} & \text{diet} & \text{bottling company} \end{matrix}$$

$$\text{store A} = \begin{bmatrix} 18 & 14 & 13 & 15 \\ 14 & 12 & 15 & 13 \\ 15 & 17 & 14 & 16 \end{bmatrix} \begin{matrix} 1 \\ 2 \\ 3 \end{matrix}$$

$$\text{store B} = \begin{bmatrix} 13 & 10 & 21 & 9 \\ 12 & 14 & 19 & 12 \\ 22 & 11 & 16 & 14 \end{bmatrix} \begin{matrix} 1 \\ 2 \\ 3 \end{matrix}$$

Solution We need to find $A + B$.

$$A + B = \begin{bmatrix} 18 & 14 & 13 & 15 \\ 14 & 12 & 15 & 13 \\ 15 & 17 & 14 & 16 \end{bmatrix} + \begin{bmatrix} 13 & 10 & 21 & 9 \\ 12 & 14 & 19 & 12 \\ 22 & 11 & 16 & 14 \end{bmatrix} = \begin{bmatrix} 31 & 24 & 34 & 24 \\ 26 & 26 & 34 & 25 \\ 37 & 28 & 30 & 30 \end{bmatrix}$$

PROBLEM 5-3 In Problem 5-2, find the total number of cartons of cola soft drinks in the combined sales of the two stores which were bottled by each of the three bottling companies.

Solution We need to add the entries in each row of $A + B$. The matrix way to do this is to multiply the matrix $A + B$ by a 4×1 matrix having all 1's as entries.

$$\begin{bmatrix} 31 & 24 & 34 & 24 \\ 26 & 26 & 34 & 25 \\ 37 & 28 & 30 & 30 \end{bmatrix} \begin{bmatrix} 1 \\ 1 \\ 1 \end{bmatrix} = \begin{bmatrix} 113 \\ 111 \\ 125 \end{bmatrix}$$

$\therefore$ 113 cartons from bottling company 1
111 cartons from bottling company 2
125 cartons from bottling company 3

PROBLEM 5-4 (Business) Use the result of Problem 5-3 to find the revenue for the week from the sale of cola soft drinks if each carton from bottler 1, 2, and 3 sold for \$2.99, \$2.69, and \$2.39, respectively.

Solution

$$R = [2.99 \quad 2.69 \quad 2.39] \begin{bmatrix} 113 \\ 111 \\ 125 \end{bmatrix} = \$935.21$$

PROBLEM 5-5 (Construction Engineering) An engineer has been asked to calculate the material and labor costs required for the renovation of 10 type A buildings, 20 type B buildings, and 30 type C buildings. The engineer estimates that 50 units of material and 200 hours of labor are required to renovate one type A structure. Each type B structure will require 20 units of material and 100 hours of labor, while a type C structure will require 30 units of material and 300 hours of labor. The costs for material is \$400 per unit and for labor is \$10 per hour.

First, show how the engineer can represent the number of buildings for each structure type as a 1×3 matrix. Next, show how the units of material and hours of labor can be expressed as a 3×2 matrix. Thirdly, express the material and labor unit costs as a 2×1 matrix. Finally, show how the engineer can calculate the total renovation cost estimate by matrix multiplication.

Solution

(a) $[10 \quad 20 \quad 30]$

(b) $\begin{bmatrix} 50 & 200 \\ 20 & 100 \\ 30 & 300 \end{bmatrix}$

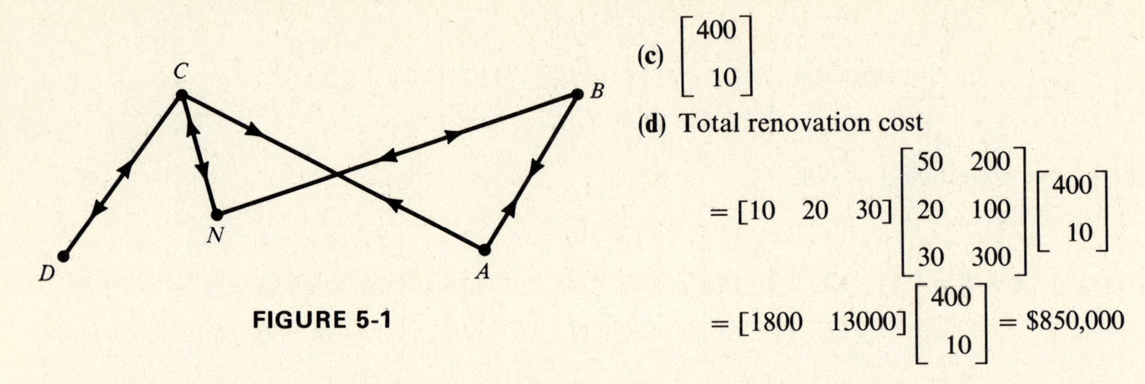

FIGURE 5-1

(c) $\begin{bmatrix} 400 \\ 10 \end{bmatrix}$

(d) Total renovation cost

$$= [10 \quad 20 \quad 30] \begin{bmatrix} 50 & 200 \\ 20 & 100 \\ 30 & 300 \end{bmatrix} \begin{bmatrix} 400 \\ 10 \end{bmatrix}$$

$$= [1800 \quad 13000] \begin{bmatrix} 400 \\ 10 \end{bmatrix} = \$850{,}000$$

PROBLEM 5-6 (Transportation, Graph Theory) Alpha airlines has daily flights between Atlanta (A), Baltimore (B), Cincinnati (C), Dallas (D), and Nashville (N) as indicated in Figure 5-1 with the arrows representing the direction of the flights. Represent the information in a matrix by using zeros and ones.

Solution Five cities are involved. We will set up a 5×5 matrix with each row representing a flight's city of origin and each of the columns representing a flight's city of terminus. For example, the row beginning with A will have 1's in the columns headed B and C because there are flights originating in Atlanta and terminating in Baltimore and Cincinnati. All other columns for row A contain zeros.

$$\therefore \quad \begin{array}{c} \\ A \\ B \\ C \\ D \\ N \end{array} \begin{array}{c} \begin{array}{ccccc} A & B & C & D & N \end{array} \\ \begin{bmatrix} 0 & 1 & 1 & 0 & 0 \\ 1 & 0 & 0 & 0 & 1 \\ 1 & 0 & 0 & 1 & 1 \\ 0 & 0 & 1 & 0 & 0 \\ 0 & 1 & 1 & 0 & 0 \end{bmatrix} \end{array}$$

note: In graph theory, a matrix of this type is called an **adjacency matrix**.

PROBLEM 5-7 (Sales) Nan Daily and Geoff Wade are the two salespersons in a specialty car agency that sells only compact cars, both imports and domestics. Gross sales for October and November are as follows:

October sales

$$\begin{array}{c} \\ \text{Daily} \\ \text{Wade} \end{array} \begin{array}{c} \begin{array}{cc} \textbf{imports} & \textbf{domestic} \end{array} \\ \begin{bmatrix} 26{,}000 & 39{,}000 \\ 39{,}000 & 13{,}000 \end{bmatrix} \end{array} = A$$

November sales

$$\begin{array}{c} \text{Daily} \\ \text{Wade} \end{array} \begin{bmatrix} 52{,}000 & 91{,}000 \\ 45{,}500 & 65{,}000 \end{bmatrix} = B$$

(a) Compute the combined dollar sales in October and November for each salesperson.
(b) If each salesperson receives a 5% commission on total sales, what commission was earned by each on the sale of imports in October?

Solution

(a)

$$A + B = \begin{bmatrix} 26{,}000 & 39{,}000 \\ 39{,}000 & 13{,}000 \end{bmatrix} + \begin{bmatrix} 52{,}000 & 91{,}000 \\ 45{,}500 & 65{,}000 \end{bmatrix}$$

$$= \begin{bmatrix} 88{,}000 & 130{,}000 \\ 84{,}500 & 78{,}000 \end{bmatrix}$$

- Daily's sales in October and November were 88,000 + 130,000 = $218,000.
- Wade's sales in October and November were 84,500 + 78,000 = $162,500.

(b)

$$[0.05]\begin{bmatrix} 26{,}000 \\ 39{,}000 \end{bmatrix} = \begin{bmatrix} 1300 \\ 1950 \end{bmatrix}$$

- Daily's commission on imports in October is $1300.
- Wade's commission on imports in October is $1950.

5-2. Matrices

PROBLEM 5-8 (Production Planning, Nutrition) A nursing home dietition is to prepare a meal consisting of three basic foods A, B, and C. The diet is to include 25 units of potassium, 22 units of zinc, and 18 units of thiamine. The number of units of potassium in one ounce of food A is 3, of food B is 3, and of food C is 2. The number of units of zinc in one ounce of food A is 3, of food B is 2, and of food C is 3. The number of units of thiamine in one ounce of food A is 2, of food B is 1, and of food C is 4. How many ounces of each food must be used to meet the diet requirements of the meal?

Solution

	units per ounce		
	food *A*	**food *B***	**food *C***
potassium	3	3	2
zinc	3	2	3
thiamine	2	1	4

Let

x = number ounces of A
y = number ounces of B
z = number ounces of C

$$\begin{bmatrix} 3 & 3 & 2 \\ 3 & 2 & 3 \\ 2 & 1 & 4 \end{bmatrix}\begin{bmatrix} x \\ y \\ z \end{bmatrix} = \begin{bmatrix} 25 \\ 22 \\ 18 \end{bmatrix}$$

Using the row reduction on the augmented matrix,

$$\begin{bmatrix} 3 & 3 & 2 & 25 \\ 3 & 2 & 3 & 22 \\ 2 & 1 & 4 & 18 \end{bmatrix} \xrightarrow{R_1 \to R_1 - R_3} \begin{bmatrix} 1 & 2 & -2 & 7 \\ 3 & 2 & 3 & 22 \\ 2 & 1 & 4 & 18 \end{bmatrix} \xrightarrow[R_3 \to R_3 - 2R_1]{R_2 \to R_2 - 3R_1} \begin{bmatrix} 1 & 2 & -2 & 7 \\ 0 & -4 & 9 & 1 \\ 0 & -3 & 8 & 4 \end{bmatrix}$$

$$\xrightarrow{R_2 \to R_2 - R_3} \begin{bmatrix} 1 & 2 & -2 & 7 \\ 0 & -1 & 1 & -3 \\ 0 & -3 & 8 & 4 \end{bmatrix} \xrightarrow[R_3 \to R_3 - 3R_2]{R_2 \to -R_2} \begin{bmatrix} 1 & 2 & -2 & 7 \\ 0 & 1 & -1 & 3 \\ 0 & 0 & 5 & 13 \end{bmatrix}$$

$$\therefore \quad \begin{aligned} x + 2y - 2z &= 7 \\ y - z &= 3 \\ 5z &= 13 \end{aligned}$$

This has solution of $x = 1$, $y = 5.6$, and $z = 2.6$. Therefore, 1 ounce of food A, 5.6 ounces of food B, and 2.6 ounces of food C are needed.

PROBLEM 5-9 (Production Planning, Tailoring) A small family-operated tailor shop in Taiwan makes ladies' jackets, skirts, and pants. Each jacket takes 1.5 hours in the cutting department, 4.0 hours in the sewing department, and 0.8 hours in the pressing and finishing department. Each skirt spends 0.6 hours in the cutting department, 0.6 hours in the sewing department, and 0.5 hours in the pressing and finishing department. Each pair of pants spends 1.0 hours in the cutting department, 1.4 hours in the sewing department, and 0.6 hours in the pressing and finishing department. If 84 hours per week is available for cutting, 146 hours per week for sewing, and 54 hours per week for pressing and finishing, how many of each piece must be produced for full production?

Solution We will set up a table of the information as follows.

	Hours spent			
Department	Jacket	Skirt	Pants	Hours/week
Cutting	1.5	0.6	1.0	84
Sewing	4.0	0.6	1.4	146
Pressing/finishing	0.8	0.5	0.6	54

Let

x = number of jackets produced

y = number of skirts

z = number of pants

$$\therefore \quad \begin{bmatrix} 1.5 & 0.6 & 1.0 \\ 4.0 & 0.6 & 1.4 \\ 0.8 & 0.5 & 0.6 \end{bmatrix} \begin{bmatrix} x \\ y \\ z \end{bmatrix} = \begin{bmatrix} 84 \\ 146 \\ 54 \end{bmatrix}$$

This is equivalent to

$$\begin{aligned} 15x + 6y + 10z &= 840 \\ 40x + 6y + 14z &= 1460 \\ 8x + 5y + 6z &= 540 \end{aligned}$$

The solution to this system is $x = 20$, $y = 40$, and $z = 30$. Therefore, to meet full production the shop must produce 20 jackets, 40 skirts, and 30 pants each week.

PROBLEM 5-10 (Civil Engineering) A civil engineer wants to use three types of concrete for his construction projects. The quantities of cement, sand, and stone required to produce a single unit of each concrete type is shown in the following table. Based on the information in the table, calculate for each concrete type the number of units that can be produced from the available supply of cement, sand, and stone.

	Concrete type			Materials
	Type 1	Type 2	Type 3	on hand
Cement (lbs)	9	10	10	3,900
Sand (lbs)	18	20	22	8,000
Stone (lbs)	28	30	34	12,200

Solution Let

$$x_1 = \text{number of units of type 1}$$
$$x_2 = \text{number of units of type 2}$$
$$x_3 = \text{number of units of type 3}$$

Therefore,

$$\begin{bmatrix} 9 & 10 & 10 \\ 18 & 20 & 22 \\ 28 & 30 & 34 \end{bmatrix} \begin{bmatrix} x_1 \\ x_2 \\ x_3 \end{bmatrix} = \begin{bmatrix} 3{,}900 \\ 8{,}000 \\ 12{,}200 \end{bmatrix}$$

or

$$\begin{aligned} 9x_1 + 10x_2 + 10x_3 &= 3{,}900 \\ 18x_1 + 20x_2 + 22x_3 &= 8{,}000 \\ 28x_1 + 30x_2 + 34x_3 &= 12{,}200 \end{aligned}$$

Using the first two of these equations, we get

$$\begin{aligned} 9x_1 + 10x_2 + 10x_3 &= 3{,}900 \\ 9x_1 + 10x_2 + 11x_3 &= 4{,}000 \end{aligned}$$

which gives $x_3 = 100$. Substituting this value in the other two equations and solving, we get $x_1 = 100$ and $x_2 = 200$. Therefore, the solution is 100 units of type 1, 200 units of type 2, and 100 units of type 3.

PROBLEM 5-11 (Cryptography) Coding and decoding messages can be done in various ways, some of which are rather simple while others are quite complex. One method uses matrices and is easy to use but difficult to break. Johnny at Pleasant State University wishes to send the message, "MOM, I LOVE YOU.", on Mother's Day by using code. Illustrate how Johnny can solve his problem.

Solution

Step 1: Assign a number to each letter of the alphabet and punctuation symbol you wish to use. Let's make the following correspondence:

A	*B*	*C*	*D*	*E*	*F*	*G*	*H*	*I*	*J*	*K*	*L*	*M*	*N*	*O*
1	2	3	4	5	6	7	8	9	10	11	12	13	14	15

P	*Q*	*R*	*S*	*T*	*U*	*V*	*W*	*X*	*Y*	*Z*	,	.	–
16	17	18	19	20	21	22	23	24	25	26	27	28	0

Step 2: Break the message into groups of three symbols with their corresponding numbers:

M	*O*	*M*	,	–	*I*	–	*L*	*O*	*V*	*E*	–	*Y*	*O*	*U*	.	–	–
13	15	13	27	0	9	9	12	15	22	5	0	25	15	21	28	0	0

Step 3: Write each block of numbers as column matrices:

$$\begin{bmatrix} 13 \\ 15 \\ 13 \end{bmatrix} \begin{bmatrix} 27 \\ 0 \\ 9 \end{bmatrix} \begin{bmatrix} 0 \\ 12 \\ 15 \end{bmatrix} \begin{bmatrix} 22 \\ 5 \\ 0 \end{bmatrix} \begin{bmatrix} 25 \\ 15 \\ 21 \end{bmatrix} \begin{bmatrix} 28 \\ 0 \\ 0 \end{bmatrix}$$

Step 4: Choose any invertible matrix in which both A and A^{-1} have integral entries. Since we used blocks of three, we need a 3×3 matrix. Let

$$A = \begin{bmatrix} 1 & -2 & 3 \\ -1 & 3 & -4 \\ 0 & 5 & -4 \end{bmatrix}$$

Then

$$A^{-1} = \begin{bmatrix} 8 & 7 & -1 \\ -4 & -4 & 1 \\ -5 & -5 & 1 \end{bmatrix}$$

Step 5: Multiply each of the column matrices in Step 3 on the left by A:

$$A\begin{bmatrix} 13 \\ 15 \\ 13 \end{bmatrix} = \begin{bmatrix} 22 \\ -20 \\ 23 \end{bmatrix} \qquad A\begin{bmatrix} 27 \\ 0 \\ 9 \end{bmatrix} = \begin{bmatrix} 54 \\ -63 \\ -36 \end{bmatrix} \qquad A\begin{bmatrix} 0 \\ 12 \\ 15 \end{bmatrix} = \begin{bmatrix} 21 \\ -24 \\ 0 \end{bmatrix}$$

$$A\begin{bmatrix} 22 \\ 5 \\ 0 \end{bmatrix} = \begin{bmatrix} 12 \\ -7 \\ 25 \end{bmatrix} \qquad A\begin{bmatrix} 25 \\ 15 \\ 21 \end{bmatrix} = \begin{bmatrix} 58 \\ -64 \\ -9 \end{bmatrix} \qquad A\begin{bmatrix} 28 \\ 0 \\ 0 \end{bmatrix} = \begin{bmatrix} 28 \\ -28 \\ 0 \end{bmatrix}$$

Step 6: Write the numbers in each result, in order, in a row:

$$22 \quad -20 \quad 23 \quad 54 \quad -63 \quad -36 \quad 21 \quad -24 \quad 0$$

$$12 \quad -7 \quad 25 \quad 58 \quad -64 \quad -9 \quad 28 \quad -28 \quad 0$$

Step 7: Reduce each number in the message by modulo 29, using only positive integers. We get

$$22 \quad 9 \quad 23 \quad 25 \quad 24 \quad 22 \quad 21 \quad 5 \quad 0$$

$$12 \quad 22 \quad 25 \quad 0 \quad 23 \quad 20 \quad 28 \quad 1 \quad 0$$

Johnny sends his mother the following message:

V I W Y X V U E – L V Y – W T . A –

Johnny's mother must decode this to get the original message. She needs A^{-1}, which was sent by her thoughtful son.

Step 8: We break the sequence of symbols into groups of three and write the transpose of their number correspondence as follows:

$$\begin{bmatrix} 22 \\ 9 \\ 23 \end{bmatrix} \begin{bmatrix} 25 \\ 24 \\ 22 \end{bmatrix} \begin{bmatrix} 21 \\ 5 \\ 0 \end{bmatrix} \begin{bmatrix} 12 \\ 22 \\ 25 \end{bmatrix} \begin{bmatrix} 0 \\ 23 \\ 20 \end{bmatrix} \begin{bmatrix} 28 \\ 1 \\ 0 \end{bmatrix}$$

Step 9: Multiply each of these on the left by A^{-1}:

$$A^{-1}\begin{bmatrix} 22 \\ 9 \\ 23 \end{bmatrix} = \begin{bmatrix} 216 \\ -101 \\ -132 \end{bmatrix} \equiv \begin{bmatrix} 13 \\ 15 \\ 13 \end{bmatrix} \qquad A^{-1}\begin{bmatrix} 25 \\ 24 \\ 22 \end{bmatrix} = \begin{bmatrix} 346 \\ -174 \\ -223 \end{bmatrix} \equiv \begin{bmatrix} 27 \\ 0 \\ 9 \end{bmatrix}$$

$$A^{-1}\begin{bmatrix}21\\5\\0\end{bmatrix}=\begin{bmatrix}203\\-104\\-130\end{bmatrix}\equiv\begin{bmatrix}0\\12\\15\end{bmatrix}\qquad A^{-1}\begin{bmatrix}12\\22\\25\end{bmatrix}=\begin{bmatrix}225\\-121\\-145\end{bmatrix}\equiv\begin{bmatrix}22\\5\\0\end{bmatrix}$$

$$A^{-1}\begin{bmatrix}0\\23\\20\end{bmatrix}=\begin{bmatrix}141\\-72\\-95\end{bmatrix}\equiv\begin{bmatrix}25\\15\\21\end{bmatrix}\qquad A^{-1}\begin{bmatrix}28\\1\\0\end{bmatrix}=\begin{bmatrix}231\\-116\\-145\end{bmatrix}\equiv\begin{bmatrix}28\\0\\0\end{bmatrix}$$

Step 10: Write the numbers obtained in Step 9 in a row and replace each by its letter correspondence or symbol to recover the original message:

MOM, I LOVE YOU.

Note! His mother is a mathematics teacher.

PROBLEM 5-12 (Food Services) Hamburger Heaven makes three types of hamburger: deluxe, regular, and special. The deluxe hamburger contains 5 ounces of meat, 0.4 ounces of lettuce, and 0.1 ounces of onion. Regular hamburgers contain 4 ounces of meat, 0.3 ounces of lettuce, and 0.1 ounces of onion, while the special hamburger contains 3 ounces of meat, 0.2 ounces of lettuce, and 0.05 ounces of onion. Determine how many ounces of each food item should be used to make 30 deluxe hamburgers, 50 regular hamburgers, and 60 special hamburgers.

Solution

	hamburger type		
	deluxe	**regular**	**special**
meat (oz.)	5	4	3
lettuce (oz.)	0.4	0.3	0.2
onion (oz.)	0.1	0.1	0.05
quantity	30	50	60

$$\begin{bmatrix}5&4&3\\0.4&0.3&0.2\\0.1&0.1&0.05\end{bmatrix}\begin{bmatrix}30\\50\\60\end{bmatrix}=\begin{bmatrix}530\\39\\11\end{bmatrix}$$

Thus, 530 ounces of meat, 39 ounces of lettuce, and 11 ounces of onion are needed.

5-3. Determinants

PROBLEM 5-13 (Geometry) In analytic geometry it is shown that the equation of a line through two given points (x_1, y_1) and (x_2, y_2) can be expressed as

$$\begin{vmatrix}x&y&1\\x_1&y_1&1\\x_2&y_2&1\end{vmatrix}=0$$

Use this to find the equation of the line through (2, 1) and (−3, −2).

Solution Let $(2, 1) = (x_1, y_1)$ and $(-3, -2) = (x_2, y_2)$; then

$$\begin{vmatrix}x&y&1\\2&1&1\\-3&-2&1\end{vmatrix}=0$$

When this is expanded, we get

$$x - 3y - 4 + 3 + 2x - 2y = 0 \qquad \text{or} \qquad 3x - 5y = 1$$

If we substitute the points into this equation,

$$3(2) - 5(1) = 1 \qquad \text{and} \qquad 3(-3) - 5(-2) = 1$$

we have verified our answer.

PROBLEM 5-14 (Geometry) A result from analytic geometry uses the determinant in Problem 5-13 to find the area of a triangle with given vertices $P_1(x_1, y_1)$, $P_2(x_2, y_2)$, and $P_3(x_3, y_3)$ chosen counterclockwise as

$$A = \frac{1}{2}\begin{vmatrix} x_1 & y_1 & 1 \\ x_2 & y_2 & 1 \\ x_3 & y_3 & 1 \end{vmatrix}$$

Use this to find the area of the triangle with vertices $(1, 4)$, $(-1, 2)$, and $(2, -4)$.

Solution

$$A = \frac{1}{2}\begin{vmatrix} 1 & 4 & 1 \\ -1 & 2 & 1 \\ 2 & -4 & 1 \end{vmatrix} = \frac{1}{2}(2 + 4 + 8 - 4 + 4 + 4) = \frac{1}{2}(18) = 9$$

PROBLEM 5-15 (Geometry) The equation of a circle through three noncollinear points $P_1(x_1, y_1)$, $P_2(x_2, y_2)$, and $P_3(x_3, y_3)$ is given in analytic geometry to be

$$\begin{vmatrix} x^2 + y^2 & x & y & 1 \\ x_1^2 + y_1^2 & x_1 & y_1 & 1 \\ x_2^2 + y_2^2 & x_2 & y_2 & 1 \\ x_3^2 + y_3^2 & x_3 & y_3 & 1 \end{vmatrix} = 0$$

Use this to find the equation of the circle through $P_1(1, 4)$, $P_2(-1, 2)$, and $P_3(2, -4)$.

Solution

$$\begin{vmatrix} x^2 + y^2 & x & y & 1 \\ 1^2 + 4^2 & 1 & 4 & 1 \\ (-1)^2 + 2^2 & -1 & 2 & 1 \\ 2^2 + (-4)^2 & 2 & -4 & 1 \end{vmatrix} = 0$$

$$\begin{vmatrix} x^2 + y^2 & x & y & 1 \\ 17 & 1 & 4 & 1 \\ 5 & -1 & 2 & 1 \\ 20 & 2 & -4 & 1 \end{vmatrix} = 0$$

The best way to expand this is to introduce some zeros by row reduction:

$$\begin{vmatrix} x^2 + y^2 - 5 & x + 1 & y - 2 & 0 \\ 12 & 2 & 2 & 0 \\ 5 & -1 & 2 & 1 \\ 15 & 3 & -6 & 0 \end{vmatrix} = -1\begin{vmatrix} x^2 + y^2 - 5 & x + 1 & y - 2 \\ 12 & 2 & 2 \\ 15 & 3 & -6 \end{vmatrix}$$

$$= -[-12(x^2 + y^2 - 5) + 36(y - 2) + 30(x + 1) - 30(y - 2) + 72(x + 1) - 6(x^2 + y^2 - 5)]$$

$$= -[-12x^2 - 12y^2 + 60 + 36y - 72 + 30x + 30 - 30y + 60 + 72x + 72 - 6x^2 - 6y^2 + 30]$$

$$= -[-18x^2 - 18y^2 + 102x + 6y + 180]$$

$$18x^2 + 18y^2 - 102x - 6y - 180 = 0$$

or

$$3x^2 + 3y^2 - 17x - y - 30 = 0$$

PROBLEM 5-16 (Geometry) The equation of a conic section through five points $P_1(x_1, y_1)$, $P_2(x_2, y_2)$, $P_3(x_3, y_3)$, $P_4(x_4, y_4)$, and $P_5(x_5, y_5)$ is given by

$$\begin{vmatrix} x^2 & xy & y^2 & x & y & 1 \\ x_1^2 & x_1y_1 & y_1^2 & x_1 & y_1 & 1 \\ x_2^2 & x_2y_2 & y_2^2 & x_2 & y_2 & 1 \\ x_3^2 & x_3y_3 & y_3^2 & x_3 & y_3 & 1 \\ x_4^2 & x_4y_4 & y_4^2 & x_4 & y_4 & 1 \\ x_5^2 & x_5y_5 & y_5^2 & x_5 & y_5 & 1 \end{vmatrix} = 0$$

Use this to find the equation of the conic through (2, 0), (2, 1), (1, 2), (−1, 0), and (−1, −1).

Solution

$$\begin{vmatrix} x^2 & xy & y^2 & x & y & 1 \\ 4 & 0 & 0 & 2 & 0 & 1 \\ 4 & 2 & 1 & 2 & 1 & 1 \\ 1 & 2 & 4 & 1 & 2 & 1 \\ 1 & 0 & 0 & -1 & 0 & 1 \\ 1 & 1 & 1 & -1 & -1 & 1 \end{vmatrix} = 0$$

This reduces to

$$\begin{vmatrix} x^2-1 & xy & y^2 & x+1 & y & 0 \\ 3 & 0 & 0 & 3 & 0 & 0 \\ 3 & 2 & 1 & 3 & 1 & 0 \\ 0 & 2 & 4 & 2 & 2 & 0 \\ 1 & 0 & 0 & -1 & 0 & 1 \\ 0 & 1 & 1 & 0 & -1 & 0 \end{vmatrix} = 0$$

$$(-1)\begin{vmatrix} x^2-1 & xy & y^2 & x+1 & y \\ 3 & 0 & 0 & 3 & 0 \\ 3 & 2 & 1 & 3 & 1 \\ 0 & 2 & 4 & 2 & 2 \\ 0 & 1 & 1 & 0 & -1 \end{vmatrix} = 0$$

$$-1\begin{vmatrix} x^2-1 & xy & y^2 & x+1 & y \\ 3 & 0 & 0 & 3 & 0 \\ 0 & 2 & 1 & 0 & 1 \\ 0 & 2 & 4 & 2 & 2 \\ 0 & 1 & 1 & 0 & -1 \end{vmatrix} = 0$$

$$-(x^2-1)\begin{vmatrix} 0 & 0 & 3 & 0 \\ 2 & 1 & 0 & 1 \\ 2 & 4 & 2 & 2 \\ 1 & 1 & 0 & -1 \end{vmatrix} + 3\begin{vmatrix} xy & y^2 & x+1 & y \\ 2 & 1 & 0 & 1 \\ 2 & 4 & 2 & 2 \\ 1 & 1 & 0 & -1 \end{vmatrix} = 0$$

$$-(x^2-1)(3)\begin{vmatrix} 2 & 1 & 1 \\ 2 & 4 & 2 \\ 1 & 1 & -1 \end{vmatrix} + 3(x+1)\begin{vmatrix} 2 & 1 & 1 \\ 2 & 4 & 2 \\ 1 & 1 & -1 \end{vmatrix} + 3(2)\begin{vmatrix} xy & y^2 & y \\ 2 & 1 & 1 \\ 1 & 1 & -1 \end{vmatrix} = 0$$

$$(-3x^2+3)[-8+2+2-4-4+2] + (3x+3)[-8+2+2-4-4+2]$$
$$+ 6[-xy+2y+y^2-y+2y^2-xy] = 0$$
$$(-3x^2+3)(-10) + (3x+3)(-10) + 6(-2xy+y+3y^2) = 0$$
$$30x^2 - 30 - 30x - 30 - 12xy + 8y + 24y^2 = 0$$
$$30x^2 - 12xy + 24y^2 - 30x + 8y - 60 = 0$$

This is the equation of an ellipse, as will be discussed in Chapter 17.

PROBLEM 5-17 (Geometry) The equation of the sphere through four noncoplanar points (x_1, y_1, z_1), (x_2, y_2, z_2), (x_3, y_3, z_3), and (x_4, y_4, z_4) is given by

$$\begin{vmatrix} x^2+y^2+z^2 & x & y & z & 1 \\ x_1^2+y_1^2+z_1^2 & x_1 & y_1 & z_1 & 1 \\ x_2^2+y_2^2+z_2^2 & x_2 & y_2 & z_2 & 1 \\ x_3^2+y_3^2+z_3^2 & x_3 & y_3 & z_3 & 1 \\ x_4^2+y_4^2+z_4^2 & x_4 & y_4 & z_4 & 1 \end{vmatrix} = 0$$

Find the equation of the sphere through the points $(1,1,0)$, $(0,2,1)$, $(-1,0,1)$, and $(-1,-1,2)$.

Solution

$$\begin{vmatrix} x^2+y^2+z^2 & x & y & z & 1 \\ 1^2+1^2 & 1 & 1 & 0 & 1 \\ 2^2+1^2 & 0 & 2 & 1 & 1 \\ (-1)^2+1^2 & -1 & 0 & 1 & 1 \\ (-1)^2+(-1)^2+2^2 & -1 & -1 & 2 & 1 \end{vmatrix} = 0$$

$$\begin{vmatrix} x^2+y^2+z^2 & x & y & z & 1 \\ 2 & 1 & 1 & 0 & 1 \\ 5 & 0 & 2 & 1 & 1 \\ 2 & -1 & 0 & 1 & 1 \\ 6 & -1 & -1 & 2 & 1 \end{vmatrix} = 0$$

$$\begin{vmatrix} x^2+y^2+z^2-2 & x-1 & y-1 & z & 0 \\ 2 & 1 & 1 & 0 & 1 \\ 3 & -1 & 1 & 1 & 0 \\ 0 & -2 & -1 & 1 & 0 \\ 4 & -2 & -2 & 2 & 0 \end{vmatrix} = 0$$

$$(-1)\begin{vmatrix} x^2+y^2+z^2-2 & x-1 & y-1 & z \\ 3 & -1 & 1 & 1 \\ 0 & -2 & -1 & 1 \\ 4 & -2 & -2 & 2 \end{vmatrix} = 0$$

$$(-1)(x^2+y^2+z^2-2)\begin{vmatrix} -1 & 1 & 1 \\ -2 & -1 & 1 \\ -2 & -2 & 2 \end{vmatrix} + (-1)(-3)\begin{vmatrix} x-1 & y-1 & z \\ -2 & -1 & 1 \\ -2 & -2 & 2 \end{vmatrix}$$

$$+(-1)(-4)\begin{vmatrix} x-1 & y-1 & z \\ -1 & 1 & 1 \\ -2 & -1 & 1 \end{vmatrix} = 0$$

$$(-x^2-y^2-z^2+2)(2+4-2-2-2+4)+3[-2(x-1)+4z-2(y-1)$$
$$-2z+2(x-1)+4(y-1)]+4[x-1+z-2(y-1)+2z+(x-1)+(y-1)]=0$$
$$(-x^2-y^2-z^2+2)(4)+3(-2x+2+4z-2y+2-2z+2x-2+4y-4)$$
$$+4(x-1+z-2y+2+2z+x-1+y-1)=0$$
$$(-x^2-y^2-z^2+2)(4)+3(2y+2z-2)+4(2x-y+3z-1)=0$$
$$-4x^2-4y^2-4z^2+8+6y+6z-6+8x-4y+12z-4=0$$
$$4x^2+4y^2+4z^2-8x-2y-18z+2=0$$
$$2x^2+2y^2+2z^2-4x-y-9z+1=0$$

This is the equation of the sphere through the given four points.

PROBLEM 5-18 (Geometry) The equation of a plane passing through the three points (x_1, y_1, z_1), (x_2, y_2, z_2), and (x_3, y_3, z_3) is given by

$$\begin{vmatrix} x & y & z & 1 \\ x_1 & y_1 & z_1 & 1 \\ x_2 & y_2 & z_2 & 1 \\ x_3 & y_3 & z_3 & 1 \end{vmatrix} = 0$$

note: This is analogous to the equation of a line through two points in R_2. Find the equation of the plane through $(1, 3, -2)$, $(0, 2, 4)$, and $(2, 1, 3)$.

Solution

$$\begin{vmatrix} x & y & z & 1 \\ 1 & 3 & -2 & 1 \\ 0 & 2 & 4 & 1 \\ 2 & 1 & 3 & 1 \end{vmatrix} = 0$$

$$\begin{vmatrix} x & y-2 & z-4 & 0 \\ 1 & 1 & -6 & 0 \\ 0 & 2 & 4 & 1 \\ 2 & -1 & -1 & 0 \end{vmatrix} = 0$$

$$(-1)\begin{vmatrix} x & y-2 & z-4 \\ 1 & 1 & -6 \\ 2 & -1 & -1 \end{vmatrix} = 0$$

$$-1[-x-12(y-2)-(z-4)-2(z-4)-6(x)+(y-2)]=0$$
$$-1[-x-12y+24-z+4-2z+8-6x+y-2]=0$$
$$-1[-7x-11y-3z+34]=0$$

Therefore, $7x+11y+3z-34=0$ is the equation of the plane.

PROBLEM 5-19 (Geometry) The equation of a parabola with a vertical axis which passes through the three noncollinear points (x_1, y_1), (x_2, y_2), and (x_3, y_3) is given by

$$\begin{vmatrix} x^2 & x & y & 1 \\ x_1^2 & x_1 & y_1 & 1 \\ x_2^2 & x_2 & y_2 & 1 \\ x_3^2 & x_3 & y_3 & 1 \end{vmatrix} = 0$$

Use this determinant equation to find the equation of the parabola passing through $(1, 2)$, $(-1, 1)$, and $(2, 0)$.

Solution

$$\begin{vmatrix} x^2 & x & y & 1 \\ 1 & 1 & 2 & 1 \\ 1 & -1 & 1 & 1 \\ 4 & 2 & 0 & 1 \end{vmatrix} = 0$$

$$\begin{vmatrix} x^2-1 & x-1 & y-2 & 0 \\ 1 & 1 & 2 & 1 \\ 0 & -2 & -1 & 0 \\ 3 & 1 & -2 & 0 \end{vmatrix} = 0$$

$$1\begin{vmatrix} x^2-1 & x-1 & y-2 \\ 0 & -2 & -1 \\ 3 & 1 & -2 \end{vmatrix} = 0$$

$$4(x^2-1) - 3(x-1) + 6(y-2) + (x^2-1) = 0$$

$$5x^2 - 3x + 6y - 14 = 0$$

is the equation of the parabola.

PROBLEM 5-20 (Differential Equations) One of the first topics discussed in differential equations is determining if two solutions are in fact different. This is usually done by evaluating a determinant called the **Wronskian** for the functions $f_1, f_2, \ldots, f_n$ which is defined by

$$W(f_1, f_2, \ldots, f_n) = \begin{vmatrix} f_1 & f_2 & \cdots & f_n \\ f_1' & f_2' & \cdots & f_1' \\ \vdots & & \cdots & \vdots \\ f_1^{(n-1)} & f_2^{(n-1)} & \cdots & f_n^{(n-1)} \end{vmatrix}$$

If $W \neq 0$ for some value of the independent variable, the functions are linearly independent. This means that the functions $f_1, f_2, \ldots, f_n$ are different solutions of the differential equation. Also, $\sin x$ and $\cos x$ are solutions of $(d^2y/dx^2) + y = 0$. Show that these are linearly independent solutions.

Solution Find the Wronskian and evaluate it. Let $f_1 = \sin x$ and $f_2 = \cos x$. Since $f_1' = \cos x$ and $f_2' = -\sin x$, then

$$W(\sin x, \cos x) = \begin{vmatrix} \sin x & \cos x \\ \cos x & -\sin x \end{vmatrix} = -\sin^2 x - \cos^2 x = -1$$

Therefore, $\sin x$ and $\cos x$ are independent solutions.

5-4. Additional Uses of Determinants

PROBLEM 5-21 (Manufacturing) A one-person furniture manufacturer makes three types of tables T1, T2, and T3. Type T1 requires 3 hours for cutting materials, 2 hours for assembly, and 2 hours for finishing. Type T2 requires 4 hours for cutting materials, 3 hours for assembly, and 3 hours for finishing. Type T3 requires 6 hours for cutting materials, 4 hours for assembly, and 3 hours for finishing. The person spends 71 hours per month cutting materials, 50 hours on assembly, and 45 hours on finishing. How many of each type of table can be made in a month? Use Cramer's rule.

Solution

Time	Type of table T1(x)	T2(y)	T3(z)	hours/month
Cutting	3	4	6	71
Assembly	2	3	4	50
Finishing	2	3	3	45

$$3x + 4y + 6z = 71$$
$$2x + 3y + 4z = 50$$
$$2x + 3y + 3z = 45$$

By Cramer's rule,

$$x = \frac{\begin{vmatrix} 71 & 4 & 6 \\ 50 & 3 & 4 \\ 45 & 3 & 3 \end{vmatrix}}{\begin{vmatrix} 3 & 4 & 6 \\ 2 & 3 & 4 \\ 2 & 3 & 3 \end{vmatrix}} = \frac{71(9) + 16(45) + 6(150) - 45(18) - 12(71) - 150(4)}{27 + 32 + 36 - 36 - 36 - 24}$$

$$= \frac{-3}{-1} = 3$$

$$y = \frac{\begin{vmatrix} 3 & 71 & 6 \\ 2 & 50 & 4 \\ 2 & 45 & 3 \end{vmatrix}}{-1} = \frac{9(50) + 8(71) + 90(6) - 100(6) - 12(45) - 6(71)}{-1}$$

$$= \frac{-8}{-1} = 8$$

$$z = \frac{\begin{vmatrix} 3 & 4 & 71 \\ 2 & 3 & 50 \\ 2 & 3 & 45 \end{vmatrix}}{-1} = \frac{9(45) + 4(100) + 71(6) - 71(6) - 9(50) - 8(45)}{-1}$$

$$= \frac{-5}{-1} = 5$$

Therefore, the custom builder can make three type T1 tables, eight type T2 tables, and five type T3 tables in a month.

PROBLEM 5-22 (Engineering, Maintenance) A maintenance engineer can use three types of bituminous paving material for highway resurfacing operations. The type 1 paving material requires 30.0 tons of aggregate, 6.0 tons of asphalt, and 1.0 tons of tack coat for each mile to be resurfaced. Each mile of resurfacing using the type 2 paving material requires 25.0 tons of aggregate, 5.0 tons of asphalt, and 1.3 tons of tack coat. A mile of highway resurfacing using the type 3 material requires 26.0 tons of aggregate, 5.0 tons of asphalt, and 1.5 tons of tack coat. If 1120 tons of aggregate, 220 tons of asphalt, and 56 tons of tack coat are available for resurfacing operations, then for each type of paving material, calculate the number of miles of highway that the maintenance engineer can expect to resurface. Use Cramer's rule.

Solution The matrix equation is

$$\begin{bmatrix} 30.0 & 25.0 & 26.0 \\ 6.0 & 5.0 & 5.0 \\ 1.0 & 1.3 & 1.5 \end{bmatrix} \begin{bmatrix} x_1 \\ x_2 \\ x_3 \end{bmatrix} = \begin{bmatrix} 1120 \\ 220 \\ 56 \end{bmatrix}$$

or

$$\begin{aligned} 30.0x_1 + 25.0x_2 + 26.0x_3 &= 1120 \\ 6.0x_1 + 5.0x_2 + 5.0x_3 &= 220 \\ 1.0x_1 + 1.3x_2 + 1.5x_3 &= 56 \end{aligned}$$

Solving by use of Cramer's rule, we get

$$x_1 = \frac{\begin{vmatrix} 1120 & 25.0 & 26.0 \\ 220 & 5.0 & 5.0 \\ 56 & 1.3 & 1.5 \end{vmatrix}}{\begin{vmatrix} 30.0 & 25.0 & 26.0 \\ 6.0 & 5.0 & 5.0 \\ 1.0 & 1.3 & 1.5 \end{vmatrix}} = \frac{\frac{1}{10}\begin{vmatrix} 1120 & 25.0 & 26.0 \\ 220 & 5.0 & 5.0 \\ 560 & 13.0 & 15.0 \end{vmatrix}}{\frac{1}{10}\begin{vmatrix} 30.0 & 25.0 & 26.0 \\ 6.0 & 5.0 & 5.0 \\ 10.0 & 13.0 & 15.0 \end{vmatrix}} = \frac{260}{28}$$

$$x_1 = 9\frac{2}{7} \doteq 9.29 \text{ miles}$$

$$x_2 = \frac{\begin{vmatrix} 30.0 & 1120 & 26.0 \\ 6.0 & 220 & 5.0 \\ 1.0 & 56 & 1.5 \end{vmatrix}}{\begin{vmatrix} 30.0 & 25.0 & 26.0 \\ 6.0 & 5.0 & 5.0 \\ 1.0 & 1.3 & 1.5 \end{vmatrix}} = \frac{\begin{vmatrix} 30 & 1120 & 26 \\ 6 & 220 & 5 \\ 10 & 560 & 15 \end{vmatrix}}{28} = \frac{360}{28} \doteq 12.86$$

$$x_3 = \frac{\begin{vmatrix} 30.0 & 25.0 & 1120 \\ 6.0 & 5.0 & 220 \\ 1.0 & 1.3 & 56 \end{vmatrix}}{\begin{vmatrix} 30.0 & 25.0 & 25.0 \\ 6.0 & 5.0 & 5.0 \\ 1.0 & 1.3 & 1.5 \end{vmatrix}} = \frac{\begin{vmatrix} 30 & 25 & 1120 \\ 6 & 5 & 220 \\ 10 & 13 & 560 \end{vmatrix}}{28} = \frac{560}{28} = 20$$

Therefore, the engineer can expect 9.29 miles from type 1 material, 12.86 miles from type 2 material, and 20.0 miles from type 3 material.

PROBLEM 5-23 (Row Echelon Form—The Lowman Method) (For the proof see the Appendices.) Find the row echelon form of

$$A = \begin{bmatrix} 1 & -1 & 2 & 1 & -2 \\ 2 & 1 & 3 & 0 & 1 \\ 3 & 2 & -1 & 2 & -1 \\ 4 & -1 & 1 & 1 & 1 \end{bmatrix}.$$

If we use elementary row operations on the matrix A, we get

$$A = \begin{bmatrix} 1 & -1 & 2 & 1 & -2 \\ 2 & 1 & 3 & 0 & 1 \\ 3 & 2 & -1 & 2 & -1 \\ 4 & -1 & 1 & 1 & 1 \end{bmatrix} \xrightarrow[\substack{R_3 \to R_3 - 3R_1 \\ R_4 \to R_4 - 4R_1}]{R_2 \to R_2 - 2R_1} \begin{bmatrix} 1 & -1 & 2 & 1 & -1 \\ 0 & 3 & -1 & -2 & 5 \\ 0 & 5 & -7 & -1 & 5 \\ 0 & 3 & -7 & -3 & 9 \end{bmatrix}$$

We get the same result by using a variation of the Sylvester eliminant (See Section 3-6).

$$\begin{bmatrix} 1 & -1 & 2 & 1 & -2 \\ 2 & 1 & 3 & 0 & 1 \\ 3 & 2 & -1 & 2 & -1 \\ 4 & -1 & 1 & 1 & 1 \end{bmatrix} \to \begin{bmatrix} 1 & -1 & 2 & 1 & -2 \\ 0 & \begin{vmatrix} 1 & -1 \\ 2 & 1 \end{vmatrix} & \begin{vmatrix} 1 & 2 \\ 2 & 3 \end{vmatrix} & \begin{vmatrix} 1 & 1 \\ 2 & 0 \end{vmatrix} & \begin{vmatrix} 1 & -2 \\ 2 & 1 \end{vmatrix} \\ 0 & \begin{vmatrix} 1 & -1 \\ 3 & 2 \end{vmatrix} & \begin{vmatrix} 1 & 2 \\ 3 & -1 \end{vmatrix} & \begin{vmatrix} 1 & 1 \\ 3 & 2 \end{vmatrix} & \begin{vmatrix} 1 & -2 \\ 3 & -1 \end{vmatrix} \\ 0 & \begin{vmatrix} 1 & -1 \\ 4 & -1 \end{vmatrix} & \begin{vmatrix} 1 & 2 \\ 4 & 1 \end{vmatrix} & \begin{vmatrix} 1 & 1 \\ 4 & 1 \end{vmatrix} & \begin{vmatrix} 1 & -2 \\ 4 & 1 \end{vmatrix} \end{bmatrix}$$

$$\to \begin{bmatrix} 1 & -1 & 2 & 1 & -2 \\ 0 & 3 & -1 & -2 & 5 \\ 0 & 5 & -7 & -1 & 5 \\ 0 & 3 & -7 & -3 & 9 \end{bmatrix}$$

Next we repeat the process with the submatrix

$$\begin{bmatrix} 1 & -1 & 2 & 1 & -2 \\ 0 & 3 & -1 & -2 & 5 \\ 0 & 5 & -7 & -1 & 5 \\ 0 & 3 & -7 & -3 & 9 \end{bmatrix} \to \begin{bmatrix} 1 & -1 & 2 & 1 & -2 \\ 0 & 3 & -1 & -2 & 5 \\ 0 & 0 & \begin{vmatrix} 3 & -1 \\ 5 & -7 \end{vmatrix} & \begin{vmatrix} 3 & -2 \\ 5 & -1 \end{vmatrix} & \begin{vmatrix} 3 & 5 \\ 5 & 5 \end{vmatrix} \\ 0 & 0 & \begin{vmatrix} 3 & -1 \\ 3 & -7 \end{vmatrix} & \begin{vmatrix} 3 & -2 \\ 3 & -3 \end{vmatrix} & \begin{vmatrix} 3 & 5 \\ 3 & 9 \end{vmatrix} \end{bmatrix}$$

$$\to \begin{bmatrix} 1 & -1 & 2 & 1 & -2 \\ 0 & 3 & -1 & -2 & 5 \\ 0 & 0 & -16 & 7 & -10 \\ 0 & 0 & -18 & -3 & 12 \end{bmatrix}$$

Now repeat with the next submatrix:

$$\begin{bmatrix} 1 & -1 & 2 & 1 & -2 \\ 0 & 3 & -1 & -2 & 5 \\ 0 & 0 & -16 & 7 & -10 \\ 0 & 0 & 0 & \begin{vmatrix} 16 & 7 \\ -18 & -3 \end{vmatrix} & \begin{vmatrix} -16 & -10 \\ -18 & 12 \end{vmatrix} \end{bmatrix} \longrightarrow \begin{bmatrix} 1 & -1 & 2 & 1 & -2 \\ 0 & 3 & -1 & -2 & 5 \\ 0 & 0 & -16 & 7 & -10 \\ 0 & 0 & 0 & 174 & -372 \end{bmatrix}$$

$$\xrightarrow[\substack{R_3 \to -1/16R_3 \\ R_4 \to 1/174R_4}]{R_2 \to 1/3R_2} \begin{bmatrix} 1 & -1 & 2 & 1 & -2 \\ 0 & 1 & -\frac{1}{3} & -\frac{2}{3} & \frac{5}{3} \\ 0 & 0 & 1 & -\frac{7}{16} & \frac{10}{16} \\ 0 & 0 & 0 & 1 & -\frac{62}{29} \end{bmatrix}$$

This is in row echelon form. With some practice the 2 × 2 determinant values can be performed mentally. Row echelon is *not* unique, so if you use the ordinary approach, your answer may be different.

PROBLEM 5-24 Solve the following system of equations using the row echelon method explained in Problem 5-23:

$$\begin{aligned} x + 2y - z &= 1 \\ 2x - y + z &= 0 \\ 3x + y + 4z &= 5 \end{aligned}$$

Solution

$$\begin{bmatrix} 1 & 2 & -1 & 1 \\ 2 & -1 & 1 & 0 \\ 3 & 1 & 4 & 5 \end{bmatrix} \to \begin{bmatrix} 1 & 2 & -1 & 1 \\ 0 & \begin{vmatrix} 1 & 2 \\ 2 & -1 \end{vmatrix} & \begin{vmatrix} 1 & -1 \\ 2 & 1 \end{vmatrix} & \begin{vmatrix} 1 & 1 \\ 2 & 0 \end{vmatrix} \\ 0 & \begin{vmatrix} 1 & 2 \\ 3 & 1 \end{vmatrix} & \begin{vmatrix} 1 & -1 \\ 3 & 4 \end{vmatrix} & \begin{vmatrix} 1 & 1 \\ 3 & 5 \end{vmatrix} \end{bmatrix} \to \begin{bmatrix} 1 & 2 & -1 & 1 \\ 0 & -5 & 3 & -2 \\ 0 & -5 & 7 & 2 \end{bmatrix}$$

$$\to \begin{bmatrix} 1 & 2 & -1 & 1 \\ 0 & -5 & 3 & -2 \\ 0 & 0 & \begin{vmatrix} -5 & 3 \\ -5 & 7 \end{vmatrix} & \begin{vmatrix} -5 & -2 \\ -5 & 2 \end{vmatrix} \end{bmatrix} \to \begin{bmatrix} 1 & 2 & -1 & 1 \\ 0 & -5 & 3 & -2 \\ 0 & 0 & -20 & -20 \end{bmatrix} \to \begin{bmatrix} 1 & 2 & -1 & 1 \\ 0 & 1 & -\frac{3}{5} & \frac{2}{5} \\ 0 & 0 & 1 & 1 \end{bmatrix}$$

Therefore, $z = 1$ and $y - 3/5z = 2/5$, which gives $y = 1$. Substituting in equation $x + 2y - z = 1$, we get $x = 0$, so the solution set is $\{(0, 1, 1)\}$

Supplementary Problems

PROBLEM 5-25 Using the correspondence between letters and numbers in Problem 5-11 and the matrix

$$\begin{bmatrix} 6 & 6 & -1 \\ -2 & 1 & 1 \\ 4 & -11 & 9 \end{bmatrix}$$

the following coded message was received: *D O G*. Find the original message.

PROBLEM 5-26 (Scheduling) The Mathematics Department at The H. Creaser Institute wishes to hire 8 teachers to teach a total of 26 sections each semester. A teacher can be hired at three ranks: Instructor, Assistant Professor, and Associate Professor, with an Instructor receiving $21,000, an Assistant Professor receiving $28,000, and an Associate Professor receiving $31,000, per year. Each Instructor will teach four classes, an Assistant Professor will teach three classes, and an Associate Professor will teach two classes. The department budget allowance contains $202,000 earmarked to pay the new teachers.

(a) Write the matrix equation for determining how many teachers should be hired at each rank.
(b) Solve the system.

PROBLEM 5-27 (Differential Equations) Three solutions of the differential equation $y''' - 2y'' - y' + 2y = 0$ are e^x, e^{-x}, and e^{2x}. Determine if these solutions are linearly independent.

Answers to Supplementary Problems

5-25 Original message was C A T.

5-26 **(a)** $$\begin{bmatrix} 1 & 1 & 1 \\ 4 & 3 & 2 \\ 21 & 28 & 31 \end{bmatrix}\begin{bmatrix} x \\ y \\ z \end{bmatrix} = \begin{bmatrix} 8 \\ 26 \\ 202 \end{bmatrix}$$

(b) Four Instructors, two Assistant Professors, two Associate Professors

5-27 Linearly independent since $W = -6e^{2x} \neq 0$.

6 VECTOR SPACES

THIS CHAPTER IS ABOUT

☑ Vector Spaces
☑ Subspaces
☑ Some Properties of Vector Spaces
☑ Linear Independence
☑ Some Properties of Linear Independence and Dependence

6-1. Vector Spaces

Definition of a vector space

A **real vector space** consists of a set V of elements called vectors on which operations of addition and scalar multiplication have been defined. The elements may be numbers, ordered pairs or n-tuples of numbers, functions, matrices, or any other objects on which the two operations $\oplus$ and $\odot$ are defined with the following properties:

(1) If $\mathbf{a}$, $\mathbf{b}$ are in V, then $\mathbf{a} \oplus \mathbf{b}$ is in V.
(2) $\mathbf{a} \oplus \mathbf{b} = \mathbf{b} \oplus \mathbf{a}$ for all $\mathbf{a}$, $\mathbf{b}$ in V.
(3) $\mathbf{a} \oplus (\mathbf{b} \oplus \mathbf{c}) = (\mathbf{a} \oplus \mathbf{b}) \oplus \mathbf{c}$ for all $\mathbf{a}$, $\mathbf{b}$, $\mathbf{c}$ in V.
(4) There exists a unique element $\mathbf{0}$ in V, called the **zero vector** or **additive identity**, such that $\mathbf{a} \oplus \mathbf{0} = \mathbf{0} \oplus \mathbf{a}$ for all $\mathbf{a}$ in V.
(5) For every $\mathbf{a}$ in V there exists a unique element denoted by $-\mathbf{a}$ called the **negative** or **additive inverse** of $\mathbf{a}$, such that $\mathbf{a} \oplus (-\mathbf{a}) = (-\mathbf{a}) \oplus \mathbf{a} = \mathbf{0}$.
(6) If $\mathbf{a}$ is in V and k is any real number called a scalar, then $k \odot \mathbf{a}$ is in V.
(7) $k \odot (\mathbf{a} \oplus \mathbf{b}) = k \odot \mathbf{a} \oplus k \odot \mathbf{b}$ for $\mathbf{a}$, $\mathbf{b}$ in V and k in R, where R is the set of real numbers.
(8) $(k + \ell) \odot \mathbf{a} = k \odot \mathbf{a} \oplus \ell \odot \mathbf{a}$ for all $\mathbf{a}$ in V and k, ℓ in R, where $+$ denotes ordinary addition of real numbers.
(9) $k \odot (\ell \odot \mathbf{a}) = (k \cdot \ell) \odot \mathbf{a}$ for all $\mathbf{a}$ in V and k, ℓ in R, where the centered dot denotes ordinary multiplication of real numbers and is usually omitted.
(10) $1 \odot \mathbf{a} = \mathbf{a}$ for all $\mathbf{a}$ in V.

The operation $\oplus$ denotes **vector addition** and $\odot$ denotes scalar multiplication. When working with real numbers, multiplication is performed before addition unless otherwise indicated by grouping symbols. Likewise, $\odot$ is performed before $\oplus$ if an expression involves both operations. In each problem, the $\oplus$ and $\odot$ operations must be defined.

EXAMPLE 6-1: Show that the set R of all real numbers with $\oplus$ ordinary addition and $\odot$ ordinary multiplication is a vector space.

Solution: In this case, real numbers are both vectors and scalars and $\oplus$ and $\odot$ are ordinary addition and multiplication, respectively. Hence a, b, c, k, ℓ are real numbers. Thus

(1) $a + b$ is in R by closure for addition of real numbers.
(2) $a + b = b + a$ by commutative property of real numbers.
(3) $a + (b + c) = (a + b) + c$ by associative property of real numbers.
(4) $0 + a = a + 0 = a$ by additive identity property of real numbers.
(5) $a + (-a) = (-a) + a = 0$ by additive inverse property of real numbers.
(6) $k \cdot a = ka$ is in R by closure for multiplication of real numbers.
(7) $k \cdot (a + b) = k \cdot a + k \cdot b = ka + kb$ by the distributive property for real numbers.
(8) $(k + \ell) \cdot a = k \cdot a + \ell \cdot a = ka + \ell a$ by the distributive property for real numbers.
(9) $k \cdot (\ell \cdot a) = (k \cdot \ell) \cdot a = (k\ell)a$ by the associative property for multiplication of real numbers.
(10) $1 \cdot a = a$ by the multiplicative identity property for real numbers.

This vector space is denoted by R, R^1, or R_1.

EXAMPLE 6-2: Show that the set of all 2×2 matrices denoted by ${}_2M_2$, with $\oplus$ defined as matrix addition and $\odot$ defined as multiplication of a matrix by a real number, is a vector space.

Solution

(1) $\mathbf{a} \oplus \mathbf{b} = \begin{bmatrix} a_1 & b_1 \\ c_1 & d_1 \end{bmatrix} + \begin{bmatrix} a_2 & b_2 \\ c_2 & d_2 \end{bmatrix} = \begin{bmatrix} a_1 + a_2 & b_1 + b_2 \\ c_1 + c_2 & d_1 + d_2 \end{bmatrix}$ is in ${}_2M_2$ and is therefore closed under matrix addition.

(2) This is satisfied because matrix addition is commutative.

(3) This is satisfied because matrix addition is associative.

(4) The zero vector in this case is the 2×2 zero matrix $\begin{bmatrix} 0 & 0 \\ 0 & 0 \end{bmatrix}$.

(5) If $a = \begin{bmatrix} a_1 & b_1 \\ c_1 & d_1 \end{bmatrix}$, then $-\mathbf{a} = \begin{bmatrix} -a_1 & -b_1 \\ -c_1 & -d_1 \end{bmatrix}$.

(6)
$$k \odot \mathbf{a} = k \begin{bmatrix} a_1 & b_1 \\ c_1 & d_1 \end{bmatrix} = \begin{bmatrix} ka_1 & kb_1 \\ kc_1 & kd_1 \end{bmatrix} \text{ is in } {}_2M_2$$

(7)
$$k \odot (\mathbf{a} \oplus \mathbf{b}) = k\left(\begin{bmatrix} a_1 & b_1 \\ c_1 & d_1 \end{bmatrix} + \begin{bmatrix} a_2 & b_2 \\ c_2 & d_2 \end{bmatrix}\right)$$
$$= k\begin{bmatrix} a_1 + a_2 & b_1 + b_2 \\ c_1 + c_2 & d_1 + d_2 \end{bmatrix} = \begin{bmatrix} ka_1 + ka_2 & kb_1 + kb_2 \\ kc_1 + kc_2 & kd_1 + kd_2 \end{bmatrix}$$
$$= \begin{bmatrix} ka_1 & kb_1 \\ kc_1 & kd_1 \end{bmatrix} + \begin{bmatrix} ka_2 & kb_2 \\ kc_2 & kd_2 \end{bmatrix}$$
$$= k\begin{bmatrix} a_1 & b_1 \\ c_1 & d_1 \end{bmatrix} + k\begin{bmatrix} a_2 & b_2 \\ c_2 & d_2 \end{bmatrix} = k \odot \mathbf{a} \oplus k \odot \mathbf{b}$$

$\therefore$ Property 7 is satisfied

(8)
$$(k + \ell) \odot \mathbf{a} = (k + \ell)\begin{bmatrix} a_1 & b_1 \\ c_1 & d_1 \end{bmatrix} = \begin{bmatrix} (k + \ell)a_1 & (k + \ell)b_1 \\ (k + \ell)c_1 & (k + \ell)d_1 \end{bmatrix}$$
$$= \begin{bmatrix} ka_1 + \ell a_1 & kb_1 + \ell b_1 \\ kc_1 + \ell c_1 & kd_1 + \ell d_1 \end{bmatrix} = \begin{bmatrix} ka_1 & kb_1 \\ kc_1 & kd_1 \end{bmatrix} + \begin{bmatrix} \ell a_1 & \ell b_1 \\ \ell c_1 & \ell d_1 \end{bmatrix}$$
$$= k\begin{bmatrix} a_1 & b_1 \\ c_1 & d_1 \end{bmatrix} + \ell\begin{bmatrix} a_1 & b_1 \\ c_1 & d_1 \end{bmatrix} = k \odot \mathbf{a} \oplus \ell \odot \mathbf{a}$$

$\therefore$ Property 8 is satisfied

(9)
$$k \odot (\ell \oplus \mathbf{a}) = k\left(\ell \begin{bmatrix} a_1 & b_1 \\ c_1 & d_1 \end{bmatrix}\right) = k \begin{bmatrix} \ell a_1 & \ell b_1 \\ \ell c_1 & \ell d_1 \end{bmatrix}$$
$$= \begin{bmatrix} k(\ell a_1) & k(\ell b_1) \\ k(\ell c_1) & k(\ell d_1) \end{bmatrix} = \begin{bmatrix} (k\ell)a_1 & (k\ell)b_1 \\ (k\ell)c_1 & (k\ell)d_1 \end{bmatrix} = (k\ell)\begin{bmatrix} a_1 & b_1 \\ c_1 & d_1 \end{bmatrix}$$
$$= (k\ell) \odot \mathbf{a}$$

$\therefore$ Property 9 is satisfied

(10)
$$1 \odot \mathbf{a} = 1 \begin{bmatrix} a_1 & b_1 \\ c_1 & d_1 \end{bmatrix} = \begin{bmatrix} 1a_1 & 1b_1 \\ 1c_1 & 1d_1 \end{bmatrix} = \begin{bmatrix} a_1 & b_1 \\ c_1 & d_1 \end{bmatrix} = \mathbf{a}$$

$\therefore$ Property 10 is satisfied

Therefore, ${}_2M_2$ is a vector space where $\oplus$ is matrix addition and $\odot$ is multiplication of a matrix by a scalar.

EXAMPLE 6-3: Verify that the set of all $m \times n$ matrices denoted by ${}_mM_n$, with $\oplus$ defined as matrix addition and $\odot$ defined as scalar multiplication, is a vector space.

Solution: Let A, B, and C be $m \times n$ matrices and a and b be real numbers. Since $\oplus$ and $\odot$ are matrix addition and multiplication, respectively, we shall use $+$ and $\cdot$ to represent them.

(1) $A \oplus B = A + B$ is an $m \times n$ matrix by definition of matrix addition.
(2) $A + B = B + A$ because the addition of matrices is commutative.
(3) $A + (B + C) = (A + B) + C$ because the addition of matrices is associative.
(4) The zero vector is the $m \times n$ zero matrix.
(5) If $A = [a_{ij}]$, then $-A = [-a_{ij}]$; that is, the additive inverse of A is the $m \times n$ matrix $-A$ in which each element of $-A$ is the negative of the corresponding element of A.
(6) aA is an $m \times n$ matrix by the definition of scalar multiplication.
(7) $a(A + B) = aA + aB$ by Property 1-5.C(**c**).
(8) $(a + b)A = aA + bA$ by Property 1-5.C(**d**).
(9) $a(bA) = (ab)A$ by Property 1-5.C(**e**).
(10) $1A = A$ by definition of scalar multiplication.

Therefore, ${}_mM_n$ with the two defined operations is a vector space.

EXAMPLE 6-4: Are the following sets with $\oplus$ and $\odot$, defined as matrix addition and scalar multiplication, respectively, vector spaces?

(**a**) The set of all $n \times 1$ matrices $\begin{bmatrix} a_1 \\ a_2 \\ \vdots \\ a_n \end{bmatrix}$ denoted by R^n.

(**b**) The set of all $1 \times n$ matrices $[a_1 \quad a_2 \quad \ldots \quad a_n]$ denoted by R_n.

Solution

(**a**) Yes, because it is a special case of Example 6-3 where $m = n$ and $n = 1$.
(**b**) Yes, because it is a special case of Example 6-3 where $m = 1$.

The individual numbers in vectors from each of these vector spaces are called **components.** If $n = 2$, R^2 denotes the set of all vectors in the plane.

EXAMPLE 6-5: Let V be the set of all polynomials of degree $\leq \mathbf{n}$ together with the zero polynomial which has no degree. If $p(x) = a_nx^n + a_{n-1}x^{n-1} + \cdots + a_1x + a_0$ and $q(x) = b_nx^n + b_{n-1}x^{n-1} + \cdots + b_1x + b_0$, define

$$p(x) \oplus q(x) = (a_n + b_n)x^n + (a_{n-1} + b_{n-1})x^{n-1} + \cdots + (a_1 + b_1)x + (a_0 + b_0)$$

and

$$k \odot p(x) = (ka_n)x^n + (ka_{n-1})x^{n-1} + \cdots + (ka_1)x + ka_0$$

Verify that V is a vector space.

Solution: Let $p(x)$ and $q(x)$ be as given, and let $h(x) = c_nx^n + c_{n-1}x^{n-1} + \cdots + c_1x + c_0$.

(1) $p(x) \oplus q(x) = p(x) + q(x)$ is either 0 or a polynomial of degree $\leq n$.

(2) $p(x) \oplus q(x) = p(x) + q(x) = (a_n + b_n)x^n + (a_{n-1} + b_{n-1})x^{n-1} + \cdots + (a_1 + b_1)x + (a_0 + b_0) = (b_n + a_n)x^n + (b_{n-1} + a_{n-1})x^{n-1} + \cdots + (b_1 + a_1)x + (b_0 + a_0)$ because the addition of real numbers is commutative. However, by the definition of $\oplus$, this last expression is equal to $q(x) \oplus p(x)$.

(3)
$$\begin{aligned} p(x) \oplus \big(q(x) \oplus h(x)\big) &= p(x) + \big(q(x) + h(x)\big) \\ &= [a_n + (b_n + c_n)]x^n + [a_{n-1} + (b_{n-1} + c_{n-1})]x^{n-1} + \cdots \\ &\quad + [a_1 + (b_1 + c_1)]x + [a_0 + (b_0 + c_0)] \\ &= [(a_n + b_n) + c_n]x^n + [(a_{n-1} + b_{n-1}) + c_{n-1}]x^{n-1} + \cdots \\ &\quad + [(a_1 + b_1) + c_1]x + [(a_0 + b_0) + c_0] \\ &= \big(p(x) + q(x)\big) + h(x) = \big(p(x) \oplus q(x)\big) \oplus h(x) \end{aligned}$$

(4) The zero polynomial is the additive identity.

(5) The additive inverse of $p(x)$ is $-p(x)$.

(6) $k \odot p(x) = kp(x)$ is either 0 or a polynomial of degree $\leq n$.

(7)
$$\begin{aligned} k \odot [p(x) \oplus q(x)] &= k[p(x) + q(x)] \\ &= [k(a_n + b_n)]x^n + [k(a_{n-1} + b_{n-1})]x^{n-1} + \cdots + k(a_1 + b_1)x + k(a_0 + b_0) \\ &= (ka_n + kb_n)x^n + (ka_{n-1} + kb_{n-1})x^{n-1} + \cdots + (ka_1 + kb_1)x + (ka_0 + kb_0) \\ &= (ka_n)x^n + (ka_{n-1})x^{n-1} + \cdots + (ka_1)x + (ka_0) \\ &\quad + (kb_n)x^n + (kb_{n-1})x^{n-1} + \cdots + (kb_1)x + (kb_0) \\ &= kp(x) + kq(x) = k \odot p(x) \oplus k \odot q(x) \end{aligned}$$

(8)
$$\begin{aligned} (k + \ell) \odot p(x) &= (k + \ell)p(x) \\ &= [(k + \ell)a_n]x^n + [(k + \ell)a_{n-1}]x^{n-1} + \cdots + [(k + \ell)a_1]x + (k + \ell)a_0 \\ &= (ka_n)x^n + (\ell a_n)x^n + (ka_{n-1})x^{n-1} + (\ell a_{n-1})x^{n-1} + \cdots \\ &\quad + (ka_1)x + (\ell a_1)x + ka_0 + \ell a_0 \\ &= kp(x) + \ell p(x) = k \odot p(x) \oplus \ell \odot p(x) \end{aligned}$$

(9)
$$\begin{aligned} k \odot \big(\ell \odot p(x)\big) &= k\big(\ell p(x)\big) \\ &= k(\ell a_nx^n) + k(\ell a_{n-1}x^{n-1}) + \cdots + k(\ell a_1x) + k(\ell a_0) \\ &= (k\ell)(a_nx^n + a_{n-1}x^{x-1} + \cdots + a_1x + a_0) \\ &= (k\ell)p(x) = (k\ell) \odot p(x) \end{aligned}$$

(10) $1 \odot p(x) = 1p(x) = p(x)$.

Therefore, V with the two defined operations is a vector space. This vector space is denoted by $\mathbf{P}_n$.

EXAMPLE 6-6: Let V be the set of all real-valued functions defined on the entire real line. If f and g are two such functions and $\oplus$ and $\odot$ are as defined by $(f \oplus g)(x) = (f + g)(x) = f(x) + g(x)$ and $(a \odot f)(x) = af(x)$, verify that V is a vector space.

Solution: Let $f(x)$, $g(x)$, and $h(x)$ be any three elements of V. Since these are real-valued, $f(x)$, $g(x)$, and $h(x)$ are real numbers. Hence, all ten properties necessary for V to be a vector space follow from properties of the real numbers. The step-by-step justifications are the same as those given in Example 6-1, where the term "real numbers" is replaced by "real-valued functions" when appropriate.

This vector space is denoted by $R(-\infty, \infty)$. If "real-valued functions" is replaced by "real-valued continuous functions," the resulting vector space is denoted by $C(-\infty, \infty)$.

EXAMPLE 6-7: If V is the set of all ordered triples of real numbers with operations $(a_1, b_1, c_1) \oplus (a_2, b_2, c_2) = (a_1 + a_2, b_1 + b_2, c_1 + c_2)$ and $k \odot (a_1, b_1, c_1) = (1, 1, 1)$, determine which of the ten properties necessary for a vector space fail to hold.

Solution: Although you must mentally go through all ten of the properties, you only need to write down those not satisfied. If, however, it is not clear to you whether a particular property holds, write it out step-by-step. Suppose, in this instance, that it is clear that Properties 1 through 6 are true but it is not clear if Properties 7 through 10 are true.

(7)
$$k \odot [(a_1, b_1, c_1) \oplus (a_2, b_2, c_2)] = k \odot (a_2 + a_2, b_1 + b_2, c_1 + c_2)$$
$$= (1, 1, 1)$$
$$k \odot (a_1, b_1, c_1) \oplus k \odot (a_2, b_2, c_2) = (1, 1, 1) \oplus (1, 1, 1) = (2, 2, 2)$$
$$\therefore \quad \text{it does not hold}$$

(8)
$$(k + \ell) \odot (a_1, b_1, c_1) = (1, 1, 1)$$
$$k \odot (a_1, b_1, c_1) \oplus \ell \odot (a_1, b_1, c_1) = (1, 1, 1) \oplus (1, 1, 1) = (2, 2, 2)$$
$$\therefore \quad \text{it does not hold}$$

(9)
$$k \odot [\ell \odot (a_1, b_1, c_1)] = k \odot (1, 1, 1) = (1, 1, 1)$$
$$(k\ell) \odot (a_1, b_1, c_1) = (1, 1, 1)$$
$$\therefore \quad \text{it holds}$$

(10)
$$1 \odot (a_1, b_1, c_1) = (1, 1, 1) \neq (a_1, b_1, c)$$
$$\text{unless} \quad a_1 = 1, \quad b_1 = 1, \quad c_1 = 1$$

In this example, Properties 7, 8, and 10 fail to hold.

6-2. Subspaces

A. Definition of subspace

If a nonempty subset W of a vector space V is itself a vector space under the same defined operations of vector addition and scalar multiplication as in V, then W is called a **subspace** of V.

As illustrated in Examples 6-1 through 6-6, to verify that a set V having defined operations $\oplus$ and $\odot$ is a vector space, you must check to see that each of the ten properties of a vector space is satisfied. However, every nonempty subset W of V inherits eight of these ten properties from its parent V. This leads to the following property.

B. Conditions for a subspace

A nonempty subset W of a vector space V having operations $\oplus$ and $\odot$ is a subspace of V if and only if it is closed under both $\oplus$ and $\odot$. This means that if **a** and **b** are elements of W and k is a scalar, then W is a subspace provided $\mathbf{a} \oplus \mathbf{b}$ is in W and $k \odot \mathbf{a}$ is in W.

For future problems you may assume that it is known that R, ${}_mM_n$, R^n, R_n, P_n, $R(-\infty, \infty)$, and $C(-\infty, \infty)$, under the defined operations of Examples 6-1 through 6-6, are vector spaces.

EXAMPLE 6-8: Find two subspaces of a vector space V having operations $\oplus$ and $\odot$.

Solution: If $\mathbf{0}$ is the zero vector of V, then sets $\{\mathbf{0}\}$ and V are nonempty subsets of V. Since V is a given vector space, it is closed under $\oplus$ and $\odot$. Likewise, for $\{\mathbf{0}\}$, $\mathbf{0} \oplus \mathbf{0} = \mathbf{0}$ and $k \odot \mathbf{0} = \mathbf{0}$. Thus $\{\mathbf{0}\}$ is closed under $\oplus$ and $\odot$. Therefore, any vector space V has at least the two trivial subspaces $\{\mathbf{0}\}$ and V. You might observe, however, that if V has only a single vector as an element, then $\{\mathbf{0}\} = V$.

EXAMPLE 6-9: Determine which of the following are vector spaces under the usual definitions of vector addition and scalar multiplication: the set W of all vectors of the form

(a) $[0 \quad 0 \quad c]$ **(b)** $\begin{bmatrix} 0 \\ b \\ 0 \end{bmatrix}$ **(c)** $[1 \quad b \quad c \quad d]$ **(d)** $\begin{bmatrix} a \\ b \\ c \\ 0 \end{bmatrix}$

Solution

(a) The set is a subset of vector space R_3. It is not empty because it contains at least $[0 \quad 0 \quad 0]$. Thus, you need to check only closure for vector addition and scalar multiplication:

$$[0 \quad 0 \quad c_1] + [0 \quad 0 \quad c_2] = [0+0 \quad 0+0 \quad c_1 + c_2]$$
$$= [0 \quad 0 \quad c_1 + c_2] \quad \text{which is in } W$$
$$k[0 \quad 0 \quad c] = [0 \quad 0 \quad kc] \qquad \text{which is in } W$$

Therefore, the set is closed under vector addition and scalar multiplication. W is, therefore, a subspace of R_3 and hence a vector space in its own right.

(b) This set is a nonempty subset of R^3. You need to check the following:

$$\begin{bmatrix} 0 \\ b_1 \\ 0 \end{bmatrix} + \begin{bmatrix} 0 \\ b_2 \\ 0 \end{bmatrix} = \begin{bmatrix} 0 \\ b_1 + b_2 \\ 0 \end{bmatrix} \qquad \text{which is in } W$$

$$k\begin{bmatrix} 0 \\ b \\ 0 \end{bmatrix} = \begin{bmatrix} 0 \\ kb \\ 0 \end{bmatrix} \qquad \text{which is in } W$$

Therefore, the set is closed under vector addition and scalar multiplication. W is, therefore, a subspace of R^3 and hence a vector space in its own right.

(c) This set is a nonempty subset of R_4.

$$[1 \quad b_1 \quad c_1 \quad d_1] + [1 \quad b_2 \quad c_2 \quad d_2] = [2 \quad b_1 + b_2 \quad c_1 + c_2 \quad d_1 + d_2]$$

which is not an element of W because the first component is 2 rather than 1. Therefore, the set is not a subspace of R_4 and hence not a vector space under the usual definitions of vector addition and scalar multiplication.

(d) This set is a nonempty subset of R^4.

$$\begin{bmatrix} a_1 \\ b_1 \\ c_1 \\ 0 \end{bmatrix} + \begin{bmatrix} a_2 \\ b_2 \\ c_2 \\ 0 \end{bmatrix} = \begin{bmatrix} a_1 + a_2 \\ b_1 + b_2 \\ c_1 + c_2 \\ 0 \end{bmatrix} \qquad \text{which is in } W$$

$$k\begin{bmatrix} a \\ b \\ c \\ 0 \end{bmatrix} = \begin{bmatrix} ka \\ kb \\ kc \\ 0 \end{bmatrix} \quad \text{which is in } W$$

Therefore, this set is a subspace of R^4 and hence a vector space in its own right.

EXAMPLE 6-10: Determine whether the set of all 2 × 2 matrices of the form $\begin{bmatrix} a & b \\ 0 & 0 \end{bmatrix}$ is a subspace of ${}_2M_2$.

note: In problems of this type you need not specify the operations because they automatically become the defined operations of ${}_2M_2$.

Solution

$$\begin{bmatrix} a_1 & b_1 \\ 0 & 0 \end{bmatrix} + \begin{bmatrix} a_2 & b_2 \\ 0 & 0 \end{bmatrix} = \begin{bmatrix} a_1 + a_2 & b_1 + b_2 \\ 0 & 0 \end{bmatrix} \quad \text{which is of form } \begin{bmatrix} a & b \\ 0 & 0 \end{bmatrix}$$

$$k\begin{bmatrix} a & b \\ 0 & 0 \end{bmatrix} = \begin{bmatrix} ka & kb \\ 0 & 0 \end{bmatrix} \quad \text{which is of form } \begin{bmatrix} a & b \\ 0 & 0 \end{bmatrix}. \quad \text{Hence a subspace.}$$

EXAMPLE 6-11: Determine which of the following subsets of ${}_3M_3$ are subspaces: the set W of all 3 × 3

(a) diagonal matrices
(b) upper triangular matrices
(c) symmetric matrices
(d) skew-symmetric matrices
(e) singular matrices
(f) nonsingular matrices.

Solution

(a) Diagonal matrices are of form

$$\begin{bmatrix} a & 0 & 0 \\ 0 & b & 0 \\ 0 & 0 & c \end{bmatrix}$$

Thus

$$\begin{bmatrix} a_1 & 0 & 0 \\ 0 & b_1 & 0 \\ 0 & 0 & c_1 \end{bmatrix} + \begin{bmatrix} a_2 & 0 & 0 \\ 0 & b_2 & 0 \\ 0 & 0 & c_2 \end{bmatrix} = \begin{bmatrix} a_1 + a_2 & 0 & 0 \\ 0 & b_1 + b_2 & 0 \\ 0 & 0 & c_1 + c_2 \end{bmatrix} \quad \text{which is diagonal}$$

and

$$k\begin{bmatrix} a & 0 & 0 \\ 0 & b & 0 \\ 0 & 0 & c \end{bmatrix} = \begin{bmatrix} ka & 0 & 0 \\ 0 & kb & 0 \\ 0 & 0 & kc \end{bmatrix} \quad \text{which is diagonal}$$

Therefore, the set of 3 × 3 diagonal matrices is closed under matrix addition and scalar multiplication and hence is a subspace of ${}_3M_3$.

(b) Upper triangular matrices are of the form

$$\begin{bmatrix} a_{11} & a_{12} & a_{13} \\ 0 & a_{22} & a_{23} \\ 0 & 0 & a_{33} \end{bmatrix}$$

Thus

$$\begin{bmatrix} a_{11} & a_{12} & a_{13} \\ 0 & a_{22} & a_{23} \\ 0 & 0 & a_{33} \end{bmatrix} + \begin{bmatrix} b_{11} & b_{12} & b_{13} \\ 0 & b_{22} & b_{23} \\ 0 & 0 & b_{33} \end{bmatrix} = \begin{bmatrix} a_{11}+b_{11} & a_{12}+b_{12} & a_{13}+b_{13} \\ 0 & a_{22}+b_{22} & a_{23}+b_{23} \\ 0 & 0 & a_{33}+b_{33} \end{bmatrix}$$

which is upper triangular

and

$$k\begin{bmatrix} a_{11} & a_{12} & a_{13} \\ 0 & a_{22} & a_{23} \\ 0 & 0 & a_{33} \end{bmatrix} = \begin{bmatrix} ka_{11} & ka_{12} & ka_{13} \\ 0 & ka_{22} & ka_{23} \\ 0 & 0 & ka_{33} \end{bmatrix} \quad \text{which is upper triangular}$$

Therefore, the set of 3×3 upper triangular matrices is closed under matrix addition and scalar multiplication and hence is a subspace of ${}_3M_3$.

(c) A matrix A is symmetric if $A^T = A$. Hence, these matrices are of the form

$$\begin{bmatrix} a & b & c \\ b & d & e \\ c & e & f \end{bmatrix}$$

Thus

$$\begin{bmatrix} a_1 & b_1 & c_1 \\ b_1 & d_1 & e_1 \\ c_1 & e_1 & f_1 \end{bmatrix} + \begin{bmatrix} a_2 & b_2 & c_2 \\ b_2 & d_2 & e_2 \\ c_2 & e_2 & f_2 \end{bmatrix}\begin{bmatrix} a_1+a_2 & b_1+b_2 & c_1+c_2 \\ b_1+b_2 & d_1+d_2 & e_1+e_2 \\ c_1+c_2 & e_1+e_2 & f_1+f_2 \end{bmatrix} \quad \text{which is symmetric}$$

and

$$k\begin{bmatrix} a & b & c \\ b & d & e \\ c & e & f \end{bmatrix} = \begin{bmatrix} ka & kb & kc \\ kb & kd & ke \\ kc & ke & kf \end{bmatrix} \quad \text{which is symmetric}$$

Therefore, the set of 3×3 symmetric matrices is closed under matrix addition and scalar multiplication and hence is a subspace of ${}_3M_3$.

(d) A matrix A is skew-symmetric if $A^T = -A$. Hence, these matrices are of form

$$\begin{bmatrix} 0 & a & b \\ -a & 0 & c \\ -b & -c & 0 \end{bmatrix}$$

Thus

$$\begin{bmatrix} 0 & a_1 & b_1 \\ -a_1 & 0 & c_1 \\ -b_1 & -c_1 & 0 \end{bmatrix} + \begin{bmatrix} 0 & a_2 & b_2 \\ -a_2 & 0 & c_2 \\ -b_2 & -c_2 & 0 \end{bmatrix} = \begin{bmatrix} 0 & a_1+a_2 & b_1+b_2 \\ -(a_1+a_2) & 0 & c_1+c_2 \\ -(b_1+b_2) & -(c_1+c_2) & 0 \end{bmatrix}$$

which is skew-symmetric

and

$$k\begin{bmatrix} 0 & a & b \\ -a & 0 & c \\ -b & -c & 0 \end{bmatrix} = \begin{bmatrix} 0 & ka & kb \\ -ka & 0 & kc \\ -kb & -kc & 0 \end{bmatrix} \quad \text{which is skew-symmetric}$$

Therefore, the set of 3×3 skew-symmetric matrices is closed under matrix addition and scalar multiplication and hence is a subspace of ${}_3M_3$.

(e) A matrix A is singular if and only if $|A| = 0$. Let

$$A = \begin{bmatrix} 1 & 0 & 0 \\ 0 & 0 & 0 \\ 0 & 0 & 0 \end{bmatrix} \quad \text{and} \quad B = \begin{bmatrix} 0 & 0 & 0 \\ 0 & 1 & 0 \\ 0 & 0 & 1 \end{bmatrix} \qquad |A| = |B| = 0$$

Therefore, A and B are 3×3 singular matrices.

$$A + B = \begin{bmatrix} 1 & 0 & 0 \\ 0 & 1 & 0 \\ 0 & 0 & 1 \end{bmatrix} \quad \text{and} \quad |A + B| = \begin{vmatrix} 1 & 0 & 0 \\ 0 & 1 & 0 \\ 0 & 0 & 1 \end{vmatrix} = 1$$

Hence, A and B are 3×3 singular matrices whose sum is a nonsingular matrix. This counterexample shows that the set of 3×3 singular matrices is not closed under matrix addition and hence is not a subspace of ${}_3M_3$.

(f) A matrix A is nonsingular or invertible if and only if $|A| \neq 0$. If

$$A = \begin{bmatrix} 1 & 0 & 0 \\ 0 & 3 & 0 \\ 0 & 0 & 5 \end{bmatrix} \quad \text{and} \quad B = \begin{bmatrix} 2 & 0 & 0 \\ 0 & -3 & 0 \\ 0 & 0 & -5 \end{bmatrix}$$

$$A + B = \begin{bmatrix} 3 & 0 & 0 \\ 0 & 0 & 0 \\ 0 & 0 & 0 \end{bmatrix}$$

$$|A| = 15, \qquad |B| = 30, \qquad |A + B| = 0$$

Hence, A and B are nonsingular, but $A + B$ is singular. This counterexample shows that the set of 3×3 nonsingular matrices is not a subspace of ${}_3M_3$.

6-3. Some Properties of Vector Spaces

Property A If V is a vector space with respect to operations $\oplus$ and $\odot$ and $\mathbf{0}$ is the zero vector of V, then

(a) $0 \odot \mathbf{a} = \mathbf{0}$ for any vector $\mathbf{a}$ in V;
(b) $k \odot \mathbf{0} = \mathbf{0}$ for any scalar k; and
(c) $(-1) \odot \mathbf{a} = -\mathbf{a}$ for any $\mathbf{a}$ in V.

EXAMPLE 6-12: Verify the properties of 6-3.A for the vector space R_4.

Solution: Let $\mathbf{a} = [a \quad b \quad c \quad d]$.

(a) $0 \odot \mathbf{a} = 0[a \quad b \quad c \quad d] = [0a \quad 0b \quad 0c \quad 0d] = [0 \quad 0 \quad 0 \quad 0] = \mathbf{0}$, the zero vector of R_4.
(b) $k \odot \mathbf{0} = k[0 \quad 0 \quad 0 \quad 0] = [0 \quad 0 \quad 0 \quad 0] = \mathbf{0}$.
(c) $(-1) \odot \mathbf{a} = (-1)[a \quad b \quad c \quad d] = [-a \quad -b \quad -c \quad -d] = -\mathbf{a}$.

Property B If V is a vector space with respect to the operations $\oplus$ and $\odot$, k is a scalar, and $\mathbf{a}$ is a vector in V such that $k \odot \mathbf{a} = \mathbf{0}$, then either $k = 0$ or $\mathbf{a} = \mathbf{0}$.

EXAMPLE 6-13: If V is the vector space ${}_2M_2$ and $a = \begin{bmatrix} 3 & 4 \\ -1 & 2 \end{bmatrix}$ such that $k\begin{bmatrix} 3 & 4 \\ -1 & 2 \end{bmatrix} = \begin{bmatrix} 0 & 0 \\ 0 & 0 \end{bmatrix}$, show that $k = 0$.

Solution: By scalar multiplication

$$k\begin{bmatrix} 3 & 4 \\ -1 & 2 \end{bmatrix} = \begin{bmatrix} 3k & 4k \\ -k & 2k \end{bmatrix}$$

Thus

$$\begin{bmatrix} 3k & 4k \\ -k & 2k \end{bmatrix} = \begin{bmatrix} 0 & 0 \\ 0 & 0 \end{bmatrix}$$

However, two matrices are equal if and only if their corresponding elements are identical. Therefore, $3k = 0$, $4k = 0$, $-k = 0$, $2k = 0$, each of which implies that $k = 0$.

EXAMPLE 6-14: If V is the vector space R_3, $k = 5$, and $\mathbf{a} = [a \quad b \quad c]$ such that $5[a \quad b \quad c] = [0 \quad 0 \quad 0]$, show that $\mathbf{a} = [0 \quad 0 \quad 0]$.

Solution: $5[a \quad b \quad c] = [0 \quad 0 \quad 0]$ implies that $5a = 0$, $5b = 0$, and $5c = 0$, which implies that $a = 0$, $b = 0$, and $c = 0$. Therefore,

$$\mathbf{a} = [0 \quad 0 \quad 0]$$

6-4. Linear Independence

A. Definition of linear combination

A vector $\mathbf{a}$ is called a linear combination of the vectors $\mathbf{a}_1, \mathbf{a}_2, \ldots, \mathbf{a}_n$ from vector space V if there exists real numbers $k_1, k_2, \ldots, k_n$ such that

$$\mathbf{a} = k_1\mathbf{a}_1 + k_1\mathbf{a}_2 + \cdots + k_n\mathbf{a}_n$$

EXAMPLE 6-15: Determine which of the following vectors in R^3 are linear combinations of $\mathbf{a}_1 = \begin{bmatrix} 1 \\ -2 \\ 3 \end{bmatrix}$ and $\mathbf{a}_2 = \begin{bmatrix} 2 \\ 3 \\ 1 \end{bmatrix}$:

(a) $\begin{bmatrix} 3 \\ 1 \\ 4 \end{bmatrix}$ **(b)** $\begin{bmatrix} 8 \\ 5 \\ 9 \end{bmatrix}$ **(c)** $\begin{bmatrix} 4 \\ 5 \\ 6 \end{bmatrix}$

Solution

(a) You must find values x_1 and x_2 such that

$$x_1\begin{bmatrix} 1 \\ -2 \\ 3 \end{bmatrix} + x_2\begin{bmatrix} 2 \\ 3 \\ 1 \end{bmatrix} = \begin{bmatrix} 3 \\ 1 \\ 4 \end{bmatrix}$$

which gives you

$$\begin{aligned} x_1 + 2x_2 &= 3 \\ -2x_1 + 3x_2 &= 1 \\ 3x_1 + x_2 &= 4. \end{aligned}$$

You need to know if this system of equations has a solution. Why not make this determination by reducing the augmented matrix to echelon form?

$$\begin{bmatrix} 1 & 2 & 3 \\ -2 & 3 & 1 \\ 3 & 1 & 4 \end{bmatrix} \xrightarrow[R_3 \to R_3 - 3R_1]{R_2 \to R_2 + 2R_1} \begin{bmatrix} 1 & 2 & 3 \\ 0 & 7 & 7 \\ 0 & -5 & -5 \end{bmatrix} \xrightarrow[R_3 \to -1/5 R_3]{R_2 \to 1/7 R_2} \begin{bmatrix} 1 & 2 & 3 \\ 0 & 1 & 1 \\ 0 & 1 & 1 \end{bmatrix}$$

$$\xrightarrow[R_3 \to R_3 - R_2]{} \begin{bmatrix} 1 & 2 & 3 \\ 0 & 1 & 1 \\ 0 & 0 & 0 \end{bmatrix} \xrightarrow{R_1 \to R_1 - 2R_2} \begin{bmatrix} 1 & 0 & 1 \\ 0 & 1 & 1 \\ 0 & 0 & 0 \end{bmatrix}$$

$$\therefore \quad x_1 = 1 \quad \text{and} \quad x_2 = 1$$

Therefore, $\begin{bmatrix} 3 \\ 1 \\ 4 \end{bmatrix}$ is a linear combination of $\mathbf{a}_1$ and $\mathbf{a}_2$.

(b) As in the **(a)** part, you must find x_1 and x_2 such that

$$x_1 \begin{bmatrix} 1 \\ -2 \\ 3 \end{bmatrix} + x_2 \begin{bmatrix} 2 \\ 3 \\ 1 \end{bmatrix} = \begin{bmatrix} 8 \\ 5 \\ 9 \end{bmatrix}$$

which gives

$$\begin{aligned} x_1 + 2x_2 &= 8 \\ -2x_1 + 3x_2 &= 5 \\ 3x_1 + x_2 &= 9 \end{aligned}$$

Let us reduce the augmented matrix:

$$\begin{bmatrix} 1 & 2 & 8 \\ -2 & 3 & 5 \\ 3 & 1 & 9 \end{bmatrix} \xrightarrow[R_3 \to R_3 - 3R_1]{R_2 \to R_2 + 2R_1} \begin{bmatrix} 1 & 2 & 8 \\ 0 & 7 & 21 \\ 0 & -5 & -15 \end{bmatrix} \xrightarrow[R_3 \to -1/5 R_3]{R_2 \to 1/7 R_2} \begin{bmatrix} 1 & 2 & 8 \\ 0 & 1 & 3 \\ 0 & 1 & 3 \end{bmatrix}$$

$$\xrightarrow[R_3 \to R_3 - R_2]{} \begin{bmatrix} 1 & 2 & 8 \\ 0 & 1 & 3 \\ 0 & 0 & 0 \end{bmatrix} \xrightarrow{R_1 \to R_1 - 2R_2} \begin{bmatrix} 1 & 0 & 2 \\ 0 & 1 & 3 \\ 0 & 0 & 0 \end{bmatrix}$$

$$\therefore \quad x_1 = 2 \quad \text{and} \quad x_2 = 3$$

Therefore, $\begin{bmatrix} 8 \\ 5 \\ 9 \end{bmatrix}$ is a linear combination of $\mathbf{a}_1$ and $\mathbf{a}_2$.

(c) Begin this problem with the linear system

$$x_1 \begin{bmatrix} 1 \\ -2 \\ 3 \end{bmatrix} + x_2 \begin{bmatrix} 2 \\ 3 \\ 1 \end{bmatrix} = \begin{bmatrix} 4 \\ 5 \\ 6 \end{bmatrix}$$

which becomes

$$\begin{aligned} x_1 + 2x_2 &= 4 \\ -2x_1 + 3x_2 &= 5 \\ 3x_1 + x_2 &= 6 \end{aligned}$$

By augmented matrix reduction,

$$\begin{bmatrix} 1 & 2 & 4 \\ -2 & 3 & 5 \\ 3 & 1 & 6 \end{bmatrix} \xrightarrow[R_3 \to R_3 - 3R_1]{R_2 \to R_2 + 2R_1} \begin{bmatrix} 1 & 2 & 4 \\ 0 & 7 & 13 \\ 0 & -5 & -6 \end{bmatrix} \xrightarrow[R_3 \to -1/5R_3]{R_2 \to 1/7R_2} \begin{bmatrix} 1 & 2 & 4 \\ 0 & 1 & \frac{13}{7} \\ 0 & 1 & \frac{6}{5} \end{bmatrix}$$

You might as well stop here, since the second row implies that $x_2 = 13/7$ and the third row implies that $x_2 = 6/5$. This is an obvious contradiction, hence, the system $x_1 \begin{bmatrix} 1 \\ -2 \\ 3 \end{bmatrix} + x_2 \begin{bmatrix} 2 \\ 3 \\ 1 \end{bmatrix} = \begin{bmatrix} 4 \\ 5 \\ 6 \end{bmatrix}$ has no solution. Therefore, $\begin{bmatrix} 4 \\ 5 \\ 6 \end{bmatrix}$ is not a linear combination of $\mathbf{a}_1$ and $\mathbf{a}_2$.

EXAMPLE 6-16: Express $6x^2 + 11x + 20$ as a linear combination of $x^2 + x + 1$, $2x^2 + x - 2$ and $3x^2 + 4x + 5$.

Solution: You need to find real numbers a_1, a_2, and a_3 such that

$$a_1(x^2 + x + 1) + a_2(2x^2 + x - 2) + a_3(3x^2 + 4x + 5) = 6x^2 + 11x + 20$$

$$(a_1 + 2a_2 + 3a_3)x^2 + (a_1 + a_2 + 4a_3)x + (a_1 - 2a_2 + 5a_3) = 6x^2 + 11x + 20$$

Setting the coefficients of like powers of x equal, you get

$$a_1 + 2a_2 + 3a_3 = 6$$

$$a_1 + a_2 + 4a_3 = 11$$

$$a_1 - 2a_2 + 5a_3 = 20.$$

Solve this system by any suitable method. Let us choose to use elementary row operations on the augmented matrix:

$$\begin{bmatrix} 1 & 2 & 3 & 6 \\ 1 & 1 & 4 & 11 \\ 1 & -2 & 5 & 20 \end{bmatrix} \xrightarrow[R_3 \to R_3 - R_1]{R_2 \to R_2 - R_1} \begin{bmatrix} 1 & 2 & 3 & 6 \\ 0 & -1 & 1 & 5 \\ 0 & -4 & 2 & 14 \end{bmatrix}$$

$$\xrightarrow[R_3 \to R_3 - 4R_2]{} \begin{bmatrix} 1 & 2 & 3 & 6 \\ 0 & -1 & 1 & 5 \\ 0 & 0 & -2 & -6 \end{bmatrix} \xrightarrow[R_3 \to -1/2R_3]{R_2 \to -R_2} \begin{bmatrix} 1 & 2 & 3 & 6 \\ 0 & 1 & -1 & -5 \\ 0 & 0 & 1 & 3 \end{bmatrix}$$

$$\xrightarrow[R_2 \to R_2 + R_3]{R_1 \to R_1 - 3R_3} \begin{bmatrix} 1 & 2 & 0 & -3 \\ 0 & 1 & 0 & -2 \\ 0 & 0 & 1 & 3 \end{bmatrix} \xrightarrow{R_1 \to R_1 - 2R_2} \begin{bmatrix} 1 & 0 & 0 & 1 \\ 0 & 1 & 0 & -2 \\ 0 & 0 & 1 & 3 \end{bmatrix}$$

$$\therefore \quad a_1 = 1, \quad a_1 = -2, \quad \text{and} \quad a_3 = 3$$

$$\therefore \quad 1(x^2 + x + 1) - 2(2x^2 + x - 2) + 3(3x^2 + 4x + 5) = 6x^2 + 11x + 20$$

B. Definition of linear independence and linear dependence

The vectors $\mathbf{a}_1, \mathbf{a}_2, \ldots, \mathbf{a}_n$ from a vector space V are **linearly independent** if the only constants k_1, $k_2, \ldots, k_n$ for which $k_1\mathbf{a}_1 + k_2\mathbf{a}_2 + \cdots + k_n\mathbf{a}_n = \mathbf{0}$ are $k_1 = k_2 = \cdots = k_n = 0$.

A set of vectors $\mathbf{a}_1, \mathbf{a}_2, \ldots, \mathbf{a}_n$ for the vector space V are said to be **linearly dependent** is they are not linearly independent. At times, it may be better to view such a set as linearly dependent if there exist constants $k_1, k_2, \ldots, k_n$, not all equal to zero, such that $k_1\mathbf{a}_1 + k_2\mathbf{a}_2 + \cdots + k_n\mathbf{a}_n = \mathbf{0}$.

EXAMPLE 6-17: Determine whether the vectors $[2, 3]$ and $[5, -6]$ from R_2 are linearly independent or dependent.

Solution: For R_2, $\mathbf{0} = [0,0]$; hence

$$k_1[2,3] + k_2[5,-6] = [0,0]$$

The resulting system from this equation is

$$2k_1 + 5k_2 = 0$$
$$3k_1 - 6k_2 = 0$$

Multiplying the first equation by 6 and the second equation by 5, you get

$$12k_1 + 30k_2 = 0$$
$$15k_1 - 30k_2 = 0$$
$$27k_1 \qquad = 0$$
$$\therefore \quad k_1 = 0 \qquad \text{which implies } k_2 = 0$$

Therefore, since the only solution is $k_1 = k_2 = 0$, the vectors are independent.

EXAMPLE 6-18: Determine whether the vectors $\mathbf{p}_1 = 2x^2 - 3x + 4$, $\mathbf{p}_2 = x^2 + 5x + 6$, and $\mathbf{p}_3 = x^2 - 21x - 10$ from P_2 are linearly independent or dependent.

Solution: For $\mathbf{P}_2$, $\mathbf{0} = 0$; hence

$$k_1(2x^2 - 3x + 4) + k_2(x^2 + 5x + 6) + k_3(x^2 - 21x - 10) = 0$$

Setting coefficients of equal powers of x equal, you get

$$2k_1 + k_2 + k_3 = 0$$
$$-3k_1 + 5k_2 - 21k_3 = 0$$
$$4k_1 + 6k_2 - 10k_3 = 0$$

A good way to solve this system is to reduce the augmented matrix to row echelon form.

$$\begin{bmatrix} 2 & 1 & 1 & 0 \\ -3 & 5 & -21 & 0 \\ 4 & 6 & -10 & 0 \end{bmatrix} \xrightarrow[R_3 \to R_3 - 2R_1]{R_1 \to R_1 + R_2} \begin{bmatrix} -1 & 6 & -20 & 0 \\ -3 & 5 & -21 & 0 \\ 0 & 4 & -12 & 0 \end{bmatrix} \xrightarrow{R_2 \to R_2 - 3R_1} \begin{bmatrix} -1 & 6 & -20 & 0 \\ 0 & -13 & 39 & 0 \\ 0 & 4 & -12 & 0 \end{bmatrix}$$

$$\xrightarrow[R_3 \to 1/4 R_3]{\substack{R_1 \to -R_1 \\ R_2 \to -1/13 R_2}} \begin{bmatrix} 1 & -6 & 20 & 0 \\ 0 & 1 & -3 & 0 \\ 0 & 1 & -3 & 0 \end{bmatrix} \xrightarrow[R_3 \to R_3 - R_2]{R_1 \to R_1 + 6R_2} \begin{bmatrix} 1 & 0 & 2 & 0 \\ 0 & 1 & -3 & 0 \\ 0 & 0 & 0 & 0 \end{bmatrix}$$

$$\therefore \quad k_1 + 2k_3 = 0$$
$$k_2 - 3k_3 = 0$$

Hence,

$$k_1 = -2k_3$$
$$k_2 = 3k_3$$

which means k_3 is arbitrary. Therefore, $k_3 = 1$, $k_1 = -2$, and $k_2 = 3$ is a solution of the system which implies that the three vectors are dependent. In other words, k_1 and k_2 depend on k_3.

In this example, you did not actually have to find the solution to the homogeneous system because you only wanted to know whether a solution exists. If the determinant of the coefficient matrix is zero, the system has a nontrivial solution:

$$\begin{vmatrix} 2 & 1 & 1 \\ -3 & 5 & -21 \\ 4 & 6 & -10 \end{vmatrix} \begin{matrix} 2 & 1 \\ -3 & 5 \\ 4 & 6 \end{matrix} = -100 - 84 - 18 - 20 + 252 - 30 = 0$$

Therefore, a solution with not all k_1, k_2, k_3 equal to zero exists, and hence the three vectors are dependent.

6-5. Some Properties of Linear Independence and Dependence

Property A If a set of vectors $\mathbf{a}_1, \mathbf{a}_2, \ldots, \mathbf{a}_k$ from a vector space V are linearly dependent, then the set obtained by adding one or more vectors from V to $\mathbf{a}_1, \mathbf{a}_2, \ldots, \mathbf{a}_k$ is also linearly dependent.

EXAMPLE 6-19: In Example 6-18, you determined that $2x^2 - 3x + 4$, $x^2 + 5x + 6$, and $x^2 - 21x - 10$ from P_2 are linearly dependent. What can be said about the linear independence of the vectors $2x^2 - 3x + 4$, $x^2 + 5x + 6$, $x^2 - 21x - 10$, and $5x^2 - 7x + 2$?

Solution: Since the first three of these vectors are linearly dependent and you are adding one vector to this linearly dependent set, the result is a set of linearly dependent vectors.

Property B If a set of vectors $\mathbf{a}_1, \mathbf{a}_2, \ldots, \mathbf{a}_k$ from a vector space V are linearly independent, then the set obtained by deleting one or more vectors from the set $\mathbf{a}_1, \mathbf{a}_2, \ldots, \mathbf{a}_k$ is also linearly independent.

EXAMPLE 6-20: Show that the vectors $[1 \quad 0 \quad 0 \quad 0]$, $[0 \quad 1 \quad 0 \quad 0]$, $[0 \quad 0 \quad 1 \quad 0]$, and $[0 \quad 0 \quad 0 \quad 1]$ from R_4 are linearly independent. What can be said about the linear independence of the vectors $[1 \quad 0 \quad 0 \quad 0]$ and $[0 \quad 0 \quad 1 \quad 0]$?

Solution: You must show that

$$a_1[1 \quad 0 \quad 0 \quad 0] + a_2[0 \quad 1 \quad 0 \quad 0] + a_3[0 \quad 0 \quad 1 \quad 0] + a_4[0 \quad 0 \quad 0 \quad 1] = [0 \quad 0 \quad 0 \quad 0]$$

implies $a_1 = a_2 = a_3 = a_4 = 0$. The sum on the left-hand side of the equation is $[a_1 \quad a_2 \quad a_3 \quad a_4]$, but $[a_1 \quad a_2 \quad a_3 \quad a_4] = [0 \quad 0 \quad 0 \quad 0]$.

Therfore, $a_1 = a_2 = a_3 = a_4 = 0$ and hence, the four vectors are linearly independent. Since the vectors $[1 \quad 0 \quad 0 \quad 0]$ and $[0 \quad 0 \quad 1 \quad 0]$ are obtained from the four given vectors by deleting $[0 \quad 1 \quad 0 \quad 0]$ and $[0 \quad 0 \quad 0 \quad 1]$, then $[1 \quad 0 \quad 0 \quad 0]$ and $[0 \quad 0 \quad 1 \quad 0]$ are also linearly independent.

Property C Linear dependence of a set of two or more nonzero vectors from a vector space V means that at least one of the vectors in the set can be written as a linear combination of the other vectors in the set.

EXAMPLE 6-21: In Example 6-18, we determined that $2x^2 - 3x + 4$, $x^2 + 5x + 6$, and $x^2 - 21x - 10$ from P_2 were linearly dependent.

(a) Show that $2x^2 - 3x + 4$ can be written as a linear combination of the other two vectors.
(b) Show that $x^2 - 21x - 10$ can be written as a linear combination of the remaining two vectors.

Solution

(a) You need to find real numbers a_1 and a_2 such that $a_1(x^2 + 5x + 6) + a_2(x^2 - 21x - 10) = 2x^2 - 3x + 4$. Setting coefficients of like powers of x equal, you get

$$a_1 + a_2 = 2$$

$$5a_1 - 21a_2 = -3$$

$$6a_1 - 10a_2 = 4$$

By solving the first and second equations, you get $a_1 = 3/2$ and $a_2 = 1/2$, which also satisfy the third equation. Therefore,

$$\tfrac{3}{2}(x^2 + 5x + 6) + \tfrac{1}{2}(x^2 - 21x - 10) = 2x^2 - 3x + 4$$

(b) You need to find real numbers a_1 and a_2 such that $a_1(2x^2 - 3x + 4) + a_2(x^2 + 5x + 6) = x^2 - 21x - 10$. Setting coefficients of like powers of x equal, you get

$$\begin{aligned} 2a_1 + a_2 &= 1 \\ -3a_1 + 5a_2 &= -21 \\ 4a_1 + 6a_2 &= -10 \end{aligned}$$

By solving the first and second equations, you get $a_1 = 2$ and $a_2 = -3$, which also satisfy the third equation. Therefore,

$$2(2x^2 - 3x + 4) - 3(x^2 + 5x + 6) = x^2 - 21x - 10$$

SUMMARY

1. A vector space V is a set of elements called vectors on which two operations $\oplus$ and $\odot$ have been defined with the properties:
   - If $\mathbf{a}, \mathbf{b}$ are in V, then $\mathbf{a} \oplus \mathbf{b}$ is in V.
   - $\mathbf{a} \oplus \mathbf{b} = \mathbf{b} \oplus \mathbf{a}$ for all $\mathbf{a}, \mathbf{b}$ in V.
   - $\mathbf{a} \oplus (\mathbf{b} \oplus \mathbf{c}) = (\mathbf{a} \oplus \mathbf{b}) \oplus \mathbf{c}$ for all $\mathbf{a}, \mathbf{b}, \mathbf{c}$ in V.
   - There exists a zero vector $\mathbf{0}$ in V such that $\mathbf{a} \oplus \mathbf{0} = \mathbf{a}$ for all $\mathbf{a}$ in V.
   - For each $\mathbf{a}$ in V there exists a unique negative $-\mathbf{a}$ such that $\mathbf{a} \oplus (-\mathbf{a}) = \mathbf{0}$.
   - If $\mathbf{a}$ is in V and k is in R, then $k \odot \mathbf{a}$ is in V.
   - $k \odot (\mathbf{a} \oplus \mathbf{b}) = k \odot \mathbf{a} \oplus k \odot \mathbf{b}$ for $\mathbf{a}, \mathbf{b}$ in V and k in R.
   - $(k + l) \odot \mathbf{a} = k \odot \mathbf{a} \oplus l \odot \mathbf{a}$ for all $\mathbf{a}$ in V and k, l in R.
   - $k \odot (l \odot \mathbf{a}) = (kl) \odot \mathbf{a}$ for all $\mathbf{a}$ in V and k, l in R.
   - $1 \odot \mathbf{a} = \mathbf{a}$ for all $\mathbf{a}$ in V.

2. The following sets with the indicated operations $\oplus$ and $\odot$ are vector spaces.

Set	$\oplus$	$\odot$	Notation
All real numbers	ordinary addition	ordinary multiplication	R
Rows of all ordered n-tuples of real numbers	matrix addition	scalar multiplication	R_n
Columns of all ordered n-tuples of real numbers	matrix addition	scalar multiplication	R^n
All $m \times n$ matrices	matrix addition	scalar multiplication	${}_mM_n$
All polynomials of degree $\leq n$ and the zero polynomial	ordinary addition of polynomials	usual multiplication of a polynomial by a scalar	P_n
All real-valued functions	function addition	multiplication of a function by a scalar	$R(-\infty, \infty)$
All real-valued continuous functions	function addition	multiplication of a function by a scalar	$C(-\infty, \infty)$

3. A subspace of a vector space V is a nonempty subset W of V which is itself a vector space under the same defined operations of vector addition and scalar multiplication as in V.
4. A nonempty subset W of a vector space V is a subspace of V if and only if it is closed under vector addition and scalar multiplication.
5. If V is a vector space with operations $\oplus$ and $\odot$ and $\mathbf{0}$ is the zero vector of V, then

- $\mathbf{0} \odot \mathbf{a} = \mathbf{0}$ for all $\mathbf{a}$ in V,
- $k \odot \mathbf{0} = \mathbf{0}$ for all k in R,
- $(-1) \odot \mathbf{a} = -\mathbf{a}$ for any $\mathbf{a}$ in V,
- if $k \odot \mathbf{a} = \mathbf{0}$, then $k = 0$ or $\mathbf{a} = \mathbf{0}$.

6. A vector $\mathbf{a}$ is a linear combination of the vectors $\mathbf{a}_1, \mathbf{a}_2, \ldots, \mathbf{a}_n$ from a vector space V if there are real numbers $k_1, k_2, \ldots, k_n$ such that $\mathbf{a} = k_1\mathbf{a}_1 + k_2\mathbf{a}_2 + \cdots + k_n\mathbf{a}_n$.
7. The vector $\mathbf{a}_1, \mathbf{a}_2, \ldots, \mathbf{a}_n$ from a vector space V are linearly independent if the only constants $k_1, k_2, \ldots, k_n$ for which $k_1\mathbf{a}_1 + k_2\mathbf{a}_2 + \cdots + k_n\mathbf{a}_n = \mathbf{0}$ are $k_1 = k_2 = \cdots = k_n = 0$.
8. A set of vectors from a vector space V are linearly dependent if they are not linearly independent.
9. If a set of vectors $\mathbf{a}_1, \mathbf{a}_2, \ldots, \mathbf{a}_k$ from a vector space V are linearly dependent, then the set obtained by adding one or more vectors from V to $\mathbf{a}_1, \mathbf{a}_2, \ldots, \mathbf{a}_k$ is also linearly dependent.
10. If a set of vectors $\mathbf{a}_1, \mathbf{a}_2, \ldots, \mathbf{a}_k$ from a vector space V are linearly independent, then the set obtained by deleting one or more vectors from the set is also linearly independent.
11. Linear dependence of a set of two or more nonzero vectors from a vector space V means that at least one of the vectors in the set can be written as a linear combination of the other vectors in the set.

RAISE YOUR GRADES

Can you ...?

☑ list all the properties that a set with two defined operations needs to be a vector space
☑ determine whether a set with specified operations is a vector space
☑ determine whether a subset of a known vector space is a subspace
☑ list four properties that are true for either any scalar k or any vector $\mathbf{a}$ of a vector space
☑ use the definition of a linear combination to determine whether a given vector is a linear combination of a set of vectors
☑ determine whether a set of vectors is linearly independent or dependent

SOLVED PROBLEMS

PROBLEM 6-1 Show that the set of all 4×1 matrices with $\oplus$ and $\odot$ defined as matrix addition and scalar multiplication, respectively, is a vector space. This set is denoted by R^4.

Solution

(1)
$$\begin{bmatrix} a_1 \\ b_1 \\ c_1 \\ d_1 \end{bmatrix} + \begin{bmatrix} a_2 \\ b_2 \\ c_2 \\ d_2 \end{bmatrix} = \begin{bmatrix} a_1 + a_2 \\ b_1 + b_2 \\ c_1 + c_2 \\ d_1 + d_2 \end{bmatrix} \quad \text{is in } R^4$$

Therefore, it is closed under matrix addition.

(2)
$$\begin{bmatrix} a_2 \\ b_2 \\ c_2 \\ d_2 \end{bmatrix} + \begin{bmatrix} a_1 \\ b_1 \\ c_1 \\ d_1 \end{bmatrix} = \begin{bmatrix} a_2 + a_1 \\ b_2 + b_1 \\ c_2 + c_1 \\ d_2 + d_1 \end{bmatrix} = \begin{bmatrix} a_1 + a_2 \\ b_1 + b_2 \\ c_1 + c_2 \\ d_1 + d_2 \end{bmatrix} = \begin{bmatrix} a_1 \\ b_1 \\ c_1 \\ d_1 \end{bmatrix} + \begin{bmatrix} a_2 \\ b_2 \\ c_2 \\ d_2 \end{bmatrix}$$

Therefore, it satisfies the commutative property for matrix addition.

(3)
$$\begin{bmatrix} a_1 \\ b_1 \\ c_1 \\ d_1 \end{bmatrix} + \left(\begin{bmatrix} a_2 \\ b_2 \\ c_2 \\ d_2 \end{bmatrix} + \begin{bmatrix} a_3 \\ b_3 \\ c_3 \\ d_3 \end{bmatrix} \right) = \begin{bmatrix} a_1 \\ b_1 \\ c_1 \\ d_1 \end{bmatrix} + \begin{bmatrix} a_2 + a_3 \\ b_2 + b_3 \\ c_2 + c_3 \\ d_2 + d_3 \end{bmatrix} = \begin{bmatrix} a_1 + a_2 + a_3 \\ b_1 + b_2 + b_3 \\ c_1 + c_2 + c_3 \\ d_1 + d_2 + d_3 \end{bmatrix}$$

$$= \begin{bmatrix} a_1 + a_2 \\ b_1 + b_2 \\ c_1 + c_2 \\ d_1 + d_2 \end{bmatrix} + \begin{bmatrix} a_3 \\ b_3 \\ c_3 \\ d_3 \end{bmatrix} = \left(\begin{bmatrix} a_1 \\ b_1 \\ c_1 \\ d_1 \end{bmatrix} + \begin{bmatrix} a_2 \\ b_2 \\ c_2 \\ d_2 \end{bmatrix} \right) + \begin{bmatrix} a_3 \\ b_3 \\ c_3 \\ d_3 \end{bmatrix}$$

Therefore, it satisfies the associative property for matrix addition.

(4)
$$\begin{bmatrix} 0 \\ 0 \\ 0 \\ 0 \end{bmatrix} + \begin{bmatrix} a_1 \\ b_1 \\ c_1 \\ d_1 \end{bmatrix} = \begin{bmatrix} a_1 \\ b_1 \\ c_1 \\ d_1 \end{bmatrix} + \begin{bmatrix} 0 \\ 0 \\ 0 \\ 0 \end{bmatrix} = \begin{bmatrix} a_1 + 0 \\ b_1 + 0 \\ c_1 + 0 \\ \mathrm{d}_1 + 0 \end{bmatrix} = \begin{bmatrix} a_1 \\ b_1 \\ c_1 \\ d_1 \end{bmatrix}$$

Therefore, there exists a zero matrix for R^4.

(5)
$$\begin{bmatrix} a_1 \\ b_1 \\ c_1 \\ d_1 \end{bmatrix} + \begin{bmatrix} -a_1 \\ -b_1 \\ -c_1 \\ -d_1 \end{bmatrix} = \begin{bmatrix} -a_1 \\ -b_1 \\ -c_1 \\ -d_1 \end{bmatrix} + \begin{bmatrix} a_1 \\ b_1 \\ c_1 \\ d_1 \end{bmatrix} = \begin{bmatrix} -a_1 + a_1 \\ -b_1 + b_1 \\ -c_1 + c_1 \\ -d_1 + d_1 \end{bmatrix} = \begin{bmatrix} 0 \\ 0 \\ 0 \\ 0 \end{bmatrix}$$

Therefore, each element in R^4 has a negative.

(6)
$$k \begin{bmatrix} a_1 \\ b_1 \\ c_1 \\ d_1 \end{bmatrix} = \begin{bmatrix} ka_1 \\ kb_1 \\ kc_1 \\ kd_1 \end{bmatrix} \quad \text{which is in } R^4$$

Therefore, multiplication by a scalar is satisfied.

(7)
$$k \left(\begin{bmatrix} a_1 \\ b_1 \\ c_1 \\ d_1 \end{bmatrix} + \begin{bmatrix} a_2 \\ b_2 \\ c_2 \\ d_2 \end{bmatrix} \right) = k \begin{bmatrix} a_1 + a_2 \\ b_1 \mid b_2 \\ c_1 + c_2 \\ d_1 + d_2 \end{bmatrix} = \begin{bmatrix} ka_1 + ka_2 \\ kb_1 + kb_2 \\ kc_1 + kc_2 \\ kd_1 + kd_2 \end{bmatrix}$$

$$= \begin{bmatrix} ka_1 \\ kb_1 \\ kc_1 \\ kd_1 \end{bmatrix} + \begin{bmatrix} ka_2 \\ kb_2 \\ kc_2 \\ kd_2 \end{bmatrix} = k \begin{bmatrix} a_1 \\ b_1 \\ c_1 \\ d_1 \end{bmatrix} + k \begin{bmatrix} a_2 \\ b_2 \\ c_2 \\ d_2 \end{bmatrix}$$

Therefore, the distributive law of multiplication by a scalar over matrix addition is satisfied.

(8)

$$(k+\ell)\begin{bmatrix} a_1 \\ b_1 \\ c_1 \\ d_1 \end{bmatrix} = \begin{bmatrix} (k+\ell)a_1 \\ (k+\ell)b_1 \\ (k+\ell)c_1 \\ (k+\ell)d_1 \end{bmatrix} = \begin{bmatrix} ka_1 + \ell a_1 \\ kb_1 + \ell b_1 \\ kc_1 + \ell c_1 \\ kd_1 + \ell d_1 \end{bmatrix} = \begin{bmatrix} ka_1 \\ kb_1 \\ kc_1 \\ kd_1 \end{bmatrix} + \begin{bmatrix} \ell a_1 \\ \ell b_1 \\ \ell c_1 \\ \ell d_1 \end{bmatrix}$$

$$= k\begin{bmatrix} a_1 \\ b_1 \\ c_1 \\ d_1 \end{bmatrix} + \ell\begin{bmatrix} a_1 \\ b_1 \\ c_1 \\ d_1 \end{bmatrix}$$

Therefore, the distributive law of scalar addition multiplied by a matrix is satisfied.

(9)

$$k\left(\ell\begin{bmatrix} a_1 \\ b_1 \\ c_1 \\ d_1 \end{bmatrix}\right) = k\begin{bmatrix} \ell a_1 \\ \ell b_1 \\ \ell c_1 \\ \ell d_1 \end{bmatrix} = \begin{bmatrix} k\ell a_1 \\ k\ell b_1 \\ k\ell c_1 \\ k\ell d_1 \end{bmatrix} = k\ell\begin{bmatrix} a_1 \\ b_1 \\ c_1 \\ d_1 \end{bmatrix}$$

(10)

$$1\begin{bmatrix} a_1 \\ b_1 \\ c_1 \\ d_1 \end{bmatrix} = \begin{bmatrix} a_1 \\ b_1 \\ c_1 \\ d_1 \end{bmatrix} \quad \text{for all vectors in } R^4$$

Therefore, R^4 is a vector space under matrix addition and scalar multiplication.

PROBLEM 6-2 Show that the set of all three component column vectors of the form $\begin{bmatrix} a \\ b \\ a+b \end{bmatrix}$ with $\oplus$ and $\odot$ defined as usual vector addition and scalar multiplication is a vector space.

Solution The set is a subset of vector space R^3.

(1)

$$\begin{bmatrix} a_1 \\ b_1 \\ a_1 + b_1 \end{bmatrix} + \begin{bmatrix} a_2 \\ b_2 \\ a_2 + b_2 \end{bmatrix} = \begin{bmatrix} a_1 + a_2 \\ b_1 + b_2 \\ (a_1 + b_1) + (a_2 + b_2) \end{bmatrix} \quad \text{is in } R^3$$

Therefore, the closure property holds.

(2)

$$\begin{bmatrix} a_2 \\ b_2 \\ a_2 + b_2 \end{bmatrix} + \begin{bmatrix} a_1 \\ b_1 \\ a_1 + b_1 \end{bmatrix} = \begin{bmatrix} a_2 + a_1 \\ b_2 + b_1 \\ a_2 + a_1 + b_2 + b_1 \end{bmatrix} = \begin{bmatrix} a_1 + a_2 \\ b_1 + b_2 \\ a_1 + b_1 + a_2 + b_2 \end{bmatrix}$$

$$= \begin{bmatrix} a_1 \\ b_1 \\ a_1 + b_1 \end{bmatrix} + \begin{bmatrix} a_2 \\ b_2 \\ a_2 + b_2 \end{bmatrix}$$

Therefore, the commutative property holds.

(3)
$$\begin{bmatrix} a_1 \\ b_1 \\ a_1 + b_1 \end{bmatrix} + \left(\begin{bmatrix} a_2 \\ b_2 \\ a_2 + b_2 \end{bmatrix} + \begin{bmatrix} a_3 \\ b_3 \\ a_3 + b_3 \end{bmatrix} \right) = \begin{bmatrix} a_1 \\ b_1 \\ a_1 + b_1 \end{bmatrix} + \begin{bmatrix} a_2 + a_3 \\ b_2 + b_3 \\ a_2 + b_2 + a_3 + b_3 \end{bmatrix}$$
$$= \begin{bmatrix} a_1 + a_2 + a_3 \\ b_1 + b_2 + b_3 \\ a_1 + b_1 + a_2 + b_2 + a_3 + b_3 \end{bmatrix} = \begin{bmatrix} a_1 + a_2 \\ b_1 + b_2 \\ a_1 + b_1 + a_2 + b_2 \end{bmatrix} + \begin{bmatrix} a_3 \\ b_3 \\ a_3 + b_3 \end{bmatrix}$$
$$= \left(\begin{bmatrix} a_1 \\ b_1 \\ a_1 + b_1 \end{bmatrix} + \begin{bmatrix} a_2 \\ b_2 \\ a_2 + b_2 \end{bmatrix} \right) + \begin{bmatrix} a_3 \\ b_3 \\ a_3 + b_3 \end{bmatrix}$$

Therefore, the associative property for vector addition holds.

(4)
$$\begin{bmatrix} 0 \\ 0 \\ 0 \end{bmatrix} + \begin{bmatrix} a \\ b \\ a + b \end{bmatrix} = \begin{bmatrix} a \\ b \\ a + b \end{bmatrix} + \begin{bmatrix} 0 \\ 0 \\ 0 \end{bmatrix} = \begin{bmatrix} a + 0 \\ b + 0 \\ a + b + 0 \end{bmatrix} = \begin{bmatrix} a \\ b \\ a + b \end{bmatrix}$$

Therefore, there exists a zero vector for the set.

(5)
$$\begin{bmatrix} a \\ b \\ a + b \end{bmatrix} + \begin{bmatrix} -a \\ -b \\ -(a + b) \end{bmatrix} = \begin{bmatrix} -a \\ -b \\ -(a + b) \end{bmatrix} + \begin{bmatrix} a \\ b \\ a + b \end{bmatrix} = \begin{bmatrix} -a + a \\ -b + b \\ -(a + b) + (a + b) \end{bmatrix} = \begin{bmatrix} 0 \\ 0 \\ 0 \end{bmatrix}$$

Therefore, each vector has a negative

(6)
$$k \begin{bmatrix} a \\ b \\ a + b \end{bmatrix} = \begin{bmatrix} ka \\ kb \\ k(a + b) \end{bmatrix} = \begin{bmatrix} ka \\ kb \\ ka + kb \end{bmatrix}$$ which is in the set

Therefore, multiplication by a scalar holds.

(7)
$$k \left(\begin{bmatrix} a_1 \\ b_1 \\ a_1 + b_1 \end{bmatrix} + \begin{bmatrix} a_2 \\ b_2 \\ a_2 + b_2 \end{bmatrix} \right) = \begin{bmatrix} ka_1 \\ kb_1 \\ ka_1 + kb_1 \end{bmatrix} + \begin{bmatrix} ka_2 \\ kb_2 \\ ka_2 + kb_2 \end{bmatrix} = k \begin{bmatrix} a_1 \\ b_1 \\ a_1 + b_1 \end{bmatrix} + k \begin{bmatrix} a_2 \\ b_2 \\ a_2 + b_2 \end{bmatrix}$$

Therefore, the distributive law of multiplication by a scalar holds.

(8)
$$(k + \ell) \begin{bmatrix} a \\ b \\ a + b \end{bmatrix} = \begin{bmatrix} (k + \ell)a \\ (k + \ell)b \\ (k + \ell)(a + b) \end{bmatrix} = \begin{bmatrix} ka + \ell a \\ kb + \ell b \\ ka + kb + \ell a + \ell b \end{bmatrix}$$
$$= \begin{bmatrix} ka \\ kb \\ k(a + b) \end{bmatrix} + \begin{bmatrix} \ell a \\ \ell b \\ \ell(a + b) \end{bmatrix} = k \begin{bmatrix} a \\ b \\ a + b \end{bmatrix} + \ell \begin{bmatrix} a \\ b \\ a + b \end{bmatrix}$$

Therefore, the distributive law of scalar addition multiplied by a vector hold.

(9)
$$k \left(\ell \begin{bmatrix} a \\ b \\ a + b \end{bmatrix} \right) = k \begin{bmatrix} \ell a \\ \ell b \\ \ell a + \ell b \end{bmatrix} = \begin{bmatrix} k\ell a \\ k\ell b \\ k\ell a + k\ell b \end{bmatrix} = k\ell \begin{bmatrix} a \\ b \\ a + b \end{bmatrix}$$ is in the set

(10)
$$1\begin{bmatrix} a \\ b \\ a+b \end{bmatrix} = \begin{bmatrix} a \\ b \\ a+b \end{bmatrix} \quad \text{for all vectors in the set}$$

Therefore, the given set is a vector space. You will notice that this is a special case of R^3.

PROBLEM 6-3 Verify that the set of all polynomials of the form $ax^2 + bx + c$ with usual addition of polynomials and multiplication by a real number is a vector space.

Solution This is a special case of Example 6-5 which has been shown to be a vector space. Hence, this is also a vector space.

PROBLEM 6-4 Show that the set of all points (x, y) in R_2 such that $x \leq 0$ and $y \geq 0$ is not a vector space under operations $(x_1, y_1) \oplus (x_2, y_2) = (x_1 + x_2, y_1 + y_2)$ and $a \odot (x, y) = (ax, ay)$ where a is any real number.

Solution We have no difficulty with the first four properties being true, but Property **(5)** gives

$$(x, y) + (a, b) = (0, 0)$$

which means

$$\begin{aligned} x + a &= 0 \qquad \text{or} \qquad & a &= -x \\ y + b &= 0 & b &= -y \end{aligned}$$

and point $(-x, -y)$ is not in the given set. Therefore, the set is not a vector space.

PROBLEM 6-5 Determine whether the following are vector spaces under the usual vector addition and scalar multiplication. The set W of all vectors of form: **(a)** $[a \quad b \quad 0]$, **(b)** $\begin{bmatrix} 1 \\ a \\ 0 \end{bmatrix}$, **(c)** $[a \quad b \quad a+b \quad 0]$.

Solution

(a) This set is a subset of R_3. We need to check only closure for vector addition and scalar multiplication since the set is obviously nonempty.

$$[a_1 \quad b_1 \quad 0] + [a_2 \quad b_2 \quad 0] = [a_1 + a_2 \quad b_1 + b_2 \quad 0] \qquad \text{is in } W$$

$$k[a \quad b \quad 0] = [ka \quad kb \quad 0] \qquad \text{is in } W$$

Therefore, the set is closed under the given operations. W is a subspace of R_3 and hence also a vector space.

(b)
$$\begin{bmatrix} 1 \\ a_1 \\ 0 \end{bmatrix} + \begin{bmatrix} 1 \\ a_2 \\ 0 \end{bmatrix} = \begin{bmatrix} 2 \\ a_1 + a_2 \\ 0 \end{bmatrix}$$

which is not an element of W, since the first component is 2. Therefore, W is not a subspace of R^3, and hence is not a vector space.

(c) The set is a nonempty subset of R_4:

$$\begin{aligned} &[a_1 \quad b_1 \quad a_1 + b_1 \quad 0] + [a_2 \quad b_2 \quad a_2 + b_2 \quad 0] \\ &\quad = [a_1 + a_2 \quad b_1 + b_2 \quad a_1 + b_1 + a_2 + b_2 \quad 0] \\ &\quad = [a_1 + a_2 \quad b_1 + b_2 \quad a_1 + a_2 + b_1 + b_2 \quad 0] \qquad \text{which is in } W \end{aligned}$$

$$k[a \quad b \quad a+b \quad 0] = [ka \quad kb \quad ka + kb \quad 0] \qquad \text{which is in } W$$

Therefore, the set is closed under vector addition and scalar multiplication. Thus W is a subspace of R_4 and is a vector space.

PROBLEM 6-6 Determine whether all 2×2 matrices of the form $\begin{bmatrix} a & b \\ b & c \end{bmatrix}$ constitute a subspace of ${}_2M_2$.

Solution

$$\begin{bmatrix} a_1 & b_1 \\ b_1 & c_1 \end{bmatrix} + \begin{bmatrix} a_2 & b_2 \\ b_2 & c_2 \end{bmatrix} = \begin{bmatrix} a_1 + a_2 & b_1 + b_2 \\ b_1 + b_2 & c_1 + c_2 \end{bmatrix} \quad \text{which is in } {}_2M_2$$

$$k\begin{bmatrix} a & b \\ b & c \end{bmatrix} = \begin{bmatrix} ka & kb \\ kb & kc \end{bmatrix} \quad \text{which is in } {}_2M_2$$

Therefore, they constitute a subspace of ${}_2M_2$.

note: This problem verifies that all 2×2 symmetric matrices are a subspace of ${}_2M_2$.

PROBLEM 6-7 Determine if the following sets are subspaces of R_3: **(a)** W is set of all triples $(a, 0, c)$. **(b)** W is set of all triples $(a, b, a - b)$, where $b > 0$. **(c)** W is set of all triples (a, b, c), where $a > 0$, $b > 0$, and $c > 0$. **(d)** W is set of all triples $(0, 0, c)$, where c is irrational.

Solution

(a) W is nonempty since at least $(0, 0, 0)$ is in W.

$$(a_1, 0, c_1) + (a_2, 0, c_2) = (a_1 + a_2, 0, c_1 + c_2) \qquad \text{is in } W$$

$$k(a, 0, c) = (ka, 0, kc) \qquad \text{is in } W$$

Therefore, W is a subspace of R_3.

(b) W is nonempty since $(1, 1, 0)$ is in W. $-1(2, 1, 1) = (-2, -1, -1)$ which is not in W, since the second component is <0. Therefore, W is not a subspace of R_3.

(c) W is nonempty and there is no zero element of W. $-1(1, 1, 1) = (-1, -1, -1)$ is not in W. Therefore, W is not a subspace of R_3.

(d) W is nonempty since at least $(0, 0, \sqrt{2})$ is in W. $\sqrt{2}(0, 0, \sqrt{2}) = (0, 0, 2)$ which is not in W since 2 is rational. Therefore, W is not a subspace of R_3.

PROBLEM 6-8 Let $W = \{f\colon f(a) = b + f(2)\}$. Show that W is not a subspace of vector space V of all functions from R into R.

Solution Let f, g be two elements of W. $f(a) = b + f(2)$ and $g(a) = b + g(2)$. Then

$$(f + g)(a) = f(a) + f(b) = 2b + f(2) + g(2) = 2b + (f + g)(2) \neq b + (f + g)(2)$$

Therefore, $f + g$ is not in W, and W is not a subspace of V.

PROBLEM 6-9 If V is a vector space of ${}_2M_2$ and $\mathbf{a} = \begin{bmatrix} 1 & 2 \\ -2 & -3 \end{bmatrix}$ such that $k\begin{bmatrix} 1 & 2 \\ -2 & -3 \end{bmatrix} = \begin{bmatrix} 0 & 0 \\ 0 & 0 \end{bmatrix}$, show that $k = 0$.

Solution

$$k\begin{bmatrix} 1 & 2 \\ -2 & -3 \end{bmatrix} = \begin{bmatrix} k & 2k \\ -2k & -3k \end{bmatrix} = \begin{bmatrix} 0 & 0 \\ 0 & 0 \end{bmatrix}$$

but two matrices are equal if and only if corresponding elements are equal. Therefore, $k = 0$, $2k = 0$, $-2k = 0$, and $-3k = 0$, each of which implies $k = 0$.

PROBLEM 6-10 If V is a vector space in R_2, $k = 3$, and $\mathbf{a} = [a \quad b]$ such that $3[a \quad b] = [0 \quad 0]$ show that $\mathbf{a} = [0 \quad 0]$.

Solution $3a = 0$ and $3b = 0$, which imply that $a = 0$ and $b = 0$. Therefore, $\mathbf{a} = [0 \quad 0]$.

PROBLEM 6-11 Determine which of the following vectors in R_3 are linear combinations of $\mathbf{a} = [-1 \quad 0 \quad 2]$ and $\mathbf{b} = [2 \quad -1 \quad 1]$: **(a)** $[3 \quad 1 \quad -11]$, **(b)** $[4 \quad -2 \quad 5]$, **(c)** $[1 \quad -2 \quad 8]$.

Solution Write the equation in each case.

(a)

$$a_1[-1 \quad 0 \quad 2] + a_2[2 \quad -1 \quad 1] = [3 \quad 1 \quad -11]$$

$$-a_1 + 2a_2 = 3$$

$$-a_2 = 1$$

$$2a_1 + a_2 = -11$$

Since we immediately know $a_2 = -1$, we will substitute this value in the other two equations: $-a_1 - 2 = 3$, implying that $a_1 = -5$; also, $2a_1 - 1 = -11$, giving $a_1 = -5$. Therefore,

$$[3 \quad 1 \quad -11] = (-5)[-1 \quad 0 \quad 2] + (-1)[2 \quad -1 \quad 1]$$

and hence is a linear combination of **a** and **b**.

(b)

$$a_1[-1 \quad 0 \quad 2] + a_2[2 \quad -1 \quad 1] = [4 \quad -2 \quad 5]$$

$$-a_1 + 2a_2 = 4$$

$$-a_2 = -2$$

$$2a_1 + a_2 = 5$$

This system gives $a_2 = 2$, but when we substitute $a_2 = 2$ in the other equations, we get

$$-a_1 + 4 = 4 \quad \text{or} \quad a_1 = 0$$

$$2a_1 + 2 = 5 \quad \text{or} \quad a_1 = 3/2$$

which are inconsistent statements. Therefore, $[4 \quad -2 \quad 5]$ is not a linear combination of $[-1 \quad 0 \quad 2]$ and $[2 \quad -1 \quad 1]$.

(c)

$$a_1[-1 \quad 0 \quad 2] + a_2[2 \quad -1 \quad 1] = [1 \quad -2 \quad 8]$$

$$-a_1 + 2a_2 = 1$$

$$-a_2 = -2$$

$$2a_1 + a_2 = 8$$

This gives $a_2 = 2$ and, by substitution,

$$-a_1 + 4 = 1 \qquad -a_1 = -3 \quad \text{or} \quad a_1 = 3$$

$$2a_1 + 2 = 8 \qquad 2a_1 = 6 \quad \text{or} \quad a_1 = 3$$

$$\therefore \quad [1 \quad -2 \quad 8] = 3[-1 \quad 0 \quad 2] + 2[2 \quad -1 \quad 1]$$

and hence is a linear combination of **a** and **b**.

PROBLEM 6-12 Express $5x^2 - 3x + 6$ as a linear combination of $x^2 - 2x + 3$, $2x^2 + x - 4$, and $2x^2 + 3x - 1$.

Solution

$$a_1(x^2 - 2x + 3) + a_2(2x^2 + x - 4) + a_3(2x^2 + 3x - 1) = 5x^2 - 3x + 6$$

$$(a_1 + 2a_2 + 2a_3)x^2 + (-2a_1 + a_2 + 3a_3)x + (3a_1 - 4a_2 - a_3) = 5x^2 - 3x + 6$$

By setting the coefficients of like powers of x equal, we get

$$a_1 + 2a_2 + 2a_3 = 5$$

$$-2a_1 + a_2 + 3a_3 = -3$$

$$3a_1 - 4a_2 - a_3 = 6$$

We will solve this system by any suitable method. In this case, we will use Gauss–Jordan elimination on the augmented matrix:

$$\begin{bmatrix} 1 & 2 & 2 & 5 \\ -2 & 1 & 3 & -3 \\ 3 & -4 & -1 & 6 \end{bmatrix} \xrightarrow[R_3 \to R_3 - 3R_1]{R_2 \to R_2 + 2R_1} \begin{bmatrix} 1 & 2 & 2 & 5 \\ 0 & 5 & 7 & 7 \\ 0 & -10 & -7 & -9 \end{bmatrix} \xrightarrow[R_3 \to R_3 + 2R_2]{} \begin{bmatrix} 1 & 2 & 2 & 5 \\ 0 & 5 & 7 & 7 \\ 0 & 0 & 7 & 5 \end{bmatrix}$$

$$\xrightarrow{R_2 \to R_2 - R_3} \begin{bmatrix} 1 & 2 & 2 & 5 \\ 0 & 5 & 0 & 2 \\ 0 & 0 & 7 & 5 \end{bmatrix} \xrightarrow[R_3 \to 1/7 R_3]{R_2 \to 1/5 R_2} \begin{bmatrix} 1 & 2 & 2 & 5 \\ 0 & 1 & 0 & \frac{2}{5} \\ 0 & 0 & 1 & \frac{5}{7} \end{bmatrix}$$

$$\xrightarrow{R_1 \to R_1 - 2R_2} \begin{bmatrix} 1 & 0 & 2 & \frac{21}{5} \\ 0 & 1 & 0 & \frac{2}{5} \\ 0 & 0 & 1 & \frac{5}{7} \end{bmatrix} \xrightarrow{R_1 \to R_1 - 2R_3} \begin{bmatrix} 1 & 0 & 0 & \frac{97}{35} \\ 0 & 1 & 0 & \frac{2}{5} \\ 0 & 0 & 1 & \frac{5}{7} \end{bmatrix}$$

$$\therefore \quad a_1 = \tfrac{97}{35} \qquad a_2 = \tfrac{14}{35} \qquad a_3 = \tfrac{25}{35}$$

$$\tfrac{97}{35}(x^2 - 2x + 3) + \tfrac{14}{35}(2x^2 + x - 4) + \tfrac{25}{35}(2x^2 + 3x - 1) = 5x^2 - 3x + 6$$

This problem demonstrates the fact that not all problems that look simple have neat answers.

PROBLEM 6-13 Determine whether the vectors $[-3 \quad 1]$ and $[5 \quad -7]$ from R_2 are linearly independent or dependent.

Solution The zero vector for R_2 is $[0 \quad 0]$; hence

$$a_1[-3 \quad 1] + a_2[5 \quad -7] = [0 \quad 0]$$
$$-3a_1 + 5a_2 = 0$$
$$a_1 - 7a_2 = 0$$

This system is equivalent to

$$-3a_1 + 5a_2 = 0$$
$$3a_1 - 21a_2 = 0$$

If we add these two equations, we get

$$-16a_2 = 0$$
$$\therefore \quad a_2 = 0 \qquad \text{which implies } a_1 = 0$$

Since the only solution is $a_1 = a_2 = 0$, the vectors are independent.

PROBLEM 6-14 Determine whether the vectors $\mathbf{a} = x^2 - 4$, $\mathbf{b} = 2x^2 + x - 5$, and $\mathbf{c} = x^2 - 7x + 4$ from P_2 are linearly independent or dependent.

Solution

$$a_1(x^2 - 4) + a_2(2x^2 + x - 5) + a_3(x^2 - 7x + 4) = 0$$
$$a_1 + 2a_2 + a_3 = 0$$
$$a_2 - 7a_3 = 0$$
$$-4a_1 - 5a_2 + 4a_3 = 0$$

If the determinant of the coefficient matrix is not zero, the vectors are linearly independent.

$$\det \begin{bmatrix} 1 & 2 & 1 \\ 0 & 1 & -7 \\ -4 & -5 & 4 \end{bmatrix} = \begin{vmatrix} 1 & 2 & 1 & 1 & 2 \\ 0 & 1 & -7 & 0 & 1 \\ -4 & -5 & 4 & -4 & -5 \end{vmatrix} = 4 + 56 + 0 + 4 - 35 - 0 = 29$$

Therefore, the vectors are linearly independent.

PROBLEM 6-15 Determine whether the set of vectors $[1 \quad 2 \quad 3]$, $[-2 \quad 3 \quad 1]$, and $[3 \quad -2 \quad 1]$ are linearly independent or dependent.

Solution If the determinant of the matrix $\begin{bmatrix} 1 & 2 & 3 \\ -2 & 3 & 1 \\ 3 & -2 & 1 \end{bmatrix}$ is zero, the set is linearly dependent:

$$\begin{vmatrix} 1 & 2 & 3 \\ -2 & 3 & 1 \\ 3 & -2 & 1 \end{vmatrix} = \frac{1}{1}\begin{vmatrix} 7 & 7 \\ -8 & -8 \end{vmatrix} = -56 + 56 = 0$$

Therefore, the set is linearly dependent.

PROBLEM 6-16 Is the set $\mathbf{a} = x^2 - 4$ and $\mathbf{b} = 2x^2 + x - 5$ linearly independent or dependent?

Solution In Problem 6-14 we showed that $x^2 - 4$, $2x^2 + x - 5$, and $x^2 - 7x + 4$ are linearly independent. If one of the vectors is deleted, we get the set **a** and **b**, which is also linearly independent by Property A of Section 6-5.

PROBLEM 6-17 Show that $[3 \quad -2 \quad 1]$ can be written as a linear combination of $[1 \quad 2 \quad 3]$ and $[-2 \quad 3 \quad 1]$.

Solution We need to determine real numbers a_1 and a_2 such that $a_1[1 \quad 2 \quad 3] + a_2[-2 \quad 3 \quad 1] = [3 \quad -2 \quad 1]$. We get

$$a_1 - 2a_2 = 3$$

$$2a_1 + 3a_2 = -2$$

$$3a_1 + a_2 = 1$$

Solving for a_1 and a_2 in the first two equations, we get $a_1 = 5/7$ and $a_2 = -8/7$. Then substituting in equation three, we get $15/7 - 8/7 = 1$. Therefore, $(5/7)[1 \quad 2 \quad 3] - (8/7)[-2 \quad 3 \quad 1] = [3 \quad -2 \quad 1]$.

PROBLEM 6-18 Show that $[-2 \quad 3 \quad 1]$ can be written as a linear combination of $[1 \quad 2 \quad 3]$ and $[3 \quad -2 \quad 1]$.

Solution

$$a_1[1 \quad 2 \quad 3] + a_2[3 \quad -2 \quad 1] = [-2 \quad 3 \quad 1]$$

$$a_1 + 3a_2 = -2$$

$$2a_1 - 2a_2 = 3$$

$$3a_1 + a_2 = 1$$

Solving for a_1 and a_2 in the first two equations, we get $a_1 = 5/8$ and $a_2 = -7/8$. Then substituting these values in the third equation, we get $15/8 - 7/8 = 1$. Therefore, $(5/8)[1 \quad 2 \quad 3] - (7/8)[3 \quad -2 \quad 1] = [-2 \quad 3 \quad 1]$.

Supplementary Problems

PROBLEM 6-19 Show that the set W of all 2×3 matrices of the form $\begin{bmatrix} a & 0 & a \\ 0 & a & 0 \end{bmatrix}$ is a vector space.

PROBLEM 6-20 Show that the set W of vectors in R_3 of the form $(a, 1, b)$ is not a subspace of R_3.

PROBLEM 6-21 Show that the set V of all polynomials of degree 2 is not a subspace of P_2.

PROBLEM 6-22 Which of the following subsets of R_4 are subspaces of R_4? **(a)** (a, b, c, d), where $a - c = d$. **(b)** (a, b, c, d), where $a + 2b = c$. **(c)** (a, b, c, d), where $c = 3d + 1$. **(d)** (a, b, c, d), where $b = 1$, $c = a - d$.

PROBLEM 6-23 Let W be the set of vectors in R_3 of form $a[2\ \ -1\ \ 3] + b[3\ \ 0\ \ -2]$. Verify that W is a subspace of R_3.

PROBLEM 6-24 Let W consist of 2×2 matrices with determinant values of zero. Show that W is not a subspace of ${}_2M_2$.

PROBLEM 6-25 Let $W = \{a_0 + a_1x + a_2x^2 | a_1 > 0\}$. Determine whether W is a subspace of P_2.

PROBLEM 6-26 Let $W = \begin{bmatrix} a & 0 \\ 0 & a \end{bmatrix}$, $a \in R$. Determine whether W is a subspace of ${}_2M_2$.

PROBLEM 6-27 Let W_1 and W_2 be subspaces of vector space V. Show that $W_1 \cap W_2$ is a subspace of V.

PROBLEM 6-28 Express $W = [8\ \ -2]$ as a linear combination of $\mathbf{a} = [4\ \ 2]$ and $\mathbf{b} = [1\ \ 2]$.

PROBLEM 6-29 Express $W = [-2\ \ 11]$ as a linear combination of $\mathbf{a} = [4\ \ 2]$ and $\mathbf{b} = [1\ \ 2]$.

PROBLEM 6-30 Express $\begin{bmatrix} 0 & 1 \\ 3 & -2 \end{bmatrix}$ as a linear combination of $\mathbf{a} = \begin{bmatrix} 1 & -1 \\ 2 & 1 \end{bmatrix}$ and $\mathbf{b} = \begin{bmatrix} 1 & -2 \\ -1 & 3 \end{bmatrix}$ if possible.

PROBLEM 6-31 Determine whether the vectors $[3\ \ -7]$ and $[-3\ \ 4]$ in R_2 are linearly independent or dependent.

PROBLEM 6-32 Determine whether the vectors $[-1\ \ 0\ \ 2]$, $[2\ \ -1\ \ 3]$ and $[4\ \ -3\ \ 13]$ are linearly independent or dependent.

PROBLEM 6-33 Show that the set $\{4\cos^2 x, \sin^2 x, -\cos 2x\}$ is a linearly dependent set.

PROBLEM 6-34 Show that the set $\{1, 2\tan^2 x, \sec^2 x\}$ is a linearly dependent set.

Answers to Supplementary Problems

6-19 Verify closure under matrix addition and multiplication by a scalar.

6-20 Show that either closure under addition does not hold or that scalar multiplication does not hold.

6-21 Counterexample: $x^2 + 4 + (-x^2 + x + 4) = x + 8$ which is not of degree 2.

6-22 (**a**) Subspace (**b**) Subspace (**c**) Not a subspace (**d**) Not a subspace

6-23 Show that closure under addition holds and that scalar multiplication holds

6-24 Let $A = \begin{bmatrix} 1 & 0 \\ 0 & 0 \end{bmatrix}$ and $B = \begin{bmatrix} 0 & 0 \\ 0 & 1 \end{bmatrix}$; then $A + B = \begin{bmatrix} 1 & 0 \\ 0 & 1 \end{bmatrix}$ but $\begin{vmatrix} 1 & 0 \\ 0 & 1 \end{vmatrix} = 1$.

6-25 Not a subspace $-1[2 + 3x + x^2] = -2 - 3x - x^2$.

6-26 W is a subspace

6-27 $W_1 \cap W_2$ is in V

6-28 $3[4\ \ 2] - 4[1\ \ 2] = [8\ \ -2]$

6-29 $-5/2[4\ \ 2] + 8[1\ \ 2] = [-2\ \ 11]$

6-30 $1\begin{bmatrix} 1 & -1 \\ 2 & 1 \end{bmatrix} + (-1)\begin{bmatrix} 1 & -2 \\ -1 & 3 \end{bmatrix} = \begin{bmatrix} 0 & 1 \\ 3 & -2 \end{bmatrix}$

6-31 Independent

6-32 Dependent; $2[-1\ \ 0\ \ 2] + 3[2\ \ -1\ \ 3] = [4\ \ -3\ \ 13]$

6-33 $-4\cos^2 x + 4\sin^2 x + 4\cos 2x = 0$; $(\cos 2x = \cos^2 x - \sin^2 x)$

6-34 $2(1) + 2\tan^2 x - 2\sec^2 x = 0$; $(1 + \tan^2 x = \sec^2 x)$

7 REPRESENTING VECTOR SPACES BY SUBSETS

THIS CHAPTER IS ABOUT

☑ Spanning Sets
☑ Bases and Dimension
☑ Properties of Bases

7-1. Spanning Sets

A. Definition of spanning set

If every vector in vector space V can be written as a linear combination of the vectors $\mathbf{a}_1, \mathbf{a}_2, \ldots, \mathbf{a}_k$, then these vectors **span** V or are said to form a **spanning set** for V. This means that V can be represented or generated by the vectors $\mathbf{a}_1, \mathbf{a}_2, \ldots, \mathbf{a}_k$.

EXAMPLE 7-1: Determine whether the vectors $[1 \quad 1 \quad 1]$, $[0 \quad 1 \quad 2]$, and $[2 \quad 3 \quad 0]$ span R_3.

Solution

Method 1: You want to know if every vector $[a \quad b \quad c]$ in R_3 can be written as a linear combination of the three given vectors:

$$a_1[1 \quad 1 \quad 1] + a_2[0 \quad 1 \quad 2] + a_3[2 \quad 3 \quad 0] = [a \quad b \quad c]$$

This results in the system

$$\begin{aligned} a_1 \qquad\quad + 2a_3 &= a \\ a_1 + \quad a_2 + 3a_3 &= b \\ a_1 + 2a_2 \qquad\quad &= c \end{aligned}$$

From the first you get $a_3 = (a - a_1)/2$, and from the third equation you get $a_2 = (c - a_1)/2$. Substituting these into the second equation, you get

$$a_1 + \frac{c - a_1}{2} + 3\frac{a - a_1}{2} = b.$$

Solving for a_1, you get $a_1 = (3a - 2b + c)/2$, which then gives

$$a_2 = \frac{2b - 3a + c}{4} \qquad a_3 = \frac{2b - a - c}{4}$$

$$\therefore \quad \frac{3a - 2b + c}{2}[1 \quad 1 \quad 1] + \frac{2b - 3a + c}{4}[0 \quad 1 \quad 2] + \frac{2b - a - c}{4}[2 \quad 3 \quad 0] = [a \quad b \quad c]$$

and thus the vectors span R_3.

Method 2: In this problem you did not actually have to find the particular linear combination of the three vectors that would give $[a \quad b \quad c]$. All you really needed to determine was the existence of such a solution for the system of equations given in Method 1. Writing that system in matrix form, you get

$$\begin{bmatrix} 1 & 0 & 2 \\ 1 & 1 & 3 \\ 1 & 2 & 0 \end{bmatrix}\begin{bmatrix} a_1 \\ a_2 \\ a_3 \end{bmatrix} = \begin{bmatrix} a \\ b \\ c \end{bmatrix}$$

Thus, you want to know whether this system has a unique solution for every 3×1 matrix $\begin{bmatrix} a \\ b \\ c \end{bmatrix}$.

From Section 2-8, you will recall that this is equivalent to determining whether the coefficient matrix which is a square matrix is invertible. You might also observe that the columns of the matrix are the transposes of the given vectors. Furthermore, a matrix A is invertible if and only if $\det(A) \neq 0$. Thus all you need to do to solve this problem is to evaluate the determinant of the coefficient matrix to see if that value is zero:

$$\begin{vmatrix} 1 & 0 & 2 \\ 1 & 1 & 3 \\ 1 & 2 & 0 \end{vmatrix} = 1\begin{vmatrix} 1 & 3 \\ 2 & 0 \end{vmatrix} - 0\begin{vmatrix} 1 & 3 \\ 1 & 0 \end{vmatrix} + 2\begin{vmatrix} 1 & 1 \\ 1 & 2 \end{vmatrix} = 1(-6) + 2(1) = -4$$

Therefore, the three vectors span R_3.

EXAMPLE 7-2: Determine which of the following sets of vectors span R^2.

(a) $\left\{\begin{bmatrix} 1 \\ 3 \end{bmatrix}, \begin{bmatrix} 2 \\ 1 \end{bmatrix}\right\}$ **(b)** $\left\{\begin{bmatrix} 1 \\ 3 \end{bmatrix}, \begin{bmatrix} 2 \\ 6 \end{bmatrix}\right\}$ **(c)** $\left\{\begin{bmatrix} -1 \\ 1 \end{bmatrix}, \begin{bmatrix} 2 \\ 3 \end{bmatrix}, \begin{bmatrix} 1 \\ 4 \end{bmatrix}\right\}$

Solution

(a)

$$a_1\begin{bmatrix} 1 \\ 3 \end{bmatrix} + a_2\begin{bmatrix} 2 \\ 1 \end{bmatrix} = \begin{bmatrix} a \\ b \end{bmatrix}$$

$$\begin{aligned} a_1 + 2a_2 &= a \\ 3a_1 + a_2 &= b \end{aligned} \quad \text{or} \quad \begin{bmatrix} 1 & 2 \\ 3 & 1 \end{bmatrix}\begin{bmatrix} a_1 \\ a_2 \end{bmatrix} = \begin{bmatrix} a \\ b \end{bmatrix}$$

$$\begin{vmatrix} 1 & 2 \\ 3 & 1 \end{vmatrix} = 1 - 6 = -5$$

Therefore, the system has a solution for every value of a and b, and hence the vectors span R^2.

(b)

$$a_1\begin{bmatrix} 1 \\ 3 \end{bmatrix} + a_2\begin{bmatrix} 2 \\ 6 \end{bmatrix} = \begin{bmatrix} a \\ b \end{bmatrix}$$

$$\begin{aligned} a_1 + 2a_2 &= a \\ 3a_1 + 6a_2 &= b \end{aligned} \quad \text{or} \quad \begin{bmatrix} 1 & 2 \\ 3 & 6 \end{bmatrix}\begin{bmatrix} a_1 \\ a_2 \end{bmatrix} = \begin{bmatrix} a \\ b \end{bmatrix}$$

$$\begin{vmatrix} 1 & 2 \\ 3 & 6 \end{vmatrix} = 0$$

Therefore, the vectors do not span R^2.

(c)

$$a_1\begin{bmatrix} -1 \\ 1 \end{bmatrix} + a_2\begin{bmatrix} 2 \\ 3 \end{bmatrix} + a_3\begin{bmatrix} 1 \\ 4 \end{bmatrix} = \begin{bmatrix} a \\ b \end{bmatrix}$$

$$\begin{aligned} -a_1 + 2a_2 + a_3 &= a \\ a_1 + 3a_2 + 4a_3 &= b \end{aligned} \quad \text{or} \quad \begin{bmatrix} -1 & 2 & 1 \\ 1 & 3 & 4 \end{bmatrix}\begin{bmatrix} a_1 \\ a_2 \\ a_3 \end{bmatrix} = \begin{bmatrix} a \\ b \end{bmatrix}$$

Since $\begin{bmatrix} -1 & 2 & 1 \\ 1 & 3 & 4 \end{bmatrix}$ is not a square matrix, its determinant is not defined and we must use some other method. Let us place the augmented matrix in reduced row echelon form:

$$\begin{bmatrix} -1 & 2 & 1 & a \\ 1 & 3 & 4 & b \end{bmatrix} \xrightarrow{R_2 \to R_2 + R_1} \begin{bmatrix} -1 & 2 & 1 & a \\ 0 & 5 & 5 & a+b \end{bmatrix} \xrightarrow[R_2 \to 1/5 R_2]{R_1 \to -R_1} \begin{bmatrix} 1 & -2 & -1 & -a \\ 0 & 1 & 1 & \dfrac{a+b}{5} \end{bmatrix}$$

$$\xrightarrow{R_1 \to R_1 + 2R_2} \begin{bmatrix} 1 & 0 & 1 & \dfrac{2b-3a}{5} \\ 0 & 1 & 1 & \dfrac{a+b}{5} \end{bmatrix}$$

$$\therefore \quad a_1 = \frac{2b-3a}{5} - a_3$$

$$a_2 = \frac{a+b}{5} - a_3$$

Since a_3 is arbitrary, let $a_3 = 0$. Therefore, $a_1 = (2b - 3a)/5$, $a_2 = (a + b)/5$, $a_3 = 0$ for one solution, although not the only solution. Therefore, the three vectors span R^2.

B. Definition of linear span

The set W of all linear combinations of a set A of vectors $\mathbf{a}_1, \mathbf{a}_2, \ldots, \mathbf{a}_k$ from a vector space V is a subspace of V called their **linear span** or, more commonly, **span**. Among the notations used to denote it are **S[A]**, **S(A)**, **span A**, **span (A)**, **span** $\{\mathbf{a}_1, \mathbf{a}_2, \ldots, \mathbf{a}_k\}$, **lin(A)**, and **lin** $\{\mathbf{a}_1, \mathbf{a}_2, \ldots, \mathbf{a}_k\}$. We shall use span **A** or span $\{\mathbf{a}_1, \mathbf{a}_2, \ldots, \mathbf{a}_k\}$.

EXAMPLE 7-3: Verify that span A where

$$A = \left\{ \begin{bmatrix} 1 \\ 1 \\ 0 \end{bmatrix}, \begin{bmatrix} 0 \\ 1 \\ 1 \end{bmatrix} \right\}$$

is a subspace of R^3.

Solution: Every vector in span A is a linear combination of $\begin{bmatrix} 1 \\ 1 \\ 0 \end{bmatrix}$ and $\begin{bmatrix} 0 \\ 1 \\ 1 \end{bmatrix}$. Let $\mathbf{a}_1$ and $\mathbf{a}_2$ be two such vectors where

$$\mathbf{a}_1 = a_1 \begin{bmatrix} 1 \\ 1 \\ 0 \end{bmatrix} + a_2 \begin{bmatrix} 0 \\ 1 \\ 1 \end{bmatrix} \quad \text{and} \quad \mathbf{a}_2 = b_1 \begin{bmatrix} 1 \\ 1 \\ 0 \end{bmatrix} + b_2 \begin{bmatrix} 0 \\ 1 \\ 1 \end{bmatrix}.$$

You must show that $\mathbf{a}_1 + \mathbf{a}_2$ and $c\mathbf{a}_1$ are in span A; that is, each is a linear combination of $\begin{bmatrix} 1 \\ 1 \\ 0 \end{bmatrix}$ and $\begin{bmatrix} 0 \\ 1 \\ 1 \end{bmatrix}$.

$$\mathbf{a}_1 + \mathbf{a}_2 = a_1 \begin{bmatrix} 1 \\ 1 \\ 0 \end{bmatrix} + a_2 \begin{bmatrix} 0 \\ 1 \\ 1 \end{bmatrix} + b_1 \begin{bmatrix} 1 \\ 1 \\ 0 \end{bmatrix} + b_2 \begin{bmatrix} 0 \\ 1 \\ 1 \end{bmatrix}$$

$$= (a_1 + b_1) \begin{bmatrix} 1 \\ 1 \\ 0 \end{bmatrix} + (a_2 + b_2) \begin{bmatrix} 0 \\ 1 \\ 1 \end{bmatrix}$$

and

$$c\mathbf{a}_1 = c\left(a_1\begin{bmatrix}1\\1\\0\end{bmatrix} + a_2\begin{bmatrix}0\\1\\1\end{bmatrix}\right) = (ca_1)\begin{bmatrix}1\\1\\0\end{bmatrix} + (ca_2)\begin{bmatrix}0\\1\\1\end{bmatrix}$$

Therefore, span A is closed under addition and scalar multiplication and thus, is a subspace of R^3.

7-2. Bases and Dimension

A. Definition of basis

A set of vectors $B = \{\mathbf{a}_1, \mathbf{a}_2, \ldots, \mathbf{a}_k\}$ from a vector space V is a **basis** for V if

- B spans V, and
- B is linearly independent.

EXAMPLE 7-4: Verify that $\{[1 \quad 0 \quad 0], [0 \quad 1 \quad 0], [0 \quad 0 \quad 1]\}$ is a basis for R_3.

Solution: Let $[a \quad b \quad c]$ be any vector in R_3: $a_1[1 \quad 0 \quad 0] + a_2[0 \quad 1 \quad 0] + a_3[0 \quad 0 \quad 1] = [a \quad b \quad c]$. Hence, $[a_1 \quad a_2 \quad a_3] = [a \quad b \quad c]$. Therefore, $a_1 = a$, $a_2 = b$, and $a_3 = c$. Thus every vector in R_3 can be written as a linear combination of the three vectors.

Furthermore, the only way this linear combination equals $[0 \quad 0 \quad 0]$ is if $a_1 = a_2 = a_3$. Therefore, the set spans R_3 and is linearly independent; hence it is a basis for R_3. This set is called the **natural** or **standard basis** for R_3. The natural or standard basis for R_n is $\{\mathbf{e}_1, \mathbf{e}_2, \ldots, \mathbf{e}_n\}$, where

$$\mathbf{e}_1 = [1 \quad 0 \quad 0 \quad \cdots \quad 0], \qquad \mathbf{e}_2 = [0 \quad 1 \quad 0 \quad \cdots \quad 0], \ldots,$$

$$\mathbf{e}_n = [0 \quad 0 \quad 0 \quad \cdots \quad 0 \quad 1].$$

In R_3, the notation that is often used is $\mathbf{i} = [1 \quad 0 \quad 0]$, $\mathbf{j} = [0 \quad 1 \quad 0]$, and $\mathbf{k} = [0 \quad 0 \quad 1]$.

EXAMPLE 7-5: Verify that $B = \{[2 \quad 1 \quad 1], [1 \quad 7 \quad 7], [4 \quad -1 \quad 0]\}$ is a basis for R_3.

Solution: To verify that B spans R_3 you must show that any vector $[a \quad b \quad c]$ in R_3 can be written as a linear combination of the vectors in B.

$$a_1[2 \quad 1 \quad 1] + a_2[1 \quad 7 \quad 7] + a_3[4 \quad -1 \quad 0] = [a \quad b \quad c]$$

$$\begin{aligned} 2a_1 + a_2 + 4a_3 &= a \\ a_1 + 7a_2 - a_3 &= b \\ a_1 + 7a_2 \quad &= c \end{aligned} \qquad \text{or} \qquad \begin{bmatrix}2 & 1 & 4\\1 & 7 & -1\\1 & 7 & 0\end{bmatrix}\begin{bmatrix}a_1\\a_2\\a_3\end{bmatrix} = \begin{bmatrix}a\\b\\c\end{bmatrix}$$

To verify linear independence, you must show that the only solution of

$$\begin{bmatrix}2 & 1 & 4\\1 & 7 & -1\\1 & 7 & 0\end{bmatrix}\begin{bmatrix}a_1\\a_2\\a_3\end{bmatrix} = \begin{bmatrix}0\\0\\0\end{bmatrix}$$

is $a_1 = a_2 = a_3 = 0$. You can, therefore, simultaneously verify that B is linearly independent and spans R_3 by showing that the matrix $\begin{bmatrix}2 & 1 & 4\\1 & 7 & -1\\1 & 7 & 0\end{bmatrix}$, which is the common coefficient matrix for the homogeneous and nonhomogenous systems, is invertible.

Since $\begin{vmatrix}2 & 1 & 4\\1 & 7 & -1\\1 & 7 & 0\end{vmatrix} = 13$, then $\begin{bmatrix}2 & 1 & 4\\1 & 7 & -1\\1 & 7 & 0\end{bmatrix}$ is invertible, and hence B is a basis for R_3.

note: the columns of the resulting matrix are the vectors of B expressed as column vectors.

B. Dimension of a vector space

In Examples 7-4 and 7-5, you may have observed that each basis of R_3 contained three vectors. This was no coincidence. If one basis for a vector space V contains some finite number, say k, of vectors, then every basis for V contains exactly k vectors. These may, however, be entirely different vectors.

The number of vectors in a basis for a vector space V having at least one nonzero vector is called the **dimension** of V and is denoted by **dim V**. In addition, we define the dimension of $\{\mathbf{0}\}$ to be zero.

A vector space that has a basis that contains a finite number of vectors is called **finite-dimensional**. If a basis has infinitely many vectors, the vector space is **infinite-dimensional**. All vector spaces in this book are finite-dimensional. The dimensions of some frequently occurring vector spaces are

Vector space	Dimension
$\{\mathbf{0}\}$	0
$\mathbf{R}$	1
$\mathbf{R}_2$	2
$\mathbf{R}_3$	3
$\mathbf{R}_n$	n
$\mathbf{R}^n$	n
${}_m\mathbf{M}_n$	mn
$\mathbf{P}_n$	$n+1$

If V is an n-dimensional vector space, then the dimension of any subspace W of V is less than or equal to n, denoted by $\dim W \le n$.

EXAMPLE 7-6: Find a basis and the dimension of the solution space of the homogeneous system:

$$\begin{aligned} x_1 + x_2 + x_3 + x_4 &= 0 \\ x_1 - x_2 + 2x_3 + 3x_4 &= 0 \\ 2x_1 \quad\quad + 3x_3 + 4x_4 &= 0 \\ 2x_2 - x_3 - 2x_4 &= 0 \end{aligned}$$

Solution: Let us use the Gauss–Jordan elimination method on the augmented matrix:

$$\begin{bmatrix} 1 & 1 & 1 & 1 & 0 \\ 1 & -1 & 2 & 3 & 0 \\ 2 & 0 & 3 & 4 & 0 \\ 0 & 2 & -1 & -2 & 0 \end{bmatrix} \xrightarrow[R_3 \to R_3 - 2R_1]{R_2 \to R_2 - R_1} \begin{bmatrix} 1 & 1 & 1 & 1 & 0 \\ 0 & -2 & 1 & 2 & 0 \\ 0 & -2 & 1 & 2 & 0 \\ 0 & 2 & -1 & -2 & 0 \end{bmatrix} \xrightarrow[R_4 \to R_4 + R_2]{R_3 \to R_3 - R_2} \begin{bmatrix} 1 & 1 & 1 & 1 & 0 \\ 0 & -2 & 1 & 2 & 0 \\ 0 & 0 & 0 & 0 & 0 \\ 0 & 0 & 0 & 0 & 0 \end{bmatrix}$$

$$\xrightarrow{R_2 \to -1/2R_2} \begin{bmatrix} 1 & 1 & 1 & 1 & 0 \\ 0 & 1 & -\frac{1}{2} & -1 & 0 \\ 0 & 0 & 0 & 0 & 0 \\ 0 & 0 & 0 & 0 & 0 \end{bmatrix} \xrightarrow{R_1 \to R_1 - R_2} \begin{bmatrix} 1 & 0 & \frac{3}{2} & 2 & 0 \\ 0 & 1 & -\frac{1}{2} & -1 & 0 \\ 0 & 0 & 0 & 0 & 0 \\ 0 & 0 & 0 & 0 & 0 \end{bmatrix}$$

$$\therefore \quad X = \begin{bmatrix} x_1 \\ x_2 \\ x_3 \\ x_4 \end{bmatrix} = \begin{bmatrix} -\frac{3}{2}s - 2t \\ \frac{1}{2}s + t \\ s \\ t \end{bmatrix} = \begin{bmatrix} -\frac{3}{2}s \\ \frac{1}{2}s \\ s \\ 0 \end{bmatrix} + \begin{bmatrix} -2t \\ t \\ 0 \\ t \end{bmatrix} = s\begin{bmatrix} -\frac{3}{2} \\ \frac{1}{2} \\ 1 \\ 0 \end{bmatrix} + t\begin{bmatrix} -2 \\ 1 \\ 0 \\ 1 \end{bmatrix}$$

Hence

$$\begin{bmatrix} -\frac{3}{2} \\ \frac{1}{2} \\ 1 \\ 0 \end{bmatrix} \quad \text{and} \quad \begin{bmatrix} -2 \\ 1 \\ 0 \\ 1 \end{bmatrix}$$

form a basis for the solution space of the given system. Since this basis contains two vectors, the dimension of the solution space is 2.

7-3. Properties of Bases

Property A Every basis of an n-dimensional vector space V contains n vectors. Any subset of V that contains fewer than n vectors does not span V, and if it contains more than n vectors, it is a linearly dependent set.

EXAMPLE 7-7: Which of the following sets of vectors are bases for R_2?

(a) $[1 \quad 3]$ **(b)** $[1 \quad 2], [3 \quad 5], [-6 \quad 4]$ **(c)** $[1 \quad 1], [0 \quad -1]$

Solution: The dimension of R_2 is 2.

(a) This is not a basis because it contains fewer than two vectors.
(b) This is not a basis because it contains more than two vectors.
(c) Since this set contains the same number of vectors as the dimension of R_2, we must check for linear independence and that they span R_2. First, let us check for linear independence:

$$a_1[1 \quad 1] + a_2[0 \quad -1] = [0 \quad 0]$$

$$a_1 = 0$$

$$a_1 - a_2 = 0$$

$$\therefore \quad a_1 = a_2 = 0$$

and hence the two vectors are linearly independent.

Now, check to see that they span R_2:

$$a_1[1 \quad 1] + a_2[0 \quad -1] = [a \quad b]$$

$$a_1 = a$$

$$a_1 - a_2 = b$$

Thus, given any a and b, there exist a_1 and a_2 such that $[a \quad b]$ is a linear combination of $[1 \quad 1]$ and $[0 \quad -1]$. Hence, they span R_2. Therefore, they are a basis for R_2 since they span R_2 and are linearly independent.

Property B A basis for a vector space V cannot contain the zero vector.

EXAMPLE 7-8: Explain why $\left\{ \begin{bmatrix} 1 \\ 1 \\ 0 \end{bmatrix}, \begin{bmatrix} 0 \\ 0 \\ 0 \end{bmatrix}, \begin{bmatrix} 1 \\ 2 \\ 3 \end{bmatrix} \right\}$ cannot be a basis for R^3.

Solution: Although we have the correct number of vectors, one is the zero vector, and hence the set is not a basis.

Property C Every vector in a vector space V can be written uniquely as a linear combination of the vectors in a given basis.

EXAMPLE 7-9: In Example 7-5, we found that $\{[2 \quad 1 \quad 1], [1 \quad 7 \quad 7], [4 \quad -1 \quad 0]\}$ is a basis for R_3.

(a) Express $[3, 24, 23]$ as a linear combination of these three vectors in the given basis.
(b) Without any computation, explain why $[3, 24, 23] = 5[2 \quad 1 \quad 1] + 4[1 \quad 7 \quad 7] - 3[4 \quad -1 \quad 0]$ cannot be correct.

Solution

(a)
$$a_1[2 \quad 1 \quad 1] + a_2[1 \quad 7 \quad 7] + a_3[4 \quad -1 \quad 0] = [3 \quad 24 \quad 23]$$
$$2a_1 + a_2 + 4a_3 = 3$$
$$a_1 + 7a_2 - a_3 = 24$$
$$a_1 + 7a_2 = 23$$

Solving by Gauss–Jordan elimination, you will find that $a_1 = 2$, $a_2 = 3$, and $a_3 = -1$. Therefore, $2[2 \quad 1 \quad 1] + 3[1 \quad 7 \quad 7] - [4 \quad -1 \quad 0] = [3 \quad 24 \quad 23]$.

(b) Since every vector in R_3 can be expressed uniquely as a linear combination of the vectors in any of its bases, then the only correct answer is the one in part **(a)**.

Property D If a subset S of an n-dimensional vector space V contains exactly n vectors, then it is either

(a) both linearly independent and spans V or
(b) neither linearly independent nor spans V.

EXAMPLE 7-10: Which of the following sets of vectors are bases for R^3?

(a) $\left\{ \begin{bmatrix} 1 \\ 1 \\ 0 \end{bmatrix}, \begin{bmatrix} 1 \\ 2 \\ 3 \end{bmatrix}, \begin{bmatrix} -1 \\ 0 \\ 1 \end{bmatrix} \right\}$ **(b)** $\left\{ \begin{bmatrix} 1 \\ 1 \\ 0 \end{bmatrix}, \begin{bmatrix} 1 \\ 2 \\ 3 \end{bmatrix}, \begin{bmatrix} -1 \\ 0 \\ 1 \end{bmatrix}, \begin{bmatrix} 3 \\ 2 \\ 1 \end{bmatrix} \right\}$

Solution

(a) Since the set contains the same number of vectors as the dimension of R^3, it suffices to check either the linear independence of the three vectors or that they span R^3. Let us check linear independence:

$$a_1 \begin{bmatrix} 1 \\ 1 \\ 0 \end{bmatrix} + a_2 \begin{bmatrix} 1 \\ 2 \\ 3 \end{bmatrix} + a_3 \begin{bmatrix} -1 \\ 0 \\ 1 \end{bmatrix} = \begin{bmatrix} 0 \\ 0 \\ 0 \end{bmatrix}$$
$$a_1 + a_2 - a_3 = 0$$
$$a_1 + 2a_2 = 0$$
$$3a_2 + a_3 = 0$$

From the second of these equations, we get $a_1 = -2a_2$. From the third equation, we get $a_3 = -3a_2$. Substituting into the first equation,

$$-2a_2 + a_2 - 3a_2 = 0$$
$$-4a_2 = 0$$
$$\therefore \quad a_1 = a_2 = a_3 = 0$$

and hence the three vectors are linearly independent. Therefore, they constitute a basis for R^3.

(b) This is not a basis because it contains four vectors.

Property E If S is a linearly independent subset of a vector space V, then there is a basis for V that contains all the vectors in S.

EXAMPLE 7-11: Find a basis for R^4 which contains the two linearly independent vectors

$$\begin{bmatrix} 2 \\ 1 \\ 0 \\ 1 \end{bmatrix} \quad \text{and} \quad \begin{bmatrix} 1 \\ 0 \\ 1 \\ 1 \end{bmatrix}$$

Solution: We shall consider the two common methods. Both of the following methods start with the natural or standard basis for R^4, which is

$$\left\{ \begin{bmatrix} 1 \\ 0 \\ 0 \\ 0 \end{bmatrix}, \begin{bmatrix} 0 \\ 1 \\ 0 \\ 0 \end{bmatrix}, \begin{bmatrix} 0 \\ 0 \\ 1 \\ 0 \end{bmatrix}, \begin{bmatrix} 0 \\ 0 \\ 0 \\ 1 \end{bmatrix} \right\}$$

Method 1: The procedure given in most books is to take e_1 of the natural basis and determine if it is a linear combination of the given vectors. If it is, discard it; otherwise, retain it. Continuing with this procedure, take the next vector in the natural basis and determine if it is a linear combination of the given vectors and any previously retained from the natural basis. When you have discarded the same number of vectors as the number given, then all remaining vectors in the natural basis along with the given vectors constitute a basis. In our problem, the procedure is

$$a_1 \begin{bmatrix} 2 \\ 1 \\ 0 \\ 1 \end{bmatrix} + a_2 \begin{bmatrix} 1 \\ 0 \\ 1 \\ 1 \end{bmatrix} = \begin{bmatrix} 1 \\ 0 \\ 0 \\ 0 \end{bmatrix}$$

$$\begin{aligned} 2a_1 + a_1 &= 1 \\ a_1 \qquad &= 0 \\ a_2 &= 0 \\ a_1 + a_2 &= 0 \end{aligned}$$

Since $a_1 = a_2 = 0$, $\begin{bmatrix} 1 \\ 0 \\ 0 \\ 0 \end{bmatrix}$ cannot be written as a linear combination of the two vectors, and therefore $\begin{bmatrix} 1 \\ 0 \\ 0 \\ 0 \end{bmatrix}$ is linearly independent of the two given vectors.

Now take the next vector

$$a_1 \begin{bmatrix} 2 \\ 1 \\ 0 \\ 1 \end{bmatrix} + a_2 \begin{bmatrix} 1 \\ 0 \\ 1 \\ 1 \end{bmatrix} + a_3 \begin{bmatrix} 1 \\ 0 \\ 0 \\ 0 \end{bmatrix} = \begin{bmatrix} 0 \\ 1 \\ 0 \\ 0 \end{bmatrix}$$

$$\begin{aligned} 2a_1 + a_2 + a_3 &= 0 \\ a_1 \qquad\qquad &= 1 \\ a_2 \qquad &= 0 \\ a_1 + a_2 \qquad &= 0 \end{aligned}$$

Since this implies $a_1 = 1$ and $a_1 = 0$, there is no solution, and thus $\begin{bmatrix} 0 \\ 1 \\ 0 \\ 0 \end{bmatrix}$ is linearly independent. Since we have four linearly independent vectors in a vector space of dimension four, then

$$\begin{bmatrix} 2 \\ 1 \\ 0 \\ 1 \end{bmatrix}, \begin{bmatrix} 1 \\ 0 \\ 1 \\ 1 \end{bmatrix}, \begin{bmatrix} 1 \\ 0 \\ 0 \\ 0 \end{bmatrix}, \text{ and } \begin{bmatrix} 0 \\ 1 \\ 0 \\ 0 \end{bmatrix} \quad \text{constitute a basis}$$

Method 2: Reduce the matrix having the given vectors as rows to echelon form. (If the given vectors are not written as n-tuples, first write them in that form.)

$$\begin{bmatrix} 2 & 1 & 0 & 1 \\ 1 & 0 & 1 & 1 \end{bmatrix} \xrightarrow{R_1 \leftrightarrow R_2} \begin{bmatrix} 1 & 0 & 1 & 1 \\ 2 & 1 & 0 & 1 \end{bmatrix} \xrightarrow{R_2 \to R_2 - 2R_1} \begin{bmatrix} 1 & 0 & 1 & 1 \\ 0 & 1 & -2 & -1 \end{bmatrix}$$

Now form the matrix with these as the first two rows and the vectors of the natural basis as the remaining rows:

$$\begin{bmatrix} 1 & 0 & 1 & 1 \\ 0 & 1 & -2 & 1 \\ \cancel{1} & \cancel{0} & \cancel{0} & \cancel{0} \\ \cancel{0} & \cancel{1} & \cancel{0} & \cancel{0} \\ 0 & 0 & 1 & 0 \\ 0 & 0 & 0 & 1 \end{bmatrix}$$

Cross out all rows in the natural basis part of this matrix in which a leading 1 is in the same column as in the echelon form of the 2×4 matrix. The given vectors along with what remains of the natural basis is a basis. That is,

$$\begin{bmatrix} 2 \\ 1 \\ 0 \\ 1 \end{bmatrix}, \begin{bmatrix} 1 \\ 0 \\ 1 \\ 1 \end{bmatrix}, \begin{bmatrix} 0 \\ 0 \\ 1 \\ 0 \end{bmatrix}, \text{ and } \begin{bmatrix} 0 \\ 0 \\ 0 \\ 1 \end{bmatrix} \quad \text{is a basis}$$

Observe that the two methods do not necessarily give the same basis, but each will contain the given linearly independent vectors. You might also notice that

$$\begin{bmatrix} 1 \\ 0 \\ 1 \\ 1 \end{bmatrix}, \begin{bmatrix} 0 \\ 1 \\ -2 \\ -1 \end{bmatrix}, \begin{bmatrix} 0 \\ 0 \\ 1 \\ 0 \end{bmatrix}, \text{ and } \begin{bmatrix} 0 \\ 0 \\ 0 \\ 1 \end{bmatrix} \quad \text{is a basis}$$

but it does not contain the two given vectors.

Property F If a set S spans a vector space V, then a subset of S is a basis for V.

EXAMPLE 7-12: The vectors [1 0 3], [1 1 0], [0 1 1], [2 1 3], and [1 2 1] span R_3. Find a subset of these five vectors that is a basis for R_3.

Solution

Step 1: Form the equation

$$a_1[1 \quad 0 \quad 3] + a_2[1 \quad 1 \quad 0] + a_3[0 \quad 1 \quad 1] + a_4[2 \quad 1 \quad 3] + a_5[1 \quad 2 \quad 1] = [0 \quad 0 \quad 0]$$

Step 2: Write the coefficient matrix resulting from the homogeneous system in Step 1, and transform to row echelon form. You get

$$\begin{bmatrix} 1 & 1 & 0 & 2 & 1 \\ 0 & 1 & 1 & 1 & 2 \\ 3 & 0 & 1 & 3 & 1 \end{bmatrix} \xrightarrow{R_3 \to R_3 - 3R_1} \begin{bmatrix} 1 & 1 & 0 & 2 & 1 \\ 0 & 1 & 1 & 1 & 2 \\ 0 & -3 & 1 & -3 & -2 \end{bmatrix}$$

$$\xrightarrow{R_3 \to R_3 + 3R_2} \begin{bmatrix} 1 & 1 & 0 & 2 & 1 \\ 0 & 1 & 1 & 1 & 2 \\ 0 & 0 & 4 & 0 & 4 \end{bmatrix} \xrightarrow[R_3 \to 1/4 R_3]{} \begin{bmatrix} 1 & 1 & 0 & 2 & 1 \\ 0 & 1 & 1 & 1 & 2 \\ 0 & 0 & 1 & 0 & 1 \end{bmatrix}$$

Step 3: There are leading 1's in columns 1, 2, and 3. The original vectors corresponding to these columns in the coefficient matrix is a basis for R_3. In this case, they are $[1 \quad 0 \quad 3]$, $[1 \quad 1 \quad 0]$, and $[0 \quad 1 \quad 1]$.

note: Steps 1 and 2 are equivalent to expressing the given vectors as n-typles (if not already in that form), writing a matrix with these as the columns and transforming to row echelon form.

Once more, there may be other subsets besides the one in Step 3 that are bases of R_3. The basis obtained is always one for **span (S)** where $S = \{[1 \quad 0 \quad 3], [1 \quad 1 \quad 0], [0 \quad 1 \quad 1], [2 \quad 1 \quad 3], [1 \quad 2 \quad 1]\}$.

SUMMARY

1. A set S spans a vector space V if every vector in V can be written as a linear combination of the vectors in S.
2. The set W of all linear combinations of the vectors in subset A of a vector space V is a subspace of V and is called span A.
3. A set of vectors B from a vector space V is a basis for V if (**a**) B spans V and (**b**) B is linearly independent.
4. The natural or standard basis for R_n is $\{\mathbf{e}_1, \mathbf{e}_2, \ldots, \mathbf{e}_n\}$ where the first component of $\mathbf{e}_1$ is 1 and all others are zero, the second component of $\mathbf{e}_2$ is 1 and all others are zero, etc.
5. The number of vectors in a basis for a nonempty vector space V is called the dimension of V. $\text{Dim}\,\mathbf{0} = 0$.
6. If V is an n-dimensional vector space, then the dimension of any subspace of V is less than or equal to n.
7. Every basis of an n-dimensional vector space V contains n vectors. Any subset of V that contains fewer than n vectors does not span V; if it contains more than n vectors, it is linearly dependent.
8. A basis for a vector space V cannot contain the zero vector.
9. Every vector in a vector space V can be written uniquely as a linear combination of the vectors in a given basis.
10. If a subset S of an n-dimensional vector space V contains exactly n vectors, then it is either (**a**) both linearly independent and spans V or (**b**) neither linearly independent nor spans V.
11. If S is a linearly independent subset of a vector space V, then there is a basis for V that contains all the vectors in S.
12. If a set S spans a vector space V, then a subset of S is a basis for V.

RAISE YOUR GRADES

Can you . . . ?

☑ determine whether a set S spans a given vector space
☑ verify that span A is a subspace of V whenever A is a subset of V
☑ find a basis for a vector space
☑ identify the natural or standard basis for R_n
☑ find the dimension of a vector space
☑ list the dimensions of the vector spaces $\{0\}$, R_n, R^n, ${}_mM_n$, and P_n
☑ express a vector uniquely as a linear combination of the vectors in a given basis
☑ find a basis for V which contains all the vectors in a subset S of linear independent vectors in V
☑ find a basis for V which is a subset of a spanning set for V

SOLVED PROBLEMS

PROBLEM 7-1 Determine whether the vectors $[0 \quad 1 \quad -2]$, $[-2 \quad 1 \quad -1]$, and $[1 \quad 2 \quad -3]$ span R_3.

Solution We shall do this problem by Method 1 discussed in Example 7-1. We are looking for a linear combination of the three vectors $a_1[0 \quad 1 \quad -2] + a_2[-2 \quad 1 \quad -1] + a_3[1 \quad 2 \quad -3] = [a \quad b \quad c]$. If this system has a solution, the given vectors span R_3:

$$\begin{aligned} -2a_2 + a_3 &= a \\ a_1 + a_2 + 2a_3 &= b \\ -2a_1 - a_2 - 3a_3 &= c \end{aligned}$$

From the first equation we get $a_2 = (a_3 - a)/2$. If we substitute this in the second equation, we get

$$a_1 + \frac{a_3 - a}{2} + 2a_3 = b$$

Solving for a_1, we get $a_1 = (2b - 5a_3 + a)/2$. Then substituting in the third equation,

$$\begin{aligned} -2\left(\frac{2b - 5a_3 + a}{2}\right) - \frac{a_3 - a}{2} - 3a_3 &= c \\ -4b + 10a_3 - 2a - a_3 + a - 6a_3 &= 2c \\ 3a_3 &= 4b + a + 2c \\ a_3 &= (4b + a + 2c)/3 \end{aligned}$$

Substituting this value in results for a_2 and a_1 gives us

$$a_1 = \frac{-7a - a - 5c}{3} \qquad \text{and} \qquad a_2 = \frac{2b - a + c}{3}$$

Therefore,

$$\frac{-7b - a - 5c}{3}[0 \quad 1 \quad -2] + \frac{2b - a + c}{3}[-2 \quad 1 \quad -1] + \frac{4b + a + 2c}{3}[1 \quad 2 \quad -3] = [a \quad b \quad c]$$

and thus the vectors span R_3.

PROBLEM 7-2 Use Method 2 in Example 7-1 to show that the vectors $[0 \;\; 1 \;\; -2]$, $[-2 \;\; 1 \;\; -1]$, and $[1 \;\; 2 \;\; -3]$ span R_3.

Solution Write the system in Problem 7-1 in matrix form:

$$\begin{bmatrix} 0 & -2 & 1 \\ 1 & 1 & 2 \\ -2 & -1 & -3 \end{bmatrix}\begin{bmatrix} a_1 \\ a_2 \\ a_3 \end{bmatrix} = \begin{bmatrix} a \\ b \\ c \end{bmatrix}$$

If the determinant of the coefficient matrix is not zero, the system of equations has a solution.

$$\begin{vmatrix} 0 & -2 & 1 \\ 1 & 1 & 2 \\ -2 & -1 & -3 \end{vmatrix} = 0 - 1 + 8 + 2 + 0 - 6 = 3$$

Therefore, the three vectors span R_3.

PROBLEM 7-3 Determine which of the following sets of vectors span R^2:

(a) $\left\{\begin{bmatrix} 1 \\ -2 \end{bmatrix}, \begin{bmatrix} 2 \\ -1 \end{bmatrix}\right\}$ **(b)** $\left\{\begin{bmatrix} 3 \\ -2 \end{bmatrix}, \begin{bmatrix} -6 \\ 4 \end{bmatrix}\right\}$ **(c)** $\left\{\begin{bmatrix} -1 \\ 2 \end{bmatrix}, \begin{bmatrix} 3 \\ -1 \end{bmatrix}, \begin{bmatrix} 4 \\ -3 \end{bmatrix}\right\}$

Solution

(a)

$$a_1\begin{bmatrix} 1 \\ -2 \end{bmatrix} + a_2\begin{bmatrix} 2 \\ -1 \end{bmatrix} = \begin{bmatrix} a \\ b \end{bmatrix}$$

$$\begin{aligned} a_1 + 2a_2 &= a \\ -2a_1 - a_2 &= b \end{aligned} \quad \text{or} \quad \begin{bmatrix} 1 & 2 \\ -2 & -1 \end{bmatrix}\begin{bmatrix} a_1 \\ a_2 \end{bmatrix} = \begin{bmatrix} a \\ b \end{bmatrix}$$

$$\begin{vmatrix} 1 & 2 \\ -2 & -1 \end{vmatrix} = -1 + 4 = 3$$

Therefore, the system has a solution for every value of a and b, and hence the vectors span R^2.

(b)

$$a_1\begin{bmatrix} 3 \\ -2 \end{bmatrix} + a_2\begin{bmatrix} -6 \\ 4 \end{bmatrix} = \begin{bmatrix} a \\ b \end{bmatrix}$$

$$\begin{aligned} 3a_1 - 6a_2 &= a \\ -2a_1 + 4a_2 &= b \end{aligned} \quad \text{or} \quad \begin{bmatrix} 3 & -6 \\ -2 & 4 \end{bmatrix}\begin{bmatrix} a_1 \\ a_2 \end{bmatrix} = \begin{bmatrix} a \\ b \end{bmatrix}$$

$$\begin{vmatrix} 3 & -6 \\ -2 & 4 \end{vmatrix} = 12 - 12 = 0$$

Therefore, the vectors do not span R^2.

(c)

$$a_1\begin{bmatrix} -1 \\ 2 \end{bmatrix} + a_2\begin{bmatrix} 3 \\ -1 \end{bmatrix} + a_3\begin{bmatrix} 4 \\ -3 \end{bmatrix} = \begin{bmatrix} a \\ b \end{bmatrix}$$

$$\begin{aligned} -a_1 + 3a_2 + 4a_3 &= a \\ 2a_1 - a_2 - 3a_3 &= b \end{aligned} \quad \text{or} \quad \begin{bmatrix} -1 & 3 & 4 \\ 2 & -1 & -3 \end{bmatrix}\begin{bmatrix} a_1 \\ a_2 \\ a_3 \end{bmatrix} = \begin{bmatrix} a \\ b \end{bmatrix}$$

Since $\begin{bmatrix} -1 & 3 & 4 \\ 2 & -1 & -3 \end{bmatrix}$ is not a square matrix, the determinant is not defined, so we must use some other method. We will reduce the augmented matrix to reduced row echelon form:

$$\begin{bmatrix} -1 & 3 & 4 & a \\ 2 & -1 & -3 & b \end{bmatrix} \xrightarrow{R_2 \to R_2 + 2R_1} \begin{bmatrix} -1 & 3 & 4 & a \\ 0 & 5 & 5 & 2a+b \end{bmatrix}$$

$$\xrightarrow[R_2 \to 1/5 R_2]{R_1 \to -R_1} \begin{bmatrix} 1 & -3 & -4 & -a \\ 0 & 1 & 1 & \dfrac{2a+b}{5} \end{bmatrix} \xrightarrow{R_1 \to R_1 + 3R_2} \begin{bmatrix} 1 & 0 & -1 & \dfrac{a+3b}{5} \\ 0 & 1 & 1 & \dfrac{2a+b}{5} \end{bmatrix}$$

$$\therefore \quad a_1 = \frac{a+3b}{5} + a_3$$

$$a_2 = \frac{2a+b}{5} - a_3$$

Since a_3 is any arbitrary number, let $a_3 = 0$. Therefore,

$$a_1 = \frac{a+3b}{5}, \qquad a_2 = \frac{2a+b}{5}, \qquad a_3 = 0$$

is one solution. For each value assigned to a_3, we get different values for a_1 and a_2. Therefore, the three vectors span R^2.

PROBLEM 7-4 Show that the vectors $[1 \quad -1 \quad 1]$, $[-1 \quad 2 \quad 3]$, and $[2 \quad -1 \quad 1]$ span R_3.

Solution Set up the system of equations

$$\begin{bmatrix} 1 & -1 & 2 \\ -1 & 2 & -1 \\ 1 & 3 & 1 \end{bmatrix} \begin{bmatrix} a_1 \\ a_2 \\ a_3 \end{bmatrix} = \begin{bmatrix} a \\ b \\ c \end{bmatrix}$$

and find

$$\det \begin{bmatrix} 1 & -1 & 2 \\ -1 & 2 & -1 \\ 1 & 3 & 1 \end{bmatrix} = \frac{1}{1} \begin{vmatrix} 1 & 1 \\ 4 & -1 \end{vmatrix} = -1 - 4 = -5$$

Since the determinant of the coefficient matrix is nonzero, the system has a solution; therefore, the vectors span R_3.

PROBLEM 7-5 Verify that span A where

$$A = \left\{ \begin{bmatrix} 1 \\ 2 \\ 0 \end{bmatrix}, \begin{bmatrix} 0 \\ 2 \\ 1 \end{bmatrix} \right\}$$

is a subspace of R^3.

Solution Every vector in span A is a linear combination of $\begin{bmatrix} 1 \\ 2 \\ 0 \end{bmatrix}$ and $\begin{bmatrix} 0 \\ 2 \\ 1 \end{bmatrix}$. Let $\mathbf{a}_1$ and $\mathbf{a}_2$ be two such vectors where

$$\mathbf{a}_1 = \mathbf{a}_1 \begin{bmatrix} 1 \\ 2 \\ 0 \end{bmatrix} + a_2 \begin{bmatrix} 0 \\ 2 \\ 1 \end{bmatrix} \qquad \mathbf{a}_2 = b_1 \begin{bmatrix} 1 \\ 2 \\ 0 \end{bmatrix} + b_2 \begin{bmatrix} 0 \\ 2 \\ 1 \end{bmatrix}$$

$$\mathbf{a}_1 + \mathbf{a}_2 = a_1\begin{bmatrix}1\\2\\0\end{bmatrix} + a_2\begin{bmatrix}0\\2\\1\end{bmatrix} + b_1\begin{bmatrix}1\\2\\0\end{bmatrix} + b_2\begin{bmatrix}0\\2\\1\end{bmatrix}$$

$$= (a_1 + b_1)\begin{bmatrix}1\\2\\0\end{bmatrix} + (a_2 + b_2)\begin{bmatrix}0\\2\\1\end{bmatrix}$$

and

$$c\mathbf{a}_1 = c\left(a_1\begin{bmatrix}1\\2\\0\end{bmatrix} + a_2\begin{bmatrix}0\\2\\1\end{bmatrix}\right) = ca_1\begin{bmatrix}1\\2\\0\end{bmatrix} + ca_2\begin{bmatrix}0\\2\\1\end{bmatrix}$$

Therefore, since span A is closed under addition and scalar multiplication, it is a subspace of R^3.

PROBLEM 7-6 Determine whether the vectors $[1\ \ -2\ \ -3]$, $[-2\ \ 3\ \ 1]$, and $[0\ \ 1\ \ -1]$ form a basis for R_3.

Solution Let $[a\ \ b\ \ c]$ be any vector in R_3

$$a_1[1\ \ -2\ \ -3] + a_2[-2\ \ 3\ \ 1] + a_3[0\ \ 1\ \ -1] = [a\ \ b\ \ c]$$

$$\begin{aligned} a_1 - 2a_2 \qquad &= a \\ -2a_1 + 3a_2 + a_3 &= b \\ -3a_1 + a_2 - a_3 &= c \end{aligned} \qquad \text{or} \qquad \begin{bmatrix}1 & -2 & 0\\ -2 & 3 & 1\\ -3 & 1 & -1\end{bmatrix}\begin{bmatrix}a_1\\a_2\\a_3\end{bmatrix} = \begin{bmatrix}a\\b\\c\end{bmatrix}$$

We can show that the three given vectors form a basis for R_3 by showing that the given vectors are linearly independent and span R_3. We shall do this by showing that $\begin{bmatrix}1 & -2 & 0\\ -2 & 3 & 1\\ -3 & 1 & -1\end{bmatrix}$ is invertible.

Since

$$\begin{vmatrix}1 & -2 & 0\\ -2 & 3 & 1\\ -3 & 1 & -1\end{vmatrix} = \frac{1}{1}\begin{vmatrix}-1 & 1\\ -5 & -1\end{vmatrix} = 1 + 5 = 6$$

which is nonzero, the matrix is invertible. Therefore, $[1\ \ -2\ \ -3]$, $[-2\ \ 3\ \ 1]$, and $[0\ \ 1\ \ -1]$ form a basis for R_3.

PROBLEM 7-7 Determine whether $\begin{bmatrix}1\\-1\\0\end{bmatrix}$, $\begin{bmatrix}1\\2\\-1\end{bmatrix}$, and $\begin{bmatrix}4\\-1\\-1\end{bmatrix}$ form a basis for R^3.

Solution Let $\begin{bmatrix}a\\b\\c\end{bmatrix}$ be any vector in R^3. Then

$$a_1\begin{bmatrix}1\\-1\\0\end{bmatrix} + a_2\begin{bmatrix}1\\2\\-1\end{bmatrix} + a_3\begin{bmatrix}4\\-1\\-1\end{bmatrix} = \begin{bmatrix}a\\b\\c\end{bmatrix}$$

$$\begin{aligned} a_1 + a_2 + 4a_3 &= a \\ -a_1 + 2a_2 - a_3 &= b \\ -a_2 - a_3 &= c \end{aligned} \quad \text{or} \quad \begin{bmatrix} 1 & 1 & 4 \\ -1 & 2 & -1 \\ 0 & -1 & -1 \end{bmatrix} \begin{bmatrix} a_1 \\ a_2 \\ a_3 \end{bmatrix} = \begin{bmatrix} a \\ b \\ c \end{bmatrix}$$

If $\det \begin{bmatrix} 1 & 1 & 4 \\ -1 & 2 & -1 \\ 0 & -1 & -1 \end{bmatrix}$ is nonzero, the vectors form a basis for R^3.

$$\begin{vmatrix} 1 & 1 & 4 \\ -1 & 2 & -1 \\ 0 & -1 & -1 \end{vmatrix} = 1 \begin{vmatrix} 2 & -1 \\ -1 & -1 \end{vmatrix} - (-1) \begin{vmatrix} 1 & 4 \\ -1 & -1 \end{vmatrix} = -2 - 1 + 1(-1 + 4)$$

$$= -3 + 3 = 0$$

Therefore, the vectors do not form a basis for R^3. This means that these vectors are not independent.

PROBLEM 7-8 Find a basis and the dimension of the solution space of the system:

$$\begin{aligned} x_1 - x_2 + 2x_3 - 3x_4 &= 0 \\ x_1 + x_2 - 3x_3 + x_4 &= 0 \\ 2x_1 - x_3 - 2x_4 &= 0 \\ 2x_2 - 5x_3 + 4x_4 &= 0 \end{aligned}$$

Solution We will use the Gauss–Jordan elimination method on the augmented matrix:

$$\begin{bmatrix} 1 & -1 & 2 & -3 & 0 \\ 1 & 1 & -3 & 1 & 0 \\ 2 & 0 & -1 & -2 & 0 \\ 0 & 2 & -5 & 4 & 0 \end{bmatrix} \xrightarrow[R_3 \to R_3 - 2R_1]{R_2 \to R_2 - R_1} \begin{bmatrix} 1 & -1 & 2 & -3 & 0 \\ 0 & 2 & -5 & 4 & 0 \\ 0 & 2 & -5 & 4 & 0 \\ 0 & 2 & -5 & 4 & 0 \end{bmatrix} \xrightarrow[R_4 \to R_4 - R_2]{R_3 \to R_3 - R_2} \begin{bmatrix} 1 & -1 & 2 & -3 & 0 \\ 0 & 2 & -5 & 4 & 0 \\ 0 & 0 & 0 & 0 & 0 \\ 0 & 0 & 0 & 0 & 0 \end{bmatrix}$$

$$\xrightarrow{R_2 \to 1/2 R_2} \begin{bmatrix} 1 & -1 & 2 & -3 & 0 \\ 0 & 1 & -\frac{5}{2} & 2 & 0 \\ 0 & 0 & 0 & 0 & 0 \\ 0 & 0 & 0 & 0 & 0 \end{bmatrix} \xrightarrow{R_1 \to R_1 + R_2} \begin{bmatrix} 1 & 0 & -\frac{1}{2} & -1 & 0 \\ 0 & 1 & -\frac{5}{2} & 2 & 0 \\ 0 & 0 & 0 & 0 & 0 \\ 0 & 0 & 0 & 0 & 0 \end{bmatrix}$$

$$\therefore \quad X = \begin{bmatrix} x_1 \\ x_2 \\ x_3 \\ x_4 \end{bmatrix} = \begin{bmatrix} \frac{1}{2}s + t \\ \frac{5}{2}s - 2t \\ s \\ t \end{bmatrix} = \begin{bmatrix} \frac{1}{2}s \\ \frac{5}{2}s \\ s \\ 0 \end{bmatrix} + \begin{bmatrix} t \\ -2t \\ 0 \\ t \end{bmatrix} = s \begin{bmatrix} \frac{1}{2} \\ \frac{5}{2} \\ 1 \\ 0 \end{bmatrix} + t \begin{bmatrix} 1 \\ -2 \\ 0 \\ 1 \end{bmatrix}$$

Hence

$$\begin{bmatrix} \frac{1}{2} \\ \frac{5}{2} \\ 1 \\ 0 \end{bmatrix} \quad \text{and} \quad \begin{bmatrix} 1 \\ -2 \\ 0 \\ 1 \end{bmatrix}$$

form a basis for the solution space. Since this basis contains two vectors, the dimension of the solution space is 2.

PROBLEM 7-9 Find the dimension and a basis of the solution space of the system

$$\begin{aligned} x + 2y - 3z + w + 2t &= 0 \\ x - 2y + 2z - w + t &= 0 \\ 3x + 2y - 4z + w + 5t &= 0 \end{aligned}$$

Solution Using the Gauss–Jordan elimination on the augmented matrix, we proceed as follows:

$$\begin{bmatrix} 1 & 2 & -3 & 1 & 2 & 0 \\ 1 & -2 & 2 & -1 & 1 & 0 \\ 3 & 2 & -4 & 1 & 5 & 0 \end{bmatrix} \xrightarrow[R_3 \to R_3 - 3R_1]{R_2 \to R_2 - R_1} \begin{bmatrix} 1 & 2 & -3 & 1 & 2 & 0 \\ 0 & -4 & 5 & -2 & -1 & 0 \\ 0 & -4 & 5 & -2 & -1 & 0 \end{bmatrix}$$

$$\xrightarrow[R_3 \to R_3 - R_2]{} \begin{bmatrix} 1 & 2 & -3 & 1 & 2 & 0 \\ 0 & -4 & 5 & -2 & -1 & 0 \\ 0 & 0 & 0 & 0 & 0 & 0 \end{bmatrix} \xrightarrow{R_2 \to -1/4 R_2} \begin{bmatrix} 1 & 2 & -3 & 1 & 2 & 0 \\ 0 & 1 & -\frac{5}{4} & \frac{1}{2} & \frac{1}{4} & 0 \\ 0 & 0 & 0 & 0 & 0 & 0 \end{bmatrix}$$

$$\xrightarrow{R_1 \to R_1 - 2R_2} \begin{bmatrix} 1 & 0 & -\frac{1}{2} & 0 & \frac{3}{2} & 0 \\ 0 & 1 & -\frac{5}{4} & \frac{1}{2} & \frac{1}{4} & 0 \\ 0 & 0 & 0 & 0 & 0 & 0 \end{bmatrix}$$

Therefore,

$$X = \begin{bmatrix} x \\ y \\ z \\ w \\ t \end{bmatrix} = \begin{bmatrix} \frac{1}{2}s - \frac{3}{2}v \\ \frac{5}{4}s - \frac{1}{2}u - \frac{1}{4}v \\ s \\ u \\ v \end{bmatrix}$$

$$= s\begin{bmatrix} \frac{1}{2} \\ \frac{5}{4} \\ 1 \\ 0 \\ 0 \end{bmatrix} + u\begin{bmatrix} 0 \\ -\frac{1}{2} \\ 0 \\ 1 \\ 0 \end{bmatrix} + v\begin{bmatrix} -\frac{3}{2} \\ -\frac{1}{4} \\ 0 \\ 0 \\ 1 \end{bmatrix}$$

Hence,

$$\begin{bmatrix} \frac{1}{2} \\ \frac{5}{4} \\ 1 \\ 0 \\ 0 \end{bmatrix}, \begin{bmatrix} 0 \\ -\frac{1}{2} \\ 0 \\ 1 \\ 0 \end{bmatrix} \quad \text{and} \quad \begin{bmatrix} -\frac{3}{2} \\ -\frac{1}{4} \\ 0 \\ 0 \\ 1 \end{bmatrix}$$

form a basis for the solution space. Since this basis has three vectors the dimension of the solution space is 3.

PROBLEM 7-10 Which of the following sets of vectors are bases for R_2? **(a)** $[-2 \quad 3]$; **(b)** $[2 \quad -1]$, $[-3 \quad 4]$, $[-1 \quad 3]$; **(c)** $[-1 \quad 2]$, $[2 \quad 0]$.

Solution The dimension of R_2 is 2.

(a) Since this set contains only one vector, it is not a basis.

(b) This set contains three vectors, and a basis for R_2 cannot contain more than two vectors; this is not a basis for R_2. In fact, $1[2 \quad -1] + 1[-3 \quad 4] = [-1 \quad 3]$.

(c) Since this set contains two vectors, we must check for linear independence and that the two vectors span R_2.

We shall check for linear independence first. The vectors will be linearly independent if there exist a_1 and a_2 such that

$$\begin{aligned} a_1[-1 \quad 2] + a_2[2 \quad 0] &= [0 \quad 0] \\ -a_1 + 2a_2 &= 0 \\ 2a_1 \qquad &= 0 \end{aligned}$$

Therefore, $a_1 = 0$ and $a_2 = 0$, and the vectors are linearly independent.

Now, we shall check to see that the vectors span R_2:

$$\begin{aligned} a_1[-1 \quad 2] + a_2[2 \quad 0] &= [a \quad b] \\ -a_1 + 2a_2 &= a \\ 2a_1 \qquad &= b \end{aligned}$$

$$\therefore \quad a_1 = \frac{b}{2} \qquad \text{and} \qquad a_2 = \frac{2a+b}{4}$$

Thus, given any a and b, there exist a_1 and a_2 such that $[a \quad b]$ is a linear combination of $[-1 \quad 2]$ and $[2 \quad 0]$. Hence, the vectors span R_2 and form a basis for R_2.

PROBLEM 7-11 Explain why $[1 \quad 1 \quad 0]$, $[0 \quad 0 \quad 0]$, and $[1 \quad 2 \quad 4]$ do not form a basis for R_3.

Solution We have three vectors given and the dim $R_3 = 3$ but a basis for a vector space cannot contain the zero vector; hence the given set is not a basis for R_3.

PROBLEM 7-12 In Problem 7-6 we found that $\{[1 \quad -2 \quad -3], [-2 \quad 3 \quad 1], [0 \quad 1 \quad -1]\}$ is a basis for R_3. **(a)** Express $[2 \quad -5 \quad 7]$ as a linear combination of these three vectors in a basis of R_3. **(b)** Explain, without any computation, why $[2 \quad -5 \quad 7] = 2[1 \quad -2 \quad -3] + 3[-2 \quad 3 \quad 1] + 2[0 \quad 1 \quad -1]$ cannot be correct.

Solution

(a)

$$a_1[1 \quad -2 \quad -3] + a_2[-2 \quad 3 \quad 1] + a_3[0 \quad 1 \quad -1] = [2 \quad -5 \quad 7]$$

$$\begin{aligned} a_1 - 2a_2 \qquad &= 2 \\ -2a_1 + 3a_2 + a_3 &= -5 \\ -3a_1 + a_2 - a_3 &= 7 \end{aligned}$$

Solving this by any method will give you $a_1 = -2$, $a_2 = -2$, and $a_3 = -3$. Therefore, $-2[1 \quad -2 \quad -3]$ $-2[-2 \quad 3 \quad 1] -3[0 \quad 1 \quad -1] = [2 \quad -5 \quad 7]$.

(b) Since any vector in R_3 can be expressed uniquely as a linear combination of the vectors in any of its bases, the only correct answer is $[2 \quad -5 \quad 7] = -2[1 \quad -2 \quad -3] - 2[-2 \quad 3 \quad 1] - 3[0 \quad 1 \quad -1]$.

PROBLEM 7-13 Find a basis for R^4 which contains the two linearly independent vectors

$$\begin{bmatrix} 2 \\ 5 \\ -4 \\ 0 \end{bmatrix} \quad \text{and} \quad \begin{bmatrix} 0 \\ 1 \\ 0 \\ 2 \end{bmatrix}$$

Solution We will do this problem by the two methods discussed in Example 7-11.

Method 1:

$$\left\{ \begin{bmatrix} 1 \\ 0 \\ 0 \\ 0 \end{bmatrix}, \begin{bmatrix} 0 \\ 1 \\ 0 \\ 0 \end{bmatrix}, \begin{bmatrix} 0 \\ 0 \\ 1 \\ 0 \end{bmatrix}, \begin{bmatrix} 0 \\ 0 \\ 0 \\ 1 \end{bmatrix} \right\} \qquad \text{standard basis for } R^4$$

$$a_1 \begin{bmatrix} 2 \\ 5 \\ -4 \\ 0 \end{bmatrix} + a_2 \begin{bmatrix} 0 \\ 1 \\ 0 \\ 2 \end{bmatrix} = \begin{bmatrix} 1 \\ 0 \\ 0 \\ 0 \end{bmatrix}$$

$$\begin{aligned} 2a_1 \qquad &= 1 \\ 5a_1 + \ a_2 &= 0 \\ -4a_1 \qquad &= 0 \\ 2a_2 &= 0 \end{aligned}$$

Since $a_1 = a_2 = 0$ from the third and fourth equations and the first equation gives $a_1 = 1/2$, $\begin{bmatrix} 1 \\ 0 \\ 0 \\ 0 \end{bmatrix}$ cannot be written as a linear combination of the two vectors and is therefore linearly independent.

Next we consider

$$a_1 \begin{bmatrix} 2 \\ 5 \\ -4 \\ 0 \end{bmatrix} + a_2 \begin{bmatrix} 0 \\ 1 \\ 0 \\ 2 \end{bmatrix} = \begin{bmatrix} 0 \\ 1 \\ 0 \\ 0 \end{bmatrix}$$

$$\begin{aligned} 2a_1 \qquad &= 0 \\ 5a_1 + \ a_2 &= 1 \\ -4a_1 \qquad &= 0 \\ 2a_2 &= 0 \end{aligned}$$

Since $a_1 = a_2 = 0$ and $5a_1 + a_2 = 1$, the vector $\begin{bmatrix} 0 \\ 1 \\ 0 \\ 0 \end{bmatrix}$ cannot be written as a linear combination of the two given vectors and is therefore linearly independent. Since R^4 has four linearly independent vectors in any basis,

$$\begin{bmatrix} 2 \\ 5 \\ -4 \\ 0 \end{bmatrix}, \begin{bmatrix} 0 \\ 1 \\ 0 \\ 2 \end{bmatrix}, \begin{bmatrix} 1 \\ 0 \\ 0 \\ 0 \end{bmatrix}, \quad \text{and} \quad \begin{bmatrix} 0 \\ 1 \\ 0 \\ 0 \end{bmatrix} \quad \text{constitute a basis}$$

Method 2: Reduce the matrix having the given vectors as rows to echelon form.

$$\begin{bmatrix} 2 & 5 & -4 & 0 \\ 0 & 1 & 0 & 2 \end{bmatrix} \xrightarrow{R_1 \to R_1 - 5R_2} \begin{bmatrix} 2 & 0 & -4 & -10 \\ 0 & 1 & 0 & 2 \end{bmatrix}$$

$$\xrightarrow{R_1 \to 1/2 R_1} \begin{bmatrix} 1 & 0 & -2 & -5 \\ 0 & 1 & 0 & 2 \end{bmatrix}$$

Now form the matrix

$$\begin{bmatrix} 1 & 0 & -2 & -5 \\ 0 & 1 & 0 & 2 \\ 1 & 0 & 0 & 0 \\ 0 & 1 & 0 & 0 \\ 0 & 0 & 1 & 0 \\ 0 & 0 & 0 & 1 \end{bmatrix}$$

Cross out the third and fourth rows since there is a leading 1 above each of these in the echelon form of the 2×4 matrix. The given vectors along with what remains of the natural basis is a basis for R^4. That is,

$$\begin{bmatrix} 2 \\ 5 \\ -4 \\ 0 \end{bmatrix}, \begin{bmatrix} 0 \\ 1 \\ 0 \\ 2 \end{bmatrix}, \begin{bmatrix} 0 \\ 0 \\ 1 \\ 0 \end{bmatrix}, \quad \text{and} \quad \begin{bmatrix} 0 \\ 0 \\ 0 \\ 1 \end{bmatrix} \quad \text{is a basis}$$

The two methods do not necessarily give the same basis. In fact,

$$\begin{bmatrix} 1 \\ 0 \\ -2 \\ -5 \end{bmatrix}, \begin{bmatrix} 0 \\ 1 \\ 0 \\ 2 \end{bmatrix}, \begin{bmatrix} 0 \\ 0 \\ 1 \\ 0 \end{bmatrix}, \quad \text{and} \quad \begin{bmatrix} 0 \\ 0 \\ 0 \\ 1 \end{bmatrix} \quad \text{is also a basis}$$

but it does not contain the two given vectors.

PROBLEM 7-14 The vectors [1 2 1], [2 3 3], [1 1 2], [2 −1 1], and [2 1 0] span R_3. Find a subset of these five vectors that is a basis for R_3.

Solution

Step 1: Form the equation

$$a_1[1 \quad 2 \quad 1] + a_2[2 \quad 3 \quad 3] + a_3[1 \quad 1 \quad 2] + a_4[2 \quad -1 \quad 1] + a_5[2 \quad 1 \quad 0] = [0 \quad 0 \quad 0]$$

Step 2:

$$\begin{aligned} a_1 + 2a_2 + a_3 + 2a_4 + 2a_5 &= 0 \\ 2a_1 + 3a_2 + a_3 - a_4 + a_5 &= 0 \\ a_1 + 3a_2 + 2a_3 + a_4 \quad &= 0 \end{aligned}$$

$$\begin{bmatrix} 1 & 2 & 1 & 2 & 2 \\ 2 & 3 & 1 & -1 & 1 \\ 1 & 3 & 2 & 1 & 0 \end{bmatrix} \xrightarrow[R_3 \to R_3 - R_1]{R_2 \to R_2 - 2R_1} \begin{bmatrix} 1 & 2 & 1 & 2 & 2 \\ 0 & -1 & -1 & -5 & -3 \\ 0 & 1 & 1 & -1 & -2 \end{bmatrix}$$

$$\xrightarrow[R_3 \to R_3 + R_2]{} \begin{bmatrix} 1 & 2 & 1 & 2 & 2 \\ 0 & -1 & -1 & -5 & -3 \\ 0 & 0 & 0 & -6 & -5 \end{bmatrix} \xrightarrow[R_3 \to -1/6R_3]{R_2 \to -R_2} \begin{bmatrix} 1 & 2 & 1 & 2 & 2 \\ 0 & 1 & 1 & 5 & 3 \\ 0 & 0 & 0 & 1 & \frac{5}{6} \end{bmatrix}$$

Step 3: There are leading ones in the first, second, and fourth columns. The original vectors corresponding to these columns in the coefficient matrix is a basis for R_3. In this case, they are [1 2 1], [2 3 3], and [2 −1 1]. If we let

$S = \{[1\ \ 2\ \ 1], [2\ \ 3\ \ 3], [1\ \ 1\ \ 2], [2\ \ {-1}\ \ 1], [2\ \ 1\ \ 0]\}$ then basis for span S is $\{[1\ \ 2\ \ 1], [2\ \ 3\ \ 3], [2\ \ {-1}\ \ 1]\}$.

PROBLEM 7-15 Do the polynomials $x^2 + 2x - 1$, $x^2 + 2$, $-x^2 + x - 5$ span P_2?

Solution We need to find a_1, a_2, and a_3 such that

$$a_1(x^2 + 2x - 1) + a_2(x^2 + 2) + a_3(-x^2 + x - 5) = ax^2 + bx + c.$$

From this equation we get the system of equations

$$\begin{aligned} a_1 + a_2 - a_3 &= a \\ 2a_1 \qquad + a_3 &= b \\ -a_1 + 2a_2 - 5a_3 &= c \end{aligned}$$

If we solve this by any method, we get

$$a_1 = \frac{-2a + 3b + c}{3}, \qquad a_2 = \frac{9a - 6b - 3c}{3} = 3a - 2b - c$$

and

$$a_3 = \frac{4a - 3b - 2c}{3}$$

Since every polynomial in P_2 can be expressed as a linear combination of given polynomials, they span P_2.

Supplementary Problems

PROBLEM 7-16 Determine whether the following sets of vectors span R_3.

(a) $\{[1\ \ 0\ \ 3], [-1\ \ 2\ \ 1], [2\ \ {-1}\ \ 3]\}$;
(b) $\{[2\ \ 1\ \ {-2}\}, [-1\ \ 3\ \ 0]\}$;
(c) $\{[1\ \ {-1}\ \ 1], [1\ \ 0\ \ {-1}], [2\ \ {-1}\ \ 0], [-2\ \ 3\ \ {-1}]\}$.

PROBLEM 7-17 Determine whether the sets of vectors form a basis for R^3:

(a) $$\left\{\begin{bmatrix}1\\0\\2\end{bmatrix}, \begin{bmatrix}1\\-2\\3\end{bmatrix}, \begin{bmatrix}2\\-2\\5\end{bmatrix}\right\}$$

(b) $$\left\{\begin{bmatrix}1\\-1\\2\end{bmatrix}, \begin{bmatrix}2\\3\\0\end{bmatrix}, \begin{bmatrix}-1\\4\\-1\end{bmatrix}\right\}$$

PROBLEM 7-18 Find a basis and the dimension of the solution space of the system

$$\begin{aligned} x_1 - 2x_2 + 3x_3 &= 0 \\ x_1 + x_2 - x_3 &= 0 \\ 2x_1 - x_2 + 2x_3 &= 0 \\ 3x_2 - 4x_3 &= 0 \end{aligned}$$

PROBLEM 7-19 Which of the following sets are bases for R_2? **(a)** $\{[1\ \ {-2}]\}$ **(b)** $\{[1\ \ 2], [-5\ \ 4]\}$ **(c)** $\{[1\ \ 1], [-1\ \ 0]\}$; **(d)** $\{[-1\ \ {-3}], [0\ \ 0], [2\ \ {-4}]\}$.

PROBLEM 7-20 Express [7 0 15] as a linear combination of [1 −1 1], [−1 2 3], and [2 −1 1].

PROBLEM 7-21 Which of the following sets form a basis for P_2? **(a)** $\{x^2 + 2x, 3x + 2, x - 4\}$; **(b)** $\{x^2 - x + 2, x + 1\}$; **(c)** $\{x^2, x^2 + 4, x^2 - 2x\}$.

PROBLEM 7-22 Find a basis for the following subspaces of R_4: **(a)** (a, b, c, d), where $a - c = d$; **(b)** (a, b, c, d), where $a + 2b = c$.

PROBLEM 7-23 Find the dimension of each subspace of R_4 as defined in Exercise 7-22.

PROBLEM 7-24 Find the dimension of the subspace W of ${}_2M_2$ spanned by

(a) $\begin{bmatrix} 1 & 3 \\ 1 & 3 \end{bmatrix}$ and $\begin{bmatrix} 1 & 1 \\ 3 & 3 \end{bmatrix}$ **(b)** $\begin{bmatrix} 1 & 2 \\ -1 & 1 \end{bmatrix}$ and $\begin{bmatrix} 2 & 4 \\ -2 & 2 \end{bmatrix}$

PROBLEM 7-25 Find a subset of the given vectors which span R_3, that is a basis for R_3. [1 −1 2], [1 0 1], [2 1 −1], and [1 0 2].

PROBLEM 7-26 Find a basis for R_4 that contains [−1 0 1 0] and [0 −2 1 0].

PROBLEM 7-27 Find the dimension of the solution space of

$$\begin{bmatrix} 1 & 1 & -1 \\ -2 & 0 & 1 \\ 3 & -1 & 2 \end{bmatrix} \begin{bmatrix} x_1 \\ x_2 \\ x_3 \end{bmatrix} = \begin{bmatrix} 0 \\ 0 \\ 0 \end{bmatrix}$$

PROBLEM 7-28 Find a basis of R_3 that includes the vectors [−1 0 −3] and [2 −1 0].

PROBLEM 7-29 Do the vectors [0 1 0], [−1 0 2], and [2 0 1] form a basis for R_3?

PROBLEM 7-30 For what values of a is $[2a^2 \;\; -a \;\; 3]$ in span {[1 −2 0], [1 −1 1], [2 −3 1]}?

Answers to Supplementary Problems

7-16 **(a)** Since determinant of the coefficient matrix is nonzero, the set spans R_3.
(b) The spanning set of R_3 must contain three vectors; hence this set does not span R_3.
(c) This set does span R_3 since it contains at least one set of three linearly independent vectors:

$$-1[1 \;\; 0 \;\; -1] + 1[2 \;\; -1 \;\; 0] + 0[-2 \;\; 3 \;\; -1] = [1 \;\; -1 \;\; 1]$$

7-17 **(a)** This is not a basis for R^3, since the third vector is the sum of the first two.
(b) This is a basis for R^3.

7-18 A basis of the solution space is $\begin{bmatrix} -\frac{1}{3} \\ \frac{4}{3} \\ 1 \end{bmatrix}$. The dimension is 1.

7-19 **(a)** Not a basis for R_2
(b) A basis for R_2
(c) A basis for R_2
(d) Not a basis for R_2. A basis cannot contain the zero vector.

7-20 2[1 −1 1] + 3[−1 2 3] + 4[2 −1 1] = [7 0 15]

7-21 **(a)** A basis for P_2 **(b)** Not a basis for P_2 **(c)** A basis for P_2

7-22 **(a)** A basis is [1 0 1 0], [0 1 0 0], and [1 0 0 1].
(b) A basis is [−2 1 0 0], [1 0 1 0], and [0 0 0 1].

7-23 **(a)** Dimension of the subspace is 3
(b) Dimension of the subspace is 3

7-24 (**a**) Dim $W = 2$, since the matrices are linearly independent
(**b**) Dim $W = 1$, since each is a multiple of the other

7-25 A basis is $[1\ \ -1\ \ 2]$, $[1\ \ 0\ \ 1]$, $[2\ \ 1\ \ -1]$.

7-26 A basis is $[-1\ \ 0\ \ 1\ \ 0]$, $[0\ \ -2\ \ 1\ \ 0]$, $[1\ \ 0\ \ 0\ \ 0]$, and $[0\ \ 0\ \ 0\ \ 1]$.

7-27 Dimension of the solution space is 0, since the system has no solution except $x_1 = x_2 = x_3 = 0$.

7-28 A basis is $[-1\ \ 0\ \ -3]$, $[2\ \ -1\ \ 0]$, and $[1\ \ 0\ \ 0]$

7-29 Yes, $\begin{vmatrix} 0 & -1 & 2 \\ 1 & 0 & 0 \\ 0 & 2 & 1 \end{vmatrix} = 5$

7-30 $a = 1$ and $a = -3/4$

ROW AND COLUMN SPACES OF A MATRIX

THIS CHAPTER IS ABOUT

☑ **Rank of a Matrix**
☑ **Properties of Row and Column Spaces**
☑ **Equivalent Statements**

8-1. Rank of a Matrix

A. Definitions of row and column spaces

Let

$$A = [a_{ij}] = \begin{bmatrix} a_{11} & a_{12} & \cdots & a_{1n} \\ a_{21} & a_{22} & \cdots & a_{2n} \\ \vdots & \cdots & \cdots & \vdots \\ a_{m1} & a_{m2} & \cdots & a_{mn} \end{bmatrix}$$

be an $m \times n$ matrix. The m rows

$$\begin{matrix} [a_{11} & a_{12} & \cdots & a_{1m}] \\ [a_{21} & a_{22} & \cdots & a_{2m}] \\ \cdots & & \vdots & \\ [a_{m1} & a_{m2} & \cdots & a_{mn}] \end{matrix}$$

span a subspace of R_n which is called the **row space** of A and is usually denoted by **row (A)**. The n columns

$$\begin{bmatrix} a_{11} \\ a_{21} \\ \vdots \\ a_{m1} \end{bmatrix}, \begin{bmatrix} a_{12} \\ a_{22} \\ \vdots \\ a_{m2} \end{bmatrix}, \ldots, \begin{bmatrix} a_{1n} \\ a_{2n} \\ \vdots \\ a_{mn} \end{bmatrix}$$

span a subspace of R^m which is called the **column space** of A and is usually denoted by **col (A)**.

B. Definitions of row and column ranks

The dimension of the row space of a matrix A is called the **row rank** of A. Likewise, the dimension of the column space of A is called the **column rank** of A.

EXAMPLE 8-1: If $A = \begin{bmatrix} 1 & 0 & 0 \\ 0 & 1 & 1 \end{bmatrix}$, find **(a)** the rank and a basis for the row space of A, and **(b)** the rank and a basis for the column space of A.

Solution

(a) We must find a maximal linearly independent set from $[1\ \ 0\ \ 0]$ and $[0\ \ 1\ \ 1]$. From $a_1[1\ \ 0\ \ 0] + a_2[0\ \ 1\ \ 1] = [0\ \ 0\ \ 0]$, $a_1 = 0$ and $a_2 = 0$; hence the two vectors are linearly independent. Therefore, a basis for the row space of A is $\{[1\ \ 0\ \ 0], [0\ \ 1\ \ 1]\}$ which has dimension 2; hence the row rank of A is 2.

(b) Let $a_1\begin{bmatrix}1\\0\end{bmatrix} + a_2\begin{bmatrix}0\\1\end{bmatrix} + a_3\begin{bmatrix}0\\1\end{bmatrix} = \begin{bmatrix}0\\0\end{bmatrix}$, with $a_1 = 0$, $a_2 + a_3 = 0$, and $a_3 = -a_2$. Thus, these are not linearly independent, but $\begin{bmatrix}1\\0\end{bmatrix}$ and $\begin{bmatrix}0\\1\end{bmatrix}$ are linearly independent. Therefore, a basis for the column space of A is $\left\{\begin{bmatrix}1\\0\end{bmatrix}, \begin{bmatrix}0\\1\end{bmatrix}\right\}$ which has dimension 2; hence the column rank is 2.

C. Definition of rank of a matrix

The row and column ranks of a matrix A are equal and are called the **rank** of A; that is, rank row (A) = rank col (A) = rank A.

EXAMPLE 8-2: Find the rank of

$$A = \begin{bmatrix} 1 & 0 & 0 \\ 0 & 1 & 1 \end{bmatrix}$$

Solution: In Example 8-1, we found rank row $(A) = 2$. Therefore, rank $A = 2$.

8-2. Properties of Row and Column Spaces

Property A If two matrices A and B are row equivalent, then they have the same row space.

EXAMPLE 8-3: Find a matrix B which has the same row space as

$$A = \begin{bmatrix} 1 & 2 & -3 & 5 \\ 2 & 0 & 4 & -2 \\ -1 & 2 & 1 & 1 \end{bmatrix}$$

Solution: Use elementary row operations to find a row equivalent matrix:

$$A = \begin{bmatrix} 1 & 2 & -3 & 5 \\ 2 & 0 & 4 & -2 \\ -1 & 2 & 1 & 1 \end{bmatrix} \xrightarrow[R_3 \to R_3 + R_1]{R_2 \to R_2 - 2R_1} \begin{bmatrix} 1 & 2 & -3 & 5 \\ 0 & -4 & 10 & -12 \\ 0 & 4 & -2 & 6 \end{bmatrix} = B$$

The row space of B is the same as the row space of A. This means that span $\{[1\ \ 2\ \ -3\ \ 5], [2\ \ 0\ \ 4\ \ -2], [-1\ \ 2\ \ 1\ \ 1]\}$ is the same vector space as span $\{[1\ \ 2\ \ -3\ \ 5], [0\ \ -4\ \ 10\ \ -12], [0\ \ 4\ \ -2\ \ 6]\}$.

Property B If two matrices A and B are column equivalent, then they have the same column space.

EXAMPLE 8-4: Find a matrix B which has the same column space as

$$A = \begin{bmatrix} 1 & 0 & 1 & 1 \\ 2 & 1 & 0 & 1 \\ 3 & 2 & 1 & 0 \end{bmatrix}$$

Solution: Use elementary column operations to find a column equivalent matrix.

$$A = \begin{bmatrix} 1 & 0 & 1 & 1 \\ 2 & 1 & 0 & 1 \\ 3 & 2 & 1 & 0 \end{bmatrix} \xrightarrow[C_4 \to C_4 - C_1]{C_3 \to C_3 - C_1} \begin{bmatrix} 1 & 0 & 0 & 0 \\ 2 & 1 & -2 & -1 \\ 3 & 2 & -2 & -3 \end{bmatrix}$$

$$\xrightarrow[C_4 \to -C_4]{C_3 \to -1/2C_3} \begin{bmatrix} 1 & 0 & 0 & 0 \\ 2 & 1 & 1 & 1 \\ 3 & 2 & 1 & 3 \end{bmatrix} = B$$

The column space of B is the same as the column space of A. Once more, this means that

$$\text{span}\left\{ \begin{bmatrix} 1 \\ 2 \\ 3 \end{bmatrix}, \begin{bmatrix} 0 \\ 1 \\ 2 \end{bmatrix}, \begin{bmatrix} 1 \\ 0 \\ 1 \end{bmatrix}, \begin{bmatrix} 1 \\ 1 \\ 0 \end{bmatrix} \right\}$$

is the same vector space as

$$\text{span}\left\{ \begin{bmatrix} 1 \\ 2 \\ 3 \end{bmatrix}, \begin{bmatrix} 0 \\ 1 \\ 2 \end{bmatrix}, \begin{bmatrix} 0 \\ 1 \\ 1 \end{bmatrix}, \begin{bmatrix} 0 \\ 1 \\ 3 \end{bmatrix} \right\}$$

Property C A^T and B^T have the same column (row) space if and only if A and B have the same row (column) space.

EXAMPLE 8-5: In Example 8-3, we found that for

$$A = \begin{bmatrix} 1 & 2 & -3 & 5 \\ 2 & 0 & 4 & -2 \\ -1 & 2 & 1 & 1 \end{bmatrix} \quad \text{and} \quad B = \begin{bmatrix} 1 & 2 & -3 & 5 \\ 0 & -4 & 10 & -12 \\ 0 & 4 & -2 & 6 \end{bmatrix}$$

row (A) = row (B). Find a matrix C different from matrix A such that col (C) = col (A^T).

Solution: Choose $C = B^T$. From this we also see that row (A) = col (A^T) and row (B) = col (B^T); that is, the row space of a matrix is the same as the column space of its transpose.

Property D The nonzero row vectors of a row echelon form of a matrix A form a basis for the row space of A.

EXAMPLE 8-6: Find a basis and the dimension of the row space of

$$A = \begin{bmatrix} 1 & 0 & 2 & 3 \\ 2 & -3 & 1 & 0 \\ 1 & -1 & 1 & 1 \end{bmatrix}$$

Solution: Reduce A to row echelon form:

$$\begin{bmatrix} 1 & 0 & 2 & 3 \\ 2 & -3 & 1 & 0 \\ 1 & -1 & 1 & 1 \end{bmatrix} \xrightarrow[R_3 \to R_3 - R_1]{R_2 \to R_2 - 2R_1} \begin{bmatrix} 1 & 0 & 2 & 3 \\ 0 & -3 & -3 & -6 \\ 0 & -1 & -1 & -2 \end{bmatrix} \xrightarrow{R_2 \to -1/3R_2} \begin{bmatrix} 1 & 0 & 2 & 3 \\ 0 & 1 & 1 & 2 \\ 0 & -1 & -1 & -2 \end{bmatrix}$$

$$\xrightarrow[R_3 \to R_3 + R_2]{} \begin{bmatrix} 1 & 0 & 2 & 3 \\ 0 & 1 & 1 & 2 \\ 0 & 0 & 0 & 0 \end{bmatrix}$$

Although the last matrix is in reduced row echelon, it is not necessary to reduce every matrix to that form unless instructed to do so. Row echelon is sufficient. In fact, the leading entries do not necessarily have to be ones. Therefore, a basis for row (A) is $\{[1 \quad 0 \quad 2 \quad 3], [0 \quad 1 \quad 1 \quad 2]\}$. The dimension of row $(A) = 2$; that is, rank row $(A) = 2$.

Property E The nonzero row vectors of a row echelon form of a matrix A^T form a basis for the column space of A.

EXAMPLE 8-7: Find a basis and the dimension of the column space of

$$A = \begin{bmatrix} 1 & 0 & 2 & 3 \\ 2 & -3 & 1 & 0 \\ 1 & -1 & 1 & 1 \end{bmatrix}$$

Solution: Find A^T and reduce it to row echelon form.

$$A^T = \begin{bmatrix} 1 & 2 & 1 \\ 0 & -3 & -1 \\ 2 & 1 & 1 \\ 3 & 0 & 1 \end{bmatrix} \xrightarrow[R_4 \to R_4 - 3R_1]{R_3 \to R_3 - 2R_1} \begin{bmatrix} 1 & 2 & 1 \\ 0 & -3 & -1 \\ 0 & -3 & -1 \\ 0 & -6 & -2 \end{bmatrix} \xrightarrow[R_4 \to R_4 - 2R_2]{R_3 \to R_3 - R_2} \begin{bmatrix} 1 & 2 & 1 \\ 0 & -3 & -1 \\ 0 & 0 & 0 \\ 0 & 0 & 0 \end{bmatrix}$$

$$\xrightarrow{R_2 \to -1/3R_2} \begin{bmatrix} 1 & 2 & 1 \\ 0 & 1 & \frac{1}{3} \\ 0 & 0 & 0 \\ 0 & 0 & 0 \end{bmatrix}$$

Therefore, a basis for the row space of A^T is $\{[1 \quad 2 \quad 1], [0 \quad 1 \quad \frac{1}{3}]\}$, and hence a basis for the column space of A is

$$\left\{ \begin{bmatrix} 1 \\ 2 \\ 1 \end{bmatrix}, \begin{bmatrix} 0 \\ 1 \\ \frac{1}{3} \end{bmatrix} \right\}$$

Since any vector in a basis can be multiplied by a nonzero scalar, then

$$\left\{\begin{bmatrix}1\\2\\1\end{bmatrix}, \begin{bmatrix}0\\3\\1\end{bmatrix}\right\}$$

is also a basis. The dimension of col $(A) = 2$.

Property F Two $m \times n$ matrices A and B have the same rank if and only if they are equivalent.

EXAMPLE 8-8: Which of the following matrices are equivalent?

$$A = \begin{bmatrix}1 & 2 & 3 & 4\\1 & -3 & 2 & 1\\2 & 3 & 4 & 5\\4 & -2 & 10 & 10\end{bmatrix} \qquad B = \begin{bmatrix}4 & 2 & 5\\1 & 1 & 2\\4 & 1 & 5\end{bmatrix} \qquad C = \begin{bmatrix}1 & -1 & 2 & 3\\0 & 3 & 3 & 2\\1 & 2 & 5 & 5\\3 & 2 & 1 & 0\end{bmatrix}$$

$$D = \begin{bmatrix}1 & 0 & 0 & 2\\0 & 2 & 1 & -4\\1 & -2 & 2 & 5\\2 & 3 & 4 & 5\end{bmatrix} \qquad E = \begin{bmatrix}1 & 0 & 1 & 1\\2 & 1 & 0 & 1\\0 & 1 & 0 & 0\\0 & 2 & 3 & 4\end{bmatrix}$$

Solution: Since B is 3×3 and the other matrices are 4×4, B is not equivalent to any of them.

By elementary row operations, transform each of the remaining matrices to a form in which the number of nonzero rows in echelon form can be determined. To make this determination, the leading entries do not have to be ones, and the zero rows do not necessarily have to be at the bottom of the matrix:

$$A = \begin{bmatrix}1 & 2 & 3 & 4\\1 & -3 & 2 & 1\\2 & 3 & 4 & 5\\4 & -2 & 10 & 10\end{bmatrix} \xrightarrow[\substack{R_3 \to R_3 - 2R_1\\ R_4 \to R_4 - 4R_1}]{R_2 \to R_2 - R_1} \begin{bmatrix}1 & 2 & 3 & 4\\0 & -5 & -1 & -3\\0 & -1 & -2 & -3\\0 & -10 & -2 & -6\end{bmatrix} \xrightarrow[R_4 \to R_4 - 2R_2]{R_2 \to R_2 - 5R_3} \begin{bmatrix}1 & 2 & 3 & 4\\0 & 0 & 9 & 12\\0 & -1 & -2 & -3\\0 & 0 & 0 & 0\end{bmatrix}$$

Therefore, rank $A = 3$ because there are three nonzero rows:

$$C = \begin{bmatrix}1 & -1 & 2 & 3\\0 & 3 & 3 & 2\\1 & 2 & 5 & 5\\3 & 2 & 1 & 0\end{bmatrix} \xrightarrow[R_4 \to R_4 - 3R_1]{R_3 \to R_3 - R_1} \begin{bmatrix}1 & -1 & 2 & 3\\0 & 3 & 3 & 2\\0 & 3 & 3 & 2\\0 & 5 & -5 & -9\end{bmatrix} \xrightarrow[R_4 \to R_4 - 5/3R_2]{R_3 \to R_3 - R_2} \begin{bmatrix}1 & -1 & 2 & 3\\0 & 3 & 3 & 2\\0 & 0 & 0 & 0\\0 & 0 & -10 & -\frac{37}{3}\end{bmatrix}$$

Therefore, rank $C = 3$ because there are three nonzero rows:

$$D = \begin{bmatrix}1 & 0 & 0 & 2\\0 & 2 & 1 & -4\\1 & -2 & 2 & 5\\2 & 3 & 4 & 5\end{bmatrix} \xrightarrow[R_4 \to R_4 - 2R_1]{R_3 \to R_3 - R_1} \begin{bmatrix}1 & 0 & 0 & 2\\0 & 2 & 1 & -4\\0 & -2 & 2 & 3\\0 & 3 & 4 & 1\end{bmatrix}$$

$$D = \xrightarrow[R_4 \to R_4 - R_2]{R_3 \to R_3 + R_2} \begin{bmatrix} 1 & 0 & 0 & 2 \\ 0 & 2 & 1 & -4 \\ 0 & 0 & 3 & -1 \\ 0 & 1 & 3 & 5 \end{bmatrix} \xrightarrow{R_2 \to R_2 - 2R_4} \begin{bmatrix} 1 & 0 & 0 & 2 \\ 0 & 0 & -5 & -14 \\ 0 & 0 & 3 & -1 \\ 0 & 1 & 3 & 5 \end{bmatrix}$$

$$\xrightarrow[R_3 \to R_3 + R_2]{R_2 \leftrightarrow R_4} \begin{bmatrix} 1 & 0 & 0 & 2 \\ 0 & 1 & 3 & 5 \\ 0 & 0 & 1 & -16 \\ 0 & 0 & -5 & -14 \end{bmatrix} \xrightarrow{R_4 \to R_4 + 5R_3} \begin{bmatrix} 1 & 0 & 0 & 2 \\ 0 & 1 & 3 & 5 \\ 0 & 0 & 1 & -16 \\ 0 & 0 & 0 & -94 \end{bmatrix}$$

Therefore, rank $D = 4$ because there are four nonzero rows:

$$E = \begin{bmatrix} 1 & 0 & 1 & 1 \\ 2 & 1 & 0 & 1 \\ 0 & 1 & 0 & 0 \\ 0 & 2 & 3 & 4 \end{bmatrix} \xrightarrow{R_2 \to R_2 - 2R_1} \begin{bmatrix} 1 & 0 & 1 & 1 \\ 0 & 1 & -2 & -1 \\ 0 & 1 & 0 & 0 \\ 0 & 2 & 3 & 4 \end{bmatrix} \xrightarrow{R_2 \leftrightarrow R_3} \begin{bmatrix} 1 & 0 & 1 & 1 \\ 0 & 1 & 0 & 0 \\ 0 & 1 & -2 & -1 \\ 0 & 2 & 3 & 4 \end{bmatrix}$$

$$\xrightarrow[R_4 \to R_4 - 2R_2]{R_3 \to R_3 - R_2} \begin{bmatrix} 1 & 0 & 1 & 1 \\ 0 & 1 & 0 & 0 \\ 0 & 0 & -2 & -1 \\ 0 & 0 & 3 & 4 \end{bmatrix} \xrightarrow{R_3 \to R_3 + R_4} \begin{bmatrix} 1 & 0 & 1 & 1 \\ 0 & 1 & 0 & 0 \\ 0 & 0 & 1 & 3 \\ 0 & 0 & 3 & 4 \end{bmatrix}$$

$$\xrightarrow{R_4 \to R_4 - 3R_3} \begin{bmatrix} 1 & 0 & 1 & 1 \\ 0 & 1 & 0 & 0 \\ 0 & 0 & 1 & 3 \\ 0 & 0 & 0 & -5 \end{bmatrix}$$

Therefore, rank $E = 4$ because there are four nonzero rows. Hence, A and C are equivalent because each has rank of 3. Likewise, D and E are equivalent because each has rank of 4.

Property G An $n \times n$ matrix A is row equivalent to I_n and hence invertible if and only if rank $A = n$.

EXAMPLE 8-9: Determine whether A^{-1} exists for

$$A = \begin{bmatrix} 1 & 0 & 0 & 0 & 0 \\ 2 & 1 & 3 & 0 & 1 \\ 0 & 1 & 2 & 0 & 1 \\ 0 & 2 & 3 & 4 & 5 \\ 0 & 0 & 0 & 1 & 2 \end{bmatrix}$$

Solution

$$A = \begin{bmatrix} 1 & 0 & 0 & 0 & 0 \\ 2 & 1 & 3 & 0 & 1 \\ 0 & 1 & 2 & 0 & 1 \\ 0 & 2 & 3 & 4 & 5 \\ 0 & 0 & 0 & 1 & 2 \end{bmatrix} \xrightarrow{R_2 \to R_2 - 2R_1} \begin{bmatrix} 1 & 0 & 0 & 0 & 0 \\ 0 & 1 & 3 & 0 & 1 \\ 0 & 1 & 2 & 0 & 1 \\ 0 & 2 & 3 & 4 & 5 \\ 0 & 0 & 0 & 1 & 2 \end{bmatrix}$$

$$A = \xrightarrow[R_4 \to R_4 - 2R_2]{R_3 \to R_3 - R_2} \begin{bmatrix} 1 & 0 & 0 & 0 & 0 \\ 0 & 1 & 3 & 0 & 1 \\ 0 & 0 & -1 & 0 & 0 \\ 0 & 0 & -3 & 4 & 3 \\ 0 & 0 & 0 & 1 & 2 \end{bmatrix} \xrightarrow[R_4 \to R_4 - 3R_3]{} \begin{bmatrix} 1 & 0 & 0 & 0 & 0 \\ 0 & 1 & 3 & 0 & 1 \\ 0 & 0 & -1 & 0 & 0 \\ 0 & 0 & 0 & 4 & 3 \\ 0 & 0 & 0 & 1 & 2 \end{bmatrix}$$

$$\xrightarrow[R_4 \to R_4 - 4R_5]{R_3 \to -R_3} \begin{bmatrix} 1 & 0 & 0 & 0 & 0 \\ 0 & 1 & 3 & 0 & 1 \\ 0 & 0 & 1 & 0 & 0 \\ 0 & 0 & 0 & 0 & -5 \\ 0 & 0 & 0 & 1 & 2 \end{bmatrix} \xrightarrow[R_4 \leftrightarrow R_5]{} \begin{bmatrix} 1 & 0 & 0 & 0 & 0 \\ 0 & 1 & 3 & 0 & 1 \\ 0 & 0 & 1 & 0 & 0 \\ 0 & 0 & 0 & 1 & 2 \\ 0 & 0 & 0 & 0 & -5 \end{bmatrix}$$

There are five nonzero rows; therefore, rank $A = 5$. Since A is 5×5 and rank $A = 5$, then A^{-1} exists—that is, A is invertible. (Observe that this tells you of the existence of an inverse but does not give you the inverse.)

Property H The row vectors in an $n \times n$ invertible matrix A form a basis for R_n and the column vectors form a basis for R^n.

EXAMPLE 8-10: Use Example 8-9 to find bases of R^5 and R_5 which are not natural bases for R^5 and R_5.

Solution: Since A is invertible, then a basis for R^5 is

$$\left\{ \begin{bmatrix} 1 \\ 2 \\ 0 \\ 0 \\ 0 \end{bmatrix}, \begin{bmatrix} 0 \\ 1 \\ 1 \\ 2 \\ 0 \end{bmatrix}, \begin{bmatrix} 0 \\ 3 \\ 2 \\ 3 \\ 0 \end{bmatrix}, \begin{bmatrix} 0 \\ 0 \\ 0 \\ 4 \\ 1 \end{bmatrix}, \begin{bmatrix} 0 \\ 1 \\ 1 \\ 5 \\ 2 \end{bmatrix} \right\}$$

Likewise, a basis for R_5 is $[1 \;\; 0 \;\; 0 \;\; 0 \;\; 0]$, $[2 \;\; 1 \;\; 3 \;\; 0 \;\; 1]$, $[0 \;\; 1 \;\; 2 \;\; 0 \;\; 1]$, $[0 \;\; 2 \;\; 3 \;\; 4 \;\; 5]$, and $[0 \;\; 0 \;\; 0 \;\; 1 \;\; 2]$.

Property I The homogeneous system $AX = 0$ where A is $n \times n$ has a nontrivial solution if and only if rank $A < n$.

EXAMPLE 8-11: Determine whether each of the following homogeneous systems has a nontrivial solution:

(a)
$$\begin{aligned} x_1 + 2x_2 &= 0 \\ x_1 + 3x_2 + 2x_3 &= 0 \\ 2x_1 + x_2 - x_3 &= 0 \end{aligned}$$

(b)
$$\begin{aligned} x_1 + 2x_2 + 3x_3 &= 0 \\ 2x_1 - x_2 + 4x_3 &= 0 \\ 3x_1 + x_2 + 7x_3 &= 0 \end{aligned}$$

Solution

(a) The coefficient matrix

$$A = \begin{bmatrix} 1 & 2 & 0 \\ 1 & 3 & 2 \\ 2 & 1 & -1 \end{bmatrix} \xrightarrow[R_3 \to R_3 - 2R_1]{R_2 \to R_2 - R_1} \begin{bmatrix} 1 & 2 & 0 \\ 0 & 1 & 2 \\ 0 & -3 & -1 \end{bmatrix} \xrightarrow[R_3 \to R_3 + 3R_2]{} \begin{bmatrix} 1 & 2 & 0 \\ 0 & 1 & 2 \\ 0 & 0 & 5 \end{bmatrix}$$

Since there are three nonzero rows, then rank $A = 3$. Therefore, the system has only the trivial solution.

(b) The coefficient matrix

$$A = \begin{bmatrix} 1 & 2 & 3 \\ 2 & -1 & 4 \\ 3 & 1 & 7 \end{bmatrix} \xrightarrow[R_3 \to R_3 - 3R_1]{R_2 \to R_2 - 2R_1} \begin{bmatrix} 1 & 2 & 3 \\ 0 & -5 & -2 \\ 0 & -5 & -2 \end{bmatrix} \xrightarrow[R_3 \to R_3 - R_2]{} \begin{bmatrix} 1 & 2 & 3 \\ 0 & -5 & -2 \\ 0 & 0 & 0 \end{bmatrix}$$

Since there are two nonzero rows, then rank $A = 2$. Therefore, the system has a nontrivial solution.

Property J A system of nonhomogeneous equations $AX = B$ is consistent; that is, it has a solution if and only if the rank of the augmented matrix equals the rank of the coefficient matrix.

EXAMPLE 8-12: Determine whether each of the following systems has a solution.

(a)
$$\begin{aligned} x + y - 3z &= 1 \\ 2x - y - 4z &= 2 \\ x + y - z &= 3 \end{aligned}$$

(b)
$$\begin{aligned} x + y - z &= 6 \\ 2x - 3y - 2z &= 2 \\ x + 6y - z &= 16 \end{aligned}$$

Solution: Compare the ranks of the coefficient and augmented matrices in each case.

(a)

$$\text{Coefficient matrix } A = \begin{bmatrix} 1 & 1 & -3 \\ 2 & -1 & -4 \\ 1 & 1 & -1 \end{bmatrix} \xrightarrow[R_3 \to R_3 - R_1]{R_2 \to R_2 - 2R_1} \begin{bmatrix} 1 & 1 & -3 \\ 0 & -3 & 2 \\ 0 & 0 & 2 \end{bmatrix}$$

Therefore, rank $A = 3$ because there are three nonzero rows in the reduced matrix which is in partial echelon form.

$$\text{Augmented matrix } [A|B] = \begin{bmatrix} 1 & 1 & -3 & 1 \\ 2 & -1 & -4 & 2 \\ 1 & 1 & -1 & 3 \end{bmatrix}$$

Rank $[A|B] = 3$ because rank $[A|B] \geq$ rank A, but rank $[A|B]$ cannot exceed the number of rows, which is 3. Therefore, the system has a solution since rank A = rank $[A|B]$.

(b)

$$A = \begin{bmatrix} 1 & 1 & -1 \\ 2 & -3 & -2 \\ 1 & 6 & -1 \end{bmatrix} \xrightarrow[R_3 \to R_3 - R_1]{R_2 \to R_2 - 2R_1} \begin{bmatrix} 1 & 1 & -1 \\ 0 & -5 & 0 \\ 0 & 5 & 0 \end{bmatrix} \xrightarrow[R_3 \to R_3 + R_2]{R_2 \to -1/5 R_2} \begin{bmatrix} 1 & 1 & -1 \\ 0 & 1 & 0 \\ 0 & 0 & 0 \end{bmatrix}$$

Therefore, rank $A = 2$ because there are two nonzero rows:

$$[A|B] = \begin{bmatrix} 1 & 1 & -1 & 6 \\ 2 & -3 & -2 & 2 \\ 1 & 6 & -1 & 16 \end{bmatrix} \xrightarrow[R_3 \to R_3 - R_1]{R_2 \to R_2 - 2R_1} \begin{bmatrix} 1 & 1 & -1 & 6 \\ 0 & -5 & 0 & -10 \\ 0 & 5 & 0 & 10 \end{bmatrix} \xrightarrow[R_3 \to R_3 + R_2]{R_2 \to -1/5 R_2} \begin{bmatrix} 1 & 1 & -1 & 6 \\ 0 & 1 & 0 & 2 \\ 0 & 0 & 0 & 0 \end{bmatrix}$$

Therefore, rank $[A|B] = 2$ because there are two nonzero rows. Hence, the system has a solution since rank A = rank $[A|B]$.

8-3. Equivalent Statements

In Section 2-8, four statements about equivalent statements were given. We can now expand that list.

If A is an $n \times n$ matrix, then the following statements are equivalent; that is, they are all true or they are all false.

(1) A is invertible.
(2) A is row equivalent to I_n.
(3) A is column equivalent to I_n.
(4) The homogeneous system $AX = 0$ has only the trivial solution.
(5) The linear system $AX = B$ has a unique solution for every $n \times 1$ matrix B.
(6) A is a product of elementary matrices.
(7) $\text{Det}(A) \neq 0$.
(8) A has rank n.
(9) The rows of A are linearly independent.
(10) The columns of A are linearly independent.
(11) In echelon form, A has no zero rows.

EXAMPLE 8-13: Determine whether the vectors

$$[1\ 0\ 1\ 0\ 0\ 1\ 0], [0\ 0\ 1\ 2\ 0\ 2\ 1],$$
$$[0\ 1\ 0\ 1\ 0\ 1\ 0], [1\ 1\ 0\ 0\ 0\ 1\ 1],$$
$$[2\ 0\ 1\ 0\ 1\ 0\ 1], [3\ 0\ 2\ 0\ 1\ 0\ 0],$$
$$[0\ 0\ 0\ 0\ 1\ 1\ 1]$$

are linearly independent.

Solution: We can form a 7×7 matrix A with the given vectors as rows. Since what we were asked to determine is now among these 11 equivalent statements, we may use any one of them. Let us determine whether in echelon form, A has any zero rows. If it does not, then the seven vectors are linearly independent. If it does, then the given vectors are linearly dependent.

$$\begin{bmatrix} 1 & 0 & 1 & 0 & 0 & 1 & 0 \\ 0 & 0 & 1 & 2 & 0 & 2 & 1 \\ 0 & 1 & 0 & 1 & 0 & 1 & 0 \\ 1 & 1 & 0 & 0 & 0 & 1 & 1 \\ 2 & 0 & 1 & 0 & 1 & 0 & 1 \\ 3 & 0 & 2 & 0 & 1 & 0 & 0 \\ 0 & 0 & 0 & 0 & 1 & 1 & 1 \end{bmatrix} \xrightarrow[\substack{R_5 \to R_5 - 2R_1 \\ R_6 \to R_6 - 3R_1}]{\substack{R_2 \leftrightarrow R_3 \\ R_4 \to R_4 - R_1}} \begin{bmatrix} 1 & 0 & 1 & 0 & 0 & 1 & 0 \\ 0 & 1 & 0 & 1 & 0 & 1 & 0 \\ 0 & 0 & 1 & 2 & 0 & 2 & 1 \\ 0 & 1 & -1 & 0 & 0 & 0 & 1 \\ 0 & 0 & -1 & 0 & 1 & -2 & 1 \\ 0 & 0 & -1 & 0 & 1 & -3 & 0 \\ 0 & 0 & 0 & 0 & 1 & 1 & 1 \end{bmatrix}$$

$$\xrightarrow[R_6 \to R_6 - R_5]{R_4 \to R_4 - R_2} \begin{bmatrix} 1 & 0 & 1 & 0 & 0 & 1 & 0 \\ 0 & 1 & 0 & 1 & 0 & 1 & 0 \\ 0 & 0 & 1 & 2 & 0 & 2 & 1 \\ 0 & 0 & -1 & -1 & 0 & -1 & 1 \\ 0 & 0 & -1 & 0 & 1 & -2 & 1 \\ 0 & 0 & 0 & 0 & 0 & -1 & -1 \\ 0 & 0 & 0 & 0 & 1 & 1 & 1 \end{bmatrix} \xrightarrow[\substack{R_5 \to R_5 + R_3 \\ R_6 \to -R_6}]{R_4 \to R_4 + R_3} \begin{bmatrix} 1 & 0 & 1 & 0 & 0 & 1 & 0 \\ 0 & 1 & 0 & 1 & 0 & 1 & 0 \\ 0 & 0 & 1 & 2 & 0 & 2 & 1 \\ 0 & 0 & 0 & 1 & 0 & 1 & 2 \\ 0 & 0 & 0 & 2 & 1 & 0 & 2 \\ 0 & 0 & 0 & 0 & 0 & 1 & 1 \\ 0 & 0 & 0 & 0 & 1 & 1 & 1 \end{bmatrix}$$

$$\xrightarrow[R_5 \to R_5 - 2R_4]{} \begin{bmatrix} 1 & 0 & 1 & 0 & 0 & 1 & 0 \\ 0 & 1 & 0 & 1 & 0 & 1 & 0 \\ 0 & 0 & 1 & 2 & 0 & 2 & 1 \\ 0 & 0 & 0 & 1 & 0 & 1 & 2 \\ 0 & 0 & 0 & 0 & 1 & -2 & -2 \\ 0 & 0 & 0 & 0 & 0 & 1 & 1 \\ 0 & 0 & 0 & 0 & 1 & 1 & 1 \end{bmatrix} \xrightarrow[R_7 \to R_7 - R_5]{} \begin{bmatrix} 1 & 0 & 1 & 0 & 0 & 1 & 0 \\ 0 & 1 & 0 & 1 & 0 & 1 & 0 \\ 0 & 0 & 1 & 2 & 0 & 2 & 1 \\ 0 & 0 & 0 & 1 & 0 & 1 & 2 \\ 0 & 0 & 0 & 0 & 1 & -2 & -2 \\ 0 & 0 & 0 & 0 & 0 & 1 & 1 \\ 0 & 0 & 0 & 0 & 0 & 3 & 3 \end{bmatrix}$$

$$\xrightarrow{R_7 \to R_7 - 3R_6} \begin{bmatrix} 1 & 0 & 1 & 0 & 0 & 1 & 0 \\ 0 & 1 & 0 & 1 & 0 & 1 & 0 \\ 0 & 0 & 1 & 2 & 0 & 2 & 1 \\ 0 & 0 & 0 & 1 & 0 & 1 & 2 \\ 0 & 0 & 0 & 0 & 1 & -2 & -2 \\ 0 & 0 & 0 & 0 & 0 & 1 & 1 \\ 0 & 0 & 0 & 0 & 0 & 0 & 0 \end{bmatrix}$$

Since there are six nonzero rows, then rank $A = 6$, which is less than the number of given vectors. Therefore, the seven given vectors are linearly dependent because statements **(8)** and **(9)** are equivalent and, hence, must both be true or both be false.

EXAMPLE 8-14: Can the following matrix A be written as a product of elementary matrices?

$$A = \begin{bmatrix} 1 & 0 & 1 & 0 & 0 & 1 & 0 \\ 0 & 0 & 1 & 2 & 0 & 2 & 1 \\ 0 & 1 & 0 & 1 & 0 & 1 & 0 \\ 1 & 1 & 0 & 0 & 0 & 1 & 1 \\ 2 & 0 & 1 & 0 & 1 & 0 & 1 \\ 3 & 0 & 2 & 0 & 1 & 0 & 0 \\ 0 & 0 & 0 & 0 & 1 & 1 & 1 \end{bmatrix}$$

Solution: The matrix is the same as that in Example 8-13. Hence, rank $A = 6$ which is less than the number of rows in A. Therefore, matrix A cannot be written as a product of elementary matrices because statement **(6)** is equivalent to statement **(8)**, and, hence, both must be true or false. Statement 8 is false, therefore A cannot be written as a product of elementary matrices.

SUMMARY

1. The span of the row vectors of a matrix A is a vector space called the row space of A.
2. The span of the column vectors of a matrix A is a vector space called the column space of A.
3. Rank row (A) = rank column (A) = rank A.
4. If two matrices are row equivalent, then they have the same row space.
5. If two matrices are column equivalent, then they have the same column space.
6. A^T and B^T have the same column (row) space if and only if A and B have the same row (column) space.
7. The nonzero row vectors of a row echelon form of a matrix A form a basis for the row space of A.
8. The nonzero row vectors of a row echelon form of a matrix A^T form a basis for the column space of A.
9. Two $m \times n$ matrices A and B have the same rank if and only if they are equivalent.
10. The row vectors in an $n \times n$ invertible matrix A form a basis for R_n and the column vectors form a basis for R^n.
11. The homogeneous system $AX = 0$ where A is $n \times n$ has a nontrivial solution if and only if rank $A < n$.
12. A system of nonhomogeneous equations $AX = B$ is consistent if and only if the rank of the augmented matrix equals the rank of the coefficient matrix.

13. If A is an $n \times n$ matrix, then the following statements are equivalent.

- A is invertible.
- A is row (column) equivalent to I_n.
- The homogeneous system $AX = 0$ has only the trivial solution.
- The linear system $AX = B$ has a unique solution for every $n \times 1$ matrix B.
- A is a product of elementary matrices.
- $\text{Det}(A) \neq 0$.
- A has rank n.
- The rows (columns) of A are linearly independent.
- In echelon form, A has no zero rows.

RAISE YOUR GRADES

Can you . . . ?

☑ find the row rank, column rank, and rank of a matrix
☑ find a basis for the row space of a matrix
☑ find a basis for the column space of a matrix
☑ determine whether two $m \times n$ matrices are equivalent by finding ranks
☑ determine whether an $n \times n$ matrix is invertible by finding its rank
☑ determine whether the homogeneous system $AX = 0$ has a nontrivial solution by using rank
☑ determine whether the nonhomogeneous system $AX = B$ is consistent by using rank
☑ use the various forms of equivalent statements

SOLVED PROBLEMS

PROBLEM 8-1 If $A = \begin{bmatrix} 1 & 0 & 1 \\ -1 & 2 & 1 \end{bmatrix}$, find **(a)** the rank and a basis for the row space of A and **(b)** the rank and a basis for the column space of A.

Solution

(a) We need to find a maximal linearly independent set from $[1 \quad 0 \quad 1]$ and $[-1 \quad 2 \quad 1]$:

$$a_1[1 \quad 0 \quad 1] + a_2[-1 \quad 2 \quad 1] = [0 \quad 0 \quad 0]$$

$a_1 = 0$ and $a_2 = 0$. Hence the two vectors are linearly independent. Therefore, a basis for the row space of A is $\{[1 \quad 0 \quad 1], [-1 \quad 2 \quad 1]\}$ which has dimension 2; hence rank (row A) = 2.

(b) Let

$$a_1\begin{bmatrix} 1 \\ -1 \end{bmatrix} + a_2\begin{bmatrix} 0 \\ 2 \end{bmatrix} + a_3\begin{bmatrix} 1 \\ 1 \end{bmatrix} = \begin{bmatrix} 0 \\ 0 \end{bmatrix} \qquad \begin{aligned} a_1 + a_3 &= 0 \\ -a_1 + 2a_2 + a_3 &= 0 \end{aligned} \quad \text{and} \quad \begin{aligned} a_1 &= -a_3 \\ a_2 &= -a_3 \end{aligned}$$

There are not linearly independent, but $\begin{bmatrix} 1 \\ -1 \end{bmatrix}$ and $\begin{bmatrix} 0 \\ 2 \end{bmatrix}$ are linearly independent. So a basis for the column space of A is $\left\{\begin{bmatrix} 1 \\ -1 \end{bmatrix}, \begin{bmatrix} 0 \\ 2 \end{bmatrix}\right\}$ which has dimension 2; hence rank (column A) = 2. Since rank (row A) = rank (column A) = rank A, rank $A = 2$.

PROBLEM 8-2 Find a matrix B that has the same row space as

$$A = \begin{bmatrix} 1 & -2 & 3 & 4 \\ 1 & 0 & 2 & -1 \\ 2 & 1 & -1 & 2 \end{bmatrix}$$

Solution Using elementary row operations on A, we can find a row equivalent matrix.

$$\begin{bmatrix} 1 & -2 & 3 & 4 \\ 1 & 0 & 2 & -1 \\ 2 & 1 & -1 & 2 \end{bmatrix} \xrightarrow[R_3 \to R_3 - 2R_1]{R_2 \to R_2 - R_1} \begin{bmatrix} 1 & -2 & 3 & 4 \\ 0 & 2 & -1 & -5 \\ 0 & 5 & -7 & -6 \end{bmatrix} = B$$

The row space of B is the same as the row space of A. Thus span $\{[1 \ -2 \ 3 \ 4], [1 \ 0 \ 2 \ -1], [2 \ 1 \ -1 \ 2]\}$ is the same vector space as span$\{[1 \ -2 \ 3 \ 4], [0 \ 2 \ -1 \ -5], [0 \ 5 \ -7 \ -6]\}$.

PROBLEM 8-3 Find a matrix B that has the same column space as

$$A = \begin{bmatrix} 1 & -2 & 3 & 4 \\ -1 & 0 & 2 & -1 \\ 2 & 1 & -1 & 2 \end{bmatrix}$$

Solution Using elementary column operations, we find

$$\begin{bmatrix} 1 & -2 & 3 & 4 \\ -1 & 0 & 2 & -1 \\ 2 & 1 & -1 & 2 \end{bmatrix} \xrightarrow[\substack{C_3 \to C_3 - 3C_1 \\ C_4 \to C_4 - 4C_1}]{C_2 \to C_2 + 2C_1} \begin{bmatrix} 1 & 0 & 0 & 0 \\ -1 & -2 & 5 & 3 \\ 2 & 5 & -7 & -6 \end{bmatrix} \xrightarrow{C_4 \to 1/3C_4} \begin{bmatrix} 1 & 0 & 0 & 0 \\ -1 & -2 & 5 & 1 \\ 2 & 5 & -7 & -2 \end{bmatrix} = B$$

The column space of B is the same as the column space of A. This means that

$$\text{span}\left\{\begin{bmatrix} 1 \\ -1 \\ 2 \end{bmatrix}, \begin{bmatrix} -2 \\ 0 \\ 1 \end{bmatrix}, \begin{bmatrix} 3 \\ 2 \\ -1 \end{bmatrix}, \begin{bmatrix} 4 \\ -1 \\ 2 \end{bmatrix}\right\}$$

is the same vector space as

$$\text{span}\left\{\begin{bmatrix} 1 \\ -1 \\ 2 \end{bmatrix}, \begin{bmatrix} 0 \\ -2 \\ 5 \end{bmatrix}, \begin{bmatrix} 0 \\ 5 \\ -7 \end{bmatrix}, \begin{bmatrix} 0 \\ 1 \\ -2 \end{bmatrix}\right\}$$

PROBLEM 8-4 For

$$A = \begin{bmatrix} 1 & -2 & 3 & 4 \\ 1 & 0 & 2 & -1 \\ 2 & 1 & -1 & 2 \end{bmatrix}$$

find a matrix C different from matrix A such that col (C) = col (A^T).

Solution In Problem 8-2, we found

$$B = \begin{bmatrix} 1 & -2 & 3 & 4 \\ 0 & 2 & -1 & -5 \\ 0 & 5 & -7 & -6 \end{bmatrix}$$

Choose

$$C = B^T = \begin{bmatrix} 1 & 0 & 0 \\ -2 & 2 & 5 \\ 3 & -1 & -7 \\ 4 & -5 & -6 \end{bmatrix}$$

Since row B = row A, then col B^T = row B = row A = col (A^T). Therefore, col (C) = col (A^T), since $B^T = C$.

PROBLEM 8-5 Find a basis and the dimension of the row space of

$$A = \begin{bmatrix} 1 & 2 & 0 & -1 \\ 3 & 2 & -1 & 0 \\ -2 & -1 & 2 & 1 \end{bmatrix}$$

Solution Reduce A to row echelon form.

$$\begin{bmatrix} 1 & 2 & 0 & -1 \\ 3 & 2 & -1 & 0 \\ -2 & -1 & 2 & 1 \end{bmatrix} \xrightarrow[R_3 \to R_3 + 2R_1]{R_2 \to R_2 - 3R_1} \begin{bmatrix} 1 & 2 & 0 & -1 \\ 0 & -4 & -1 & 3 \\ 0 & 3 & 2 & -1 \end{bmatrix} \xrightarrow{R_2 \to R_2 + R_3} \begin{bmatrix} 1 & 2 & 0 & -1 \\ 0 & -1 & 1 & 2 \\ 0 & 3 & 2 & -1 \end{bmatrix}$$

$$\xrightarrow[R_3 \to R_3 + 3R_2]{} \begin{bmatrix} 1 & 2 & 0 & -1 \\ 0 & -1 & 1 & 2 \\ 0 & 0 & 5 & 5 \end{bmatrix} \xrightarrow[R_3 \to 1/5 R_3]{R_2 \to -R_2} \begin{bmatrix} 1 & 2 & 0 & -1 \\ 0 & 1 & -1 & -2 \\ 0 & 0 & 1 & 1 \end{bmatrix}$$

Therefore, a basis for row (A) is $\{[1 \quad 2 \quad 0 \quad -1], [0 \quad 1 \quad -1 \quad -2], [0 \quad 0 \quad 1 \quad 1]\}$. The dim row $(A) = 3$; that is rank row $(A) = 3$.

PROBLEM 8-6 Find a basis and the dimension of the column space of

$$A = \begin{bmatrix} 1 & 2 & 0 & -1 \\ 3 & 2 & -1 & 0 \\ -2 & -1 & 2 & 1 \end{bmatrix}$$

Solution Find A^T and reduce A^T to row echelon form.

$$A^T = \begin{bmatrix} 1 & 3 & -2 \\ 2 & 2 & -1 \\ 0 & -1 & 2 \\ -1 & 0 & 1 \end{bmatrix} \xrightarrow[R_4 \to R_4 + R_1]{R_2 \to R_2 - 2R_1} \begin{bmatrix} 1 & 3 & -2 \\ 0 & -4 & 3 \\ 0 & -1 & 2 \\ 0 & 3 & -1 \end{bmatrix} \xrightarrow[R_4 \to R_4 + 3R_3]{R_2 \to R_2 + R_4} \begin{bmatrix} 1 & 3 & -2 \\ 0 & -1 & 2 \\ 0 & -1 & 2 \\ 0 & 0 & 5 \end{bmatrix}$$

$$\xrightarrow[\substack{R_3 \to R_3 - R_2 \\ R_4 \to 1/5 R_4}]{R_2 \to -R_2} \begin{bmatrix} 1 & 3 & -2 \\ 0 & 1 & -2 \\ 0 & 0 & 0 \\ 0 & 0 & 1 \end{bmatrix} \xrightarrow{R_3 \leftrightarrow R_4} \begin{bmatrix} 1 & 3 & -2 \\ 0 & 1 & -2 \\ 0 & 0 & 1 \\ 0 & 0 & 0 \end{bmatrix}$$

Therefore, a basis for row space of A^T is $\{[1 \quad 3 \quad -2], [0 \quad 1 \quad -2], [0 \quad 0 \quad 1]\}$, and, hence, a basis for column space of A is

$$\left\{ \begin{bmatrix} 1 \\ 3 \\ -2 \end{bmatrix}, \begin{bmatrix} 0 \\ 1 \\ -2 \end{bmatrix}, \begin{bmatrix} 0 \\ 0 \\ 1 \end{bmatrix} \right\}$$

and dim col $(A) = 3$.

PROBLEM 8-7 Which of the following matrices are equivalent?

$$A = \begin{bmatrix} 1 & 1 & 2 & -1 \\ 2 & -1 & -1 & 2 \\ 3 & 1 & -1 & -3 \\ 2 & 4 & -3 & -1 \end{bmatrix} \quad B = \begin{bmatrix} 1 & 2 & 3 & 1 \\ 2 & 1 & 1 & 3 \\ 0 & 2 & -1 & -1 \\ 2 & -1 & 2 & 4 \end{bmatrix} \quad C = \begin{bmatrix} 1 & -2 & 3 & -1 \\ 2 & -1 & 3 & -1 \\ 0 & -2 & 1 & 1 \\ -1 & 1 & 2 & 3 \end{bmatrix}$$

$$D = \begin{bmatrix} 1 & 2 & 0 & 0 \\ 0 & 1 & -1 & 2 \\ 1 & -1 & 2 & 1 \\ 2 & -3 & 1 & -1 \end{bmatrix} \quad E = \begin{bmatrix} 1 & 1 & 2 & -1 \\ -2 & 1 & 1 & -2 \\ 1 & -2 & -3 & 3 \\ 2 & 4 & -2 & 1 \end{bmatrix}$$

Solution

(1)

$$A = \begin{bmatrix} 1 & 1 & 2 & -1 \\ 2 & -1 & -1 & 2 \\ 3 & 1 & -1 & -3 \\ 2 & 4 & -3 & -1 \end{bmatrix} \xrightarrow[\substack{R_3 \to R_3 - 3R_1 \\ R_4 \to R_4 - 2R_1}]{R_2 \to R_2 - 2R_1} \begin{bmatrix} 1 & 1 & 2 & -1 \\ 0 & -3 & -5 & 4 \\ 0 & -2 & -7 & 0 \\ 0 & 2 & -7 & 1 \end{bmatrix} \xrightarrow[R_4 \to R_4 + R_3]{R_2 \to R_2 - R_3} \begin{bmatrix} 1 & 1 & 2 & -1 \\ 0 & -1 & 2 & 4 \\ 0 & -2 & -7 & 0 \\ 0 & 0 & -14 & 1 \end{bmatrix}$$

$$\xrightarrow[R_3 \to R_3 - 2R_2]{} \begin{bmatrix} 1 & 1 & 2 & -1 \\ 0 & -1 & 2 & 4 \\ 0 & 0 & -11 & -8 \\ 0 & 0 & -14 & 1 \end{bmatrix} \xrightarrow[R_4 \to R_4 - R_3]{R_2 \to -R_2} \begin{bmatrix} 1 & 0 & 2 & -1 \\ 0 & 1 & -2 & -4 \\ 0 & 0 & -11 & -8 \\ 0 & 0 & -3 & 9 \end{bmatrix}$$

$$\xrightarrow[R_4 \to -1/3R_4]{} \begin{bmatrix} 1 & 1 & 2 & -1 \\ 0 & 1 & -2 & -4 \\ 0 & 0 & -11 & -8 \\ 0 & 0 & 1 & -3 \end{bmatrix} \xrightarrow[R_3 \to R_3 + 11R_4]{} \begin{bmatrix} 1 & 1 & 2 & -1 \\ 0 & 1 & -2 & -4 \\ 0 & 0 & 0 & -41 \\ 0 & 0 & 1 & -3 \end{bmatrix}$$

$$\xrightarrow[R_3 \leftrightarrow R_4]{} \begin{bmatrix} 1 & 1 & 2 & -1 \\ 0 & 1 & -2 & -4 \\ 0 & 0 & 1 & -3 \\ 0 & 0 & 0 & -41 \end{bmatrix} \xrightarrow[R_4 \to -1/41R_4]{} \begin{bmatrix} 1 & 1 & 2 & -1 \\ 0 & 1 & -2 & -4 \\ 0 & 0 & 1 & -3 \\ 0 & 0 & 0 & 1 \end{bmatrix}$$

Therefore, rank $A = 4$.

(2)

$$B = \begin{bmatrix} 1 & 2 & 3 & 1 \\ 2 & 1 & 1 & 3 \\ 0 & 2 & -1 & -1 \\ 2 & -1 & 2 & 4 \end{bmatrix} \xrightarrow[R_4 \to R_4 - 2R_1]{R_2 \to R_2 - 2R_1} \begin{bmatrix} 1 & 2 & 3 & 1 \\ 0 & -3 & -5 & 1 \\ 0 & 2 & -1 & -1 \\ 0 & -5 & -4 & 2 \end{bmatrix} \xrightarrow[R_4 \to R_4 + 2R_3]{R_2 \to R_2 + R_3} \begin{bmatrix} 1 & 2 & 3 & 1 \\ 0 & -1 & -6 & 0 \\ 0 & 2 & -1 & -1 \\ 0 & -1 & -6 & 0 \end{bmatrix}$$

$$\xrightarrow[R_4 \to R_4 - R_2]{} \begin{bmatrix} 1 & 2 & 3 & 1 \\ 0 & -1 & -6 & 0 \\ 0 & 2 & -1 & -1 \\ 0 & 0 & 0 & 0 \end{bmatrix} \xrightarrow[R_3 \to R_3 + 2R_2]{} \begin{bmatrix} 1 & 2 & 3 & 1 \\ 0 & -1 & -6 & 0 \\ 0 & 0 & -13 & -1 \\ 0 & 0 & 0 & 0 \end{bmatrix}$$

$$\xrightarrow[R_3 \to -1/13R_3]{R_2 \to -R_2} \begin{bmatrix} 1 & 2 & 3 & 1 \\ 0 & 1 & 6 & 0 \\ 0 & 0 & 1 & \frac{1}{13} \\ 0 & 0 & 0 & 0 \end{bmatrix}$$

Therefore, rank $B = 3$.

(3)
$$C = \begin{bmatrix} 1 & -2 & 3 & -1 \\ 2 & -1 & 3 & -1 \\ 0 & -2 & 1 & 1 \\ -1 & 1 & 2 & 3 \end{bmatrix} \xrightarrow[R_4 \to R_4 + R_1]{R_2 \to R_2 - 2R_1} \begin{bmatrix} 1 & -2 & 3 & -1 \\ 0 & 3 & -3 & 1 \\ 0 & -2 & 1 & 1 \\ 0 & -1 & 5 & 2 \end{bmatrix} \xrightarrow[R_3 \to R_3 - 2R_4]{R_2 \to R_2 + R_3} \begin{bmatrix} 1 & -2 & 3 & -1 \\ 0 & 1 & -2 & 2 \\ 0 & 0 & -9 & -3 \\ 0 & -1 & 5 & 2 \end{bmatrix}$$

$$\xrightarrow[R_4 \to R_4 + R_2]{R_3 \to -1/3 R_3} \begin{bmatrix} 1 & -2 & 3 & -1 \\ 0 & 1 & -2 & 2 \\ 0 & 0 & 3 & 1 \\ 0 & 0 & 3 & 4 \end{bmatrix} \xrightarrow{R_4 \to R_4 - R_3} \begin{bmatrix} 1 & -2 & 3 & -1 \\ 0 & 1 & -2 & 2 \\ 0 & 0 & 3 & 1 \\ 0 & 0 & 0 & 3 \end{bmatrix}$$

Therefore, rank $C = 4$.

(4)
$$D = \begin{bmatrix} 1 & 2 & 0 & 0 \\ 0 & 1 & -1 & 2 \\ 1 & -1 & 2 & 1 \\ 2 & -3 & 1 & -1 \end{bmatrix} \xrightarrow[R_4 \to R_4 - 2R_1]{R_3 \to R_3 - R_1} \begin{bmatrix} 1 & 2 & 0 & 0 \\ 0 & 1 & -1 & 2 \\ 0 & -3 & 2 & 1 \\ 0 & -7 & 1 & -1 \end{bmatrix} \xrightarrow[R_4 \to R_4 + 7R_2]{R_3 \to R_3 + 3R_2} \begin{bmatrix} 1 & 2 & 0 & 0 \\ 0 & 1 & -1 & 2 \\ 0 & 0 & -1 & 7 \\ 0 & 0 & -6 & 13 \end{bmatrix}$$

$$\xrightarrow[R_4 \to R_4 - 6R_3]{R_3 \to -R_3} \begin{bmatrix} 1 & 2 & 0 & 0 \\ 0 & 1 & -1 & 2 \\ 0 & 0 & 1 & -7 \\ 0 & 0 & 0 & -29 \end{bmatrix}$$

Therefore, rank $D = 4$.

(5)
$$E = \begin{bmatrix} 1 & 1 & 2 & -1 \\ -2 & 1 & 1 & -2 \\ 1 & -2 & -3 & 3 \\ 2 & 4 & -2 & 1 \end{bmatrix} \xrightarrow[\substack{R_3 \to R_3 - R_1 \\ R_4 \to R_4 - 2R_1}]{R_2 \to R_2 + 2R_1} \begin{bmatrix} 1 & 1 & 2 & -1 \\ 0 & 3 & 5 & -4 \\ 0 & -3 & -5 & 4 \\ 0 & 2 & -6 & 3 \end{bmatrix} \xrightarrow{R_3 \to R_3 + R_2} \begin{bmatrix} 1 & 1 & 2 & -1 \\ 0 & 3 & 5 & -4 \\ 0 & 0 & 0 & 0 \\ 0 & 2 & -6 & 3 \end{bmatrix}$$

$$\xrightarrow[R_3 \leftrightarrow R_4]{R_2 \to R_2 - R_4} \begin{bmatrix} 1 & 1 & 2 & -1 \\ 0 & 1 & 11 & -7 \\ 0 & 2 & -6 & 3 \\ 0 & 0 & 0 & 0 \end{bmatrix} \xrightarrow{R_3 \to R_3 - 2R_2} \begin{bmatrix} 1 & 1 & 2 & -1 \\ 0 & 1 & 11 & -7 \\ 0 & 0 & -28 & 17 \\ 0 & 0 & 0 & 0 \end{bmatrix}$$

Therefore, rank $E = 3$. Therefore, A, C, and D are equivalent, and B and E are equivalent.

PROBLEM 8-8 Determine whether A^{-1} exists for

$$A = \begin{bmatrix} 1 & -1 & 2 & 1 \\ 1 & 2 & -2 & 3 \\ 0 & 1 & -1 & 0 \\ 1 & 0 & 1 & 1 \end{bmatrix}$$

Solution A^{-1} exists for A if rank $A = 4$. We will find the rank of A.

$$A = \begin{bmatrix} 1 & -1 & 2 & 1 \\ 1 & 2 & -2 & 3 \\ 0 & 1 & -1 & 0 \\ 1 & 0 & 1 & 1 \end{bmatrix} \xrightarrow[R_4 \to R_4 - R_1]{R_2 \to R_2 - R_1} \begin{bmatrix} 1 & -1 & 2 & 1 \\ 0 & 3 & -4 & 2 \\ 0 & 1 & -1 & 0 \\ 0 & 1 & -1 & 0 \end{bmatrix} \xrightarrow[R_4 \to R_4 - R_3]{R_2 \to R_2 - 2R_3} \begin{bmatrix} 1 & -1 & 2 & 1 \\ 0 & 1 & -2 & 2 \\ 0 & 1 & -1 & 0 \\ 0 & 0 & 0 & 0 \end{bmatrix}$$

Since there is a row of zeros rank $A \neq 4$. Therefore, A^{-1} does not exist.

PROBLEM 8-9 Determine whether A^{-1} exists for

$$A = \begin{bmatrix} 1 & 2 & -1 \\ 2 & 1 & 1 \\ 1 & -1 & -2 \end{bmatrix}$$

Solution A^{-1} exists for A if rank $A = 3$. The determinant of A is nonzero if A is of rank 3. We will find det A.

$$\det A = \begin{vmatrix} 1 & 2 & -1 \\ 2 & 1 & 1 \\ 1 & -1 & -2 \end{vmatrix} = \frac{1}{1}\begin{vmatrix} -3 & 3 \\ -3 & -1 \end{vmatrix} = 3 + 9 = 12 \neq 0$$

Therefore, A^{-1} exists, but we have not found A^{-1}. We have only established its existence.

PROBLEM 8-10 Use Problem 8-9 to find bases for R^3 and R_3 which are not natural bases for R^3 and R_3.

Solution Since A is invertible, then a basis for R^3 is

$$\left\{ \begin{bmatrix} 1 \\ 2 \\ 1 \end{bmatrix}, \begin{bmatrix} 2 \\ 1 \\ -1 \end{bmatrix}, \begin{bmatrix} -1 \\ 1 \\ -2 \end{bmatrix} \right\}$$

Likewise, a basis for R_3 is $\{[1 \quad 2 \quad -1], [2 \quad 1 \quad 1], [1 \quad -1 \quad -2]\}$.

PROBLEM 8-11 Determine whether each homogeneous system has a nontrivial solution.

(a)
$$\begin{aligned} x_1 - x_2 &= 0 \\ 2x_1 + 3x_2 - x_3 &= 0 \\ x_1 - 2x_2 + 2x_3 &= 0 \end{aligned}$$

(b)
$$\begin{aligned} x_1 - x_2 + 3x_3 &= 0 \\ 2x_1 - x_2 + x_3 &= 0 \\ 3x_1 - 2x_2 + 4x_3 &= 0 \end{aligned}$$

Solution

(a) We will reduce the coefficient matrix A.

$$A = \begin{bmatrix} 1 & -1 & 0 \\ 2 & 3 & -1 \\ 1 & -2 & 2 \end{bmatrix} \xrightarrow[R_3 \to R_3 - R_1]{R_2 \to R_2 - 2R_1} \begin{bmatrix} 1 & -1 & 0 \\ 0 & 5 & -1 \\ 0 & -1 & 2 \end{bmatrix} \xrightarrow{R_2 \leftrightarrow R_3} \begin{bmatrix} 1 & -1 & 0 \\ 0 & -1 & 2 \\ 0 & 5 & -1 \end{bmatrix}$$

$$\xrightarrow[R_3 \to R_3 + 5R_2]{R_2 \to -R_2} \begin{bmatrix} 1 & -1 & 0 \\ 0 & 1 & -2 \\ 0 & 0 & 9 \end{bmatrix}$$

Since there are three nonzero rows, rank $A = 3$. Therefore, the system has only the trivial solution.

(b)
$$B = \begin{bmatrix} 1 & -1 & 3 \\ 2 & -1 & 1 \\ 3 & -2 & 4 \end{bmatrix} \xrightarrow[R_3 \to R_3 - 3R_1]{R_2 \to R_2 - 2R_1} \begin{bmatrix} 1 & -1 & 3 \\ 0 & 1 & -5 \\ 0 & 1 & -5 \end{bmatrix} \xrightarrow[R_3 \to R_3 - R_2]{} \begin{bmatrix} 1 & -1 & 3 \\ 0 & 1 & -5 \\ 0 & 0 & 0 \end{bmatrix}$$

Since there are only two nonzero rows, rank $B = 2$. Therefore, the system has a nontrivial solution.

PROBLEM 8-12 Determine whether the following nonhomogeneous systems have solutions.

(a)
$$\begin{aligned} x + y + 2z &= 4 \\ 2x - y - z &= 0 \\ x + 2y + 3z &= 6 \end{aligned}$$

(b)
$$\begin{aligned} x + y + z &= 3 \\ 2x + 3y + 2z &= 3 \\ x + 2y + z &= 1 \end{aligned}$$

Solution Since we only want to know if the systems have solutions, we'll compare the ranks of the coefficient and augmented matrices for each system.

(a) $$\begin{bmatrix} 1 & 1 & 2 \\ 2 & -1 & -1 \\ 1 & 2 & 3 \end{bmatrix} \xrightarrow[R_3 \to R_3 - R_1]{R_2 \to R_2 - 2R_1} \begin{bmatrix} 1 & 1 & 2 \\ 0 & -3 & -5 \\ 0 & 1 & 1 \end{bmatrix} \xrightarrow{R_2 \leftrightarrow R_3} \begin{bmatrix} 1 & 1 & 2 \\ 0 & 1 & 1 \\ 0 & -3 & -5 \end{bmatrix} \xrightarrow[R_3 \to R_3 + 3R_2]{} \begin{bmatrix} 1 & 1 & 2 \\ 0 & 1 & 1 \\ 0 & 0 & -2 \end{bmatrix}$$

The coefficient matrix has rank 3.

$$\begin{bmatrix} 1 & 1 & 2 & 4 \\ 2 & -1 & -1 & 0 \\ 1 & 2 & 3 & 6 \end{bmatrix} \xrightarrow[R_3 \to R_3 - R_1]{R_2 \to R_2 - 2R_1} \begin{bmatrix} 1 & 1 & 2 & 4 \\ 0 & -3 & -5 & -8 \\ 0 & 1 & 1 & 2 \end{bmatrix} \xrightarrow{R_2 \leftrightarrow R_3} \begin{bmatrix} 1 & 1 & 2 & 4 \\ 0 & 1 & 1 & 2 \\ 0 & -3 & -5 & -8 \end{bmatrix}$$

$$\xrightarrow[R_3 \to R_3 + 3R_2]{} \begin{bmatrix} 1 & 1 & 2 & 4 \\ 0 & 1 & 1 & 2 \\ 0 & 0 & -2 & -2 \end{bmatrix}$$

The augmented matrix has rank 3. Therefore, the system has a solution, since ranks are the same.

(b) $$\begin{bmatrix} 1 & 1 & 1 \\ 2 & 3 & 2 \\ 1 & 2 & 1 \end{bmatrix} \xrightarrow[R_3 \to R_3 - R_1]{R_2 \to R_2 - 2R_1} \begin{bmatrix} 1 & 1 & 1 \\ 0 & 1 & 0 \\ 0 & 1 & 0 \end{bmatrix} \xrightarrow[R_3 \to R_3 - R_2]{} \begin{bmatrix} 1 & 1 & 1 \\ 0 & 1 & 0 \\ 0 & 0 & 0 \end{bmatrix}$$

The coefficient matrix has rank 2.

$$\begin{bmatrix} 1 & 1 & 1 & 3 \\ 2 & 3 & 2 & 3 \\ 1 & 2 & 1 & 1 \end{bmatrix} \xrightarrow[R_3 \to R_3 - R_1]{R_2 \to R_2 - 2R_1} \begin{bmatrix} 1 & 1 & 1 & 3 \\ 0 & 1 & 0 & -3 \\ 0 & 1 & 0 & -2 \end{bmatrix} \xrightarrow[R_3 \to R_3 - R_2]{} \begin{bmatrix} 1 & 1 & 1 & 3 \\ 0 & 1 & 0 & -3 \\ 0 & 0 & 0 & 1 \end{bmatrix}$$

The augmented matrix has rank 3. Therefore, the system does not have a solution since the rank of the coefficient matrix does not equal the rank of the augmented matrix.

PROBLEM 8-13 Determine whether the set of vectors is linearly independent or linearly dependent.

$$\{[1 \quad 1 \quad 0 \quad 1], [0 \quad 1 \quad -1 \quad 2], [-2 \quad 1 \quad -2 \quad -1], [2 \quad 1 \quad -1 \quad -3]\}$$

Solution We shall form a 4×4 matrix A with the given vectors as rows, and then reduce the matrix to echelon form.

$$A = \begin{bmatrix} 1 & 1 & 0 & 1 \\ 0 & 1 & -1 & 2 \\ -2 & 1 & -2 & -1 \\ 2 & 1 & -1 & -3 \end{bmatrix} \xrightarrow[\substack{R_3 \to R_3 + 2R_1 \\ R_4 \to R_4 - 2R_1}]{} \begin{bmatrix} 1 & 1 & 0 & 1 \\ 0 & 1 & -1 & 2 \\ 0 & 3 & -2 & 1 \\ 0 & -1 & -1 & -5 \end{bmatrix} \xrightarrow[\substack{R_3 \to R_3 - 3R_2 \\ R_4 \to R_4 + R_2}]{} \begin{bmatrix} 1 & 1 & 0 & 1 \\ 0 & 1 & -1 & 2 \\ 0 & 0 & 1 & -5 \\ 0 & 0 & -2 & -3 \end{bmatrix}$$

$$\xrightarrow[R_4 \to R_4 + 2R_3]{} \begin{bmatrix} 1 & 1 & 0 & 1 \\ 0 & 1 & -1 & 2 \\ 0 & 0 & 1 & -5 \\ 0 & 0 & 0 & -13 \end{bmatrix} \xrightarrow[R_4 \to -1/13 R_4]{} \begin{bmatrix} 1 & 1 & 0 & 1 \\ 0 & 1 & -1 & 2 \\ 0 & 0 & 1 & -5 \\ 0 & 0 & 0 & 1 \end{bmatrix}$$

Rank $A = 4$. Therefore, the set is linearly independent, by statements **(9)** and **(11)** of Section 8.3.

PROBLEM 8-14 Show that the following set of vectors is linearly independent in R_5: $\{[1 \quad k \quad 0 \quad 0 \quad t], [0 \quad 0 \quad 1 \quad 0 \quad s], [0 \quad 0 \quad 0 \quad 1 \quad u]\}$.

Solution We shall form a matrix A with the vectors as rows:

$$A = \begin{bmatrix} 1 & k & 0 & 0 & t \\ 0 & 0 & 1 & 0 & s \\ 0 & 0 & 0 & 1 & u \end{bmatrix}$$

This matrix is in echelon form and contains no zero rows. The number of rows is 3, so the given set forms a linearly independent set of vectors in R_5.

Supplementary Problems

PROBLEM 8-15 Given $A = \begin{bmatrix} 1 & 1 & 4 \\ 2 & 3 & -1 \end{bmatrix}$, find the rank and a basis for (**a**) the row space of A, and (**b**) the column space of A.

PROBLEM 8-16 Determine which of the following matrices have the same row space.

$$A = \begin{bmatrix} 1 & 0 & 2 \\ -1 & 3 & -1 \end{bmatrix} \qquad B = \begin{bmatrix} 1 & -2 & -1 \\ 3 & -5 & -1 \end{bmatrix} \qquad C = \begin{bmatrix} 1 & -2 & -1 \\ 2 & 3 & -4 \\ -1 & 1 & 3 \end{bmatrix}$$

PROBLEM 8-17 Find a matrix that has the same row space as

$$A = \begin{bmatrix} 1 & 3 & -2 & 4 \\ -1 & 2 & 5 & -1 \\ 0 & -1 & 2 & 3 \end{bmatrix}$$

PROBLEM 8-18 Given

$$A = \begin{bmatrix} 1 & -2 & 3 & 4 \\ -1 & 0 & 2 & -1 \\ 2 & 1 & -1 & 2 \end{bmatrix}$$

find a basis and the dimension of the (**a**) column space of A and (**b**) row space of A.

PROBLEM 8-19 Which of the following matrices are equivalent?

$$A = \begin{bmatrix} 1 & 2 & -3 \\ -1 & 1 & 2 \end{bmatrix} \qquad B = \begin{bmatrix} -1 & 0 & 2 \\ 2 & 1 & -3 \end{bmatrix} \qquad C = \begin{bmatrix} 1 & -2 & 0 & 3 \\ -1 & 1 & 2 & -4 \\ 1 & -3 & 2 & 2 \end{bmatrix}$$

$$D = \begin{bmatrix} 1 & 0 & 2 & 1 \\ -1 & 2 & -3 & 5 \\ 2 & 3 & -1 & 4 \end{bmatrix} \qquad E = \begin{bmatrix} 2 & -1 & 3 \\ 1 & 2 & -1 \\ 4 & 3 & -2 \end{bmatrix}$$

PROBLEM 8-20 Determine whether A^{-1} exists for

$$A = \begin{bmatrix} 2 & -1 & 3 \\ 1 & 2 & -1 \\ 1 & -3 & 4 \end{bmatrix}$$

PROBLEM 8-21 Determine whether the set of vectors $\{[1 \quad -1 \quad 1], [-1 \quad 2 \quad 3], [3 \quad -5 \quad -5]\}$ is linearly independent.

PROBLEM 8-22 Determine whether the homogeneous systems have nontrivial solutions.

(a) $\begin{aligned} 2x_1 - x_2 \quad &= 0 \\ x_1 + x_2 - 2x_3 &= 0 \\ x_1 - 2x_2 + 2x_3 &= 0 \end{aligned}$ **(b)** $\begin{aligned} x_1 + 2x_2 - x_3 &= 0 \\ 2x_1 - 3x_2 + x_3 &= 0 \\ x_1 - x_2 - 2x_3 &= 0 \end{aligned}$

PROBLEM 8-23 Determine whether the nonhomogeneous systems have solutions.

(a) $\begin{aligned} 2x - y \quad &= 1 \\ x + y - 2z &= 2 \\ x - y + 2z &= -1 \end{aligned}$ **(b)** $\begin{aligned} x - 2y + z &= 3 \\ 2x + y - 2z &= 2 \\ x + y + z &= -1 \end{aligned}$

PROBLEM 8-24 Determine if $[1\ \ 4\ \ -1\ \ 2]$, $[2\ \ -1\ \ 3\ \ 1]$ and $[2\ \ 0\ \ 1\ \ -4]$ are linearly independent or linearly dependent.

PROBLEM 8-25 Find a basis for the set of vectors given in Problem 8-24.

PROBLEM 8-26 Let W be a subspace of R_5 spanned by $[1\ \ 0\ \ 1\ \ -1\ \ 0]$, $[2\ \ -2\ \ 4\ \ -6\ \ 8]$, $[1\ \ -2\ \ 3\ \ 3\ \ 4]$, $[1\ \ 5\ \ -3\ \ 2\ \ 6]$, and $[2\ \ -2\ \ 4\ \ 2\ \ 4]$. Find a subset of the vectors which form a basis for W.

PROBLEM 8-27 Let W be a subspace of R_4 defined by $W = \{[a\ \ b\ \ c\ \ d]: a = 2b \text{ and } d = 3c\}$. Find the dimension and a basis for W.

PROBLEM 8-28 Let W be a subspace of R_4 such that $W = \text{span}\{[1\ \ 2\ \ 0\ \ -1], [1\ \ 3\ \ 4\ \ 2], [1\ \ 4\ \ 8\ \ 5]\}$. Find a homogeneous system whose solution space is W.

Answers to Supplementary Problems

8-15 **(a)** Rank 2, basis $\{[1\ \ 0\ \ 13], [0\ \ 1\ \ -9]\}$
(b) Rank 2, basis $\left\{\begin{bmatrix} 1 \\ 0 \end{bmatrix}, \begin{bmatrix} 0 \\ -1 \end{bmatrix}\right\}$

8-16 A and B have rank 2, but not the same row space. C has rank 3.

8-17 There are several answers. One is $\begin{bmatrix} 1 & 3 & -2 & 4 \\ 0 & 5 & 3 & 3 \\ 0 & -1 & 2 & 3 \end{bmatrix}$.

8-18 **(a)** Dim col $(A) = 3$; a basis is
$$\left\{\begin{bmatrix} 1 \\ -1 \\ 2 \end{bmatrix}, \begin{bmatrix} 0 \\ 1 \\ 3 \end{bmatrix}, \begin{bmatrix} 0 \\ 0 \\ 1 \end{bmatrix}\right\}$$
(b) Dim row $(A) = 3$; a basis is $\{[1\ \ -2\ \ 3\ \ 4], [0\ \ 1\ \ 3\ \ 0], [0\ \ 0\ \ 11\ \ 3]\}$.

8-19 A and B are the only equivalent matrices. D and E both have rank 3, but D is a 3×4 matrix and E is a 3×3 matrix.

8-20 Det $A = 0$; therefore A^{-1} does not exist.

8-21 The set is not linearly independent.

8-22 **(a)** This has a nontrivial solution.
(b) This has only the trivial solution.

8-23 **(a)** This has a solution.
(b) This has a solution.

8-24 Linearly independent

8-25 A basis is $\{[1\ \ 4\ \ -1\ \ 2], [0\ \ 1\ \ -2\ \ -5], [0\ \ 0\ \ 13\ \ 48]\}$.

8-26 $[1\ \ 0\ \ 1\ \ -1\ \ 0]$, $[2\ \ -2\ \ 4\ \ -6\ \ 8]$, $[1\ \ -2\ \ 3\ \ 3\ \ 4]$, and $[1\ \ 5\ \ -3\ \ 2\ \ 6]$ form a basis for W.

8-27 A basis for W is $\{[2\ \ 1\ \ 0\ \ 0], [0\ \ 0\ \ 1\ \ 3]\}$. Dim $W = 2$.

8-28 $8x - 4y + z = 0$, $7x - 3y + w = 0$
$$\left(\textit{Hint: } \text{reduce} \begin{bmatrix} 1 & 2 & 0 & -1 \\ 1 & 3 & 4 & 2 \\ 1 & 4 & 8 & 5 \\ x & y & z & w \end{bmatrix}\right)$$

EXAM 2 (Chapters 6–8)

1. Determine whether all matrices of the form $\begin{bmatrix} a & b \\ 0 & 0 \end{bmatrix}$ is a subspace of the vector space of 2×2 matrices.

2. Identify each of the following as either a linearly independent or linearly dependent set of vectors. Give reasons for your answers.

   (a) $\mathbf{v}_1 = [3 \quad 4], \mathbf{v}_2 = [-6 \quad -8]$
   (b) $\mathbf{v}_1, \mathbf{v}_2, \mathbf{v}_3, \mathbf{v}_4$ in a vector space of dimension 3
   (c) $\mathbf{v}_1 = [2 \quad 1 \quad 0], \mathbf{v}_2 = [1 \quad 0 \quad -1]$ in R_3.

3. Do $\mathbf{a} = [1 \quad 3 \quad 1]$ and $\mathbf{b} = [0 \quad 1 \quad 1]$ span R_3? Why or why not?

4. If $\mathbf{a}$ and $\mathbf{b}$ are as defined in exam question 3, determine whether $\mathbf{x} = [5 \quad 19 \quad 9]$ belongs to span $\{\mathbf{a}, \mathbf{b}\}$.

5. Determine whether the three vectors $x^2 - 1$, $2x^2 + x + 1$, and $7x^2 + 3x + 2$ are linearly independent in P_2.

6. Find a basis for the row space of

$$A = \begin{bmatrix} 1 & 0 & 1 & 2 \\ 0 & 1 & 2 & 0 \\ 1 & 0 & 0 & 1 \end{bmatrix}$$

7. Find a basis for the column space of the matrix in exam question 6.

8. If A is the matrix in question 6, what is (a) the row rank of A? (b) the column rank of A? (c) the rank of A?

9. Determine the dimension and find a basis for the solution space of the following homogeneous system:

$$\begin{aligned} x - y + 2z &= 0 \\ x + y + 4z &= 0 \\ x - 2y + z &= 0 \end{aligned}$$

10. Express $[-10 \quad 3 \quad -1]$ as a linear combination of $[1 \quad 2 \quad 3]$, $[4 \quad -1 \quad 5]$, and $[0 \quad 1 \quad -2]$.

Solutions to Exam 2

1. $$\begin{bmatrix} a & b \\ 0 & 0 \end{bmatrix} + \begin{bmatrix} c & d \\ 0 & 0 \end{bmatrix} = \begin{bmatrix} a+c & b+d \\ 0 & 0 \end{bmatrix}$$

$\therefore$ closed under addition

$$k\begin{bmatrix} a & b \\ 0 & 0 \end{bmatrix} = \begin{bmatrix} ka & kb \\ 0 & 0 \end{bmatrix}$$

$\therefore$ closed under scalar multiplication

Therefore, matrices of form $\begin{bmatrix} a & b \\ 0 & 0 \end{bmatrix}$ form a subspace.

2. (a) Dependent because $\mathbf{v}_2 = -2\mathbf{v}_1$
   (b) Dependent because there are four vectors in the vector space of dimension 3

(c) $c_1[2 \quad 1 \quad 0] + c_2[1 \quad 0 \quad -1] = [0 \quad 0 \quad 0]$

$$\begin{aligned} 2c_1 + c_2 &= 0 \\ c_1 \quad &= 0 \\ -c_2 &= 0 \end{aligned}$$

$\therefore$ $\mathbf{v}_1$ and $\mathbf{v}_2$ are independent

3. No. Since the dimension of R_3 is 3, the minimum number of vectors in a spanning set for R_3 is 3.

4. We must determine whether $\mathbf{x}$ can be written as a linear combination of $\mathbf{a}$ and $\mathbf{b}$.

$$c_1[1 \quad 3 \quad 1] + c_2[0 \quad 1 \quad 1] = [5 \quad 19 \quad 9]$$

$$\begin{aligned} c_1 \quad &= 5 \\ 3c_1 + c_2 &= 19 \\ c_1 + c_2 &= 9 \end{aligned}$$

Hence $c_1 = 5$ and $c_2 = 4$ and therefore $[5 \quad 19 \quad 9]$ belongs to span $\{\mathbf{a}, \mathbf{b}\}$.

5. Express $x^2 - 1$, $2x^2 + x + 1$, and $7x^2 + 3x + 2$ as $[1 \quad 0 \quad -1]$, $[2 \quad 1 \quad 1]$, and $[7 \quad 3 \quad 2]$, respectively, and consider the matrix with these matrices as rows. Find the determinant of this matrix.

$$\det \begin{bmatrix} 1 & 0 & -1 \\ 2 & 1 & 1 \\ 7 & 3 & 2 \end{bmatrix} = 1 \begin{vmatrix} 1 & 1 \\ 3 & 2 \end{vmatrix} - 1 \begin{vmatrix} 2 & 1 \\ 7 & 3 \end{vmatrix}$$

$$= -1 + 1 = 0$$

Since the determinant equals zero, the rank of the matrix is less than 3 and the vectors are dependent.

6. Change to echelon form.

$$\begin{bmatrix} 1 & 0 & 1 & 2 \\ 0 & 1 & 2 & 0 \\ 1 & 0 & 0 & 1 \end{bmatrix} \xrightarrow{R_3 \to R_3 - R_1} \begin{bmatrix} 1 & 0 & 1 & 2 \\ 0 & 1 & 2 & 0 \\ 0 & 0 & -1 & -1 \end{bmatrix}$$

$$\xrightarrow{R_3 \to -R_3} \begin{bmatrix} 1 & 0 & 1 & 2 \\ 0 & 1 & 2 & 0 \\ 0 & 0 & 1 & 1 \end{bmatrix}$$

Therefore, a basis for the row space of A is $\{[1 \quad 0 \quad 1 \quad 2], [0 \quad 1 \quad 2 \quad 0], [0 \quad 0 \quad 1 \quad 1]\}$.

7. Find the transpose and change to echelon form:

$$\begin{bmatrix} 1 & 0 & 1 \\ 0 & 1 & 0 \\ 1 & 2 & 0 \\ 2 & 0 & 1 \end{bmatrix} \xrightarrow[R_4 \to R_4 - 2R_1]{R_3 \to R_3 - R_1} \begin{bmatrix} 1 & 0 & 1 \\ 0 & 1 & 0 \\ 0 & 2 & -1 \\ 0 & 0 & -1 \end{bmatrix}$$

$$\xrightarrow[\substack{R_3 \to R_3 - R_4 \\ R_4 \to -R_4}]{R_1 \to R_1 + R_4} \begin{bmatrix} 1 & 0 & 0 \\ 0 & 1 & 0 \\ 0 & 2 & 0 \\ 0 & 0 & 1 \end{bmatrix} \xrightarrow{R_3 \to R_3 - 2R_2} \begin{bmatrix} 1 & 0 & 0 \\ 0 & 1 & 0 \\ 0 & 0 & 0 \\ 0 & 0 & 1 \end{bmatrix}$$

$$\xrightarrow{R_3 \leftrightarrow R_4} \begin{bmatrix} 1 & 0 & 0 \\ 0 & 1 & 0 \\ 0 & 0 & 1 \\ 0 & 0 & 0 \end{bmatrix}$$

Therefore, a basis for the column space of A is

$$\left\{ \begin{bmatrix} 1 \\ 0 \\ 0 \end{bmatrix}, \begin{bmatrix} 0 \\ 1 \\ 0 \end{bmatrix}, \begin{bmatrix} 0 \\ 0 \\ 1 \end{bmatrix} \right\}$$

8. (a) From question 6, the dimension of the row space is 3. Therefore, the row rank is 3.
   (b) From question 7, the dimension of the column space is 3. Therefore, the column rank is 3.
   (c) Row rank = column rank = rank = 3

9. Use Gauss–Jordan elimination on the augmented matrix:

$$\begin{bmatrix} 1 & -1 & 2 & 0 \\ 1 & 1 & 4 & 0 \\ 1 & -2 & 1 & 0 \end{bmatrix} \xrightarrow[R_3 \to R_3 - R_1]{R_2 \to R_2 - R_1} \begin{bmatrix} 1 & -1 & 2 & 0 \\ 0 & 2 & 2 & 0 \\ 0 & -1 & -1 & 0 \end{bmatrix}$$

$$\xrightarrow[R_3 \to -R_3]{R_2 \to R_2 + 2R_3} \begin{bmatrix} 1 & -1 & 2 & 0 \\ 0 & 0 & 0 & 0 \\ 0 & 1 & 1 & 0 \end{bmatrix}$$

$$\xrightarrow[R_2 \leftrightarrow R_3]{R_1 \to R_1 + R_3} \begin{bmatrix} 1 & 0 & 3 & 0 \\ 0 & 1 & 1 & 0 \\ 0 & 0 & 0 & 0 \end{bmatrix}$$

$$\therefore \quad x = -3z$$
$$y = -z$$

Let $z = t$, where t is arbitrary.

$$\therefore \begin{bmatrix} x \\ y \\ z \end{bmatrix} = \begin{bmatrix} -3t \\ -t \\ t \end{bmatrix} = -t \begin{bmatrix} 3 \\ 1 \\ -1 \end{bmatrix}$$

Therefore, $\begin{bmatrix} 3 \\ 1 \\ -1 \end{bmatrix}$ is a basis and 1 is the dimension of the solution space.

10. $a[1 \quad 2 \quad 3] + b[4 \quad -1 \quad 5] + c[0 \quad 1 \quad -2]$
$= [-10 \quad 3 \quad -1]$

$$\begin{aligned} a + 4b \quad &= -10 \\ 2a - b + c &= 3 \\ 3a + 5b - 2c &= -1 \end{aligned}$$

Using Gauss–Jordan elimination, we get:

$$\begin{bmatrix} 1 & 4 & 0 & -10 \\ 2 & -1 & 1 & 3 \\ 3 & 5 & -2 & -1 \end{bmatrix}$$

$$\xrightarrow[\substack{R_2 \to R_2 - 2R_1 \\ R_3 \to R_3 - 3R_1}]{} \begin{bmatrix} 1 & 4 & 0 & -10 \\ 0 & -9 & 1 & 23 \\ 0 & -7 & -2 & 29 \end{bmatrix}$$

$$\xrightarrow[R_3 \to -4R_3]{} \begin{bmatrix} 1 & 4 & 0 & -10 \\ 0 & -9 & 1 & 23 \\ 0 & 28 & 8 & -116 \end{bmatrix}$$

$$\xrightarrow[R_3 \to R_3 + 3R_2]{} \begin{bmatrix} 1 & 4 & 0 & -10 \\ 0 & -9 & 1 & 23 \\ 0 & 1 & 11 & -47 \end{bmatrix}$$

$$\xrightarrow{R_2 \leftrightarrow R_3} \begin{bmatrix} 1 & 4 & 0 & -10 \\ 0 & 1 & 11 & -47 \\ 0 & -9 & 1 & 23 \end{bmatrix}$$

$$\xrightarrow[R_3 \to R_3 + 9R_2]{} \begin{bmatrix} 1 & 4 & 0 & -10 \\ 0 & 1 & 11 & -47 \\ 0 & 0 & 100 & -400 \end{bmatrix}$$

$$\xrightarrow[R_3 \to 1/100R_3]{} \begin{bmatrix} 1 & 4 & 0 & -10 \\ 0 & 1 & 11 & -47 \\ 0 & 0 & 1 & -4 \end{bmatrix}$$

$$\xrightarrow{R_2 \to R_2 - 11R_3} \begin{bmatrix} 1 & 4 & 0 & -10 \\ 0 & 1 & 0 & -3 \\ 0 & 0 & 1 & -4 \end{bmatrix}$$

$$\xrightarrow{R_1 \to R_1 - 4R_2} \begin{bmatrix} 1 & 0 & 0 & 2 \\ 0 & 1 & 0 & -3 \\ 0 & 0 & 1 & -4 \end{bmatrix}$$

$\therefore \quad a = 2, \qquad b = -3, \qquad c = -4$

$$\therefore \quad [-10 \quad 3 \quad -1] = 2[1 \quad 2 \quad 3] - 3[4 \quad -1 \quad 5] - 4[0 \quad 1 \quad -2]$$

9 INNER PRODUCTS

THIS CHAPTER IS ABOUT

- ☑ **Inner Product**
- ☑ **Length of a Vector**
- ☑ **Distance between Vectors**
- ☑ **Cauchy–Schwarz Inequality**
- ☑ **Triangle Inequality**
- ☑ **Unit Vectors**

9-1. Inner Product

A. Definition of inner product and inner product space

An **inner product** on a real vector space V is a function that assigns to each ordered pair of vectors $\mathbf{a}$ and $\mathbf{b}$ in V a real number denoted by $\langle \mathbf{a}, \mathbf{b} \rangle$ such that if $\mathbf{a}$, $\mathbf{b}$, and $\mathbf{c}$ are in V and k is a scalar, then

(a) $\langle \mathbf{a}, \mathbf{a} \rangle > 0$ for all $\mathbf{a}$ not the zero element of V
(b) $\langle \mathbf{a}, \mathbf{b} \rangle = \langle \mathbf{b}, \mathbf{a} \rangle$
(c) $\langle \mathbf{a} + \mathbf{b}, \mathbf{c} \rangle = \langle \mathbf{a}, \mathbf{c} \rangle + \langle \mathbf{b}, \mathbf{c} \rangle$
(d) $\langle k\mathbf{a}, \mathbf{b} \rangle = k\langle \mathbf{a}, \mathbf{b} \rangle$

A vector space with an inner product is called an **inner product space**.

EXAMPLE 9-1: Verify that if $\mathbf{a} = [a_1, a_2]$ and $\mathbf{b} = [b_1, b_2]$, then $\langle \mathbf{a}, \mathbf{b} \rangle = a_1 b_1 + a_2 b_2$ defines an inner product on R_2.

Solution

Step 1: $$\langle \mathbf{a}, \mathbf{a} \rangle = a_1 a_1 + a_2 a_2 = a_1^2 + a_2^2 > 0 \qquad \text{for} \qquad a \neq [0,0]$$

Step 2: $$\langle \mathbf{a}, \mathbf{b} \rangle = a_1 b_1 + a_2 b_2 \qquad \text{and}$$
$$\langle \mathbf{b}, \mathbf{a} \rangle = b_1 a_1 + b_2 a_2 = a_1 b_1 + a_2 b_2$$
$$\langle \mathbf{a}, \mathbf{b} \rangle = \langle \mathbf{b}, \mathbf{a} \rangle$$

Step 3: Let

$$\mathbf{c} = [c_1, c_2]$$
$$\begin{aligned}\langle \mathbf{a} + \mathbf{b}, \mathbf{c} \rangle &= \langle [a_1 + b_1, a_2 + b_2], [c_1, c_2] \rangle \\ &= (a_1 + b_1)c_1 + (a_2 + b_2)c_2 \\ &= a_1 c_1 + b_1 c_1 + a_2 c_2 + b_2 c_2 \\ &= (a_1 c_1 + a_2 c_2) + (b_1 c_1 + b_2 c_2) \\ &= \langle \mathbf{a}, \mathbf{c} \rangle + \langle \mathbf{b}, \mathbf{c} \rangle\end{aligned}$$

Step 4:

$$\begin{aligned}\langle k\mathbf{a}, \mathbf{b}\rangle &= \langle [ka_1, ka_2], [b_1, b_2]\rangle \\ &= (ka_1)b_1 + (ka_2)b_2 \\ &= k(a_1b_1 + a_2b_2) \\ &= k\langle \mathbf{a}, \mathbf{b}\rangle\end{aligned}$$

Therefore, it defines an inner product on R_2.

EXAMPLE 9-2: Verify that $\langle \mathbf{a}, \mathbf{b}\rangle = 3a_1b_1 + 4a_2b_2$ defines an inner product on R_2.

Solution

Step 1: $\langle \mathbf{a}, \mathbf{a}\rangle = 3a_1a_1 + 4a_2a_2 = 3a_1^2 + 4a_2^2 > 0 \quad \text{for} \quad \mathbf{a} \neq (0,0)$

Step 2: $\langle \mathbf{a}, \mathbf{b}\rangle = 3a_1b_1 + 4a_2b_2 = 3b_1a_1 + 4b_2a_2 = \langle \mathbf{b}, \mathbf{a}\rangle$

Step 3:

$$\begin{aligned}\langle \mathbf{a} + \mathbf{b}, \mathbf{c}\rangle &= \langle [a_1 + b_1, a_2 + b_2], [c_1, c_2]\rangle \\ &= 3(a_1 + b_1)c_1 + 4(a_2 + b_2)c_2 \\ &= (3a_1c_1 + 4a_2c_2) + (3b_1c_1 + 4b_2c_2) \\ &= \langle \mathbf{a}, \mathbf{c}\rangle + \langle \mathbf{b}, \mathbf{c}\rangle\end{aligned}$$

Step 4:

$$\begin{aligned}\langle k\mathbf{a}, \mathbf{b}\rangle &= 3(ka_1)b_1 + 4(ka_2)b_2 \\ &= k(3a_1b_1 + 4a_2b_2) \\ &= k\langle \mathbf{a}, \mathbf{b}\rangle\end{aligned}$$

Therefore, it defines an inner product.

EXAMPLE 9-3: Verify that $\langle \mathbf{a}, \mathbf{b}\rangle = 2a_1b_1 + 3a_2b_2 - a_1b_2 - a_2b_1$ defines an inner product on R_2.

Solution

Step 1:

$$\begin{aligned}\langle \mathbf{a}, \mathbf{a}\rangle &= 2a_1a_1 + 3a_2a_2 - a_1a_2 - a_2a_1 \\ &= (a_1^2 - 2a_1a_2 + a_2^2) + a_1^2 + 2a_2^2 \\ &= (a_1 - a_2)^2 + a_1^2 + 2a_2^2 > 0 \quad \text{for} \quad \mathbf{a} \neq (0,0)\end{aligned}$$

Step 2:

$$\begin{aligned}\langle \mathbf{a}, \mathbf{b}\rangle &= 2a_1b_1 + 3a_2b_2 - a_1b_2 - a_2b_1 \\ &= 2b_1a_1 + 3b_2a_2 - b_1a_2 - b_2a_1 \\ &= \langle \mathbf{b}, \mathbf{a}\rangle\end{aligned}$$

Step 3:

$$\begin{aligned}\langle \mathbf{a} + \mathbf{b}, \mathbf{c}\rangle &= \langle [a_1 + b_1, a_2 + b_2], [c_1, c_2]\rangle \\ &= 2(a_1 + b_1)c_1 + 3(a_2 + b_2)c_2 - (a_1 + b_1)c_2 - (a_2 + b_2)c_1 \\ &= (2a_1c_1 + 3a_2c_2 - a_1c_1 - a_2c_1) \\ &\quad + (2b_1c_1 + 3b_2c_2 - b_1c_2 - b_2c_1) \\ &= \langle \mathbf{a}, \mathbf{c}\rangle + \langle \mathbf{b}, \mathbf{c}\rangle\end{aligned}$$

Step 4:

$$\begin{aligned}\langle k\mathbf{a}, \mathbf{b}\rangle &= 2(ka_1)b_1 + 3(ka_2)b_2 - (ka_1)b_2 - (ka_2)b_1 \\ &= k(2a_1b_1 + 3a_2b_2 - a_1b_2 - a_2b_1) \\ &= k\langle \mathbf{a}, \mathbf{b}\rangle\end{aligned}$$

Therefore, it defines an inner product.

EXAMPLE 9-4: Show that $\langle \mathbf{A}, \mathbf{B}\rangle = \det(\mathbf{A})\det(\mathbf{B})$ does not define an inner product on the vector space of all 2×2 matrices.

Solution: Let $A = \begin{vmatrix} 0 & 2 \\ 0 & 4 \end{vmatrix}$ and $B = \begin{vmatrix} 2 & 1 \\ 2 & 1 \end{vmatrix}$. Then

$$\langle \mathbf{A}, \mathbf{B}\rangle = \det(\mathbf{A})\det(\mathbf{B}) = \begin{vmatrix} 0 & 2 \\ 0 & 4 \end{vmatrix}\begin{vmatrix} 2 & 1 \\ 2 & 1 \end{vmatrix} = 0(0) = 0$$

But neither **A** nor **B** is the zero matrix. Hence, this counterexample shows that the given definition violates property **(a)** of the inner product definition and thus does not define an inner product.

B. Definition of standard inner product

The **standard inner product** on R_n is defined by $\langle \mathbf{a}, \mathbf{b} \rangle = a_1b_1 + a_2b_2 + a_3b_3 + \cdots + a_nb_n$. On R_2 and R_3 the standard inner product is often called the **dot product** and is frequently denoted by $\mathbf{a} \cdot \mathbf{b}$ rather than $\langle \mathbf{a}, \mathbf{b} \rangle$.

EXAMPLE 9-5: Find the standard inner product for the following pairs of vectors: **(a)** $\mathbf{a} = [2 \quad 1 \quad 3]$, $\mathbf{b} = [0 \quad 5 \quad 6]$; **(b)** $\mathbf{a} = [5 \quad 0 \quad 4]$, $\mathbf{b} = [4 \quad -3 \quad -2]$; **(c)** $\mathbf{a} = [3 \quad -4 \quad 5 \quad 7]$, $\mathbf{b} = [2 \quad 4 \quad -1 \quad 6]$.

Solution

(a) $\langle \mathbf{a}, \mathbf{b} \rangle = \mathbf{a} \cdot \mathbf{b} = 2(0) + 1(5) + 3(6) = 23$
(b) $\langle \mathbf{a}, \mathbf{b} \rangle = \mathbf{a} \cdot \mathbf{b} = 5(4) + 0(-3) + 4(-2) = 12$
(c) $\langle \mathbf{a}, \mathbf{b} \rangle = 3(2) - 4(4) + 5(-1) + 7(6) = 27$

EXAMPLE 9-6: Use $\langle \mathbf{a}, \mathbf{b} \rangle = 2a_1b_1 + 3a_2b_2 + 4a_3b_3$ on R_3 to find the inner product of the following pairs of vectors: **(a)** $\mathbf{a} = [2 \quad 1 \quad 3]$, $\mathbf{b} = [0 \quad 5 \quad 6]$; **(b)** $\mathbf{a} = [5 \quad 0 \quad 4]$, $\mathbf{b} = [4 \quad -3 \quad -2]$.

Solution

(a) $\langle \mathbf{a}, \mathbf{b} \rangle = 2(2)(0) + 3(1)(5) + 4(3)(6) = 87$
(b) $\langle \mathbf{a}, \mathbf{b} \rangle = 2(5)(4) + 3(0)(-3) + 4(4)(-2) = 8$

EXAMPLE 9-7: Use $\langle \mathbf{a}, \mathbf{b} \rangle = 2a_1b_1 + 3a_2b_2 - a_1b_2 - a_2b_1$ on R_2 to find the inner product of $\mathbf{a} = [2 \quad 5]$ and $\mathbf{b} = [3 \quad 4]$.

Solution

$$\langle \mathbf{a}, \mathbf{b} \rangle = 2(2)(3) + 3(5)(4) - 2(4) - 5(3) = 49$$

9-2. Length of a Vector

Definition of length of a vector

In an inner product space V, the **length, norm,** or **magnitude** of a vector **a** is $\|\mathbf{a}\| = \sqrt{\langle \mathbf{a}, \mathbf{a} \rangle}$.

Observe that in R_2 with the standard inner product the length of $\mathbf{a} = [a_1 \quad a_2]$ is $\|\mathbf{a}\| = \|[a_1 \quad a_2]\| = \sqrt{\langle \mathbf{a}, \mathbf{a} \rangle} = \sqrt{a_1^2 + a_2^2}$, which is the square root of the sum of the squares of the components of the vector **a**. Thus, in R_2 with the standard inner product the length of a vector $[a_1 \quad a_2]$ is the distance from the origin to the point (a_1, a_2).

Likewise in R_3 with the standard inner product the length of $\mathbf{a} = [a_1 \quad a_2 \quad a_3]$ is $\|[a_1 \quad a_2 \quad a_3]\| = \sqrt{\langle \mathbf{a}, \mathbf{a} \rangle} = \sqrt{a_1^2 + a_2^2 + a_3^2}$ which is the distance from the origin to the point (a_1, a_2, a_3).

EXAMPLE 9-8: Find the length of the vector $\mathbf{a} = [3 \quad 4]$ in R_2 with the standard inner product.

Solution

$$\|[3 \quad 4]\| = \sqrt{3^2 + 4^2} = \sqrt{25} = 5$$

EXAMPLE 9-9: Find the length of the vector $\mathbf{a} = [3 \quad 4]$ in R_2 with inner product $\langle \mathbf{a}, \mathbf{b} \rangle = 3a_1b_1 + 4a_2b_2$.

Solution

$$\|\mathbf{a}\| = \sqrt{\langle \mathbf{a}, \mathbf{a} \rangle} = \sqrt{3a_1^2 + 4a_2^2} = \sqrt{3(3)^2 + 4(4)^2} = \sqrt{27 + 64} = \sqrt{91}$$

EXAMPLE 9-10: Find the length of the vector $\mathbf{a} = [3 \quad 4]$ in R_2 with the inner product $\langle \mathbf{a}, \mathbf{b} \rangle = 2a_1b_1 + 3a_2b_2 - a_1b_2 - a_2b_1$.

Solution

$$\|\mathbf{a}\| = \sqrt{\langle \mathbf{a}, \mathbf{a} \rangle} = \sqrt{2(3)(3) + 3(4)(4) - 3(4) - 4(3)} = \sqrt{18 + 48 - 12 - 12} = \sqrt{42}$$

You have observed in Examples 9-8, 9-9, and 9-10 that the length of a vector is dependent upon the defined inner product.

9-3. Distance between Vectors

If $\mathbf{a}$ and $\mathbf{b}$ are two vectors in an inner product space, then the **distance** between $\mathbf{a}$ and $\mathbf{b}$ is $d(\mathbf{a}, \mathbf{b}) = \|\mathbf{a} - \mathbf{b}\|$.

EXAMPLE 9-11: Find the distance between the vectors $\mathbf{a} = [2 \quad 3]$ and $\mathbf{b} = [6 \quad 2]$ in R_2 if R_2 has the standard inner product.

Solution

$$d(\mathbf{a}, \mathbf{b}) = \|\mathbf{a} - \mathbf{b}\| = \sqrt{\langle \mathbf{a} - \mathbf{b}, \mathbf{a} - \mathbf{b} \rangle} = \sqrt{(2 - 6)^2 + (3 - 2)^2} = \sqrt{16 + 1} = \sqrt{17}$$

Observe that this is merely the distance between the points (2, 3) and (6, 2).

EXAMPLE 9-12: Find the distance between the vectors $\mathbf{a} = [2 \quad 3]$ and $\mathbf{b} = [6 \quad 2]$ in R_2 if R_2 has the inner product $\langle \mathbf{a}, \mathbf{b} \rangle = 3a_1b_1 + 5a_2b_2$.

Solution

$$d(\mathbf{a}, \mathbf{b}) = \|\mathbf{a} - \mathbf{b}\| = \sqrt{\langle \mathbf{a} - \mathbf{b}, \mathbf{a} - \mathbf{b} \rangle} = \sqrt{3(2 - 6)^2 + 5(3 - 2)^2} = \sqrt{3(16) + 5(1)} = \sqrt{53}$$

Observe that Examples 9-11 and 9-12 illustrate that the distance between two vectors is dependent upon the inner product.

9-4. Cauchy–Schwarz Inequality

If $\mathbf{a}$ and $\mathbf{b}$ are vectors in an inner product space V, then $\langle \mathbf{a}, \mathbf{b} \rangle^2 \leq \|\mathbf{a}\|^2 \|\mathbf{b}\|^2$.

EXAMPLE 9-13: Verify the Cauchy–Schwarz inequality for $\mathbf{a} = [2 \quad 3 \quad 5]$ and $\mathbf{b} = [1 \quad 2 \quad 3]$ in R_3 with the standard inner product.

Solution

$$\langle \mathbf{a}, \mathbf{b} \rangle^2 = [2(1) + 3(2) + 5(3)]^2 = 23^2 = 529$$

$$\|\mathbf{a}\|^2 = \sqrt{\langle \mathbf{a}, \mathbf{a} \rangle}^2 = \langle \mathbf{a}, \mathbf{a} \rangle = 2^2 + 3^2 + 5^2 = 38$$

$$\|\mathbf{b}\|^2 = \sqrt{\langle \mathbf{b}, \mathbf{b} \rangle}^2 = \langle \mathbf{b}, \mathbf{b} \rangle = 1^2 + 2^2 + 3^3 = 14$$

$$\|\mathbf{a}\|^2 \|\mathbf{b}\|^2 = 38(14) = 532 \qquad \text{and} \qquad 529 < 532$$

Thus $\langle \mathbf{a}, \mathbf{b} \rangle^2 \le \|\mathbf{a}\|^2 \|\mathbf{b}\|^2$ for the given **a** and **b**. Often the Cauchy–Schwarz inequality occurs in the form

$$[2(1) + 3(2) + 5(3)]^2 \le (2^2 + 3^2 + 5^2)(1^2 + 2^2 + 3^2)$$

EXAMPLE 9-14: Verify the Cauchy–Schwarz inequality for $\mathbf{a} = [2 \quad 3 \quad 5]$ and $\mathbf{b} = [1 \quad 2 \quad 3]$ in R_3 with the inner product $\langle \mathbf{a}, \mathbf{b} \rangle = 2a_1b_1 + 3a_2b_2 + a_3b_3 - a_1b_2 - a_2b_1$.

Solution

$$\langle \mathbf{a}, \mathbf{b} \rangle^2 = [2(2)(1) + 3(3)(2) + 5(3) - 2(2) - 3(1)]^2 = 900$$

$$\|\mathbf{a}\|^2 = \langle \mathbf{a}, \mathbf{a} \rangle = 2(2)(2) + 3(3)(3) + 5(5) - 2(3) - 3(2) = 48$$

$$\|\mathbf{b}\|^2 = \langle \mathbf{b}, \mathbf{b} \rangle = 2(1)(1) + 3(2)(2) + 3(3) - 1(2) - 2(1) = 19$$

$$\|\mathbf{a}\|^2 \|\mathbf{b}\|^2 = 48(19) = 912 \qquad 900 < 912$$

Thus $\langle \mathbf{a}, \mathbf{b} \rangle \le \|\mathbf{a}\|^2 \|\mathbf{b}\|^2$ for the given **a** and **b**.

9-5. Triangle Inequality

If **a** and **b** are vectors in an inner product space V, then $\|\mathbf{a} + \mathbf{b}\| \le \|\mathbf{a}\| + \|\mathbf{b}\|$.

EXAMPLE 9-15: Verify the triangle inequality for $\mathbf{a} = [2 \quad 3 \quad 5]$ and $\mathbf{b} = [1 \quad 2 \quad 3]$ in R_3 with the standard inner product.

Solution

$$\|\mathbf{a} + \mathbf{b}\| = \|[3 \quad 5 \quad 8]\| = \sqrt{\langle [3 \quad 5 \quad 8], [3 \quad 5 \quad 8] \rangle}$$

$$= \sqrt{9 + 25 + 64} = \sqrt{98} = 7\sqrt{2} \doteq 9.899$$

$$\|\mathbf{a}\| + \|\mathbf{b}\| = \sqrt{\langle \mathbf{a}, \mathbf{a} \rangle} + \sqrt{\langle \mathbf{b}, \mathbf{b} \rangle} = \sqrt{4 + 9 + 25} + \sqrt{1 + 4 + 9} = \sqrt{38} + \sqrt{14} \doteq 9.906$$

Hence $\sqrt{98} < \sqrt{38} + \sqrt{14}$ and $\|\mathbf{a} + \mathbf{b}\| \le \|\mathbf{a}\| + \|\mathbf{b}\|$ for the given **a** and **b**.

9-6. Unit Vectors

If **a** is a nonzero vector in an inner product space V and $\mathbf{b} = \mathbf{a}/\|\mathbf{a}\|$, then $\|\mathbf{b}\| = 1$. Such vectors are called **unit vectors** and are in the direction of **a**.

EXAMPLE 9-16: Transform $[2 \quad -3 \quad 4]$ in R_3 with standard inner product into a unit vector.

Solution

$$b = \frac{\mathbf{a}}{\|\mathbf{a}\|} = \frac{[2 \quad -3 \quad 4]}{\|[2 \quad -3 \quad 4]\|} = \frac{[2 \quad -3 \quad 4]}{\sqrt{4 + 9 + 16}} = \begin{bmatrix} \frac{2}{\sqrt{29}} & -\frac{3}{\sqrt{29}} & \frac{4}{\sqrt{29}} \end{bmatrix}$$

EXAMPLE 9-17: Transform $[2 \quad -3 \quad 4]$ in R_3 with inner product $\langle \mathbf{a}, \mathbf{b} \rangle = 3a_1b_1 + 2a_2b_2 + a_3b_3$ into a unit vector.

Solution

$$\mathbf{b} = \frac{\mathbf{a}}{\|\mathbf{a}\|} = \frac{\mathbf{a}}{\langle \mathbf{a}, \mathbf{a} \rangle} = \frac{[2 \quad -3 \quad 4]}{\sqrt{3(2)(2) + 2(-3)(-3) + 4(4)}} = \begin{bmatrix} \frac{2}{\sqrt{46}} & -\frac{3}{\sqrt{46}} & \frac{4}{\sqrt{46}} \end{bmatrix}$$

caution: When the inner product is not defined in working with unit vectors, it is assumed to be the standard inner product.

SUMMARY

1. An inner product is a function that assigns to each ordered pair of vectors in a vector space V a real number denoted by $\langle \mathbf{a}, \mathbf{b} \rangle$ such that if **a**, **b**, and **c** are in V and k is a scalar, then

$$\langle \mathbf{a}, \mathbf{a} \rangle > 0 \text{ for } \mathbf{a} \text{ not the zero element of } V$$

$$\langle \mathbf{a}, \mathbf{b} \rangle = \langle \mathbf{b}, \mathbf{a} \rangle$$

$$\langle \mathbf{a} + \mathbf{b}, \mathbf{c} \rangle = \langle \mathbf{a}, \mathbf{c} \rangle + \langle \mathbf{b}, \mathbf{c} \rangle$$

$$\langle k\mathbf{a}, \mathbf{b} \rangle = k\langle \mathbf{a}, \mathbf{b} \rangle$$

2. The standard inner product on R_n is $\langle \mathbf{a}, \mathbf{b} \rangle = a_1b_1 + a_2b_2 + \cdots + a_nb_n$.
3. In R_2 and R_3 the standard inner product is usually denoted by $\mathbf{a} \cdot \mathbf{b}$ and called dot product.
4. In an inner product space, the length, norm or magnitude of a vector **a** is $\|\mathbf{a}\| = \sqrt{\langle \mathbf{a}, \mathbf{a} \rangle}$.
5. In R_2 and R_3 with the standard inner product the lengths of the vectors $[a_1 \quad a_2]$ and $[a_1 \quad a_2 \quad a_3]$ are $\sqrt{a_1^2 + a_2^2}$ and $\sqrt{a_1^2 + a_2^2 + a_3^2}$ which are the distances from the origin to the points (a_1, a_2) and (a_1, a_2, a_3), respectively.
6. The length of a vector is dependent upon the defined inner product.
7. The distance between vectors **a** and **b** in an inner product space is $d(\mathbf{a}, \mathbf{b}) = \|\mathbf{a} - \mathbf{b}\|$.
8. The distance between two vectors is dependent upon the inner product.
9. If **a** and **b** are vectors in an inner product space, then the Cauchy–Schwarz inequality is $\langle \mathbf{a}, \mathbf{b} \rangle^2 \leq \|\mathbf{a}\|^2\|\mathbf{b}\|^2$.
10. If **a** and **b** are vectors in an inner product space V, then the triangle inequality is $\|\mathbf{a} + \mathbf{b}\| \leq \|\mathbf{a}\| + \|\mathbf{b}\|$.
11. If **a** is a nonzero vector in an inner product space then $\mathbf{a}/\|\mathbf{a}\|$ is a unit vector in the direction of **a**.

RAISE YOUR GRADES

Can you ...?

☑ determine if a function is an inner product
☑ find the inner product of two vectors when the inner product is defined
☑ find the standard inner product in R_2
☑ find the length of a vector
☑ find the distance between two vectors
☑ apply the Cauchy–Schwarz inequality
☑ apply thc triangle inequality
☑ find a unit vector in the direction of **a**

SOLVED PROBLEMS

PROBLEM 9-1 Verify that if $\mathbf{a} = [a_1, a_2]$ and $\mathbf{b} = [b_1, b_2]$, then $\langle \mathbf{a}, \mathbf{b} \rangle = 2a_1b_1 + a_2b_2$ defines an inner produced on R_2.

Solution

Step 1: $\langle \mathbf{a}, \mathbf{a} \rangle = 2a_1a_1 + a_2a_2 = 2a_1^2 + a_2^2 > 0 \quad \text{for} \quad \mathbf{a} \neq [0,0]$

Step 2:

$$\langle \mathbf{a}, \mathbf{b} \rangle = 2a_1b_1 + a_2b_2$$

$$\langle \mathbf{b}, \mathbf{a} \rangle = 2b_1a_1 + b_2a_2 = 2a_1b_1 + a_2b_2$$

$$\langle \mathbf{a}, \mathbf{b} \rangle = \langle \mathbf{b}, \mathbf{a} \rangle$$

Step 3: Let $\mathbf{c} = [c_1, c_2]$.

$$\begin{aligned} \langle \mathbf{a} + \mathbf{b}, \mathbf{c} \rangle &= \langle [a_1 + b_1, a_2 + b_2], [c_1, c_2] \rangle \\ &= 2(a_1 + b_1)c_1 + (a_2 + b_2)c_2 \\ &= 2a_1c_1 + 2b_1c_1 + a_2c_2 + b_2c_2 \\ &= (2a_1c_1 + a_2c_2) + (2b_1c_1 + b_2c_2) \\ &= \langle \mathbf{a}, \mathbf{c} \rangle + \langle \mathbf{b}, \mathbf{c} \rangle \end{aligned}$$

Step 4:

$$\begin{aligned} \langle k\mathbf{a}, \mathbf{b} \rangle &= \langle [ka_1, ka_2], [b_1, b_2] \rangle \\ &= 2(ka_1)b_1 + (ka_2)b_2 \\ &= k(2a_1b_1 + a_2b_2) \\ &= k\langle \mathbf{a}, \mathbf{b} \rangle \end{aligned}$$

It defines an inner product on R_2.

PROBLEM 9-2 Verify that $\langle \mathbf{a}, \mathbf{b} \rangle = 5a_1b_1 + 2a_2b_2$ defines an inner product on R_2.

Solution

Step 1: $\langle \mathbf{a}, \mathbf{a} \rangle = 5a_1a_1 + 2a_2a_2 = 5a_1^2 + 2a_2^2 > 0 \quad \text{for} \quad \mathbf{a} \neq [0,0]$

Step 2:

$$\langle \mathbf{a}, \mathbf{b} \rangle = 5a_1b_1 + 2a_2b_2$$

$$\langle \mathbf{b}, \mathbf{a} \rangle = 5b_1a_1 + 2b_2a_2 = 5a_1b_1 + 2a_2b_2$$

$$\langle \mathbf{a}, \mathbf{b} \rangle = \langle \mathbf{b}, \mathbf{a} \rangle$$

Step 3: Let $\mathbf{c} = [c_1, c_2]$.

$$\begin{aligned} \langle \mathbf{a} + \mathbf{b}, \mathbf{c} \rangle &= \langle [a_1 + b_1, a_2 + b_2], [c_1, c_2] \rangle \\ &= 5(a_1 + b_1)c_1 + 2(a_2 + b_2)c_2 \\ &= 5a_1c_1 + 5b_1c_1 + 2a_2c_2 + 2b_2c_2 \\ &= (5a_1c_1 + 2a_2c_2) + (5b_1c_1 + 2b_2c_2) \\ &= \langle \mathbf{a}, \mathbf{c} \rangle + \langle \mathbf{b}, \mathbf{c} \rangle \end{aligned}$$

Step 4:

$$\begin{aligned} \langle k\mathbf{a}, \mathbf{b} \rangle &= \langle ka_1, ka_2], [b_1, b_2] \rangle \\ &= 5(ka_1)b_1 + 2(ka_2)b_2 \\ &= k(5a_1b_1 + 2a_2b_2) \\ &= k\langle \mathbf{a}, \mathbf{b} \rangle \end{aligned}$$

It defines an inner product on R_2.

PROBLEM 9-3 Verify that $\langle \mathbf{a}, \mathbf{b} \rangle = 3a_1b_1 - 2a_2b_2$ does not define an inner product on R_2.

Solution

Step 1: $\langle \mathbf{a}, \mathbf{a} \rangle = 3a_1a_1 - 2a_2a_2 = 3a_1^2 - 2a_2^2$, but this expression is not always greater than zero for $\mathbf{a} \neq [0,0]$.

It does not define an inner product on R_2.

PROBLEM 9-4 Does $\langle \mathbf{a}, \mathbf{b} \rangle = 3a_1b_1 + 2a_2b_2 - 4a_1b_2$ define an inner product on R_2?

Solution

Step 1:
$$\begin{aligned}\langle \mathbf{a}, \mathbf{a} \rangle &= 3a_1^2 + 2a_2^2 - 4a_1a_2 = 2a_1^2 - 4a_1a_2 + 2a_2^2 + a_1^2 \\ &= 2(a_1^2 - 2a_1a_2 + a_2^2) + a_1^2 \\ &= 2(a_1 - a_2)^2 + a_1^2 > 0 \qquad \text{for} \qquad \mathbf{a} \neq [0,0]\end{aligned}$$

Step 2:
$$\begin{aligned}\langle \mathbf{a}, \mathbf{b} \rangle &= 3a_1b_1 + 2a_2b_2 - 4a_1b_2 \\ &= 3b_1a_1 + 2b_2a_2 - 4b_2a_1 \neq \langle \mathbf{b}, \mathbf{a} \rangle\end{aligned}$$

Step 3:
$$\begin{aligned}\langle \mathbf{a} + \mathbf{b}, \mathbf{c} \rangle &= \langle [a_1 + b_1, a_2 + b_2], [c_1, c_2] \rangle \\ &= 3(a_1 + b_1)c_1 + 2(a_2 + b_2)c_2 - 4(a_1 + b_1)c_2 \\ &= 3a_1c_1 + 3b_1c_1 + 2a_2c_2 + 2b_2c_2 - 4a_1c_2 - 4b_1c_2 \\ &= (3a_1c_1 + 2a_2c_2 - 4a_1c_2) + (3b_1c_1 + 2b_2c_2 - 4b_1c_2) \\ &= \langle \mathbf{a}, \mathbf{c} \rangle + \langle \mathbf{b}, \mathbf{c} \rangle\end{aligned}$$

Step 4:
$$\begin{aligned}\langle k\mathbf{a}, \mathbf{b} \rangle &= 3(ka_1)b_1 + 2(ka_2)b_2 - 4(ka_1)b_2 \\ &= k(3a_1b_1 + 2a_2b_2 - 4a_1b_2) \\ &= k\langle \mathbf{a}, \mathbf{b} \rangle.\end{aligned}$$

It is not an inner product because $\langle \mathbf{a}, \mathbf{b} \rangle \neq \langle \mathbf{b}, \mathbf{a} \rangle$.

PROBLEM 9-5 Find the standard inner product for the following pairs of vectors: **(a)** $\mathbf{a} = [-1 \quad 2 \quad -4]$, $\mathbf{b} = [2 \quad 3 \quad -1]$; **(b)** $\mathbf{a} = [5 \quad -3 \quad 1]$, $\mathbf{b} = [0 \quad -3 \quad 7]$; **(c)** $\mathbf{a} = [2 \quad -3 \quad 1 \quad -6]$, $\mathbf{b} = [1 \quad 2 \quad -1 \quad 5]$.

Solution

(a) $\langle \mathbf{a}, \mathbf{b} \rangle = \mathbf{a} \cdot \mathbf{b} = (-1)2 + (2)(3) + (-4)(-1) = -2 + 6 + 4 = 8$
(b) $\langle \mathbf{a}, \mathbf{b} \rangle = \mathbf{a} \cdot \mathbf{b} = 5(0) + (-3)(-3) + 1(7) = 9 + 7 = 16$
(c) $\langle \mathbf{a}, \mathbf{b} \rangle = \mathbf{a} \cdot \mathbf{b} = 2(1) + (-3)(2) + (1)(-1) + (-6)(5) = 2 - 6 - 1 - 30 = -35$

PROBLEM 9-6 Use $\langle \mathbf{a}, \mathbf{b} \rangle = 3a_1b_1 + 2a_2b_2 + 4a_3b_3$ on R_3 to find the inner product of the following pairs of vectors: **(a)** $\mathbf{a} = [-1 \quad 2 \quad -4]$, $\mathbf{b} = [2 \quad 3 \quad -1]$; **(b)** $\mathbf{a} = [5 \quad -3 \quad 1]$, $\mathbf{b} = [0 \quad -3 \quad 7]$.

Solution

(a) $\langle \mathbf{a}, \mathbf{b} \rangle = 3(-1)(2) + 2(2)(3) + 4(-4)(-1) = -6 + 12 + 16 = 22$
(b) $\langle \mathbf{a}, \mathbf{b} \rangle = 3(5)(0) + 2(-3)(-3) + 4(1)(7) = 0 + 18 + 28 = 46$

PROBLEM 9-7 Use $\langle \mathbf{a}, \mathbf{b} \rangle = a_1b_1 + 2a_2b_2 + 3a_3b_3 + 4a_4b_4$ on R_4 to find the inner product of the vectors, $\mathbf{a} = [2 \quad -3 \quad 1 \quad -6]$, $\mathbf{b} = [1 \quad 2 \quad -1 \quad 5]$.

Solution

$$\begin{aligned}\langle \mathbf{a}, \mathbf{b} \rangle &= 2(1) + 2(-3)(2) + 3(1)(-1) + 4(-6)(5) \\ &= 2 - 12 - 3 - 120 = -133\end{aligned}$$

PROBLEM 9-8 Find the length of the vector $\mathbf{a} = [1 \quad 2 \quad 3]$ in R_3 with the standard inner product.

Solution

$$\|[1 \quad 2 \quad 3]\| = \sqrt{1^2 + 2^2 + 3^2} = \sqrt{1 + 4 + 9} = \sqrt{14}$$

PROBLEM 9-9 Find the length of the vector $\mathbf{a} = [1 \quad 2 \quad 3]$ in R_3 with inner product $\langle \mathbf{a}, \mathbf{b} \rangle = 2a_1b_1 + 3a_2b_2 + a_3b_3 - a_1b_2 + a_2b_3 + a_1b_3$

Solution

$$\begin{aligned}\|a\| &= \sqrt{\langle \mathbf{a}, \mathbf{a} \rangle} = \sqrt{2(1)(1) + 3(2)(2) + (3)(3) - 1(2) + 2(3) + 1(3)} \\ &= \sqrt{2 + 12 + 9 - 2 + 6 + 3} = \sqrt{30}\end{aligned}$$

PROBLEM 9-10 Find the length of the vector $\mathbf{a} = [1 \quad 2 \quad 3]$ in R_3 with inner product $\langle \mathbf{a}, \mathbf{b} \rangle = a_1b_1 + a_2b_2 + a_3b_3 + 1a_1b_2 + a_2b_1$.

Solution

$$\|\mathbf{a}\| = \sqrt{\langle \mathbf{a}, \mathbf{a} \rangle} = \sqrt{1(1) + 2(2) + 3(3) + (1)(2) + 2(1)}$$
$$= \sqrt{1 + 4 + 9 + 2} = \sqrt{18}$$

PROBLEM 9-11 Find the distance between the vectors $\mathbf{a} = [1 \quad -4]$ and $\mathbf{b} = [-2 \quad 5]$ in R_2 if R_2 has the standard inner product.

Solution

$$d(\mathbf{a}, \mathbf{b}) = \|\mathbf{a} - \mathbf{b}\| = \sqrt{\langle \mathbf{a} - \mathbf{b}, \mathbf{a} - \mathbf{b} \rangle} = \sqrt{(1 - (-2))^2 + (-4 - 5)^2}$$
$$= \sqrt{9 + 81} = \sqrt{90} = 3\sqrt{10}$$

PROBLEM 9-12 Find the distance between the vectors $\mathbf{a} = [1 \quad -4]$ and $\mathbf{b} = [-2 \quad 5]$ in R_2 if R_2 has the inner product $\langle \mathbf{a}, \mathbf{b} \rangle = 2a_1b_1 + 3a_2b_2$.

Solution

$$d(\mathbf{a}, \mathbf{b}) = \|\mathbf{a} - \mathbf{b}\| = \sqrt{\langle \mathbf{a} - \mathbf{b}, \mathbf{a} - \mathbf{b} \rangle} = \sqrt{2(1 + 2)^2 + 3(-4 - 5)^2}$$
$$= \sqrt{2(9) + 3(81)} = \sqrt{18 + 243} = \sqrt{261} = 3\sqrt{29}$$

PROBLEM 9-13 Verify the Cauchy–Schwarz inequality for $\mathbf{a} = [1 \quad -2 \quad 3]$ and $\mathbf{b} = [2 \quad 3 \quad -4]$ in R_3 with the standard inner product.

Solution

$$\langle \mathbf{a}, \mathbf{b} \rangle^2 = [1(2) + (-2)(3) + 3(-4)]^2 = (2 - 6 - 12)^2 = (-16)^2 = 256$$
$$\|\mathbf{a}\|^2 = (\sqrt{\langle \mathbf{a}, \mathbf{a} \rangle})^2 = \langle \mathbf{a}, \mathbf{a} \rangle = 1^2 + (-2)^2 + 3^2 = 1 + 4 + 9 = 14$$
$$\|\mathbf{b}\|^2 = (\sqrt{\langle \mathbf{b}, \mathbf{b} \rangle})^2 = \langle \mathbf{b}, \mathbf{b} \rangle = 2^2 + 3^2 + (-4)^2 = 4 + 9 + 16 = 29$$
$$\|\mathbf{a}\|^2\|\mathbf{b}\|^2 = 14(29) = 406 \qquad \text{and} \qquad 256 < 406$$

Therefore, $\langle \mathbf{a}, \mathbf{b} \rangle^2 \leq \|\mathbf{a}\|^2\|\mathbf{b}\|^2$ for the given **a** and **b**.

PROBLEM 9-14 Verify the Cauchy–Schwarz inequality for $\mathbf{a} = [1 \quad -2 \quad 3]$ and $\mathbf{b} = [2 \quad 3 \quad -4]$ in R_3 with the inner product $\langle \mathbf{a}, \mathbf{b} \rangle = 2a_1b_1 + a_2b_2 + 3a_3b_3 + a_1b_2 + a_2b_1$.

Solution

$$\langle \mathbf{a}, \mathbf{b} \rangle^2 = [2(1)(2) + (-2)(3) + 3(3)(-4) + 1(3) + (-2)(2)]^2$$
$$= [4 - 6 - 36 + 3 \quad -4]^2 = (-39)^2 = 1521$$
$$\|\mathbf{a}\|^2 = \langle \mathbf{a}, \mathbf{a} \rangle = 2(1)(1) + (-2)(-2) + 3(3)(3) + (1)(-2) + (-2)(1)$$
$$= 2 + 4 + 27 - 2 - 2 = 29$$
$$\|\mathbf{b}\|^2 = \langle \mathbf{b}, \mathbf{b} \rangle = 2(2)(2) + (3)(3) + 3(-4)(-4) + (2)(3) + (3)(2)$$
$$= 8 + 9 + 48 + 6 + 6 = 77.$$
$$\|\mathbf{a}\|^2\|\mathbf{b}\|^2 = (29)(77) = 2233 \qquad \text{and} \qquad 1521 < 2233$$

Therefore, $\langle \mathbf{a}, \mathbf{b} \rangle^2 \leq \|\mathbf{a}\|^2\|\mathbf{b}\|^2$ for the given vectors.

PROBLEM 9-15 Verify the triangle inequality for $\mathbf{a} = [2 \quad -3]$ and $\mathbf{b} = [3 \quad 5]$ in R_2 with the standard inner product.

Solution

$$\|\mathbf{a} + \mathbf{b}\| = \|[5 \quad 2]\| = \sqrt{\langle [5 \quad 2], [5 \quad 2] \rangle} = \sqrt{25 + 4} = \sqrt{29}$$
$$\|\mathbf{a}\| + \|\mathbf{b}\| = \sqrt{\langle \mathbf{a}, \mathbf{a} \rangle} + \sqrt{\langle \mathbf{b}, \mathbf{b} \rangle} = \sqrt{4 + 9} + \sqrt{9 + 25} = \sqrt{13} + \sqrt{34}$$

Hence $\sqrt{29} < \sqrt{13} + \sqrt{34}$, and $\|\mathbf{a} + \mathbf{b}\| \leq \|\mathbf{a}\| + \|\mathbf{b}\|$ for the given vectors.

PROBLEM 9-16 Verify the triangle inequality for $\mathbf{a} = [1 \quad 2 \quad 3]$ and $\mathbf{b} = [2 \quad 1 \quad 5]$ in R_3 with the standard inner product.

Solution

$$\|\mathbf{a} + \mathbf{b}\| = \|[3 \quad 3 \quad 8]\| = \sqrt{\langle [3 \quad 3 \quad 8], [3 \quad 3 \quad 8] \rangle} = \sqrt{9 + 9 + 64} = \sqrt{82}$$

$$\|\mathbf{a}\| + \|\mathbf{b}\| = \sqrt{\langle \mathbf{a}, \mathbf{a} \rangle} + \sqrt{\langle \mathbf{b}, \mathbf{b} \rangle} = \sqrt{1 + 4 + 9} + \sqrt{4 + 1 + 25} = \sqrt{14} + \sqrt{30}$$

$$\sqrt{82} \doteq 9.06 \qquad \sqrt{14} \doteq 3.74 \qquad \sqrt{30} \doteq 5.47$$

$$9.06 < 3.74 + 5.47 = 9.21$$

$$\|\mathbf{a} + \mathbf{b}\| \leq \|\mathbf{a}\| + \|\mathbf{b}\| \qquad \text{for the given vectors.}$$

PROBLEM 9-17 Transform $[-1 \quad 4 \quad 6]$ in R_3 with the standard inner product into a unit vector.

Solution

$$\mathbf{b} = \frac{\mathbf{a}}{\|\mathbf{a}\|} = \frac{[-1 \quad 4 \quad 6]}{\sqrt{1 + 16 + 36}} = \begin{bmatrix} \frac{-1}{\sqrt{53}} & \frac{4}{\sqrt{53}} & \frac{6}{\sqrt{53}} \end{bmatrix}$$

PROBLEM 9-18 Transform $[-1 \quad 4 \quad 6]$ in R_3 with inner product $\langle \mathbf{a}, \mathbf{b} \rangle = 3a_1b_1 + a_2b_2 + 2a_3b_3$ into a unit vector.

Solution

$$\mathbf{b} = \frac{\mathbf{a}}{\|\mathbf{a}\|} = \frac{\mathbf{a}}{\sqrt{\langle \mathbf{a}, \mathbf{a} \rangle}} = \frac{[-1 \quad 4 \quad 6]}{\sqrt{3(-1)(-1) + 4(4) + 2(6)(6)}} = \begin{bmatrix} \frac{-1}{\sqrt{91}} & \frac{4}{\sqrt{91}} & \frac{6}{\sqrt{91}} \end{bmatrix}$$

PROBLEM 9-19 Let $A = \begin{bmatrix} a_{11} & a_{12} \\ a_{21} & a_{22} \end{bmatrix}$ and $B = \begin{bmatrix} b_{11} & b_{12} \\ b_{21} & b_{22} \end{bmatrix}$ with the standard inner product on ${}_2M_2$ defined as $\langle A, B \rangle = a_{11}b_{11} + a_{12}b_{12} + a_{21}b_{21} + a_{22}b_{22}$. For the given pair $\mathbf{a} = \begin{bmatrix} -1 & 3 \\ 2 & -1 \end{bmatrix}$ and $\mathbf{b} = \begin{bmatrix} 2 & 4 \\ -3 & -2 \end{bmatrix}$, find $\langle \mathbf{a}, \mathbf{b} \rangle$ and $\|\mathbf{a}\|$.

Solution

$$\langle \mathbf{a}, \mathbf{b} \rangle = (-1)(2) + 3(4) + 2(-3) + (-1)(-2) = -2 + 12 - 6 + 2 = 6$$

$$\|\mathbf{a}\| = \sqrt{(-1)^2 + 3^2 + 2^2 + (-1)^2} = \sqrt{1 + 9 + 4 + 1} = \sqrt{15}$$

Supplementary Problems

PROBLEM 9-20 Verify that if $\mathbf{a} = [a_1, a_2]$ and $\mathbf{b} = [b_1, b_2]$, then $\langle \mathbf{a}, \mathbf{b} \rangle = 2a_1b_1 - a_2b_2$ does not define an inner product on R_2.

PROBLEM 9-21 Verify that $\langle \mathbf{a}, \mathbf{b} \rangle = 3a_1b_1 + 2a_2b_2 + 2a_1b_2 + 2a_2b_1$ defines an inner product on R_2.

PROBLEM 9-22 Find the standard inner product for the following pairs of vectors: **(a)** $\mathbf{a} = [2 \quad -3]$, $\mathbf{b} = [4 \quad 7]$ in R_2; **(b)** $\mathbf{a} = [2 \quad -5 \quad 4]$, $\mathbf{b} = [1 \quad -3 \quad 8]$ in R_3; **(c)** $\mathbf{a} = [1 \quad 2 \quad -3 \quad 4]$, $\mathbf{b} = [0 \quad 3 \quad 5 \quad -7]$ in R_4.

PROBLEM 9-23 Use $\langle \mathbf{a}, \mathbf{b} \rangle = a_1b_1 + 3a_2b_2 - a_1b_2 - a_2b_1$ on R_2 to find the inner product of the following pairs: **(a)** $\mathbf{a} = [2 \quad -3]$, $\mathbf{b} = [4 \quad 7]$; **(b)** $\mathbf{a} = [-1 \quad 4]$, $\mathbf{b} = [-3 \quad -5]$.

PROBLEM 9-24 Use $\langle \mathbf{a}, \mathbf{b} \rangle = 3a_1b_1 + 2a_2b_2 + a_1b_2 + a_2b_1$ on R_2 to find the inner product of $\mathbf{a} = [-3 \quad 4]$, $\mathbf{b} = [7 \quad 8]$.

PROBLEM 9-25 Find the length of $\mathbf{a} = [-3 \quad 2 \quad -4]$ in R_3 with the standard inner product.

PROBLEM 9-26 Find the length of $\mathbf{a} = [-3 \quad 2 \quad -4]$ in R_3 with inner product $\langle \mathbf{a}, \mathbf{b} \rangle = 2a_1b_1 + a_2b_2 + 3a_3b_3$.

PROBLEM 9-27 Find the distance between $\mathbf{a} = [5 \quad 7]$ and $\mathbf{b} = [9 \quad 4]$ in R_2 with standard inner product.

PROBLEM 9-28 Find the distance between $\mathbf{a} = [5 \quad 7]$ and $\mathbf{b} = [9 \quad 4]$ in R_2 with inner product $\langle \mathbf{a}, \mathbf{b} \rangle = 2a_1b_1 + 3a_2b_2$.

PROBLEM 9-29 Verify the Cauchy–Schwarz inequality for $\mathbf{a} = [2 \quad 5 \quad -4]$ and $\mathbf{b} = [1 \quad 3 \quad 8]$ in R_3 with standard inner product.

PROBLEM 9-30 Verify the Cauchy–Schwarz inequality for $\mathbf{a} = [2 \quad 5 \quad -4]$ and $\mathbf{b} = [1 \quad 3 \quad 8]$ in R_3 with inner product $\langle \mathbf{a}, \mathbf{b} \rangle = 2a_1b_1 + a_2b_2 + 3a_3b_3$.

PROBLEM 9-31 Verify the triangle inequality for $\mathbf{a} = [2 \quad 5 \quad -4]$ and $\mathbf{b} = [1 \quad 3 \quad 8]$ in R_3 with standard inner product.

PROBLEM 9-32 Transform $[2 \quad 5 \quad -4]$ in R_3 with standard inner product into a unit vector.

PROBLEM 9-33 Transform $[2 \quad 5 \quad -4]$ in R_3 with inner product $\langle \mathbf{a}, \mathbf{b} \rangle = 2a_1b_1 + a_2b_2 + 3a_3b_3$ into a unit vector.

PROBLEM 9-34 Let ${}_2M_2$ be defined with the standard inner product, and let $\mathbf{a} = \begin{bmatrix} 1 & 2 \\ -1 & 4 \end{bmatrix}$ and $\mathbf{b} = \begin{bmatrix} 3 & -2 \\ 2 & -3 \end{bmatrix}$. Find $\langle \mathbf{a}, \mathbf{b} \rangle$ and $\|\mathbf{b}\|$.

PROBLEM 9-35 If $\mathbf{a} = a_0 + a_1 + a_2x^2$ and $\mathbf{b} = b_0 + b_1x + b_2x^2$ in P_2 with standard inner product defined as $\langle \mathbf{a}, \mathbf{b} \rangle = a_0b_0 + a_1b_1 + a_2b_2$, and given pair in P_2 of $\mathbf{a} = 2 - x + 3x^2$ and $\mathbf{b} = 1 + 3x + 4x^2$, find **(a)** $\langle \mathbf{a}, \mathbf{b} \rangle$ and **(b)** $\mathbf{b}/\|\mathbf{b}\|$.

PROBLEM 9-36 In Problem 9-34 find a unit vector in the direction of $\mathbf{b} = \begin{bmatrix} 3 & -2 \\ 2 & -3 \end{bmatrix}$.

Answers to Supplementary Problems

9-20 $\langle \mathbf{a}, \mathbf{a} \rangle = 2a_1^2 - a_2^2$ is not always positive.

9-21 Show that all properties of inner product hold.

9-22 **(a)** -13 **(b)** 49 **(c)** -37

9-23 **(a)** -57 **(b)** -50

9-24 5

9-25 $\sqrt{29}$

9-26 $\sqrt{70}$

9-27 $\sqrt{25} = 5$

9-28 $\sqrt{59}$

9-29 $\langle \mathbf{a}, \mathbf{b} \rangle^2 = 225$, $\|\mathbf{a}\|^2 = 45$, $\|\mathbf{b}\|^2 = 74$, $225 < 45(74)$

9-30 $\langle \mathbf{a}, \mathbf{b} \rangle^2 = 5929$, $\|\mathbf{a}\|^2 = 81$, $\|\mathbf{b}\|^2 = 203$, $5929 < 81(203)$

9-31 $\sqrt{89} < \sqrt{45} + \sqrt{74}$

9-32 $(1/\sqrt{45})[2 \quad 5 \quad -4]$

9-33 $(1/9)[2 \quad 5 \quad -4]$

9-34 $\langle \mathbf{a}, \mathbf{b} \rangle = -15$, $\|\mathbf{b}\| = \sqrt{26}$

9-35 **(a)** 11
(b) $(1/\sqrt{26})[1 + 3x + 4x^2]$

9-36 $\dfrac{1}{\sqrt{26}} \begin{bmatrix} 3 & -2 \\ 2 & -3 \end{bmatrix}$

10 GEOMETRY OF VECTORS

THIS CHAPTER IS ABOUT

☑ Geometric Representation of Vectors in R_2 and R_3
☑ Geometric Interpretations of Operations on Vectors in R_2 and R_3
☑ Angle between Two Vectors
☑ Vector Equation of a Plane
☑ Lines in R_3
☑ Cross Product

10-1. Geometric Representation of Vectors in R_2 and R_3

Until now, our concept of a vector has been an abstraction that satisfies certain conditions. In the physical sciences, however, such quantities as force, displacement, velocity, and acceleration are usually represented geometrically by arrows or directed line segments with the length of each proportional to the magnitude of the quantity and pointing in the appropriate direction. The point from which the arrow begins is called the **initial point**, and the tip of the arrow is called the **terminal point**.

It is important to understand that an arrow or directed line segment is merely a graphical way of representing the mathematical objects we have called vectors. As long as an arrow is moved parallel to its initial position and its length remains unchanged, the vectors represented are all equal. If the initial point of an arrow is at the origin of a coordinate system, the vector represented is called a **position vector**. The vector $[3 \quad 4]$ can thus be represented by an arrow with the initial point at $(0, 0)$ and terminal point at $(3, 4)$ or any arrow of length five parallel to the one described.

10-2. Geometric Interpretations of Operations on Vectors in R_2 and R_3

A. Multiplication by a scalar

To multiply a vector $\mathbf{a}$ by a positive scalar k, you merely multiply the length or magnitude of the vector by k. The process is called a **dilation** if $k > 1$ and a **contraction** if $0 < k < 1$. If $k < 0$, the direction of $\mathbf{a}$ is reversed and the length is multiplied by the absolute value of k. This process is sometimes called an **opposition**.

EXAMPLE 10-1: Describe the arrow with its initial point at the origin that would be used to represent the vector $2[3 \quad 4]$.

Solution: The vector $2[3 \quad 4]$ has the same direction as $[3 \quad 4]$, but it is twice as long. Since $2[3 \quad 4] = [6 \quad 8]$ the arrow would be drawn from the origin to $(6, 8)$.

note: Observe that the length of $[3 \quad 4]$ is $\|[3 \quad 4]\| = \sqrt{3^2 + 4^2} = 5$. Likewise, the length of $[6 \quad 8]$ is $\|[6 \quad 8]\| = \sqrt{6^2 + 8^2} = 10$ which is 2 times the length of $[3 \quad 4]$.

B. Addition of vectors (parallelogram law)

If $\mathbf{a}$ and $\mathbf{b}$ are adjacent sides of a parallelogram, then $\mathbf{a} + \mathbf{b}$ is the diagonal of the parallelogram that has as its initial point the common initial point of $\mathbf{a}$ and $\mathbf{b}$.

EXAMPLE 10-2: Add, geometrically, the vectors $\mathbf{a} = [2 \quad 3]$ and $\mathbf{b} = [6 \quad 2]$.

Solution: Plot the points (2, 3) and (6, 2). Then draw the arrows from (0, 0) to these two points. Draw the arrow from (2, 3) parallel to vector **b** with length $\|\mathbf{b}\|$ and the arrow from (6, 2) parallel to vector **a** with length $\|\mathbf{a}\|$, as shown in Figure 10-1. The diagonal from (0, 0) of the resulting parallelogram represents $\mathbf{a} + \mathbf{b} = [8 \quad 5]$.

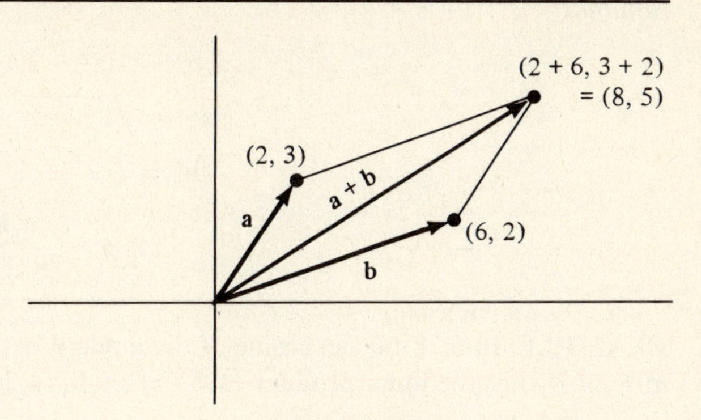

FIGURE 10-1

C. Subtraction of vectors

If **a** and **b** are positioned so that their initial points coincide, then $\mathbf{a} - \mathbf{b}$ is represented by the arrow with initial point at the terminal point of **b** and its terminal point at the terminal point of **a**.

EXAMPLE 10-3: If $\mathbf{a} = [2 \quad 3]$ and $\mathbf{b} = [6 \quad 2]$, find, geometrically, the vector $\mathbf{a} - \mathbf{b}$.

Solution: Plot the points (2, 3) and (6, 2). Then draw the arrows from the origin to these two points, as in Figure 10-2. $\mathbf{a} - \mathbf{b}$ is the arrow from (6, 2) to (2, 3).

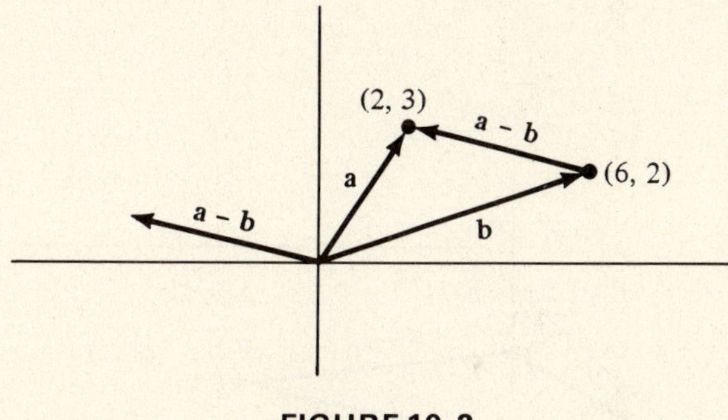

FIGURE 10-2

10-3. Angle between Two Vectors

If ϕ is the angle between two nonzero vectors **a** and **b** in an inner product space V; then $\cos\phi = \langle \mathbf{a}, \mathbf{b} \rangle / \|\mathbf{a}\| \, \|\mathbf{b}\|$, $0 \le \phi \le \pi$.

The two nonzero vectors are **orthogonal** or **perpendicular** if and only if $\langle \mathbf{a}, \mathbf{b} \rangle = 0$, or $\cos\phi = 0$. With the standard inner product or dot product, $\mathbf{a} \cdot \mathbf{b} = 0$.

EXAMPLE 10-4: Find the cosine of the angle ϕ between the two vectors $\mathbf{a} = [3 \quad 4]$ and $\mathbf{b} = [2 \quad 7]$ in R_2 if R_2 has the standard inner product.

Solution

$$\langle \mathbf{a}, \mathbf{b} \rangle = 3(2) + 4(7) = 34$$

$$\|\mathbf{a}\| = \sqrt{\langle \mathbf{a}, \mathbf{a} \rangle} = \sqrt{3^2 + 4^2} = \sqrt{25} = 5$$

$$\|\mathbf{b}\| = \sqrt{\langle \mathbf{b}, \mathbf{b} \rangle} = \sqrt{2^2 + 7^2} = \sqrt{53}$$

$$\therefore \quad \cos\phi = \frac{\langle \mathbf{a}, \mathbf{b} \rangle}{\|\mathbf{a}\| \, \|\mathbf{b}\|} = \frac{34}{5\sqrt{53}}$$

EXAMPLE 10-5: Find the cosine of the angle ϕ between the two vectors $\mathbf{a} = [1 \quad 2 \quad -1]$ and $\mathbf{b} = [2 \quad 1 \quad 3]$ in R_3 if R_3 has the standard inner product.

Solution

$$\langle \mathbf{a}, \mathbf{b} \rangle = 1(2) + 2(1) + (-1)(3) = 1$$
$$\|\mathbf{a}\| = \sqrt{1^2 + 2^2 + (-1)^2} = \sqrt{6}$$
$$\|\mathbf{b}\| = \sqrt{2^2 + 1^2 + 3^2} = \sqrt{14}$$
$$\therefore \quad \cos\phi = \frac{\langle \mathbf{a}, \mathbf{b} \rangle}{\|\mathbf{a}\|\,\|\mathbf{b}\|} = \frac{1}{\sqrt{6}\sqrt{14}}$$

EXAMPLE 10-6: Find the cosine of the angle ϕ between the two vectors $\mathbf{a} = [3 \quad 4]$ and $\mathbf{b} = [2 \quad 7]$ in R_2 if R_2 has the inner product $\langle \mathbf{a}, \mathbf{b} \rangle = 5\mathbf{a}_1\mathbf{b}_1 + 2\mathbf{a}_2\mathbf{b}_2$.

Solution

$$\langle \mathbf{a}, \mathbf{b} \rangle = 5(3)(2) + 2(4)(7) = 86$$
$$\|\mathbf{a}\| = \sqrt{\langle \mathbf{a}, \mathbf{a} \rangle} = \sqrt{5(3)^2 + 2(4)^2} = \sqrt{77}$$
$$\|\mathbf{b}\| = \sqrt{\langle \mathbf{b}, \mathbf{b} \rangle} = \sqrt{5(2)^2 + 2(7)^2} = \sqrt{118}$$
$$\cos\phi = \frac{\langle \mathbf{a}, \mathbf{b} \rangle}{\|\mathbf{a}\|\,\|\mathbf{b}\|} = \frac{86}{\sqrt{77}\sqrt{118}}$$

10-4. Vector Equation of a Plane

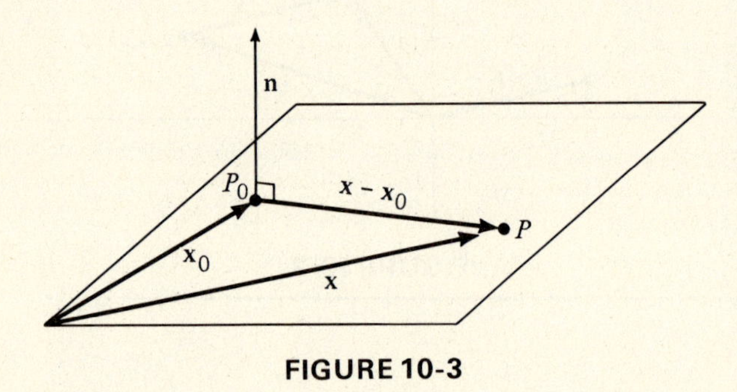

FIGURE 10-3

Let $\mathbf{x}$ be the position vector of any point P in a plane and $\mathbf{x}_0$ be the position vector of any fixed point P_0 in the plane. If $\mathbf{n}$ is a nonzero vector that is **perpendicular** to the plane at P_0, as shown in Figure 10-3, then $\mathbf{n}$ is called a **normal**, and in terms of the dot product, the equation of the plane is $\mathbf{n}\cdot(\mathbf{x} - \mathbf{x}_0) = 0$. This is often called the **point-normal** or **vector form** of the equation of a plane. If $\mathbf{n} = [a \quad b \quad c]$, $\mathbf{x} = [x \quad y \quad z]$, and $\mathbf{x}_0 = [x_0 \quad y_0 \quad z_0]$, then the equation in expanded point-normal form becomes $a(x - x_0) + b(y + y_0) + c(z - z_0) = 0$, which can be simplified to $ax + by + cz = d$. The latter is called the **general equation** of a plane. Observe that the coefficients of x, y, and z are the components of the vector $\mathbf{n} = [a \quad b \quad c]$ which is normal to the plane.

EXAMPLE 10-7: Find the equation of the plane passing through $(4, -5, 3)$ and perpendicular to the vector $[2 \quad 3 \quad -2]$ in both point-normal and general form.

Solution

$$\mathbf{n}\cdot(\mathbf{x} - \mathbf{x}_0) = 0$$
$$[2 \quad 3 \quad -2]\cdot[x - 4, y + 5, z - 3] = 0$$

or

$$2(x - 4) + 3(y + 5) - 2(z - 3) = 0 \qquad \text{in point-normal form}$$
$$2x + 3y - 2z = -13 \qquad \text{in general form}$$

EXAMPLE 10-8: Find the general form of the equation of a plane passing through the point $(1, 3, 2)$ and parallel to the plane $4x - y + 3z = 5$.

Solution: A normal to the given plane is $[4 \quad -1 \quad 3]$ that will also be a normal to the desired plane since the planes must be parallel. Hence,

$$[4 \quad -1 \quad 3]\cdot[x-1, y-3, z-2] = 0$$

or

$$4(x-1)-(y-3)+3(z-2)=0$$
$$4x - y + 3z = 7$$

10-5. Lines in R_3

A. Direction cosines

The angles a vector makes with the positive x-, y-, and z-axes are called **direction angles**, and the cosines of these angles are called **direction cosines**. In Figure 10-4, the direction angles are denoted by α, β, and γ. The direction of a vector is determined by its direction cosines. If $[a \quad b \quad c] = k[\cos\alpha \quad \cos\beta \quad \cos\gamma]$ then a, b, and c are called **direction numbers**.

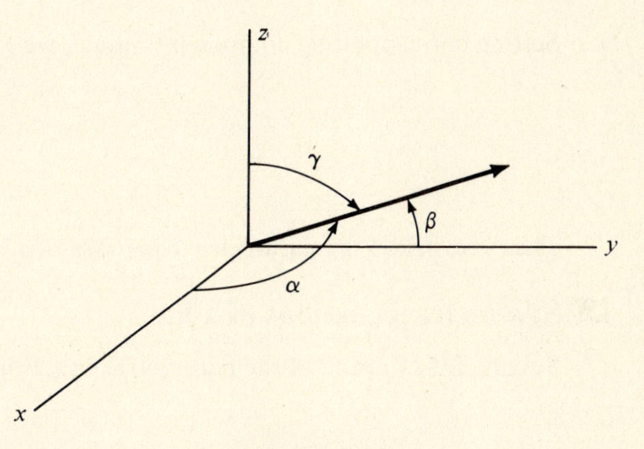

FIGURE 10-4

EXAMPLE 10-9: Find the direction cosines of the vector $[1 \quad 2 \quad 3]$.

Solution: Find vectors in the directions of the positive x-, y-, and z-axes. Since $(0, 0, 0)$ is a point on all three, we merely need to find one other point on each axis. We might as well use $(1, 0, 0)$, $(0, 1, 0)$, and $(0, 0, 1)$, respectively.

Therefore, $[1 \quad 0 \quad 0]$, $[0 \quad 1 \quad 0]$, and $[0 \quad 0 \quad 1]$ are vectors in the directions of the positive x-, y-, and z-axes, respectively. These are the natural basis vectors for R_3 and are often denoted by **i**, **j**, and **k**.

$$\cos\alpha = \frac{[1 \quad 2 \quad 3]\cdot[1 \quad 0 \quad 0]}{\|[1 \quad 2 \quad 3]\|\,\|[1 \quad 0 \quad 0]\|} = \frac{1}{\sqrt{14}}$$

$$\cos\beta = \frac{[1 \quad 2 \quad 3]\cdot[0 \quad 1 \quad 0]}{\|[1 \quad 2 \quad 3]\|\,\|[0 \quad 1 \quad 0]\|} = \frac{2}{\sqrt{14}}$$

$$\cos\gamma = \frac{[1 \quad 2 \quad 3]\cdot[0 \quad 0 \quad 1]}{\|[1 \quad 2 \quad 3]\|\,\|[0 \quad 0 \quad 1]\|} = \frac{3}{\sqrt{14}}$$

B. Vector equation of a line

Let $\mathbf{x}$ be the position vector of any point P on a line in R_2 or R_3 and $\mathbf{x}_0$ be the position vector of a fixed point on the line. If $\mathbf{a}$ is a vector parallel to the line, the vector equation of the line is $\mathbf{x} - \mathbf{x}_0 = t\mathbf{a}$ where t is a scalar.

If we are in R_3 with $\mathbf{x} = [x \quad y \quad z]$, $\mathbf{x}_0 = [x_0 \quad y_0 \quad z_0]$, and $\mathbf{a} = [a \quad b \quad c]$, then in terms of components the equation of the line becomes

$$[x \quad y \quad z] - [x_0 \quad y_0 \quad z_0] = t[a \quad b \quad c]$$

or

$$[x - x_0 \quad y - y_0 \quad z - z_0] = t[a \quad b \quad c]$$

EXAMPLE 10-10: In terms of components, find the vector equation of the line passing through the point $(4, 3, -2)$ and parallel to $[2 \quad 3 \quad 4]$.

Solution

$$[x \quad y \quad z] - [4 \quad 3 \quad -2] = t[2 \quad 3 \quad 4]$$

or

$$[x - 4, y - 3, z + 2] = t[2 \quad 3 \quad 4]$$

C. Parametric equations of a line

Consider the vector form in terms of components of the equation of the line given in Section B:

$$[x - x_0 \quad y - y_0 \quad z - z_0] = t[a \quad b \quad c]$$

Setting corresponding components equal, we get

$$x = x_0 + ta$$
$$y = y_0 + tb$$
$$z = z_0 + tc$$

These are called the **parametric equations** of a line.

D. Symmetric equations of a line

Solving for t in each of the parametric equations,

$$x = x_0 + ta$$
$$y = y_0 + tb$$
$$z = z_0 + tc$$

we get

$$t = \frac{x - x_0}{a} \qquad t = \frac{y - y_0}{b} \qquad t = \frac{z - z_0}{c}$$

Since t is the same in each of these equations, we get

$$\frac{x - x_0}{a} = \frac{y - y_0}{b} = \frac{z - z_0}{c}$$

which are called the **symmetric equations** of the line. It is assumed that $a \neq 0$, $b \neq 0$, and $c \neq 0$. If a denominator in any ratio is zero, the numerator in that ratio is also zero. The a, b, and c in the vector, parametric, and symmetric forms of the equation of a line are the components of a vector to which the line is parallel. Therefore, they are direction numbers of the vector and the line.

EXAMPLE 10-11: Find the vector, parametric and symmetric forms of the line passing through $(-3, 5, 2)$ and parallel to $[6 \quad -7 \quad 3]$.

Solution: Let (x, y, z) by any point on the line. Hence, $[x \quad y \quad z] - [-3 \quad 5 \quad 2] = t[6 \quad -7 \quad 3]$ or $[x + 3, y - 5, z - 2] = t[6 \quad -7 \quad 3]$ is the vector form of the equation of the line.

Setting corresponding components equal, we get

$$x = -3 + 6t$$
$$y = 5 - 7t$$
$$z = 2 + 3t$$

which are parametric equations. Solving for t and setting the expressions equal, we get

$$\frac{x + 3}{6} = \frac{y - 5}{-7} = \frac{z - 2}{3}$$

which are symmetric equations.

EXAMPLE 10-12: Find the symmetric equations of a line perpendicular to the plane $3x - 4y - 6z = 10$ and containing the point $(1, 2, 3)$.

Solution: The coefficients of x, y, and z are the components of a normal to the plane:

$$\frac{x - x_0}{a} = \frac{y - y_0}{b} = \frac{z - z_0}{c} \quad \text{becomes} \quad \frac{x-1}{3} = \frac{y-2}{-4} = \frac{z-3}{-6}$$

EXAMPLE 10-13: Find the cosine of the angle ϕ between the lines

$$\frac{x-1}{3} = \frac{y-2}{-4} = \frac{z-3}{-6} \quad \text{and} \quad \frac{x+2}{2} = \frac{y+3}{3} = \frac{z-5}{-4}$$

Solution: Vectors parallel to the lines are $\mathbf{a} = [3 \quad -4 \quad -6]$ and $\mathbf{b} = [2 \quad 3 \quad -4]$. The angle between the lines is the same as the angle between the vectors **a** and **b**. Since no inner product is indicated, we assume it to be the standard inner product which is the dot product:

$$\cos\phi = \frac{\mathbf{a}\cdot\mathbf{b}}{\|\mathbf{a}\|\,\|\mathbf{b}\|} = \frac{3(2) + (-4)(3) + (-6)(-4)}{\sqrt{3^2 + (-4)^2 + (-6)^2}\sqrt{2^2 + 3^2 + (-4)^2}} = \frac{18}{\sqrt{61}\sqrt{29}}$$

10-6. Cross Product

A. Definition of cross product

If we restrict our consideration to R_3 with $\mathbf{a} = [a_1 \quad a_2 \quad a_3]$ and $\mathbf{b} = [b_1 \quad b_2 \quad b_3]$, then the **cross product** of **a** and **b** is denoted by $\mathbf{a} \times \mathbf{b}$ and defined by $\mathbf{a} \times \mathbf{b} = [a_2b_3 - a_3b_2, a_3b_1 - a_1b_3, a_1b_2 - a_2b_1]$.

An easy way to obtain the subscripts is to draw a circle with the numbers indicated in a clockwise fashion, as shown in Figure 10-5. Beginning with 2 and always going clockwise, write down 23, 31, and 12. These are the subscripts of the first terms in the first, second, and third components, respectively. Interchange the order of each of these numbers to obtain 32, 13, and 21. These are the subscripts of the second terms in the first, second, and third components, respectively.

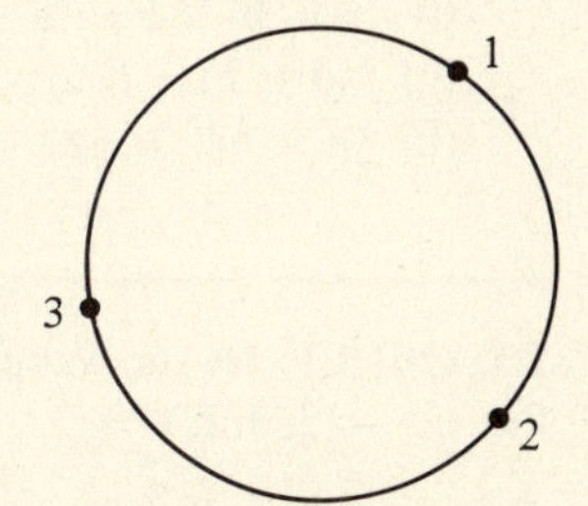

FIGURE 10-5

EXAMPLE 10-14: If $\mathbf{a} = [3 \quad -4 \quad -6]$ and $\mathbf{b} = [2 \quad 3 \quad -4]$, find $\mathbf{a} \times \mathbf{b}$.

Solution

$$\begin{aligned}\mathbf{a} \times \mathbf{b} &= [a_2b_3 - a_3b_2, a_3b_1 - a_1b_3, a_1b_2 - a_2b_1]\\ &= [(-4)(-4) - (-6)(3), (-6)(2) - 3(-4), (3)(3) - (-4)(2)]\\ &= [34 \quad 0 \quad 17]\end{aligned}$$

B. Determinant representation of cross product

Recall that in R_3, $\mathbf{i} = [1 \quad 0 \quad 0]$, $\mathbf{j} = [0 \quad 1 \quad 0]$, and $\mathbf{k} = [0 \quad 0 \quad 1]$. Consequently, $\mathbf{a} = a_1\mathbf{i} + a_2\mathbf{j} + a_3\mathbf{k}$ and $\mathbf{b} = b_1\mathbf{i} + b_2\mathbf{j} + b_3\mathbf{k}$. The cross product $\mathbf{a} \times \mathbf{b}$ may be obtained by expanding the following determinant by the elements in the first row and the cofactors of those elements.

$$\begin{aligned}\mathbf{a} \times \mathbf{b} &= \begin{vmatrix} \mathbf{i} & \mathbf{j} & \mathbf{k} \\ a_1 & a_2 & a_3 \\ b_1 & b_2 & b_3 \end{vmatrix}\\ &= \begin{vmatrix} a_2 & a_3 \\ b_2 & b_3 \end{vmatrix}\mathbf{i} - \begin{vmatrix} a_1 & a_3 \\ b_1 & b_3 \end{vmatrix}\mathbf{j} + \begin{vmatrix} a_1 & a_2 \\ b_1 & b_2 \end{vmatrix}\mathbf{k}\\ &= (a_2b_3 - a_3b_2)\mathbf{i} + (a_3b_1 - a_1b_3)\mathbf{j} + (a_1b_2 - a_2b_1)\mathbf{k}\end{aligned}$$

EXAMPLE 10-15: Use determinant representation to find the cross product $\mathbf{a} \times \mathbf{b}$ for the vectors in Example 10-14.

Solution

$$\mathbf{a} \times \mathbf{b} = \begin{vmatrix} \mathbf{i} & \mathbf{j} & \mathbf{k} \\ 3 & -4 & -6 \\ 2 & 3 & -4 \end{vmatrix} = \begin{vmatrix} -4 & -6 \\ 3 & -4 \end{vmatrix}\mathbf{i} - \begin{vmatrix} 3 & -6 \\ 2 & -4 \end{vmatrix}\mathbf{j} + \begin{vmatrix} 3 & -4 \\ 2 & 3 \end{vmatrix}\mathbf{k}$$

$$= 34\mathbf{i} + 0\mathbf{j} + 17\mathbf{k} = [34 \quad 0 \quad 17],$$

which is same answer as obtained in Example 10-14.

C. Properties of cross product

If **a**, **b**, and **c** are vectors in R_3 with standard inner product and k is a scalar, then

(a) $\mathbf{a} \times \mathbf{b} = -(\mathbf{b} \times \mathbf{a})$
(b) $\mathbf{a} \times \mathbf{b}$ is orthogonal to both **a** and **b**
(c) $\mathbf{a} \times \mathbf{0} = \mathbf{0} \times \mathbf{a} = \mathbf{0}$
(d) $\mathbf{a} \times \mathbf{a} = \mathbf{0}$
(e) $\mathbf{a} \times (\mathbf{b} + \mathbf{c}) = (\mathbf{a} \times \mathbf{b}) + (\mathbf{a} \times \mathbf{c})$
(f) $(\mathbf{a} + \mathbf{b}) \times \mathbf{c} = (\mathbf{a} \times \mathbf{c}) + (\mathbf{b} \times \mathbf{c})$
(g) $k(\mathbf{a} \times \mathbf{b}) = (k\mathbf{a}) \times \mathbf{b} = \mathbf{a} \times (k\mathbf{b})$
(h) $\|\mathbf{a} \times \mathbf{b}\|^2 = \|\mathbf{a}\|^2\|\mathbf{b}\|^2 - (\mathbf{a} \cdot \mathbf{b})^2$, which is called **Lagrange's identity**.

EXAMPLE 10-16: Verify each of the properties in Section C for $\mathbf{a} = [1 \quad 2 \quad 3]$, $\mathbf{b} = [2 \quad -1 \quad 2]$, $\mathbf{c} = [3 \quad 2 \quad -1]$, and $k = 2$.

Solution

(a)

$$\mathbf{a} \times \mathbf{b} = \begin{vmatrix} \mathbf{i} & \mathbf{j} & \mathbf{k} \\ 1 & 2 & 3 \\ 2 & -1 & 2 \end{vmatrix} = \begin{vmatrix} 2 & 3 \\ -1 & 2 \end{vmatrix}\mathbf{i} - \begin{vmatrix} 1 & 3 \\ 2 & 2 \end{vmatrix}\mathbf{j} + \begin{vmatrix} 1 & 2 \\ 2 & -1 \end{vmatrix}\mathbf{k} = 7\mathbf{i} + 4\mathbf{j} - 5\mathbf{k}$$

$$\mathbf{b} \times \mathbf{a} = \begin{vmatrix} \mathbf{i} & \mathbf{j} & \mathbf{k} \\ 2 & -1 & 2 \\ 1 & 2 & 3 \end{vmatrix} = \begin{vmatrix} -1 & 2 \\ 2 & 3 \end{vmatrix}\mathbf{i} - \begin{vmatrix} 2 & 2 \\ 1 & 3 \end{vmatrix}\mathbf{j} + \begin{vmatrix} 2 & -1 \\ 1 & 2 \end{vmatrix}\mathbf{k}$$

$$= -7\mathbf{i} - 4\mathbf{j} + 5\mathbf{k} = [-7 \quad -4 \quad 5]$$

$$\therefore \quad \mathbf{a} \times \mathbf{b} = -(\mathbf{b} \times \mathbf{a}) \text{ for the given } \mathbf{a} \text{ and } \mathbf{b}$$

(b) $\mathbf{a} \times \mathbf{b}$ is orthogonal to **a** and also to **b** if and only if $(\mathbf{a} \times \mathbf{b}) \cdot \mathbf{a} = 0$ and $(\mathbf{a} \times \mathbf{b}) \cdot \mathbf{b} = 0$. From part **(a)**, $\mathbf{a} \times \mathbf{b} = [7 \quad 4 \quad -5]$. Therefore,

$$(\mathbf{a} \times \mathbf{b}) \cdot \mathbf{a} = [7 \quad 4 \quad -5] \cdot [1 \quad 2 \quad 3] = 7(1) + 4(2) - 5(3) = 0$$

and

$$(\mathbf{a} \times \mathbf{b}) \cdot \mathbf{b} = [7 \quad 4 \quad -5] \cdot [2 \quad -1 \quad 2] = 7(2) + 4(-1) - 5(2) = 0$$

$$\therefore \quad \mathbf{a} \times \mathbf{b} \text{ is orthogonal to both } \mathbf{a} \text{ and } \mathbf{b}$$

(c)

$$\mathbf{a} \times \mathbf{0} = \begin{vmatrix} \mathbf{i} & \mathbf{j} & \mathbf{k} \\ 1 & 2 & 3 \\ 0 & 0 & 0 \end{vmatrix} = -\begin{vmatrix} \mathbf{i} & \mathbf{j} & \mathbf{k} \\ 0 & 0 & 0 \\ 1 & 2 & 3 \end{vmatrix} = 0\mathbf{i} + 0\mathbf{j} + 0\mathbf{k} = [0 \quad 0 \quad 0] = \mathbf{0}$$

$$\therefore \quad \mathbf{a} \times \mathbf{0} = \mathbf{0} \times \mathbf{a} = \mathbf{0} \text{ for the given } \mathbf{a}$$

(d)

$$\mathbf{a} \times \mathbf{a} = \begin{vmatrix} \mathbf{i} & \mathbf{j} & \mathbf{k} \\ 1 & 2 & 3 \\ 1 & 2 & 3 \end{vmatrix} = 0\mathbf{i} + 0\mathbf{j} + 0\mathbf{k} = [0 \quad 0 \quad 0] = \mathbf{0}$$

$$\therefore \quad \mathbf{a} \times \mathbf{a} = \mathbf{0} \text{ for the given } \mathbf{a}$$

(e)

$$\mathbf{b} + \mathbf{c} = [2 \quad -1 \quad 2] + [3 \quad 2 \quad -1] = [5 \quad 1 \quad 1]$$

$$\therefore \quad \mathbf{a} \times (\mathbf{b} + \mathbf{c}) = [1 \quad 2 \quad 3] \times [5 \quad 1 \quad 1]$$

$$= \begin{vmatrix} \mathbf{i} & \mathbf{j} & \mathbf{k} \\ 1 & 2 & 3 \\ 5 & 1 & 1 \end{vmatrix} = -\mathbf{i} + 14\mathbf{j} - 9\mathbf{k} = [-1 \quad 14 \quad -9]$$

$$(\mathbf{a} \times \mathbf{b}) + (\mathbf{a} \times \mathbf{c}) = \begin{vmatrix} \mathbf{i} & \mathbf{j} & \mathbf{k} \\ 1 & 2 & 3 \\ 2 & -1 & 2 \end{vmatrix} + \begin{vmatrix} \mathbf{i} & \mathbf{j} & \mathbf{k} \\ 1 & 2 & 3 \\ 3 & 2 & -1 \end{vmatrix} = (7\mathbf{i} + 4\mathbf{j} - 5\mathbf{k}) + (-8\mathbf{i} + 10\mathbf{j} - 4\mathbf{k})$$

$$= -\mathbf{i} + 14\mathbf{k} - 9\mathbf{k} = [-1 \quad 14 \quad -9]$$

$$\therefore \quad \mathbf{a} \times (\mathbf{b} + \mathbf{c}) = (\mathbf{a} \times \mathbf{b}) + (\mathbf{a} \times \mathbf{c}) \text{ for the given vectors}$$

(f)

$$\mathbf{a} + \mathbf{b} = [1 \quad 2 \quad 3] + [2 \quad -1 \quad 2] = [3 \quad 1 \quad 5]$$

$$(\mathbf{a} + \mathbf{b}) \times \mathbf{c} = [3 \quad 1 \quad 5] \times [3 \quad 2 \quad -1] = \begin{vmatrix} \mathbf{i} & \mathbf{j} & \mathbf{k} \\ 3 & 1 & 5 \\ 3 & 2 & -1 \end{vmatrix}$$

$$= -11\mathbf{i} + 18\mathbf{j} + 3\mathbf{k} = [-11 \quad 18 \quad 3]$$

$$(\mathbf{a} \times \mathbf{c}) + (\mathbf{b} \times \mathbf{c}) = \begin{vmatrix} \mathbf{i} & \mathbf{j} & \mathbf{k} \\ 1 & 2 & 3 \\ 3 & 2 & -1 \end{vmatrix} + \begin{vmatrix} \mathbf{i} & \mathbf{j} & \mathbf{k} \\ 2 & -1 & 2 \\ 3 & 2 & -1 \end{vmatrix} = (-8\mathbf{i} + 10\mathbf{j} - 4\mathbf{k}) + (-3\mathbf{i} + 8\mathbf{j} + 7\mathbf{k})$$

$$= -11\mathbf{i} + 18\mathbf{j} + 3\mathbf{k} = [-11 \quad 18 \quad 3]$$

$$\therefore \quad (\mathbf{a} + \mathbf{b}) \times \mathbf{c} = (\mathbf{a} \times \mathbf{c}) + (\mathbf{b} \times \mathbf{c}) \text{ for the given vectors}$$

(g) From part **(a)**, $\mathbf{a} \times \mathbf{b} = [7 \quad 4 \quad -5]$. Therefore,

$$k(\mathbf{a} \times \mathbf{b}) = 2([1 \quad 2 \quad 3] \times [2 \quad -1 \quad 2]) = 2[7 \quad 4 \quad -5] = [14 \quad 8 \quad -10]$$

$$(k\mathbf{a}) \times \mathbf{b} = \begin{vmatrix} \mathbf{i} & \mathbf{j} & \mathbf{k} \\ 2 & 4 & 6 \\ 2 & -1 & 2 \end{vmatrix} = 14\mathbf{i} + 8\mathbf{j} - 10\mathbf{k} = [14 \quad 8 \quad -10]$$

$$\mathbf{a} \times (k\mathbf{b}) = \begin{vmatrix} \mathbf{i} & \mathbf{j} & \mathbf{k} \\ 1 & 2 & 3 \\ 4 & -2 & 4 \end{vmatrix} = 14\mathbf{i} + 8\mathbf{j} - 10\mathbf{k} = [14 \quad 8 \quad -10]$$

$$\therefore \quad k(\mathbf{a} \times \mathbf{b}) = (k\mathbf{a}) \times \mathbf{b} = \mathbf{a} \times (k\mathbf{b}) \text{ for the given } \mathbf{a} \text{ and } \mathbf{b}$$

(h) From part **(a)**, $\mathbf{a} \times \mathbf{b} = [7 \quad 4 \quad -5]$. Therefore,

$$\|\mathbf{a} \times \mathbf{b}\|^2 = 7^2 + 4^2 + (-5)^2 = 90$$

$$\|\mathbf{a}\|^2 \|\mathbf{b}\|^2 - (\mathbf{a} \cdot \mathbf{b})^2 = [1^2 + 2^2 + 3^2][2^2 + (-1)^2 + 2^2] - [1(2) + 2(-1) + 3(2)]^2$$

$$= 14(9) - 36 = 90$$

Therefore, Lagrange's identity holds for the given **a** and **b**.

EXAMPLE 10-17: Find the general form of the equation of the plane which passes through the three points $P_1 = (3, 4, 5)$, $P_2 = (-1, 2, 3)$, and $P_3 = (2, 1, 3)$.

Solution: Since $\mathbf{a} \times \mathbf{b}$ is orthogonal to both $\mathbf{a}$ and $\mathbf{b}$; a normal to the plane is given by $\overrightarrow{P_1P_2} \times \overrightarrow{P_1P_3}$.

$$n = \overrightarrow{P_1P_2} \times \overrightarrow{P_1P_3} = [-4 \;\; -2 \;\; -2] \times [-1 \;\; -3 \;\; -2] = \begin{vmatrix} \mathbf{i} & \mathbf{j} & \mathbf{k} \\ -4 & -2 & -2 \\ -1 & -3 & -2 \end{vmatrix}$$

$$= -2\mathbf{i} - 6\mathbf{j} + 10\mathbf{k} = [-2 \;\; -6 \;\; 10]$$

Let (x, y, z) be any point in the plane and choose any of the three points as a point through which the plane passes. Let us use P_1.

$$\mathbf{n} \cdot [x - 3, y - 4, z - 5] = 0$$

or

$$[-2 \;\; -6 \;\; 10] \cdot [x - 3, y - 4, z - 5] = 0$$

$$-2(x - 3) - 6(y - 4) + 10(z - 5) = 0$$

Hence, in general form the equation simplifies to

$$x + 3y - 5z = -10$$

SUMMARY

1. A vector may be represented geometrically by an arrow with the length proportional to magnitude and pointing in the appropriate direction.
2. The point from which the arrow begins is called the initial point, and the tip of the arrow is called the terminal point.
3. As long as an arrow is moved parallel to its initial position and its length is unchanged, the vectors represented are all equal.
4. If the initial point of an arrow is at the origin of a coordinate system, then the vector represented is a position vector.
5. If a vector is lengthened, you have a dilation.
6. If a vector is shortened, you have a contraction.
7. If a vector is multiplied by -1, its direction is reversed; an opposition.
8. Geometrically, vectors may be added by the parallelogram law.
9. $\mathbf{a} - \mathbf{b}$ is represented by the arrow with initial point at the terminal point of $\mathbf{b}$ and its terminal point at the terminal point of $\mathbf{a}$.
10. The cosine of the angle ϕ between two nonzero vectors $\mathbf{a}$ and $\mathbf{b}$ in an inner product space V is
$$\cos \phi = \frac{\langle \mathbf{a}, \mathbf{b} \rangle}{\|\mathbf{a}\| \, \|\mathbf{b}\|}$$
11. Two nonzero vectors $\mathbf{a}$ and $\mathbf{b}$ are orthogonal if and only if $\langle \mathbf{a}, \mathbf{b} \rangle = 0$. For the standard inner product, $\langle \mathbf{a}, \mathbf{b} \rangle = \mathbf{a} \cdot \mathbf{b} = 0$.
12. The point-normal form of the equation of a plane is $[a \;\; b \;\; c] \cdot [x - x_0, y - y_0, z - z_0] = 0$ where $[a \;\; b \;\; c]$ is normal to the plane and (x, y, z) is any variable point on the plane with (x_0, y_0, z_0) a fixed point.
13. The general form of the equation of a plane is $ax + by + cz = d$.
14. The angles that a vector makes with the positive x-, y-, and z-axes are called direction angles.
15. The cosines of the direction angles are called direction cosines.
16. Scalar multiples of direction cosines are called direction numbers; that is, $k \cos \alpha$, $k \cos \beta$, $k \cos \gamma$ are direction numbers.
17. The natural basis vectors for R_3, $[1 \;\; 0 \;\; 0]$, $[0 \;\; 1 \;\; 0]$, and $[0 \;\; 0 \;\; 1]$, are often denoted by $\mathbf{i}$, $\mathbf{j}$, and $\mathbf{k}$, respectively.
18. In terms of components, the vector equation of a line is $[x - x_0, y - y_0, z - z_0] = t[a \;\; b \;\; c]$ where

$[a \quad b \quad c]$ is a vector parallel to the line and (x, y, z) is any variable point on the line, (x_0, y_0, z_0) any fixed point, and t is a scalar.

19. Parametric equations of a line are

$$x = x_0 + ta$$
$$y = y_0 + tb$$
$$z = z_0 + tc$$

20. Symmetric equations of a line are

$$\frac{x - x_0}{a} = \frac{y - y_0}{b} = \frac{z - z_0}{c}$$

21. The cross product $\mathbf{a} \times \mathbf{b}$ is defined by

$$\mathbf{a} \times \mathbf{b} = [a_2b_3 - a_3b_2, a_3b_1 - a_1b_3, a_1b_2 - a_2b_1] = \begin{vmatrix} \mathbf{i} & \mathbf{j} & \mathbf{k} \\ a_1 & a_2 & a_3 \\ b_1 & b_2 & b_3 \end{vmatrix}$$

22. If $\mathbf{a}$, $\mathbf{b}$, and $\mathbf{c}$ are vectors in R_3 with standard inner product and k is a scalar, then

- $\mathbf{a} \times \mathbf{b} = -(\mathbf{b} \times \mathbf{a})$
- $\mathbf{a} \times \mathbf{b}$ is orthogonal to both $\mathbf{a}$ and $\mathbf{b}$
- $\mathbf{a} \times \mathbf{0} = \mathbf{0} \times \mathbf{a} = \mathbf{0}$
- $\mathbf{a} \times \mathbf{a} = \mathbf{0}$
- $\mathbf{a} \times (\mathbf{b} + \mathbf{c}) = (\mathbf{a} \times \mathbf{b}) + (\mathbf{a} \times \mathbf{c})$
- $(\mathbf{a} + \mathbf{b}) \times \mathbf{c} = (\mathbf{a} \times \mathbf{c}) + (\mathbf{b} \times \mathbf{c})$
- $k(\mathbf{a} \times \mathbf{b}) = (k\mathbf{a}) \times \mathbf{b} = \mathbf{a} \times (k\mathbf{b})$
- $\|\mathbf{a} \times \mathbf{b}\|^2 = \|\mathbf{a}\|^2 \|\mathbf{b}\|^2 - (\mathbf{a} \cdot \mathbf{b})^2$, which is Lagrange's identity.

RAISE YOUR GRADES

Can you ...?

☑ represent a vector by an arrow
☑ identify what happens if a vector is multiplied by a constant k
☑ add vectors by the parallelogram law
☑ subtract vectors
☑ find the cosine of the angle between two vectors
☑ find the point-normal and general forms of the equation of a plane
☑ find the direction cosines of a vector or a line
☑ use the i, j, k notation
☑ find the vector, parametric, and symmetric forms of the equations of a line
☑ find the cross product of two vectors
☑ identify the various properties of the cross product

SOLVED PROBLEMS

PROBLEM 10-1 Describe the arrow with initial point at the origin which would be used to represent the following vectors: **(a)** $2[1 \quad -4]$. **(b)** $1/2[1 \quad -4]$. **(c)** $-3[1 \quad -4]$.

Solution

(**a**) This vector has the same direction as $[1 \;\; -4]$ but is twice as long. Since $2[1 \;\; -4] = [2 \;\; -8]$, the arrow would be drawn from $(0,0)$ to the point $(2, -8)$. The length of $[1 \;\; -4]$ is $\|[1 \;\; -4]\| = \sqrt{1^2 + (-4)^2} = \sqrt{17}$, so the length of $2[1 \;\; -4]$ is $\|[2 \;\; -8]\| = \sqrt{2^2 + (-8)^2} = \sqrt{68} = 2\sqrt{17}$.

(**b**) This vector has the same direction as $[1 \;\; -4]$ but is only half as long. Since $\frac{1}{2}[1 \;\; -4] = [\frac{1}{2} \;\; -2]$, the arrow would be drawn from $(0,0)$ to the point $(\frac{1}{2}, -2)$. The length of $[\frac{1}{2} \;\; -2]$ is $\|[\frac{1}{2} \;\; -2]\| = \sqrt{\frac{1}{4} + 4} = \sqrt{\frac{17}{4}} = \frac{1}{2}\sqrt{17}$, which is half the length of $[1 \;\; -4]$.

(**c**) This vector has the opposite direction as that of $[1 \;\; -4]$ and is three times as long. Since $-3[1 \;\; -4] = [-3 \;\; 12]$, the arrow would be drawn from $(0,0)$ to $(-3, 12)$. The length of $[-3 \;\; 12]$ is $\|[-3 \;\; 12]\| = \sqrt{9 + 144} = \sqrt{153} = 3\sqrt{17}$, which is three times the length of $[1 \;\; -4]$.

Part (**a**) is a dilation, part (**b**) a contraction, and part (**c**) an opposition.

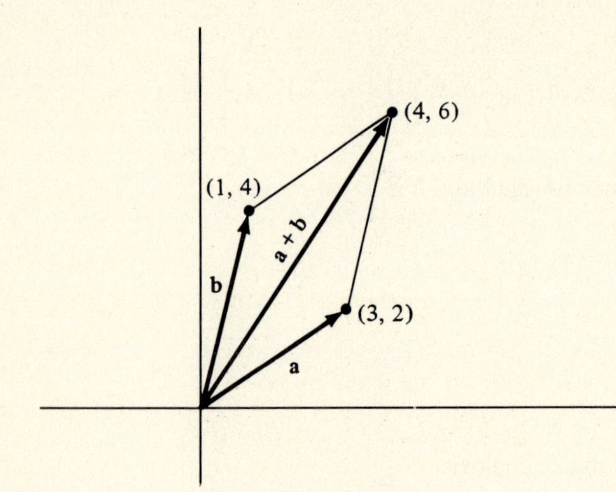

FIGURE 10-6

PROBLEM 10-2 Add, geometrically, the vectors $\mathbf{a} = [3 \;\; 2]$ and $\mathbf{b} = [1 \;\; 4]$.

Solution Plot points $(3,2)$ and $(1,4)$, then draw the arrows from $(0,0)$ to each of these points. Complete the parallelogram with these vectors as adjacent sides, and draw the diagonal from the origin of the parallelogram, as shown in Figure 10-6. This diagonal is the geometric representation of the sum $\mathbf{a} + \mathbf{b}$.

PROBLEM 10-3 Given $\mathbf{a} = [3 \;\; 2]$ and $\mathbf{b} = [1 \;\; 4]$, find, geometrically, the vector $\mathbf{a} - \mathbf{b}$.

Solution Plot points $(3,2)$ and $(1,4)$, then draw the arrows from $(0,0)$ to each of these points. $\mathbf{a} - \mathbf{b}$ is the arrow drawn from $(1,4)$ to $(3,2)$, as in Figure 10-7. It is the other diagonal of the parallelogram in Problem 10-2.

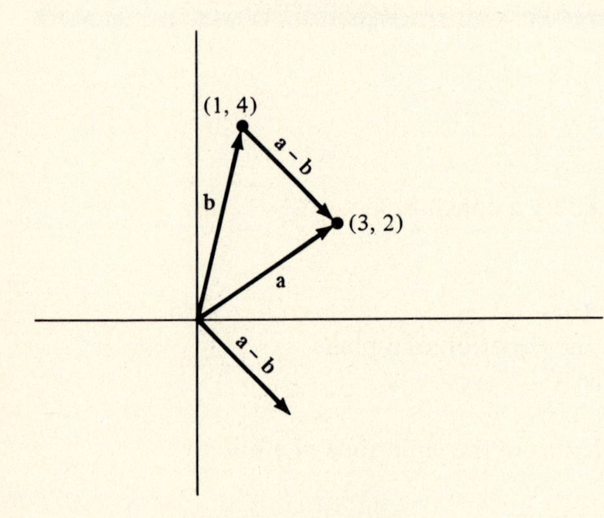

FIGURE 10-7

PROBLEM 10-4 Find the cosine of the angle ϕ between the two vectors $\mathbf{a} = [3 \;\; 4]$ and $\mathbf{b} = [1 \;\; 2]$ in R_2 if R_2 has the standard inner product.

Solution

$$\langle \mathbf{a}, \mathbf{b} \rangle = 3(1) + 4(2) = 11$$

$$\|\mathbf{a}\| = \sqrt{\langle \mathbf{a}, \mathbf{a} \rangle} = \sqrt{3^2 + 4^2} = 5$$

$$\|\mathbf{b}\| = \sqrt{\langle \mathbf{b}, \mathbf{b} \rangle} = \sqrt{1^2 + 2^2} = \sqrt{5}$$

$$\therefore \quad \cos\phi = \frac{\langle \mathbf{a}, \mathbf{b} \rangle}{\|\mathbf{a}\| \, \|\mathbf{b}\|} = \frac{11}{5\sqrt{5}}$$

PROBLEM 10-5 Find the cosine of the angle ϕ between the two vectors $\mathbf{a} = [-2 \;\; 3 \;\; 4]$ and $\mathbf{b} = [1 \;\; 2 \;\; 5]$ in R_3 if R_3 has the standard inner product.

Solution

$$\langle \mathbf{a}, \mathbf{b} \rangle = (-2)(1) + 3(2) + 4(5) = 24$$

$$\|\mathbf{a}\| = \sqrt{\langle \mathbf{a}, \mathbf{a} \rangle} = \sqrt{(-2)^2 + 3^2 + 4^2} = \sqrt{29}$$

$$\|\mathbf{b}\| = \sqrt{\langle \mathbf{b}, \mathbf{b} \rangle} = \sqrt{1^2 + 2^2 + 5^2} = \sqrt{30}$$

$$\therefore \quad \cos\phi = \frac{\langle \mathbf{a}, \mathbf{b} \rangle}{\|\mathbf{a}\|\,\|\mathbf{b}\|} = \frac{24}{\sqrt{29}\sqrt{30}}$$

PROBLEM 10-6 Find the cosine of the angle ϕ between the two vectors $\mathbf{a} = [3 \quad 4]$ and $\mathbf{b} = [1 \quad 2]$ in R_2 if R_2 has the inner product $\langle \mathbf{a}, \mathbf{b} \rangle = 3a_1 b_1 + 4a_2 b_2$.

Solution

$$\langle \mathbf{a}, \mathbf{b} \rangle = 3(3)(1) + 4(4)(2) = 9 + 32 = 41$$

$$\|\mathbf{a}\| = \sqrt{3(3)^2 + 4(4)^2} = \sqrt{27 + 64} = \sqrt{91}$$

$$\|\mathbf{b}\| = \sqrt{3(1)^2 + 4(2)^2} = \sqrt{3 + 16} = \sqrt{19}$$

$$\therefore \quad \cos\phi = \frac{\langle \mathbf{a}, \mathbf{b} \rangle}{\|\mathbf{a}\|\,\|\mathbf{b}\|} = \frac{41}{\sqrt{91}\sqrt{19}}$$

PROBLEM 10-7 Find the equation of the plane passing through $(-2, 3, 4)$ and perpendicular to the vector $[3 \quad -4 \quad 2]$ in both point-normal and general forms.

Solution

$$\mathbf{n} \cdot (\mathbf{x} - \mathbf{x}_0) = 0$$

$$\therefore \quad [3 \quad -4 \quad 2] \cdot [x + 2, y - 3, z - 4] = 0$$

or

$$3(x + 2) - 4(y - 3) + 2(z - 4) = 0 \qquad \text{in point-normal form}$$

Therefore, $3x - 4y + 2z = -10$ in general form.

PROBLEM 10-8 Find the equation of the plane containing point $(-2, 1, -4)$ and parallel to plane $2x + 3y - z = 6$.

Solution A normal to the plane $2x + 3y - z = 6$ is $[2 \quad 3 \quad -1]$ which will also be normal to the requested plane. Hence,

$$[2 \quad 3 \quad -1] \cdot [x + 2, y - 1, z + 4] = 0$$

or

$$2(x + 2) + 3(y - 1) - 1(z + 4) = 0$$

$$\therefore \quad 2x + 3y - z = 3$$

PROBLEM 10-9 Find the direction cosines of the vector $[2 \quad 1 \quad 4]$.

Solution Using any three vectors in the directions of the positive x-, y-, and z-axes, we will find the cosines of the angles between the given vector and the three vectors $[1 \quad 0 \quad 0]$, $[0 \quad 1 \quad 0]$, and $[0 \quad 0 \quad 1]$ in the direction of the positive x, y, and z-axes, respectively.

$$\therefore \quad \cos\alpha = \frac{[2 \quad 1 \quad 4] \cdot [1 \quad 0 \quad 0]}{\|[2 \quad 1 \quad 4]\|\,\|[1 \quad 0 \quad 0]\|} = \frac{2}{\sqrt{21}}$$

$$\cos\beta = \frac{[2 \quad 1 \quad 4] \cdot [0 \quad 1 \quad 0]}{\|[2 \quad 1 \quad 4]\|\,\|[1 \quad 0 \quad 0]\|} = \frac{1}{\sqrt{21}}$$

$$\cos\gamma = \frac{[2 \quad 1 \quad 4] \cdot [0 \quad 0 \quad 1]}{\|[2 \quad 1 \quad 4]\|\,\|[1 \quad 0 \quad 0]\|} = \frac{4}{\sqrt{21}}$$

PROBLEM 10-10 Find the vector equation of the line passing through the point $(-3, 2, 4)$ and parallel to the vector $[1 \quad 2 \quad -2]$, in terms of components.

Solution Let (x, y, z) be any point on the line, then $[x, y, z] - [-3 \quad 2 \quad 4] = t[1 \quad 2 \quad -2]$ or $[x + 3, y - 2, z - 4] = t[1 \quad 2 \quad -2]$.

PROBLEM 10-11 Find the parametric equations of the line in Problem 10-10.

Solution We set corresponding components equal to get

$$x = t - 3$$

$$y = 2t + 2$$

$$z = -2t + 4$$

PROBLEM 10-12 Find the symmetric equations of the line in Problem 10-11.

Solution Solving the three equations in Problem 10-11 for t, we get

$$t = x + 3$$

$$t = \frac{y - 2}{2}$$

$$t = \frac{z - 4}{-2}$$

Therefore, the symmetric equations of the line are

$$\frac{x + 3}{1} = \frac{y - 2}{2} = \frac{z - 4}{-2}$$

Observe that the denominators are the components of the given vector.

PROBLEM 10-13 Find the vector, parametric, and symmetric forms of the line passing through $(4, -3, 6)$ and parallel to the vector $[3 \quad -5 \quad -4]$.

Solution Let (x, y, z) be any point on the line. Hence, $[x \quad y \quad z] - [4 \quad -3 \quad 6] = t[3 \quad -5 \quad -4]$ or $[x - 4, \quad y + 3, \quad z - 6] = t[3 \quad -5 \quad -4]$, the vector form of the line.

Setting corresponding components equal, we get

$$x = 3t + 4$$

$$y = -5t - 3$$

$$z = -4t + 6$$

which are parametric equations. Solving each of the parametric equations for t, we get

$$t = \frac{x - 4}{3}, \qquad t = \frac{y + 3}{-5}, \qquad t = \frac{z - 6}{-4}$$

Therefore, $(x - 4)/3 = (y + 3)/-5 = (z - 6)/-4$ are the symmetric equations.

PROBLEM 10-14 Find the symmetric equations of the line perpendicular to the plane $5x + 2y - 3z = 7$ and containing the point $(-2, 1, -3)$.

Solution The coefficients of x, y, and z are the components of a normal to the plane. Therefore,

$$\frac{x - x_0}{a} = \frac{y - y_0}{b} = \frac{z - z_0}{c}$$

becomes

$$\frac{x + 2}{5} = \frac{y - 1}{2} = \frac{z + 3}{-3}$$

PROBLEM 10-15 Find the cosine of the angle ϕ between the lines

$$\frac{x+2}{2} = \frac{y-3}{3} = \frac{z-1}{4} \quad \text{and} \quad \frac{x-1}{5} = \frac{y+4}{3} = \frac{z-3}{-2}$$

Solution Vectors parallel to the given lines are $\mathbf{a} = [2 \quad 3 \quad 4]$ and $\mathbf{b} = [5 \quad 3 \quad -2]$. The angle between the lines is the same angle as that between vectors **a** and **b**. We will use the standard inner product which is the dot product. Therefore,

$$\cos\phi = \frac{\mathbf{a}\cdot\mathbf{b}}{\|\mathbf{a}\|\,\|\mathbf{b}\|} = \frac{2\cdot 5 + 3\cdot 3 + 4(-2)}{\sqrt{2^2+3^2+4^2}\sqrt{5^2+3^2+(-2)^2}} = \frac{11}{\sqrt{29}\sqrt{38}}$$

PROBLEM 10-16 Find the cosine of the angle ϕ between the planes $x + 2y - z = 3$ and $2x - 3y + 4z = 5$.

Solution The normal vectors to the given planes are $[1 \quad 2 \quad -1]$ and $[2 \quad -3 \quad 4]$. The angle between these two vectors is the same as angle ϕ. Thus,

$$\cos\phi = \frac{[1 \quad 2 \quad -1]\cdot[2 \quad -3 \quad 4]}{\|[1 \quad 2 \quad -1]\|\,\|[2 \quad -3 \quad 4]\|} = \frac{2-6-4}{\sqrt{6}\sqrt{29}} = \frac{-8}{\sqrt{6}\sqrt{29}}$$

PROBLEM 10-17 If $\mathbf{a} = [3 \quad -2 \quad 5]$ and $\mathbf{b} = [1 \quad -3 \quad 4]$, find $\mathbf{a}\times\mathbf{b}$.

Solution

$$\begin{aligned}\mathbf{a}\times\mathbf{b} &= [a_2b_3 - a_3b_2, a_3b_1 - a_1b_3, a_1b_2 - a_2b_1]\\ &= [(-2)(4) - (5)(-3), (5)(1) - (3)(4), (3)(-3) - (-2)(1)]\\ &= [7 \quad -7 \quad -7]\end{aligned}$$

PROBLEM 10-18 Use the determinant representation to find the cross product $\mathbf{a}\times\mathbf{b}$ for the vectors in Problem 10-17.

Solution

$$\begin{aligned}\mathbf{a}\times\mathbf{b} &= \begin{vmatrix}\mathbf{i} & \mathbf{j} & \mathbf{k}\\ 3 & -2 & 5\\ 1 & -3 & 4\end{vmatrix} = \mathbf{i}\begin{vmatrix}-2 & 5\\ -3 & 4\end{vmatrix} - \mathbf{j}\begin{vmatrix}3 & 5\\ 1 & 4\end{vmatrix} + \mathbf{k}\begin{vmatrix}3 & -2\\ 1 & -3\end{vmatrix}\\ &= 7\mathbf{i} - 7\mathbf{j} - 7\mathbf{k} = [7 \quad -7 \quad -7]\end{aligned}$$

PROBLEM 10-19 Verify each of the properties of a cross product given in Section 10-6C for $\mathbf{a} = [-1 \quad 2 \quad -4]$, $\mathbf{b} = [3 \quad 1 \quad -2]$, $\mathbf{c} = [3 \quad -2 \quad -3]$, and $k = 3$.

Solution

(a)

$$\begin{aligned}\mathbf{a}\times\mathbf{b} &= \begin{vmatrix}\mathbf{i} & \mathbf{j} & \mathbf{k}\\ -1 & 2 & -4\\ 3 & 1 & -2\end{vmatrix} = \begin{vmatrix}2 & -4\\ 1 & -2\end{vmatrix}\mathbf{i} - \begin{vmatrix}-1 & -4\\ 3 & -2\end{vmatrix}\mathbf{j} + \begin{vmatrix}-1 & 2\\ 3 & 1\end{vmatrix}\mathbf{k}\\ &= 0\mathbf{i} - 14\mathbf{j} - 7\mathbf{k}\\ &= [0 \quad -14 \quad -7]\end{aligned}$$

$$\begin{aligned}\mathbf{b}\times\mathbf{a} &= \begin{vmatrix}\mathbf{i} & \mathbf{j} & \mathbf{k}\\ 3 & 1 & -2\\ -1 & 2 & -4\end{vmatrix} = \begin{vmatrix}1 & -2\\ 2 & -4\end{vmatrix}\mathbf{i} - \begin{vmatrix}3 & -2\\ -1 & -4\end{vmatrix}\mathbf{j} + \begin{vmatrix}3 & 1\\ -1 & 2\end{vmatrix}\mathbf{k}\\ &= 0\mathbf{i} + 14\mathbf{j} + 7\mathbf{k}\\ &= [0 \quad 14 \quad 7]\end{aligned}$$

$$\therefore \quad \mathbf{a}\times\mathbf{b} = -(\mathbf{b}\times\mathbf{a})$$

(b) $\mathbf{a} \times \mathbf{b}$ is orthogonal to $\mathbf{a}$ and to $\mathbf{b}$ if and only if $(\mathbf{a} \times \mathbf{b})\cdot\mathbf{a} = 0$ and $(\mathbf{a} \times \mathbf{b})\cdot\mathbf{b} = 0$. Thus,

$$(\mathbf{a} \times \mathbf{b})\cdot\mathbf{a} = [0 \quad -14 \quad -7]\cdot[-1 \quad 2 \quad -4] = 0(-1) + (-14)(2) + (-7)(-4) = 0$$

and

$$(\mathbf{a} \times \mathbf{b})\cdot\mathbf{b} = [0 \quad -14 \quad -7]\cdot[3 \quad 1 \quad -2] = 0(3) + (-14)(1) + (-7)(-2) = 0$$

$$\therefore \quad \mathbf{a} \times \mathbf{b} \text{ is orthogonal to both } \mathbf{a} \text{ and } \mathbf{b}$$

(c)
$$\mathbf{a} \times \mathbf{0} = \begin{vmatrix} \mathbf{i} & \mathbf{j} & \mathbf{k} \\ -1 & 2 & -4 \\ 0 & 0 & 0 \end{vmatrix} = -\begin{vmatrix} \mathbf{i} & \mathbf{j} & \mathbf{k} \\ 0 & 0 & 0 \\ -1 & 2 & -4 \end{vmatrix} = 0\mathbf{i} + 0\mathbf{j} + 0\mathbf{k} = [0 \quad 0 \quad 0]$$

$$\therefore \quad \mathbf{a} \times \mathbf{0} = \mathbf{0} \times \mathbf{a} = \mathbf{0} \text{ for the given } \mathbf{a}$$

(d)
$$\mathbf{a} \times \mathbf{a} = \begin{vmatrix} \mathbf{i} & \mathbf{j} & \mathbf{k} \\ -1 & 2 & -4 \\ -1 & 2 & -4 \end{vmatrix} = 0\mathbf{i} + 0\mathbf{j} + 0\mathbf{k} = [0 \quad 0 \quad 0]$$

$$\therefore \quad \mathbf{a} \times \mathbf{a} = \mathbf{0} \text{ for the given } \mathbf{a}$$

(e)
$$\mathbf{b} + \mathbf{c} = [3 \quad 1 \quad -2] + [3 \quad -2 \quad -3] = [6 \quad -1 \quad -5]$$

$$\mathbf{a} \times (\mathbf{b} + \mathbf{c}) = [-1 \quad 2 \quad -4] \times [6 \quad -1 \quad -5] = \begin{vmatrix} \mathbf{i} & \mathbf{j} & \mathbf{k} \\ -1 & 2 & -4 \\ 6 & -1 & -5 \end{vmatrix} = -14\mathbf{i} - 29\mathbf{j} - 11\mathbf{k}$$

$$= [-14 \quad -29 \quad -11]$$

$$(\mathbf{a} \times \mathbf{b}) + (\mathbf{a} \times \mathbf{c}) = \begin{vmatrix} \mathbf{i} & \mathbf{j} & \mathbf{k} \\ -1 & 2 & -4 \\ 3 & 1 & -2 \end{vmatrix} + \begin{vmatrix} \mathbf{i} & \mathbf{j} & \mathbf{k} \\ -1 & 2 & -4 \\ 3 & -2 & -3 \end{vmatrix} = (0\mathbf{i} - 14\mathbf{j} - 7\mathbf{k}) + (-14\mathbf{i} - 15\mathbf{j} - 4\mathbf{k})$$

$$= -14\mathbf{i} - 29\mathbf{j} - 11\mathbf{k} = [-14 \quad -29 \quad -11]$$

$$\therefore \quad \mathbf{a} \times (\mathbf{b} + \mathbf{c}) = (\mathbf{a} \times \mathbf{b}) + (\mathbf{a} \times \mathbf{c}) \text{ for the given vectors}$$

(f)
$$\mathbf{a} + \mathbf{b} = [-1 \quad 2 \quad -4] + [3 \quad 1 \quad -2] = [2 \quad 3 \quad -6]$$

$$(\mathbf{a} + \mathbf{b}) \times \mathbf{c} = [2 \quad 3 \quad -6] \times [3 \quad -2 \quad -3] = \begin{vmatrix} \mathbf{i} & \mathbf{j} & \mathbf{k} \\ 2 & 3 & -6 \\ 3 & -2 & -3 \end{vmatrix} = -21\mathbf{i} - 12\mathbf{j} - 13\mathbf{k}$$

$$= [-21 \quad -12 \quad -13]$$

$$(\mathbf{a} \times \mathbf{c}) + (\mathbf{b} \times \mathbf{c}) = \begin{vmatrix} \mathbf{i} & \mathbf{j} & \mathbf{k} \\ -1 & 2 & -4 \\ 3 & -2 & -3 \end{vmatrix} + \begin{vmatrix} \mathbf{i} & \mathbf{j} & \mathbf{k} \\ 3 & 1 & -2 \\ 3 & -2 & -3 \end{vmatrix} = (-14\mathbf{i} - 15\mathbf{j} - 4\mathbf{k}) + (-7\mathbf{i} + 3\mathbf{j} - 9\mathbf{k})$$

$$= -21\mathbf{i} - 12\mathbf{j} - 13\mathbf{k} = [-21 \quad -12 \quad -13]$$

$$\therefore \quad (\mathbf{a} + \mathbf{b}) \times \mathbf{c} = (\mathbf{a} \times \mathbf{c}) + (\mathbf{b} \times \mathbf{c}) \text{ for the given vector}$$

(g) From part **(a)**, $\mathbf{a} \times \mathbf{b} = [0 \quad -14 \quad -7]$. Therefore,

$$k[\mathbf{a} \times \mathbf{b}] = 3[0 \quad -14 \quad -7] = [0 \quad -42 \quad -21]$$

$$(k\mathbf{a}) \times \mathbf{b} = \begin{vmatrix} \mathbf{i} & \mathbf{j} & \mathbf{k} \\ -3 & 6 & -12 \\ 3 & 1 & -2 \end{vmatrix} = 0\mathbf{i} - 42\mathbf{j} - 21\mathbf{k} = [0 \quad -42 \quad -21]$$

$$\mathbf{a} \times (k\mathbf{b}) = \begin{vmatrix} \mathbf{i} & \mathbf{j} & \mathbf{k} \\ -1 & 2 & -4 \\ 9 & 3 & -6 \end{vmatrix} = 0\mathbf{i} - 42\mathbf{j} - 21\mathbf{k} = [0 \quad -42 \quad -21]$$

$$\therefore \quad k(\mathbf{a} \times \mathbf{b}) = (k\mathbf{a}) \times \mathbf{b} = \mathbf{a} \times (k\mathbf{b}) \text{ for the given } \mathbf{a} \text{ and } \mathbf{b} \text{ and } k$$

(h) From part **(a)**, $\mathbf{a} \times \mathbf{b} = [0 \quad -14 \quad -7]$. Therefore,

$$\|\mathbf{a} \times \mathbf{b}\|^2 = 0^2 + (-14)^2 + (-7)^2 = 196 + 49 = 245$$

$$\|\mathbf{a}\|^2 \|\mathbf{b}\|^2 - (\mathbf{a} \cdot \mathbf{b})^2 = [(-1)^2 + 2^2 + (-4)^2][3^2 + 1^2 + (-2)^2] - [(-1)(3) + 2(1) + (-4)(-2)]^2$$
$$= (21)(14) - (7)^2 = 294 - 49 = 245$$

Therefore, Lagrange's identity holds for the given **a** and **b**.

PROBLEM 10-20 Find the general form of the equation of a plane that passes through the three points $P_1 = (1, -2, 3)$, $P_2 = (-3, 1, 4)$, and $P_3 = (-1, 3, 5)$.

Solution Since $\mathbf{a} \times \mathbf{b}$ is orthogonal to both **a** and **b**, a normal to the plane is given by

$$\overrightarrow{P_1P_2} \times \overrightarrow{P_1P_3}$$

$$\therefore \quad \mathbf{n} = \overrightarrow{P_1P_2} \times \overrightarrow{P_1P_3} = [-4 \quad 3 \quad 1] \times [-2 \quad 5 \quad 2]$$

$$= \begin{vmatrix} \mathbf{i} & \mathbf{j} & \mathbf{k} \\ -4 & 3 & 1 \\ -2 & 5 & 2 \end{vmatrix} = \mathbf{i} + 6\mathbf{j} - 14\mathbf{k} = [1 \quad 6 \quad -14]$$

Let (x, y, z) be any point in the plane and use any one of the given points on the plane. Let us choose P_3. Therefore,

$$n \cdot [x + 1, y - 3, z - 5] = 0$$

or

$$[1 \quad 6 \quad -14] \cdot [x + 1, y - 3, z - 5] = 0$$

$$\therefore \quad 1(x + 1) + 6(y - 3) - 14(z - 5) = 0$$

Thus, in general form the equation becomes $x + 6y - 14z = -53$.

Supplementary Problems

PROBLEM 10-21 Describe the arrow with initial point at (0, 0) that would represent each of the following in R_2: **(a)** $3[2 \quad 3]$, **(b)** $-2[2 \quad 3]$.

PROBLEM 10-22 If $\mathbf{a} = [1 \quad 2]$ and $\mathbf{b} = [5 \quad 2]$, describe, geometrically, the following: **(a)** $\mathbf{a} + \mathbf{b}$. **(b)** $\mathbf{a} - \mathbf{b}$.

PROBLEM 10-23 Find the cosine of the angle ϕ between the two vectors in Problem 10-22.

PROBLEM 10-24 Find the cosine of the angle ϕ between each of the following pairs of vectors: **(a)** $\mathbf{a} = [2 \quad -3]$, $\mathbf{b} = [4 \quad 7]$ in R_2. **(b)** $\mathbf{a} = [3 \quad 1 \quad -2]$, $\mathbf{b} = [-5 \quad 2 \quad 1]$ in R_3.

PROBLEM 10-25 Find the cosine of the angle ϕ between the two vectors $\mathbf{a} = [4 \quad -5]$ and $\mathbf{b} = [1 \quad 3]$ in R_2 with inner product $\langle \mathbf{a}, \mathbf{b} \rangle = 4a_1b_1 + a_2b_2$.

PROBLEM 10-26 Find the cosine of the angle ϕ between $\mathbf{a} = [-1 \quad 3]$ and $\mathbf{b} = [1 \quad -3]$ in R_2.

PROBLEM 10-27 Find the equation of the plane through (1, 2, 3) and parallel to plane $x - 3y + 4z = 6$.

PROBLEM 10-28 Find the equation of the plane passing through (1, 2, 3) and perpendicular to the vector $[-2 \quad 1 \quad 3]$ in point-normal form.

PROBLEM 10-29 Find the equation of the plane passing through the point (2, 1, −3) and having normal direction $\mathbf{n} = 3\mathbf{i} + 4\mathbf{j} - 2\mathbf{k}$.

PROBLEM 10-30 Find the direction cosines of each of the following vectors: **(a)** $\mathbf{a} = [2 \quad -1 \quad 3]$. **(b)** $\mathbf{b} = [0 \quad 4 \quad -3]$.

PROBLEM 10-31 Find the vector equation, in component form, of the line passing through point (6, 7, −3) and parallel to vector $[3 \quad -2 \quad 5]$

PROBLEM 10-32 Find the vector equation, parametric equations, and symmetric equations of the line going through (−4, 2, −6) and parallel to $[2 \quad -3 \quad 5]$.

PROBLEM 10-33 Find the symmetric equations of the line containing point (2, −1, 4) and perpendicular to the plane $3x + y - 5z = 11$.

PROBLEM 10-34 Find the cosine of the angle ϕ between the planes $5x + 2y - 3z = 4$ and $x - y + 4z = 6$.

PROBLEM 10-35 Find the cosine of the angle ϕ between the lines

$$\frac{x-3}{1} = \frac{y+4}{-2} = \frac{z-1}{3} \qquad \text{and} \qquad \frac{x+4}{3} = \frac{y-5}{-4} = \frac{z+2}{6}$$

PROBLEM 10-36 Given the vectors $\mathbf{a} = [4 \quad -3 \quad 7]$, $\mathbf{b} = [2 \quad 4 \quad -1]$, $\mathbf{c} = [1 \quad 2 \quad -1]$ find the following:

(a) $\mathbf{a} \times \mathbf{b}$
(b) $\mathbf{a} \times \mathbf{c}$
(c) $\mathbf{b} \times \mathbf{c}$
(d) $(\mathbf{a} + \mathbf{b}) \times \mathbf{c}$ and $\mathbf{a} \times (\mathbf{b} + \mathbf{c})$
(e) $(\mathbf{a} \times \mathbf{b}) + (\mathbf{a} \times \mathbf{c})$
(f) $(\mathbf{a} \times \mathbf{c}) + (\mathbf{b} \times \mathbf{c})$

PROBLEM 10-37 Find the general form of the equation of the plane passing through the points $P_1 = (2, -1, -2)$, $P_2 = (-1, 2, 4)$, and $P_3 = (3, -4, 3)$.

PROBLEM 10-38 Find the parametric equations of the line through (2, 4, −1) and perpendicular to the plane $3x - 2y + 5z = 8$.

Answers to Supplementary Problems

10-21 (a) Arrow begins at (0, 0) and end at (6, 9).
(b) Arrow begins at (0, 0) and ends at (−4, −6).

10-22 (a) $\mathbf{a} + \mathbf{b}$ begins at (0, 0) and ends at (6, 4).
(b) $\mathbf{a} - \mathbf{b}$ begins at (5, 2) and ends at (1, 2).

10-23 $\cos\phi = 9/\sqrt{5}\sqrt{29}$

10-24 (a) $\cos\phi = -1/\sqrt{5}$
(b) $\cos\phi = -15/2\sqrt{105}$

10-25 $\cos\phi = 1/\sqrt{89}\sqrt{13}$

10-26 $\cos\phi = -10/\sqrt{10}\sqrt{10} = -1$. The vectors are opposites.

10-27 $x - 3y + 4z = 7$

10-28 $-2(x - 1) + 1(y - 2) + 3(z - 3) = 0$

10-29 $3(x - 1) + 4(y - 2) - 2(z + 3) = 0$

10-30 **(a)** $2/\sqrt{14}, -1/\sqrt{14}, 3/\sqrt{14}$
(b) $0, 4/5, -3/5$

10-31 $[x-6, y-7, z+3] = t[3 \quad -2 \quad 5]$

10-32 $[x+4, y-2, z+6] = t[2 \quad -3 \quad 5]$

$x = -4 + 2t, y = 2 - 3t, z = -6 + 5t$

$$\frac{x+4}{2} = \frac{y-2}{-3} = \frac{z+6}{5}$$

10-33 $(x-2)/3 = (y+1)/1 = (z-4)/-5$

10-34 $\cos\phi = -9/\sqrt{38}\sqrt{18}$

10-35 $\cos\phi = 29/\sqrt{14}\sqrt{61}$

10-36 **(a)** $[-25 \quad 18 \quad 22]$
(b) $[-11 \quad 11 \quad 11]$
(c) $[-2 \quad 1 \quad 0]$
(d) $[-13 \quad 12 \quad 11], [-36 \quad 29 \quad 33]$
(e) $[-36 \quad 29 \quad 33]$
(f) $[-13 \quad 12 \quad 11]$

10-37 $11x + 7y + 2z = 11$

10-38 $x = 2 + 3t, y = 4 - 2t, z = -1 + 5t$

11 ORTHOGONALITY AND GRAM–SCHMIDT

THIS CHAPTER IS ABOUT

☑ **Orthonormal Basis**
☑ **Orthogonal Projection**
☑ **Gram–Schmidt Process**

11-1. Orthonormal Basis

A. Definitions of orthogonal and orthonormal sets

Two vectors **a** and **b** in an inner product space V are **orthogonal** if $\langle \mathbf{a}, \mathbf{b} \rangle = 0$. A set S of vectors in V is an **orthogonal set** if every pair of distinct vectors in S are orthogonal. If, in addition, each vector in S is of length one, then S is an **orthonormal set**. Observe that the zero vector cannot be in an orthonormal set.

EXAMPLE 11-1: Determine whether the following sets of vectors from R_3 with the standard inner product is orthonormal. **(a)** $[1\ \ 0\ \ 0], [0\ \ 1\ \ 0], [0\ \ 0\ \ 1]$; **(b)** $[1\ \ 0\ \ 1], [2\ \ 0\ \ -2], [0\ \ 1\ \ 0]$; **(c)** $[1\ \ 1\ \ 0], [2\ \ 0\ \ 1], [1\ \ 1\ \ 1]$

Solution

(a)

$$\langle [1\ \ 0\ \ 0], [0\ \ 1\ \ 0] \rangle = 0 + 0 + 0 = 0$$

$$\langle [1\ \ 0\ \ 0], [0\ \ 0\ \ 1] \rangle = 0 + 0 + 0 = 0$$

$$\langle [0\ \ 1\ \ 0], [0\ \ 0\ \ 1] \rangle = 0 + 0 + 0 = 0$$

Therefore, the set of vectors is orthogonal.

$$\|[1\ \ 0\ \ 0]\| = \sqrt{1^2 + 0^2 + 0^2} = 1$$

$$\|[0\ \ 1\ \ 0]\| = \sqrt{0^2 + 1^2 + 0^2} = 1$$

$$\|[0\ \ 0\ \ 1]\| = \sqrt{0^2 + 0^2 + 1^2} = 1$$

Therefore, each vector is of length one; that is, they are unit vectors. Hence, the set is orthonormal.

(b)

$$\langle [1\ \ 0\ \ 1], [2\ \ 0\ \ -2] \rangle = 2 + 0 - 2 = 0$$

$$\langle [1\ \ 0\ \ 1], [0\ \ 1\ \ 0] \rangle = 0 + 0 + 0 = 0$$

$$\langle [2\ \ 0\ \ -2], [0\ \ 1\ \ 0] \rangle = 0 + 0 + 0 = 0$$

Therefore, the set of vectors is orthogonal.

$$\|[1\ \ 0\ \ 1]\| = \sqrt{1^2 + 0^2 + 1^2} = \sqrt{2}$$

$$\|[2\ \ 0\ \ -2]\| = \sqrt{2^2 + 0^2 + (-2)^2} = \sqrt{8} = 2\sqrt{2}$$

$$\|[0\ \ 1\ \ 0]\| = \sqrt{0^2 + 1^2 + 0^2} = 1$$

Only one of these vectors is of length one. Hence, the set is orthogonal, but not orthonormal.

(c)
$$\langle [1 \quad 1 \quad 0], [2 \quad 0 \quad 1] \rangle = 2 + 0 + 0 = 2 \neq 0$$

Since these two vectors are not orthogonal, we can conclude that the set is neither orthogonal nor orthonormal.

EXAMPLE 11-2: Consider the vector space ${}_2M_2$ of all 2×2 matrices with inner product defined by $\langle A, B \rangle = a_{11}b_{11} + a_{12}b_{12} + a_{21}b_{21} + a_{22}b_{22}$. Show that the matrices

$$A = \begin{bmatrix} 2 & 2 \\ -2 & 0 \end{bmatrix} \qquad B = \begin{bmatrix} -1 & 0 \\ -1 & 2 \end{bmatrix} \qquad C = \begin{bmatrix} 0 & 4 \\ 4 & 2 \end{bmatrix}$$

form an orthogonal, but not orthonormal, set.

Solution

$$\langle A, B \rangle = 2(-1) + 2(0) + (-2)(-1) + 0(2) = 0$$
$$\langle A, C \rangle = 2(0) + 2(4) + (-2)(4) + 0(2) = 0$$
$$\langle B, C \rangle = -1(0) + 0(4) + (-1)(4) + 2(2) = 0$$

Therefore, it is an orthogonal set.

$$\|A\| = \sqrt{\langle A, A \rangle} = \sqrt{2^2 + 2^2 + (-2)^2 + 0^2} = \sqrt{12} = 2\sqrt{3}$$

Since A is not of length one, the set is orthogonal but not orthonormal.

B. Linear independence of an orthogonal set

If $S = \{\mathbf{a}_1, \mathbf{a}_2, \ldots, \mathbf{a}_n\}$ is a finite orthogonal set of nonzero vectors in an inner product space V, then S is **linearly independent**.

EXAMPLE 11-3: Show that the three vectors in Example 11-2 are linearly independent.

Solution: Let a, b, and c be scalars such that

$$a\begin{bmatrix} 2 & 2 \\ -2 & 0 \end{bmatrix} + b\begin{bmatrix} -1 & 0 \\ -1 & 2 \end{bmatrix} + c\begin{bmatrix} 0 & 4 \\ 4 & 2 \end{bmatrix} = \begin{bmatrix} 0 & 0 \\ 0 & 0 \end{bmatrix}$$

$$\begin{aligned} 2a - b \quad\quad &= 0 \\ 2a \quad\quad + 4c &= 0 \\ -2a - b + 4c &= 0 \\ 2b + 2c &= 0 \end{aligned}$$

From the first two equations,

$$b = 2a$$
$$c = -\tfrac{1}{2}a$$

Substituting into the fourth equation we find

$$4a - a = 0 \qquad \text{or} \qquad 3a = 0$$

Hence $a = 0$ and therefore $b = 0$ and $c = 0$. Hence, the three vectors are linearly independent.

C. Definition of orthogonal and orthonormal basis

If B is a basis for an inner product space V and B is an orthogonal set, then B is called an **orthogonal basis** for V. If B is an orthonormal set, it is an **orthonormal basis** for V.

EXAMPLE 11-4: The standard basis for R_5 with standard inner product is $\{\mathbf{e}_1, \mathbf{e}_2, \mathbf{e}_3, \mathbf{e}_4, \mathbf{e}_5\}$, where

$$\mathbf{e}_1 = [1 \quad 0 \quad 0 \quad 0 \quad 0]$$
$$\mathbf{e}_2 = [0 \quad 1 \quad 0 \quad 0 \quad 0]$$
$$\mathbf{e}_3 = [0 \quad 0 \quad 1 \quad 0 \quad 0]$$
$$\mathbf{e}_4 = [0 \quad 0 \quad 0 \quad 1 \quad 0]$$
$$\mathbf{e}_5 = [0 \quad 0 \quad 0 \quad 0 \quad 1]$$

Verify that this is an orthonormal basis.

Solution

$$\langle \mathbf{e}_1, \mathbf{e}_2 \rangle = 0, \qquad \langle \mathbf{e}_1, \mathbf{e}_3 \rangle = 0, \qquad \langle \mathbf{e}_1, \mathbf{e}_4 \rangle = 0, \qquad \langle \mathbf{e}_1, \mathbf{e}_5 \rangle = 0,$$
$$\langle \mathbf{e}_2, \mathbf{e}_3 \rangle = 0, \qquad \langle \mathbf{e}_2, \mathbf{e}_4 \rangle = 0, \qquad \langle \mathbf{e}_2, \mathbf{e}_5 \rangle = 0, \qquad \langle \mathbf{e}_3, \mathbf{e}_4 \rangle = 0,$$
$$\langle \mathbf{e}_3, \mathbf{e}_5 \rangle = 0, \qquad \langle \mathbf{e}_4, \mathbf{e}_5 \rangle = 0$$

This shows that it is at least an orthogonal basis.

$$\|\mathbf{e}_1\| = 1, \|\mathbf{e}_2\| = 1, \qquad \|\mathbf{e}_3\| = 1, \qquad \|\mathbf{e}_4\| = 1, \qquad \|\mathbf{e}_5\| = 1$$

Hence, it is an orthonormal basis.

EXAMPLE 11-5: Let R_3 have the standard inner product. Show that

$$\mathbf{b}_1 = \left[\frac{1}{\sqrt{2}} \quad \frac{1}{\sqrt{2}} \quad 0\right], \qquad \mathbf{b}_2 = \left[\frac{1}{\sqrt{2}} \quad -\frac{1}{\sqrt{2}} \quad 0\right], \qquad \text{and} \qquad \mathbf{b}_3 = [0 \quad 0 \quad 1]$$

is an orthonormal basis of R_3.

Solution: Since we have three vectors and the dimension of R_3 is 3, we have to show linear independence of these vectors or that they span R_3 to conclude that they form a basis for R_3. If the determinant having these three vectors as rows is nonzero, then they are linearly independent.

$$\begin{vmatrix} \frac{1}{\sqrt{2}} & \frac{1}{\sqrt{2}} & 0 \\ \frac{1}{\sqrt{2}} & -\frac{1}{\sqrt{2}} & 0 \\ 0 & 0 & 1 \end{vmatrix} = 1 \begin{vmatrix} \frac{1}{\sqrt{2}} & \frac{1}{\sqrt{2}} \\ \frac{1}{\sqrt{2}} & -\frac{1}{\sqrt{2}} \end{vmatrix} = -\frac{1}{2} - \frac{1}{2} = -1$$

Therefore, the vectors are linearly independent and hence, form a basis for R_3.

We now must show that the vectors are orthogonal.

$$\langle \mathbf{b}_1, \mathbf{b}_2 \rangle = \tfrac{1}{2} - \tfrac{1}{2} + 0 = 0$$
$$\langle \mathbf{b}_1, \mathbf{b}_3 \rangle = 0 + 0 + 0 = 0$$
$$\langle \mathbf{b}_2, \mathbf{b}_3 \rangle = 0 + 0 + 0 = 0$$

Therefore, they are orthogonal. Now we must show that they are unit vectors.

$$\|\mathbf{b}_1\| = \sqrt{\langle \mathbf{b}_1, \mathbf{b}_1 \rangle} = \sqrt{\tfrac{1}{2} + \tfrac{1}{2} + 0} = 1$$
$$\|\mathbf{b}_2\| = \sqrt{\langle \mathbf{b}_2, \mathbf{b}_2 \rangle} = \sqrt{\tfrac{1}{2} + \tfrac{1}{2} + 0} = 1$$
$$\|\mathbf{b}_3\| = \sqrt{\langle \mathbf{b}_3, \mathbf{b}_3 \rangle} = \sqrt{0 + 0 + 1} = 1$$

Each of the vectors has length one; hence, the set of the three vectors is an orthonormal basis or R_3.

D. Linear combination of orthonormal basis vectors

If $B = \{\mathbf{b}_1, \mathbf{b}_2, \mathbf{b}_3, \ldots, \mathbf{b}_n\}$ is an orthonormal basis for an inner product space V and $\mathbf{b}$ is any vector in V, then the unique linear representation of $\mathbf{b}$ in terms of B is

$$\mathbf{b} = \langle \mathbf{b}, \mathbf{b}_1 \rangle \mathbf{b}_1 + \langle \mathbf{b}, \mathbf{b}_2 \rangle \mathbf{b}_2 + \langle \mathbf{b}, \mathbf{b}_3 \rangle \mathbf{b}_3 + \cdots + \langle \mathbf{b}, \mathbf{b}_n \rangle \mathbf{b}_n.$$

EXAMPLE 11-6: Express the vector $\mathbf{b} = [1 \quad 2 \quad 3]$ as a linear combination of the basis vectors in Example 11-5.

Solution

$$\langle \mathbf{b}, \mathbf{b}_1 \rangle = 1\left(\frac{1}{\sqrt{2}}\right) + 2\left(\frac{1}{\sqrt{2}}\right) + 3(0) = \frac{3}{\sqrt{2}}$$

$$\langle \mathbf{b}, \mathbf{b}_2 \rangle = 1\left(\frac{1}{\sqrt{2}}\right) + 2\left(-\frac{1}{\sqrt{2}}\right) + 3(0) = -\frac{1}{\sqrt{2}}$$

$$\langle \mathbf{b}, \mathbf{b}_3 \rangle = 1(0) + 2(0) + 3(1) = 3$$

$$\therefore \quad [1 \quad 2 \quad 3] = \frac{3}{\sqrt{2}}\begin{bmatrix} \frac{1}{\sqrt{2}} & \frac{1}{\sqrt{2}} & 0 \end{bmatrix} - \frac{1}{\sqrt{2}}\begin{bmatrix} \frac{1}{\sqrt{2}} & -\frac{1}{\sqrt{2}} & 0 \end{bmatrix} + 3[0 \quad 0 \quad 1]$$

11-2. Orthogonal Projection

If V is an inner product space and $B = \{\mathbf{b}_1, \mathbf{b}_2, \ldots, \mathbf{b}_n\}$ is an orthogonal basis for a subspace W of V, then the **orthogonal projection** of any vector $\mathbf{a}$ in V onto W is given by

$$\text{proj}_W \mathbf{a} = \frac{\langle \mathbf{b}_1, \mathbf{a} \rangle}{\langle \mathbf{b}_1, \mathbf{b}_1 \rangle}\mathbf{b}_1 + \frac{\langle \mathbf{b}_2, \mathbf{a} \rangle}{\langle \mathbf{b}_2, \mathbf{b}_2 \rangle}\mathbf{b}_2 + \cdots + \frac{\langle \mathbf{b}_n, \mathbf{a} \rangle}{\langle \mathbf{b}_n, \mathbf{b}_n \rangle}\mathbf{b}_n$$

If B is an orthonormal basis, the orthogonal projection is given by

$$\text{proj}_W \mathbf{a} = \langle \mathbf{b}_1, \mathbf{a} \rangle \mathbf{b}_1 + \langle \mathbf{b}_2, \mathbf{a} \rangle \mathbf{b}_2 + \cdots + \langle \mathbf{b}_n, \mathbf{a} \rangle \mathbf{b}_n$$

The vector $\mathbf{a} - \text{proj}_W \mathbf{a}$ is orthogonal to each vector in W.

EXAMPLE 11-7: Let V be R_2 with the standard inner project. Let $\mathbf{a}$ be the vector from point 0 to point A and $\mathbf{b}$ the vector from point 0 to point B. If W is the subspace of R_2 consisting of all scalar multiples of $\mathbf{b}$, find the orthogonal projection of $\mathbf{a}$ onto W. (If, as in this case, W is the span of a single nonzero vector, we usually say the orthogonal projection of $\mathbf{a}$ onto $\mathbf{b}$ and denote it by $\text{proj}_\mathbf{b}\, \mathbf{a}$.)

Solution

Method 1 (*trigonometric method*) See Figure 11.1. Let C be the foot of the perpendicular from point A to line OB, and ϕ the angle between vectors $\mathbf{a}$ and $\mathbf{b}$. $\mathbf{OC} = \mathbf{c}$ is the orthogonal projection of $\mathbf{a}$ onto $\mathbf{b}$. From right-triangle trigonometry,

$$\cos\phi = \frac{\|\mathbf{c}\|}{\|\mathbf{a}\|} \quad \text{or} \quad \|\mathbf{c}\| = \|\mathbf{a}\| \cos\phi$$

Since a unit vector in the direction of $\mathbf{c}$ is $\mathbf{b}/\|\mathbf{b}\|$, then

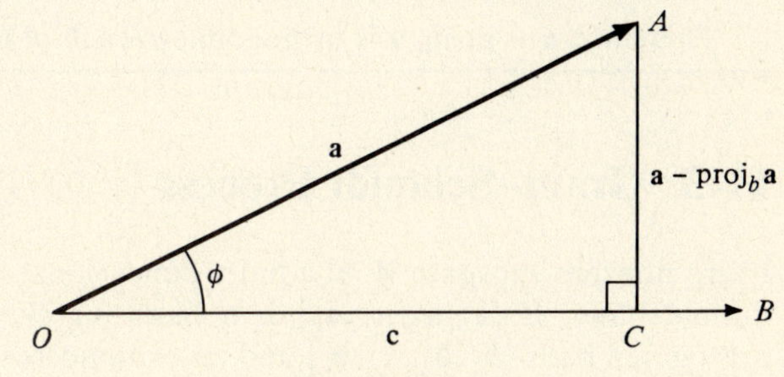

FIGURE 11-1

$$\mathbf{c} = \text{proj}_{\mathbf{b}}\,\mathbf{a} = (\|\mathbf{a}\| \cos \phi)\frac{\mathbf{b}}{\|\mathbf{b}\|}$$

This can be rewritten as

$$\mathbf{c} = \text{proj}_{\mathbf{b}}\,\mathbf{a} = \frac{\|\mathbf{a}\|\,\|\mathbf{b}\| \cos \phi}{\|\mathbf{b}\|^2}\mathbf{b}$$

Method 2

$$\text{proj}_{\mathbf{b}}\,\mathbf{a} = \frac{\langle \mathbf{b}, \mathbf{a} \rangle}{\langle \mathbf{b}, \mathbf{b} \rangle}\mathbf{b}$$

If you recall that $\langle \mathbf{b}, \mathbf{a} \rangle = \|\mathbf{b}\|\,\|\mathbf{a}\| \cos \phi$ and $\|\mathbf{b}\| = \sqrt{\langle \mathbf{b}, \mathbf{b} \rangle}$, you will see that this becomes

$$\text{proj}_{\mathbf{b}}\,\mathbf{a} = \frac{\|\mathbf{b}\|\,\|\mathbf{a}\| \cos \phi}{\|\mathbf{b}\|^2}\mathbf{b} = \mathbf{c}$$

and the two methods are equivalent.

EXAMPLE 11-8: Let V be R_4 with the standard inner product and W be the subspace of V of which $\mathbf{b}_1 = [1 \;\; 1 \;\; 0 \;\; 0]$, $\mathbf{b}_2 = [1 \;\; -1 \;\; 0 \;\; 0]$, $\mathbf{b}_3 = [0 \;\; 0 \;\; 1 \;\; 0]$ is an orthogonal basis. **(a)** Find the orthogonal projection of $\mathbf{a} = [1 \;\; 2 \;\; 3 \;\; -1]$ onto W. **(b)** Show that $\mathbf{a} - \text{proj}_W\,\mathbf{a}$ is orthogonal to each of the vectors $\mathbf{b}_1, \mathbf{b}_2, \mathbf{b}_3$.

Solution

(a)

$$\text{proj}_W\,\mathbf{a} = \frac{\langle \mathbf{b}_1, \mathbf{a} \rangle}{\langle \mathbf{b}_1, \mathbf{b}_1 \rangle}\mathbf{b}_1 + \frac{\langle \mathbf{b}_2, \mathbf{a} \rangle}{\langle \mathbf{b}_2, \mathbf{b}_2 \rangle}\mathbf{b}_2 + \frac{\langle \mathbf{b}_3, \mathbf{a} \rangle}{\langle \mathbf{b}_3, \mathbf{b}_3 \rangle}\mathbf{b}_3$$

$$= \frac{[1(1) + 1(2) + 0(3) + 0(-1)]}{1^2 + 1^2 + 0^2 + 0^2}\mathbf{b}_1 + \frac{[1(1) + (-1)(2) + 0(3) + 0(-1)}{1^2 + (-1)^2 + 0^2 + 0^2}\mathbf{b}_2$$

$$+ \frac{[0(1) + 0(2) + 1(3) + 0(-1)]}{0^2 + 0^2 + 1^2 + 0^2}\mathbf{b}_3$$

$$= \tfrac{3}{2}\mathbf{b}_1 - \tfrac{1}{2}\mathbf{b}_2 + 3\mathbf{b}_3$$

$$= \tfrac{3}{2}[1 \;\; 1 \;\; 0 \;\; 0] - \tfrac{1}{2}[1 \;\; -1 \;\; 0 \;\; 0] + 3[0 \;\; 0 \;\; 1 \;\; 0] = [1 \;\; 2 \;\; 3 \;\; 0]$$

(b)

$$\mathbf{a} - \text{proj}_W\,\mathbf{a} = [1 \;\; 2 \;\; 3 \;\; -1] - [1 \;\; 2 \;\; 3 \;\; 0]$$

$$= [0 \;\; 0 \;\; 0 \;\; -1]$$

$$\langle (\mathbf{a} - \text{proj}_W\,\mathbf{a}), \mathbf{b}_1 \rangle = 0$$

$$\langle (\mathbf{a} - \text{proj}_W\,\mathbf{a}), \mathbf{b}_2 \rangle = 0$$

$$\langle (\mathbf{a} - \text{proj}_W\,\mathbf{a}), \mathbf{b}_3 \rangle = 0$$

Therefore, $\mathbf{a} - \text{proj}_W\,\mathbf{a}$ is orthogonal to each of the vectors $\mathbf{b}_1, \mathbf{b}_2, \mathbf{b}_3$.

11-3. Gram–Schmidt Process

Every nonzero subspace W of a finite dimensional inner product space V has orthogonal and orthonormal bases. If $\{\mathbf{a}_1, \mathbf{a}_2, \ldots, \mathbf{a}_n\}$ is a basis for W, then the **Gram–Schmidt process** for finding an orthogonal basis $\{\mathbf{b}_1, \mathbf{b}_2, \ldots, \mathbf{b}_n\}$ and an orthonormal basis $\{\mathbf{c}_1, \mathbf{c}_2, \ldots, \mathbf{c}_n\}$ is as follows:

$$\mathbf{b}_1 = \mathbf{a}_1$$

$$\mathbf{b}_2 = \mathbf{a}_2 - \frac{\langle \mathbf{a}_2, \mathbf{b}_1 \rangle}{\langle \mathbf{b}_1, \mathbf{b}_1 \rangle}\mathbf{b}_1$$

$$\mathbf{b}_3 = \mathbf{a}_3 - \frac{\langle \mathbf{a}_3, \mathbf{b}_1 \rangle}{\langle \mathbf{b}_1, \mathbf{b}_1 \rangle}\mathbf{b}_1 - \frac{\langle \mathbf{a}_3, \mathbf{b}_2 \rangle}{\langle \mathbf{b}_2, \mathbf{b}_2 \rangle}\mathbf{b}_2$$

$$\vdots$$

$$\mathbf{b}_n = \mathbf{a}_n - \frac{\langle \mathbf{a}_n, \mathbf{b}_1 \rangle}{\langle \mathbf{b}_1, \mathbf{b}_1 \rangle}\mathbf{b}_1 - \frac{\langle \mathbf{a}_n, \mathbf{b}_2 \rangle}{\langle \mathbf{b}_2, \mathbf{b}_2 \rangle}\mathbf{b}_2 - \cdots - \frac{\langle \mathbf{a}_n, \mathbf{b}_{n-1} \rangle}{\langle \mathbf{b}_{n-1}, \mathbf{b}_{n-1} \rangle}\mathbf{b}_{n-1}$$

$$\mathbf{c}_1 = \frac{\mathbf{b}_1}{\|\mathbf{b}_1\|}, \mathbf{c}_2 = \frac{\mathbf{b}_2}{\|\mathbf{b}_2\|}, \ldots, \mathbf{c}_n = \frac{\mathbf{b}_n}{\|\mathbf{b}_n\|}$$

EXAMPLE 11-9: Let R_2 have the standard inner product. Use the Gram–Schmidt process to transform the basis $\{\mathbf{a}_1, \mathbf{a}_2\}$ where $\mathbf{a}_1 = [1 \quad 2]$ and $\mathbf{a}_2 = [-3 \quad 4]$ into an (**a**) orthogonal basis and (**b**) into an orthonormal basis.

Solution

(**a**) Let $\mathbf{b}_1 = \mathbf{a}_1 = [1 \quad 2]$. Then

$$\begin{aligned}\mathbf{b}_2 &= \mathbf{a}_2 - \frac{\langle \mathbf{a}_2, \mathbf{b}_1 \rangle}{\langle \mathbf{b}_1, \mathbf{b}_1 \rangle}\mathbf{b}_1 \\ &= [-3 \quad 4] - \frac{\langle [-3 \quad 4], [1 \quad 2] \rangle}{\langle [1 \quad 2], [1 \quad 2] \rangle}[1 \quad 2] \\ &= [-3 \quad 4] - \frac{(-3)(1) + 4(2)}{1^2 + 2^2}[1 \quad 2] \\ &= [-3 \quad 4] - 5/5[1 \quad 2] = [-4 \quad 2]\end{aligned}$$

Therefore, $\{[1 \quad 2], [-4 \quad 2]\}$ is an orthogonal basis of R_2. Since $[-4 \quad 2] = 2[-2 \quad 1]$, then $\{[1 \quad 2], [-2 \quad 1]\}$ is also an orthogonal basis.

(**b**)

$$\mathbf{c}_1 = \frac{\mathbf{b}_1}{\|\mathbf{b}_1\|} = \frac{[1 \quad 2]}{\|[1 \quad 2]\|} = \left[\frac{1}{\sqrt{5}} \quad \frac{2}{\sqrt{5}}\right]$$

$$\mathbf{c}_2 = \frac{\mathbf{b}_2}{\|\mathbf{b}_2\|} = \frac{[-2 \quad 1]}{\|[-2 \quad 1]\|} = \left[-\frac{2}{\sqrt{5}} \quad \frac{1}{\sqrt{5}}\right]$$

Therefore, $\{[1/\sqrt{5} \quad 2/\sqrt{5}], [-2/\sqrt{5} \quad 1/\sqrt{5}]\}$ is an orthonormal basis of R_2.

EXAMPLE 11-10: Let R_3 have the standard inner product. Let $\{\mathbf{a}_1, \mathbf{a}_2\}$ where $\mathbf{a}_1 = [1 \quad 1 \quad 0]$ and $\mathbf{a}_2 = [2 \quad -1 \quad 3]$ be a basis of a subspace W of R_3. Use the Gram–Schmidt process to transform the basis into an (**a**) orthogonal basis and (**b**) into an orthonormal basis.

Solution

(**a**) Let $\mathbf{b}_1 = \mathbf{a}_1 = [1 \quad 1 \quad 0]$. Then

$$\begin{aligned}\mathbf{b}_2 &= \mathbf{a}_2 - \frac{\langle \mathbf{a}_2, \mathbf{b}_1 \rangle}{\langle \mathbf{b}_1, \mathbf{b}_1 \rangle}\mathbf{b}_1 \\ &= [2 \quad -1 \quad 3] - \frac{2(1) + (-1)(1) + 3(0)}{1^2 + 1^2 + 0^2}[1 \quad 1 \quad 0] \\ &= [2 \quad -1 \quad 3] - \tfrac{1}{2}[1 \quad 1 \quad 0] = [\tfrac{3}{2} \quad -\tfrac{3}{2} \quad 3]\end{aligned}$$

We can choose $\mathbf{b}_2$ as $[\frac{1}{2} \quad -\frac{1}{2} \quad 1]$ or $[1 \quad -1 \quad 2]$ since each of these is a scalar multiple of $[\frac{3}{2} \quad -\frac{3}{2} \quad 3]$. Therefore, $\{[1 \quad 1 \quad 0], [1 \quad -1 \quad 2]\}$ is an orthogonal basis of W.

(**b**)

$$\mathbf{c}_1 = \frac{\mathbf{b}_1}{\|\mathbf{b}_1\|} = \frac{[1 \quad 1 \quad 0]}{\|[1 \quad 1 \quad 0]\|} = \left[\frac{1}{\sqrt{2}} \quad \frac{1}{\sqrt{2}} \quad 0\right]$$

$$\mathbf{c}_2 = \frac{\mathbf{b}_2}{\|\mathbf{b}_2\|} = \frac{[1 \quad -1 \quad 2]}{\|[1 \quad -1 \quad 2]\|} = \left[\frac{1}{\sqrt{6}} \quad -\frac{1}{\sqrt{6}} \quad \frac{2}{\sqrt{6}}\right]$$

Therefore, $\{[1/\sqrt{2} \quad 1/\sqrt{2} \quad 0], [1/\sqrt{6} \quad -1/\sqrt{6} \quad 2/\sqrt{6}]\}$ is an orthonormal basis of W.

EXAMPLE 11-11: Let R_4 have the standard inner product and $\{\mathbf{a}_1, \mathbf{a}_2, \mathbf{a}_3, \mathbf{a}_4\}$ where $\mathbf{a}_1 = [1\ \ 1\ \ 0\ \ 0]$, $\mathbf{a}_2 = [0\ \ 1\ \ 1\ \ 0]$, $\mathbf{a}_3 = [0\ \ 0\ \ 1\ \ 1]$, and $\mathbf{a}_4 = [1\ \ 0\ \ 0\ \ 2]$ be a basis for R_4. Use the Gram–Schmidt process to transform the given basis into an orthonormal basis for R_4.

Solution: Let $\mathbf{b}_1 = \mathbf{a}_1$. Then

$$\mathbf{b}_1 = \mathbf{a}_1 = [1\ \ 1\ \ 0\ \ 0]$$

$$\begin{aligned}
\mathbf{b}_2 &= \mathbf{a}_2 - \frac{\langle \mathbf{a}_2, \mathbf{b}_1 \rangle}{\langle \mathbf{b}_1, \mathbf{b}_1 \rangle}\mathbf{b}_1 \\
&= [0\ \ 1\ \ 1\ \ 0] - \frac{\langle [0\ \ 1\ \ 1\ \ 0], [1\ \ 1\ \ 0\ \ 0] \rangle}{\langle [1\ \ 1\ \ 0\ \ 0], [1\ \ 1\ \ 0\ \ 0] \rangle}[1\ \ 1\ \ 0\ \ 0] \\
&= [0\ \ 1\ \ 1\ \ 0] - \frac{0(1) + 1(1) + 1(0) + 0(0)}{1^2 + 1^2 + 0^2 + 0^2}[1\ \ 1\ \ 0\ \ 0] \\
&= [0\ \ 1\ \ 1\ \ 0] - \tfrac{1}{2}[1\ \ 1\ \ 0\ \ 0] \\
&= [-\tfrac{1}{2}\ \ \tfrac{1}{2}\ \ 1\ \ 0]
\end{aligned}$$

Since all we need at this stage in the process is a vector orthogonal to $\mathbf{b}_1$, and most of us would prefer to work with integers, let us choose $\mathbf{b}_2 = [-1\ \ 1\ \ 2\ \ 0]$, which is a scalar multiple of $[-\tfrac{1}{2}\ \ \tfrac{1}{2}\ \ 1\ \ 0]$.

$$\begin{aligned}
\mathbf{b}_3 &= \mathbf{a}_3 - \frac{\langle \mathbf{a}_3, \mathbf{b}_1 \rangle}{\langle \mathbf{b}_1, \mathbf{b}_1 \rangle}\mathbf{b}_1 - \frac{\langle \mathbf{a}_3, \mathbf{b}_2 \rangle}{\langle \mathbf{b}_2, \mathbf{b}_2 \rangle}\mathbf{b}_2 \\
&= [0\ \ 0\ \ 1\ \ 1] - \frac{\langle [0\ \ 0\ \ 1\ \ 1], [1\ \ 1\ \ 0\ \ 0] \rangle}{\langle [1\ \ 1\ \ 0\ \ 0], [1\ \ 1\ \ 0\ \ 0] \rangle}[1\ \ 1\ \ 0\ \ 0] \\
&\quad - \frac{\langle [0\ \ 0\ \ 1\ \ 1], [-1\ \ 1\ \ 2\ \ 0] \rangle}{\langle [-1\ \ 1\ \ 2\ \ 0], [-1\ \ 1\ \ 2\ \ 0] \rangle}[-1\ \ 1\ \ 2\ \ 0] \\
&= [0\ \ 0\ \ 1\ \ 1] - 0[1\ \ 1\ \ 0\ \ 0] - \tfrac{2}{6}[-1\ \ 1\ \ 2\ \ 0] \\
&= [\tfrac{1}{3}\ \ -\tfrac{1}{3}\ \ \tfrac{1}{3}\ \ 1]
\end{aligned}$$

This time, choose $\mathbf{b}_3 = [1\ \ -1\ \ 1\ \ 3]$ which is a scalar multiple of $[\tfrac{1}{3}\ \ -\tfrac{1}{3}\ \ \tfrac{1}{3}\ \ 1]$.

$$\begin{aligned}
\mathbf{b}_4 &= \mathbf{a}_4 - \frac{\langle \mathbf{a}_4, \mathbf{b}_1 \rangle}{\langle \mathbf{b}_1, \mathbf{b}_1 \rangle}\mathbf{b}_1 - \frac{\langle \mathbf{a}_4, \mathbf{b}_2 \rangle}{\langle \mathbf{b}_2, \mathbf{b}_2 \rangle}\mathbf{b}_2 - \frac{\langle \mathbf{a}_4, \mathbf{b}_3 \rangle}{\langle \mathbf{b}_3, \mathbf{b}_3 \rangle}\mathbf{b}_3 \\
&= [1\ \ 0\ \ 0\ \ 2] - \frac{\langle [1\ \ 0\ \ 0\ \ 2], [1\ \ 1\ \ 0\ \ 0] \rangle}{\langle [1\ \ 1\ \ 0\ \ 0], [1\ \ 1\ \ 0\ \ 0] \rangle}[1\ \ 1\ \ 0\ \ 0] \\
&\quad - \frac{\langle [1\ \ 0\ \ 0\ \ 2], [-1\ \ 1\ \ 2\ \ 0] \rangle}{\langle [-1\ \ 1\ \ 2\ \ 0], [-1\ \ 1\ \ 2\ \ 0] \rangle}[-1\ \ 1\ \ 2\ \ 0] \\
&\quad - \frac{\langle [1\ \ 0\ \ 0\ \ 2], [1\ \ -1\ \ 1\ \ 3] \rangle}{\langle [1\ \ -1\ \ 1\ \ 3], [1\ \ -1\ \ 1\ \ 3] \rangle}[1\ \ -1\ \ 1\ \ 3] \\
&= [1\ \ 0\ \ 0\ \ 2] - \tfrac{1}{2}[1\ \ 1\ \ 0\ \ 0] + \tfrac{1}{6}[-1\ \ 1\ \ 2\ \ 0] - \tfrac{7}{12}[1\ \ -1\ \ 1\ \ 3] \\
&= [-\tfrac{3}{12}\ \ \tfrac{3}{12}\ \ -\tfrac{3}{12}\ \ \tfrac{3}{12}] = [-\tfrac{1}{4}\ \ \tfrac{1}{4}\ \ -\tfrac{1}{4}\ \ \tfrac{1}{4}]
\end{aligned}$$

This time, choose $\mathbf{b}_4 = [-1\ \ 1\ \ -1\ \ 1]$. Therefore $\{[1\ \ 1\ \ 0\ \ 0], [-1\ \ 1\ \ 2\ \ 0], [1\ \ -1\ \ 1\ \ 3], [-1\ \ 1\ \ -1\ \ 1]\}$ is an orthogonal basis for R_4.

$$\mathbf{c}_1 = \frac{[1\ \ 1\ \ 0\ \ 0]}{\|[1\ \ 1\ \ 0\ \ 0]\|} = \left[\frac{1}{\sqrt{2}}\ \ \frac{1}{\sqrt{2}}\ \ 0\ \ 0\right]$$

$$\mathbf{c}_2 = \frac{[-1\ \ 1\ \ 2\ \ 0]}{\|[-1\ \ 1\ \ 2\ \ 0]\|} = \left[-\frac{1}{\sqrt{6}}\ \ \frac{1}{\sqrt{6}}\ \ \frac{2}{\sqrt{6}}\ \ 0\right]$$

$$\mathbf{c}_3 = \frac{[1\ \ -1\ \ 1\ \ 3]}{\|[1\ \ -1\ \ 1\ \ 3]\|} = \left[\frac{1}{\sqrt{12}}\ \ -\frac{1}{\sqrt{12}}\ \ \frac{1}{\sqrt{12}}\ \ \frac{3}{\sqrt{12}}\right]$$

$$\mathbf{c}_4 = \frac{[-1 \quad 1 \quad -1 \quad 1]}{\|[-1 \quad 1 \quad -1 \quad 1]\|} = [-\tfrac{1}{2} \quad \tfrac{1}{2} \quad -\tfrac{1}{2} \quad \tfrac{1}{2}]$$

Therefore, $\{\mathbf{c}_1, \mathbf{c}_2, \mathbf{c}_3, \mathbf{c}_4\}$ is an orthonormal basis for R_4.

note: $\{[1 \quad 0 \quad 0 \quad 0], [0 \quad 1 \quad 0 \quad 0], [0 \quad 0 \quad 1 \quad 0], [0 \quad 0 \quad 0 \quad 1]\}$ is an orthonormal basis for R_4. Thus, an orthonormal basis is not unique.

SUMMARY

1. Two vectors **a** and **b** in an inner product space are orthogonal if $\langle \mathbf{a}, \mathbf{b} \rangle = 0$.
2. A set S of vectors in an inner product space V is orthogonal if every pair of distinct vectors in S are orthogonal.
3. An orthogonal set S of vectors in an inner product space are orthonormal if every vector in S is of length one.
4. A finite orthogonal set of nonzero vectors in an inner product space is linearly independent.
5. An orthogonal basis for an inner product space is a basis which is also orthogonal.
6. An orthonormal basis for an inner product space is a basis which is also orthonormal.
7. If $B = \{\mathbf{b}_1, \mathbf{b}_2, \ldots, \mathbf{b}_n\}$ is an orthonormal basis for an inner product space V and **b** is any vector in V, then the unique linear representation of **b** in terms of B is $\mathbf{b} = \langle \mathbf{b}, \mathbf{b}_1 \rangle \mathbf{b}_1 + \langle \mathbf{b}, \mathbf{b}_2 \rangle \mathbf{b}_2 + \cdots + \langle \mathbf{b}, \mathbf{b}_n \rangle \mathbf{b}_n$.
8. If V is an inner product space and $B = \{\mathbf{b}_1, \mathbf{b}_2, \ldots, \mathbf{b}_n\}$ is an orthogonal basis for a subspace W of V, then the orthogonal projection of any vector **a** in V onto W is given by

   $$\text{proj}_W \mathbf{a} = \frac{\langle \mathbf{b}_1, \mathbf{a} \rangle}{\langle \mathbf{b}_1, \mathbf{b}_1 \rangle} \mathbf{b}_1 + \frac{\langle \mathbf{b}_2, \mathbf{a} \rangle}{\langle \mathbf{b}_2, \mathbf{b}_2 \rangle} \mathbf{b}_2 + \cdots + \frac{\langle \mathbf{b}_n, \mathbf{a} \rangle}{\langle \mathbf{b}_n, \mathbf{b}_n \rangle} \mathbf{b}_n$$

   The vector $\mathbf{a} - \text{proj}_W \mathbf{a}$ is orthogonal to each of the vectors in W.
9. Every nonzero subspace W of a finite dimensional inner product space V has orthogonal and orthonormal bases.
10. If $\{\mathbf{a}_1, \mathbf{a}_2, \ldots, \mathbf{a}_n\}$ is a basis for a subspace W of an inner product space V, then the Gram–Schmidt process for finding an orthogonal basis $\{\mathbf{b}_1, \mathbf{b}_2, \ldots, \mathbf{b}_n\}$ and an orthonormal basis $\{\mathbf{c}_1, \mathbf{c}_2, \ldots, \mathbf{c}_n\}$ is

    $$\mathbf{b}_1 = \mathbf{a}_1$$

    $$\mathbf{b}_2 = \mathbf{a}_2 - \frac{\langle \mathbf{a}_2, \mathbf{b}_1 \rangle}{\langle \mathbf{b}_1, \mathbf{b}_1 \rangle} \mathbf{b}_1$$

    $$\mathbf{b}_3 = \mathbf{a}_3 - \frac{\langle \mathbf{a}_3, \mathbf{b}_1 \rangle}{\langle \mathbf{b}_1, \mathbf{b}_1 \rangle} \mathbf{b}_1 - \frac{\langle \mathbf{a}_3, \mathbf{b}_2 \rangle}{\langle \mathbf{b}_2, \mathbf{b}_2 \rangle} \mathbf{b}_2$$

    $$\vdots$$

    $$\mathbf{b}_n = \mathbf{a}_n - \frac{\langle \mathbf{a}_n, \mathbf{b}_1 \rangle}{\langle \mathbf{b}_1, \mathbf{b}_1 \rangle} \mathbf{b}_1 - \frac{\langle \mathbf{a}_n, \mathbf{b}_2 \rangle}{\langle \mathbf{b}_2, \mathbf{b}_2 \rangle} \mathbf{b}_2 - \cdots - \frac{\langle \mathbf{a}_n, \mathbf{b}_{n-1} \rangle}{\langle \mathbf{b}_{n-1}, \mathbf{b}_{n-1} \rangle} \mathbf{b}_{n-1}$$

    $$\mathbf{c}_1 = \frac{\mathbf{b}_1}{\|\mathbf{b}_1\|}, \mathbf{c}_2 = \frac{\mathbf{b}_2}{\|\mathbf{b}_2\|}, \ldots, \mathbf{c}_n = \frac{\mathbf{b}_n}{\|\mathbf{b}_n\|}$$
11. Orthogonal and orthonormal bases for an inner product space are not unique.

RAISE YOUR GRADES

Can you ...?

☑ determine if two vectors are orthogonal

☑ determine if two vectors are orthonormal

☑ determine whether a set of vectors is orthogonal or orthonormal

☑ find the orthogonal projection of any vector **a** in an inner product space V onto a subspace W
☑ find orthogonal and orthonormal bases by the Gram–Schmidt process

SOLVED PROBLEMS

PROBLEM 11-1 Determine whether each of the following pairs of vectors from R_2 with the standard inner product is orthonormal: **(a)** $[1\ \ -2]$, $[-2\ \ -1]$; **(b)** $[1\ \ 0]$, $[0\ \ -1]$.

Solution

(a) $\langle[1\ \ -2],[-2\ \ -1]\rangle = -2+2=0$. Therefore, the set is orthogonal. $\|[1\ \ -2]\| = \sqrt{1^2+2^2} = \sqrt{5}$. Therefore, the set is not orthonormal.

(b) $\langle[1\ \ 0],[0\ \ -1]\rangle = 0+0$. Therefore, the set is orthogonal.

$$\|[1\ \ 0]\| = \sqrt{1^2+0^2} = 1$$

$$\|[0\ \ -1]\| = \sqrt{0^2+(-1)^2} = 1$$

Therefore, the set is also orthonormal.

PROBLEM 11-2 Let R_3 have the standard inner product. Classify each of the following sets as orthogonal, orthonormal, or nonorthogonal.

(a)
$$\left\{\left[\frac{1}{\sqrt{3}}\ \ 0\ \ -\frac{1}{\sqrt{3}}\right], [0\ \ 1\ \ 0], \left[\frac{\sqrt{2}}{\sqrt{5}}\ \ 0\ \ \frac{\sqrt{2}}{\sqrt{5}}\right]\right\}$$

(b)
$$\left\{\left[-\frac{1}{\sqrt{5}}\ \ 0\ \ \frac{2}{\sqrt{5}}\right], [0\ \ 1\ \ 0], \left[-\frac{2}{\sqrt{5}}\ \ 0\ \ -\frac{1}{\sqrt{5}}\right]\right\}$$

Solution

(a)
$$\left\langle\left[\frac{1}{\sqrt{3}}\ \ 0\ \ -\frac{1}{\sqrt{3}}\right],\quad [0\ \ 1\ \ 0]\right\rangle = 0+0+0=0$$

$$\left\langle\left[\frac{1}{\sqrt{3}}\ \ 0\ \ -\frac{1}{\sqrt{3}}\right],\quad \left[\frac{\sqrt{2}}{\sqrt{5}}\ \ 0\ \ \frac{\sqrt{2}}{\sqrt{5}}\right]\right\rangle = \frac{\sqrt{2}}{\sqrt{15}} - \frac{\sqrt{2}}{\sqrt{15}} = 0$$

$$\left\langle[0\ \ 1\ \ 0],\quad \left[\frac{\sqrt{2}}{\sqrt{5}}\ \ 0\ \ \frac{\sqrt{2}}{\sqrt{5}}\right]\right\rangle = 0+0+0=0$$

Therefore, the set is orthogonal.

$$\left\|\left[\frac{1}{\sqrt{3}}\ \ 0\ \ -\frac{1}{\sqrt{3}}\right]\right\| = \sqrt{\left(\frac{1}{\sqrt{3}}\right)^2 + 0^2 + \left(-\frac{1}{\sqrt{3}}\right)^2} = \sqrt{\frac{1}{3}+\frac{1}{3}} = \sqrt{\frac{2}{3}} \neq 1$$

$$\|[0\ \ 1\ \ 0]\| = \sqrt{0^2+1^2+0^2} = 1$$

$$\left\|\left[\frac{\sqrt{2}}{\sqrt{5}}\ \ 0\ \ \frac{\sqrt{2}}{\sqrt{5}}\right]\right\| = \sqrt{\frac{4}{5}+\frac{4}{5}} = \sqrt{\frac{8}{5}} \neq 1$$

Therefore, the set is not orthonormal.

(b)
$$\left\langle\left[-\frac{1}{\sqrt{5}}\ \ 0\ \ \frac{2}{\sqrt{5}}\right],\quad [0\ \ 1\ \ 0]\right\rangle = 0+0+0=0$$

$$\left\langle\left[-\frac{1}{\sqrt{5}}\ \ 0\ \ \frac{2}{\sqrt{5}}\right],\quad \left[-\frac{2}{\sqrt{5}}\ \ 0\ \ -\frac{1}{\sqrt{5}}\right]\right\rangle = \frac{2}{5}+0-\frac{2}{5}=0$$

$$\left\langle [0 \quad 1 \quad 0], \quad \left[-\frac{2}{\sqrt{5}} \quad 0 \quad -\frac{1}{\sqrt{5}}\right]\right\rangle = 0 + 0 + 0 = 0$$

$$\left\|\left[-\frac{1}{\sqrt{5}} \quad 0 \quad \frac{2}{\sqrt{5}}\right]\right\| = \sqrt{\frac{1}{5} + 0 + \frac{4}{5}} = 1$$

$$\|[0 \quad 1 \quad 0]\| = \sqrt{0 + 1 + 0} = 1$$

$$\left\|\left[-\frac{2}{\sqrt{5}} \quad 0 \quad -\frac{1}{\sqrt{5}}\right]\right\| = \sqrt{\frac{4}{5} + 0 + \frac{1}{5}} = 1$$

Therefore, the set is orthogonal and orthonormal.

PROBLEM 11-3 Consider the vector space ${}_2M_2$ of all 2×2 matrices with inner product defined by $\langle A, B\rangle = a_{11}b_{11} + a_{12}b_{12} + a_{21}b_{21} + a_{22}b_{22}$. Show that the matrices

$$A = \begin{bmatrix} \frac{1}{\sqrt{3}} & \frac{1}{\sqrt{3}} \\ \frac{-1}{\sqrt{3}} & 0 \end{bmatrix}, \qquad B = \begin{bmatrix} \frac{1}{\sqrt{6}} & 0 \\ \frac{1}{\sqrt{6}} & -\frac{2}{\sqrt{6}} \end{bmatrix}, \qquad C = \begin{bmatrix} 0 & \frac{2}{3} \\ \frac{2}{3} & \frac{1}{3} \end{bmatrix}$$

form an orthonormal set.

Solution

$$\langle A, B\rangle = \frac{1}{\sqrt{3}}\left(\frac{1}{\sqrt{6}}\right) + \frac{1}{\sqrt{3}}(0) - \frac{1}{\sqrt{3}}\left(\frac{1}{\sqrt{6}}\right) + 0\left(-\frac{2}{\sqrt{6}}\right) = 0$$

$$\langle A, C\rangle = \frac{1}{\sqrt{3}}(0) + \frac{1}{\sqrt{3}}\left(\frac{2}{3}\right) - \frac{1}{\sqrt{3}}\left(\frac{2}{3}\right) + 0\left(\frac{1}{3}\right) = 0$$

$$\langle B, C\rangle = \frac{1}{\sqrt{6}}(0) + 0\left(\frac{2}{3}\right) + \frac{1}{\sqrt{6}}\left(\frac{2}{3}\right) - \frac{2}{\sqrt{6}}\left(\frac{1}{3}\right) = 0$$

$$\|A\| = \sqrt{\left(\frac{1}{\sqrt{3}}\right)^2 + \left(\frac{1}{\sqrt{3}}\right)^2 + \left(-\frac{1}{\sqrt{3}}\right)^2} = \sqrt{\frac{3}{3}} = 1$$

$$\|B\| = \sqrt{\left(\frac{1}{\sqrt{6}}\right)^2 + \left(\frac{1}{\sqrt{6}}\right)^2 + \left(-\frac{2}{\sqrt{6}}\right)^2} = \sqrt{\frac{6}{6}} = 1$$

$$\|C\| = \sqrt{\left(\frac{2}{3}\right)^2 + \left(\frac{2}{3}\right)^2 + \left(\frac{1}{3}\right)^2} = \sqrt{\frac{9}{9}} = 1$$

Therefore, the set is orthonormal.

PROBLEM 11-4 Let P_2 have the standard inner product $\langle \mathbf{p}_1, \mathbf{p}_2\rangle = a_0b_0 + a_1b_1 + a_2b_2$. Show that the set of polynomials $\mathbf{p}_1 = \frac{2}{3} + \frac{2}{3}x + \frac{1}{3}x^2$, $\mathbf{p}_2 = \frac{1}{3} - \frac{2}{3}x + \frac{2}{3}x^2$, $\mathbf{p}_3 = -\frac{2}{3} + \frac{1}{3}x + \frac{2}{3}x^2$ form an orthonormal set.

Solution

$$\langle \mathbf{p}_1, \mathbf{p}_2\rangle = \tfrac{2}{3}(\tfrac{1}{3}) + \tfrac{2}{3}(-\tfrac{2}{3}) + \tfrac{1}{3}(\tfrac{2}{3}) = 0$$

$$\langle \mathbf{p}_1, \mathbf{p}_3\rangle = \tfrac{2}{3}(-\tfrac{2}{3}) + \tfrac{2}{3}(\tfrac{1}{3}) + \tfrac{1}{3}(\tfrac{2}{3}) = 0$$

$$\langle \mathbf{p}_2, \mathbf{p}_3\rangle = \tfrac{1}{3}(-\tfrac{2}{3}) - \tfrac{2}{3}(\tfrac{1}{3}) + \tfrac{2}{3}(\tfrac{2}{3}) = 0$$

$$\|\langle \mathbf{p}_1, \mathbf{p}_1\rangle\| = \sqrt{(\tfrac{2}{3})^2 + (\tfrac{2}{3})^2 + (\tfrac{1}{3})^2} = \sqrt{\tfrac{9}{9}} = 1$$

$$\|\langle \mathbf{p}_2, \mathbf{p}_2\rangle\| = \sqrt{(\tfrac{1}{3})^2 + (-\tfrac{2}{3})^2 + (\tfrac{2}{3})^2} = \sqrt{\tfrac{9}{9}} = 1$$

$$\|\langle \mathbf{p}_3, \mathbf{p}_3\rangle\| = \sqrt{(-\tfrac{2}{3})^2 + (\tfrac{1}{3})^2 + (\tfrac{2}{3})^2} = 1$$

Therefore, the set forms an orthonormal basis of P_2.

PROBLEM 11-5 Show that

$$\begin{bmatrix} -\frac{1}{\sqrt{5}} & 0 & \frac{2}{\sqrt{5}} \end{bmatrix}, \qquad [0 \quad 1 \quad 0], \qquad \begin{bmatrix} -\frac{2}{\sqrt{5}} & 0 & -\frac{1}{\sqrt{5}} \end{bmatrix}$$

are linearly independent.

Solution Choose scalars a, b, c such that

$$a\begin{bmatrix} -\frac{1}{\sqrt{5}} & 0 & \frac{2}{\sqrt{5}} \end{bmatrix} + b[0 \quad 1 \quad 0] + c\begin{bmatrix} -\frac{2}{\sqrt{5}} & 0 & -\frac{1}{\sqrt{5}} \end{bmatrix} = [0 \quad 0 \quad 0]$$

$$\begin{aligned} -\frac{1}{\sqrt{5}}a \qquad -\frac{2}{\sqrt{5}}c &= 0 \\ b \qquad\qquad &= 0 \\ \frac{2}{\sqrt{5}}a \qquad -\frac{1}{\sqrt{5}}c &= 0 \end{aligned}$$

This system has solution $a = 0$, $b = 0$, $c = 0$. Hence, the three vectors are linearly independent.

PROBLEM 11-6 Let R_3 have the standard inner product. Show that $\mathbf{b}_1 = [-1/\sqrt{5} \quad 0 \quad 2/\sqrt{5}]$, $\mathbf{b}_2 = [0 \quad 1 \quad 0]$, and $\mathbf{b}_3 = [-2/\sqrt{5} \quad 0 \quad -1/\sqrt{5}]$ is an orthonormal basis for R_3.

Solution Since the dimension of R_3 is 3 and we have three vectors, we can show linear independence by evaluating the determinant

$$\begin{vmatrix} -\frac{1}{\sqrt{5}} & 0 & \frac{2}{\sqrt{5}} \\ 0 & 1 & 0 \\ -\frac{2}{\sqrt{5}} & 0 & -\frac{1}{\sqrt{5}} \end{vmatrix} = \frac{1}{5} + \frac{4}{5} = 1$$

Since this is different from zero the vectors form a basis. From Problem 11-2, part (**b**), the set of vectors is also orthonormal. Hence, they form an orthonormal basis for R_3.

PROBLEM 11-7 Express $\mathbf{b} = [2 \quad 1 \quad 4]$ as a linear combination of the basis vectors in Problem 11-6.

Solution

$$\langle \mathbf{b}, \mathbf{b}_1 \rangle = 2\left(-\frac{1}{\sqrt{5}}\right) + 1(0) + 4\left(\frac{2}{\sqrt{5}}\right) = \frac{6}{\sqrt{5}}$$

$$\langle \mathbf{b}, \mathbf{b}_2 \rangle = 2(0) + 1(1) + 4(0) = 1$$

$$\langle \mathbf{b}, \mathbf{b}_3 \rangle = 2\left(-\frac{2}{\sqrt{5}}\right) + 1(0) + 4\left(-\frac{1}{\sqrt{5}}\right) = -\frac{8}{\sqrt{5}}$$

$$\therefore \quad [2 \quad 1 \quad 4] = \frac{6}{\sqrt{5}}\begin{bmatrix} -\frac{1}{\sqrt{5}} & 0 & \frac{2}{\sqrt{5}} \end{bmatrix} + 1[0 \quad 1 \quad 0] - \frac{8}{\sqrt{5}}\begin{bmatrix} -\frac{2}{\sqrt{5}} & 0 & -\frac{1}{\sqrt{5}} \end{bmatrix}$$

PROBLEM 11-8 Let V be R_4 with the standard inner product and W be the subspace of V of which $\mathbf{b}_1 = [1 \quad 2 \quad -1 \quad 1]$, $\mathbf{b}_2 = [0 \quad 1 \quad 2 \quad 0]$, $\mathbf{b}_3 = [1 \quad 0 \quad 0 \quad -1]$ is an orthogonal basis. (**a**) Find $\text{proj}_W \mathbf{a}$ if $\mathbf{a} = [1 \quad -2 \quad 1 \quad 3]$. (**b**) Show that $\mathbf{a} - \text{proj}_W \mathbf{a}$ is orthogonal to each of the vectors $\mathbf{b}_1$, $\mathbf{b}_2$, $\mathbf{b}_3$.

Solution

(a)
$$\text{proj}_W \mathbf{a} = \frac{\langle \mathbf{b}_1, \mathbf{a} \rangle}{\langle \mathbf{b}_1, \mathbf{b}_1 \rangle}\mathbf{b}_1 + \frac{\langle \mathbf{b}_2, \mathbf{a} \rangle}{\langle \mathbf{b}_2, \mathbf{b}_2 \rangle}\mathbf{b}_2 + \frac{\langle \mathbf{b}_3, \mathbf{a} \rangle}{\langle \mathbf{b}_3, \mathbf{b}_3 \rangle}\mathbf{b}_3$$
$$= \frac{[1(1) + 2(-2) + (-1)(1) + 1(3)]}{1^2 + 2^2 + (-1)^2 + 1^2}\mathbf{b}_1 + \frac{[0(1) + 1(-2) + 2(1) + 0(3)]}{0^2 + 1^2 + 2^2 + 0^2}\mathbf{b}_2$$
$$+ \frac{[1(1) + 0(-2) + 0(1) + (-1)(3)]}{1^2 + 0^2 + 0^2 + (-1)^2}\mathbf{b}_3$$
$$= -\tfrac{1}{7}\mathbf{b}_1 + \tfrac{0}{5}\mathbf{b}_2 + -\tfrac{2}{2}\mathbf{b}_3$$
$$= -\tfrac{1}{7}[1 \quad 2 \quad -1 \quad 1] + 0[0 \quad 1 \quad 2 \quad 0] - 1[1 \quad 0 \quad 0 \quad -1]$$
$$= [-\tfrac{8}{7} \quad -\tfrac{2}{7} \quad \tfrac{1}{7} \quad \tfrac{6}{7}]$$

(b)
$$\mathbf{a} - \text{proj}_W \mathbf{a} = [1 \quad -2 \quad 1 \quad 3] - [-\tfrac{8}{7} \quad -\tfrac{2}{7} \quad \tfrac{1}{7} \quad \tfrac{6}{7}]$$
$$= [\tfrac{15}{7} \quad -\tfrac{12}{7} \quad \tfrac{6}{7} \quad \tfrac{15}{7}]$$
$$\langle (\mathbf{a} - \text{proj}_W \mathbf{a}), \mathbf{b}_1 \rangle = 0$$
$$\langle (\mathbf{a} - \text{proj}_W \mathbf{a}), \mathbf{b}_2 \rangle = 0$$
$$\langle (\mathbf{a} - \text{proj}_W \mathbf{a}), \mathbf{b}_3 \rangle = 0$$

Therefore, $\mathbf{a} - \text{proj}_W \mathbf{a}$ is orthogonal to each vector $\mathbf{b}_1$, $\mathbf{b}_2$, $\mathbf{b}_3$.

PROBLEM 11-9 Let R_2 have the standard inner product. Use the Gram–Schmidt process to transform the basis $\{\mathbf{a}_1, \mathbf{a}_2\}$ where $\mathbf{a}_1 = [5 \quad -2]$ and $\mathbf{a}_2 = [-4 \quad 3]$ into an **(a)** orthogonal basis and **(b)** into an orthonormal basis.

Solution

(a) Let $\mathbf{b}_1 = \mathbf{a}_1 = [5 \quad -2]$. Then

$$\mathbf{b}_2 = \mathbf{a}_2 - \frac{\langle \mathbf{a}_2, \mathbf{b}_1 \rangle}{\langle \mathbf{b}_1, \mathbf{b}_1 \rangle}\mathbf{b}_1$$
$$= [-4 \quad 3] - \frac{\langle [-4 \quad 3], [5 \quad -2] \rangle}{\langle [5 \quad -2], [5 \quad -2] \rangle}[5 \quad -2]$$
$$= [-4 \quad 3] - \frac{(-4)(5) + 3(-2)}{5^2 + (-2)^2}[5 \quad -2]$$
$$= [-4 \quad 3] + \tfrac{26}{29}[5 \quad -2] = [\tfrac{14}{29} \quad \tfrac{35}{29}]$$

Therefore, $\{[5 \quad -2], [\tfrac{14}{29} \quad \tfrac{35}{29}]\}$ is an orthogonal basis of R_2. Since $[\tfrac{14}{29} \quad \tfrac{35}{29}] = \tfrac{7}{29}[2 \quad 5]$, then $\{[5 \quad -2], [2 \quad 5]\}$ is also an orthogonal basis.

(b)
$$\mathbf{c}_1 = \frac{\mathbf{b}_1}{\|\mathbf{b}_1\|} = \frac{[5 \quad -2]}{\|[5 \quad -2]\|} = \left[\frac{5}{\sqrt{29}} \quad -\frac{2}{\sqrt{29}}\right]$$
$$\mathbf{c}_2 = \frac{\mathbf{b}_2}{\|\mathbf{b}_2\|} = \frac{[2 \quad 5]}{\|[2 \quad 5]\|} = \left[\frac{2}{\sqrt{29}} \quad \frac{5}{\sqrt{29}}\right]$$

Therefore, $\{[5/\sqrt{29} \quad -2/\sqrt{29}], [2/\sqrt{29} \quad 5/\sqrt{29}]\}$ is an orthonormal basis of R_2.

PROBLEM 11-10 Let R_3 have the standard inner product. Let $\{\mathbf{a}_1, \mathbf{a}_2\}$ where $\mathbf{a}_1 = [1 \quad -1 \quad 1]$ and $\mathbf{a} = [2 \quad 0 \quad -3]$ be a basis of the subspace W of R_3. Use the Gram–Schmidt process to transform the basis into an **(a)** orthogonal basis and **(b)** into an orthonormal basis.

Solution

(a) Let $\mathbf{b}_1 = \mathbf{a}_1 = [1 \quad -1 \quad 1]$. Then

$$\begin{aligned}\mathbf{b}_2 &= \mathbf{a}_2 - \frac{\langle \mathbf{a}_2, \mathbf{b}_1 \rangle}{\langle \mathbf{b}_1, \mathbf{b}_1 \rangle}\mathbf{b}_1 \\ &= [2 \quad 0 \quad -3] - \frac{(2)(1) + 0(-1) + (-3)(1)}{1^2 + (-1)^2 + 1^2}[1 \quad -1 \quad 1] \\ &= [2 \quad 0 \quad -3] + \tfrac{1}{3}[1 \quad -1 \quad 1] = [\tfrac{7}{3} \quad -\tfrac{1}{3} \quad -\tfrac{8}{3}]\end{aligned}$$

We can choose $\mathbf{b}_2$ as $[7 \quad -1 \quad -8]$. Therefore, $\{[1 \quad -1 \quad 1], [7 \quad -1 \quad -8]\}$ is an orthogonal basis of W.

(b)

$$\mathbf{c}_1 = \frac{\mathbf{b}_1}{\|\mathbf{b}_1\|} = \frac{[1 \quad -1 \quad 1]}{\|[1 \quad -1 \quad 1]\|} = \left[\frac{1}{\sqrt{3}} \quad -\frac{1}{\sqrt{3}} \quad \frac{1}{\sqrt{3}}\right]$$

$$\mathbf{c}_2 = \frac{\mathbf{b}_2}{\|\mathbf{b}_2\|} = \frac{[7 \quad -1 \quad -8]}{\|[7 \quad -1 \quad -8]\|} = \left[\frac{7}{\sqrt{114}} \quad -\frac{1}{\sqrt{114}} \quad -\frac{8}{\sqrt{114}}\right]$$

Therefore,

$$\left\{\left[\frac{1}{\sqrt{3}} \quad -\frac{1}{\sqrt{3}} \quad \frac{1}{\sqrt{3}}\right], \left[\frac{7}{\sqrt{114}} \quad -\frac{1}{\sqrt{114}} \quad -\frac{8}{\sqrt{114}}\right]\right\}$$

is an orthonormal basis of W.

PROBLEM 11-11 Let R_3 have the standard inner product and $\{\mathbf{a}_1, \mathbf{a}_2, \mathbf{a}_3\}$ where $\mathbf{a}_1 = [1 \quad 1 \quad 1]$, $\mathbf{a}_2 = [1 \quad 1 \quad 0]$, and $\mathbf{a}_3 = [0 \quad 1 \quad 0]$ be a basis for R_3. Use the Gram–Schmidt process to transform the given basis into an orthonormal basis for R_3.

Solution Let $\mathbf{b}_1 = \mathbf{a}_1$. Then

$$\begin{aligned}\mathbf{b}_1 &= \mathbf{a}_1 = [1 \quad 1 \quad 1] \\ \mathbf{b}_2 &= \mathbf{a}_2 - \frac{\langle \mathbf{a}_2, \mathbf{b}_1 \rangle}{\langle \mathbf{b}_1, \mathbf{b}_1 \rangle}\mathbf{b}_1 \\ &= [1 \quad 1 \quad 0] - \frac{1(1) + 1(1) + 0(1)}{1^2 + 1^2 + 1^2}[1 \quad 1 \quad 1] \\ &= [1 \quad 1 \quad 0] - \tfrac{2}{3}[1 \quad 1 \quad 1] = [\tfrac{1}{3} \quad \tfrac{1}{3} \quad -\tfrac{2}{3}]\end{aligned}$$

Let us choose $\mathbf{b}_2 = [1 \quad 1 \quad -2]$ which is a scalar multiple of $[\tfrac{1}{3} \quad \tfrac{1}{3} \quad -\tfrac{2}{3}]$.

$$\begin{aligned}\mathbf{b}_3 &= \mathbf{a}_3 - \frac{\langle \mathbf{a}_3, \mathbf{b}_1 \rangle}{\langle \mathbf{b}_1, \mathbf{b}_1 \rangle}\mathbf{b}_1 - \frac{\langle \mathbf{a}_3, \mathbf{b}_2 \rangle}{\langle \mathbf{b}_2, \mathbf{b}_2 \rangle}\mathbf{b}_2 \\ &= [0 \quad 1 \quad 0] - \frac{0(1) + 1(1) + 0(1)}{1^2 + 1^2 + 1^2}[1 \quad 1 \quad 1] \\ &\quad - \frac{0(1) + 1(1) + 0(-2)}{1^2 + 1^2 + (-2)^2}[1 \quad 1 \quad -2] \\ &= [0 \quad 1 \quad 0] - \tfrac{1}{3}[1 \quad 1 \quad 1] - \tfrac{1}{6}[1 \quad 1 \quad -2] \\ &= [-\tfrac{3}{6} \quad \tfrac{3}{6} \quad 0]\end{aligned}$$

Choose $\mathbf{b}_3 = [-1 \quad 1 \quad 0]$, a scalar multiple of $[-\tfrac{3}{6} \quad \tfrac{3}{6} \quad 0]$.

$$\mathbf{c}_1 = \frac{\mathbf{b}_1}{\|\mathbf{b}_1\|} = \frac{[1 \quad 1 \quad 1]}{\sqrt{1^2 + 1^2 + 1^2}} = \left[\frac{1}{\sqrt{3}} \quad \frac{1}{\sqrt{3}} \quad \frac{1}{\sqrt{3}}\right]$$

$$\mathbf{c}_2 = \frac{\mathbf{b}_2}{\|\mathbf{b}_2\|} = \frac{[1 \quad 1 \quad -2]}{\sqrt{1^2 + 1^2 + (-2)^2}} = \left[\frac{1}{\sqrt{6}} \quad \frac{1}{\sqrt{6}} \quad -\frac{2}{\sqrt{6}}\right]$$

$$\mathbf{c}_3 = \frac{\mathbf{b}_3}{\|\mathbf{b}_3\|} = \frac{[-1 \quad 1 \quad 0]}{\sqrt{(-1)^2 + 1^2 + 0^2}} = \left[-\frac{1}{\sqrt{2}} \quad \frac{1}{\sqrt{2}} \quad 0\right]$$

Therefore, $\{\mathbf{c}_1, \mathbf{c}_2, \mathbf{c}_3\}$ is an orthonormal basis for R_3.

PROBLEM 11-12 Let R_4 have standard inner product and $\{\mathbf{a}_1, \mathbf{a}_2, \mathbf{a}_3, \mathbf{a}_4\}$ where $\mathbf{a}_1 = [1 \quad -1 \quad 0 \quad 1]$, $\mathbf{a}_2 = [0 \quad -1 \quad 1 \quad 0]$, $\mathbf{a}_3 = [1 \quad 0 \quad 1 \quad 1]$, and $\mathbf{a}_4 = [0 \quad 1 \quad -2 \quad 1]$ be a basis for R_4. Use the Gram–Schmidt process to transform the given basis into an orthonormal basis for R_4.

Solution Let $\mathbf{b}_1 = \mathbf{a}_1$. Then

$$\mathbf{b}_1 = \mathbf{a}_1 = [1 \quad -1 \quad 0 \quad 1]$$

$$\mathbf{b}_2 = \mathbf{a}_2 - \frac{\langle \mathbf{a}_2, \mathbf{b}_1 \rangle}{\langle \mathbf{b}_1, \mathbf{b}_1 \rangle}\mathbf{b}_1$$

$$= [0 \quad -1 \quad 1 \quad 0] - \frac{0(1) + (-1)(-1) + 1(0) + 0(1)}{1^2 + (-1)^2 + 0^2 + 1^2}[1 \quad -1 \quad 0 \quad 1]$$

$$= [0 \quad -1 \quad 1 \quad 0] - \tfrac{1}{3}[1 \quad -1 \quad 0 \quad 1] = [-\tfrac{1}{3} \quad -\tfrac{2}{3} \quad 1 \quad -\tfrac{1}{3}]$$

Choose $\mathbf{b}_2 = [-1 \quad -2 \quad 3 \quad -1]$. Then

$$\mathbf{b}_3 = \mathbf{a}_3 - \frac{\langle \mathbf{a}_3, \mathbf{b}_1 \rangle}{\langle \mathbf{b}_1, \mathbf{b}_1 \rangle}\mathbf{b}_1 - \frac{\langle \mathbf{a}_3, \mathbf{b}_2 \rangle}{\langle \mathbf{b}_2, \mathbf{b}_2 \rangle}\mathbf{b}_2$$

$$= [1 \quad 0 \quad 1 \quad 1] - \frac{1(1) + 0(-1) + 1(0) + (1)(1)}{1^2 + (-1)^2 + 0^2 + 1^2}[1 \quad -1 \quad 0 \quad 1]$$

$$- \frac{1(-1) + 0(-2) + 1(3) + 1(-1)}{(-1)^2 + (-2)^2 + 3^2 + (-1)^2}[-1 \quad -2 \quad 3 \quad -1]$$

$$= [1 \quad 0 \quad 1 \quad 1] - \tfrac{2}{3}[1 \quad -1 \quad 0 \quad 1] - \tfrac{1}{15}[-1 \quad -2 \quad 3 \quad -1]$$

$$= [\tfrac{6}{15} \quad \tfrac{12}{15} \quad \tfrac{12}{15} \quad \tfrac{6}{15}]$$

Choose $\mathbf{b}_3 = [1 \quad 2 \quad 2 \quad 1]$. Then

$$\mathbf{b}_4 = \mathbf{a}_4 - \frac{\langle \mathbf{a}_4, \mathbf{b}_1 \rangle}{\langle \mathbf{b}_1, \mathbf{b}_1 \rangle}\mathbf{b}_1 - \frac{\langle \mathbf{a}_4, \mathbf{b}_2 \rangle}{\langle \mathbf{b}_2, \mathbf{b}_2 \rangle}\mathbf{b}_2 - \frac{\langle \mathbf{a}_4, \mathbf{b}_3 \rangle}{\langle \mathbf{b}_3, \mathbf{b}_3 \rangle}\mathbf{b}_3$$

$$= [0 \quad 1 \quad -2 \quad 1] - \frac{0(1) + 1(-1) + (-2)(0) + 1(1)}{1^2 + (-1)^2 + 0^2 + 1^2}[1 \quad -1 \quad 0 \quad 1]$$

$$- \frac{0(-1) + 1(-2) + (-2)(3) + 1(-1)}{(-1)^2 + (-2)^2 + 3^2 + (-1)^2}[-1 \quad -2 \quad 3 \quad -1]$$

$$- \frac{0(1) + 1(2) + (-2)(2) + 1(1)}{1^2 + 2^2 + 2^2 + 1^2}[1 \quad 2 \quad 2 \quad 1]$$

$$= [0 \quad 1 \quad -2 \quad 1] - \tfrac{0}{3}[1 \quad -1 \quad 0 \quad 1] + \tfrac{9}{15}[-1 \quad -2 \quad 3 \quad -1] + \tfrac{1}{10}[1 \quad 2 \quad 2 \quad 1]$$

$$= [-\tfrac{5}{10} \quad 0 \quad 0 \quad \tfrac{5}{10}] = [-\tfrac{1}{2} \quad 0 \quad 0 \quad \tfrac{1}{2}]$$

Choose $\mathbf{b}_4 = [-1 \quad 0 \quad 0 \quad 1]$. Then

$$\mathbf{c}_1 = \frac{\mathbf{b}_1}{\|\mathbf{b}_1\|} = \frac{[1 \ -1 \ 0 \ 1]}{\sqrt{1^2 + (-1)^2 + 1^2}} = \left[\frac{1}{\sqrt{3}} \ -\frac{1}{\sqrt{3}} \ 0 \ \frac{1}{\sqrt{3}}\right]$$

$$\mathbf{c}_2 = \frac{\mathbf{b}_2}{\|\mathbf{b}_2\|} = \frac{[-1 \ -2 \ 3 \ 1]}{\sqrt{(-1)^2 + (-2)^2 + 3^2 + 1^2}} = \left[-\frac{1}{\sqrt{15}} \ -\frac{2}{\sqrt{15}} \ \frac{3}{\sqrt{15}} \ \frac{1}{\sqrt{15}}\right]$$

$$\mathbf{c}_3 = \frac{\mathbf{b}_3}{\|\mathbf{b}_3\|} = \frac{[1 \ 2 \ 2 \ 1]}{\sqrt{1^2 + 2^2 + 2^2 + 1^2}} = \left[\frac{1}{\sqrt{10}} \ \frac{2}{\sqrt{10}} \ \frac{2}{\sqrt{10}} \ \frac{1}{\sqrt{10}}\right]$$

$$\mathbf{c}_4 = \frac{\mathbf{b}_4}{\|\mathbf{b}_4\|} = \frac{[-1 \ 0 \ 0 \ 1]}{\sqrt{(-1)^2 + 0^2 + 0^2 + 1^2}} = \left[-\frac{1}{\sqrt{2}} \ 0 \ 0 \ \frac{1}{\sqrt{2}}\right]$$

Therefore, $\{\mathbf{c}_1, \mathbf{c}_2, \mathbf{c}_3, \mathbf{c}_4\}$ is an orthonormal basis for R_4.

PROBLEM 11-13 Let R_3 have the standard inner product. Let $\{\mathbf{a}_1, \mathbf{a}_2\}$ where $\mathbf{a}_1 = [1 \quad 1 \quad 0]$ and $\mathbf{a}_2 = [1 \quad 0 \quad 1]$ be a basis for the subspace W of R_3. Use the Gram–Schmidt process to find an orthonormal basis of W.

Solution Let $\mathbf{b}_1 = \mathbf{a}_1 = [1 \quad 1 \quad 0]$.

$$\begin{aligned}\mathbf{b}_2 &= \mathbf{a}_2 - \frac{\langle \mathbf{a}_2, \mathbf{b}_1 \rangle}{\langle \mathbf{b}_1, \mathbf{b}_1 \rangle}\mathbf{b}_1 \\ &= [1 \ 0 \ 1] - \frac{1(1) + 0(1) + 1(0)}{1^2 + 1^2 + 0^2}[1 \ 1 \ 0] \\ &= [1 \ 0 \ 1] - \tfrac{1}{2}[1 \ 1 \ 0] = [\tfrac{1}{2} \ -\tfrac{1}{2} \ 1]\end{aligned}$$

Choose $\mathbf{b}_2 = [1 \quad -1 \quad 2]$. Then

$$\mathbf{c}_1 = \frac{\mathbf{b}_1}{\|\mathbf{b}_1\|} = \frac{[1 \ 1 \ 0]}{\sqrt{1^2 + 1^2}} = \left[\frac{1}{\sqrt{2}} \ \frac{1}{\sqrt{2}} \ 0\right]$$

$$\mathbf{c}_2 = \frac{\mathbf{b}_2}{\|\mathbf{b}_2\|} = \frac{[1 \ -1 \ 2]}{\sqrt{1^2 + (-1)^2 + 2^2}} = \left[\frac{1}{\sqrt{6}} \ -\frac{1}{\sqrt{6}} \ \frac{2}{\sqrt{6}}\right]$$

Therefore, $\{\mathbf{c}_1, \mathbf{c}_2\}$ form an orthonormal basis of W.

Supplementary Problems

PROBLEM 11-14 Let R_2 have the standard inner product and determine which of the following sets form orthogonal sets and orthonormal sets.

(a) $\{[-1 \quad 0], [0 \quad 2]\}$ **(b)** $\left\{\left[\frac{2}{\sqrt{5}} \ -\frac{1}{\sqrt{5}}\right], \left[\frac{1}{\sqrt{5}} \ \frac{2}{\sqrt{5}}\right]\right\}$

PROBLEM 11-15 Let R_3 have the standard inner product and determine which of the following sets form orthogonal sets. If a set is orthogonal is it also orthonormal?

(a) $$\left\{[1 \ 0 \ -1], \left[\frac{1}{\sqrt{2}} \ 1 \ \frac{1}{\sqrt{2}}\right], \left[-\frac{1}{\sqrt{2}} \ 1 \ -\frac{1}{\sqrt{2}}\right]\right\}$$

(b) $$\left\{\left[\frac{2}{\sqrt{6}} \ -\frac{1}{\sqrt{6}} \ -\frac{1}{\sqrt{6}}\right], \left[0 \ -\frac{1}{\sqrt{2}} \ \frac{1}{\sqrt{2}}\right]\right\}$$

PROBLEM 11-16 Let R_2 have the inner product $\langle \mathbf{a}, \mathbf{b} \rangle = a_1 b_1 + 2a_2 b_2$ and $\mathbf{a} = [2/\sqrt{6} \quad -1/\sqrt{6}]$ and $\mathbf{b} = [1/\sqrt{3} \quad 1/\sqrt{3}]$. Show that $\{\mathbf{a}, \mathbf{b}\}$ is an orthonormal set of R_2.

PROBLEM 11-17 Show that the following vectors in P_2 are linearly independent:

$$\mathbf{p}_1 = \tfrac{2}{3} + \tfrac{2}{3}x + \tfrac{1}{3}x^2, \qquad \mathbf{p}_2 = \tfrac{1}{3} - \tfrac{2}{3}x + \tfrac{2}{3}x^2, \qquad \mathbf{p}_3 = -\tfrac{2}{3} + \tfrac{1}{3}x + \tfrac{2}{3}x^2$$

PROBLEM 11-18 Express $\mathbf{p} = 4 + x - 3x^2$ as a linear combination of the vectors $\mathbf{p}_1$, $\mathbf{p}_2$, and $\mathbf{p}_3$ in Problem 11-17.

PROBLEM 11-19 Show that the set of vectors $\{1, \sin x, \sin 2x\}$ in vector space V are linearly independent.

PROBLEM 11-20 Express the vector $D = \begin{bmatrix} 1 & 2 \\ -\frac{1}{3} & 1 \end{bmatrix}$ as a linear combination of

$$A = \begin{bmatrix} 2 & 2 \\ -1 & 0 \end{bmatrix} \qquad B = \begin{bmatrix} -1 & 0 \\ -1 & 2 \end{bmatrix} \qquad C = \begin{bmatrix} 0 & 4 \\ 4 & 2 \end{bmatrix}$$

PROBLEM 11-21 The set $\{\mathbf{b}_1, \mathbf{b}_2, \mathbf{b}_2\}$ where $\mathbf{b}_1 = [-1/\sqrt{5} \quad 0 \quad 2/\sqrt{5}]$, $\mathbf{b}_2 = [0 \quad 1 \quad 0]$, and $\mathbf{b}_3 = [-2/\sqrt{5} \quad 0 \quad -1/\sqrt{5}]$ is an orthonormal basis for R_3. Express the vector $\mathbf{b} = [1 \quad -1 \quad 3]$ as a linear combination of $\mathbf{b}_1$, $\mathbf{b}_2$, and $\mathbf{b}_3$.

PROBLEM 11-22 The set $\{\mathbf{b}_1, \mathbf{b}_2\}$ where $\mathbf{b}_1 = [1 \quad 1 \quad 1]$, and $\mathbf{b}_2 = [-1 \quad 1 \quad 0]$ is an orthogonal basis for subspace W of R_3 with standard inner product. Find **(a)** $\text{proj}_W \mathbf{a}$ if $\mathbf{a} = [1 \quad -2 \quad 3]$ and **(b)** $\text{proj}_W \mathbf{a}$ if $\mathbf{a} = [2 \quad 1 \quad 1]$.

PROBLEM 11-23 Let R_2 have the standard inner product. Use the Gram–Schmidt process to transform the basis $\{\mathbf{a}_1, \mathbf{a}_2\}$ where $\mathbf{a}_1 = [2 \quad -3]$ and $\mathbf{a}_2 = [1 \quad 4]$ into an orthogonal basis.

PROBLEM 11-24 Let R_3 have the standard inner product. Find an orthonormal basis for the subspace spanned by $\mathbf{a}_1 = [0 \quad 2 \quad -1]$ and $\mathbf{a}_2 = [-2 \quad 0 \quad 1]$.

PROBLEM 11-25 Let R_4 have the standard inner product and $\{\mathbf{a}_1, \mathbf{a}_2, \mathbf{a}_3, \mathbf{a}_4\}$ where $\mathbf{a}_1 = [1 \quad 1 \quad 0 \quad 1]$, $\mathbf{a}_2 = [0 \quad 1 \quad 1 \quad -1]$, $\mathbf{a}_3 = [1 \quad -1 \quad 0 \quad 1]$, and $\mathbf{a}_4 = [1 \quad 0 \quad -1 \quad 1]$ be a basis for R_4. Find an orthonormal basis for R_4.

PROBLEM 11-26 Let $A = \{[-1 \quad -1 \quad 1], [4 \quad -3 \quad 1], [2 \quad 5 \quad 7]\}$ be a subspace of R_3. Is set A an independent set? Is set A orthogonal? Is A a basis for R_3?

PROBLEM 11-27 Given $A = \{[-1 \quad -1 \quad 1], [4 \quad -3 \quad 1], [2 \quad 5 \quad 7]\}$, normalize each of the vectors. Is the result an orthonormal set?

PROBLEM 11-28 Write $\mathbf{b} = [7 \quad -10 \quad -6]$ as a linear combination of the vectors in Problem 11-26.

PROBLEM 11-29 Given $\mathbf{a} = [2 \quad 6]$ and $\mathbf{b} = [3 \quad 2]$, find **(a)** $\text{proj}_b \mathbf{a}$ and **(b)** $\text{proj}_a \mathbf{b}$.

PROBLEM 11-30 Given $\mathbf{a} = [2 \quad -5]$, $\mathbf{b} = [1 \quad 3t]$, find value(s) of t so that $\mathbf{a}$ and $\mathbf{b}$ are orthogonal.

PROBLEM 11-31 Given $\mathbf{a} = [t \quad 1 \quad -1]$, $\mathbf{b} = [t \quad 2t \quad 3]$, find values of t so that $\mathbf{a}$ and $\mathbf{b}$ are orthogonal.

Answers to Supplementary Problems

11-14 **(a)** Orthogonal
(b) Orthogonal and orthonormal

11-15 **(a)** Orthogonal but not orthonormal
(b) Orthogonal and orthonormal

11-16 $\langle \mathbf{a}, \mathbf{b} \rangle = 0$ and $\|\mathbf{a}\| = 1$, $\|\mathbf{b}\| = 1$

11-17

$$\begin{vmatrix} \frac{2}{3} & \frac{2}{3} & \frac{1}{3} \\ \frac{1}{3} & -\frac{2}{3} & \frac{2}{3} \\ -\frac{2}{3} & \frac{1}{3} & \frac{2}{3} \end{vmatrix} \neq 0$$

Therefore, the vectors are linearly independent.

11-18 $\mathbf{p} = (7/3)\mathbf{p}_1 - (4/3)\mathbf{p}_2 - (13/3)\mathbf{p}_3$

11-19 Sin $2x = 2 \sin x \cos x$ and cannot be expressed in terms of 1 and $\sin x$.

11-20 $\begin{bmatrix} 1 & 2 \\ -\frac{1}{3} & 1 \end{bmatrix} = \frac{2}{3}\begin{bmatrix} 2 & 2 \\ -1 & 0 \end{bmatrix} + \frac{1}{3}\begin{bmatrix} -1 & 0 \\ -1 & 2 \end{bmatrix} + \frac{1}{6}\begin{bmatrix} 0 & 4 \\ 4 & 2 \end{bmatrix}$

11-21 $[1 \quad -1 \quad 3] = \sqrt{5}\left[-\frac{1}{\sqrt{5}} \quad 0 \quad \frac{2}{\sqrt{5}}\right] + (-1)[0 \quad 1 \quad 0] + (-\sqrt{5})\left[-\frac{2}{\sqrt{5}} \quad 0 \quad -\frac{1}{\sqrt{5}}\right]$

11-22 **(a)** $[13/6 \quad -5/6 \quad 2/3]$
(b) $[11/6 \quad 5/6 \quad 4/3]$

11-23 $\{[2 \quad -3], [33/13 \quad 22/13]\}$

11-24 $\left\{\left[0 \quad \frac{2}{\sqrt{5}} \quad -\frac{1}{\sqrt{5}}\right], \left[-\frac{5}{\sqrt{30}} \quad \frac{1}{\sqrt{30}} \quad \frac{2}{\sqrt{30}}\right]\right\}$

11-25
$$\left\{\left[\frac{1}{\sqrt{3}} \quad \frac{1}{\sqrt{3}} \quad 0 \quad \frac{1}{\sqrt{3}}\right], \left[0 \quad \frac{1}{\sqrt{3}} \quad \frac{1}{\sqrt{3}} \quad -\frac{1}{\sqrt{3}}\right], \left[\frac{1}{\sqrt{3}} \quad -\frac{1}{\sqrt{3}} \quad \frac{1}{\sqrt{3}} \quad 0\right], \left[\frac{1}{\sqrt{3}} \quad 0 \quad -\frac{1}{\sqrt{3}} \quad -\frac{1}{\sqrt{3}}\right]\right\}$$

11-26 Yes, A is an independent set. Yes, A is orthogonal. Yes, A is a basis for R_3.

11-27
$$\left\{\left[-\frac{1}{\sqrt{3}} \quad -\frac{1}{\sqrt{3}} \quad \frac{1}{\sqrt{3}}\right], \left[\frac{4}{\sqrt{26}} \quad -\frac{3}{\sqrt{26}} \quad \frac{1}{\sqrt{26}}\right], \left[\frac{2}{\sqrt{78}} \quad \frac{5}{\sqrt{78}} \quad \frac{7}{\sqrt{78}}\right]\right\}$$

This is an orthonormal set.

11-28 $[7 \quad -10 \quad -6] = -1[-1 \quad -1 \quad 1] + 2[4 \quad -3 \quad 1] - 1[2 \quad 5 \quad 7]$

11-29 **(a)** $[54/13 \quad 36/13]$
(b) $[36/40 \quad 108/40]$

11-30 $t = 2/15$

11-31 $t = 1, t = -3$

EXAM 3 (Chapters 9–11)

1. Find the standard inner product for the vectors $\mathbf{a} = [2 \quad 1 \quad -3]$ and $\mathbf{b} = [3 \quad 2 \quad 1]$.

2. Use $\langle \mathbf{a}, \mathbf{b} \rangle = 4a_1b_1 + 3a_2b_2 + 2a_3b_3$ on R_3 to find the inner product of the vectors $\mathbf{a} = [2 \quad 1 \quad -3]$ and $\mathbf{b} = [3 \quad 2 \quad 1]$.

3. Find the length of the vector $\mathbf{a} = [5 \quad 6]$ in R_2 with the standard inner product.

4. Find the length of the vector $\mathbf{a} = [5 \quad 6]$ in R_2 if the inner product is $\langle \mathbf{a}, \mathbf{b} \rangle = 4a_1b_1 + 3a_2b_2$.

5. Find the distance between the vectors $\mathbf{a} = [3 \quad 4]$ and $\mathbf{b} = [5 \quad 1]$ in R_2 if R_2 has the standard inner product.

6. Find the distance between the vectors $\mathbf{a} = [3 \quad 4]$ and $\mathbf{b} = [5 \quad 1]$ in R_2 if R_2 has the inner product $\langle \mathbf{a}, \mathbf{b} \rangle = 2a_1b_1 + 3a_2b_2$.

7. Transform $\mathbf{a} = [1 \quad 2 \quad -3]$ in R_3 with standard inner product into a unit vector $\mathbf{b}$.

8. Find the cosine of the angle ϕ between the two vectors $\mathbf{a} = [1 \quad 1]$ and $\mathbf{b} = [2 \quad 3]$ in R_2 if R_2 has the standard inner product.

9. Find symmetric equations for the line passing through the point $(2, -3, 4)$ and parallel to the vector $[2 \quad 1 \quad 2]$.

10. Find the parametric equations of the line in question 9.

11. If $\mathbf{a} = [2 \quad -1 \quad 1]$ and $\mathbf{b} = [1 \quad 2 \quad 3]$, find $\mathbf{a} \times \mathbf{b}$.

12. If $\mathbf{a} = [2 \quad -1 \quad 1]$, $\mathbf{b} = [1 \quad 2 \quad 3]$, and $\mathbf{c} = [1 \quad -2 \quad 3]$, find $\mathbf{a} \cdot (\mathbf{b} \times \mathbf{c})$.

13. If R_2 has the standard inner product, find two vectors of norm one that are orthogonal to $[3 \quad -4]$.

14. If R_3 has the standard inner product, find the orthogonal projection of $\mathbf{a} = [1 \quad -2 \quad 1]$ onto $\mathbf{b} = [2 \quad -1 \quad 3]$.

15. Let R_2 have the standard inner product. Use the Gram–Schmidt process to transform the basis $\{\mathbf{a}_1, \mathbf{a}_2\}$ where $\mathbf{a}_1 = [-1 \quad 3]$ and $\mathbf{a}_2 = [2 \quad 1]$ into an **(a)** orthogonal basis and **(b)** into an orthonormal basis.

Solutions to Exam 3

1. $\langle \mathbf{a}, \mathbf{b} \rangle = \mathbf{a} \cdot \mathbf{b} = 2(3) + 1(2) + (-3)(1) = 5$

2. $\langle \mathbf{a}, \mathbf{b} \rangle = 4(2)(3) + 3(1)(2) + 2(-3)(1) = 24$

3. $\|\mathbf{a}\| = \sqrt{\langle \mathbf{a}, \mathbf{a} \rangle} = \sqrt{5^2 + 6^2} = \sqrt{61}$

4. $\|\mathbf{a}\| = \sqrt{\langle \mathbf{a}, \mathbf{a} \rangle} = \sqrt{4(5)^2 + 3(6)^2} = \sqrt{208}$
 $= 4\sqrt{13}$

5. $d(\mathbf{a}, \mathbf{b} \rangle = \|\mathbf{a} - \mathbf{b}\| = \sqrt{\langle \mathbf{a} - \mathbf{b}, \mathbf{a} - \mathbf{b} \rangle}$
 $= \sqrt{(3-5)^2 + (4-1)^2}$
 $= \sqrt{4 + 9} = \sqrt{13}$

6. $d(\mathbf{a}, \mathbf{b}) = \|\mathbf{a} - \mathbf{b}\| = \sqrt{\langle \mathbf{a} - \mathbf{b}, \mathbf{a} - \mathbf{b} \rangle}$
 $= \sqrt{2(3-5)^2 + 3(4-1)^2}$
 $= \sqrt{8 + 27} = \sqrt{35}$

7. $\mathbf{b} = \mathbf{a}/\|\mathbf{a}\| = [1 \quad 2 \quad -3]/\|[1 \quad 2 \quad -3]\|$

$= [1 \quad 2 \quad -3]/\sqrt{1^2 + 2^2 + (-3)^2}$

$= [1/\sqrt{14} \quad 2/\sqrt{14} \quad -3/\sqrt{14}]$

8. $$\cos\phi = \frac{\langle \mathbf{a}, \mathbf{b}\rangle}{\|\mathbf{a}\|\,\|\mathbf{b}\|} = \frac{\langle \mathbf{a}, \mathbf{b}\rangle}{\sqrt{\langle \mathbf{a}, \mathbf{a}\rangle}\sqrt{\langle \mathbf{b}, \mathbf{b}\rangle}}$$

$$= \frac{1(2) + 1(3)}{\sqrt{1^2 + 1^2}\sqrt{2^2 + 3^2}} = \frac{5}{\sqrt{2}\sqrt{13}} = \frac{5}{\sqrt{26}}$$

9. The forms of the equations are $(x - x_0)/a = (y - y_0)/b = (z - z_0)/c$. Therefore $(x - 2)/2 = (y + 3)/1 = (z - 4)/2$.

10. The forms of the equations are $x = x_0 + ta$, $y = y_0 + tb$, $z = z_0 + tc$. Therefore, $x = 2 + 2t$, $y = -3 + t$, and $z = 4 + 2t$.

11. $$\mathbf{a} \times \mathbf{b} = \begin{vmatrix} \mathbf{i} & \mathbf{j} & \mathbf{k} \\ 2 & -1 & 1 \\ 1 & 2 & 3 \end{vmatrix} = -5\mathbf{i} - 5\mathbf{j} + 5\mathbf{k}$$

$= [-5 \quad -5 \quad 5]$

12. $$\mathbf{a}\cdot(\mathbf{b} \times \mathbf{c}) = \begin{vmatrix} 2 & -1 & 1 \\ 1 & 2 & 3 \\ 1 & -2 & 3 \end{vmatrix} = 12 - 2 - 3 - 2$$

$+ 12 + 3 = 20$

13. $$\frac{\pm[3 \quad -4]}{\|[3 \quad -4]\|} = \frac{\pm[3 \quad -4]}{\sqrt{3^2 + (-4)^2}} = \pm\begin{bmatrix} \frac{3}{5} & -\frac{4}{5} \end{bmatrix}$$

14. $$\text{proj}_b\,\mathbf{a} = \frac{\|\mathbf{a}\|\cos\phi}{\|\mathbf{b}\|}\mathbf{b} = \frac{\|\mathbf{a}\|\,\|\mathbf{b}\|\cos\phi}{\|\mathbf{b}\|^2}\mathbf{b}$$

$$= \frac{\langle \mathbf{a}, \mathbf{b}\rangle}{\|\mathbf{b}\|^2}\mathbf{b} = \frac{2 + 2 + 3}{4 + 1 + 9}\mathbf{b}$$

$$= \tfrac{7}{14}[2 \quad -1 \quad 3] = [1 \quad -\tfrac{1}{2} \quad \tfrac{3}{2}]$$

15. **(a)** Let $\mathbf{b}_1 = \mathbf{a}_1 = [-1 \quad 3]$. Then

$$\mathbf{b}_2 = \mathbf{a}_2 - \frac{\langle \mathbf{a}_2, \mathbf{b}_1\rangle}{\langle \mathbf{b}_1, \mathbf{b}_1\rangle}\mathbf{b}_1$$

$$= [2 \quad 1] - \frac{(-1)(2) + 3(1)}{(-1)^2 + (3)^2}[-1 \quad 3]$$

$$= [2 \quad 1] - \tfrac{1}{10}[-1 \quad 3] = [\tfrac{21}{10} \quad \tfrac{7}{10}]$$

or we could use $[3 \quad 1]$ which is a scalar multiple of $\mathbf{b}_2 = [\tfrac{21}{10} \quad \tfrac{7}{10}]$. Therefore, $[-1 \quad 3]$ and $[3 \quad 1]$ form an orthogonal basis for R_2.

(b) Changing each of these to unit vectors, we get $[-1/\sqrt{10} \quad 3/\sqrt{10}]$ and $[3/\sqrt{10} \quad 1/\sqrt{10}]$ which form an orthonormal basis for R_2.

12 LINEAR TRANSFORMATIONS

THIS CHAPTER IS ABOUT

☑ Linear Transformation
☑ Properties of Linear Transformations
☑ Isomorphism
☑ Properties of Isomorphic Vector Spaces
☑ Kernel and Range

12-1. Linear Transformation

Definition of linear transformation

A function from a vector space V into a vector space W denoted by $L: V \to W$ is called a **linear transformation** provided that

(1) $L(\mathbf{a} + \mathbf{b}) = L(\mathbf{a}) + L(\mathbf{b})$ and
(2) $L(k\mathbf{a}) = kL(\mathbf{a})$

for all vectors $\mathbf{a}$ and $\mathbf{b}$ in V and for every scalar k.

EXAMPLE 12-1: Determine if the function $L: {}_2M_2 \to R$ defined by $L\left(\begin{bmatrix} a & b \\ c & d \end{bmatrix}\right) = a + 2d$ is linear.

Solution

$$L\left(\begin{bmatrix} a_1 & b_1 \\ c_1 & d_1 \end{bmatrix} + \begin{bmatrix} a_2 & b_2 \\ c_2 & d_2 \end{bmatrix}\right) = L\left(\begin{bmatrix} a_1 + a_2 & b_1 + b_2 \\ c_1 + c_2 & d_1 + d_2 \end{bmatrix}\right) = (a_1 + a_2) + 2(d_1 + d_2)$$

$$= (a_1 + 2d_1) + (a_2 + 2d_2)$$

$$= L\left(\begin{bmatrix} a_1 & b_1 \\ c_1 & d_1 \end{bmatrix}\right) + L\left(\begin{bmatrix} a_2 & b_2 \\ c_2 & d_2 \end{bmatrix}\right).$$

$$L\left(k\begin{bmatrix} a & b \\ c & d \end{bmatrix}\right) = L\left(\begin{bmatrix} ka & kb \\ kc & kd \end{bmatrix}\right) = ka + 2kd = k(a + 2d) = kL\left(\begin{bmatrix} a & b \\ c & d \end{bmatrix}\right)$$

Therefore, the function is linear.

EXAMPLE 12-2: Determine if each of the following functions is a linear transformation:

(a) $L: R_3 \to R_2$ defined by $L([a_1 \quad a_2 \quad a_3]) = [a_1 + a_2 \quad a_3]$
(b) $L: R_3 \to R_3$ defined by $L([a_1 \quad a_2 \quad a_3]) = [a_2 \quad a_1 \quad 1]$
(c) $L: R_3 \to R_3$ defined by $L([a_1 \quad a_2 \quad a_3]) = [a_2 \quad a_1 \quad 0]$
(d) $L: R^2 \to R^3$ defined by $L\left(\begin{bmatrix} a \\ b \end{bmatrix}\right) = \begin{bmatrix} 1 & 3 \\ 2 & -1 \\ 3 & 2 \end{bmatrix}\begin{bmatrix} a \\ b \end{bmatrix}$

Solution

(a)

$$\begin{aligned} L([a_1 \quad a_2 \quad a_3] + [b_1 \quad b_2 \quad b_3]) &= L([a_1 + b_1 \quad a_2 + b_2 \quad a_3 + b_3]) \\ &= [(a_1 + b_1) + (a_2 + b_2) \quad a_3 + b_3] \\ &= [(a_1 + a_2) + (b_1 + b_2) \quad a_3 + b_3] \\ &= [a_1 + a_2 \quad a_3] + [b_1 + b_2 \quad b_3] \\ &= L([a_1 \quad a_2 \quad a_3]) + L([b_1 \quad b_2 \quad b_3]) \end{aligned}$$

$$\begin{aligned} L(k[a_1 \quad a_2 \quad a_3]) &= L([ka_1 \quad ka_2 \quad ka_3]) \\ &= [ka_1 + ka_2 \quad ka_3] \\ &= k[a_1 + a_2 \quad a_3] \\ &= kL([a_1 \quad a_2 \quad a_3]) \end{aligned}$$

Therefore, the transformation in part (**a**) is linear.

(b)

$$\begin{aligned} L([a_1 \quad a_2 \quad a_3] + [b_1 \quad b_2 \quad b_3]) &= L([a_1 + b_1 \quad a_2 + b_2 \quad a_3 + b_3]) \\ &= [a_2 + b_2 \quad a_1 + b_1 \quad 1] \end{aligned}$$

$$\begin{aligned} L([a_1 \quad a_2 \quad a_3]) + L([b_1 \quad b_2 \quad b_3]) &= [a_2 \quad a_1 \quad 1] + [b_2 \quad b_1 \quad 1] \\ &= [a_2 + b_2 \quad a_1 + b_1 \quad 2] \\ &\neq [a_2 + b_2 \quad a_1 + b_1 \quad 1] \end{aligned}$$

Therefore, the transformation in part (**b**) is not linear.

(c)

$$\begin{aligned} L([a_1 \quad a_2 \quad a_3] + [b_1 \quad b_2 \quad b_3]) &= L([a_1 + b_1 \quad a_2 + b_2 \quad a_3 + b_3]) \\ &= [a_2 + b_2 \quad a_1 + b_1 \quad 0] \\ &= [a_2 \quad a_1 \quad 0] + [b_2 \quad b_1 \quad 0] \\ &= L([a_1 \quad a_2 \quad a_3]) + L([b_1 \quad b_2 \quad b_3]) \end{aligned}$$

$$\begin{aligned} L(k[a_1 \quad a_2 \quad a_3]) &= L([ka_1 \quad ka_2 \quad ka_3]) \\ &= [ka_2 \quad ka_1 \quad 0] \\ &= k[a_2 \quad a_1 \quad 0] \\ &= kL([a_1 \quad a_2 \quad a_3]) \end{aligned}$$

Therefore, the transformation in part (**c**) is linear.

(d)

$$\begin{aligned} L\left(\begin{bmatrix} a \\ b \end{bmatrix} + \begin{bmatrix} c \\ d \end{bmatrix}\right) = L\left(\begin{bmatrix} a + c \\ b + d \end{bmatrix}\right) &= \begin{bmatrix} 1 & 3 \\ 2 & -1 \\ 3 & 2 \end{bmatrix}\begin{bmatrix} a + c \\ b + d \end{bmatrix} = \begin{bmatrix} a + c + 3(b + d) \\ 2(a + c) - (b + d) \\ 3(a + c) + 2(b + d) \end{bmatrix} \\ &= \begin{bmatrix} a + 3b \\ 2a - b \\ 3a + 2b \end{bmatrix} + \begin{bmatrix} c + 3d \\ 2c - d \\ 3c + 2d \end{bmatrix} \\ &= \begin{bmatrix} 1 & 3 \\ 2 & -1 \\ 3 & 2 \end{bmatrix}\begin{bmatrix} a \\ b \end{bmatrix} + \begin{bmatrix} 1 & 3 \\ 2 & -1 \\ 3 & 2 \end{bmatrix}\begin{bmatrix} c \\ d \end{bmatrix} = L\left(\begin{bmatrix} a \\ b \end{bmatrix}\right) + L\left(\begin{bmatrix} c \\ d \end{bmatrix}\right) \end{aligned}$$

Therefore, the transformation in part (**d**) is linear.

12-2. Properties of Linear Transformations

Property A Let $L: V \to W$ be a linear transformation of an n-dimensional vector space V into an m-dimensional vector space W and $A = \{\mathbf{a}_1, \mathbf{a}_2, \ldots, \mathbf{a}_n\}$ be a basis for V. If $\mathbf{a}$ is any vector in V, then its image $L(\mathbf{a})$ is a linear combination of the vectors in the set $\{L(\mathbf{a}_1), L(\mathbf{a}_2), \ldots, L(\mathbf{a}_n)\}$; that is, $L(\mathbf{a}) = c_1L(\mathbf{a}_1) + c_2L(\mathbf{a}_2) + \cdots + c_nL(\mathbf{a}_n)$ for some scalars $c_1, c_2, \ldots, c_n$.

note: This property tells you that if $\mathbf{a}$ is in V, then $L(\mathbf{a})$ is a linear combination of the images of the basis vectors for V. It does not say that every vector in W is a linear combination of the images of the basis vectors for V.

EXAMPLE 12-3: Consider the function $L: R_3 \to R_4$ defined by $L([a_1 \quad a_2 \quad a_3]) = [a_1 - a_2 \quad a_1 + a_2 \quad a_3 \quad 0]$. **(a)** Show that the function is a linear transformation. **(b)** If $\mathbf{e}_1$, $\mathbf{e}_2$, and $\mathbf{e}_3$ are the natural basis vectors for R_3, show that $[0 \quad 2 \quad 1 \quad 0]$ is a linear combination of $\{L(\mathbf{e}_1), L(\mathbf{e}_2), L(\mathbf{e}_3)\}$ but that $[1 \quad 1 \quad 1 \quad 1]$ is not a linear combination of $\{L(\mathbf{e}_1), L(\mathbf{e}_2), L(\mathbf{e}_3)\}$.

Solution

(a)

$$\begin{aligned}
L([a_1 \quad a_2 \quad a_3] + [b_1 \quad b_2 \quad b_3]) &= L([a_1 + b_1 \quad a_2 + b_2 \quad a_3 + b_3]) \\
&= [(a_1 + b_1) - (a_2 + b_2) \quad (a_1 + b_1) + (a_2 + b_2) \quad a_3 + b_3 \quad 0] \\
&= [a_1 - a_2 \quad a_1 + a_2 \quad a_3 \quad 0] + [b_1 - b_2 \quad b_1 + b_2 \quad b_3 \quad 0] \\
&= L([a_1 \quad a_2 \quad a_3]) + L([b_1 \quad b_2 \quad b_3]) \\
L(k[a_1 \quad a_2 \quad a_3]) &= L([ka_1 \quad ka_2 \quad ka_3]) \\
&= [ka_1 - ka_2 \quad ka_1 + ka_2 \quad ka_3 \quad 0] \\
&= k[a_1 - a_2 \quad a_1 + a_2 \quad a_3 \quad 0] \\
&= kL([a_1 \quad a_2 \quad a_3])
\end{aligned}$$

Therefore, it is a linear transformation.

(b)

$$\begin{aligned}
L(\mathbf{e}_1) &= L([1 \quad 0 \quad 0]) = [1 \quad 1 \quad 0 \quad 0] \\
L(\mathbf{e}_2) &= L([0 \quad 1 \quad 0]) = [-1 \quad 1 \quad 0 \quad 0] \\
L(\mathbf{e}_3) &= L([0 \quad 0 \quad 1]) = [0 \quad 0 \quad 1 \quad 0]
\end{aligned}$$

$$\begin{aligned}
[0 \quad 2 \quad 1 \quad 0] &= c_1L(\mathbf{e}_1) + c_2L(\mathbf{e}_2) + c_3L(\mathbf{e}_3) \\
&= c_1[1 \quad 1 \quad 0 \quad 0] + c_2[-1 \quad 1 \quad 0 \quad 0] + c_3[0 \quad 0 \quad 1 \quad 0]
\end{aligned}$$

Thus

$$\begin{aligned}
c_1 - c_2 &= 0 \\
c_1 + c_2 &= 2 \\
c_3 &= 1
\end{aligned}$$

This system has $c_1 = 1$, $c_2 = 1$, $c_3 = 1$ as its solution. Therefore, $[0 \quad 2 \quad 1 \quad 0] = [1 \quad 1 \quad 0 \quad 0] + [-1 \quad 1 \quad 0 \quad 0] + [0 \quad 0 \quad 1 \quad 0]$. Now suppose that $[1 \quad 1 \quad 1 \quad 1]$ is a linear combination of $L(\mathbf{e}_1)$, $L(\mathbf{e}_2)$, and $L(\mathbf{e}_3)$. Thus,

$$c_1[1 \quad 1 \quad 0 \quad 0] + c_2[-1 \quad 1 \quad 0 \quad 0] + c_3[0 \quad 0 \quad 1 \quad 0] = [1 \quad 1 \quad 1 \quad 1]$$

$$\begin{aligned}
c_1 - c_2 &= 1 \\
c_1 + c_2 &= 2 \\
c_3 &= 0 \\
0 &= 1
\end{aligned}$$

which is a contradiction. Hence, $[1 \quad 1 \quad 1 \quad 1]$ cannot be expressed as a linear combination of $L(\mathbf{e}_1)$, $L(\mathbf{e}_2)$, and $L(\mathbf{e}_3)$. The trouble is that $[1 \quad 1 \quad 1 \quad 1]$ is not the image of any vector in R_3 under the given linear transformation.

Property B Let $L: V \to W$ be a linear transformation of an n-dimensional vector space V into an m-dimensional vector space W, and $\mathbf{a}$ and $\mathbf{b}$ are vectors in V. Then

(a) $L(\mathbf{0}_V) = \mathbf{0}_W$,
(b) $L(-\mathbf{a}) = -L(\mathbf{a})$, and
(c) $L(\mathbf{a} - \mathbf{b}) = L(\mathbf{a}) - L(\mathbf{b})$.

EXAMPLE 12-4: Verify Property B for the linear transformation $L: R_3 \to R_4$ defined by $L([a_1 \quad a_2 \quad a_3]) = [a_1 - a_2 \quad a_1 + a_2 \quad a_3 \quad 0]$ where $\mathbf{a} = [1 \quad 2 \quad 3]$ and $\mathbf{b} = [4 \quad -3 \quad 2]$.

Solution

(a)

$$\mathbf{0}_{R_3} = [0 \quad 0 \quad 0]$$
$$L(\mathbf{0}_{R_3}) = L([0 \quad 0 \quad 0]) = [0 - 0 \quad 0 + 0 \quad 0 \quad 0]$$
$$= [0 \quad 0 \quad 0 \quad 0] = \mathbf{0}_W$$

(b)

$$L(-\mathbf{a}) = L(-[1 \quad 2 \quad 3]) = L([-1 \quad -2 \quad -3])$$
$$= [1 \quad -3 \quad -3 \quad 0]$$
$$= -L(\mathbf{a})$$

(c)

$$L(\mathbf{a} - \mathbf{b}) = L([1 \quad 2 \quad 3] - [4 \quad -3 \quad 2]) = L([-3 \quad 5 \quad 1])$$
$$= [-8 \quad 2 \quad 1 \quad 0]$$
$$L(\mathbf{a}) - L(\mathbf{b}) = L([1 \quad 2 \quad 3]) - L([4 \quad -3 \quad 2])$$
$$= [-1 \quad 3 \quad 3 \quad 0] - [7 \quad 1 \quad 2 \quad 0]$$
$$= [-8 \quad 2 \quad 1 \quad 0] = L(\mathbf{a} - \mathbf{b})$$

Hence, all three parts of the property are true for the given vectors and linear transformation.

12-3. Isomorphism

A. Definition of one-to-one and onto

A linear transformation $L: V \to W$ is called **one-to-one** or **injective** if, for all $\mathbf{a}$, $\mathbf{b}$ in V, $L(\mathbf{a}) = L(\mathbf{b})$ implies $\mathbf{a} = \mathbf{b}$. A linear transformation $L: V \to W$ is called **onto** or **surjective** if every vector in W is the image of at least one vector in V.

EXAMPLE 12-5: Determine if each of the following linear transformations is one-to-one and onto. **(a)** $L: R_3 \to R_2$ defined by $L([a_1 \quad a_2 \quad a_3]) = [a_1 + a_2 \quad a_3]$. **(b)** $L: R_3 \to R_3$ defined by $L([a_1 \quad a_2 \quad a_3]) = [a_2 \quad a_1 \quad 0]$.

Solution

(a) Let $L([a_1 \quad a_2 \quad a_3]) = L([b_1 \quad b_2 \quad b_3])$. Thus, $[a_1 + a_2 \quad a_3] = [b_1 + b_2 \quad b_3]$. Therefore, $a_1 + a_2 = b_1 + b_2$ and $a_3 = b_3$. Although this implies $a_3 = b_3$, it does not imply that $a_1 = b_1$ and $a_2 = b_2$ since it is possible that $a_1 = b_2$ and $a_2 = b_1$. Hence, the linear transformation is not one-to-one.

To determine whether it is onto, let $[a_1 \quad a_2]$ be any vector in R_2. Hence,

$$L([x \quad y \quad z]) = [a_1 \quad a_2]$$
$$x + y = a_1$$
$$z = a_2$$

Thus, to find a vector in V whose image is $[a_1 \quad a_2]$, we choose any two real numbers x and y whose sum is a_1 for the first two components in V and let the third component be a_2. Therefore, this linear transformation is onto but not one-to-one.

(b) Let $L([a_1 \quad a_2 \quad a_3]) = L([b_1 \quad b_2 \quad b_3])$. Thus, $[a_2 \quad a_1 \quad 0] = [b_2 \quad b_1 \quad 0]$. Therefore, $a_2 = b_2$, $a_1 = b_1$, but a_3 and b_3 are arbitrary. Hence, the linear transformation is not one-to-one. Neither is it onto because $[a_1 \quad a_2 \quad 2]$ is not the image of any vector in R_3. Therefore, the linear transformation in this case is neither one-to-one nor onto.

B. Definition of isomorphic vector spaces

Two vector spaces V and W are **isomorphic** if a linear transformation $L: V \to W$ exists that is one-to-one and onto. Since functions that are one-to-one and onto possess inverses, an inverse linear transformation $L^{-1}: W \to V$ exists that is also one-to-one and onto.

EXAMPLE 12-6: Show that the linear transformation $L: {}_2M_2 \to R_4$ defined by

$$L\left(\begin{bmatrix} a & b \\ c & d \end{bmatrix}\right) = [a \quad b \quad c \quad d]$$

is an isomorphism and, thus, that ${}_2M_2$ is isomorphic to R_4.

Solution: Let

$$L\left(\begin{bmatrix} a_1 & b_1 \\ c_1 & d_1 \end{bmatrix}\right) = L\left(\begin{bmatrix} a_2 & b_2 \\ c_2 & d_2 \end{bmatrix}\right)$$

Thus $[a_1 \quad b_1 \quad c_1 \quad d] = [a_2 \quad b_2 \quad c_2 \quad d_2]$ which is true only if $a_1 = a_2$, $b_1 = b_2$, $c_1 = c_2$, and $d_1 = d_2$. Thus, the transformation is one-to-one. Furthermore, it is onto because $[a \quad b \quad c \quad d]$ is the image of $\begin{bmatrix} a & b \\ c & d \end{bmatrix}$.

Therefore, the linear transformation in part **(c)** is both one-to-one and onto and, hence, is an isomorphism. Therefore, ${}_2M_2$ is isomorphic to R_4.

12-4. Properties of Isomorphic Vector Spaces

You may be wondering why so many of the examples for vector spaces have involved R_n. The reason is that algebraic properties of isomorphic vector spaces are identical. The only difference is in the notation used to denote their vectors. This is indicated more specifically in the following properties.

Property A Two finite-dimensional vector spaces are isomorphic if and only if their dimensions are equal.

EXAMPLE 12-7: Indicate a vector space from R_n which is isomorphic to each of the following vector spaces: **(a)** ${}_3M_4$, **(b)** R^5, **(c)** P_3, **(d)** P_6.

Solution

(a) Since $\dim {}_3M_4 = 3(4) = 12$, then it is isomorphic to R_{12}.
(b) Since $\dim R^5 = 5$, then it is isomorphic to R_5.
(c) Since $\dim P_3 = 3 + 1 = 4$, then it is isomorphic to R_4.
(d) Since $\dim P_6 = 6 + 1 = 7$, then it is isomorphic to R_7.

Property B Isomorphism of vector spaces is an equivalence relation. That is, all of the following are true.

(a) (Reflexive) Every vector space V is isomorphic to itself.
(b) (Symmetric) If V is isomorphic to W, then W is isomorphic to V.
(c) (Transitive) If V is isomorphic to W and W is isomorphic to U, then V is isomorphic to U.

EXAMPLE 12-8: If a vector space V is isomorphic to R_8, explain why V is isomorphic to ${}_2M_4$.

Solution: R_8 is isomorphic to ${}_2M_4$ since $\dim R_8 = \dim {}_2M_4 = 8$. Therefore, by the transitive property of an equivalence relation, V is isomorphic to ${}_2M_4$.

12-5. Kernel and Range

A. Definition of kernel

If $L: V \to W$ is a linear transformation of a vector space V into a vector space W, then the subset of V consisting of all vectors $\mathbf{a}$ in V having the zero element $\mathbf{0}_W$ of W as image is the **kernel** of L and is denoted by $\ker L$. This means that $\ker L = \{\mathbf{a} \mid \mathbf{a} \text{ is in } V \text{ and } L(\mathbf{a}) = \mathbf{0}_W\}$. The kernel of L is a subspace of V.

EXAMPLE 12-9: Determine whether each of the following vectors is in the kernel of the linear transformation $L: R_3 \to R_2$ defined by $L([a_1 \quad a_2 \quad a_3]) = [a_1 + a_2 \quad a_3]$: **(a)** $[0 \quad 0 \quad 0]$, **(b)** $[3 \quad -3 \quad 0]$, **(c)** $[-5 \quad 5 \quad 0]$, **(d)** $[-5 \quad 5 \quad 3]$, **(e)** $[1 \quad 2 \quad 0]$.

Solution

(a) $L([0 \quad 0 \quad 0]) = [0 + 0 \quad 0] = [0 \quad 0]$, $\therefore$ $[0 \quad 0 \quad 0]$ is in $\ker L$.
(b) $L([3 \quad -3 \quad 0]) = [3 + (-3) \quad 0] = [0 \quad 0]$, $\therefore$ $[3 \quad -3 \quad 0]$ is in $\ker L$.
(c) $L([-5 \quad 5 \quad 0]) = [-5 + 5 \quad 0] = [0 \quad 0]$, $\therefore$ $[-5 \quad 5 \quad 0]$ is in $\ker L$.
(d) $L([-5 \quad 5 \quad 3]) = [-5 + 5 \quad 3] = [0 \quad 3]$, $\therefore$ $[-5 \quad 5 \quad 3]$ is not in $\ker L$.
(e) $L([1 \quad 2 \quad 0]) = [1 + 2 \quad 0] = [3 \quad 0]$, $\therefore$ $[1 \quad 2 \quad 0]$ is not in $\ker L$.

EXAMPLE 12-10: For the linear transformation given in Example 12-9, find **(a)** $\ker L$ and **(b)** a basis for $\ker L$.

Solution

(a) We need to find all vectors $[a_1 \quad a_2 \quad a_3]$ in V such that $L([a_1 \quad a_2 \quad a_3]) = [0 \quad 0]$, but $L([a_1 \quad a_2 \quad a_3]) = [a_1 + a_2 \quad a_3]$. Hence, $[a_1 + a_2 \quad a_3] = [0 \quad 0]$. Thus

$$a_1 + a_2 = 0 \qquad \text{and} \qquad a_3 = 0$$

or

$$a_2 = -a_1 \qquad \text{and} \qquad a_3 = 0$$

Therefore, $\ker L = [a_1 \quad -a_1 \quad 0]$, which is more often written as $[t \quad -t \quad 0]$.

(b) $[t \quad -t \quad 0] = t[1 \quad -1 \quad 0]$. Therefore, $\{[1 \quad -1 \quad 0]\}$ is a basis for $\ker L$.

B. Definition of range

If $L: V \to W$ is a linear transformation of a vector space V into a vector space W, then the **range** of L, denoted by **range L**, is the subset of W which consists of all vectors in W that are images of vectors in V. The range of L is a subspace of W.

EXAMPLE 12-11: Determine whether each of the following vectors is in the range of the linear transformation $L: R_3 \to R_2$ defined by $L([a_1 \quad a_2 \quad a_3]) = [a_1 + a_2 \quad a_3]$: **(a)** $[1 \quad 2]$; **(b)** $[5 \quad 3]$.

Solution

(a) We must determine whether there is some $[a_1 \quad a_2 \quad a_3]$ such that

$$L([a_1 \quad a_2 \quad a_3]) = [1 \quad 2]$$
$$L([a_1 \quad a_2 \quad a_3]) = [a_1 + a_2 \quad a_3] = [1 \quad 2]$$
$$\therefore \quad a_1 + a_2 = 1$$
$$a_3 = 2$$

Thus, one of the many solutions of the first of these equations is $a_1 = 2$, $a_2 = -1$. Hence, $L([2 \quad -1 \quad 2] = [1 \quad 2]$. Therefore, $[1 \quad 2]$ is the image of $[2 \quad -1 \quad 2]$ and is, thus, in range L.

(b) We must determine if some $[a_1 \quad a_2 \quad a_3]$ exists such that $L([a_1 \quad a_2 \quad a_3]) = [5 \quad 3]$.

$$L([a_1 \quad a_2 \quad a_3]) = [a_1 + a_2 \quad a_3] = [5 \quad 3]$$
$$\therefore \quad a_1 + a_2 = 5$$
$$a_3 = 3$$

Thus, one of the many solutions of the first of these equations is $a_1 = 2$, $a_2 = 3$. Hence, $L([2 \quad 3 \quad 3]) = [5 \quad 3]$. Therefore, $[5 \quad 3]$ is the image of $[2 \quad 3 \quad 3]$ and is, thus, in range L.

EXAMPLE 12-12: Find a basis for the range of the linear transformation in Example 12-11.

Solution: We first need to find a set of vectors that spans range L. We then need to find a maximal linearly independent subset of these spanning vectors. To find a spanning set, we start with L.

$$L([a_1 \quad a_2 \quad a_3]) = [a_1 + a_2 \quad a_3]$$

The right-hand side of this equation can be written as

$$[a_1 + a_2 \quad 0] + [0 \quad a_3] = [a_1 \quad 0] + [a_2 \quad 0] + [0 \quad a_3]$$
$$= a_1[1 \quad 0] + a_2[1 \quad 0] + a_3[0 \quad 1]$$

Thus $\{[1 \quad 0], [1 \quad 0], [0 \quad 1]\}$ spans range L. Since these are the natural basis vectors of R_2, we see that $\{[1 \quad 0], [0 \quad 1]\}$ is a maximal linearly independent subset and, hence, a basis for range L.

EXAMPLE 12-13: Find a basis for the range of the linear transformation $L: R_4 \to R_4$ defined by

$$L([a_1 \quad a_2 \quad a_3 \quad a_4]) = [a_1 + a_2 + 2a_3 \quad a_1 - a_2 - 2a_4 \quad a_1 + 2a_2 + 3a_3 + a_4 \quad a_1 + a_3 - a_4]$$

Solution

$$L([a_1 \quad a_2 \quad a_3 \quad a_4)] = [a_1 + a_2 + 2a_3 \quad a_1 - a_2 - 2a_4 \quad a_1 + 2a_2 + 3a_3 + a_4 \quad a_1 + a_3 - a_4]$$
$$= [a_1 \quad a_1 \quad a_1 \quad a_1] + [a_2 \quad -a_2 \quad 2a_2 \quad 0]$$
$$+ [2a_3 \quad 0 \quad 3a_3 \quad a_3] + [0 \quad -2a_4 \quad a_4 \quad -a_4]$$
$$= a_1[1 \quad 1 \quad 1 \quad 1] + a_2[1 \quad -1 \quad 2 \quad 0]$$
$$+ a_3[2 \quad 0 \quad 3 \quad 1] + a_4[0 \quad -2 \quad 1 \quad -1]$$

Thus, $\{[1 \quad 1 \quad 1 \quad 1], [1 \quad -1 \quad 2 \quad 0], [2 \quad 0 \quad 3 \quad 1], [0 \quad -2 \quad 1 \quad -1]\}$ spans range L.

One of the easiest ways to find a maximal linearly independent subset of these vectors is, first, to form a matrix with these four vectors as rows. Then reduce this matrix by elementary row operations to a form where the nonzero rows in echelon form can be determined. These nonzero rows constitute a basis for range L.

$$\begin{bmatrix} 1 & 1 & 1 & 1 \\ 1 & -1 & 2 & 0 \\ 2 & 0 & 3 & 1 \\ 0 & -2 & 1 & -1 \end{bmatrix} \xrightarrow[R_3 \to R_3 - 2R_1]{R_2 \to R_2 - R_1} \begin{bmatrix} 1 & 1 & 1 & 1 \\ 0 & -2 & 1 & -1 \\ 0 & -2 & 1 & -1 \\ 0 & -2 & 1 & -1 \end{bmatrix} \xrightarrow[R_4 \to R_4 - R_2]{R_3 \to R_3 - R_2} \begin{bmatrix} 1 & 1 & 1 & 1 \\ 0 & -2 & 1 & -1 \\ 0 & 0 & 0 & 0 \\ 0 & 0 & 0 & 0 \end{bmatrix}$$

Therefore, $\{[1 \quad 1 \quad 1 \quad 1], [0 \quad -2 \quad 1 \quad -1]\}$ is a basis for range L.

C. Kernel, range, and dimension

If $L: V \to W$ is a linear transformation of a vector space V into a vector space W, then dim (ker L) + dim (range L) = dim V.

EXAMPLE 12-14: Verify dim (ker L) + dim (range L) = dim V for the linear transformation in Example 12-13.

Solution: In Example 12-13, we found that a basis for range L consists of two vectors. Thus, dim (range L) = 2. Also, $V = R_4$, thus dim V = dim $R_4 = 4$. We now must find ker L and its dimension. Kernel L consists of all vectors $[a_1 \quad a_2 \quad a_3 \quad a_4]$ such that

$$L([a_1 \quad a_2 \quad a_3 \quad a_4]) = [0 \quad 0 \quad 0 \quad 0]$$

but

$$L([a_1 \quad a_2 \quad a_3 \quad a_4]) = [a_1 + a_2 + 2a_3 \quad a_1 - a_2 - 2a_4 \quad a_1 + 2a_2 + 3a_3 + a_4 \quad a_1 + a_3 - a_4]$$

Hence,

$$\begin{aligned} a_1 + a_2 + 2a_3 \qquad &= 0 \\ a_1 - a_2 \qquad - 2a_4 &= 0 \\ a_1 + 2a_2 + 3a_3 + a_4 &= 0 \\ a_1 \qquad + a_3 - a_4 &= 0 \end{aligned}$$

Solving by the Gauss–Jordan elimination method, we have

$$\begin{bmatrix} 1 & 1 & 2 & 0 \\ 1 & -1 & 0 & -2 \\ 1 & 2 & 3 & 1 \\ 1 & 0 & 1 & -1 \end{bmatrix} \xrightarrow[\substack{R_3 \to R_3 - R_1 \\ R_4 \to R_4 - R_1}]{R_2 \to R_2 - R_1} \begin{bmatrix} 1 & 1 & 2 & 0 \\ 0 & -2 & -2 & -2 \\ 0 & 1 & 1 & 1 \\ 0 & -1 & -1 & -1 \end{bmatrix} \xrightarrow{R_2 \leftrightarrow R_3} \begin{bmatrix} 1 & 1 & 2 & 0 \\ 0 & 1 & 1 & 1 \\ 0 & -2 & -2 & -2 \\ 0 & -1 & -1 & -1 \end{bmatrix}$$

$$\xrightarrow[\substack{R_3 \to R_3 + 2R_2 \\ R_4 \to R_4 + R_2}]{} \begin{bmatrix} 1 & 1 & 2 & 0 \\ 0 & 1 & 1 & 1 \\ 0 & 0 & 0 & 0 \\ 0 & 0 & 0 & 0 \end{bmatrix}$$

$$\therefore \quad \begin{aligned} a_1 + a_2 + 2a_3 &= 0 \\ a_2 + a_3 + a_4 &= 0 \end{aligned}$$

Thus

$$\begin{aligned} a_1 &= -a_2 - 2a_3 \\ a_4 &= -a_2 - a_3 \end{aligned}$$

Let $a_2 = s$ and $a_3 = t$ where s and t are arbitrary real numbers, then $a_1 = -s - 2t$ and $a_4 = -s - t$. Therefore, the kernel is all vectors of the form $[-s-2t \quad s \quad t \quad -s-t]$. However, this equals

$$[-s \quad s \quad 0 \quad -s] + [-2t \quad 0 \quad t \quad -t] = s[-1 \quad 1 \quad 0 \quad -1] + t[-2 \quad 0 \quad 1 \quad -1]$$

Hence, a basis for $\ker L$ is $\{[-1 \quad 1 \quad 0 \quad -1], [-2 \quad 0 \quad 1 \quad -1]\}$ which has dimension 2. Therefore, $\dim(\ker L) + \dim(\text{range } L) = 2 + 2 = 4 = \dim V$.

SUMMARY

1. A function $L: V \to W$ from a vector space V into a vector space W is called a linear transformation provided that
   - $L(\mathbf{a} + \mathbf{b}) = L(\mathbf{a}) + L(\mathbf{b})$ and
   - $L(k\mathbf{a}) = kL(\mathbf{a})$ for all vectors $\mathbf{a}$ and $\mathbf{b}$ in V and for every scalar k.
2. If $L: V \to W$ is a linear transformation of an n-dimensional vector space V into an m-dimensional vector space W and $A = \{\mathbf{a}_1, \mathbf{a}_2, \ldots, \mathbf{a}_n\}$ is a basis for V, then for any vector $\mathbf{a}$ in V its image $L(\mathbf{a}) = c_1L(\mathbf{a}_1) + c_2L(\mathbf{a}_2) + \cdots + c_nL(\mathbf{a}_n)$ for some scalars $c_1, c_2, \ldots, c_n$.
3. The last property tells us that if $\mathbf{a}$ is in V, then $L(\mathbf{a})$ is a linear combination of the images of the basis vectors for V, but there may be vectors in W that are not linear combinations of the images of the basis vectors for V.
4. If $L: V \to W$ is a linear transformation of an n-dimensional vector space V into an m-dimensional vector space W and $\mathbf{a}$ and $\mathbf{b}$ are vectors in V, then
   - $L(\mathbf{0}_V) = \mathbf{0}_W$;
   - $L(-\mathbf{a}) = -L(\mathbf{a})$;
   - $L(\mathbf{a} - \mathbf{b}) = L(\mathbf{a}) - L(\mathbf{b})$.
5. A linear transformation $L: V \to W$ is called
   - one-to-one or injective if for all $\mathbf{a}, \mathbf{b}$ in V, $L(\mathbf{a}) = L(\mathbf{b})$ implies $\mathbf{a} = \mathbf{b}$;
   - onto or surjective if every vector in W is the image of at least one vector in V.
6. Two vector spaces V and W are isomorphic if there exists a linear transformation $L: V \to W$ that is one-to-one and onto.
7. Algebraic properties of isomorphic vector spaces are identical.
8. Two finite-dimensional vector spaces are isomorphic if and only if their dimensions are equal.
9. Isomorphism of vector spaces is an equivalence relation.
10. If $L: V \to W$ is a linear transformation of a vector space V into a vector space W, then
    - the subset of V consisting of all vectors $\mathbf{a}$ in V having the zero element $\mathbf{0}_W$ of W as image is the kernel of L and is denoted by $\ker L$;
    - the range of L denoted by range L is the subset of W which consists of all vectors in W that are images of vectors in V;
    - $\dim(\ker L) + \dim(\text{range } L) = \dim V$;
    - $\ker L$ is a subspace of V;
    - range L is a subspace of W.

RAISE YOUR GRADES

Can you ...?

☑ determine if a function $L: V \to W$ is a linear transformation
☑ express the image of $\mathbf{a}$ in V as a linear combination of vectors in W
☑ when given $L: V \to W$ and $\mathbf{0}_V$, find $\mathbf{0}_W$
☑ determine if the linear transformation $L: V \to W$ is one-to-one
☑ determine if the linear transformation $L: V \to W$ is onto

☑ determine if two vector spaces are isomorphic
☑ when given a finite-dimensional vector space V, find a vector space from Euclidean space E_m isomorphic to it
☑ find the kernel of a linear transformation $L\colon V \to W$
☑ find the range of a linear transformation $L\colon V \to W$
☑ when given the dimensions of any two of $\ker L$, range L, and V of a linear transformation $L\colon V \to W$, find the dimension of the third vector space

SOLVED PROBLEMS

PROBLEM 12-1 Determine if the function $L\colon R_2 \to R_2$ defined by $L([a_1 \quad a_2]) = [2a_1 + a_2 \quad a_2]$ is linear.

Solution

$$\begin{aligned} L([a_1 \quad a_2] + [b_1 \quad b_2]) &= L([a_1 + b_1 \quad a_2 + b_2]) \\ &= [2(a_1 + b_1) + (a_2 + b_2) \quad a_2 + b_2] \\ &= [(2a_1 + a_2) + (2b_1 + b_2) \quad a_2 + b_2] \\ &= [2a_1 + a_2 \quad a_2] + [2b_1 + b_2 \quad b_2] \\ &= L([a_1 \quad a_2]) + L([b_1 \quad b_2]) \end{aligned}$$

$$\begin{aligned} L(k[a_1 \quad a_2]) &= L([ka_1 \quad ka_2]) \\ &= [2(ka_1) + ka_2 \quad ka_2] \\ &= k[2a_1 + a_2 \quad a_2] \\ &= kL([a_1 \quad a_2]) \end{aligned}$$

Therefore, the function is linear.

PROBLEM 12-2 Determine if each of the following functions is a linear transformation: **(a)** $L\colon R_3 \to R_3$ defined by $L([a_1 \quad a_2 \quad a_3]) = [a_1 - a_2 \quad 0 \quad a_3]$; **(b)** $L\colon R_3 \to R_2$ defined by $L([a_1 \quad a_2 \quad a_3]) = [a_1 - 3a_2 \quad a_3]$; and **(c)** $L\colon R_3 \to R_3$ defined by $L([a_1 \quad a_2 \quad a_3]) = [a_1 + a_2 \quad a_3 \quad 1]$.

Solution

(a)

$$\begin{aligned} L([a_1 \quad a_2 \quad a_3] + [b_1 \quad b_2 \quad b_3]) &= L([a_1 + b_1 \quad a_2 + b_2 \quad a_3 + b_3]) \\ &= [(a_1 + b_1) - (a_2 + b_2) \quad 0 \quad a_3 + b_3] \\ &= [(a_1 - a_2) + (b_1 - b_2) \quad 0 \quad a_3 + b_3] \\ &= [a_1 - a_2 \quad 0 \quad a_3] + [b_1 - b_2 \quad 0 \quad b_3] \\ &= L([a_1 \quad a_2 \quad a_3)] + L([b_1 \quad b_2 \quad b_3]) \end{aligned}$$

$$\begin{aligned} L(k[a_1 \quad a_2 \quad a_3]) &= L([ka_1 \quad ka_2 \quad ka_3]) \\ &= [ka_1 - ka_2 \quad 0 \quad ka_3] \\ &= k[a_1 - a_2 \quad 0 \quad a_3] \\ &= kL([a_1 \quad a_2 \quad a_3]) \end{aligned}$$

Therefore, the transformation is linear.

(b)

$$\begin{aligned} L([a_1 \quad a_2 \quad a_3] + [b_1 \quad b_2 \quad b_3]) &= L([a_1 + b_1 \quad a_2 + b_2 \quad a_3 + b_3]) \\ &= [(a_1 + b_1) - 3(a_2 + b_2) \quad a_3 + b_3] \\ &= [(a_1 - 3a_2) + (b_1 - 3b_2) \quad a_3 + b_3] \\ &= [a_1 - 3a_2 \quad a_3] + [b_1 - 3b_2 \quad b_3] \\ &= L([a_1 \quad a_2 \quad a_3]) + L([b_1 \quad b_2 \quad b_3]) \\ L(k[a_1 \quad a_2 \quad a_3]) &= L([ka_1 \quad ka_2 \quad ka_3]) \\ &= [ka_1 - 3ka_2 \quad ka_3] \\ &= k[a_1 - 3a_2 \quad a_3] \\ &= kL([a_1 \quad a_2 \quad a_3]) \end{aligned}$$

Therefore, the transformation is linear.

(c)

$$\begin{aligned} L([a_1 \quad a_2 \quad a_3] + [b_1 \quad b_2 \quad b_3]) &= L([a_1 + b_1 \quad a_2 + b_2 \quad a_3 + b_3]) \\ &= [(a_1 + b_1) + (a_2 + b_2) \quad a_3 + b_3 \quad 1] \\ &= [(a_1 + a_2) + (b_1 + b_2) \quad a_3 + b_3 \quad 1] \end{aligned}$$

but

$$\begin{aligned} L([a_1 \quad a_2 \quad a_3]) + L([b_1 \quad b_2 \quad b_3]) &= [a_1 + a_2 \quad a_3 \quad 1] + [b_1 + b_2 \quad b_3 \quad 1] \\ &= [(a_1 + a_2) + (b_1 + b_2) \quad a_3 + b_3 \quad 2] \\ &\neq [(a_1 + a_2) + (b_1 + b_2) \quad a_3 + b_3 \quad 1] \end{aligned}$$

Therefore, the transformation is not linear.

PROBLEM 12-3 Show that the following function is a linear transformation. $L\colon R^2 \to R^3$ defined by

$$L\left(\begin{bmatrix} a \\ b \end{bmatrix}\right) = \begin{bmatrix} 1 & 2 \\ -1 & 3 \\ 2 & -1 \end{bmatrix}\begin{bmatrix} a \\ b \end{bmatrix}$$

Solution

$$\begin{aligned} L\left(\begin{bmatrix} a \\ b \end{bmatrix} + \begin{bmatrix} c \\ d \end{bmatrix}\right) &= L\left(\begin{bmatrix} a + c \\ b + d \end{bmatrix}\right) = \begin{bmatrix} 1 & 2 \\ -1 & 3 \\ 2 & -1 \end{bmatrix}\begin{bmatrix} a + c \\ b + d \end{bmatrix} \\ &= \begin{bmatrix} a + c + 2(b + d) \\ -(a + c) + 3(b + d) \\ 2(a + c) - 1(b + d) \end{bmatrix} = \begin{bmatrix} a + 2b \\ -a + 3b \\ 2a - b \end{bmatrix} + \begin{bmatrix} c + 2d \\ -c + 3d \\ 2c - d \end{bmatrix} \\ &= \begin{bmatrix} 1 & 2 \\ -1 & 3 \\ 2 & -1 \end{bmatrix}\begin{bmatrix} a \\ b \end{bmatrix} + \begin{bmatrix} 1 & 2 \\ -1 & 3 \\ 2 & -1 \end{bmatrix}\begin{bmatrix} c \\ d \end{bmatrix} = L\left(\begin{bmatrix} a \\ b \end{bmatrix}\right) + L\left(\begin{bmatrix} c \\ d \end{bmatrix}\right) \end{aligned}$$

$$\begin{aligned} L\left(k\begin{bmatrix} a \\ b \end{bmatrix}\right) &= L\left(\begin{bmatrix} ka \\ kb \end{bmatrix}\right) = \begin{bmatrix} 1 & 2 \\ -1 & 3 \\ 2 & -1 \end{bmatrix}\begin{bmatrix} ka \\ kb \end{bmatrix} = \begin{bmatrix} ka + 2kb \\ -ka + 3kb \\ 2ka - kb \end{bmatrix} \\ &= k\left(\begin{bmatrix} a + 2b \\ -a + 3b \\ 2a - b \end{bmatrix}\right) = k\left(\begin{bmatrix} 1 & 2 \\ -1 & 3 \\ 2 & -1 \end{bmatrix}\begin{bmatrix} a \\ b \end{bmatrix}\right) = kL\left(\begin{bmatrix} a \\ b \end{bmatrix}\right) \end{aligned}$$

PROBLEM 12-4 Consider the function $L: R_3 \to R$ defined by $L([a_1 \quad a_2 \quad a_3]) = [2a_1 + a_2 + 3a_3]$. Show that this function is linear.

Solution

$$\begin{aligned} L([a_1 \quad a_2 \quad a_3] + [b_1 \quad b_2 \quad b_3]) &= L([a_1 + b_1 \quad a_2 + b_2 \quad a_3 + b_3]) \\ &= [2(a_1 + b_1) + (a_2 + b_2) + 3(a_3 + b_3)] \\ &= [2a_1 + a_2 + 3a_3] + [2b_1 + b_2 + 3b_3] \\ &= L([a_1 \quad a_2 \quad a_3]) + L([b_1 \quad b_2 \quad b_3]) \\ L(k[a_1 \quad a_2 \quad a_3]) &= L([ka_1 \quad ka_2 \quad ka_3]) \\ &= [2ka_1 + ka_2 + 3ka_3] \\ &= k[2a_1 + a_2 + 3a_3] = kL([a_1 \quad a_2 \quad a_3]) \end{aligned}$$

Therefore, the function is linear.

PROBLEM 12-5 Let $L: R_3 \to R_4$ be defined by $L([a_1 \quad a_2 \quad a_3]) = [a_1 - a_2 \quad a_1 + a_3 \quad a_2 + 2a_3 \quad a_1 + a_2 + a_3]$. Show that the function is a linear transformation.

Solution

$$\begin{aligned} L([a_1 \quad a_2 \quad a_3] + [b_1 \quad b_2 \quad b_3]) &= L([a_1 + b_1 \quad a_2 + b_2 \quad a_3 + b_3]) \\ &= [(a_1 + b_1) - (a_2 + b_2) \quad (a_1 + b_1) + (a_3 + b_3) \\ &\qquad (a_2 + b_2) + 2(a_3 + b_3) \quad (a_1 + b_1) + (a_2 + b_2) + (a_3 + b_3)] \\ &= [(a_1 - a_2) + (b_1 - b_2) \quad (a_1 + a_3) + (b_1 + b_3) \\ &\qquad (a_2 + 2a_3) + (b_2 + 2b_3) \quad (a_1 + a_2 + a_3) + (b_1 + b_2 + b_3)] \\ &= [a_1 - a_2 \quad a_1 + a_3 \quad a_2 + 2a_3 \quad a_1 + a_2 + a_3] \\ &\quad + [b_1 - b_2 \quad b_1 + b_3 \quad b_2 + 2b_3 \quad b_1 + b_2 + b_3] \\ &= L([a_1 \quad a_2 \quad a_3]) + L([b_1 \quad b_2 \quad b_3]) \\ L(k[a_1 \quad a_2 \quad a_3]) &= L([ka_1 \quad ka_2 \quad ka_3]) \\ &= [ka_1 - ka_2 \quad ka_1 + ka_3 \quad ka_2 + 2ka_3 \quad ka_1 + ka_2 + ka_3] \\ &= k[a_1 - a_2 \quad a_1 + a_3 \quad a_2 + 2a_3 \quad a_1 + a_2 + a_3] \\ &= kL([a_1 \quad a_2 \quad a_3]) \end{aligned}$$

Therefore, the function is linear.

PROBLEM 12-6 For the function defined in Problem 12-5 and given $\mathbf{e}_1$, $\mathbf{e}_2$, $\mathbf{e}_3$ the natural basis vectors for R_3, show that $[1 \quad 1 \quad 2 \quad -1]$ is a linear combination of $\{L(\mathbf{e}_1), L(\mathbf{e}_2), L(\mathbf{e}_3)\}$ but that $[1 \quad 1 \quad 2 \quad 1]$ is not a linear combination of $\{L(\mathbf{e}_1), L(\mathbf{e}_2), L(\mathbf{e}_3)\}$.

Solution

$$\begin{aligned} L([1 \quad 0 \quad 0]) &= [1 \quad 1 \quad 0 \quad 1] \\ L([0 \quad 1 \quad 0]) &= [-1 \quad 0 \quad 1 \quad 1] \\ L([0 \quad 0 \quad 1]) &= [0 \quad 1 \quad 2 \quad 1] \end{aligned}$$

$$[1 \quad 1 \quad 2 \quad -1] = c_1[1 \quad 1 \quad 0 \quad 1] + c_2[-1 \quad 0 \quad 1 \quad 1] + c_3[0 \quad 1 \quad 2 \quad 1]$$

Thus,

$$\begin{aligned} c_1 - c_2 \qquad &= 1 \\ c_1 \qquad + c_3 &= 1 \\ c_2 + 2c_3 &= 2 \\ c_1 + c_2 + c_3 &= -1 \end{aligned}$$

This system has $c_1 = -1$, $c_2 = -2$, and $c_3 = 2$ as its solution. Therefore, $[1 \quad 1 \quad 2 \quad -1] = -[1 \quad 1 \quad 0 \quad 1] - 2[-1 \quad 0 \quad 1 \quad 1] + 2[0 \quad 1 \quad 2 \quad 1]$. Now suppose that $[1 \quad 1 \quad 2 \quad 1]$ is a linear combination of $L(\mathbf{e}_1)$, $L(\mathbf{e}_2)$, and $L(\mathbf{e}_3)$. Thus,

$$c_1[1 \quad 1 \quad 0 \quad 1] + c_2[-1 \quad 0 \quad 1 \quad 1] + c_3[0 \quad 1 \quad 2 \quad 1] = [1 \quad 1 \quad 2 \quad 1]$$

$$\begin{aligned} c_1 - c_2 \qquad &= 1 \\ c_1 \qquad + c_3 &= 1 \\ c_2 + 2c_3 &= 2 \\ c_1 + c_2 + c_3 &= 1 \end{aligned}$$

The first three of these equations give $c_1 = -1$, $c_2 = -2$, and $c_3 = 2$, but substituting in the last equation gives us $-1 - 2 + 2 = 1$ which is a contradiction. Hence, $[1 \quad 1 \quad 2 \quad 1]$ cannot be expressed as a linear combination of $L(\mathbf{e}_1)$, $L(\mathbf{e}_2)$, and $L(\mathbf{e}_3)$.

PROBLEM 12-7 Given the linear transformation $L\colon R_3 \to R_4$ defined by $L([a_1 \quad a_2 \quad a_3]) = [a_1 \quad a_2 - a_3 \quad a_1 + a_3 \quad 0]$, where $\mathbf{a} = [2 \quad -1 \quad 4]$ and $\mathbf{b} = [3 \quad 2 \quad 4]$, verify the following: **(a)** $L(-\mathbf{a}) = -L(\mathbf{a})$; **(b)** $L(\mathbf{a} - \mathbf{b}) = L(\mathbf{a}) - L(\mathbf{b})$.

Solution

(a)

$$\begin{aligned} L(-\mathbf{a}) &= L(-[2 \quad -1 \quad 4]) = L([-2 \quad 1 \quad -4]) \\ &= [-2 \quad 5 \quad -6 \quad 0] = -[2 \quad -5 \quad 6 \quad 0] = -L(\mathbf{a}) \end{aligned}$$

(b)

$$\begin{aligned} L(\mathbf{a} - \mathbf{b}) &= L([2 \quad -1 \quad 4] - [3 \quad 2 \quad 4]) = L([-1 \quad -3 \quad 0]) \\ &= [-1 \quad -3 \quad -1 \quad 0] \\ L(\mathbf{a}) - L(\mathbf{b}) &= L([2 \quad -1 \quad 4]) - L([3 \quad 2 \quad 4]) \\ &= [2 \quad -5 \quad 6 \quad 0] - [3 \quad -2 \quad 7 \quad 0] \\ &= [-1 \quad -3 \quad -1 \quad 0] \end{aligned}$$

Therefore, $L(\mathbf{a} - \mathbf{b}) = L(\mathbf{a}) - L(\mathbf{b})$.

PROBLEM 12-8 Determine if each of the following linear transformation is one-to-one and onto: **(a)** $L\colon R_3 \to R_2$ defined by $L([a_1 \quad a_2 \quad a_3]) = [a_1 + 2a_2 \quad a_3]$, **(b)** $L\colon R_3 \to R_3$ defined by $L([a_1 \quad a_2 \quad a_3]) = [a_3 \quad -a_2 \quad 0]$.

Solution

(a) Let $L([a_1 \quad a_2 \quad a_3]) = L([b_1 \quad b_2 \quad b_3])$. Thus, $[a_1 + 2a_2 \quad a_3] = [b_1 + 2b_2 \quad b_3]$, and therefore,

$$\begin{aligned} a_1 + 2a_2 &= b_1 + 2b_2 \\ a_3 &= b_3 \end{aligned}$$

Although this implies $a_3 = b_3$, it does not imply $a_1 = b_1$ and $a_2 = b_2$, since $a_1 = 2b_2$ and $a_2 = (\frac{1}{2})b_1$ is also possible. Hence, the transformation is not one-to-one.

To determine if the transformation is onto, let $[a_1 \quad a_2]$ be any vector in R_2. Hence,

$$\begin{aligned} L([x \quad y \quad z]) &= [a_1 \quad a_2] \\ x + 2y &= a_1 \\ z &= a_2 \end{aligned}$$

Thus, to find a vector in V whose image is $[a_1 \quad a_2]$, we will choose any two real numbers x and y such that $x + 2y = a_1$ for the first two components, and let the third component be a_2. For example, if $[3 \quad 2]$ is any vector in R_2, $x = 1$, $y = 1$, and $z = 2$ would give vector $[1 \quad 1 \quad 2]$ in V. Therefore, this linear transformation is onto but not one-to-one.

(b) Let $L([a_1 \quad a_2 \quad a_3]) = L([b_1 \quad b_2 \quad b_3])$. Thus, $[a_3 \quad -a_2 \quad 0] = [b_3 \quad -b_2 \quad 0]$. Therefore, $a_3 = b_3$, $a_2 = b_2$, but a_1 and b_1 are arbitrary. Hence, the transformation is not one-to-one.

Since $[a_1 \quad a_2 \quad 1]$ is not the image of any vector in R_3, the transformation is not onto.

PROBLEM 12-9 Show that the linear transformation $L: P_3 \to {}_2M_2$ defined by

$$L(a_0 + a_1x + a_2x^2 + a_3x^3) = \begin{bmatrix} a_0 & a_1 \\ a_2 & a_3 \end{bmatrix}$$

is an isomorphism.

Solution Let

$$L(a_0 + a_1x + a_2x^2 + a_3x^3) = a_0\begin{bmatrix} 1 & 0 \\ 0 & 0 \end{bmatrix} + a_1\begin{bmatrix} 0 & 1 \\ 0 & 0 \end{bmatrix} + a_2\begin{bmatrix} 0 & 0 \\ 1 & 0 \end{bmatrix} + a_3\begin{bmatrix} 0 & 0 \\ 0 & 1 \end{bmatrix} = \begin{bmatrix} a_0 & a_1 \\ a_2 & a_3 \end{bmatrix}$$

Then

$${}_2M_2 = \begin{bmatrix} a & b \\ c & d \end{bmatrix} = \begin{bmatrix} a_0 & a_1 \\ a_2 & a_3 \end{bmatrix}$$

is true only if $a_0 = a$, $a_1 = b$, $a_2 = c$, and $a_3 = d$. Thus, the transformation is one-to-one.

Furthermore, the transformation is onto because

$$L(a + bx + cx^2 + dx^3) = \begin{bmatrix} a & b \\ c & d \end{bmatrix} = {}_2M_2$$

Therefore, P_3 is isomorphic to ${}_2M_2$.

PROBLEM 12-10 Indicate a vector space from R_n which is isomorphic to each of the following vector spaces: **(a)** ${}_2M_4$, **(b)** R^6, **(c)** P_4, **(d)** $\mathbf{P}_7$.

Solution

(a) Since $\dim {}_2M_4 = 2(4) = 8$, ${}_2M_4$ is isomorphic to R_8.
(b) Since $\dim R^6 = 6$, R^6 is isomorphic to R_6.
(c) Since $\dim P_4 = 4 + 1 = 5$, P_4 is isomorphic to R_5.
(d) Since $\dim P_7 = 7 + 1 = 8$, P_7 is isomorphic to R_8.

PROBLEM 12-11 If a vector space V is isomorphic to R_9, explain why V is isomorphic to ${}_3M_3$.

Solution R_9 is isomorphic to ${}_3M_3$ since $\dim R_9 = \dim {}_3M_3 = 9$; and by the transitive property of an equivalence relation V is isomorphic to V.

PROBLEM 12-12 Determine which of the following vectors are in the kernel of a linear transformation $L: R_3 \to R_2$ defined by $L([a_1 \quad a_2 \quad a_3]) = [a_1 + 2a_3 \quad a_2]$: **(a)** $[2 \quad 0 \quad -1]$, **(b)** $[3 \quad 0 \quad -\frac{3}{2}]$, **(c)** $[0 \quad 0 \quad 0]$, **(d)** $[2 \quad 1 \quad -1]$, **(e)** $[3 \quad 1 \quad 0]$.

Solution

(a) $L([2 \quad 0 \quad -1]) = [2 - 2 \quad 0] = [0 \quad 0] \therefore [2 \quad 0 \quad -1]$ is in $\ker L$.
(b) $L([3 \quad 0 \quad -\frac{3}{2}]) = [3 - 3 \quad 0] = [0 \quad 0] \therefore [3 \quad 0 \quad -\frac{3}{2}]$ is in $\ker L$.
(c) $L([0 \quad 0 \quad 0]) = [0 + 0 \quad 0] = [0 \quad 0] \therefore [0 \quad 0 \quad 0]$ is in $\ker L$.
(d) $L([2 \quad 1 \quad -1]) = [2 - 2 \quad 1] = [0 \quad 1] \therefore [2 \quad 1 \quad -1]$ is not in $\ker L$.
(e) $L([3 \quad 1 \quad 0]) = [3 + 0 \quad 1] = [3 \quad 0] \therefore [3 \quad 1 \quad 0]$ is not in $\ker L$.

PROBLEM 12-13 For the linear transformation $L: R_3 \to R_2$ defined by $L([a_1 \quad a_2 \quad a_3]) = [a_1 + 2a_3 \quad a_2]$ find **(a)** $\ker L$ and **(b)** a basis for $\ker L$.

Solution

(a) We must find all the vectors $[a_1 \quad a_2 \quad a_3]$ in V such that $L([a_1 \quad a_2 \quad a_3]) = [0 \quad 0]$, but $L([a_1 \quad a_2 \quad a_3]) = [a_1 + 2a_3 \quad a_2]$. Hence, $[a_1 + 2a_3 \quad a_2] = [0 \quad 0]$. Thus, $a_1 + 2a_3 = 0$ and

$a_2 = 0$ or $a_1 = -2a_3$ and $a_2 = 0$. Therefore, $\ker L = [-2a_3 \quad 0 \quad a_3]$ or $[-2t \quad 0 \quad t]$ as usually written.

(b) $[-2t \quad 0 \quad t] = t[-2 \quad 0 \quad 1]$. Therefore, $\{[-2 \quad 0 \quad 1]\}$ is a basis for $\ker L$.

PROBLEM 12-14 Determine whether each of the given vectors is in the range of the linear transformation $L: R_3 \to R_2$ defined by $L([a_1 \quad a_2 \quad a_3]) = [a_1 + 2a_3 \quad a_2]$: **(a)** $[7 \quad -2]$, **(b)** $[2 \quad 3]$.

Solution

(a) We must determine if there is some $[a_1 \quad a_2 \quad a_3]$ such that

$$L([a_1 \quad a_2 \quad a_3]) = [7 \quad -2]$$

$$L([a_1 \quad a_2 \quad a_3]) = [a_1 + 2a_3 \quad a_2] = [7 \quad -2]$$

$$\therefore \quad a_1 + 2a_3 = 7$$

$$a_2 = -2$$

Among the many solutions of the first of these equations is $a_1 = 1, a_3 = 3$. Hence, $L([1 \quad -2 \quad 3]) = [7 \quad -2]$. Therefore, $[7 \quad -2]$ is the image of $[1 \quad -2 \quad 3]$ and is in range L.

(b) We need to find some $[a_1 \quad a_2 \quad a_3]$ such that

$$L([a_1 \quad a_2 \quad a_3]) = [2 \quad 3]$$

$$L([a_1 \quad a_2 \quad a_3]) = [a_1 + 2a_3 \quad a_2] = [2 \quad 3]$$

$$\therefore \quad a_1 + 2a_3 = 2$$

$$a_2 = 3$$

Thus, among the solutions to the first equation is $a_1 = 4$ and $a_3 = -1$. Hence, $L([4 \quad 3 \quad -1]) = [2 \quad 3]$. Therefore, $[2 \quad 3]$ is the image of $[4 \quad 3 \quad -1]$ and is, thus, in range L.

PROBLEM 12-15 Find a basis for the range of the linear transformation in Problem 12-14.

Solution We first must find a set of vectors that spans range L. We then need to find a maximal linearly independent subset of these spanning vectors. We begin with

$$L([a_1 \quad a_2 \quad a_3]) = [a_1 + 2a_3 \quad a_2]$$

We can rewrite the right side as

$$[a_1 + 2a_3 \quad 0] + [0 \quad a_2] = [a_1 \quad 0] + [2a_3 \quad 0] + [0 \quad a_3] = a_1[1 \quad 0] + a_3[2 \quad 0] + a_2[0 \quad 1]$$

Thus $\{[1 \quad 0], [2 \quad 0], [0 \quad 1]\}$ spans range L. Since $[2 \quad 0] = 2[1 \quad 0]$, we write this as $\{[1 \quad 0], [1 \quad 0], [0 \quad 1]\}$. These are the natural basis vectors of R_2, and the maximal linearly independent subset is $\{[1 \quad 0], [0 \quad 1]\}$. Therefore, $\{[1 \quad 0], [0 \quad 1]\}$ is a basis for range L.

PROBLEM 12-16 Find a basis for the range of the linear transformation $L: R_4 \to R_4$ defined by $L([a_1 \quad a_2 \quad a_3 \quad a_4]) = [a_1 + 2a_2 + a_4 \quad a_2 - a_3 - a_4 \quad a_1 + 3a_3 + a_4 \quad a_1 + a_2 + 2a_3]$.

Solution

$$\begin{aligned} L([a_1 \quad a_2 \quad a_3 \quad a_4]) &= [a_1 + 2a_2 + a_4 \quad a_2 - a_3 - a_4 \quad a_1 + 3a_3 + a_4 \quad a_1 + a_2 + 2a_3] \\ &= [a_1 \quad 0 \quad a_1 \quad a_1] + [2a_2 \quad a_2 \quad 0 \quad a_2] \\ &\quad + [0 \quad -a_3 \quad 3a_3 \quad 2a_3] + [a_4 \quad -a_4 \quad a_4 \quad 0] \\ &= a_1[1 \quad 0 \quad 1 \quad 1] + a_2[2 \quad 1 \quad 0 \quad 1] + a_3[0 \quad -1 \quad 3 \quad 2] + a_4[1 \quad -1 \quad 1 \quad 0] \end{aligned}$$

Thus, $\{[1 \quad 0 \quad 1 \quad 1], [2 \quad 1 \quad 0 \quad 1], [0 \quad -1 \quad 3 \quad 2], [1 \quad -1 \quad 1 \quad 0]\}$ spans range L.

To find a maximal linearly independent subset of these vectors we will form a matrix with these four vectors as rows. After reducing the matrix to echelon form, we will write the nonzero rows to get a basis for range L.

$$\begin{bmatrix} 1 & 0 & 1 & 1 \\ 2 & 1 & 0 & 1 \\ 0 & -1 & 3 & 2 \\ 1 & -1 & 1 & 0 \end{bmatrix} \xrightarrow[R_4 \to R_4 - R_1]{R_2 \to R_2 - 2R_1} \begin{bmatrix} 1 & 0 & 1 & 1 \\ 0 & 1 & -2 & -1 \\ 0 & -1 & 3 & 2 \\ 0 & -1 & 0 & -1 \end{bmatrix} \xrightarrow[R_4 \to R_4 + R_2]{R_3 \to R_3 + R_2} \begin{bmatrix} 1 & 0 & 1 & 1 \\ 0 & 1 & -2 & -1 \\ 0 & 0 & 1 & 1 \\ 0 & 0 & -2 & -2 \end{bmatrix}$$

$$\xrightarrow[R_4 \to R_4 + 2R_3]{} \begin{bmatrix} 1 & 0 & 1 & 1 \\ 0 & 1 & -2 & -1 \\ 0 & 0 & 1 & 1 \\ 0 & 0 & 0 & 0 \end{bmatrix}$$

Therefore, $\{[1 \quad 0 \quad 1 \quad 1], [0 \quad 1 \quad -2 \quad -1], [0 \quad 0 \quad 1 \quad 1]\}$ is a basis for range L.

PROBLEM 12-17 Verify $\dim(\ker L) + \dim(\text{range } L) = \dim V$ for the linear transformation in Problem 12-16.

Solution In Problem 12-16 we found that a basis for range L consists of three vectors. Thus, $\dim(\text{range } L) = 3$. Also, $V = R_4$, so $\dim V = \dim R_4 = 4$. We must find ker L and its dimension. Kernel L consists of all vectors $[a_1 \quad a_2 \quad a_3 \quad a_4]$ such that

$$L([a_1 \quad a_2 \quad a_3 \quad a_4]) = [0 \quad 0 \quad 0 \quad 0]$$

but

$$L([a_1 \quad a_2 \quad a_3 \quad a_4]) = [a_1 + 2a_2 + a_4 \quad a_2 - a_3 - a_4 \quad a_1 + 3a_3 + a_4 \quad a_1 + a_2 + 2a_3]$$

Thus,

$$\begin{aligned} a_1 + 2a_2 \qquad\quad + a_4 &= 0 \\ a_2 - \ a_3 - a_4 &= 0 \\ a_1 \qquad\quad + 3a_3 + a_4 &= 0 \\ a_1 + \ a_2 + 2a_3 \qquad &= 0 \end{aligned}$$

Solving by the Gauss–Jordan elimination, we get

$$\begin{bmatrix} 1 & 2 & 0 & 1 \\ 0 & 1 & -1 & -1 \\ 1 & 0 & 3 & 1 \\ 1 & 1 & 2 & 0 \end{bmatrix} \xrightarrow[R_4 \to R_4 - R_1]{R_3 \to R_3 - R_1} \begin{bmatrix} 1 & 2 & 0 & 1 \\ 0 & 1 & -1 & -1 \\ 0 & -2 & 3 & 0 \\ 0 & -1 & 2 & -1 \end{bmatrix} \xrightarrow[R_4 \to R_4 + R_2]{R_3 \to R_3 + 2R_2} \begin{bmatrix} 1 & 2 & 0 & 1 \\ 0 & 1 & -1 & -1 \\ 0 & 0 & 1 & -2 \\ 0 & 0 & 1 & -2 \end{bmatrix}$$

$$\xrightarrow[R_4 \to R_4 - R_3]{} \begin{bmatrix} 1 & 2 & 0 & 1 \\ 0 & 1 & -1 & -1 \\ 0 & 0 & 1 & -2 \\ 0 & 0 & 0 & 0 \end{bmatrix} \xrightarrow{R_2 \to R_2 + R_3} \begin{bmatrix} 1 & 2 & 0 & 1 \\ 0 & 1 & 0 & -3 \\ 0 & 0 & 1 & -2 \\ 0 & 0 & 0 & 0 \end{bmatrix}$$

$$\xrightarrow{R_1 \to R_1 - 2R_2} \begin{bmatrix} 1 & 0 & 0 & 7 \\ 0 & 1 & 0 & -3 \\ 0 & 0 & 1 & -2 \\ 0 & 0 & 0 & 0 \end{bmatrix}$$

$$\begin{aligned} \therefore \quad a_1 \qquad\quad + 7a_4 &= 0 \\ a_2 \qquad - 3a_4 &= 0 \\ a_3 - 2a_4 &= 0 \end{aligned}$$

Thus

$$a_1 = -7a_4$$
$$a_2 = 3a_4$$
$$a_3 = 2a_4$$

Let $a_4 = t$ where t is any real number, then $a_1 = -7t$, $a_2 = 3t$, and $a_3 = 2t$. Therefore, the kernel is all vectors of form $[-7t \quad 3t \quad 2t \quad t]$. However, this is equivalent to $[-7 \quad 3 \quad 2 \quad 1]$. Thus, a basis for $\ker L$ is $\{[-7 \quad 3 \quad 2 \quad 1]\}$ which has dimension 1. Therefore, $\dim(\ker L) + \dim(\text{range } L) = 1 + 3 = 4 = \dim V$.

Supplementary Problems

PROBLEM 12-18 Determine if each of the following functions is a linear function: **(a)** $L: R_2 \to R_2$ defined by $L([a_1 \quad a_2]) = [a_1 - a_2 \quad 0]$. **(b)** $L: R_3 \to R_3$ defined by $L([a_1 \quad a_2 \quad a_3]) = [a_1 + a_2 \quad a_2 \quad a_2 + a_3]$. **(c)** $L: R_3 \to R_2$ defined by $L([a_1 \quad a_2 \quad a_3]) = [a_1 + a_2 + a_3 \quad 1]$. **(d)** $L: R_2 \to R_2$ defined by $L([a_1 \quad a_2]) = [a_1 + a_2 \quad a_1 - a_2]$. **(e)** $L: R^2 \to R^3$ defined by $L\left(\begin{bmatrix} a_1 \\ a_2 \end{bmatrix}\right) = \begin{bmatrix} a_1 + a_2 \\ a_1 - a_2 \\ 0 \end{bmatrix}$. **(f)** $L: R^3 \to R$ defined by $L\left(\begin{bmatrix} a_1 \\ a_2 \\ a_3 \end{bmatrix}\right) = [a_1 + a_2 - 2a_3]$.

PROBLEM 12-19 Determine if each of the following linear transformations is one-to-one and onto: **(a)** $L: R_3 \to R_2$ defined by $L([a_1 \quad a_2 \quad a_3]) = [a_1 \quad a_2 + a_3]$. **(b)** $L: R_3 \to R_3$ defined by $L([a_1 \quad a_2 \quad a_3]) = [a_1 - a_2 \quad a_2 \quad a_3 + a_1]$. **(c)** $L: R_3 \to R_3$ defined by $L([a_1 \quad a_2 \quad a_3]) = [a_1 - a_2 \quad a_3 \quad 0]$.

PROBLEM 12-20 Determine if each of the following linear transformations is an isomorphism: **(a)** $L: R^2 \to R^3$ defined by $L\left(\begin{bmatrix} a \\ b \end{bmatrix}\right) = \begin{bmatrix} a \\ a + b \\ b \end{bmatrix}$. **(b)** $L: R_4 \to {}_2M_2$ defined by $L([a_1 \quad a_2 \quad a_3 \quad a_4]) = \begin{bmatrix} a_1 & a_2 \\ a_3 & a_4 \end{bmatrix}$.

PROBLEM 12-21 Give three vector spaces that are isomorphic to ${}_2M_3$.

PROBLEM 12-22 Determine which of the following vectors are in the kernel of a linear transformation $L: R_3 \to R_3$ defined by $L([a_1 \quad a_2 \quad a_3]) = [a_1 + a_2 \quad a_2 + a_3 \quad a_1 - a_3]$: **(a)** $[1 \quad -1 \quad 1]$, **(b)** $[0 \quad -1 \quad 0]$, **(c)** $[-2 \quad 2 \quad -2]$, **(d)** $[0 \quad 0 \quad 0]$, **(e)** $[\frac{1}{3} \quad -\frac{1}{3} \quad \frac{1}{3}]$.

PROBLEM 12-23 For the linear transformation $L: R_3 \to R_2$ defined by $L([a_1 \quad a_2 \quad a_3]) = [a_1 - a_2 \quad 2a_3]$ find: **(a)** $\ker L$; **(b)** a basis for $\ker L$.

PROBLEM 12-24 Determine whether each of the following given vectors is in the range of the linear transformation $L: R_3 \to R_2$ defined by $L([a_1 \quad a_2 \quad a_3]) = [a_1 - a_2 \quad 2a_3]$: **(a)** $[-1 \quad -10]$, **(b)** $[-2 \quad 6]$, **(c)** $[1 \quad 4]$.

PROBLEM 12-25 Find a basis for the range of $L: R_3 \to R_2$ defined by $L([a_1 \quad a_2 \quad a_3]) = [a_1 - a_2 \quad 2a_3]$.

PROBLEM 12-26 Find a basis for the range of $L: R_4 \to R_4$ defined by

$$L([a_1 \quad a_2 \quad a_3 \quad a_4]) = [a_1 \quad -a_2 + 2a_3 \quad a_1 + a_3 - a_4 \quad a_1 + 2a_2 - 2a_4 \quad a_2 + 2a_3 + 2a_4]$$

PROBLEM 12-27 Find a basis of the kernel of the linear transformation in Problem 12-26. What is dim (ker L)?

PROBLEM 12-28 Let $L: R^3 \to R^3$ be defined by $L([a \quad b \quad c]) = [a - b \quad b + c \quad a + c]$. Find a basis for range L.

PROBLEM 12-29 Let $L: P_2 \to P_1$ be defined by $L([a_0 \quad a_1x \quad a_2x^2]) = [a_0 + a_2 \quad (a_0 + a_1)x \quad 0]$. (**a**) Is $1 - x - x^2$ in ker L? (**b**) Is $2 + x + x^2$ in range L?

PROBLEM 12-30 In Problem 12-29 find a basis for ker L.

PROBLEM 12-31 Let $L: R_3 \to R_2$ be defined by $L([a_1 \quad a_2 \quad a_3]) = [a_1 - a_2 + a_3 \quad a_1 + 2a_2 - 3a_3]$. Find (**a**) ker L and (**b**) the nullity of L.

Answers to Supplementary Problems

12-18 (**a**), (**b**), (**d**), (**e**), (**f**) Linear (**c**) Nonlinear

12-19 (**a**) Onto but not one-to-one
(**b**) Onto and one-to-one
(**c**) Not onto and not one-to-one

12-20 (**a**) Not an isomorphism
(**b**) An isomorphism

12-21 P_5, R^6, R_6

12-22 (**a**), (**c**), (**d**), (**e**) In ker L

12-23 (**a**) $[t \quad t \quad 0]$ (**b**) $\{[1 \quad 1 \quad 0]\}$

12-24 (**a**), (**b**), (**c**) in range L

12-25 $\{[1 \quad 0], [0 \quad 1]\}$

12-26 $\{[1 \quad 1 \quad 1 \quad 0], [0 \quad 1 \quad 3 \quad 1], [0 \quad 0 \quad 1 \quad 3]\}$

12-27 $\{[2 \quad 0 \quad -1 \quad 1]\}$; dim (ker L) = 1

12-28 $\{[1 \quad 0 \quad 1], [0 \quad 1 \quad 1]\}$

12-29 (**a**) Yes (**b**) No

12-30 $\{-1 + x + x^2\}$

12-31 (**a**) $[t \quad 4t \quad 3t]$
(**b**) Nullity L = dim (ker L) = 1

13 MATRICES OF LINEAR TRANSFORMATIONS

THIS CHAPTER IS ABOUT

- ☑ Matrix Transformation
- ☑ Coordinates
- ☑ Transition Matrices
- ☑ Matrix of a Linear Transformation
- ☑ Change of Bases for Matrix Transformation
- ☑ Partitioned-Matrix Method for Change of Bases

13-1. Matrix Transformation

Definition of matrix transformation

If A is any $m \times n$ matrix, the function $L: R^n \to R^m$ defined by $L(\mathbf{a}) = A\mathbf{a}$ for $\mathbf{a}$ in R^n is called a **matrix transformation**.

Property A matrix transformation is a linear transformation.

EXAMPLE 13-1: Verify that the matrix transformation $L: R^2 \to R^4$ defined by

$$L\left(\begin{bmatrix} a \\ b \end{bmatrix}\right) = \begin{bmatrix} 2 & -1 \\ 0 & 3 \\ 1 & 5 \\ 4 & 2 \end{bmatrix}\begin{bmatrix} a \\ b \end{bmatrix}$$

is a linear transformation.

Solution

$$L\left(\begin{bmatrix} a \\ b \end{bmatrix} + \begin{bmatrix} c \\ d \end{bmatrix}\right) = L\left(\begin{bmatrix} a + c \\ b + d \end{bmatrix}\right) = \begin{bmatrix} 2 & -1 \\ 0 & 3 \\ 1 & 5 \\ 4 & 2 \end{bmatrix}\begin{bmatrix} a + c \\ b + d \end{bmatrix} = \begin{bmatrix} 2(a + c) - 1(b + d) \\ 0(a + c) + 3(b + d) \\ 1(a + c) + 5(b + d) \\ 4(a + c) + 2(b + d) \end{bmatrix}$$

$$= \begin{bmatrix} 2a - b \\ 0 + 3b \\ a + 5b \\ 4a + 2b \end{bmatrix} + \begin{bmatrix} 2c - d \\ 0c + 3d \\ c + 5d \\ 4c + 2d \end{bmatrix} = \begin{bmatrix} 2 & -1 \\ 0 & 3 \\ 1 & 5 \\ 4 & 2 \end{bmatrix}\begin{bmatrix} a \\ b \end{bmatrix} + \begin{bmatrix} 2 & -1 \\ 0 & 3 \\ 1 & 5 \\ 4 & 2 \end{bmatrix}\begin{bmatrix} c \\ d \end{bmatrix} = L\left(\begin{bmatrix} a \\ b \end{bmatrix}\right) + L\left(\begin{bmatrix} c \\ d \end{bmatrix}\right)$$

$$L\left(k\begin{bmatrix} a \\ b \end{bmatrix}\right) = L\left(\begin{bmatrix} ka \\ kb \end{bmatrix}\right) = \begin{bmatrix} 2 & -1 \\ 0 & 3 \\ 1 & 5 \\ 4 & 2 \end{bmatrix}\begin{bmatrix} ka \\ kb \end{bmatrix} = \begin{bmatrix} 2(ka) - 1(kb) \\ 0(ka) + 3(kb) \\ 1(ka) + 5(kb) \\ 4(ka) + 2(kb) \end{bmatrix}$$

$$= k\begin{bmatrix} 2a - b \\ 0a + 3b \\ a + 5b \\ 4a + 2b \end{bmatrix} = k\begin{bmatrix} 2 & -1 \\ 0 & 3 \\ 1 & 5 \\ 4 & 2 \end{bmatrix}\begin{bmatrix} a \\ b \end{bmatrix} = kL\left(\begin{bmatrix} a \\ b \end{bmatrix}\right)$$

Therefore, the given matrix transformation is a linear transformation.

note: After we have developed some needed background, you will see later in this chapter that every linear transformation between finite-dimensional vector spaces can be represented as a matrix transformation.

13-2. Coordinates

A. Definition of ordered basis

A basis $B = \{\mathbf{b}_1, \mathbf{b}_2, \ldots, \mathbf{b}_n\}$ for a vector space V is an **ordered basis** if the vectors in B are assigned a first position, a second position, and so on. In this instance, $\mathbf{b}_1$ is the first vector, $\mathbf{b}_2$ the second vector, and $\mathbf{b}_n$ the last vector.

EXAMPLE 13-2: An ordered basis for R_3 is $\{[1\ \ 0\ \ 1], [0\ \ 1\ \ 1], [1\ \ 1\ \ 0]\}$. Write five other ordered bases for R_3 and denote them by B_1, B_2, B_3, B_4, and B_5.

Solution

$$B_1 = \{[1\ \ 0\ \ 1], [1\ \ 1\ \ 0], [0\ \ 1\ \ 1]\}$$
$$B_2 = \{[1\ \ 1\ \ 0], [1\ \ 0\ \ 1], [0\ \ 1\ \ 1]\}$$
$$B_3 = \{[1\ \ 1\ \ 0], [0\ \ 1\ \ 1], [1\ \ 0\ \ 1]\}$$
$$B_4 = \{[0\ \ 1\ \ 1], [1\ \ 1\ \ 0], [1\ \ 0\ \ 1]\}$$
$$B_5 = \{[0\ \ 1\ \ 1], [1\ \ 0\ \ 1], [1\ \ 1\ \ 0]\}$$

B. Definition of coordinate vector

If $B = \{\mathbf{b}_1, \mathbf{b}_2, \ldots, \mathbf{b}_n\}$ is an ordered basis for the n-dimensional vector space V and $\mathbf{a}$ is any vector in V, then

$$[\mathbf{a}]_B = \begin{bmatrix} c_1 \\ c_2 \\ \vdots \\ c_n \end{bmatrix}$$

is the coordinate vector of $\mathbf{a}$ with respect to the ordered basis B where $c_1, c_2, \ldots, c_n$ are the unique coefficients for $\mathbf{a} = c_1\mathbf{b}_1 + c_2\mathbf{b}_2 + \cdots + c_n\mathbf{b}_n$.

EXAMPLE 13-3: Find the coordinate vector of $[1 \quad 2 \quad 3]$ with respect to each of the following ordered bases for R_3: **(a)** $B = \{[1 \quad 0 \quad 0], [0 \quad 1 \quad 0], [0 \quad 0 \quad 1]\}$, **(b)** $T = \{[1 \quad 0 \quad 1], [0 \quad 1 \quad 1], [1 \quad 1 \quad 0]\}$, and **(c)** $S = \{[0 \quad 1 \quad 1], [1 \quad 1 \quad 0], [1 \quad 0 \quad 1]\}$.

Solution

(a)

$$c_1[1 \quad 0 \quad 0] + c_2[0 \quad 1 \quad 0] + c_3[0 \quad 0 \quad 1] = [1 \quad 2 \quad 3]$$

$$[c_1 \quad c_2 \quad c_3] = [1 \quad 2 \quad 3]$$

$$\therefore \quad c_1 = 1, c_2 = 2, c_3 = 3$$

Hence,

$$[1 \quad 2 \quad 3]_B = \begin{bmatrix} 1 \\ 2 \\ 3 \end{bmatrix}$$

Observe that a vector is its own coordinate vector with respect to the ordered natural basis.

(b)

$$c_1[1 \quad 0 \quad 1] + c_2[0 \quad 1 \quad 1] + c_3[1 \quad 1 \quad 0] = [1 \quad 2 \quad 3]$$

$$\begin{aligned} \therefore \quad c_1 \quad\quad\; + c_3 &= 1 \\ c_2 + c_3 &= 2 \\ c_1 + c_2 \quad\quad &= 3 \end{aligned}$$

which has $c_1 = 1$, $c_2 = 2$, and $c_3 = 0$ as its solution. Therefore,

$$[1 \quad 2 \quad 3]_T = \begin{bmatrix} 1 \\ 2 \\ 0 \end{bmatrix}$$

(c)

$$c_1[0 \quad 1 \quad 1] + c_2[1 \quad 1 \quad 0] + c_3[1 \quad 0 \quad 1] = [1 \quad 2 \quad 3]$$

$$\begin{aligned} \therefore \quad c_2 + c_3 &= 1 \\ c_1 + c_2 \quad\quad &= 2 \\ c_1 \quad\quad\; + c_3 &= 3 \end{aligned}$$

which has $c_1 = 2$, $c_2 = 0$, and $c_3 = 1$ as solution. Therefore,

$$[1 \quad 2 \quad 3]_S = \begin{bmatrix} 2 \\ 0 \\ 1 \end{bmatrix}$$

Observe that the coordinate vector of a vector depends not only on the basis but also on the ordering of the elements in the basis.

13-3. Transition Matrices

A. Definition of transition matrix

If B and S are two ordered bases and $\mathbf{a}$ any vector in an n-dimensional vector space V, then P is the **transition matrix** from basis B to basis S provided $[\mathbf{a}]_S = P[\mathbf{a}]_B$ where $[\mathbf{a}]_B$ is the coordinate vector of $\mathbf{a}$ with respect to the basis B and $[\mathbf{a}]_S$ is the coordinate vector of $\mathbf{a}$ with respect to the basis S.

EXAMPLE 13-4: For Example 13-3, verify that the matrix

$$P = \begin{bmatrix} -\frac{1}{2} & \frac{1}{2} & \frac{1}{2} \\ \frac{1}{2} & \frac{1}{2} & -\frac{1}{2} \\ \frac{1}{2} & -\frac{1}{2} & \frac{1}{2} \end{bmatrix}$$

transforms the coordinate vector [1 2 3] with respect to basis B into its coordinate vector with respect to basis S.

Solution: From Example 13-3,

$$[1 \quad 2 \quad 3]_B = \begin{bmatrix} 1 \\ 2 \\ 3 \end{bmatrix} \quad \text{and} \quad [1 \quad 2 \quad 3]_S = \begin{bmatrix} 2 \\ 0 \\ 1 \end{bmatrix}$$

Therefore,

$$P[1 \quad 2 \quad 3]_B = \begin{bmatrix} -\frac{1}{2} & \frac{1}{2} & \frac{1}{2} \\ \frac{1}{2} & \frac{1}{2} & -\frac{1}{2} \\ \frac{1}{2} & -\frac{1}{2} & \frac{1}{2} \end{bmatrix} \begin{bmatrix} 1 \\ 2 \\ 3 \end{bmatrix} = \begin{bmatrix} 2 \\ 0 \\ 1 \end{bmatrix} = [1 \quad 2 \quad 3]_S$$

B. Nonsingularity of transition matrices

A transition matrix P from basis B to basis S is **nonsingular**, and its inverse P^{-1} is the transition matrix from basis S to basis B.

EXAMPLE 13-5: Find the inverse matrix P^{-1} in Example 13-4, and verify that it transforms the coordinate vector of [1 2 3] with respect to basis S into its coordinate vector with respect to basis B.

Solution

$$\left[\begin{array}{ccc|ccc} -\frac{1}{2} & \frac{1}{2} & \frac{1}{2} & 1 & 0 & 0 \\ \frac{1}{2} & \frac{1}{2} & -\frac{1}{2} & 0 & 1 & 0 \\ \frac{1}{2} & -\frac{1}{2} & \frac{1}{2} & 0 & 0 & 1 \end{array}\right] \xrightarrow[R_3 \to R_3 + R_1]{R_2 \to R_2 + R_1} \left[\begin{array}{ccc|ccc} -\frac{1}{2} & \frac{1}{2} & \frac{1}{2} & 1 & 0 & 0 \\ 0 & 1 & 0 & 1 & 1 & 0 \\ 0 & 0 & 1 & 1 & 0 & 1 \end{array}\right]$$

$$\xrightarrow{R_1 \to R_1 - 1/2 R_3} \left[\begin{array}{ccc|ccc} -\frac{1}{2} & \frac{1}{2} & 0 & \frac{1}{2} & 0 & -\frac{1}{2} \\ 0 & 1 & 0 & 1 & 1 & 0 \\ 0 & 0 & 1 & 1 & 0 & 1 \end{array}\right]$$

$$\xrightarrow{R_1 \to R_1 - 1/2 R_2} \left[\begin{array}{ccc|ccc} -\frac{1}{2} & 0 & 0 & 0 & -\frac{1}{2} & -\frac{1}{2} \\ 0 & 1 & 0 & 1 & 1 & 0 \\ 0 & 0 & 1 & 1 & 0 & 1 \end{array}\right]$$

$$\xrightarrow{R_1 \to -2R_1} \left[\begin{array}{ccc|ccc} 1 & 0 & 0 & 0 & 1 & 1 \\ 0 & 1 & 0 & 1 & 1 & 0 \\ 0 & 0 & 1 & 1 & 0 & 1 \end{array}\right]$$

$$\therefore \quad P^{-1} = \begin{bmatrix} 0 & 1 & 1 \\ 1 & 1 & 0 \\ 1 & 0 & 1 \end{bmatrix}$$

$$P^{-1}[1 \quad 2 \quad 3]_S = \begin{bmatrix} 0 & 1 & 1 \\ 1 & 1 & 0 \\ 1 & 0 & 1 \end{bmatrix} \begin{bmatrix} 2 \\ 0 \\ 1 \end{bmatrix} = \begin{bmatrix} 1 \\ 2 \\ 3 \end{bmatrix} = [1 \quad 2 \quad 3]_B$$

C. Finding a transition matrix

If $B = \{\mathbf{b}_1, \mathbf{b}_2, \ldots, \mathbf{b}_n\}$ and $S = \{\mathbf{s}_1, \mathbf{s}_2, \ldots, \mathbf{s}_n\}$ are two ordered bases of an n-dimensional vector space V, the transition matrix P from basis B to basis S can be found as follows.

Step 1: Find $[\mathbf{b}_1]_S, [\mathbf{b}_2]_S, \ldots, [\mathbf{b}_n]_S$.
Step 2: Form the matrix P having $[\mathbf{b}_1]_S$ as its first column, $[\mathbf{b}_2]_S$ as its second column, and so on.

EXAMPLE 13-6: Let

$$B = \left\{ \begin{bmatrix} 2 \\ 1 \\ 1 \end{bmatrix}, \begin{bmatrix} 1 \\ 7 \\ 7 \end{bmatrix}, \begin{bmatrix} 4 \\ -1 \\ 0 \end{bmatrix} \right\} \quad \text{and} \quad S = \left\{ \begin{bmatrix} 0 \\ 1 \\ 1 \end{bmatrix}, \begin{bmatrix} 1 \\ 1 \\ 0 \end{bmatrix}, \begin{bmatrix} 1 \\ 0 \\ 1 \end{bmatrix} \right\}$$

be ordered bases for R^3. If $\mathbf{a} = \begin{bmatrix} 3 \\ 24 \\ 23 \end{bmatrix}$, find **(a)** the coordinate vector of **a** with respect to basis B; **(b)** the coordinate vector of **a** with respect to basis S; **(c)** the transition matrix P from basis B to basis S; **(d)** the transition matrix P^{-1} from basis S to basis B; **(e)** the product $P[\mathbf{a}]_B$ to show that it is the same as the answer in part **(b)**; **(f)** the product $P^{-1}[\mathbf{a}]_S$ to show that it is the same as the answer in part **(a)**.

Solution

(a)

$$c_1 \begin{bmatrix} 2 \\ 1 \\ 1 \end{bmatrix} + c_2 \begin{bmatrix} 1 \\ 7 \\ 7 \end{bmatrix} + c_3 \begin{bmatrix} 4 \\ -1 \\ 0 \end{bmatrix} = \begin{bmatrix} 3 \\ 24 \\ 23 \end{bmatrix}$$

$$\therefore \quad \begin{aligned} 2c_1 + c_2 + 4c_3 &= 3 \\ c_1 + 7c_2 - c_3 &= 24 \\ c_1 + 7c_2 \qquad &= 23 \end{aligned}$$

Solving by Gauss–Jordan elimination, you find $c_1 = 2$, $c_2 = 3$, and $c_3 = -1$. Therefore,

$$\begin{bmatrix} 3 \\ 24 \\ 23 \end{bmatrix}_B = \begin{bmatrix} 2 \\ 3 \\ -1 \end{bmatrix}$$

(b)

$$c_1 \begin{bmatrix} 0 \\ 1 \\ 1 \end{bmatrix} + c_2 \begin{bmatrix} 1 \\ 1 \\ 0 \end{bmatrix} + c_3 \begin{bmatrix} 1 \\ 0 \\ 1 \end{bmatrix} = \begin{bmatrix} 3 \\ 24 \\ 23 \end{bmatrix}$$

$$\therefore \quad \begin{aligned} c_2 + c_3 &= 3 \\ c_1 + c_2 \qquad &= 24 \\ c_1 \qquad + c_3 &= 23 \end{aligned}$$

from which $c_1 = 22$, $c_2 = 2$, and $c_3 = 1$. Therefore,

$$\begin{bmatrix} 3 \\ 24 \\ 23 \end{bmatrix}_S = \begin{bmatrix} 22 \\ 2 \\ 1 \end{bmatrix}$$

(c)

$$c_1 \begin{bmatrix} 0 \\ 1 \\ 1 \end{bmatrix} + c_2 \begin{bmatrix} 1 \\ 1 \\ 0 \end{bmatrix} + c_3 \begin{bmatrix} 1 \\ 0 \\ 1 \end{bmatrix} = \begin{bmatrix} 2 \\ 1 \\ 1 \end{bmatrix}$$

$$\therefore \quad \begin{aligned} c_2 + c_3 &= 2 \\ c_1 + c_2 \quad &= 1 \\ c_1 \quad + c_3 &= 1 \end{aligned}$$

from which $c_1 = 0$, $c_2 = 1$, and $c_3 = 1$. Therefore,

$$\begin{bmatrix} 2 \\ 1 \\ 1 \end{bmatrix}_S = \begin{bmatrix} 0 \\ 1 \\ 1 \end{bmatrix}$$

$$c_1 \begin{bmatrix} 0 \\ 1 \\ 1 \end{bmatrix} + c_2 \begin{bmatrix} 1 \\ 1 \\ 0 \end{bmatrix} + c_3 \begin{bmatrix} 1 \\ 0 \\ 1 \end{bmatrix} = \begin{bmatrix} 1 \\ 7 \\ 7 \end{bmatrix}$$

$$\therefore \quad \begin{aligned} c_2 + c_3 &= 1 \\ c_1 + c_2 \quad &= 7 \\ c_1 \quad + c_3 &= 7 \end{aligned}$$

from which $c_1 = 13/2$, $c_2 = 1/2$, and $c_3 = 1/2$. Therefore,

$$\begin{bmatrix} 1 \\ 7 \\ 7 \end{bmatrix}_S = \begin{bmatrix} \frac{13}{2} \\ \frac{1}{2} \\ \frac{1}{2} \end{bmatrix}$$

$$c_1 \begin{bmatrix} 0 \\ 1 \\ 1 \end{bmatrix} + c_2 \begin{bmatrix} 1 \\ 1 \\ 0 \end{bmatrix} + c_3 \begin{bmatrix} 1 \\ 0 \\ 1 \end{bmatrix} = \begin{bmatrix} 4 \\ -1 \\ 0 \end{bmatrix}$$

$$\therefore \quad \begin{aligned} c_2 + c_3 &= 4 \\ c_1 + c_2 \quad &= -1 \\ c_1 \quad + c_3 &= 0 \end{aligned}$$

from which $c_1 = -5/2$, $c_2 = 3/2$, and $c_3 = 5/2$.

$$\therefore \quad \begin{bmatrix} 4 \\ -1 \\ 0 \end{bmatrix}_S = \begin{bmatrix} -\frac{5}{2} \\ \frac{3}{2} \\ \frac{5}{2} \end{bmatrix}$$

Therefore,

$$P = \begin{bmatrix} 0 & \frac{13}{2} & -\frac{5}{2} \\ 1 & \frac{1}{2} & \frac{3}{2} \\ 1 & \frac{1}{2} & \frac{5}{2} \end{bmatrix}$$

(d) To find the transition matrix from basis S to basis B, we may do directly as in part **(c)**. However, the easiest method is to find P^{-1}.

$$\left[\begin{array}{ccc|ccc} 0 & \frac{13}{2} & -\frac{5}{2} & 1 & 0 & 0 \\ 1 & \frac{1}{2} & \frac{3}{2} & 0 & 1 & 0 \\ 1 & \frac{1}{2} & \frac{5}{2} & 0 & 0 & 1 \end{array}\right] \xrightarrow{R_1 \leftrightarrow R_2} \left[\begin{array}{ccc|ccc} 1 & \frac{1}{2} & \frac{3}{2} & 0 & 1 & 0 \\ 0 & \frac{13}{2} & -\frac{5}{2} & 1 & 0 & 0 \\ 1 & \frac{1}{2} & \frac{5}{2} & 0 & 0 & 1 \end{array}\right]$$

$$\xrightarrow[R_3 \to R_3 - R_1]{R_2 \to 2/13 R_2} \left[\begin{array}{ccc|ccc} 1 & \frac{1}{2} & \frac{3}{2} & 0 & 1 & 0 \\ 0 & 1 & -\frac{5}{13} & \frac{2}{13} & 0 & 0 \\ 0 & 0 & 1 & 0 & -1 & 1 \end{array}\right]$$

$$\xrightarrow[R_2 \to R_2 + 5/13R_3]{R_1 \to R_1 - 3/2R_3} \left[\begin{array}{ccc|ccc} 1 & \frac{1}{2} & 0 & 0 & \frac{5}{2} & -\frac{3}{2} \\ 0 & 1 & 0 & \frac{2}{13} & -\frac{5}{13} & \frac{5}{13} \\ 0 & 0 & 1 & 0 & -1 & 1 \end{array}\right]$$

$$\xrightarrow{R_1 \to R_1 - 1/2R_2} \left[\begin{array}{ccc|ccc} 1 & 0 & 0 & -\frac{1}{13} & \frac{35}{13} & -\frac{22}{13} \\ 0 & 1 & 0 & \frac{2}{13} & -\frac{5}{13} & \frac{5}{13} \\ 0 & 0 & 1 & 0 & -1 & 1 \end{array}\right]$$

$$\therefore \quad P^{-1} = \begin{bmatrix} -\frac{1}{13} & \frac{35}{13} & -\frac{22}{13} \\ \frac{2}{13} & -\frac{5}{13} & \frac{5}{13} \\ 0 & -1 & 1 \end{bmatrix}$$

(e)
$$P[\mathbf{a}]_B = \begin{bmatrix} 0 & \frac{13}{2} & -\frac{5}{2} \\ 1 & \frac{1}{2} & \frac{3}{2} \\ 1 & \frac{1}{2} & \frac{5}{2} \end{bmatrix} \begin{bmatrix} 2 \\ 3 \\ -1 \end{bmatrix} = \begin{bmatrix} 22 \\ 2 \\ 1 \end{bmatrix}$$

(f)
$$P^{-1}[\mathbf{a}]_S = \begin{bmatrix} -\frac{1}{13} & \frac{35}{13} & -\frac{22}{13} \\ \frac{2}{13} & -\frac{5}{13} & \frac{5}{13} \\ 0 & -1 & 1 \end{bmatrix} \begin{bmatrix} 22 \\ 2 \\ 1 \end{bmatrix} = \begin{bmatrix} 2 \\ 3 \\ -1 \end{bmatrix}$$

13-4. Matrix of a Linear Transformation

A. Definition of the matrix of a linear transformation

If B is an ordered basis of an n-dimensional vector space V and S is an ordered basis of an m-dimensional vector space W, then A is the **matrix of the linear transformation** $L: V \to W$ with respect to the ordered bases B and S provided $[\mathbf{w}]_S = A[\mathbf{v}]_B$ where $\mathbf{w} = L(\mathbf{v})$ for any vector $\mathbf{v}$ in V and $[\mathbf{v}]_B$ and $[\mathbf{w}]_S$ are the coordinate vectors of $\mathbf{v}$ and $\mathbf{w}$ with respect to the bases B and S, respectively.

EXAMPLE 13-7: If $L: R_3 \to R_2$ is the linear transformation defined by $L([a_1 \quad a_2 \quad a_3]) = [a_1 + a_2 \quad a_3]$, verify that the matrix of L with respect to the natural ordered bases of R_3 and R_2 is

$$A = \begin{bmatrix} 1 & 1 & 0 \\ 0 & 0 & 1 \end{bmatrix}$$

Solution: Let B and S be the natural ordered bases of R_3 and R_2, respectively. Now find the coordinate vector of $[a_1 \quad a_2 \quad a_3]$ with respect to B and the coordinate vector of $[a_1 + a_2 \quad a_3]$ with respect to S.

$$[a_1 \quad a_2 \quad a_3] = a_1[1 \quad 0 \quad 0] + a_2[0 \quad 1 \quad 0] + a_3[0 \quad 0 \quad 1]$$

$$[a_1 \quad a_2 \quad a_3]_B = \begin{bmatrix} a_1 \\ a_2 \\ a_3 \end{bmatrix}$$

$$[a_1 + a_2 \quad a_3] = (a_1 + a_2)[1 \quad 0] + a_3[0 \quad 1]$$

$$[a_1 + a_2 \quad a_3]_S = \begin{bmatrix} a_1 + a_2 \\ a_3 \end{bmatrix}$$

Hence,

$$A[a_1 \quad a_2 \quad a_3]_B = \begin{bmatrix} 1 & 1 & 0 \\ 0 & 0 & 1 \end{bmatrix} \begin{bmatrix} a_1 \\ a_2 \\ a_3 \end{bmatrix} = \begin{bmatrix} a_1 + a_2 \\ a_3 \end{bmatrix} = [a_1 + a_2 \quad a_3]_S$$

B. Finding the matrix of a linear transformation

If $B = \{\mathbf{b}_1, \mathbf{b}_2, \ldots, \mathbf{b}_n\}$ is an ordered basis of an n-dimensional vector space V and $S = \{\mathbf{s}_1, \mathbf{s}_2, \ldots, s_m\}$ is an ordered basis of an m-dimensional vector space W, then matrix A of the linear transformation $L: V \to W$ with respect to the ordered bases B and S can be found as follows:

Step 1: Find $L(\mathbf{b}_1)$, $L(\mathbf{b}_2)$, ..., $L(\mathbf{b}_n)$.
Step 2: Find the coordinate vectors $[L(\mathbf{b}_1)]_S, [L(\mathbf{b}_2)]_S, \ldots, [L(\mathbf{b}_n)]_S$.
Step 3: Form the matrix A having $[L(\mathbf{b}_1)]_S$ as its first column, $[L(\mathbf{b}_2)]_S$ as its second column, and so on.

EXAMPLE 13-8: If $L: R_3 \to R_2$ is the linear transformation defined by $L([a_1 \quad a_2 \quad a_3]) = [a_1 + a_2 \quad a_3]$, $B = \{[0 \quad 1 \quad 1], [1 \quad 1 \quad 0], [1 \quad 0 \quad 1]\}$ is an ordered basis of R_3 and $S = \{[1 \quad 2], [3 \quad 4]\}$ is an ordered basis of R_2, find the matrix A of L with respect to the bases B and S.

Solution

Step 1:

$$L([0 \quad 1 \quad 1]) = [0 + 1 \quad 1] = [1 \quad 1]$$
$$L([1 \quad 1 \quad 0]) = [1 + 1 \quad 0] = [2 \quad 0]$$
$$L([1 \quad 0 \quad 1]) = [1 + 0 \quad 1] = [1 \quad 1]$$

Step 2:

$$[1 \quad 1] = c_1[1 \quad 2] + c_2[3 \quad 4]$$
$$c_1 + 3c_2 = 1$$
$$\therefore \quad 2c_1 + 4c_2 = 1$$

which has $c_1 = -1/2$, $c_2 = 1/2$ as its solution. Therefore,

$$[1 \quad 1]_s = \begin{bmatrix} -\frac{1}{2} \\ \frac{1}{2} \end{bmatrix}$$

$$[2 \quad 0] = c_1[1 \quad 2] + c_2[3 \quad 4]$$
$$\therefore \quad c_1 + 3c_2 = 2$$
$$2c_1 + 4c_2 = 0$$

which has $c_1 = -4$, $c_2 = 2$ as its solution.

$$\therefore \quad [2 \quad 0]_S = \begin{bmatrix} -4 \\ 2 \end{bmatrix}$$

$$L([1 \quad 0 \quad 1]) = L([0 \quad 1 \quad 1]) = [1 \quad 1]$$

$$\therefore \quad [L([1 \quad 0 \quad 1])]_S = [1 \quad 1]_S = \begin{bmatrix} -\frac{1}{2} \\ \frac{1}{2} \end{bmatrix}$$

Step 3:

$$A = \begin{bmatrix} -\frac{1}{2} & -4 & -\frac{1}{2} \\ \frac{1}{2} & 2 & \frac{1}{2} \end{bmatrix}$$

EXAMPLE 13-9: For the linear transformation in Example 13-8, find $L([4 \quad 2 \quad 5])$ by **(a)** direct calculation using the definition of L, and **(b)** by using matrix A.

Solution

(a) $L([4 \quad 2 \quad 5]) = [4 + 2 \quad 5] = [6 \quad 5]$

(b) To use matrix A, we must first find the coordinate vector of $[4 \quad 2 \quad 5]$ with respect to basis B.

$$[4 \quad 2 \quad 5] = c_1[0 \quad 1 \quad 1] + c_2[1 \quad 1 \quad 0] + c_3[1 \quad 0 \quad 1]$$

$$\begin{aligned} c_2 + c_3 &= 4 \\ c_1 + c_2 \quad\quad &= 2 \\ c_1 \quad\quad + c_3 &= 5 \end{aligned}$$

which has $c_1 = 3/2$, $c_2 = 1/2$, $c_3 = 7/2$ as solution. Therefore,

$$[4 \quad 2 \quad 5]_B = \begin{bmatrix} \frac{3}{2} \\ \frac{1}{2} \\ \frac{7}{2} \end{bmatrix}$$

Hence,

$$\begin{bmatrix} -\frac{1}{2} & -4 & -\frac{1}{2} \\ \frac{1}{2} & 2 & \frac{1}{2} \end{bmatrix} \begin{bmatrix} \frac{3}{2} \\ \frac{1}{2} \\ \frac{7}{2} \end{bmatrix} = \begin{bmatrix} -\frac{9}{2} \\ \frac{7}{2} \end{bmatrix}$$

The last vector is the coordinate vector of the desired answer.

$$-\tfrac{9}{2}[1 \quad 2] + \tfrac{7}{2}[3 \quad 4] = [6 \quad 5]$$

$$L([4 \quad 2 \quad 5]) = [6 \quad 5]$$

note: In many instances you will have only the matrix of the linear transformation; thus, your only method of solution would be that given in part (**b**).

13-5. Change of Bases for Matrix Transformation

If A is the matrix of the linear transformation $L: V \to W$ from an n-dimensional vector space V into an m-dimensional vector space W with respect to the ordered bases $B = \{\mathbf{b}_1, \mathbf{b}_2, \ldots, \mathbf{b}_n\}$ and $S = \{\mathbf{s}_1, \mathbf{s}_2, \ldots, \mathbf{s}_m\}$, P the transition matrix from $B' = \{\mathbf{b}'_1, \mathbf{b}'_2, \ldots, \mathbf{b}'_n\}$ to B and Q the transition matrix from $S' = \{\mathbf{s}'_1, \mathbf{s}'_2, \ldots, \mathbf{s}'_m\}$ to S, then $Q^{-1}AP$ is the matrix representation of L with respect to the ordered bases B' and S'.

EXAMPLE 13-10: Let $L: R^2 \to R^3$ be defined by

$$L\left(\begin{bmatrix} a_1 \\ a_2 \end{bmatrix}\right) = \begin{bmatrix} a_1 - a_2 \\ a_1 + a_2 \\ 2a_1 + 3a_2 \end{bmatrix}$$

Consider the order bases

$$B = \left\{\begin{bmatrix} 1 \\ 1 \end{bmatrix}, \begin{bmatrix} 0 \\ 1 \end{bmatrix}\right\} \quad \text{and} \quad B' = \left\{\begin{bmatrix} 2 \\ 3 \end{bmatrix}, \begin{bmatrix} -1 \\ 1 \end{bmatrix}\right\} \quad \text{for } R^2$$

and

$$S = \left\{\begin{bmatrix} 1 \\ 0 \\ 1 \end{bmatrix}, \begin{bmatrix} 0 \\ 1 \\ 1 \end{bmatrix}, \begin{bmatrix} 1 \\ 1 \\ 0 \end{bmatrix}\right\} \quad \text{and} \quad S' = \left\{\begin{bmatrix} 1 \\ 0 \\ 2 \end{bmatrix}, \begin{bmatrix} 1 \\ 0 \\ 0 \end{bmatrix}, \begin{bmatrix} 0 \\ 1 \\ 0 \end{bmatrix}\right\} \quad \text{for } R^3$$

(**a**) Find the matrix representation A of L with respect to bases B and S by direct computation. (**b**) Find the matrix representation A' of L with respect to bases B' and S' by direct computation. (**c**) Find the transition matrix P from B' to B. (**d**) Find the transition matrix Q from S' to S. (**e**) Verify that $A' = Q^{-1}AP$.

Solution

(**a**)

$$L\left(\begin{bmatrix}1\\1\end{bmatrix}\right) = \begin{bmatrix}1-1\\1+1\\2+3\end{bmatrix} = \begin{bmatrix}0\\2\\5\end{bmatrix}$$

$$L\left(\begin{bmatrix}0\\1\end{bmatrix}\right) = \begin{bmatrix}0-1\\0+1\\0+3\end{bmatrix} = \begin{bmatrix}-1\\1\\3\end{bmatrix}$$

$$\begin{bmatrix}0\\2\\5\end{bmatrix} = c_1\begin{bmatrix}1\\0\\1\end{bmatrix} + c_2\begin{bmatrix}0\\1\\1\end{bmatrix} + c_3\begin{bmatrix}1\\1\\0\end{bmatrix}$$

$$\therefore \quad \begin{aligned} c_1 \qquad + c_3 &= 0 \\ c_2 + c_3 &= 2 \\ c_1 + c_2 \qquad &= 5 \end{aligned}$$

which has as its solution $c_1 = 3/2$, $c_2 = 7/2$, $c_3 = -3/2$. Therefore,

$$\begin{bmatrix}0\\2\\5\end{bmatrix}_S = \begin{bmatrix}\frac{3}{2}\\\frac{7}{2}\\-\frac{3}{2}\end{bmatrix}$$

$$\begin{bmatrix}-1\\1\\3\end{bmatrix} = c_1\begin{bmatrix}1\\0\\1\end{bmatrix} + c_2\begin{bmatrix}0\\1\\1\end{bmatrix} + c_3\begin{bmatrix}1\\1\\0\end{bmatrix}$$

$$\therefore \quad \begin{aligned} c_1 \qquad + c_3 &= -1 \\ c_2 + c_3 &= 1 \\ c_1 + c_2 \qquad &= 3 \end{aligned}$$

which has as its solution $c_1 = 1/2$, $c_2 = 5/2$, $c_3 = -3/2$. Therefore,

$$\begin{bmatrix}-1\\1\\3\end{bmatrix}_S = \begin{bmatrix}\frac{1}{2}\\\frac{5}{2}\\-\frac{3}{2}\end{bmatrix}$$

Therefore,

$$A = \begin{bmatrix}\frac{3}{2} & \frac{1}{2}\\\frac{7}{2} & \frac{5}{2}\\-\frac{3}{2} & -\frac{3}{2}\end{bmatrix}$$

(**b**)

$$L\left(\begin{bmatrix}2\\3\end{bmatrix}\right) = \begin{bmatrix}2-3\\2+3\\4+9\end{bmatrix} = \begin{bmatrix}-1\\5\\13\end{bmatrix}$$

$$L\left(\begin{bmatrix}-1\\1\end{bmatrix}\right) = \begin{bmatrix}-1-1\\-1+1\\-2+3\end{bmatrix} = \begin{bmatrix}-2\\0\\1\end{bmatrix}$$

$$\begin{bmatrix} -1 \\ 5 \\ 13 \end{bmatrix} = c_1 \begin{bmatrix} 1 \\ 0 \\ 2 \end{bmatrix} + c_2 \begin{bmatrix} 1 \\ 0 \\ 0 \end{bmatrix} + c_3 \begin{bmatrix} 0 \\ 1 \\ 0 \end{bmatrix}$$

$$\begin{aligned} \therefore \quad c_1 + c_2 \quad &= -1 \\ c_3 &= \ \ 5 \\ 2c_1 \quad &= \ \ 13 \end{aligned}$$

which has $c_1 = 13/2$, $c_2 = -15/2$, $c_3 = 5$ as the solution. Therefore,

$$\begin{bmatrix} -1 \\ 5 \\ 13 \end{bmatrix}_{S'} = \begin{bmatrix} \frac{13}{2} \\ -\frac{15}{2} \\ 5 \end{bmatrix}$$

$$\begin{bmatrix} -2 \\ 0 \\ 1 \end{bmatrix} = c_1 \begin{bmatrix} 1 \\ 0 \\ 2 \end{bmatrix} + c_2 \begin{bmatrix} 1 \\ 0 \\ 0 \end{bmatrix} + c_3 \begin{bmatrix} 0 \\ 1 \\ 0 \end{bmatrix}$$

$$\begin{aligned} \therefore \quad c_1 + c_2 \quad &= -2 \\ c_3 &= \ \ 0 \\ 2c_1 \quad &= \ \ 1 \end{aligned}$$

which has $c_1 = 1/2$, $c_2 = -5/2$, $c_3 = 0$ as the solution. Therefore,

$$\begin{bmatrix} -2 \\ 0 \\ 1 \end{bmatrix}_{S'} = \begin{bmatrix} \frac{1}{2} \\ -\frac{5}{2} \\ 0 \end{bmatrix}$$

$$A' = \begin{bmatrix} \frac{13}{2} & \frac{1}{2} \\ -\frac{15}{2} & -\frac{5}{2} \\ 5 & 0 \end{bmatrix}$$

(c)

$$\begin{bmatrix} 2 \\ 3 \end{bmatrix} = c_1 \begin{bmatrix} 1 \\ 1 \end{bmatrix} + c_2 \begin{bmatrix} 0 \\ 1 \end{bmatrix}$$

$$\begin{aligned} \therefore \quad c_1 \quad &= 2 \\ c_1 + c_2 &= 3 \end{aligned}$$

which has as the solution $c_1 = 2$, $c_2 = 1$. Therefore,

$$\begin{bmatrix} 2 \\ 3 \end{bmatrix}_B = \begin{bmatrix} 2 \\ 1 \end{bmatrix}$$

$$\begin{bmatrix} -1 \\ 1 \end{bmatrix} = c_1 \begin{bmatrix} 1 \\ 1 \end{bmatrix} + c_2 \begin{bmatrix} 0 \\ 1 \end{bmatrix}$$

$$\begin{aligned} \therefore \quad c_1 \quad &= -1 \\ c_1 + c_2 &= \ \ 1 \end{aligned}$$

which has as a solution $c_1 = -1$, $c_2 = 2$. Therefore,

$$\begin{bmatrix} -1 \\ 1 \end{bmatrix}_B = \begin{bmatrix} -1 \\ 2 \end{bmatrix}$$

Therefore,

$$P_{B' \text{ to } B} = \begin{bmatrix} 2 & -1 \\ 1 & 2 \end{bmatrix}$$

(d)

$$\begin{bmatrix}1\\0\\2\end{bmatrix} = c_1\begin{bmatrix}1\\0\\1\end{bmatrix} + c_2\begin{bmatrix}0\\1\\1\end{bmatrix} + c_3\begin{bmatrix}1\\1\\0\end{bmatrix}$$

$$\begin{aligned}\therefore\quad c_1 \qquad\quad + c_3 &= 1\\ c_2 + c_3 &= 0\\ c_1 + c_2 \qquad &= 2\end{aligned}$$

which has as the solution $c_1 = 3/2$, $c_2 = 1/2$, $c_3 = -1/2$. Therefore,

$$\begin{bmatrix}1\\0\\2\end{bmatrix}_S = \begin{bmatrix}\frac{3}{2}\\ \frac{1}{2}\\ -\frac{1}{2}\end{bmatrix}$$

$$\begin{bmatrix}1\\0\\0\end{bmatrix} = c_1\begin{bmatrix}1\\0\\1\end{bmatrix} + c_2\begin{bmatrix}0\\1\\1\end{bmatrix} + c_3\begin{bmatrix}1\\1\\0\end{bmatrix}$$

$$\begin{aligned}\therefore\quad c_1 \qquad\quad + c_3 &= 1\\ c_2 + c_3 &= 0\\ c_1 + c_2 \qquad &= 0\end{aligned}$$

which has as the solution $c_1 = 1/2$, $c_2 = -1/2$, $c_3 = 1/2$. Therefore,

$$\begin{bmatrix}1\\0\\0\end{bmatrix}_S = \begin{bmatrix}\frac{1}{2}\\ -\frac{1}{2}\\ \frac{1}{2}\end{bmatrix}$$

$$\begin{bmatrix}0\\1\\0\end{bmatrix} = c_1\begin{bmatrix}1\\0\\1\end{bmatrix} + c_2\begin{bmatrix}0\\1\\1\end{bmatrix} + c_3\begin{bmatrix}1\\1\\0\end{bmatrix}$$

$$\begin{aligned}\therefore\quad c_1 \qquad\quad + c_3 &= 0\\ c_2 + c_3 &= 1\\ c_1 + c_2 \qquad &= 0\end{aligned}$$

which has as the solution $c_1 = -1/2$, $c_2 = 1/2$, $c_3 = 1/2$. Therefore,

$$\begin{bmatrix}0\\1\\0\end{bmatrix}_S = \begin{bmatrix}-\frac{1}{2}\\ \frac{1}{2}\\ \frac{1}{2}\end{bmatrix}$$

Therefore,

$$Q_{S' \text{ to } S} = \begin{bmatrix}\frac{3}{2} & \frac{1}{2} & -\frac{1}{2}\\ \frac{1}{2} & -\frac{1}{2} & \frac{1}{2}\\ -\frac{1}{2} & \frac{1}{2} & \frac{1}{2}\end{bmatrix}$$

(e) First find Q^{-1}

$$\left[\begin{array}{ccc|ccc} \frac{3}{2} & \frac{1}{2} & -\frac{1}{2} & 1 & 0 & 0 \\ \frac{1}{2} & -\frac{1}{2} & \frac{1}{2} & 0 & 1 & 0 \\ -\frac{1}{2} & \frac{1}{2} & \frac{1}{2} & 0 & 0 & 1 \end{array}\right] \xrightarrow{R_1 \leftrightarrow R_2} \left[\begin{array}{ccc|ccc} \frac{1}{2} & -\frac{1}{2} & \frac{1}{2} & 0 & 1 & 0 \\ \frac{3}{2} & \frac{1}{2} & -\frac{1}{2} & 1 & 0 & 0 \\ -\frac{1}{2} & \frac{1}{2} & \frac{1}{2} & 0 & 0 & 1 \end{array}\right]$$

$$\xrightarrow[\substack{R_2 \to 2R_2 \\ R_3 \to 2R_3}]{R_1 \to 2R_1} \left[\begin{array}{ccc|ccc} 1 & -1 & 1 & 0 & 2 & 0 \\ 3 & 1 & -1 & 2 & 0 & 0 \\ -1 & 1 & 1 & 0 & 0 & 2 \end{array}\right] \xrightarrow[R_3 \to R_3 + R_1]{R_2 \to R_2 - 3R_1} \left[\begin{array}{ccc|ccc} 1 & -1 & 1 & 0 & 2 & 0 \\ 0 & 4 & -4 & 2 & -6 & 0 \\ 0 & 0 & 2 & 0 & 2 & 2 \end{array}\right]$$

$$\xrightarrow[R_3 \to 1/2R_3]{} \left[\begin{array}{ccc|ccc} 1 & -1 & 1 & 0 & 2 & 0 \\ 0 & 4 & -4 & 2 & -6 & 0 \\ 0 & 0 & 1 & 0 & 1 & 1 \end{array}\right] \xrightarrow[R_2 \to R_2 + 4R_3]{R_1 \to R_1 - R_3} \left[\begin{array}{ccc|ccc} 1 & -1 & 0 & 0 & 1 & -1 \\ 0 & 4 & 0 & 2 & -2 & 4 \\ 0 & 0 & 1 & 0 & 1 & 1 \end{array}\right]$$

$$\xrightarrow{R_2 \to 1/4R_2} \left[\begin{array}{ccc|ccc} 1 & -1 & 0 & 0 & 1 & -1 \\ 0 & 1 & 0 & \frac{1}{2} & -\frac{1}{2} & 1 \\ 0 & 0 & 1 & 0 & 1 & 1 \end{array}\right] \xrightarrow{R_1 \to R_1 + R_2} \left[\begin{array}{ccc|ccc} 1 & 0 & 0 & \frac{1}{2} & \frac{1}{2} & 0 \\ 0 & 1 & 0 & \frac{1}{2} & -\frac{1}{2} & 1 \\ 0 & 0 & 1 & 0 & 1 & 1 \end{array}\right]$$

$$\therefore \quad Q^{-1} = \begin{bmatrix} \frac{1}{2} & \frac{1}{2} & 0 \\ \frac{1}{2} & -\frac{1}{2} & 1 \\ 0 & 1 & 1 \end{bmatrix}$$

$$Q^{-1}AP = \begin{bmatrix} \frac{1}{2} & \frac{1}{2} & 0 \\ \frac{1}{2} & -\frac{1}{2} & 1 \\ 0 & 1 & 1 \end{bmatrix} \begin{bmatrix} \frac{3}{2} & \frac{1}{2} \\ \frac{7}{2} & \frac{5}{2} \\ -\frac{3}{2} & -\frac{3}{2} \end{bmatrix} \begin{bmatrix} 2 & -1 \\ 1 & 2 \end{bmatrix} = \begin{bmatrix} \frac{5}{2} & \frac{3}{2} \\ -\frac{5}{2} & -\frac{5}{2} \\ 2 & 1 \end{bmatrix} \begin{bmatrix} 2 & -1 \\ 1 & 2 \end{bmatrix} = \begin{bmatrix} \frac{13}{2} & \frac{1}{2} \\ -\frac{15}{2} & -\frac{5}{2} \\ 5 & 0 \end{bmatrix} = A'$$

by comparing with answer in part (**b**). Therefore, $A' = Q^{-1}AP$.

13-6. Partitioned-Matrix Method for Change of Bases

The authors have found the following procedure for change of bases to be very useful for computational purposes.

If $B = \{\mathbf{b}_1, \mathbf{b}_2, \ldots, \mathbf{b}_n\}$ and $S = \{\mathbf{s}_1, \mathbf{s}_2, \ldots, \mathbf{s}_n\}$ are two ordered bases of an n-dimensional vector space V, the transition matrix P from basis B to basis S can be found as follows:

Step 1: Form the partitioned matrix $[S|B]$.

Step 2: By elementary row operations change $[S|B]$ to a form where the left part is the $n \times n$ identity matrix. The right part of this result is the transition matrix P from basis B to basis S.

EXAMPLE 13-11: Let $B = \left\{\begin{bmatrix} 2 \\ 3 \end{bmatrix}, \begin{bmatrix} -1 \\ 1 \end{bmatrix}\right\}$ and $S = \left\{\begin{bmatrix} 1 \\ 1 \end{bmatrix}, \begin{bmatrix} 0 \\ 1 \end{bmatrix}\right\}$ be ordered bases for R^2. Find the transition matrix P from B to S.

Solution

$$\left[\begin{array}{cc|cc} 1 & 0 & 2 & -1 \\ 1 & 1 & 3 & 1 \end{array}\right] \xrightarrow[R_2 \to R_2 - R_1]{} \left[\begin{array}{cc|cc} 1 & 0 & 2 & -1 \\ 0 & 1 & 1 & 2 \end{array}\right]$$

$$\therefore \quad P_{B \text{ to } S} = \begin{bmatrix} 2 & -1 \\ 1 & 2 \end{bmatrix}$$

EXAMPLE 13-12: Let

$$B = \left\{ \begin{bmatrix} 1 \\ 0 \\ 2 \end{bmatrix}, \begin{bmatrix} 1 \\ 0 \\ 0 \end{bmatrix}, \begin{bmatrix} 0 \\ 1 \\ 0 \end{bmatrix} \right\} \quad \text{and} \quad S = \left\{ \begin{bmatrix} 1 \\ 0 \\ 1 \end{bmatrix}, \begin{bmatrix} 0 \\ 1 \\ 1 \end{bmatrix}, \begin{bmatrix} 1 \\ 1 \\ 0 \end{bmatrix} \right\}$$

be ordered bases for R^3. Find the transition matrix from B to S.

Solution

$$\left[\begin{array}{ccc|ccc} 1 & 0 & 1 & 1 & 1 & 0 \\ 0 & 1 & 1 & 0 & 0 & 1 \\ 1 & 1 & 0 & 2 & 0 & 0 \end{array}\right] \xrightarrow[R_3 \to R_3 - R_1]{} \left[\begin{array}{ccc|ccc} 1 & 0 & 1 & 1 & 1 & 0 \\ 0 & 1 & 1 & 0 & 0 & 1 \\ 0 & 1 & -1 & 1 & -1 & 0 \end{array}\right] \xrightarrow[R_3 \to R_3 - R_2]{} \left[\begin{array}{ccc|ccc} 1 & 0 & 1 & 1 & 1 & 0 \\ 0 & 1 & 1 & 0 & 0 & 1 \\ 0 & 0 & -2 & 1 & -1 & -1 \end{array}\right]$$

$$\xrightarrow[R_3 \to -1/2 R_3]{} \left[\begin{array}{ccc|ccc} 1 & 0 & 1 & 1 & 1 & 0 \\ 0 & 1 & 1 & 0 & 0 & 1 \\ 0 & 0 & 1 & -\frac{1}{2} & \frac{1}{2} & \frac{1}{2} \end{array}\right] \xrightarrow[R_2 \to R_2 - R_3]{R_1 \to R_1 - R_3} \left[\begin{array}{ccc|ccc} 1 & 0 & 0 & \frac{3}{2} & \frac{1}{2} & -\frac{1}{2} \\ 0 & 1 & 0 & \frac{1}{2} & -\frac{1}{2} & \frac{1}{2} \\ 0 & 0 & 1 & -\frac{1}{2} & \frac{1}{2} & \frac{1}{2} \end{array}\right]$$

$$\therefore \quad P_{B\,\text{to}\,S} = \begin{bmatrix} \frac{3}{2} & \frac{1}{2} & -\frac{1}{2} \\ \frac{1}{2} & -\frac{1}{2} & \frac{1}{2} \\ -\frac{1}{2} & \frac{1}{2} & \frac{1}{2} \end{bmatrix}$$

EXAMPLE 13-13: Let $L: R^2 \to R^3$ be defined by

$$L\left(\begin{bmatrix} a_1 \\ a_2 \end{bmatrix}\right) = \begin{bmatrix} a_1 - a_2 \\ a_1 + a_2 \\ 2a_1 + 3a_2 \end{bmatrix}$$

with ordered basis $B = \left\{ \begin{bmatrix} 1 \\ 1 \end{bmatrix}, \begin{bmatrix} 0 \\ 1 \end{bmatrix} \right\}$ for R^2 and ordered basis $S = \left\{ \begin{bmatrix} 1 \\ 0 \\ 1 \end{bmatrix}, \begin{bmatrix} 0 \\ 1 \\ 1 \end{bmatrix}, \begin{bmatrix} 1 \\ 1 \\ 0 \end{bmatrix} \right\}$ for R^3. Find the matrix representation A of L with respect to bases B and S.

Solution We must first find $L\left(\begin{bmatrix} 1 \\ 1 \end{bmatrix}\right)$ and $L\left(\begin{bmatrix} 0 \\ 1 \end{bmatrix}\right)$.

$$L\left(\begin{bmatrix} 1 \\ 1 \end{bmatrix}\right) = \begin{bmatrix} 1 - 1 \\ 1 + 1 \\ 2 + 3 \end{bmatrix} = \begin{bmatrix} 0 \\ 2 \\ 5 \end{bmatrix}$$

$$L\left(\begin{bmatrix} 0 \\ 1 \end{bmatrix}\right) = \begin{bmatrix} 0 - 1 \\ 0 + 1 \\ 0 + 3 \end{bmatrix} = \begin{bmatrix} -1 \\ 1 \\ 3 \end{bmatrix}$$

Then we form the partitioned matrix $\left[\begin{array}{ccc|cc} 1 & 0 & 1 & 0 & -1 \\ 0 & 1 & 1 & 2 & 1 \\ 1 & 1 & 0 & 5 & 3 \end{array}\right]$. Performing elementary row operations on this result, we get

$$\left[\begin{array}{ccc|cc} 1 & 0 & 1 & 0 & -1 \\ 0 & 1 & 1 & 2 & 1 \\ 1 & 1 & 0 & 5 & 3 \end{array}\right] \xrightarrow[R_3 \to R_3 - R_1]{} \left[\begin{array}{ccc|cc} 1 & 0 & 1 & 0 & -1 \\ 0 & 1 & 1 & 2 & 1 \\ 0 & 1 & -1 & 5 & 4 \end{array}\right] \xrightarrow[R_3 \to R_3 - R_2]{} \left[\begin{array}{ccc|cc} 1 & 0 & 1 & 0 & -1 \\ 0 & 1 & 1 & 2 & 1 \\ 0 & 0 & -2 & 3 & 3 \end{array}\right]$$

$$\xrightarrow[R_3 \to -1/2 R_3]{} \left[\begin{array}{ccc|cc} 1 & 0 & 1 & 0 & -1 \\ 0 & 1 & 1 & 2 & 1 \\ 0 & 0 & 1 & -\frac{3}{2} & -\frac{3}{2} \end{array}\right] \xrightarrow[R_2 \to R_2 - R_3]{R_1 \to R_1 - R_3} \left[\begin{array}{ccc|cc} 1 & 0 & 0 & \frac{3}{2} & \frac{1}{2} \\ 0 & 1 & 0 & \frac{7}{2} & \frac{5}{2} \\ 0 & 0 & 0 & -\frac{3}{2} & -\frac{3}{2} \end{array}\right]$$

$$\therefore \quad A = \begin{bmatrix} \frac{3}{2} & \frac{1}{2} \\ \frac{7}{2} & \frac{5}{2} \\ -\frac{3}{2} & -\frac{3}{2} \end{bmatrix}$$

Compare this result with that of Example 13-10, part (**a**).

SUMMARY

1. If A is any $m \times n$ matrix, then the function $L: R^n \to R^m$ defined by $L(\mathbf{a}) = A\mathbf{a}$ for $\mathbf{a}$ in R^n is a matrix transformation.
2. Every matrix transformation is a linear transformation.
3. Every linear transformation between finite-dimensional vector spaces can be represented as a matrix transformation.
4. A basis B of a vector space V is an ordered basis if the vectors of B are assigned a first position, second position, and so on.
5. If $B = \{\mathbf{b}_1, \mathbf{b}_2, \ldots, \mathbf{b}_n\}$ is an ordered basis for the n-dimensional vector space V and $\mathbf{a}$ is any vector in V, then

$$[\mathbf{a}]_B = \begin{bmatrix} c_1 \\ c_2 \\ \vdots \\ c_n \end{bmatrix}$$

is the coordinate vector of $\mathbf{a}$ with respect to the ordered basis B where $c_1, c_2, \ldots, c_n$ are the unique coefficients for $\mathbf{a} = c_1\mathbf{b}_1 + c_2\mathbf{b}_2 + \cdots + c_n\mathbf{b}_n$.
6. With respect to the ordered natural basis for R^n, a vector is its own coordinate vector.
7. A coordinate vector depends on both the basis and the ordering of the vectors in the basis.
8. The transition matrix P from an ordered basis $B = \{\mathbf{b}_1, \mathbf{b}_2, \ldots, \mathbf{b}_n\}$ to an ordered basis $S = \{\mathbf{s}_1, \mathbf{s}_2, \ldots, s_m\}$ of a vector space V
   - transforms a coordinate vector with respect to basis B into a coordinate vector with respect to basis S—that is, $[\mathbf{a}]_s = P[\mathbf{a}]_B$ for any $\mathbf{a}$ in V;
   - is always invertible;
   - can be found as follows:

     Step 1: compute $[\mathbf{b}_1]_s, [\mathbf{b}_2]_s, \ldots, [\mathbf{b}_n]_s$;
     Step 2: form the matrix P having $[\mathbf{b}_1]_s$ as its first column, $[\mathbf{b}_2]_s$ as its second column, and so on.
9. The matrix A of the linear transformation $L: V \to W$ with respect to the ordered bases B and S where $B = \{\mathbf{b}_1, \mathbf{b}_2, \ldots, \mathbf{b}_n\}$ is an ordered basis of the n-dimensional vector space V and $S = \{\mathbf{s}_1, \mathbf{s}_2, \ldots, \mathbf{s}_m\}$ is an ordered basis of the m-dimensional vector space V

- transforms a coordinate vector with respect to basis B of V into a coordinate vector with respect to basis S of W—that is, $[\mathbf{w}]_s = A[\mathbf{v}]_B$;
- is not necessarily invertible;
- can be found as follows:

  Step 1: Find $L(\mathbf{b}_1), L(\mathbf{b}_2), \ldots, L(\mathbf{b}_n)$;
  Step 2: Find the coordinate vectors $[L(\mathbf{b}_1)]_s, [L(\mathbf{b}_2)]_s, \ldots, [L(\mathbf{b}_n)]_s$;
  Step 3: Form the matrix A having $[L(\mathbf{b}_1)]_s$ as its first column, $[L(\mathbf{b}_2)]_s$ as its second column, and so on;

- can be changed to the matrix $Q^{-1}AP$ of the linear transformation L with respect to the ordered bases B' and S' of V and W, respectively, where P is the transition matrix from B' to B and Q is the transition matrix from S' to S.

RAISE YOUR GRADES

Can you ...?

☑ find the coordinate vector with respect to an ordered basis for any given vector in a vector space V
☑ find the transition matrix from a basis B to a basis S for a vector space
☑ find the transition matrix from a basis S to a basis B when given the transition matrix from basis B to basis S
☑ find the matrix of a linear transformation
☑ find the matrix of a linear transformation when the bases of the vector space are changed
☑ use the partitioned-matrix method to find the transition matrix from a basis B to a basis S

SOLVED PROBLEMS

PROBLEM 13-1 Verify that the matrix transformation $L: R^2 \to R^4$ defined by

$$L\left(\begin{bmatrix} a \\ b \end{bmatrix}\right) = \begin{bmatrix} 1 & -2 \\ 2 & 1 \\ 0 & 2 \\ 3 & 1 \end{bmatrix}\begin{bmatrix} a \\ b \end{bmatrix}$$

is a linear transformation.

Solution

$$L\left(\begin{bmatrix} a \\ b \end{bmatrix} + \begin{bmatrix} c \\ d \end{bmatrix}\right) = L\left(\begin{bmatrix} a+c \\ b+d \end{bmatrix}\right) = \begin{bmatrix} 1 & -2 \\ 2 & 1 \\ 0 & 2 \\ 3 & 1 \end{bmatrix}\begin{bmatrix} a+c \\ b+d \end{bmatrix}$$

$$L\left(\begin{bmatrix} a \\ b \end{bmatrix} + \begin{bmatrix} c \\ d \end{bmatrix}\right) = \begin{bmatrix} (a+c) - 2(b+d) \\ 2(a+c) + 1(b+d) \\ 0(a+c) + 2(b+d) \\ 3(a+c) + 1(b+d) \end{bmatrix} = \begin{bmatrix} a - 2b \\ 2a + b \\ 2b \\ 3a + b \end{bmatrix} + \begin{bmatrix} c - 2d \\ 2c + d \\ 2d \\ 3c + d \end{bmatrix}$$

$$= \begin{bmatrix} 1 & -2 \\ 2 & 1 \\ 0 & 2 \\ 3 & 1 \end{bmatrix}\begin{bmatrix} a \\ b \end{bmatrix} + \begin{bmatrix} 1 & -2 \\ 2 & 1 \\ 0 & 2 \\ 3 & 1 \end{bmatrix}\begin{bmatrix} c \\ d \end{bmatrix} = L\left(\begin{bmatrix} a \\ b \end{bmatrix}\right) + L\left(\begin{bmatrix} c \\ d \end{bmatrix}\right).$$

$$L\left(k\begin{bmatrix} a \\ b \end{bmatrix}\right) = L\left(\begin{bmatrix} ka \\ kb \end{bmatrix}\right) = \begin{bmatrix} 1 & -2 \\ 2 & 1 \\ 0 & 2 \\ 3 & 1 \end{bmatrix}\begin{bmatrix} ka \\ kb \end{bmatrix} = \begin{bmatrix} ka - 2kb \\ 2ka + kb \\ 2kb \\ 3ka + kb \end{bmatrix}$$

$$= k\begin{bmatrix} a - 2b \\ 2a + b \\ 2b \\ 3a + b \end{bmatrix} = k\left(\begin{bmatrix} 1 & -2 \\ 2 & 1 \\ 0 & 2 \\ 3 & 1 \end{bmatrix}\begin{bmatrix} a \\ b \end{bmatrix}\right) = kL\left(\begin{bmatrix} a \\ b \end{bmatrix}\right).$$

PROBLEM 13-2 An ordered basis for R_3 is $\{[1\ \ -2\ \ 0], [0\ \ -1\ \ 1], [1\ \ 1\ \ 2]\}$. Write the five other ordered bases for R_3 with these same vectors, and denote them by B_1, B_2, B_3, B_4, and B_5.

Solution

$$B_1 = \{[1\ \ -2\ \ 0], [1\ \ 1\ \ 2], [0\ \ -1\ \ 1]\}$$
$$B_2 = \{[1\ \ 1\ \ 2], [1\ \ -2\ \ 0], [0\ \ -1\ \ 1]\}$$
$$B_3 = \{[1\ \ 1\ \ 2], [0\ \ -1\ \ 1], [1\ \ -2\ \ 0]\}$$
$$B_4 = \{[0\ \ -1\ \ 1], [1\ \ 1\ \ 2], [1\ \ -2\ \ 0]\}$$
$$B_5 = \{[0\ \ -1\ \ 1], [1\ \ -2\ \ 0], [1\ \ 1\ \ 2]\}$$

PROBLEM 13-3 Find the coordinate vector of $[2\ \ -3\ \ 4]$ with respect to each of the following ordered bases for R_3: **(a)** $B = \{[1\ \ 0\ \ 0], [0\ \ 1\ \ 0], [0\ \ 0\ \ 1]\}$; **(b)** $T = \{[1\ \ -2\ \ 0], [0\ \ 1\ \ 1], [1\ \ 1\ \ 2]\}$; and **(c)** $S = \{[0\ \ -1\ \ 1], [1\ \ -2\ \ 0], [1\ \ 1\ \ 2]\}$.

Solution

(a)

$$c_1[1\ \ 0\ \ 0] + c_2[0\ \ 1\ \ 0] + c_3[0\ \ 0\ \ 1] = [2\ \ -3\ \ 4]$$
$$[c_1\ \ c_2\ \ c_3] = [2\ \ -3\ \ 4]$$
$$\therefore\ \ c_1 = 2, c_2 = -3, c_3 = 4$$

Hence,

$$[2\ \ -3\ \ 4]_B = \begin{bmatrix} 2 \\ -3 \\ 4 \end{bmatrix}$$

(b)

$$c_1[1\ \ -2\ \ 0] + c_2[0\ \ 1\ \ 1] + c_3[1\ \ 1\ \ 2] = [2\ \ -3\ \ 4]$$

$$\begin{aligned} c_1 \quad\quad + c_3 &= 2 \\ -2c_1 + c_2 + c_3 &= -3 \\ c_2 + 2c_3 &= 4 \end{aligned}$$

which has $c_1 = 5, c_2 = 10, c_3 = -3$ as its solution. Hence,

$$[2 \ -3 \ \ 4]_T = \begin{bmatrix} 5 \\ 10 \\ -3 \end{bmatrix}$$

(c) $$c_1[0 \ -1 \ \ 1] + c_2[1 \ -2 \ \ 0] + c_3[1 \ \ 1 \ \ 2] = [2 \ -3 \ \ 4]$$

$$\begin{aligned} c_2 + c_3 &= 2 \\ -c_1 - 2c_2 + c_3 &= -3 \\ c_1 \quad\quad + 2c_3 &= 4 \end{aligned}$$

which has $c_1 = 2, c_2 = 1, c_3 = 1$ as its solution. Hence,

$$[2 \ -3 \ \ 4]_S = \begin{bmatrix} 2 \\ 1 \\ 1 \end{bmatrix}$$

PROBLEM 13-4 For Problem 13-3, verify that

$$P = \begin{bmatrix} \frac{3}{5} & -1 & \frac{1}{5} \\ \frac{4}{5} & -2 & \frac{3}{5} \\ \frac{2}{5} & 1 & -\frac{1}{5} \end{bmatrix}$$

transforms $[2 \ -3 \ \ 4]_B$ into $[2 \ -3 \ \ 4]_T$.

Solution

$$[2 \ -3 \ \ 4]_B = \begin{bmatrix} 2 \\ -3 \\ 4 \end{bmatrix} \quad \text{and} \quad [2 \ -3 \ \ 4]_T = \begin{bmatrix} 5 \\ 10 \\ -3 \end{bmatrix}$$

Therefore,

$$P[2 \ -3 \ \ 4]_B = \begin{bmatrix} \frac{3}{5} & -1 & \frac{1}{5} \\ \frac{4}{5} & -2 & \frac{3}{5} \\ \frac{2}{5} & 1 & -\frac{1}{5} \end{bmatrix} \begin{bmatrix} 2 \\ -3 \\ 4 \end{bmatrix} = \begin{bmatrix} 5 \\ 10 \\ -3 \end{bmatrix}$$

$$\therefore \quad P[2 \ -3 \ \ 4]_B = [2 \ -3 \ \ 4]_T$$

PROBLEM 13-5 Find the inverse matrix P^{-1} of Problem 13-4 and verify that P^{-1} transforms $[2 \ -3 \ \ 4]_T$ into $[2 \ -3 \ \ 4]_B$.

Solution

$$\left[\begin{array}{ccc|ccc} \frac{3}{5} & -1 & \frac{1}{5} & 1 & 0 & 0 \\ \frac{4}{5} & -2 & \frac{3}{5} & 0 & 1 & 0 \\ \frac{2}{5} & 1 & -\frac{1}{5} & 0 & 0 & 1 \end{array}\right] \xrightarrow[R_2 \to R_2 - 2R_3]{R_1 \to R_1 + R_3} \left[\begin{array}{ccc|ccc} 1 & 0 & 0 & 1 & 0 & 1 \\ 0 & -4 & 1 & 0 & 1 & -2 \\ \frac{2}{5} & 1 & -\frac{1}{5} & 0 & 0 & 1 \end{array}\right]$$

$$\xrightarrow[R_3 \to 5R_3]{} \left[\begin{array}{ccc|ccc} 1 & 0 & 0 & 1 & 0 & 1 \\ 0 & -4 & 1 & 0 & 1 & -2 \\ 2 & 5 & -1 & 0 & 0 & 5 \end{array}\right] \xrightarrow[R_3 \to R_3 - 2R_1]{} \left[\begin{array}{ccc|ccc} 1 & 0 & 0 & 1 & 0 & 1 \\ 0 & -4 & 1 & 0 & 1 & -2 \\ 0 & 5 & -1 & -2 & 0 & 3 \end{array}\right]$$

$$\xrightarrow{R_2 \to R_2 + R_3} \left[\begin{array}{ccc|ccc} 1 & 0 & 0 & 1 & 0 & 1 \\ 0 & 1 & 0 & -2 & 1 & 1 \\ 0 & 5 & -1 & -2 & 0 & 3 \end{array}\right] \xrightarrow{R_3 \to R_3 - 5R_2} \left[\begin{array}{ccc|ccc} 1 & 0 & 0 & 1 & 0 & 1 \\ 0 & 1 & 0 & -2 & 1 & 1 \\ 0 & 0 & -1 & 8 & -5 & -2 \end{array}\right]$$

$$\xrightarrow{R_3 \to -R_3} \left[\begin{array}{ccc|ccc} 1 & 0 & 0 & 1 & 0 & 1 \\ 0 & 1 & 0 & -2 & 1 & 1 \\ 0 & 0 & 1 & -8 & 5 & 2 \end{array}\right]$$

$$\therefore \quad P^{-1} = \begin{bmatrix} 1 & 0 & 1 \\ -2 & 1 & 1 \\ -8 & 5 & 2 \end{bmatrix}$$

Then

$$P^{-1}[2 \ \ -3 \ \ 4]_T = \begin{bmatrix} 1 & 0 & 1 \\ -2 & 1 & 1 \\ -8 & 5 & 2 \end{bmatrix} \begin{bmatrix} 5 \\ 10 \\ -3 \end{bmatrix} = \begin{bmatrix} 2 \\ -3 \\ 4 \end{bmatrix}$$

Therefore, $P^{-1}[2 \ \ -3 \ \ 4]_T = [2 \ \ -3 \ \ 4]_B$.

PROBLEM 13-6 Let

$$B = \left\{ \begin{bmatrix} 2 \\ 1 \\ -1 \end{bmatrix}, \begin{bmatrix} 1 \\ 0 \\ 3 \end{bmatrix}, \begin{bmatrix} 2 \\ -1 \\ 4 \end{bmatrix} \right\} \quad \text{and} \quad S = \left\{ \begin{bmatrix} 1 \\ 0 \\ 1 \end{bmatrix}, \begin{bmatrix} 1 \\ 1 \\ 0 \end{bmatrix}, \begin{bmatrix} 0 \\ 1 \\ 1 \end{bmatrix} \right\}$$

be ordered bases for R^3. If $\mathbf{a} = \begin{bmatrix} 5 \\ -1 \\ 4 \end{bmatrix}$, find **(a)** the coordinate vector of **a** with respect to basis B; **(b)** the coordinate vector of **a** with respect to basis S; **(c)** the transition matrix P from basis S to basis B; **(d)** the transition matrix P^{-1} from basis B to basis S; **(e)** the product $P[\mathbf{a}]_S$ to show that it is the same as the answer in part **(a)**; **(f)** the product $P^{-1}[\mathbf{a}]_B$ to show that it is the same as the answer in part **(b)**.

Solution

(a)

$$c_1 \begin{bmatrix} 2 \\ 1 \\ -1 \end{bmatrix} + c_2 \begin{bmatrix} 1 \\ 0 \\ 3 \end{bmatrix} + c_3 \begin{bmatrix} 2 \\ -1 \\ 4 \end{bmatrix} = \begin{bmatrix} 5 \\ -1 \\ 4 \end{bmatrix}$$

$$\begin{aligned} 2c_1 + c_2 + 2c_3 &= 5 \\ c_1 \qquad - c_3 &= -1 \\ -c_1 + 3c_2 + 4c_3 &= 4 \end{aligned}$$

which has the solution $c_1 = 1$, $c_2 = -1$, $c_3 = 2$. Therefore,

$$\begin{bmatrix} 5 \\ -1 \\ 4 \end{bmatrix}_B = \begin{bmatrix} 1 \\ -1 \\ 2 \end{bmatrix}$$

(b)

$$c_1 \begin{bmatrix} 1 \\ 0 \\ 1 \end{bmatrix} + c_2 \begin{bmatrix} 1 \\ 1 \\ 0 \end{bmatrix} + c_3 \begin{bmatrix} 0 \\ 1 \\ 1 \end{bmatrix} = \begin{bmatrix} 5 \\ -1 \\ 4 \end{bmatrix}$$

$$\begin{aligned} c_1 + c_2 \quad\quad &= 5 \\ c_2 + c_3 &= -1 \\ c_1 \quad\quad + c_3 &= 4 \end{aligned}$$

which has as its solution $c_1 = 5, c_2 = 0, c_3 = -1$. Therefore,

$$\begin{bmatrix} 5 \\ -1 \\ 4 \end{bmatrix}_S = \begin{bmatrix} 5 \\ 0 \\ -1 \end{bmatrix}$$

(c)

$$c_1\begin{bmatrix} 2 \\ 1 \\ -1 \end{bmatrix} + c_2\begin{bmatrix} 1 \\ 0 \\ 3 \end{bmatrix} + c_3\begin{bmatrix} 2 \\ -1 \\ 4 \end{bmatrix} = \begin{bmatrix} 1 \\ 0 \\ 1 \end{bmatrix}$$

$$\therefore \quad \begin{aligned} 2c_1 + c_2 + 2c_3 &= 1 \\ c_1 \quad\quad - c_3 &= 0 \\ -c_1 + 3c_2 + 4c_3 &= 1 \end{aligned}$$

which has as its solution $c_1 = 2/9, c_2 = 1/9, c_3 = 2/9$;

$$c_1\begin{bmatrix} 2 \\ 1 \\ -1 \end{bmatrix} + c_2\begin{bmatrix} 1 \\ 0 \\ 3 \end{bmatrix} + c_3\begin{bmatrix} 2 \\ -1 \\ 4 \end{bmatrix} = \begin{bmatrix} 1 \\ 1 \\ 0 \end{bmatrix}$$

$$\therefore \quad \begin{aligned} 2c_1 + c_2 + 2c_3 &= 1 \\ c_1 \quad\quad - c_3 &= 1 \\ -c_1 + 3c_2 + 4c_3 &= 0 \end{aligned}$$

which has as its solution $c_1 = 5/9, c_2 = 7/9, c_3 = -4/9$;

$$c_1\begin{bmatrix} 2 \\ 1 \\ -1 \end{bmatrix} + c_2\begin{bmatrix} 1 \\ 0 \\ 3 \end{bmatrix} + c_3\begin{bmatrix} 2 \\ -1 \\ 4 \end{bmatrix} = \begin{bmatrix} 0 \\ 1 \\ 1 \end{bmatrix}$$

$$\therefore \quad \begin{aligned} 2c_1 + c_2 + 2c_3 &= 0 \\ c_1 \quad\quad - c_3 &= 1 \\ -c_1 + 3c_2 + 4c_3 &= 1 \end{aligned}$$

which has as its solution $c_1 = 1/9, c_2 = 14/9, c_3 = -8/9$. Therefore,

$$P = \begin{bmatrix} \frac{2}{9} & \frac{5}{9} & \frac{1}{9} \\ \frac{1}{9} & \frac{7}{9} & \frac{14}{9} \\ \frac{2}{9} & -\frac{4}{9} & -\frac{8}{9} \end{bmatrix}$$

is the transition matrix from S to B.

(d) To find the transition matrix from B to S we will find P^{-1}.

$$\left[\begin{array}{ccc|ccc} \frac{2}{9} & \frac{5}{9} & \frac{1}{9} & 1 & 0 & 0 \\ \frac{1}{9} & \frac{7}{9} & \frac{14}{9} & 0 & 1 & 0 \\ \frac{2}{9} & -\frac{4}{9} & -\frac{8}{9} & 0 & 0 & 1 \end{array}\right] \xrightarrow{R_1 \leftrightarrow R_2} \left[\begin{array}{ccc|ccc} \frac{1}{9} & \frac{7}{9} & \frac{14}{9} & 0 & 1 & 0 \\ \frac{2}{9} & \frac{5}{9} & \frac{1}{9} & 1 & 0 & 0 \\ \frac{2}{9} & -\frac{4}{9} & -\frac{8}{9} & 0 & 0 & 1 \end{array}\right]$$

$$\xrightarrow[R_3 \to R_3 - 2R_1]{R_2 \to R_2 - 2R_1} \left[\begin{array}{ccc|ccc} \frac{1}{9} & \frac{7}{9} & \frac{14}{9} & 0 & 1 & 0 \\ 0 & -1 & -3 & 1 & -2 & 0 \\ 0 & -2 & -4 & 0 & -2 & 1 \end{array}\right] \xrightarrow[R_3 \to R_3 - 2R_2]{R_2 \to -R_2} \left[\begin{array}{ccc|ccc} \frac{1}{9} & \frac{7}{9} & \frac{14}{9} & 0 & 1 & 0 \\ 0 & 1 & 3 & -1 & 2 & 0 \\ 0 & 0 & 2 & -2 & 2 & 1 \end{array}\right]$$

$$\xrightarrow[R_3\to 1/2R_3]{R_1\to 9R_1}\left[\begin{array}{ccc|ccc}1&7&14&0&9&0\\0&1&3&-1&2&0\\0&0&1&-1&1&\frac{1}{2}\end{array}\right]\xrightarrow{R_2\to R_2-3R_3}\left[\begin{array}{ccc|ccc}1&7&14&0&9&0\\0&1&0&2&-1&-\frac{3}{2}\\0&0&1&-1&1&\frac{1}{2}\end{array}\right]$$

$$\xrightarrow{R_1\to R_1-7R_2}\left[\begin{array}{ccc|ccc}1&0&14&-14&16&\frac{21}{2}\\0&1&0&2&-1&-\frac{3}{2}\\0&0&1&-1&1&\frac{1}{2}\end{array}\right]\xrightarrow{R_1\to R_1-14R_3}\left[\begin{array}{ccc|ccc}1&0&0&0&2&\frac{7}{2}\\0&1&0&2&-1&-\frac{3}{2}\\0&0&1&-1&1&\frac{1}{2}\end{array}\right]$$

$$\therefore\quad P^{-1}=\begin{bmatrix}0&2&\frac{7}{2}\\2&-1&-\frac{3}{2}\\-1&1&\frac{1}{2}\end{bmatrix}$$

is the transition matrix from B to S.

(e) To find the product $P[\mathbf{a}]_S$, we multiply the matrix P by $[\mathbf{a}]_S$. Therefore,

$$\begin{bmatrix}\frac{2}{9}&\frac{5}{9}&\frac{1}{9}\\\frac{1}{9}&\frac{7}{9}&\frac{14}{9}\\\frac{2}{9}&-\frac{4}{9}&-\frac{8}{9}\end{bmatrix}\begin{bmatrix}5\\0\\-1\end{bmatrix}=\begin{bmatrix}1\\-1\\2\end{bmatrix}$$

but this is the same as in part **(a)**.

(f) To find product $P^{-1}[\mathbf{a}]_B$, we multiply the matrix P^{-1} by $[\mathbf{a}]_B$. Therefore,

$$\begin{bmatrix}0&2&\frac{7}{2}\\2&-1&-\frac{3}{2}\\-1&1&\frac{1}{2}\end{bmatrix}\begin{bmatrix}1\\-1\\2\end{bmatrix}=\begin{bmatrix}5\\0\\-1\end{bmatrix}$$

which is same as in part **(b)**.

PROBLEM 13-7 If $L: R_3\to R_2$ is the linear transformation defined by $L([a_1\ \ a_2\ \ a_3])=[a_1+2a_2\ \ a_3]$, verify that the matrix of L with respect to the natural ordered bases of R_3 and R_2 is $A=\begin{bmatrix}1&2&0\\0&0&1\end{bmatrix}$.

Solution Let $B=\{[1\ \ 0\ \ 0],[0\ \ 1\ \ 0],[0\ \ 0\ \ 1]\}$, and $S=\{[1\ \ 0],[0\ \ 1]\}$. We must find the coordinate vector of $[a_1\ \ a_2\ \ a_3]$ with respect to B and the coordinate vector of $[a_1+2a_2\ \ a_3]$ with respect to S.

$$[a_1\ \ a_2\ \ a_3]=a_1[1\ \ 0\ \ 0]+a_2[0\ \ 1\ \ 0]+a_3[0\ \ 0\ \ 1]$$

$$\therefore\quad [a_1\ \ a_2\ \ a_3]_B=\begin{bmatrix}a_1\\a_2\\a_3\end{bmatrix}$$

$$[a_1+2a_2\ \ a_3]=(a_1+2a_2)[1\ \ 0]+a_3[0\ \ 1]$$

$$\therefore\quad [a_1+2a_2]_S=\begin{bmatrix}a_1+2a_2\\a_3\end{bmatrix}$$

Hence,

$$A[a_1\ \ a_2\ \ a_3]_B=\begin{bmatrix}1&2&0\\0&0&1\end{bmatrix}\begin{bmatrix}a_1\\a_2\\a_3\end{bmatrix}=\begin{bmatrix}a_1+2a_2\\a_3\end{bmatrix}=[a_1+2a_2]_S$$

PROBLEM 13-8 Let $L: R_3\to R_2$ be the linear transformation defined by $L([a_1\ \ a_2\ \ a_3])=[a_1+2a_2\ \ a_3]$ and $B=\{[1\ \ 0\ \ 1],[0\ \ 1\ \ 1],[1\ \ 1\ \ 0]\}$ is an ordered basis of R_3 and $S=\{[2\ \ -1],[4\ \ 3]\}$ is an ordered basis of R_2, find the matrix A of L with respect to bases B and S.

Solution

Step 1:

$$L([1\quad 0\quad 1]) = [1+0\quad 1] = [1\quad 1]$$

$$L([0\quad 1\quad 1]) = [0+2\quad 1] = [2\quad 1]$$

$$L([1\quad 1\quad 0]) = [1+2\quad 0] = [3\quad 0]$$

Step 2:

$$[1\quad 1] = c_1[2\quad -1] + c_2[4\quad 3]$$

$$\therefore\quad 2c_1 + 4c_2 = 1$$

$$-c_1 + 3c_2 = 1$$

which has $c_1 = -1/10$ and $c_2 = 3/10$ as its solution.

$$[2\quad 1] = c_1[2\quad -1] + c_2[4\quad 3]$$

$$\therefore\quad 2c_1 + 4c_2 = 2$$

$$-c_1 + 3c_2 = 1$$

which has $c_1 = 1/5$ and $c_2 = 2/5$ as its solution.

$$[3\quad 0] = c_1[2\quad -1] + c_2[4\quad 3]$$

$$\therefore\quad 2c_1 + 4c_2 = 3$$

$$-c_1 + 3c_2 = 0$$

which has $c_1 = 9/10$ and $c_2 = 3/10$ as solution. Therefore,

$$[1\quad 1]_S = \begin{bmatrix} -\frac{1}{10} \\ \frac{3}{10} \end{bmatrix} \qquad [2\quad 1]_S = \begin{bmatrix} \frac{1}{5} \\ \frac{2}{5} \end{bmatrix}, \qquad [3\quad 0]_S = \begin{bmatrix} \frac{9}{10} \\ \frac{3}{10} \end{bmatrix}$$

Step 3:

$$A = \begin{bmatrix} -\frac{1}{10} & \frac{1}{5} & \frac{9}{10} \\ \frac{3}{10} & \frac{2}{5} & \frac{3}{10} \end{bmatrix}$$

PROBLEM 13-9 For the linear transformation in Problem 13-8, find L: ([2 1 −3]) by (**a**) direct calculation using the definition of L and (**b**) using matrix A.

Solution

(**a**) $L([2\quad 1\quad -3]) = [2+2\quad -3] = [4\quad -3]$.
(**b**) To use matrix A, we must first find the coordinate vector of [2 1 −3] with respect to basis B.

$$[2\quad 1\quad -3] = c_1[1\quad 0\quad 1] + c_2[0\quad 1\quad 1] + c_3[1\quad 1\quad 0]$$

$$\therefore\quad c_1 \qquad + c_3 = 2$$

$$c_2 + c_3 = 1$$

$$c_1 + c_2 \qquad = -3$$

which has as its solution $c_1 = -1$, $c_2 = -2$, $c_3 = 3$. Therefore, $[2\quad 1\quad -3]_B = \begin{bmatrix} -1 \\ -2 \\ 3 \end{bmatrix}$. Therefore,

$$A[2\quad 1\quad -3]_B = \begin{bmatrix} -\frac{1}{10} & \frac{1}{5} & \frac{9}{10} \\ \frac{3}{10} & \frac{2}{5} & \frac{3}{10} \end{bmatrix} \begin{bmatrix} -1 \\ -2 \\ 3 \end{bmatrix} = \begin{bmatrix} \frac{12}{5} \\ -\frac{1}{5} \end{bmatrix}$$

The last vector is the coordinate vector of the desired answer. Therefore, $(12/5)[2\quad -1] - (1/5)[4\quad 3] = [4\quad -3]$, which is equal to $L([2\quad 1\quad -3])$ in part (**a**).

PROBLEM 13-10 Let $L: R^2 \to R^3$ be defined by

$$L\begin{bmatrix} a_1 \\ a_2 \end{bmatrix} = \begin{bmatrix} 2a_1 - a_2 \\ a_1 + a_2 \\ -a_1 + 3a_2 \end{bmatrix}$$

Consider the ordered bases

$$B = \left\{ \begin{bmatrix} 1 \\ 1 \end{bmatrix}, \begin{bmatrix} 0 \\ 2 \end{bmatrix} \right\} \quad \text{and} \quad B' = \left\{ \begin{bmatrix} 2 \\ 1 \end{bmatrix}, \begin{bmatrix} -1 \\ 1 \end{bmatrix} \right\} \quad \text{for } R^2$$

$$S = \left\{ \begin{bmatrix} 1 \\ 1 \\ 0 \end{bmatrix}, \begin{bmatrix} 0 \\ 1 \\ 1 \end{bmatrix}, \begin{bmatrix} 1 \\ 0 \\ 1 \end{bmatrix} \right\} \quad \text{and} \quad S' = \left\{ \begin{bmatrix} 1 \\ 0 \\ -2 \end{bmatrix}, \begin{bmatrix} 0 \\ 1 \\ 0 \end{bmatrix}, \begin{bmatrix} 0 \\ 0 \\ 1 \end{bmatrix} \right\} \quad \text{for } R^3$$

(**a**) Find the matrix representation A of L with respect to bases B and S by direct computation. (**b**) Find the matrix representation A' of L with respect to bases B' and S' by direct computation. (**c**) Find the transition matrix P from B' to B. (**d**) Find the transition matrix Q from S' to S. (**e**) Verify that $A' = Q^{-1}AP$.

Solution

(**a**)

$$L\left(\begin{bmatrix} 1 \\ 1 \end{bmatrix}\right) = \begin{bmatrix} 2-1 \\ 1+1 \\ -1+3 \end{bmatrix} = \begin{bmatrix} 1 \\ 2 \\ 2 \end{bmatrix}$$

$$L\left(\begin{bmatrix} 0 \\ 2 \end{bmatrix}\right) = \begin{bmatrix} 0-2 \\ 0+2 \\ -0+6 \end{bmatrix} = \begin{bmatrix} -2 \\ 2 \\ 6 \end{bmatrix}$$

$$\begin{bmatrix} 1 \\ 2 \\ 2 \end{bmatrix} = c_1 \begin{bmatrix} 1 \\ 1 \\ 0 \end{bmatrix} + c_2 \begin{bmatrix} 0 \\ 1 \\ 1 \end{bmatrix} + c_3 \begin{bmatrix} 1 \\ 0 \\ 1 \end{bmatrix}$$

$$\begin{aligned} \therefore\quad c_1 \phantom{{}+c_2} + c_3 &= 1 \\ c_1 + c_2 \phantom{{}+c_3} &= 2 \\ c_2 + c_3 &= 2 \end{aligned}$$

which has as its solution $c_1 = 1/2$, $c_2 = 3/2$, $c_3 = 1/2$. Therefore,

$$\begin{bmatrix} 1 \\ 2 \\ 2 \end{bmatrix}_S = \begin{bmatrix} \frac{1}{2} \\ \frac{3}{2} \\ \frac{1}{2} \end{bmatrix}$$

$$\begin{bmatrix} -2 \\ 2 \\ 6 \end{bmatrix} = c_1 \begin{bmatrix} 1 \\ 1 \\ 0 \end{bmatrix} + c_2 \begin{bmatrix} 0 \\ 1 \\ 1 \end{bmatrix} + c_3 \begin{bmatrix} 1 \\ 0 \\ 1 \end{bmatrix}$$

$$\begin{aligned} \therefore\quad c_1 \phantom{{}+c_2} + c_3 &= -2 \\ c_1 + c_2 \phantom{{}+c_3} &= 2 \\ c_2 + c_3 &= 6 \end{aligned}$$

which has as its solution $c_1 = -3, c_2 = 5, c_3 = 1$. Therefore,

$$\begin{bmatrix} -2 \\ 2 \\ 6 \end{bmatrix}_S = \begin{bmatrix} -3 \\ 5 \\ 1 \end{bmatrix}$$

Therefore,

$$A = \begin{bmatrix} \frac{1}{2} & -3 \\ \frac{3}{2} & 5 \\ \frac{1}{2} & 1 \end{bmatrix}$$

(b)

$$L\left(\begin{bmatrix} 2 \\ 1 \end{bmatrix}\right) = \begin{bmatrix} 4-1 \\ 2+1 \\ -2+3 \end{bmatrix} = \begin{bmatrix} 3 \\ 3 \\ 1 \end{bmatrix}$$

$$L\left(\begin{bmatrix} -1 \\ 1 \end{bmatrix}\right) = \begin{bmatrix} -2-1 \\ -1+1 \\ 1+3 \end{bmatrix} = \begin{bmatrix} -3 \\ 0 \\ 4 \end{bmatrix}$$

$$\begin{bmatrix} 3 \\ 3 \\ 1 \end{bmatrix} = c_1 \begin{bmatrix} 1 \\ 0 \\ -2 \end{bmatrix} + c_2 \begin{bmatrix} 0 \\ 1 \\ 0 \end{bmatrix} + c_3 \begin{bmatrix} 0 \\ 0 \\ 1 \end{bmatrix}$$

$$\begin{aligned} \therefore \quad c_1 \qquad\qquad &= 3 \\ c_2 \quad &= 3 \\ -2c_1 \qquad + c_3 &= 1 \end{aligned}$$

which has as its solution $c_1 = 3, c_2 = 3, c_3 = 7$. Therefore,

$$\begin{bmatrix} 3 \\ 3 \\ 1 \end{bmatrix}_{S'} = \begin{bmatrix} 3 \\ 3 \\ 7 \end{bmatrix}$$

$$\begin{bmatrix} -3 \\ 0 \\ 4 \end{bmatrix} = c_1 \begin{bmatrix} 1 \\ 0 \\ -2 \end{bmatrix} + c_2 \begin{bmatrix} 0 \\ 1 \\ 0 \end{bmatrix} + c_3 \begin{bmatrix} 0 \\ 0 \\ 1 \end{bmatrix}$$

$$\begin{aligned} \therefore \quad c_1 \qquad\qquad &= -3 \\ c_2 \quad &= 0 \\ -2c_1 \qquad + c_3 &= 4 \end{aligned}$$

which has as its solution $c_1 = -3, c_2 = 0, c_3 = -2$. Therefore,

$$\begin{bmatrix} -3 \\ 0 \\ 4 \end{bmatrix}_{S'} = \begin{bmatrix} -3 \\ 0 \\ -2 \end{bmatrix}$$

Therefore,

$$A' = \begin{bmatrix} 3 & -3 \\ 3 & 0 \\ 7 & -2 \end{bmatrix}$$

(c)

$$\begin{bmatrix} 2 \\ 1 \end{bmatrix} = c_1 \begin{bmatrix} 1 \\ 1 \end{bmatrix} + c_2 \begin{bmatrix} 0 \\ 2 \end{bmatrix}$$

$$\begin{aligned} \therefore \quad c_1 \qquad &= 2 \\ c_1 + 2c_2 &= 1 \end{aligned}$$

which has as its solution $c_1 = 2$, $c_2 = -1/2$. Therefore,

$$\begin{bmatrix} 2 \\ 1 \end{bmatrix}_B = \begin{bmatrix} 2 \\ -\frac{1}{2} \end{bmatrix}$$

$$\begin{bmatrix} -1 \\ 1 \end{bmatrix} = c_1 \begin{bmatrix} 1 \\ 1 \end{bmatrix} + c_2 \begin{bmatrix} 0 \\ 2 \end{bmatrix}$$

$$\begin{aligned} \therefore \quad c_1 \qquad &= -1 \\ c_1 + 2c_2 &= 1 \end{aligned}$$

which has $c_1 = -1$ and $c_2 = 1$ as its solution. Therefore,

$$\begin{bmatrix} -1 \\ 1 \end{bmatrix}_B = \begin{bmatrix} -1 \\ 1 \end{bmatrix}$$

Therefore,

$$P_{B' \text{ to } B} = \begin{bmatrix} 2 & -1 \\ -\frac{1}{2} & 1 \end{bmatrix}$$

(d)

$$\begin{bmatrix} 1 \\ 0 \\ -2 \end{bmatrix} = c_1 \begin{bmatrix} 1 \\ 1 \\ 0 \end{bmatrix} + c_2 \begin{bmatrix} 0 \\ 1 \\ 1 \end{bmatrix} + c_3 \begin{bmatrix} 1 \\ 0 \\ 1 \end{bmatrix}$$

$$\begin{aligned} \therefore \quad c_1 \qquad + c_3 &= 1 \\ c_1 + c_2 \qquad &= 0 \\ c_2 + c_3 &= -2 \end{aligned}$$

which has $c_1 = 3/2$, $c_2 = -3/2$, $c_3 = -1/2$ as its solution. Therefore,

$$\begin{bmatrix} 1 \\ 0 \\ -2 \end{bmatrix}_S = \begin{bmatrix} \frac{3}{2} \\ -\frac{3}{2} \\ -\frac{1}{2} \end{bmatrix}$$

$$\begin{bmatrix} 0 \\ 1 \\ 0 \end{bmatrix} = c_1 \begin{bmatrix} 1 \\ 1 \\ 0 \end{bmatrix} + c_2 \begin{bmatrix} 0 \\ 1 \\ 1 \end{bmatrix} + c_3 \begin{bmatrix} 1 \\ 0 \\ 1 \end{bmatrix}$$

$$\begin{aligned} \therefore \quad c_1 \qquad + c_3 &= 0 \\ c_1 + c_2 \qquad &= 1 \\ c_2 + c_3 &= 0 \end{aligned}$$

which has $c_1 = 1/2$, $c_2 = 1/2$, $c_3 = -1/2$ as its solution. Therefore,

$$\begin{bmatrix} 0 \\ 1 \\ 0 \end{bmatrix}_S = \begin{bmatrix} \frac{1}{2} \\ \frac{1}{2} \\ -\frac{1}{2} \end{bmatrix}$$

$$\begin{bmatrix} 0 \\ 0 \\ 1 \end{bmatrix} = c_1 \begin{bmatrix} 1 \\ 1 \\ 0 \end{bmatrix} + c_2 \begin{bmatrix} 0 \\ 1 \\ 1 \end{bmatrix} + c_3 \begin{bmatrix} 1 \\ 0 \\ 1 \end{bmatrix}$$

$$\begin{aligned} \therefore \quad c_1 \qquad + c_3 &= 0 \\ c_1 + c_2 \qquad &= 0 \\ c_2 + c_3 &= 1 \end{aligned}$$

which has as its solution $c_1 = -1/2$, $c_2 = 1/2$, $c_3 = 1/2$. Therefore,

$$\begin{bmatrix} 0 \\ 0 \\ 1 \end{bmatrix}_S = \begin{bmatrix} -\frac{1}{2} \\ \frac{1}{2} \\ \frac{1}{2} \end{bmatrix}$$

Therefore,

$$Q_{S' \text{ to } S} = \begin{bmatrix} \frac{3}{2} & \frac{1}{2} & -\frac{1}{2} \\ -\frac{3}{2} & \frac{1}{2} & \frac{1}{2} \\ -\frac{1}{2} & -\frac{1}{2} & \frac{1}{2} \end{bmatrix}$$

(e) To verify $A' = Q^{-1}AP$, we must first find Q^{-1}.

$$\left[\begin{array}{ccc|ccc} \frac{3}{2} & \frac{1}{2} & -\frac{1}{2} & 1 & 0 & 0 \\ -\frac{3}{2} & \frac{1}{2} & \frac{1}{2} & 0 & 1 & 0 \\ -\frac{1}{2} & -\frac{1}{2} & \frac{1}{2} & 0 & 0 & 1 \end{array}\right] \xrightarrow[R_2 \to R_2 - R_3]{R_1 \to R_1 + R_3} \left[\begin{array}{ccc|ccc} 1 & 0 & 0 & 1 & 0 & 1 \\ -1 & 1 & 0 & 0 & 1 & -1 \\ -\frac{1}{2} & -\frac{1}{2} & \frac{1}{2} & 0 & 0 & 1 \end{array}\right]$$

$$\xrightarrow[R_3 \to 2R_3]{R_2 \to R_2 + R_1} \left[\begin{array}{ccc|ccc} 1 & 0 & 0 & 1 & 0 & 1 \\ 0 & 1 & 0 & 1 & 1 & 0 \\ -1 & -1 & 1 & 0 & 0 & 2 \end{array}\right]$$

$$\xrightarrow[R_3 \to R_3 + R_1]{} \left[\begin{array}{ccc|ccc} 1 & 0 & 0 & 1 & 0 & 1 \\ 0 & 1 & 0 & 1 & 1 & 0 \\ 0 & -1 & 1 & 1 & 0 & 3 \end{array}\right]$$

$$\xrightarrow[R_3 \to R_3 + R_2]{} \left[\begin{array}{ccc|ccc} 1 & 0 & 0 & 1 & 0 & 1 \\ 0 & 1 & 0 & 1 & 1 & 0 \\ 0 & 0 & 1 & 2 & 1 & 3 \end{array}\right]$$

$$\therefore \quad Q^{-1} = \begin{bmatrix} 1 & 0 & 1 \\ 1 & 1 & 0 \\ 2 & 1 & 3 \end{bmatrix}$$

So

$$Q^{-1}AP = \begin{bmatrix} 1 & 0 & 1 \\ 1 & 1 & 0 \\ 2 & 1 & 3 \end{bmatrix} \begin{bmatrix} \frac{1}{2} & -3 \\ \frac{3}{2} & 5 \\ \frac{1}{2} & 1 \end{bmatrix} \begin{bmatrix} 2 & -1 \\ -\frac{1}{2} & 1 \end{bmatrix} = \begin{bmatrix} 3 & -3 \\ 3 & 0 \\ 7 & -2 \end{bmatrix} = A'$$

$$A' = Q^{-1}AP$$

PROBLEM 13-11 Let $L: R^2 \to R^2$ be a linear transformation defined by $L\left(\begin{bmatrix} a \\ b \end{bmatrix}\right) = \begin{bmatrix} 2a - 3b \\ 3a + 5b \end{bmatrix}$. Find the standard matrix of the linear transformation.

Solution

$$L\left(\begin{bmatrix}1\\0\end{bmatrix}\right)=\begin{bmatrix}2\\3\end{bmatrix},\qquad L\left(\begin{bmatrix}0\\1\end{bmatrix}\right)=\begin{bmatrix}-3\\5\end{bmatrix}$$

Thus, $A=\begin{bmatrix}2 & -3\\3 & 5\end{bmatrix}$ is the matrix used to define L.

note: This matrix is simply the matrix of the coefficients in the definition of the transformation.

PROBLEM 13-12 Let $B=\left\{\begin{bmatrix}1\\-1\end{bmatrix},\begin{bmatrix}2\\0\end{bmatrix}\right\}$ be a basis for R^2, and let $\mathbf{a}=\begin{bmatrix}4\\-1\end{bmatrix}$. Using the linear transformation given in Problem 13-11, find the matrix of L with respect to B, and $[\mathbf{a}]_B$.

Solution

$$L\left(\begin{bmatrix}1\\-1\end{bmatrix}\right)=\begin{bmatrix}5\\-2\end{bmatrix}\quad\text{and}\quad L\left(\begin{bmatrix}2\\0\end{bmatrix}\right)=\begin{bmatrix}4\\6\end{bmatrix}$$

$$\begin{bmatrix}5\\-2\end{bmatrix}=c_1\begin{bmatrix}1\\-1\end{bmatrix}+c_2\begin{bmatrix}2\\0\end{bmatrix}$$

$$\therefore\quad \begin{aligned} c_1+2c_2&=5\\ -c_1\qquad&=-2\end{aligned}$$

which has as its solution $c_1=2$, $c_2=3/2$. Therefore,

$$\begin{bmatrix}5\\-2\end{bmatrix}_B=\begin{bmatrix}2\\\frac{3}{2}\end{bmatrix}$$

$$\begin{bmatrix}4\\6\end{bmatrix}=c_1\begin{bmatrix}1\\-1\end{bmatrix}+c_2\begin{bmatrix}2\\0\end{bmatrix}$$

$$\therefore\quad \begin{aligned} c_1+2c_2&=4\\ -c_1\qquad&=6\end{aligned}$$

which has as its solution $c_1=-6$, $c_2=5$. Therefore,

$$\begin{bmatrix}4\\6\end{bmatrix}_B=\begin{bmatrix}-6\\5\end{bmatrix}$$

Therefore, $[L]_B=\begin{bmatrix}2 & -6\\\frac{3}{2} & 5\end{bmatrix}$. Now we will find $[\mathbf{a}]_B$.

$$\begin{bmatrix}4\\-1\end{bmatrix}=c_1\begin{bmatrix}1\\-1\end{bmatrix}+c_2\begin{bmatrix}2\\0\end{bmatrix}$$

$$\therefore\quad \begin{aligned} c_1+2c_2&=4\\ -c_1\qquad&=-1\end{aligned}$$

which has as its solution $c_1=1$, $c_2=3/2$. Thus $[\mathbf{a}]_B=\begin{bmatrix}1\\\frac{3}{2}\end{bmatrix}$.

Supplementary Problems

PROBLEM 13-13 Find the coordinate vector of $[2\ \ -3\ \ 5]$ with respect to each of the following ordered bases for R_3: **(a)** $B=\{[1\ \ 0\ \ 2],[0\ \ -1\ \ 1],[1\ \ 0\ \ 1]\}$, **(b)** $T=\{[1\ \ 0\ \ 1],[1\ \ 0\ \ 2],[0\ \ -1\ \ 1]\}$, and **(c)** $S=\{[0\ \ -1\ \ 1],[1\ \ 0\ \ 2],[1\ \ 0\ \ 1]\}$.

PROBLEM 13-14 Let

$$B = \left\{ \begin{bmatrix} 1 \\ 0 \\ 2 \end{bmatrix}, \begin{bmatrix} 0 \\ -1 \\ 1 \end{bmatrix}, \begin{bmatrix} 2 \\ -1 \\ 4 \end{bmatrix} \right\} \quad \text{and} \quad S = \left\{ \begin{bmatrix} 1 \\ 1 \\ 0 \end{bmatrix}, \begin{bmatrix} 0 \\ 1 \\ 1 \end{bmatrix}, \begin{bmatrix} 1 \\ 0 \\ 1 \end{bmatrix} \right\}$$

be ordered bases for R^3. If $\mathbf{a} = \begin{bmatrix} 2 \\ -3 \\ 1 \end{bmatrix}$, find **(a)** the coordinate vector of **a** with respect to basis B and **(b)** the coordinate vector of **a** with respect to basis S.

PROBLEM 13-15 Using the information given in Problem 13-14, find the following: **(a)** the transition matrix P from basis S to basis B, **(b)** the transition matrix P^{-1} from basis B to basis S, **(c)** the product $P[\mathbf{a}]_S$, and **(d)** the product $P^{-1}[\mathbf{a}]_B$.

PROBLEM 13-16 If $L: R_3 \to R_2$ is the linear transformation defined by $L[(a_1 \quad a_2 \quad a_3]) = [2a_1 + 3a_2 \quad a_1 + 2a_2 - a_3]$, find the matrix of L with respect to the natural ordered bases of R_3 and R_2.

PROBLEM 13-17 If $L: R^4 \to R^3$ is the linear transformation defined by

$$L \begin{bmatrix} a_1 \\ a_2 \\ a_3 \\ a_4 \end{bmatrix} = \begin{bmatrix} a_1 + 2a_3 \\ 2a_2 - a_4 \\ a_1 + a_2 - a_3 \\ 2a_1 + a_3 + 3a_4 \end{bmatrix}$$

find the matrix of L with respect to the natural ordered bases of R^4 and R^3.

PROBLEM 13-18 Let $L: R_3 \to R_2$ as defined in Problem 13-16. $B = \{[1 \quad 1 \quad 0], [1 \quad 0 \quad 1], [0 \quad 1 \quad 1]\}$ is an ordered basis of R_3, and $S = \{[1 \quad 2], [-1 \quad 0]\}$ is an ordered basis of R_2. Find the matrix A of L with respect to bases B and S.

PROBLEM 13-19 For the linear transformation in Problem 13-18, find $L([2 \quad 0 \quad -1])$ by **(a)** direct calculation using the definition of L, and **(b)** using the matrix A found in Problem 13-18.

PROBLEM 13-20 Let $L: R^2 \to R^3$ be defined by

$$L\left(\begin{bmatrix} a_1 \\ a_2 \end{bmatrix} \right) = \begin{bmatrix} a_1 + a_2 \\ 2a_1 + 3a_2 \\ a_2 - a_1 \end{bmatrix}$$

Consider the ordered bases

$$B = \left\{ \begin{bmatrix} 1 \\ 1 \end{bmatrix}, \begin{bmatrix} 0 \\ -1 \end{bmatrix} \right\} \quad \text{and} \quad B' = \left\{ \begin{bmatrix} 2 \\ -1 \end{bmatrix}, \begin{bmatrix} -1 \\ 1 \end{bmatrix} \right\} \quad \text{for } R^2$$

$$S = \left\{ \begin{bmatrix} 1 \\ 0 \\ 1 \end{bmatrix}, \begin{bmatrix} 0 \\ 1 \\ 1 \end{bmatrix}, \begin{bmatrix} 1 \\ 1 \\ 0 \end{bmatrix} \right\} \quad \text{and} \quad S' = \left\{ \begin{bmatrix} 0 \\ 1 \\ 0 \end{bmatrix}, \begin{bmatrix} 1 \\ 0 \\ 1 \end{bmatrix}, \begin{bmatrix} 1 \\ 0 \\ 0 \end{bmatrix} \right\} \quad \text{for } R^3$$

(a) Find the matrix representation A of L with respect to bases B and S by direct calculation. **(b)** Find the matrix representation A' of L with respect to bases B' and S' by direct calculation.

PROBLEM 13-21 For the linear transformation in Problem 13-20, find **(a)** the transition matrix P from B' to B and **(b)** the transition matrix Q from S' to S.

PROBLEM 13-22 Let $L: R^2 \to R^2$ be a linear transformation defined by $L\left(\begin{bmatrix} a \\ b \end{bmatrix}\right) = \begin{bmatrix} a + 2b \\ 2a + b \end{bmatrix}$, let $B = \left\{\begin{bmatrix} 1 \\ -2 \end{bmatrix}, \begin{bmatrix} 1 \\ 0 \end{bmatrix}\right\}$ be a basis for R^2, and let $\mathbf{a} = \begin{bmatrix} 2 \\ -3 \end{bmatrix}$. Find

(a) the matrix of L with respect to B,
(b) $[\mathbf{a}]_B$.

PROBLEM 13-23 Let $L: P_2 \to P_2$ be the linear transformation defined by $L([1]) = 1 - x + x^2$, $L([x]) = 2 + 3x - x^2$ and $L([x^2]) = 3x^2$. Find the matrix of L with respect to basis $B = \{[1 \quad x \quad x^2]\}$.

Answers to Supplementary Problems

13-13 **(a)** $\begin{bmatrix} 0 \\ 3 \\ 2 \end{bmatrix}$ **(b)** $\begin{bmatrix} 2 \\ 0 \\ 3 \end{bmatrix}$ **(c)** $\begin{bmatrix} 3 \\ 0 \\ 2 \end{bmatrix}$

13-14 **(a)** $\begin{bmatrix} 2 \\ -3 \\ 1 \end{bmatrix}_B = \begin{bmatrix} -10 \\ -3 \\ 6 \end{bmatrix}$

(b) $\begin{bmatrix} 2 \\ -3 \\ 1 \end{bmatrix}_S = \begin{bmatrix} -1 \\ -2 \\ 3 \end{bmatrix}$

13-15 **(a)** $P = \begin{bmatrix} -1 & 4 & -1 \\ -2 & 1 & -1 \\ 1 & -2 & 1 \end{bmatrix}$

(b) $P^{-1} \begin{bmatrix} -\frac{1}{2} & -1 & -\frac{3}{2} \\ \frac{1}{2} & 0 & \frac{1}{2} \\ \frac{3}{2} & 1 & \frac{7}{2} \end{bmatrix}$

(c) $P[a]_S = \begin{bmatrix} -10 \\ -3 \\ 6 \end{bmatrix}$

(d) $P^{-1}[a]_B = \begin{bmatrix} -1 \\ -2 \\ 3 \end{bmatrix}$

13-16 $\begin{bmatrix} 2 & 3 & 0 \\ 1 & 2 & -1 \end{bmatrix}$

13-17 $\begin{bmatrix} 1 & 0 & 2 & 0 \\ 0 & 2 & 0 & -1 \\ 1 & 1 & -1 & 0 \\ 2 & 0 & 1 & 3 \end{bmatrix}$

13-18 $A = \begin{bmatrix} \frac{3}{2} & 0 & \frac{1}{2} \\ -\frac{7}{2} & -2 & -\frac{5}{2} \end{bmatrix}$

13-19 **(a)** $[4 \quad 3]$

(b) $\begin{bmatrix} \frac{3}{2} \\ -\frac{5}{2} \end{bmatrix} \to \frac{3}{2}[1 \quad 2] - \frac{5}{2}[-1 \quad 0] = [4 \quad 3]$

13-20 **(a)** $A = \begin{bmatrix} -\frac{3}{2} & \frac{1}{2} \\ \frac{3}{2} & -\frac{3}{2} \\ \frac{7}{2} & -\frac{3}{2} \end{bmatrix}$

(b) $A' = \begin{bmatrix} 1 & 1 \\ -3 & 2 \\ 4 & -2 \end{bmatrix}$

13-21 **(a)** $P = \begin{bmatrix} 2 & -1 \\ 3 & -2 \end{bmatrix}$

(b) $Q = \begin{bmatrix} -\frac{1}{2} & 1 & \frac{1}{2} \\ \frac{1}{2} & 0 & -\frac{1}{2} \\ \frac{1}{2} & 0 & \frac{1}{2} \end{bmatrix}$

13-22 **(a)** $[L]_B = \begin{bmatrix} 0 & -1 \\ -3 & 2 \end{bmatrix}$

(b) $[a]_B = \begin{bmatrix} \frac{3}{2} \\ \frac{1}{2} \end{bmatrix}$

13-23 $\begin{bmatrix} 1 & 2 & 0 \\ -1 & 3 & 0 \\ 1 & -1 & 3 \end{bmatrix}$

14 SIMILARITY OF MATRICES

THIS CHAPTER IS ABOUT

- ☑ **Similarity**
- ☑ **Properties of Similarity**
- ☑ **Similarity of Matrices of a Linear Transformation**
- ☑ **Rank and Nullity of a Linear Transformation**

14-1. Similarity

A. Definition of similar matrices

If A and B are $n \times n$ matrices and P is an invertible matrix such that $B = P^{-1}AP$, then B is **similar** to A.

EXAMPLE 14-1: Use $P = \begin{bmatrix} 1 & 2 \\ -3 & 1 \end{bmatrix}$ to verify that $B = \begin{bmatrix} 0 & 0 \\ -3 & 1 \end{bmatrix}$ is similar to $A = \begin{bmatrix} 0 & 2 \\ 0 & 1 \end{bmatrix}$.

Solution: P^{-1} can be found to be $\begin{bmatrix} \frac{1}{7} & -\frac{2}{7} \\ \frac{3}{7} & \frac{1}{7} \end{bmatrix}$.

$$P^{-1}AP = \begin{bmatrix} \frac{1}{7} & -\frac{2}{7} \\ \frac{3}{7} & \frac{1}{7} \end{bmatrix}\begin{bmatrix} 0 & 2 \\ 0 & 1 \end{bmatrix}\begin{bmatrix} 1 & 2 \\ -3 & 1 \end{bmatrix} = \begin{bmatrix} 0 & 0 \\ -3 & 1 \end{bmatrix} = B$$

B. Finding a matrix similar to a given matrix

If all you need is any matrix B similar to a given $n \times n$ matrix A, then a suitable B can be found as follows:

Step 1: Find any $n \times n$ invertible matrix P.
Step 2: Find P^{-1}.
Step 3: $B = P^{-1}AP$.

EXAMPLE 14-2: Find a matrix B that is similar to the matrix $A = \begin{bmatrix} 1 & 2 \\ 3 & 4 \end{bmatrix}$.

Solution: Since $\begin{vmatrix} 1 & 2 \\ 1 & -1 \end{vmatrix} = -3$, then $\begin{bmatrix} 1 & 2 \\ 1 & -1 \end{bmatrix}$ is invertible. Thus, let $P = \begin{bmatrix} 1 & 2 \\ 1 & -1 \end{bmatrix}$. Calculating P^{-1}, we get $P^{-1} = \begin{bmatrix} \frac{1}{3} & \frac{2}{3} \\ \frac{1}{3} & -\frac{1}{3} \end{bmatrix}$. Therefore,

$$B = P^{-1}AP = \begin{bmatrix} \frac{1}{3} & \frac{2}{3} \\ \frac{1}{3} & -\frac{1}{3} \end{bmatrix}\begin{bmatrix} 1 & 2 \\ 3 & 4 \end{bmatrix}\begin{bmatrix} 1 & 2 \\ 1 & -1 \end{bmatrix} = \begin{bmatrix} \frac{17}{3} & \frac{4}{3} \\ -\frac{4}{3} & -\frac{2}{3} \end{bmatrix}$$

14-2. Properties of Similarity

Property A Similarity of $n \times n$ matrices is an equivalence relation. That is, all of the following statements are true.

(a) **(Reflexive)** Every matrix A is similar to itself.
(b) **(Symmetric)** If A is similar to B, then B is similar to A.
(c) **(Transitive)** If A is similar to B and B is similar to C, then A is similar to C.

EXAMPLE 14-3: If matrices A, B, and C satisfy the hypotheses of Property A, show that each of the conclusions follows.

Solution

(a) Let $P = I$, the $n \times n$ identity. Hence, $A = IAI = I^{-1}AI$, which means that A is similar to A.
(b) A similar to B implies that an invertible matrix P exists such that $A = P^{-1}BP$. Multiplying on the left by P and on the right by P^{-1}, we get

$$\begin{aligned} PAP^{-1} &= P(P^{-1}BP)P^{-1} \\ &= (PP^{-1})B(PP^{-1}) \\ &= IBI \\ &= B \end{aligned}$$

which means that B is similar to A.
(c) A similar to B implies an invertible matrix P exists such that $A = P^{-1}BP$. B is similar to C implies an invertible matrix Q exists such that $B = Q^{-1}CQ$. Hence,

$$\begin{aligned} A = P^{-1}(Q^{-1}CQ)P &= (P^{-1}Q^{-1})C(QP) \\ &= (QP)^{-1}C(QP) \end{aligned}$$

which means that A is similar to C. Because similarity is an equivalence relation, we no longer have to say that A is similar to B. We simply say that matrices A and B are similar.

Property B If A and B are similar matrices, then

(a) $\det(A) = \det(B)$;
(b) $\operatorname{trace}(A) = \operatorname{trace}(B)$ where **trace A** is the sum of all elements on the main diagonal of A; this is usually given as $\operatorname{tr}(A) = \operatorname{tr}(B)$;
(c) A^k and B^k are similar where k is any positive integer;
(d) A is invertible if and only if B is invertible, in which case A^{-1} and B^{-1} are similar;
(e) $\operatorname{rank} A = \operatorname{rank} B$.

EXAMPLE 14-4: Let

$$A = \begin{bmatrix} 1 & 1 & 0 \\ 0 & 2 & 1 \\ 0 & 0 & 3 \end{bmatrix} \quad \text{and} \quad P = \begin{bmatrix} 1 & 0 & 1 \\ 0 & 1 & 0 \\ 0 & 0 & 1 \end{bmatrix}$$

(a) Find $B = P^{-1}AP$. **(b)** Find $\det(A)$ and $\det(B)$ to show they are equal. **(c)** Find A^2 and B^2 and verify that they are similar. **(d)** Verify that A^{-1} and B^{-1} are similar. **(e)** Verify that $\operatorname{rank} A = \operatorname{rank} B$.

Solution

(a)

$$P^{-1} = \begin{bmatrix} 1 & 0 & -1 \\ 0 & 1 & 0 \\ 0 & 0 & 1 \end{bmatrix}$$

$$\therefore \quad B = P^{-1}AP = \begin{bmatrix} 1 & 0 & -1 \\ 0 & 1 & 0 \\ 0 & 0 & 1 \end{bmatrix} \begin{bmatrix} 1 & 1 & 0 \\ 0 & 2 & 1 \\ 0 & 0 & 3 \end{bmatrix} \begin{bmatrix} 1 & 0 & 1 \\ 0 & 1 & 0 \\ 0 & 0 & 1 \end{bmatrix} = \begin{bmatrix} 1 & 1 & 1 \\ 0 & 2 & 1 \\ 0 & 0 & 3 \end{bmatrix}$$

(b)
$$\det(A) = \begin{vmatrix} 1 & 1 & 0 \\ 0 & 2 & 1 \\ 0 & 0 & 3 \end{vmatrix} = 6 \qquad \det(B) = \begin{vmatrix} 1 & 1 & 1 \\ 0 & 2 & 1 \\ 0 & 0 & 3 \end{vmatrix} = 6$$

$$\therefore \quad \det(A) = \det(B)$$

(c)
$$A^2 = \begin{bmatrix} 1 & 1 & 0 \\ 0 & 2 & 1 \\ 0 & 0 & 3 \end{bmatrix} \begin{bmatrix} 1 & 1 & 0 \\ 0 & 2 & 1 \\ 0 & 0 & 3 \end{bmatrix} = \begin{bmatrix} 1 & 3 & 1 \\ 0 & 4 & 5 \\ 0 & 0 & 9 \end{bmatrix}$$

$$B^2 = \begin{bmatrix} 1 & 1 & 1 \\ 0 & 2 & 1 \\ 0 & 0 & 3 \end{bmatrix} \begin{bmatrix} 1 & 1 & 1 \\ 0 & 2 & 1 \\ 0 & 0 & 3 \end{bmatrix} = \begin{bmatrix} 1 & 3 & 5 \\ 0 & 4 & 5 \\ 0 & 0 & 9 \end{bmatrix}$$

Since $B^2 = (P^{-1}AP)^2 = (P^{-1}AP)(P^{-1}AP) = (P^{-1}A^2P)$ then A^2 and B^2 are similar when A and B are similar.

(d) $B^{-1} = (P^{-1}AP)^{-1} = P^{-1}A^{-1}(P^{-1})^{-1} = P^{-1}A^{-1}P$. Therefore, A^{-1} and B^{-1} are similar when A and B are similar and invertible.

(e)
$$A = \begin{bmatrix} 1 & 1 & 0 \\ 0 & 2 & 1 \\ 0 & 0 & 3 \end{bmatrix} \xrightarrow{R_1 \to R_1 + 1/3R_3} \begin{bmatrix} 1 & 1 & 1 \\ 0 & 2 & 1 \\ 0 & 0 & 3 \end{bmatrix} = B$$

Therefore, A and B are row equivalent and, hence, must have the same rank. For a second approach, you can observe that for both A and B the number of nonzero rows in echelon form is three. Thus, rank A = rank B = 3.

14-3. Similarity of Matrices of a Linear Transformation

Two $n \times n$ matrices A and B are similar if and only if they are matrix representations of the same linear transformation $L: V \to V$ with respect to different ordered bases for the n-dimensional vector space V. If A is the matrix representation of L with respect to basis S and P is the transition matrix from basis T to basis S, then $B = P^{-1}AP$ is the matrix representation of L with respect to T.

It is helpful to understand that if A and B are similar $n \times n$ matrices, they are equivalent. However, if A and B are equivalent, they are not necessarily similar.

EXAMPLE 14-5: Let $L: R^3 \to R^3$ be defined by

$$L\left(\begin{bmatrix} a_1 \\ a_2 \\ a_3 \end{bmatrix}\right) = \begin{bmatrix} a_1 - a_2 \\ a_2 + a_3 \\ a_1 + a_3 \end{bmatrix}$$

Consider the ordered bases

$$S = \left\{ \begin{bmatrix} 1 \\ 0 \\ 1 \end{bmatrix}, \begin{bmatrix} 0 \\ 1 \\ 1 \end{bmatrix}, \begin{bmatrix} 1 \\ 1 \\ 0 \end{bmatrix} \right\} \quad \text{and} \quad T = \left\{ \begin{bmatrix} 1 \\ 1 \\ 1 \end{bmatrix}, \begin{bmatrix} 1 \\ 0 \\ 2 \end{bmatrix}, \begin{bmatrix} 0 \\ 2 \\ 1 \end{bmatrix} \right\} \quad \text{for } R^3$$

(**a**) Find A, the matrix representation of L with respect to S. (**b**) Find B, the matrix representation of L with respect to T without finding the transition matrix from T to S. (**c**) Find the transition matrix P from T to S. (**d**) Find the transition matrix P^{-1} from S to T. (**e**) Show that $B = P^{-1}AP$.

Solution

(**a**)

$$L\left(\begin{bmatrix}1\\0\\1\end{bmatrix}\right) = \begin{bmatrix}1-0\\0+1\\1+1\end{bmatrix} = \begin{bmatrix}1\\1\\2\end{bmatrix}$$

$$L\left(\begin{bmatrix}0\\1\\1\end{bmatrix}\right) = \begin{bmatrix}0-1\\1+1\\0+1\end{bmatrix} = \begin{bmatrix}-1\\2\\1\end{bmatrix}$$

$$L\left(\begin{bmatrix}1\\1\\0\end{bmatrix}\right) = \begin{bmatrix}1-1\\1+0\\1+0\end{bmatrix} = \begin{bmatrix}0\\1\\1\end{bmatrix}$$

$$\begin{bmatrix}1\\1\\2\end{bmatrix} = c_1\begin{bmatrix}1\\0\\1\end{bmatrix} + c_2\begin{bmatrix}0\\1\\1\end{bmatrix} + c_3\begin{bmatrix}1\\1\\0\end{bmatrix}$$

$$\begin{aligned} c_1 \quad\quad + c_3 &= 1 \\ c_2 + c_3 &= 1 \\ c_1 + c_2 \quad\quad &= 2 \end{aligned}$$

which has as its solution $c_1 = 1, c_2 = 1, c_3 = 0$. Therefore,

$$\begin{bmatrix}1\\1\\2\end{bmatrix}_s = \begin{bmatrix}1\\1\\0\end{bmatrix}$$

$$\begin{bmatrix}-1\\2\\1\end{bmatrix} = c_1\begin{bmatrix}1\\0\\1\end{bmatrix} + c_2\begin{bmatrix}0\\1\\1\end{bmatrix} + c_3\begin{bmatrix}1\\1\\0\end{bmatrix}$$

$$\begin{aligned} c_1 \quad\quad + c_3 &= -1 \\ c_2 + c_3 &= 2 \\ c_1 + c_2 \quad\quad &= 1 \end{aligned}$$

which has as its solution $c_1 = -1, c_2 = 2, c_3 = 0$. Therefore,

$$\begin{bmatrix}-1\\2\\1\end{bmatrix}_s = \begin{bmatrix}-1\\2\\0\end{bmatrix}$$

Likewise,

$$\begin{bmatrix}0\\1\\1\end{bmatrix} = c_1\begin{bmatrix}1\\0\\1\end{bmatrix} + c_2\begin{bmatrix}0\\1\\1\end{bmatrix} + c_3\begin{bmatrix}1\\1\\0\end{bmatrix}$$

which has as its solution $c_1 = 0, c_2 = 1, c_3 = 0$. Therefore,

$$\begin{bmatrix}0\\1\\1\end{bmatrix}_S = \begin{bmatrix}0\\1\\0\end{bmatrix}$$

Therefore,

$$A = \begin{bmatrix}1 & -1 & 0\\1 & 2 & 1\\0 & 0 & 0\end{bmatrix}$$

(b)

$$L\left(\begin{bmatrix}1\\1\\1\end{bmatrix}\right) = \begin{bmatrix}1-1\\1+1\\1+1\end{bmatrix} = \begin{bmatrix}0\\2\\2\end{bmatrix}$$

$$L\left(\begin{bmatrix}1\\0\\2\end{bmatrix}\right) = \begin{bmatrix}1-0\\0+2\\1+2\end{bmatrix} = \begin{bmatrix}1\\2\\3\end{bmatrix}$$

$$L\left(\begin{bmatrix}0\\2\\1\end{bmatrix}\right) = \begin{bmatrix}0-2\\2+1\\0+1\end{bmatrix} = \begin{bmatrix}-2\\3\\1\end{bmatrix}$$

$$\begin{bmatrix}0\\2\\2\end{bmatrix} = c_1\begin{bmatrix}1\\1\\1\end{bmatrix} + c_2\begin{bmatrix}1\\0\\2\end{bmatrix} + c_3\begin{bmatrix}0\\2\\1\end{bmatrix}$$

$$\begin{aligned} c_1 + c_2 &= 0\\ c_1 + 2c_3 &= 2\\ c_1 + 2c_2 + c_3 &= 2 \end{aligned}$$

which has as its solution $c_1 = -2/3$, $c_2 = 2/3$, $c_3 = 4/3$. Therefore,

$$\begin{bmatrix}0\\2\\2\end{bmatrix}_T = \begin{bmatrix}-\frac{2}{3}\\\frac{2}{3}\\\frac{4}{3}\end{bmatrix}$$

$$\begin{bmatrix}1\\2\\3\end{bmatrix} = c_1\begin{bmatrix}1\\1\\1\end{bmatrix} + c_2\begin{bmatrix}1\\0\\2\end{bmatrix} + c_3\begin{bmatrix}0\\2\\1\end{bmatrix}$$

$$\begin{aligned} c_1 + c_2 &= 1\\ c_1 + 2c_3 &= 2\\ c_1 + 2c_2 + c_3 &= 3 \end{aligned}$$

which has as its solution $c_1 = 0$, $c_2 = 1$, $c_3 = 1$. Therefore,

$$\begin{bmatrix}1\\2\\3\end{bmatrix}_T = \begin{bmatrix}0\\1\\1\end{bmatrix}$$

$$\begin{bmatrix}-2\\3\\1\end{bmatrix} = c_1\begin{bmatrix}1\\1\\1\end{bmatrix} + c_2\begin{bmatrix}1\\0\\2\end{bmatrix} + c_3\begin{bmatrix}0\\2\\1\end{bmatrix}$$

which has as its solution $c_1 = -7/3$, $c_2 = 1/3$, $c_3 = 8/3$. Therefore,

$$\begin{bmatrix} -2 \\ 3 \\ 1 \end{bmatrix}_T = \begin{bmatrix} -\frac{7}{3} \\ \frac{1}{3} \\ \frac{8}{3} \end{bmatrix}$$

Therefore,

$$B = \begin{bmatrix} -\frac{2}{3} & 0 & -\frac{7}{3} \\ \frac{2}{3} & 1 & \frac{1}{3} \\ \frac{4}{3} & 1 & \frac{8}{3} \end{bmatrix}$$

(c)

$$\begin{bmatrix} 1 \\ 1 \\ 1 \end{bmatrix} = c_1 \begin{bmatrix} 1 \\ 0 \\ 1 \end{bmatrix} + c_2 \begin{bmatrix} 0 \\ 1 \\ 1 \end{bmatrix} + c_3 \begin{bmatrix} 1 \\ 1 \\ 0 \end{bmatrix}$$

$$\begin{aligned} c_1 \quad\quad + c_3 &= 1 \\ c_2 + c_3 &= 1 \\ c_1 + c_2 \quad\quad &= 1 \end{aligned}$$

whose solution is $c_1 = c_2 = c_3 = 1/2$. Therefore,

$$\begin{bmatrix} 1 \\ 1 \\ 1 \end{bmatrix}_s = \begin{bmatrix} \frac{1}{2} \\ \frac{1}{2} \\ \frac{1}{2} \end{bmatrix}$$

$$\begin{bmatrix} 1 \\ 0 \\ 2 \end{bmatrix} = c_1 \begin{bmatrix} 1 \\ 0 \\ 1 \end{bmatrix} + c_2 \begin{bmatrix} 0 \\ 1 \\ 1 \end{bmatrix} + c_3 \begin{bmatrix} 1 \\ 1 \\ 0 \end{bmatrix}$$

$$\begin{aligned} c_1 \quad\quad + c_3 &= 1 \\ c_2 + c_3 &= 0 \\ c_1 + c_2 \quad\quad &= 2 \end{aligned}$$

whose solution is $c_1 = 3/2$, $c_2 = 1/2$, $c_3 = -1/2$. Therefore,

$$\begin{bmatrix} 1 \\ 0 \\ 2 \end{bmatrix}_s = \begin{bmatrix} \frac{3}{2} \\ \frac{1}{2} \\ -\frac{1}{2} \end{bmatrix}$$

$$\begin{bmatrix} 0 \\ 2 \\ 1 \end{bmatrix} = c_1 \begin{bmatrix} 1 \\ 0 \\ 1 \end{bmatrix} + c_2 \begin{bmatrix} 0 \\ 1 \\ 1 \end{bmatrix} + c_3 \begin{bmatrix} 1 \\ 1 \\ 0 \end{bmatrix}$$

This system has the solution $c_1 = -1/2$, $c_2 = 3/2$, $c_3 = 1/2$. Therefore,

$$\begin{bmatrix} 0 \\ 2 \\ 1 \end{bmatrix}_s = \begin{bmatrix} -\frac{1}{2} \\ \frac{3}{2} \\ \frac{1}{2} \end{bmatrix}$$

Therefore,

$$P = \begin{bmatrix} \frac{1}{2} & \frac{3}{2} & -\frac{1}{2} \\ \frac{1}{2} & \frac{1}{2} & \frac{3}{2} \\ \frac{1}{2} & -\frac{1}{2} & \frac{1}{2} \end{bmatrix}$$

(d)

$$\left[\begin{array}{rrr|rrr} \frac{1}{2} & \frac{3}{2} & -\frac{1}{2} & 1 & 0 & 0 \\ \frac{1}{2} & \frac{1}{2} & \frac{3}{2} & 0 & 1 & 0 \\ \frac{1}{2} & -\frac{1}{2} & \frac{1}{2} & 0 & 0 & 1 \end{array}\right] \xrightarrow[R_3 \to R_3 - R_1]{R_2 \to R_2 - R_1} \left[\begin{array}{rrr|rrr} \frac{1}{2} & \frac{3}{2} & -\frac{1}{2} & 1 & 0 & 0 \\ 0 & -1 & 2 & -1 & 1 & 0 \\ 0 & -2 & 1 & -1 & 0 & 1 \end{array}\right]$$

$$\xrightarrow[\substack{R_2 \to -R_2 \\ R_3 \to R_3 - 2R_2}]{R_1 \to 2R_1} \left[\begin{array}{rrr|rrr} 1 & 3 & -1 & 2 & 0 & 0 \\ 0 & 1 & -2 & 1 & -1 & 0 \\ 0 & 0 & -3 & 1 & -2 & 1 \end{array}\right]$$

$$\xrightarrow[R_3 \to -1/3 R_3]{} \left[\begin{array}{rrr|rrr} 1 & 3 & -1 & 2 & 0 & 0 \\ 0 & 1 & -2 & 1 & -1 & 0 \\ 0 & 0 & 1 & -\frac{1}{3} & \frac{2}{3} & -\frac{1}{3} \end{array}\right]$$

$$\xrightarrow[R_2 \to R_2 + 2R_3]{R_1 \to R_1 + R_3} \left[\begin{array}{rrr|rrr} 1 & 3 & 0 & \frac{5}{3} & \frac{2}{3} & -\frac{1}{3} \\ 0 & 1 & 0 & \frac{1}{3} & \frac{1}{3} & -\frac{2}{3} \\ 0 & 0 & 1 & -\frac{1}{3} & \frac{2}{3} & -\frac{1}{3} \end{array}\right]$$

$$\xrightarrow{R_1 \to R_1 - 3R_2} \left[\begin{array}{rrr|rrr} 1 & 0 & 0 & \frac{2}{3} & -\frac{1}{3} & \frac{5}{3} \\ 0 & 1 & 0 & \frac{1}{3} & \frac{1}{3} & -\frac{2}{3} \\ 0 & 0 & 1 & -\frac{1}{3} & \frac{2}{3} & -\frac{1}{3} \end{array}\right]$$

Therefore,

$$P^{-1} = \begin{bmatrix} \frac{2}{3} & -\frac{1}{3} & \frac{5}{3} \\ \frac{1}{3} & \frac{1}{3} & -\frac{2}{3} \\ -\frac{1}{3} & \frac{2}{3} & -\frac{1}{3} \end{bmatrix}$$

(e)

$$P^{-1}AP = \begin{bmatrix} \frac{2}{3} & -\frac{1}{3} & \frac{5}{3} \\ \frac{1}{3} & \frac{1}{3} & -\frac{2}{3} \\ -\frac{1}{3} & \frac{2}{3} & -\frac{1}{3} \end{bmatrix} \begin{bmatrix} 1 & -1 & 0 \\ 1 & 2 & 1 \\ 0 & 0 & 0 \end{bmatrix} \begin{bmatrix} \frac{1}{2} & \frac{3}{2} & -\frac{1}{2} \\ \frac{1}{2} & \frac{1}{2} & \frac{3}{2} \\ \frac{1}{2} & -\frac{1}{2} & \frac{1}{2} \end{bmatrix}$$

$$= \begin{bmatrix} -\frac{2}{3} & 0 & -\frac{7}{3} \\ \frac{2}{3} & 1 & \frac{1}{3} \\ \frac{4}{3} & 1 & \frac{8}{3} \end{bmatrix} = B$$

14-4. Rank and Nullity of a Linear Transformation

In Chapter 12 we found that if $L: V \to W$ is a linear transformation of a vector space V into a vector space W, then $\dim \ker L + \dim \text{range } L = \dim V$. To find $\dim \ker L$ and $\dim \text{range } L$, it was not necessary first to find a matrix representation of L. Consequently, it seems natural to consider whether having the matrix representation of L with respect to the ordered bases $B = \{\mathbf{b}_1, \mathbf{b}_2, \ldots, \mathbf{b}_n\}$ for V and $S = \{\mathbf{s}_1, \mathbf{s}_2, \ldots, \mathbf{s}_m\}$ for W would simplify the procedure. If the matrix representation is given or can be quickly obtained, the procedure is, in fact, simplified because both the $\dim \ker L$ and $\dim \text{range } L$ can be found at the same time. The computation gives coordinate vectors for kernel L with respect to basis B and coordinate vectors for range L with respect to basis S. We want to remember that kernel L is the subset of V consisting of all vectors that map into the zero vector of W and that range L is the set of all vectors in W that are images of vectors in V. Because of this, $\dim \ker L$ is often called the **nullity of L** and $\dim \text{range } L$ is called the **rank of L**.

The procedure for finding bases and dimensions of $\ker L$ and range L is as follows.

Step 1: Find A, the matrix representation of L with respect to the ordered bases B and S.

Step 2: Set up the augmented matrix of the homogeneous system $AX = 0$, and use only elementary row operations to change it to echelon form.

Step 3: The columns in A corresponding to the columns in echelon form with leading ones are the coordinate vector representations with respect to the ordered basis S of the basis vectors for range L. The number of leading ones is the rank of L or dim range L.

Step 4: The solution of the homogeneous system in Step 2 gives the coordinate vector representations with respect to the ordered basis B of the basis vectors for kernel L. The number of vectors in the kernel is the nullity of L or $\dim \ker L$.

EXAMPLE 14-6: In Example 14-5 we found that the matrix representation of the linear transformation $L: R^3 \to R^3$ defined by $L\left(\begin{bmatrix} a_1 \\ a_2 \\ a_3 \end{bmatrix}\right) = \begin{bmatrix} a_1 - a_2 \\ a_2 + a_3 \\ a_1 + a_3 \end{bmatrix}$ is $A = \begin{bmatrix} 1 & -1 & 0 \\ 1 & 2 & 1 \\ 0 & 0 & 0 \end{bmatrix}$ with respect to the ordered basis

$$S = \left\{ \begin{bmatrix} 1 \\ 0 \\ 1 \end{bmatrix}, \begin{bmatrix} 0 \\ 1 \\ 1 \end{bmatrix}, \begin{bmatrix} 1 \\ 1 \\ 0 \end{bmatrix} \right\} \quad \text{and} \quad B = \begin{bmatrix} -\frac{2}{3} & 0 & -\frac{7}{3} \\ \frac{2}{3} & 1 & \frac{1}{3} \\ \frac{4}{3} & 1 & \frac{8}{3} \end{bmatrix}$$

with respect to the ordered basis

$$T = \left\{ \begin{bmatrix} 1 \\ 1 \\ 1 \end{bmatrix}, \begin{bmatrix} 1 \\ 0 \\ 2 \end{bmatrix}, \begin{bmatrix} 0 \\ 2 \\ 1 \end{bmatrix} \right\}$$

(**a**) Use A to find the dimension and a basis for range L. (**b**) Use the result in part (**a**) to find the dimension and a basis for $\ker L$. (**c**) Use B to find the dimension and a basis for range L. (**d**) Use the result in part (**c**) to find the dimension and a basis for $\ker L$. (**e**) Show that parts (**a**) and (**c**) give the same range and parts (**b**) and (**d**) give the same kernel of L.

Solution

(**a**) We begin by reducing the augmented matrix of A to echelon form.

$$\begin{bmatrix} 1 & -1 & 0 & 0 \\ 1 & 2 & 1 & 0 \\ 0 & 0 & 0 & 0 \end{bmatrix} \xrightarrow{R_2 \to R_2 - R_1} \begin{bmatrix} 1 & -1 & 0 & 0 \\ 0 & 3 & 1 & 0 \\ 0 & 0 & 0 & 0 \end{bmatrix} \xrightarrow{R_2 \to 1/3 R_2} \begin{bmatrix} 1 & -1 & 0 & 0 \\ 0 & 1 & \frac{1}{3} & 0 \\ 0 & 0 & 0 & 0 \end{bmatrix}$$

$$\xrightarrow{R_1 \to R_1 + R_2} \begin{bmatrix} 1 & 0 & \frac{1}{3} & 0 \\ 0 & 1 & \frac{1}{3} & 0 \\ 0 & 0 & 0 & 0 \end{bmatrix}$$

Since we used only row operations and the first and second columns have leading ones, the coordinate vector representations of the basis vectors for range L are $\begin{bmatrix} 1 \\ 1 \\ 0 \end{bmatrix}$ and $\begin{bmatrix} -1 \\ 2 \\ 0 \end{bmatrix}$ which are the first two columns of A. Thus, dim range L or the rank of L is 2. To find a basis transform these coordinate vectors into their corresponding vectors in R^3:

$$\mathbf{s}_1 = 1\begin{bmatrix}1\\0\\1\end{bmatrix} + 1\begin{bmatrix}0\\1\\1\end{bmatrix} + 0\begin{bmatrix}1\\1\\0\end{bmatrix} = \begin{bmatrix}1\\1\\2\end{bmatrix}$$

$$\mathbf{s}_2 = -1\begin{bmatrix}1\\0\\1\end{bmatrix} + 2\begin{bmatrix}0\\1\\1\end{bmatrix} + 0\begin{bmatrix}1\\1\\0\end{bmatrix} = \begin{bmatrix}-1\\2\\1\end{bmatrix}$$

Therefore, a basis for range L is $\left\{\begin{bmatrix}1\\1\\2\end{bmatrix}, \begin{bmatrix}-1\\2\\1\end{bmatrix}\right\}$.

(b) From the reduced row echelon form of the augmented matrix in part (**a**), we find $a_1 = -(1/3)a_3$ and $a_2 = -(1/3)a_3$. Therefore, let $a_3 = t$ where t is arbitrary. Hence, the solution is $\begin{bmatrix}-1/3t\\-1/3t\\t\end{bmatrix} = t\begin{bmatrix}-1/3\\-1/3\\1\end{bmatrix}$. Thus, $\begin{bmatrix}-1/3\\-1/3\\1\end{bmatrix}$ or more appropriately $\begin{bmatrix}1\\1\\-3\end{bmatrix}$ is the coordinate vector representation for kernel L. Transform this vector into its corresponding vector in R^3.

$$1\begin{bmatrix}1\\0\\1\end{bmatrix} + 1\begin{bmatrix}0\\1\\1\end{bmatrix} - 3\begin{bmatrix}1\\1\\0\end{bmatrix} = \begin{bmatrix}-2\\-2\\2\end{bmatrix}$$

Therefore, a basis for kernel L is $\left\{\begin{bmatrix}1\\1\\-1\end{bmatrix}\right\}$ and dim ker L or the nullity of L is 1.

(c) We begin this with the augmented matrix of B.

$$\begin{bmatrix}-\frac{2}{3} & 0 & -\frac{7}{3} & 0\\ \frac{2}{3} & 1 & \frac{1}{3} & 0\\ \frac{4}{3} & 1 & \frac{8}{3} & 0\end{bmatrix} \xrightarrow[R_3 \to R_3 + 2R_1]{R_2 \to R_2 + R_1} \begin{bmatrix}-\frac{2}{3} & 0 & -\frac{7}{3} & 0\\ 0 & 1 & -2 & 0\\ 0 & 1 & -2 & 0\end{bmatrix} \xrightarrow[R_3 \to R_3 - R_2]{R_1 \to -3/2R_1} \begin{bmatrix}1 & 0 & \frac{7}{2} & 0\\ 0 & 1 & -2 & 0\\ 0 & 0 & 0 & 0\end{bmatrix}$$

Since we used only row operations and the first and second columns have leading ones, the coordinate vector representations of the basis vectors for range L are $\begin{bmatrix}-2/3\\2/3\\4/3\end{bmatrix}$ and $\begin{bmatrix}0\\1\\1\end{bmatrix}$. From this we see that dim range L or rank L is 2. To find a basis transform these coordinate vectors into their corresponding vectors in R^3.

$$\mathbf{t}_1 = -\frac{2}{3}\begin{bmatrix}1\\1\\1\end{bmatrix} + \frac{2}{3}\begin{bmatrix}1\\0\\2\end{bmatrix} + \frac{4}{3}\begin{bmatrix}0\\2\\1\end{bmatrix} = \begin{bmatrix}0\\2\\2\end{bmatrix} \quad \text{or} \quad \begin{bmatrix}0\\1\\1\end{bmatrix}$$

$$\mathbf{t}_2 = 0\begin{bmatrix}1\\1\\1\end{bmatrix} + 1\begin{bmatrix}1\\0\\2\end{bmatrix} + 1\begin{bmatrix}0\\2\\1\end{bmatrix} = \begin{bmatrix}1\\2\\3\end{bmatrix}$$

Therefore, a basis for range L is $\left\{\begin{bmatrix}0\\1\\1\end{bmatrix}, \begin{bmatrix}1\\2\\3\end{bmatrix}\right\}$.

(d) From the reduced row echelon form of the augmented matrix in part (**c**), we find $a_1 = -(7/2)a_3$ and $a_2 = 2a_3$. Therefore, let $a_3 = t$. Hence, the solution is $\begin{bmatrix}-7/2t\\2t\\t\end{bmatrix} = t\begin{bmatrix}-7/2\\2\\1\end{bmatrix}$. Thus, $\begin{bmatrix}-7/2\\2\\1\end{bmatrix}$ or

$\begin{bmatrix} -7 \\ 4 \\ 2 \end{bmatrix}$ is the coordinate vector representation of the basis vector for kernel L. Transform this vector into its corresponding vector in R^3:

$$-7\begin{bmatrix} 1 \\ 1 \\ 1 \end{bmatrix} + 4\begin{bmatrix} 1 \\ 0 \\ 2 \end{bmatrix} + 2\begin{bmatrix} 0 \\ 2 \\ 1 \end{bmatrix} = \begin{bmatrix} -3 \\ -3 \\ 3 \end{bmatrix}$$

Therefore, a basis for kernel L is $\left\{\begin{bmatrix} 1 \\ 1 \\ -1 \end{bmatrix}\right\}$ and dim ker L or the nullity of L is 1.

(e) To show that parts **(a)** and **(c)** give the same range of L, we must show that each of the vectors in part **(c)** can be expressed as a linear combination of the basis vectors in part **(a)**. This results in

$$\begin{bmatrix} 0 \\ 1 \\ 1 \end{bmatrix} = \frac{1}{3}\begin{bmatrix} 1 \\ 1 \\ 2 \end{bmatrix} + \frac{1}{3}\begin{bmatrix} -1 \\ 2 \\ 1 \end{bmatrix}$$

$$\begin{bmatrix} 1 \\ 2 \\ 3 \end{bmatrix} = \frac{4}{3}\begin{bmatrix} 1 \\ 1 \\ 2 \end{bmatrix} + \frac{1}{3}\begin{bmatrix} -1 \\ 2 \\ 1 \end{bmatrix}$$

Therefore, range L is the same in both parts. To show that parts **(b)** and **(d)** give the same kernel of L, we must show that each of the basis vectors in part **(d)** can be expressed as a linear combination of the basis vectors in part **(b)**. This is obvious since the same vector $\begin{bmatrix} 1 \\ 1 \\ -1 \end{bmatrix}$ occurs in both parts **(b)** and **(d)**.

EXAMPLE 14-7: In Example 13-8 we found that the matrix representation of the linear transformation $L: R_3 \to R_2$ defined by $L([a_1 \quad a_2 \quad a_3]) = [a_1 + a_2 \quad a_3]$ is $A = \begin{bmatrix} -1/2 & -4 & -1/2 \\ 1/2 & 2 & 1/2 \end{bmatrix}$ with respect to the ordered bases $B = \{[0 \quad 1 \quad 1], [1 \quad 1 \quad 0], [1 \quad 0 \quad 1]\}$ and $S = \{[1 \quad 2], [3 \quad 4]\}$. **(a)** Use A to find the dimension and a basis for range L. **(b)** Use the result in part **(a)** to find the dimension and a basis for ker L.

Solution

(a) Augmented matrix $= \begin{bmatrix} -\frac{1}{2} & -4 & -\frac{1}{2} & 0 \\ \frac{1}{2} & 2 & \frac{1}{2} & 0 \end{bmatrix} \xrightarrow{R_2 \to R_2 + R_1} \begin{bmatrix} -\frac{1}{2} & -4 & -\frac{1}{2} & 0 \\ 0 & -2 & 0 & 0 \end{bmatrix}$

$$\xrightarrow[R_2 \to -1/2R_2]{R_1 \to -2R_1} \begin{bmatrix} 1 & 8 & 1 & 0 \\ 0 & 1 & 0 & 0 \end{bmatrix} \xrightarrow{R_1 \to R_1 - 8R_2} \begin{bmatrix} 1 & 0 & 1 & 0 \\ 0 & 1 & 0 & 0 \end{bmatrix}$$

Since we used only row operations and the first and second columns have leading ones, the coordinate vector representations of the basis vectors for range L are $\begin{bmatrix} -1/2 \\ 1/2 \end{bmatrix}$ and $\begin{bmatrix} -4 \\ 2 \end{bmatrix}$. Thus, dim range L or the rank of L is 2. Although we can conclude that range $L = R_2$, let us continue the computation for the practice. We must transform each of the last two coordinate vectors into their corresponding vectors in R_2. We get

$$\mathbf{s}_1 = -\tfrac{1}{2}[1 \quad 2] + \tfrac{1}{2}[3 \quad 4] = [1 \quad 1]$$

$$\mathbf{s}_2 = -4[1 \quad 2] + 2[3 \quad 4] = [2 \quad 0]$$

Therefore, $\{[1\ \ 1], [2\ \ 0]\}$ or $\{[1\ \ 1], [1\ \ 0]\}$ is a basis for range L. We also know that this basis can be simplified to $\{[1\ \ 0], [0\ \ 1]\}$.

(b) From the reduced row echelon form of the augmented matrix in part **(a)**, we find $a_1 = -a_3$ and $a_2 = 0$. If $a_3 = t$, the solution is $\begin{bmatrix} -t \\ 0 \\ t \end{bmatrix} = t\begin{bmatrix} -1 \\ 0 \\ 1 \end{bmatrix}$. Thus, $\begin{bmatrix} -1 \\ 0 \\ 1 \end{bmatrix}$ is the coordinate vector representation of the basis vector for kernel L. Transforming this into its corresponding vector in R_3, we get $-1[0\ \ 1\ \ 1] + 0[1\ \ 1\ \ 0] + 1[1\ \ 0\ \ 1] = [1\ \ -1\ \ 0]$. Therefore, a basis for kernel L is $\{[1\ \ -1\ \ 0]\}$ and dim ker L or the nullity of L is 1. Observe that dim range L + dim ker L = dim R_3 or rank of L + nullity of L = dim R_3.

SUMMARY

1. If A and B are $n \times n$ matrices and P is an invertible matrix such that $B = P^{-1}AP$, then B is similar to A.
2. A matrix B similar to an $n \times n$ matrix A can be obtained by finding any $n \times n$ invertible matrix P and computing $B = P^{-1}AP$.
3. Similarity of $n \times n$ matrices is an equivalence relation, and hence, we can say that matrices A and B are similar.
4. If matrices A and B are similar, then
   - $\det(A) = \det(B)$,
   - $\operatorname{tr}(A) = \operatorname{tr}(B)$,
   - A^k and B^k are similar where k is any positive integer,
   - A is invertible if and only if B is invertible, in which case A^{-1} and B^{-1} are similar,
   - rank A = rank B.
5. If A and B are similar matrices, then A and B are equivalent.
6. If A and B are equivalent matrices, then they are not necessarily similar.
7. Two matrices are similar if and only if they are matrix representations of the same linear transformation with respect to different ordered bases.
8. If A is the matrix representation of the linear transformation $L: V \to V$ with respect to basis S and P is the transition matrix from basis T to basis S, then $B = P^{-1}AP$ is the matrix representation of L with respect to T.
9. Dim ker L is often called nullity of L.
10. Dim range L is often called rank of L.
11. Bases and dimensions of ker L and range L can be found as follows:

    Step 1: Find A, the matrix representation of L with respect to the ordered bases B and S.
    Step 2: Change the augmented matrix of the homogeneous system $AX = \mathbf{0}$ by elementary row operations only into echelon form.
    Step 3: The columns in A corresponding to the columns in echelon form having leading ones are the coordinate vector representations with respect to the ordered basis S of the basis vectors for range L.
    Step 4: The solution of the homogeneous system gives the coordinate vector representations with respect to the ordered basis B of the basis vectors for kernel L.

12. If $L: V \to W$ is a linear transformation, then dim range L + dim ker 1 = dim V or, stated differently, rank of L + nullity of L = dim V.

RAISE YOUR GRADES

Can you ...?

☑ find a matrix similar to an $n \times n$ matrix
☑ determine if two $n \times n$ matrices are similar
☑ verify that the similarity of $n \times n$ matrices is an equivalence relationship

☑ determine det(B), tr(B), and rank B if it is known that B is similar to a given matrix A
☑ when given the matrix representation of a linear transformation with respect to basis S and P the transition matrix from basis T to basis S, find the matrix representation of L with respect to T
☑ find a basis and the dimension of ker L
☑ find a basis and the dimension of range L

SOLVED PROBLEMS

PROBLEM 14-1 Use $P = \begin{bmatrix} 2 & -1 \\ 1 & -1 \end{bmatrix}$ to verify that $B = \begin{bmatrix} 1 & -1 \\ -1 & 1 \end{bmatrix}$ is similar to $A = \begin{bmatrix} 0 & 3 \\ 0 & 2 \end{bmatrix}$.

Solution P^{-1} is found to be $\begin{bmatrix} 1 & -1 \\ 1 & -2 \end{bmatrix}$.

$$P^{-1}AP = \begin{bmatrix} 1 & -1 \\ 1 & -2 \end{bmatrix}\begin{bmatrix} 0 & 3 \\ 0 & 2 \end{bmatrix}\begin{bmatrix} 2 & -1 \\ 1 & -1 \end{bmatrix} = \begin{bmatrix} 1 & -1 \\ -1 & 1 \end{bmatrix} = B$$

Therefore, B is similar to A.

PROBLEM 14-2 Use $P = \begin{bmatrix} 2 & -1 \\ 1 & -1 \end{bmatrix}$ to verify that $B = \begin{bmatrix} -2 & 1 \\ -11 & 5 \end{bmatrix}$ is similar to $A = \begin{bmatrix} 4 & -1 \\ 5 & -1 \end{bmatrix}$.

Solution $P^{-1} = \begin{bmatrix} 1 & -1 \\ 1 & -2 \end{bmatrix}$ from Problem 14-1.

$$P^{-1}AP = \begin{bmatrix} 1 & -1 \\ 1 & -2 \end{bmatrix}\begin{bmatrix} 4 & -1 \\ 5 & -1 \end{bmatrix}\begin{bmatrix} 2 & -1 \\ 1 & -1 \end{bmatrix} = \begin{bmatrix} -2 & 1 \\ -11 & 5 \end{bmatrix} = B$$

Therefore, B is similar to A.

PROBLEM 14-3 Find a matrix B that is similar to the matrix $A = \begin{bmatrix} 1 & 2 \\ -2 & -1 \end{bmatrix}$.

Solution We will choose any 2×2 invertible matrix for P. Let

$$P = \begin{bmatrix} 2 & 2 \\ 1 & -1 \end{bmatrix} \qquad P^{-1} = \begin{bmatrix} \frac{1}{4} & \frac{1}{2} \\ \frac{1}{4} & -\frac{1}{2} \end{bmatrix}$$

Therefore,

$$B = P^{-1}AP = \begin{bmatrix} \frac{1}{4} & \frac{1}{2} \\ \frac{1}{4} & -\frac{1}{2} \end{bmatrix}\begin{bmatrix} 1 & 2 \\ -2 & -1 \end{bmatrix}\begin{bmatrix} 2 & 2 \\ 1 & -1 \end{bmatrix} = \begin{bmatrix} -\frac{3}{2} & -\frac{3}{2} \\ \frac{7}{2} & \frac{3}{2} \end{bmatrix}$$

PROBLEM 14-4 Verify that if A and B are two similar matrices, then A^{-1} and B^{-1} are similar.

Solution B similar to A means an invertible matrix P exists such that $B = P^{-1}AP$. Then $B^{-1} = (P^{-1}AP)^{-1} = P^{-1}A^{-1}(P^{-1})^{-1} = P^{-1}A^{-1}P$. Therefore, B^{-1} is similar to A^{-1}.

Likewise, A is similar to B means there exists an invertible matrix Q such that $A = Q^{-1}BQ$. Then $A^{-1} = (Q^{-1}BQ)^{-1} = Q^{-1}B^{-1}(Q^{-1})^{-1} = Q^{-1}B^{-1}Q$. Therefore, A^{-1} is similar to B^{-1}.

PROBLEM 14-5 Let

$$A = \begin{bmatrix} 1 & 0 & 1 \\ 1 & 1 & 0 \\ 0 & 0 & 2 \end{bmatrix} \quad \text{and} \quad P = \begin{bmatrix} 1 & 0 & 0 \\ 0 & 1 & 0 \\ 1 & 0 & 1 \end{bmatrix}$$

(a) Find $B = P^{-1}AP$. **(b)** Find $\det(A)$ and $\det(B)$. **(c)** Find $\operatorname{tr}(A)$ and $\operatorname{tr}(B)$. **(d)** Verify that rank A = rank B.

Solution

(a)

$$\left[\begin{array}{ccc|ccc} 1 & 0 & 0 & 1 & 0 & 0 \\ 0 & 1 & 0 & 0 & 1 & 0 \\ 1 & 0 & 1 & 0 & 0 & 1 \end{array}\right] \xrightarrow{R_3 \to R_3 - R_1} \left[\begin{array}{ccc|ccc} 1 & 0 & 0 & 1 & 0 & 0 \\ 0 & 1 & 0 & 0 & 1 & 0 \\ 0 & 0 & 1 & -1 & 0 & 1 \end{array}\right]$$

$$P^{-1} = \begin{bmatrix} 1 & 0 & 0 \\ 0 & 1 & 0 \\ -1 & 0 & 1 \end{bmatrix}$$

$$\therefore \quad B = P^{-1}AP = \begin{bmatrix} 1 & 0 & 0 \\ 0 & 1 & 0 \\ -1 & 0 & 1 \end{bmatrix} \begin{bmatrix} 1 & 0 & 1 \\ 1 & 1 & 0 \\ 0 & 0 & 2 \end{bmatrix} \begin{bmatrix} 1 & 0 & 0 \\ 0 & 1 & 0 \\ 1 & 0 & 1 \end{bmatrix}$$

$$= \begin{bmatrix} 2 & 0 & 1 \\ 1 & 1 & 0 \\ 0 & 0 & 1 \end{bmatrix}$$

(b)

$$\det \begin{bmatrix} 1 & 0 & 1 \\ 1 & 1 & 0 \\ 0 & 0 & 2 \end{bmatrix} = 2 \qquad \det \begin{bmatrix} 2 & 0 & 1 \\ 1 & 1 & 0 \\ 0 & 0 & 1 \end{bmatrix} = 2$$

$$\therefore \quad \det(A) = \det(B)$$

(c)

$$\operatorname{tr} \begin{bmatrix} 1 & 0 & 1 \\ 1 & 1 & 0 \\ 0 & 0 & 2 \end{bmatrix} = 1 + 1 + 2 = 4 \qquad \operatorname{tr} \begin{bmatrix} 2 & 0 & 1 \\ 1 & 1 & 0 \\ 0 & 0 & 1 \end{bmatrix} = 2 + 1 + 4 = 4$$

$$\therefore \quad \operatorname{tr}(A) = \operatorname{tr}(B)$$

(d)

$$A = \begin{bmatrix} 1 & 0 & 1 \\ 1 & 1 & 0 \\ 0 & 0 & 2 \end{bmatrix} \xrightarrow[R_3 \to 1/2 R_3]{R_2 \to R_2 - R_1} \begin{bmatrix} 1 & 0 & 1 \\ 0 & 1 & -1 \\ 0 & 0 & 1 \end{bmatrix}$$

$$B = \begin{bmatrix} 2 & 0 & 1 \\ 1 & 1 & 0 \\ 0 & 0 & 1 \end{bmatrix} \xrightarrow{R_1 \leftrightarrow R_2} \begin{bmatrix} 1 & 1 & 0 \\ 2 & 0 & 1 \\ 0 & 0 & 1 \end{bmatrix} \xrightarrow{R_2 \to R_2 - 2R_1} \begin{bmatrix} 1 & 1 & 0 \\ 0 & -2 & 1 \\ 0 & 0 & 1 \end{bmatrix} \xrightarrow{R_2 \to -1/2 R_2} \begin{bmatrix} 1 & 1 & 0 \\ 0 & 1 & -\frac{1}{2} \\ 0 & 0 & 1 \end{bmatrix}$$

Since the number of nonzero rows of each reduction in echelon is 3, rank A = rank B = 3.

PROBLEM 14-6 Let $L: R^2 \to R^2$ be defined by $L\left(\begin{bmatrix} a \\ b \end{bmatrix}\right) = \begin{bmatrix} 2a - 3b \\ 3a + 5b \end{bmatrix}$. Consider the ordered bases $S = \left\{\begin{bmatrix} 1 \\ 1 \end{bmatrix}, \begin{bmatrix} 0 \\ 1 \end{bmatrix}\right\}$ and $T = \left\{\begin{bmatrix} 1 \\ 0 \end{bmatrix}, \begin{bmatrix} 2 \\ 1 \end{bmatrix}\right\}$ for R^2. **(a)** Find A, the matrix representation of L with respect to S. **(b)** Find B, the matrix representation of L with respect to T without finding the transition matrix from T to S.

Solution

(a)

$$L\left(\begin{bmatrix} 1 \\ 1 \end{bmatrix}\right) = \begin{bmatrix} 2 - 3 \\ 3 + 5 \end{bmatrix} = \begin{bmatrix} -1 \\ 8 \end{bmatrix}$$

$$L\left(\begin{bmatrix} 0 \\ 1 \end{bmatrix}\right) = \begin{bmatrix} 0 - 3 \\ 0 + 5 \end{bmatrix} = \begin{bmatrix} -3 \\ 5 \end{bmatrix}$$

$$\begin{bmatrix} -1 \\ 8 \end{bmatrix} = c_1 \begin{bmatrix} 1 \\ 1 \end{bmatrix} + c_2 \begin{bmatrix} 0 \\ 1 \end{bmatrix}$$

$$c_1 \qquad = -1$$

$$c_1 + c_2 = \quad 8$$

$$\therefore \quad c_1 = -1, \qquad c_2 = 9$$

Therefore, $\begin{bmatrix} -1 \\ 8 \end{bmatrix}_S = \begin{bmatrix} -1 \\ 9 \end{bmatrix}$.

$$\begin{bmatrix} -3 \\ 5 \end{bmatrix} = c_1 \begin{bmatrix} 1 \\ 1 \end{bmatrix} + c_2 \begin{bmatrix} 0 \\ 1 \end{bmatrix}$$

$$c_1 \qquad = -3$$

$$c_1 + c_2 = \quad 5$$

$$\therefore \quad c_1 = -3, \qquad c_2 = 8$$

Therefore, $\begin{bmatrix} -3 \\ 5 \end{bmatrix}_S = \begin{bmatrix} -3 \\ 8 \end{bmatrix}$, and $A = \begin{bmatrix} -1 & -3 \\ 9 & 8 \end{bmatrix}$.

(b)

$$L\left(\begin{bmatrix} 1 \\ 0 \end{bmatrix}\right) = \begin{bmatrix} 2 - 0 \\ 3 + 0 \end{bmatrix} = \begin{bmatrix} 2 \\ 3 \end{bmatrix}$$

$$L\left(\begin{bmatrix} 2 \\ 1 \end{bmatrix}\right) = \begin{bmatrix} 4 - 3 \\ 6 + 5 \end{bmatrix} = \begin{bmatrix} 1 \\ 11 \end{bmatrix}$$

$$\begin{bmatrix} 2 \\ 3 \end{bmatrix} = c_1 \begin{bmatrix} 1 \\ 0 \end{bmatrix} + c_2 \begin{bmatrix} 2 \\ 1 \end{bmatrix}$$

$$c_1 + 2c_2 = 2$$

$$c_2 = 3$$

$$\therefore \quad c_1 = -4, \qquad c_2 = 3$$

Therefore, $\begin{bmatrix} 2 \\ 3 \end{bmatrix}_T = \begin{bmatrix} -4 \\ 3 \end{bmatrix}$.

$$\begin{bmatrix} 1 \\ 11 \end{bmatrix} = c_1 \begin{bmatrix} 1 \\ 0 \end{bmatrix} + c_2 \begin{bmatrix} 2 \\ 1 \end{bmatrix}$$

$$c_1 + 2c_2 = \ 1$$

$$c_2 = 11$$

$$\therefore \quad c_1 = -21, \qquad c_2 = 11$$

Therefore, $\begin{bmatrix} 1 \\ 11 \end{bmatrix}_T = \begin{bmatrix} -21 \\ 11 \end{bmatrix}$ and $B = \begin{bmatrix} -4 & -21 \\ 3 & 11 \end{bmatrix}$.

PROBLEM 14-7 Using the linear transformation and the ordered bases given in Problem 14-6 find **(a)** the transition matrix P from T to S, **(b)** the transition matrix P^{-1} from S to T, and **(c)** show that $B = P^{-1}AP$.

Solution

(a)

$$\begin{bmatrix} 1 \\ 0 \end{bmatrix} = c_1 \begin{bmatrix} 1 \\ 1 \end{bmatrix} + c_2 \begin{bmatrix} 0 \\ 1 \end{bmatrix}$$

$$c_1 \qquad = 1$$

$$c_1 + c_2 = 0$$

$$\therefore \quad c_1 = 1, \qquad c_2 = -1$$

$$\therefore \quad \begin{bmatrix} 1 \\ 0 \end{bmatrix}_T = \begin{bmatrix} 1 \\ -1 \end{bmatrix}$$

$$\begin{bmatrix} 2 \\ 1 \end{bmatrix} = c_1 \begin{bmatrix} 1 \\ 1 \end{bmatrix} + c_2 \begin{bmatrix} 0 \\ 1 \end{bmatrix}$$

$$\begin{aligned} c_1 \qquad &= 2 \\ c_1 + c_2 &= 1 \end{aligned}$$

$$\therefore \quad c_1 = 2, \qquad c_2 = -1$$

$$\therefore \quad \begin{bmatrix} 2 \\ 1 \end{bmatrix}_T = \begin{bmatrix} 2 \\ -1 \end{bmatrix}$$

Therefore, $P = \begin{bmatrix} 1 & 2 \\ -1 & -1 \end{bmatrix}$.

(b)

$$\left[\begin{array}{rr|rr} 1 & 2 & 1 & 0 \\ -1 & -1 & 0 & 1 \end{array}\right] \xrightarrow{R_2 \to R_2 + R_1} \left[\begin{array}{rr|rr} 1 & 2 & 1 & 0 \\ 0 & 1 & 1 & 1 \end{array}\right] \xrightarrow{R_1 \to R_1 - 2R_2} \left[\begin{array}{rr|rr} 1 & 0 & -1 & -2 \\ 0 & 1 & 1 & 1 \end{array}\right]$$

$$\therefore \quad P^{-1} = \begin{bmatrix} -1 & -2 \\ 1 & 1 \end{bmatrix}$$

(c)

$$P^{-1}AP = \begin{bmatrix} -1 & -2 \\ 1 & 1 \end{bmatrix} \begin{bmatrix} -1 & -3 \\ 9 & 8 \end{bmatrix} \begin{bmatrix} 1 & 2 \\ -1 & -1 \end{bmatrix} = \begin{bmatrix} -4 & -21 \\ 3 & 11 \end{bmatrix} = B$$

PROBLEM 14-8 Let $L: R^3 \to R^3$ be defined by

$$L\left(\begin{bmatrix} a_1 \\ a_2 \\ a_3 \end{bmatrix}\right) = \begin{bmatrix} a_1 + 2a_2 \\ a_2 - 3a_3 \\ a_1 + a_2 \end{bmatrix}$$

Consider the ordered bases

$$S = \left\{ \begin{bmatrix} 1 \\ 1 \\ 0 \end{bmatrix}, \begin{bmatrix} 0 \\ 1 \\ 1 \end{bmatrix}, \begin{bmatrix} 1 \\ 0 \\ 1 \end{bmatrix} \right\} \quad \text{and} \quad T = \left\{ \begin{bmatrix} 1 \\ 0 \\ 2 \end{bmatrix}, \begin{bmatrix} 0 \\ 2 \\ 1 \end{bmatrix}, \begin{bmatrix} 1 \\ 1 \\ 1 \end{bmatrix} \right\} \quad \text{for } R^3$$

(a) Find A, the matrix representation of L with respect to S. **(b)** Find the transition matrix P from T to S. **(c)** Find the transition matrix P^{-1} from S to T. **(d)** Find B, the matrix representation of L with respect to T by using parts **(a)**, **(b)**, and **(c)**.

Solution

(a)

$$L\left(\begin{bmatrix} 1 \\ 1 \\ 0 \end{bmatrix}\right) = \begin{bmatrix} 1 + 2 \\ 1 - 0 \\ 1 + 1 \end{bmatrix} = \begin{bmatrix} 3 \\ 1 \\ 2 \end{bmatrix}$$

$$L\left(\begin{bmatrix} 0 \\ 1 \\ 1 \end{bmatrix}\right) = \begin{bmatrix} 0 + 2 \\ 1 - 3 \\ 0 + 1 \end{bmatrix} = \begin{bmatrix} 2 \\ -2 \\ 1 \end{bmatrix}$$

$$L\left(\begin{bmatrix}1\\0\\1\end{bmatrix}\right)=\begin{bmatrix}1+0\\0-3\\1+0\end{bmatrix}=\begin{bmatrix}1\\-3\\1\end{bmatrix}$$

$$\begin{bmatrix}3\\1\\2\end{bmatrix}=c_1\begin{bmatrix}1\\1\\0\end{bmatrix}+c_2\begin{bmatrix}0\\1\\1\end{bmatrix}+c_3\begin{bmatrix}1\\0\\1\end{bmatrix}$$

$$\begin{aligned}c_1 \qquad + c_3 &= 3\\ c_1 + c_2 \qquad &= 1\\ c_2 + c_3 &= 2\end{aligned}$$

$$\therefore \quad c_1 = 1, \qquad c_2 = 0, \qquad c_3 = 2$$

$$\therefore \quad \begin{bmatrix}3\\1\\2\end{bmatrix}_s=\begin{bmatrix}1\\0\\2\end{bmatrix}$$

$$\begin{bmatrix}2\\-2\\1\end{bmatrix}=c_1\begin{bmatrix}1\\1\\0\end{bmatrix}+c_2\begin{bmatrix}0\\1\\1\end{bmatrix}+c_3\begin{bmatrix}1\\0\\1\end{bmatrix}$$

$$\begin{aligned}c_1 \qquad + c_3 &= 2\\ c_1 + c_2 \qquad &= -2\\ c_2 + c_3 &= 1\end{aligned}$$

$$\therefore \quad c_1 = -\tfrac{1}{2}, \qquad c_2 = -\tfrac{3}{2}, \qquad c_3 = \tfrac{5}{2}$$

$$\therefore \quad \begin{bmatrix}2\\-2\\1\end{bmatrix}_s=\begin{bmatrix}-1/2\\-3/2\\5/2\end{bmatrix}$$

$$\begin{bmatrix}1\\-3\\1\end{bmatrix}=c_1\begin{bmatrix}1\\1\\0\end{bmatrix}+c_2\begin{bmatrix}0\\1\\1\end{bmatrix}+c_3\begin{bmatrix}1\\0\\1\end{bmatrix}$$

$$\begin{aligned}c_1 \qquad c_3 &= 1\\ c_1 + c_2 \qquad &= -3\\ c_2 + c_3 &= 1\end{aligned}$$

$$\therefore \quad c_1 = -\tfrac{3}{2}, \qquad c_2 = -\tfrac{3}{2}, \qquad c_3 = \tfrac{5}{2}$$

$$\therefore \quad \begin{bmatrix}1\\-3\\1\end{bmatrix}_s=\begin{bmatrix}-\frac{3}{2}\\-\frac{3}{2}\\\frac{5}{2}\end{bmatrix}$$

Therefore,

$$A = \begin{bmatrix} 1 & -\frac{1}{2} & -\frac{3}{2} \\ 0 & -\frac{3}{2} & -\frac{3}{2} \\ 2 & \frac{5}{2} & \frac{5}{2} \end{bmatrix}$$

(b)

$$\begin{bmatrix} 1 \\ 0 \\ 2 \end{bmatrix} = c_1 \begin{bmatrix} 1 \\ 1 \\ 0 \end{bmatrix} + c_2 \begin{bmatrix} 0 \\ 1 \\ 1 \end{bmatrix} + c_3 \begin{bmatrix} 1 \\ 0 \\ 1 \end{bmatrix}$$

$$\begin{aligned} c_1 \qquad + c_3 &= 1 \\ c_1 + c_2 \qquad &= 0 \\ c_2 + c_3 &= 2 \end{aligned}$$

$$\therefore \quad c_1 = -\tfrac{1}{2}, \qquad c_2 = \tfrac{1}{2}, \qquad c_3 = \tfrac{3}{2}$$

$$\therefore \quad \begin{bmatrix} 1 \\ 0 \\ 2 \end{bmatrix}_T = \begin{bmatrix} -\frac{1}{2} \\ \frac{1}{2} \\ \frac{3}{2} \end{bmatrix}$$

$$\begin{bmatrix} 0 \\ 2 \\ 1 \end{bmatrix} = c_1 \begin{bmatrix} 1 \\ 1 \\ 0 \end{bmatrix} + c_2 \begin{bmatrix} 0 \\ 1 \\ 1 \end{bmatrix} + c_3 \begin{bmatrix} 1 \\ 0 \\ 1 \end{bmatrix}$$

$$\begin{aligned} c_1 \qquad + c_3 &= 0 \\ c_1 + c_2 \qquad &= 2 \\ c_2 + c_3 &= 1 \end{aligned}$$

$$\therefore \quad c_1 = \tfrac{1}{2}, \qquad c_2 = \tfrac{3}{2}, \qquad c_3 = -\tfrac{1}{2}$$

$$\therefore \quad \begin{bmatrix} 0 \\ 2 \\ 1 \end{bmatrix}_T = \begin{bmatrix} \frac{1}{2} \\ \frac{3}{2} \\ -\frac{1}{2} \end{bmatrix}$$

$$\begin{bmatrix} 1 \\ 1 \\ 1 \end{bmatrix} = c_1 \begin{bmatrix} 1 \\ 1 \\ 0 \end{bmatrix} + c_2 \begin{bmatrix} 0 \\ 1 \\ 1 \end{bmatrix} + c_3 \begin{bmatrix} 1 \\ 0 \\ 1 \end{bmatrix}$$

$$\begin{aligned} c_1 \qquad + c_3 &= 1 \\ c_1 + c_2 \qquad &= 1 \\ c_2 + c_3 &= 1 \end{aligned}$$

$$\therefore \quad c_1 = \tfrac{1}{2}, \qquad c_2 = \tfrac{1}{2}, \qquad c_3 = \tfrac{1}{2}$$

$$\therefore \quad \begin{bmatrix} 1 \\ 1 \\ 1 \end{bmatrix}_T = \begin{bmatrix} \frac{1}{2} \\ \frac{1}{2} \\ \frac{1}{2} \end{bmatrix}$$

Therefore, the transition matrix P from T to S is

$$P = \begin{bmatrix} -\frac{1}{2} & \frac{1}{2} & \frac{1}{2} \\ \frac{1}{2} & \frac{3}{2} & \frac{1}{2} \\ \frac{3}{2} & -\frac{1}{2} & \frac{1}{2} \end{bmatrix}$$

(c)

$$\left[\begin{array}{ccc|ccc} -\frac{1}{2} & \frac{1}{2} & \frac{1}{2} & 1 & 0 & 0 \\ \frac{1}{2} & \frac{3}{2} & \frac{1}{2} & 0 & 1 & 0 \\ \frac{3}{2} & -\frac{1}{2} & \frac{1}{2} & 0 & 0 & 1 \end{array}\right] \xrightarrow[R_3 \to R_3 + 3R_1]{R_2 \to R_2 + R_1} \left[\begin{array}{ccc|ccc} -\frac{1}{2} & \frac{1}{2} & \frac{1}{2} & 1 & 0 & 0 \\ 0 & 2 & 1 & 1 & 1 & 0 \\ 0 & 1 & 2 & 3 & 0 & 1 \end{array}\right]$$

$$\xrightarrow[R_2 \to R_2 - R_3]{R_1 \to -2R_1} \left[\begin{array}{ccc|ccc} 1 & -1 & -1 & -2 & 0 & 0 \\ 0 & 1 & -1 & -2 & 1 & -1 \\ 0 & 1 & 2 & 3 & 0 & 1 \end{array}\right] \xrightarrow[R_3 \to R_3 - R_2]{R_1 \to R_1 + R_2} \left[\begin{array}{ccc|ccc} 1 & 0 & -2 & -4 & 1 & -1 \\ 0 & 1 & -1 & -2 & 1 & -1 \\ 0 & 0 & 3 & 5 & -1 & 2 \end{array}\right]$$

$$\xrightarrow[R_3 \to 1/3R_3]{} \left[\begin{array}{ccc|ccc} 1 & 0 & -2 & -4 & 1 & -1 \\ 0 & 1 & -1 & -2 & 1 & -1 \\ 0 & 0 & 1 & \frac{5}{3} & -\frac{1}{3} & \frac{2}{3} \end{array}\right] \xrightarrow[R_2 \to R_2 + R_3]{R_1 \to R_1 + 2R_3} \left[\begin{array}{ccc|ccc} 1 & 0 & 0 & -\frac{2}{3} & \frac{1}{3} & \frac{1}{3} \\ 0 & 1 & 0 & -\frac{1}{3} & \frac{2}{3} & -\frac{1}{3} \\ 0 & 0 & 1 & \frac{5}{3} & -\frac{1}{3} & \frac{2}{3} \end{array}\right]$$

$$\therefore \quad P^{-1} = \begin{bmatrix} -\frac{2}{3} & \frac{1}{3} & \frac{1}{3} \\ -\frac{1}{3} & \frac{2}{3} & -\frac{1}{3} \\ \frac{5}{3} & -\frac{1}{3} & \frac{2}{3} \end{bmatrix}$$

(d)

$$B = P^{-1}AP = \begin{bmatrix} -\frac{2}{3} & \frac{1}{3} & \frac{1}{3} \\ -\frac{1}{3} & \frac{2}{3} & -\frac{1}{3} \\ \frac{5}{3} & -\frac{1}{3} & \frac{2}{3} \end{bmatrix} \begin{bmatrix} 1 & -\frac{1}{2} & -\frac{3}{2} \\ 0 & -\frac{3}{2} & -\frac{3}{2} \\ 2 & \frac{5}{2} & \frac{5}{2} \end{bmatrix} \begin{bmatrix} -\frac{1}{2} & \frac{1}{2} & \frac{1}{2} \\ \frac{1}{2} & \frac{3}{2} & \frac{1}{2} \\ \frac{3}{2} & -\frac{1}{2} & \frac{1}{2} \end{bmatrix}$$

$$= \begin{bmatrix} 0 & \frac{2}{3} & \frac{4}{3} \\ -1 & -\frac{5}{3} & -\frac{4}{3} \\ 3 & \frac{4}{3} & -\frac{1}{3} \end{bmatrix} \begin{bmatrix} -\frac{1}{2} & \frac{1}{2} & \frac{1}{2} \\ \frac{1}{2} & \frac{3}{2} & \frac{1}{2} \\ \frac{3}{2} & -\frac{1}{2} & \frac{1}{2} \end{bmatrix} = \begin{bmatrix} \frac{7}{3} & \frac{1}{3} & 1 \\ -\frac{7}{3} & -\frac{7}{3} & -2 \\ -\frac{4}{3} & \frac{11}{3} & 2 \end{bmatrix}$$

Therefore,

$$B = \begin{bmatrix} \frac{7}{3} & \frac{1}{3} & 1 \\ -\frac{7}{3} & -\frac{7}{3} & -2 \\ -\frac{4}{3} & \frac{11}{3} & 2 \end{bmatrix}$$

PROBLEM 14-9 In Problem 14-8 we found the matrix representation of L with respect to the basis

$$S = \left\{ \begin{bmatrix} 1 \\ 1 \\ 0 \end{bmatrix}, \begin{bmatrix} 0 \\ 1 \\ 1 \end{bmatrix}, \begin{bmatrix} 1 \\ 0 \\ 1 \end{bmatrix} \right\}$$

is

$$A = \begin{bmatrix} 1 & -\frac{1}{2} & -\frac{3}{2} \\ 0 & -\frac{3}{2} & -\frac{3}{2} \\ 2 & \frac{5}{2} & \frac{5}{2} \end{bmatrix}$$

(a) Use A to find the dimension and a basis for range L. **(b)** Use the result in part **(a)** to find the dimension and a basis for ker L.

Solution

(a)

$$\text{Augmented matrix} = \begin{bmatrix} 1 & -\frac{1}{2} & -\frac{3}{2} & 0 \\ 0 & -\frac{3}{2} & -\frac{3}{2} & 0 \\ 2 & \frac{5}{2} & \frac{5}{2} & 0 \end{bmatrix} \xrightarrow[R_3 \to R_3 - 2R_1]{R_2 \to -2/3R_2} \begin{bmatrix} 1 & -\frac{1}{2} & -\frac{3}{2} & 0 \\ 0 & 1 & 1 & 0 \\ 0 & \frac{7}{2} & \frac{11}{2} & 0 \end{bmatrix}$$

$$\xrightarrow[R_3 \to R_3 - 7/2R_2]{R_1 \to R_1 + 1/2R_2} \begin{bmatrix} 1 & 0 & -1 & 0 \\ 0 & 1 & 1 & 0 \\ 0 & 0 & 2 & 0 \end{bmatrix}$$

$$\xrightarrow[R_3 \to 1/2R_3]{} \begin{bmatrix} 1 & 0 & -1 & 0 \\ 0 & 1 & 1 & 0 \\ 0 & 0 & 1 & 0 \end{bmatrix}$$

Since there are leading ones in the first, second, and third columns, the coordinate vector representation of the basis for range L are $\begin{bmatrix} 1 \\ 0 \\ 2 \end{bmatrix}$, $\begin{bmatrix} -\frac{1}{2} \\ -\frac{3}{2} \\ \frac{5}{2} \end{bmatrix}$, and $\begin{bmatrix} -\frac{3}{2} \\ -\frac{3}{2} \\ \frac{5}{2} \end{bmatrix}$. Thus, dim range $L = 3$. To find a basis we transform these coordinate vectors into their corresponding vectors in R^3.

$$\mathbf{s}_1 = 1\begin{bmatrix} 1 \\ 1 \\ 0 \end{bmatrix} + 0\begin{bmatrix} 0 \\ 1 \\ 1 \end{bmatrix} + 2\begin{bmatrix} 1 \\ 0 \\ 1 \end{bmatrix} = \begin{bmatrix} 3 \\ 1 \\ 2 \end{bmatrix}$$

$$\mathbf{s}_2 = -\frac{1}{2}\begin{bmatrix} 1 \\ 1 \\ 0 \end{bmatrix} - \frac{3}{2}\begin{bmatrix} 0 \\ 1 \\ 1 \end{bmatrix} + \frac{5}{2}\begin{bmatrix} 1 \\ 0 \\ 1 \end{bmatrix} = \begin{bmatrix} 2 \\ -2 \\ 1 \end{bmatrix}$$

$$\mathbf{s}_3 = -\frac{3}{2}\begin{bmatrix} 1 \\ 1 \\ 0 \end{bmatrix} - \frac{3}{2}\begin{bmatrix} 0 \\ 1 \\ 1 \end{bmatrix} + \frac{5}{2}\begin{bmatrix} 1 \\ 0 \\ 1 \end{bmatrix} = \begin{bmatrix} 1 \\ -3 \\ 1 \end{bmatrix}$$

Therefore, a basis for range L is $\left\{ \begin{bmatrix} 3 \\ 1 \\ 2 \end{bmatrix}, \begin{bmatrix} 2 \\ -2 \\ 1 \end{bmatrix}, \begin{bmatrix} 1 \\ -3 \\ 1 \end{bmatrix} \right\}$.

(b) Since dim range $L = 3$ and dim range L + dim ker $L = 3$, then dim ker $L = 0$ and the only vector in ker L is $\begin{bmatrix} 0 \\ 0 \\ 0 \end{bmatrix}$.

PROBLEM 14-10 In Problem 14-8, we found that the matrix representation of L with respect to basis

$$T = \left\{ \begin{bmatrix} 1 \\ 0 \\ 2 \end{bmatrix}, \begin{bmatrix} 0 \\ 2 \\ 1 \end{bmatrix}, \begin{bmatrix} 1 \\ 1 \\ 1 \end{bmatrix} \right\}$$

is

$$B = \begin{bmatrix} \frac{7}{3} & \frac{1}{3} & 1 \\ -\frac{7}{3} & -\frac{7}{3} & -2 \\ -\frac{4}{3} & \frac{11}{3} & 2 \end{bmatrix}$$

(a) Use B to find the dimension and a basis for range L. **(b)** Show that the basis for range L in part **(a)** gives the same range as the basis in part **(a)** of Problem 14-9.

Solution

(a) Augmented matrix $= \begin{bmatrix} \frac{7}{3} & \frac{1}{3} & 1 & 0 \\ -\frac{7}{3} & -\frac{7}{3} & -2 & 0 \\ -\frac{4}{3} & \frac{11}{3} & 2 & 0 \end{bmatrix} \xrightarrow[R_2 \to R_2 + R_1]{R_1 \to R_1 + R_3} \begin{bmatrix} 1 & 4 & 3 & 0 \\ 0 & -2 & -1 & 0 \\ -\frac{4}{3} & \frac{11}{3} & 2 & 0 \end{bmatrix}$

$$\xrightarrow[R_3 \to R_3 + 4/3R_1]{} \begin{bmatrix} 1 & 4 & 3 & 0 \\ 0 & -2 & -1 & 0 \\ 0 & 9 & 6 & 0 \end{bmatrix}$$

$$\xrightarrow[R_3 \to 1/3R_3]{R_2 \to -R_2} \begin{bmatrix} 1 & 4 & 3 & 0 \\ 0 & 2 & 1 & 0 \\ 0 & 3 & 2 & 0 \end{bmatrix} \xrightarrow[R_3 \to R_3 - R_2]{} \begin{bmatrix} 1 & 4 & 3 & 0 \\ 0 & 2 & 1 & 0 \\ 0 & 1 & 1 & 0 \end{bmatrix}$$

$$\xrightarrow[R_2 \to R_2 - R_3]{R_1 \to R_1 - 2R_2} \begin{bmatrix} 1 & 0 & 1 & 0 \\ 0 & 1 & 0 & 0 \\ 0 & 1 & 1 & 0 \end{bmatrix} \xrightarrow[R_3 \to R_3 - R_2]{} \begin{bmatrix} 1 & 0 & 1 & 0 \\ 0 & 1 & 0 & 0 \\ 0 & 0 & 1 & 0 \end{bmatrix}$$

Since there are leading ones in the first, second, and third columns, the coordinate vector representations of the basis for range L are $\begin{bmatrix} 7/3 \\ -7/3 \\ -4/3 \end{bmatrix}$, $\begin{bmatrix} 1/3 \\ -7/3 \\ 11/3 \end{bmatrix}$, and $\begin{bmatrix} 1 \\ -2 \\ 2 \end{bmatrix}$. Thus, dim range $L = 3$. Now we proceed to find a basis.

$$\mathbf{t}_1 = \frac{7}{3}\begin{bmatrix} 1 \\ 0 \\ 2 \end{bmatrix} - \frac{7}{3}\begin{bmatrix} 0 \\ 2 \\ 1 \end{bmatrix} - \frac{4}{3}\begin{bmatrix} 1 \\ 1 \\ 1 \end{bmatrix} = \begin{bmatrix} 1 \\ -6 \\ 1 \end{bmatrix}$$

$$\mathbf{t}_2 = \frac{1}{3}\begin{bmatrix} 1 \\ 0 \\ 2 \end{bmatrix} - \frac{7}{3}\begin{bmatrix} 0 \\ 2 \\ 1 \end{bmatrix} + \frac{11}{3}\begin{bmatrix} 1 \\ 1 \\ 1 \end{bmatrix} = \begin{bmatrix} 4 \\ -1 \\ 2 \end{bmatrix}$$

$$\mathbf{t}_3 = 1\begin{bmatrix} 1 \\ 0 \\ 2 \end{bmatrix} - 2\begin{bmatrix} 0 \\ 2 \\ 1 \end{bmatrix} + 2\begin{bmatrix} 1 \\ 1 \\ 1 \end{bmatrix} = \begin{bmatrix} 3 \\ -2 \\ 2 \end{bmatrix}$$

Therefore, a basis for range L is $\left\{ \begin{bmatrix} 1 \\ -6 \\ 1 \end{bmatrix}, \begin{bmatrix} 4 \\ -1 \\ 2 \end{bmatrix}, \begin{bmatrix} 3 \\ -2 \\ 2 \end{bmatrix} \right\}$.

(b) To show that the basis for range L in part **(a)** gives the same range as the basis in part **(a)** of Problem 14-9, we must show that each vector in $\left\{ \begin{bmatrix} 1 \\ -6 \\ 1 \end{bmatrix}, \begin{bmatrix} 4 \\ -1 \\ 2 \end{bmatrix}, \begin{bmatrix} 3 \\ -2 \\ 2 \end{bmatrix} \right\}$ can be expressed as a linear combination of the basis vectors in $\left\{ \begin{bmatrix} 3 \\ 1 \\ 2 \end{bmatrix}, \begin{bmatrix} 2 \\ -2 \\ 1 \end{bmatrix}, \begin{bmatrix} 1 \\ -3 \\ 1 \end{bmatrix} \right\}$. This results in the following:

(1) $$\begin{bmatrix} 1 \\ -6 \\ 1 \end{bmatrix} = -\frac{1}{2}\begin{bmatrix} 3 \\ 1 \\ 2 \end{bmatrix} + \frac{1}{2}\begin{bmatrix} 2 \\ -2 \\ 1 \end{bmatrix} + \frac{3}{2}\begin{bmatrix} 1 \\ -3 \\ 1 \end{bmatrix}$$

(2) $$\begin{bmatrix} 4 \\ -1 \\ 2 \end{bmatrix} = \frac{1}{2}\begin{bmatrix} 3 \\ 1 \\ 2 \end{bmatrix} + \frac{3}{2}\begin{bmatrix} 2 \\ -2 \\ 1 \end{bmatrix} - \frac{1}{2}\begin{bmatrix} 1 \\ -3 \\ 1 \end{bmatrix}$$

(3) $$\begin{bmatrix} 3 \\ -2 \\ 2 \end{bmatrix} = \frac{1}{2}\begin{bmatrix} 3 \\ 1 \\ 2 \end{bmatrix} + \frac{1}{2}\begin{bmatrix} 2 \\ -2 \\ 1 \end{bmatrix} + \frac{1}{2}\begin{bmatrix} 1 \\ -3 \\ 1 \end{bmatrix}$$

Therefore, range L is the same in both parts.

PROBLEM 14-11 In Problem 13-18 we found that the matrix representation of the linear transformation $L: R_3 \to R_2$ defined by $L([a_1 \quad a_2 \quad a_3]) = [2a_1 + 3a_2 \quad a_1 + 2a_2 - a_3]$ is $A = \begin{bmatrix} 3/2 & 0 & 1/2 \\ -7/2 & -2 & -5/2 \end{bmatrix}$ with respect to the ordered bases $B = \{[1 \quad 1 \quad 0], [1 \quad 0 \quad 1], [0 \quad 1 \quad 1]\}$ and $S = \{[1 \quad 2], [-1 \quad 0]\}$. **(a)** Use A to find the dimension and a basis for range L. **(b)** Use the result in part **(a)** to find the dimension and a basis for ker L.

Solution

(a) Augmented matrix $= \begin{bmatrix} \frac{3}{2} & 0 & \frac{1}{2} & 0 \\ -\frac{7}{2} & -2 & -\frac{5}{2} & 0 \end{bmatrix} \xrightarrow{R_2 \to R_2 + R_1} \begin{bmatrix} \frac{3}{2} & 0 & \frac{1}{2} & 0 \\ -2 & -2 & -2 & 0 \end{bmatrix}$

$$\xrightarrow[R_2 \to -1/2 R_2]{R_1 \to 2R_1} \begin{bmatrix} 3 & 0 & 1 & 0 \\ 1 & 1 & 1 & 0 \end{bmatrix} \xrightarrow{R_1 \to R_1 - 2R_2} \begin{bmatrix} 1 & -2 & -1 & 0 \\ 1 & 1 & 1 & 0 \end{bmatrix}$$

$$\xrightarrow{R_2 \to R_2 - R_1} \begin{bmatrix} 1 & -2 & -1 & 0 \\ 0 & 3 & 2 & 0 \end{bmatrix} \xrightarrow{R_2 \to 1/3 R_2} \begin{bmatrix} 1 & -2 & -1 & 0 \\ 0 & 1 & \frac{2}{3} & 0 \end{bmatrix}$$

$$\xrightarrow{R_1 \to R_1 + 2R_2} \begin{bmatrix} 1 & 0 & \frac{1}{3} & 0 \\ 0 & 1 & \frac{2}{3} & 0 \end{bmatrix}$$

Since we used only row operations and there are leading ones in the first, and second columns, the coordinate vector representations of the basis vectors for range L are $\begin{bmatrix} 3/2 \\ -7/2 \end{bmatrix}$ and $\begin{bmatrix} 0 \\ -2 \end{bmatrix}$. Thus dim range $L = 2$.

We must transform each of these coordinate vectors into their corresponding vectors in R_2. We get

$$\mathbf{s}_1 = \tfrac{3}{2}[1 \quad 2] - \tfrac{7}{2}[-1 \quad 0] = [5 \quad 3]$$

$$\mathbf{s}_2 = 0[1 \quad 2] - 2[-1 \quad 0] = [2 \quad 0]$$

Therefore, $\{[5 \quad 3], [2 \quad 0]\}$ is a basis for range L.

(b) From the reduced echelon form in part **(a)** we find $a_1 = -(1/3)a_3$ and $a_2 = -(2/3)a_3$. If $a_3 = t$, the solution is

$$\begin{bmatrix} -\frac{1}{3}t \\ -\frac{2}{3}t \\ t \end{bmatrix} = \frac{1}{3}t\begin{bmatrix} -1 \\ -2 \\ 3 \end{bmatrix}$$

Thus $\begin{bmatrix} -1 \\ -2 \\ 3 \end{bmatrix}$ is the coordinate vector representation of the basis vector for kernel L. Therefore, $-1[1 \quad 1 \quad 0] - 2[1 \quad 0 \quad 1] + 3[0 \quad 1 \quad 1] = [-3 \quad 2 \quad 1]$. Thus, a basis for kernel L is $\{[-3 \quad 2 \quad 1]\}$ and dim ker $L = 1$.

PROBLEM 14-12 Let $L: R_3 \to R_3$ be defined by $L([a_0 \quad a_1 \quad a_2]) = [a_0 - 2a_1 \quad 2a_2 + a_1 \quad 2a_0 + 3a_2]$. Consider the ordered bases $S = \{[1 \quad 0 \quad 0], [0 \quad 1 \quad 0], [0 \quad 0 \quad 1]\}$ and $T = \{[1 \quad 1 \quad 0], [0 \quad 1 \quad 1],$

[1 0 1]} for R_3. **(a)** Find A, the matrix representation of L with respect to S. **(b)** Find B, the matrix representation of L with respect to T.

Solution

(a)

$$L([1 \quad 0 \quad 0]) = [1 - 0 \quad 0 + 0 \quad 2 + 0] = [1 \quad 0 \quad 2]$$

$$L([0 \quad 1 \quad 0]) = [0 - 2 \quad 0 + 1 \quad 0 + 0] = [-2 \quad 1 \quad 0]$$

$$L([0 \quad 0 \quad 0]) = [0 + 0 \quad 2 + 0 \quad 0 + 3] = [0 \quad 2 \quad 3]$$

Since S is the standard basis, these are the columns of A. Therefore,

$$A = \begin{bmatrix} 1 & -2 & 0 \\ 0 & 1 & 2 \\ 2 & 0 & 3 \end{bmatrix}$$

(b)

$$L([1 \quad 1 \quad 0]) = [-1 \quad 1 \quad 2]$$

$$L([0 \quad 1 \quad 1]) = [-2 \quad 3 \quad 3]$$

$$L([1 \quad 0 \quad 1]) = [1 \quad 2 \quad 5]$$

$$[-1 \quad 1 \quad 2] = c_1[1 \quad 1 \quad 0] + c_2[0 \quad 1 \quad 1] + c_3[1 \quad 0 \quad 1]$$

$$\begin{aligned} c_1 \quad\quad + c_3 &= -1 \\ c_1 + c_2 \quad\quad &= 1 \\ c_2 + c_3 &= 2 \end{aligned}$$

$$\therefore \quad c_1 = -1, \qquad c_2 = 2, \qquad c_3 = 0$$

$$\therefore \quad [-1 \quad 1 \quad 2]_T = [-1 \quad 2 \quad 0]$$

$$[-2 \quad 3 \quad 3] = c_1[1 \quad 1 \quad 0] + c_2[0 \quad 1 \quad 1] + c_3[1 \quad 0 \quad 1]$$

$$\begin{aligned} c_1 \quad\quad + c_3 &= -2 \\ c_1 + c_2 \quad\quad &= 3 \\ c_2 + c_3 &= 3 \end{aligned}$$

$$\therefore \quad c_1 = -1, \qquad c_2 = 4, \qquad c_3 = -1$$

$$\therefore \quad [-2 \quad 3 \quad 3]_T = [-1 \quad 4 \quad -1]$$

$$[1 \quad 2 \quad 5] = c_1[1 \quad 1 \quad 0] + c_2[0 \quad 1 \quad 1] + c_3[1 \quad 0 \quad 1]$$

$$\begin{aligned} c_1 \quad\quad + c_3 &= 1 \\ c_1 + c_2 \quad\quad &= 2 \\ c_2 + c_3 &= 5 \end{aligned}$$

$$\therefore \quad c_1 = -1, \qquad c_2 = 3, \qquad c_3 = 2$$

$$\therefore \quad [1 \quad 2 \quad 5]_T = [-1 \quad 3 \quad 2]$$

Therefore, $B = \begin{bmatrix} -1 & -1 & -1 \\ 2 & 4 & 3 \\ 0 & -1 & 2 \end{bmatrix}$ where columns are formed from the vectors $[-1 \quad 2 \quad 0]$, $[-1 \quad 4 \quad -1]$, $[-1 \quad 3 \quad 2]$.

PROBLEM 14-13 In Problem 14-12, find **(a)** the transition matrix P from T to S, and **(b)** the transition matrix P^{-1} from S to T.

Solution

(a) Since S is the standard basis, the transition matrix from T to S is

$$P = \begin{bmatrix} 1 & 0 & 1 \\ 1 & 1 & 0 \\ 0 & 1 & 1 \end{bmatrix}$$

whose columns are the vectors in the basis T.

(b)

$$\left[\begin{array}{ccc|ccc} 1 & 0 & 1 & 1 & 0 & 0 \\ 1 & 1 & 0 & 0 & 1 & 0 \\ 0 & 1 & 1 & 0 & 0 & 1 \end{array}\right] \xrightarrow[R_2 \to R_2 - R_1]{} \left[\begin{array}{ccc|ccc} 1 & 0 & 1 & 1 & 0 & 0 \\ 0 & 1 & -1 & -1 & 1 & 0 \\ 0 & 1 & 1 & 0 & 0 & 1 \end{array}\right]$$

$$\xrightarrow[R_3 \to R_3 - R_2]{} \left[\begin{array}{ccc|ccc} 1 & 0 & 1 & 1 & 0 & 0 \\ 0 & 1 & -1 & -1 & 1 & 0 \\ 0 & 0 & 2 & 1 & -1 & 1 \end{array}\right] \xrightarrow[R_3 \to 1/2 R_3]{} \left[\begin{array}{ccc|ccc} 1 & 1 & 0 & 1 & 0 & 0 \\ 0 & 1 & -1 & -1 & 1 & 0 \\ 0 & 0 & 1 & \frac{1}{2} & -\frac{1}{2} & \frac{1}{2} \end{array}\right]$$

$$\xrightarrow[R_2 \to R_2 + R_3]{} \left[\begin{array}{ccc|ccc} 1 & 1 & 0 & 1 & 0 & 0 \\ 0 & 1 & 0 & -\frac{1}{2} & \frac{1}{2} & \frac{1}{2} \\ 0 & 0 & 1 & \frac{1}{2} & -\frac{1}{2} & \frac{1}{2} \end{array}\right] \xrightarrow[]{R_1 \to R_1 - R_2} \left[\begin{array}{ccc|ccc} 1 & 0 & 0 & \frac{1}{2} & \frac{1}{2} & -\frac{1}{2} \\ 0 & 1 & 0 & -\frac{1}{2} & \frac{1}{2} & \frac{1}{2} \\ 0 & 0 & 1 & \frac{1}{2} & -\frac{1}{2} & \frac{1}{2} \end{array}\right]$$

$$\therefore \quad P^{-1} = \begin{bmatrix} \frac{1}{2} & \frac{1}{2} & -\frac{1}{2} \\ -\frac{1}{2} & \frac{1}{2} & \frac{1}{2} \\ \frac{1}{2} & -\frac{1}{2} & \frac{1}{2} \end{bmatrix}$$

PROBLEM 14-14 Show that $B = P^{-1}AP$ using the results of Problems 14-12 and 14-13.

Solution

$$P^{-1}AP = \begin{bmatrix} \frac{1}{2} & \frac{1}{2} & -\frac{1}{2} \\ -\frac{1}{2} & \frac{1}{2} & \frac{1}{2} \\ \frac{1}{2} & -\frac{1}{2} & \frac{1}{2} \end{bmatrix} \begin{bmatrix} 1 & -2 & 0 \\ 0 & 1 & 2 \\ 2 & 0 & 3 \end{bmatrix} \begin{bmatrix} 1 & 0 & 1 \\ 1 & 1 & 0 \\ 0 & 1 & 1 \end{bmatrix}$$

$$= \begin{bmatrix} -\frac{1}{2} & -\frac{1}{2} & -\frac{1}{2} \\ \frac{1}{2} & \frac{3}{2} & \frac{5}{2} \\ \frac{3}{2} & -\frac{3}{2} & \frac{1}{2} \end{bmatrix} \begin{bmatrix} 1 & 0 & 1 \\ 1 & 1 & 0 \\ 0 & 1 & 1 \end{bmatrix} = \begin{bmatrix} -1 & -1 & -1 \\ 2 & 4 & 3 \\ 0 & -1 & 2 \end{bmatrix} = B$$

Supplementary Problems

PROBLEM 14-15 Use $P = \begin{bmatrix} 1 & -1 \\ 2 & 3 \end{bmatrix}$ to verify that $B = \begin{bmatrix} 3 & -4 \\ 1 & 1 \end{bmatrix}$ is similar to $A = \begin{bmatrix} 0 & 1 \\ 1 & 4 \end{bmatrix}$.

PROBLEM 14-16 Find a matrix B that is similar to the matrix $A = \begin{bmatrix} 0 & 1 \\ 1 & -1 \end{bmatrix}$.

PROBLEM 14-17 Let $A = \begin{bmatrix} 1 & 0 & 1 \\ 0 & 2 & -1 \\ 0 & 1 & 2 \end{bmatrix}$ and $P = \begin{bmatrix} 1 & 0 & 0 \\ 0 & 1 & 0 \\ 1 & 0 & 1 \end{bmatrix}$. **(a)** Find $B = P^{-1}AP$. **(b)** Find $\det(A)$ and $\det(B)$.

PROBLEM 14-18 Let $L: R^2 \to R^2$ be defined by

$$L\left(\begin{bmatrix} a_1 \\ a_2 \end{bmatrix}\right) = \begin{bmatrix} a_1 + 5a_2 \\ 2a_1 + 3a_2 \end{bmatrix}$$

Consider the ordered bases

$$S = \left\{\begin{bmatrix} 1 \\ 0 \end{bmatrix}, \begin{bmatrix} 3 \\ -1 \end{bmatrix}\right\} \quad \text{and} \quad T = \left\{\begin{bmatrix} 1 \\ -3 \end{bmatrix}, \begin{bmatrix} 0 \\ -1 \end{bmatrix}\right\}$$

(a) Find A, the matrix of L with respect to S. **(b)** Find B, the matrix of L with respect to T. **(c)** Find the transition matrix P from T to S. **(d)** Find the transition matrix P^{-1} from S to T.

PROBLEM 14-19 Let $L: P_1 \to P_1$ be defined by $L(a_0 + a_1x) = a_0 + a_1(x - 2)$ with ordered bases $S = \{6 + 3x, 8 - 2x\}$ and $T = \{1, 5 + x\}$. **(a)** Find A, the matrix of L with respect to S. **(b)** Find B, the matrix of L with respect to T.

PROBLEM 14-20 In Problem 14-19 use A to find the dimension and a basis for range L.

PROBLEM 14-21 In Problem 14-19 use B to find the dimension and a basis for ker L.

PROBLEM 14-22 Let $A = \begin{bmatrix} 1 & 1 & -1 \\ 0 & 1 & 1 \\ 0 & 0 & 0 \end{bmatrix}$ be the matrix representation of a linear transformation L with respect to the ordered basis $S = \left\{\begin{bmatrix} 1 \\ 1 \\ 0 \end{bmatrix}, \begin{bmatrix} 0 \\ 1 \\ 1 \end{bmatrix}, \begin{bmatrix} 1 \\ 0 \\ 1 \end{bmatrix}\right\}$. **(a)** Find the dimension and a basis for range L. **(b)** Find the dimension and a basis for ker L.

PROBLEM 14-23 Write the polynomial $-3 + 4x - 10x^2$ in P_2 in terms of the given basis **(a)** $\{5, 3 - 2x, 2 + x - 5x^2\}$ and **(b)** $\{1 - x, 1 + x, 1 - x^2\}$.

PROBLEM 14-24 Find the kernel and the nullity of $A = \begin{bmatrix} 1 & -1 & 3 \\ 2 & 0 & 4 \\ 0 & -2 & 2 \end{bmatrix}$.

Answers to Supplementary Problems

14-15 $P^{-1} = \frac{1}{5}\begin{bmatrix} 3 & 1 \\ -2 & 1 \end{bmatrix}$

14-16 $B = \begin{bmatrix} 1 & 1 \\ -1 & -2 \end{bmatrix}$

14-17 $B = \begin{bmatrix} 2 & 0 & 1 \\ -1 & 2 & -1 \\ 0 & 1 & 1 \end{bmatrix}$; $\det A = 5$, $\det B = 5$

14-18 **(a)** $A = \begin{bmatrix} 7 & 7 \\ -2 & -3 \end{bmatrix}$

(b) $B = \begin{bmatrix} -14 & -5 \\ 49 & 18 \end{bmatrix}$

(c) $P = \begin{bmatrix} -8 & -3 \\ 3 & 1 \end{bmatrix}$

(d) $P^{-1} = \begin{bmatrix} 1 & 3 \\ -3 & -8 \end{bmatrix}$

14-19 **(a)** $A = \begin{bmatrix} 2/3 & 2/9 \\ -1/2 & 4/3 \end{bmatrix}$ **(b)** $B = \begin{bmatrix} 1 & -2 \\ 0 & 1 \end{bmatrix}$

14-20 Dim range $L = 2$, basis is $\{x, 6 - x\}$

14-21 Dim ker $L = 0$, basis is $\{0 + 0x\}$

14-22 (**a**) Dim range $L = 2$, basis is $\left\{\begin{bmatrix}1\\1\\0\end{bmatrix}, \begin{bmatrix}1\\2\\1\end{bmatrix}\right\}$

(**b**) Dim ker $L = 1$, basis is $\left\{\begin{bmatrix}3\\1\\0\end{bmatrix}\right\}$

14-23 (**a**) $-(4/5)(5) - 1(3 - 2x) + 2(2 + x - 5x^2)$
(**b**) $-(17/2)(1 - x) - (9/2)(1 + x) + 10(1 - x^2)$

14-24 Ker $A = \text{span}\left\{\begin{bmatrix}-2\\1\\1\end{bmatrix}\right\}$; nullity is 1

EXAM 4 (Chapters 12–14)

1. Determine whether the transformation $L: R_3 \to R_2$ defined by $L([a_1 \ \ a_2 \ \ a_3]) = [a_1 + a_2 \ \ a_2 + a_3]$ is linear.

2. Find a basis for the kernal of L if $L: R^3 \to R^4$ is defined by

$$L\left(\begin{bmatrix} a_1 \\ a_2 \\ a_3 \end{bmatrix}\right) = \begin{bmatrix} 1 & 0 & 2 \\ 0 & -1 & 1 \\ 1 & 1 & 1 \\ 2 & 4 & 0 \end{bmatrix}\begin{bmatrix} a_1 \\ a_2 \\ a_3 \end{bmatrix}$$

3. If L is the linear transformation in question 2, find the range of L.

4. What is the rank and nullity of the transformation L in question 2 and 3? Give reasons for your answers.

5. Let $B = \left\{\begin{bmatrix} 1 \\ 0 \end{bmatrix}, \begin{bmatrix} 2 \\ 1 \end{bmatrix}\right\}$ and $S = \left\{\begin{bmatrix} 1 \\ 1 \end{bmatrix}, \begin{bmatrix} 3 \\ 2 \end{bmatrix}\right\}$ be ordered bases for R^2. If $\mathbf{a} = \begin{bmatrix} 1 \\ 4 \end{bmatrix}$ and $\mathbf{b} = \begin{bmatrix} 3 \\ 5 \end{bmatrix}$, find the coordinate vectors of **a** and **b** with respect to B.

6. Find the transition matrix P from the basis B to the basis S in question 5.

7. Use the result in question 6 to find the coordinate vector of **a** with respect to S for the **a** and S of question 5.

8. Find the transition matrix Q from the S to the B basis for question 5.

9. $L: R^2 \to R^2$ defined by $L\left(\begin{bmatrix} a_1 \\ a_2 \end{bmatrix}\right) = \begin{bmatrix} a_1 - 3a_2 \\ a_1 + \ a_2 \end{bmatrix}$ is a linear transformation with ordered bases $B = \left\{\begin{bmatrix} 1 \\ 3 \end{bmatrix}, \begin{bmatrix} 1 \\ 1 \end{bmatrix}\right\}$ and $S = \left\{\begin{bmatrix} 2 \\ 1 \end{bmatrix}, \begin{bmatrix} -1 \\ 0 \end{bmatrix}\right\}$ and transition matrix $P = \begin{bmatrix} -1/2 & 1/2 \\ 5/2 & -3/2 \end{bmatrix}$ from S to B. If the matrix representation of L is $A = \begin{bmatrix} 6 & 2 \\ -14 & -4 \end{bmatrix}$ with respect to B, find the matrix representation B of L with respect to S.

10. Find any matrix B different from A which is similar to $A = \begin{bmatrix} 1 & 1 & 2 \\ -1 & 0 & -1 \\ 2 & 3 & 7 \end{bmatrix}$.

Solutions to Exam 4

1. $L(\mathbf{a} + \mathbf{b})$

$$= L([a_1 \ \ a_2 \ \ a_3] + [b_1 \ \ b_2 \ \ b_3])$$
$$= L([a_1 + b_1 \ \ a_2 + b_2 \ \ a_3 + b_3])$$
$$= [a_1 + b_1 + a_2 + b_2 \ \ a_2 + b_2 + a_3 + b_3]$$
$$= [(a_1 + a_2) + (b_1 + b_2) \ \ (a_2 + a_3) + (b_2 + b_3)]$$
$$= [a_1 + a_2 \ \ a_2 + a_3] + [b_1 + b_2 \ \ b_2 + b_3]$$
$$= L([a_1 \ \ a_2 \ \ a_3]) + L([b_1 \ \ b_2 \ \ b_3])$$

$$\begin{aligned} L(k[a_1 \quad a_2 \quad a_3]) &= L([ka_1 \quad ka_2 \quad ka_3]) \\ &= [ka_1 + ka_2 \quad ka_2 + ka_3] \\ &= k[a_1 + a_2 \quad a_2 + a_3] \\ &= kL([a_1 \quad a_2 \quad a_3]) \end{aligned}$$

Therefore, it is linear.

2. Transform the augmented matrix to row echelon form.

$$\begin{bmatrix} 1 & 0 & 2 & 0 \\ 0 & -1 & 1 & 0 \\ 1 & 1 & 1 & 0 \\ 2 & 4 & 0 & 0 \end{bmatrix} \xrightarrow[R_4 \to R_4 - 2R_1]{R_3 \to R_3 - R_1} \begin{bmatrix} 1 & 0 & 2 & 0 \\ 0 & -1 & 1 & 0 \\ 0 & 1 & -1 & 0 \\ 0 & 4 & -4 & 0 \end{bmatrix}$$

$$\xrightarrow[\substack{R_3 \to R_3 + R_2 \\ R_4 \to R_4 + 4R_2}]{R_2 \to -R_2} \begin{bmatrix} 1 & 0 & 2 & 0 \\ 0 & 1 & -1 & 0 \\ 0 & 0 & 0 & 0 \\ 0 & 0 & 0 & 0 \end{bmatrix}$$

$$\therefore \quad a_1 + 2a_3 = 0$$

$$a_2 - \ a_3 = 0$$

If we let $a_3 = t$ where t is arbitrary, the solution is $\begin{bmatrix} -2t \\ t \\ t \end{bmatrix} = t\begin{bmatrix} -2 \\ 1 \\ 1 \end{bmatrix}$. Therefore, a basis for $\ker L$ is $\left\{\begin{bmatrix} -2 \\ 1 \\ 1 \end{bmatrix}\right\}$.

3. Since only row operations were used in question 2 and there are leading ones in only columns one and two, the vectors in columns one and two of the coefficient matrix form a basis for range L. Therefore,

a basis for range $L = \left\{\begin{bmatrix} 1 \\ 0 \\ 1 \\ 2 \end{bmatrix}, \begin{bmatrix} 0 \\ -1 \\ 1 \\ 4 \end{bmatrix}\right\}$.

4. The rank of L is 2 because dim (range L) = 2. The nullity of L is 1 because dim (ker L) = 1.

5. $\mathbf{a} = \begin{bmatrix} 1 \\ 4 \end{bmatrix} = a_1\begin{bmatrix} 1 \\ 0 \end{bmatrix} + a_2\begin{bmatrix} 2 \\ 1 \end{bmatrix}$ which gives $a_1 = -7$ and $a_2 = 4$. Therefore, $[\mathbf{a}]_B = \begin{bmatrix} -7 \\ 4 \end{bmatrix}$ and $\mathbf{b} = \begin{bmatrix} 3 \\ 5 \end{bmatrix} = a_1\begin{bmatrix} 1 \\ 0 \end{bmatrix} + a_2\begin{bmatrix} 2 \\ 1 \end{bmatrix}$ which gives $a_1 = -7$ and $a_2 = 5$. Therefore, $[\mathbf{b}]_B = \begin{bmatrix} -7 \\ 5 \end{bmatrix}$.

6. $\begin{bmatrix} 1 \\ 0 \end{bmatrix} = a_1\begin{bmatrix} 1 \\ 1 \end{bmatrix} + a_2\begin{bmatrix} 3 \\ 2 \end{bmatrix}$ which gives $a_1 = -2$ and $a_2 = 1$. $\begin{bmatrix} 2 \\ 1 \end{bmatrix} = a_1\begin{bmatrix} 1 \\ 1 \end{bmatrix} + a_2\begin{bmatrix} 3 \\ 2 \end{bmatrix}$ which gives $a_1 = -1$ and $a_2 = 1$. Hence, $\begin{bmatrix} 1 \\ 0 \end{bmatrix}_S = \begin{bmatrix} -2 \\ 1 \end{bmatrix}$ and $\begin{bmatrix} 2 \\ 1 \end{bmatrix}_S = \begin{bmatrix} -1 \\ 1 \end{bmatrix}$. Therefore, $P = \begin{bmatrix} -2 & -1 \\ 1 & 1 \end{bmatrix}$.

7.
$$\begin{bmatrix} 1 \\ 4 \end{bmatrix}_S = P\begin{bmatrix} 1 \\ 4 \end{bmatrix}_B = \begin{bmatrix} -2 & -1 \\ 1 & 1 \end{bmatrix}\begin{bmatrix} -7 \\ 4 \end{bmatrix} = \begin{bmatrix} 10 \\ -3 \end{bmatrix}$$

8. Q is the inverse of P in question 6. Find P^{-1}.

$$\left[\begin{array}{cc|cc} -2 & -1 & 1 & 0 \\ 1 & 1 & 0 & 1 \end{array}\right] \xrightarrow{R_1 \leftrightarrow R_2} \left[\begin{array}{cc|cc} 1 & 1 & 0 & 1 \\ -2 & -1 & 1 & 0 \end{array}\right]$$

$$\xrightarrow{R_2 \to R_2 + 2R_1} \left[\begin{array}{cc|cc} 1 & 1 & 0 & 1 \\ 0 & 1 & 1 & 2 \end{array}\right] \xrightarrow{R_1 \to R_1 - R_2} \left[\begin{array}{cc|cc} 1 & 0 & -1 & -1 \\ 0 & 1 & 1 & 2 \end{array}\right]$$

$$\therefore \quad Q = P^{-1} = \begin{bmatrix} -1 & -1 \\ 1 & 2 \end{bmatrix}$$

9. Find P^{-1}.

$$\left[\begin{array}{cc|cc} -\frac{1}{2} & \frac{1}{2} & 1 & 0 \\ \frac{5}{2} & -\frac{3}{2} & 0 & 1 \end{array}\right] \xrightarrow[R_2 \to 2R_2]{R_1 \to -2R_1} \left[\begin{array}{cc|cc} 1 & -1 & -2 & 0 \\ 5 & -3 & 0 & 2 \end{array}\right]$$

$$\xrightarrow{R_2 \to R_2 - 5R_1} \left[\begin{array}{cc|cc} 1 & -1 & -2 & 0 \\ 0 & 2 & 10 & 2 \end{array}\right] \xrightarrow{R_2 \to 1/2R_2} \left[\begin{array}{cc|cc} 1 & -1 & -2 & 0 \\ 0 & 1 & 5 & 1 \end{array}\right] \xrightarrow{R_1 \to R_1 + R_2} \left[\begin{array}{cc|cc} 1 & 0 & 3 & 1 \\ 0 & 1 & 5 & 1 \end{array}\right]$$

$$\therefore \quad P^{-1} = \begin{bmatrix} 3 & 1 \\ 5 & 1 \end{bmatrix}$$

$$\begin{aligned} \therefore \quad B_S &= P^{-1}AP \\ &= \begin{bmatrix} 3 & 1 \\ 5 & 1 \end{bmatrix}\begin{bmatrix} 6 & 2 \\ -14 & -4 \end{bmatrix}\begin{bmatrix} -\frac{1}{2} & \frac{1}{2} \\ \frac{5}{2} & -\frac{3}{2} \end{bmatrix} \\ &= \begin{bmatrix} 3 & -1 \\ 7 & -1 \end{bmatrix} \end{aligned}$$

10. Choose P to be any 3×3 invertible matrix, other than I_3. Thus, $B = P^{-1}AP$. Let $P = \begin{bmatrix} 1 & 0 & 0 \\ 0 & 1 & 0 \\ 2 & 0 & 1 \end{bmatrix}$.

Since P is an elementary matrix, P^{-1} is the elementary matrix $\begin{bmatrix} 1 & 0 & 0 \\ 0 & 1 & 0 \\ -2 & 0 & 1 \end{bmatrix}$.

$$B = P^{-1}AP$$

$$= \begin{bmatrix} 1 & 0 & 0 \\ 0 & 1 & 0 \\ -2 & 0 & 1 \end{bmatrix} \begin{bmatrix} 1 & 1 & 2 \\ -1 & 0 & -1 \\ 2 & 3 & 7 \end{bmatrix} \begin{bmatrix} 1 & 0 & 0 \\ 0 & 1 & 0 \\ 2 & 0 & 1 \end{bmatrix}$$

$$= \begin{bmatrix} 1 & 0 & 0 \\ 0 & 1 & 0 \\ -2 & 0 & 1 \end{bmatrix} \begin{bmatrix} 5 & 1 & 2 \\ -3 & 0 & -1 \\ 16 & 3 & 7 \end{bmatrix}$$

$$= \begin{bmatrix} 5 & 1 & 2 \\ -3 & 0 & -1 \\ 6 & 1 & 3 \end{bmatrix}$$

15 EIGENVALUES AND EIGENVECTORS

THIS CHAPTER IS ABOUT

- ☑ The Matrix Eigenproblem
- ☑ The Linear Transformation Eigenproblem
- ☑ Properties of Eigenvalues and Eigenvectors

15-1. The Matrix Eigenproblem

A. Definition of eigenvalue and eigenvector of a matrix

If A is an $n \times n$ matrix, then a nonzero vector $\mathbf{a}$ in R^n is called an **eigenvector** corresponding to the **eigenvalue** provided $A\mathbf{a} = \lambda\mathbf{a}$ and λ is a real number. Eigenvalues are also called **characteristic values, proper values**, or **latent roots**.

EXAMPLE 15-1: Verify that $\begin{bmatrix} 1 \\ -1 \end{bmatrix}$ and $\begin{bmatrix} 1 \\ -2 \end{bmatrix}$ are eigenvectors of $A = \begin{bmatrix} 3 & 1 \\ -2 & 0 \end{bmatrix}$.

Solution

$$\begin{bmatrix} 3 & 1 \\ -2 & 0 \end{bmatrix}\begin{bmatrix} 1 \\ -1 \end{bmatrix} = \begin{bmatrix} 2 \\ -2 \end{bmatrix} = 2\begin{bmatrix} 1 \\ -1 \end{bmatrix}$$

Therefore, $\begin{bmatrix} 1 \\ -1 \end{bmatrix}$ is an eigenvector corresponding to the eigenvalue 2.

$$\begin{bmatrix} 3 & 1 \\ -2 & 0 \end{bmatrix}\begin{bmatrix} 1 \\ -2 \end{bmatrix} = \begin{bmatrix} 1 \\ -2 \end{bmatrix} = 1\begin{bmatrix} 1 \\ -2 \end{bmatrix}$$

Therefore, $\begin{bmatrix} 1 \\ -2 \end{bmatrix}$ is an eigenvector corresponding to the eigenvalue 1.

B. Definition of characteristic determinant, polynomial, and equation

If A is an $n \times n$ matrix, then $\det(\lambda I - A)$ is called the **characteristic determinant** of A. When expanded, this determinant is a polynomial of degree n in λ and is called the **characteristic polynomial** of A. The corresponding equation $\det(\lambda I - A) = 0$ is called the **characteristic equation** of A.

note: Some texts use the matrix $(A - \lambda I)$ rather than $(\lambda I - A)$.

EXAMPLE 15-2: If $A = \begin{bmatrix} 1 & 0 & 2 \\ 3 & 0 & 1 \\ 1 & 2 & 3 \end{bmatrix}$, find **(a)** the characteristic determinant, **(b)** the characteristic polynomial, and **(c)** the characteristic equation.

Solution

(a) $$\det(\lambda I - A) = \left| \begin{bmatrix} \lambda & 0 & 0 \\ 0 & \lambda & 0 \\ 0 & 0 & \lambda \end{bmatrix} - \begin{bmatrix} 1 & 0 & 2 \\ 3 & 0 & 1 \\ 1 & 2 & 3 \end{bmatrix} \right| = \begin{vmatrix} \lambda - 1 & 0 & -2 \\ -3 & \lambda & -1 \\ -1 & -2 & \lambda - 3 \end{vmatrix}$$

which is the characteristic determinant.

(b) $$\begin{vmatrix} \lambda - 1 & 0 & -2 \\ -3 & \lambda & -1 \\ -1 & -2 & \lambda - 3 \end{vmatrix} = (\lambda - 1)\lambda(\lambda - 3) + (-2)(-3)(-2) - (-1)\lambda(-2) - (\lambda - 1)(-2)(-1)$$

$$= \lambda^3 - 4\lambda^2 + 3\lambda - 12 - 2\lambda - 2\lambda + 2$$
$$= \lambda^3 - 4\lambda^2 - \lambda - 10$$

which is the characteristic polynomial.

(c) $\lambda^3 - 4\lambda^2 - \lambda - 10 = 0$ is the characteristic equation.

C. Procedure for finding eigenvalues and eigenvectors

Step 1: Find the real solutions of the characteristic equation. These solutions λ_1, $\lambda_2, \ldots, \lambda_n$ are the eigenvalues of matrix A.

Step 2: To find the eigenvector $\mathbf{a}$ corresponding to an eigenvalue λ_i, solve the homogeneous equations $[(\lambda_i I - A)\mathbf{a} = \mathbf{0}]$.

EXAMPLE 15-3: Find the characteristic equation, the eigenvalues, and their associated eigenvectors for each of the following matrices.

(a) $A = \begin{bmatrix} 3 & 1 \\ -2 & 0 \end{bmatrix}$ **(b)** $B = \begin{bmatrix} 1 & 0 & 0 \\ 3 & 2 & 0 \\ 1 & 2 & 3 \end{bmatrix}$

Solution

(a) $$\det(\lambda I - A) = \begin{vmatrix} \lambda - 3 & -1 \\ 2 & \lambda \end{vmatrix} = \lambda(\lambda - 3) + 2 = 0$$

Therefore, $\lambda^2 - 3\lambda + 2 = 0$ is the characteristic equation.

$$(\lambda - 2)(\lambda - 1) = 0$$

Therefore, $\lambda = 2$ and $\lambda = 1$ are the eigenvalues.

$$\left(1\begin{bmatrix} 1 & 0 \\ 0 & 1 \end{bmatrix} - \begin{bmatrix} 3 & 1 \\ -2 & 0 \end{bmatrix}\right)\begin{bmatrix} a_1 \\ a_2 \end{bmatrix} = \begin{bmatrix} 0 \\ 0 \end{bmatrix}$$

$$\begin{bmatrix} -2 & -1 \\ 2 & 1 \end{bmatrix}\begin{bmatrix} a_1 \\ a_2 \end{bmatrix} = \begin{bmatrix} 0 \\ 0 \end{bmatrix}$$

$$-2a_1 - a_2 = 0$$

$$2a_1 + a_2 = 0$$

Hence, $a_2 = -2a_1$. Let $a_1 = t$. Therefore, $\begin{bmatrix} t \\ -2t \end{bmatrix} = t\begin{bmatrix} 1 \\ -2 \end{bmatrix}$. Thus $\begin{bmatrix} 1 \\ -2 \end{bmatrix}$ is a basis for the eigenvectors associated with the eigenvalue 1 of the matrix $\begin{bmatrix} 3 & 1 \\ -2 & 0 \end{bmatrix}$.

$$\left(2\begin{bmatrix}1 & 0\\ 0 & 1\end{bmatrix} - \begin{bmatrix}3 & 1\\ -2 & 0\end{bmatrix}\right)\begin{bmatrix}a_1\\ a_2\end{bmatrix} = \begin{bmatrix}0\\ 0\end{bmatrix}$$

$$\begin{bmatrix}-1 & -1\\ 2 & 2\end{bmatrix}\begin{bmatrix}a_1\\ a_2\end{bmatrix} = \begin{bmatrix}0\\ 0\end{bmatrix}$$

$$-a_1 - a_2 = 0$$

$$2a_1 + 2a_2 = 0$$

Hence, $a_2 = -a_1$. Let $a_1 = t$. Therefore, $\begin{bmatrix}t\\ -t\end{bmatrix} = t\begin{bmatrix}1\\ -1\end{bmatrix}$. Thus $\begin{bmatrix}1\\ -1\end{bmatrix}$ is a basis for the eigenvectors associated with the eigenvalue 2 of the matrix $\begin{bmatrix}3 & 1\\ -2 & 0\end{bmatrix}$.

(b)

$$\det(\lambda I - B) = \begin{vmatrix}\lambda - 1 & 0 & 0\\ -3 & \lambda - 2 & 0\\ -1 & -2 & \lambda - 3\end{vmatrix} = (\lambda - 1)(\lambda - 2)(\lambda - 3) = 0$$

is the characteristic equation. Therefore, the eigenvalues are 1, 2, and 3.

$$(1I - B)\mathbf{a} = \begin{bmatrix}1 - 1 & 0 & 0\\ -3 & 1 - 2 & 0\\ -1 & -2 & 1 - 3\end{bmatrix}\begin{bmatrix}a_1\\ a_2\\ a_3\end{bmatrix} = \begin{bmatrix}0\\ 0\\ 0\end{bmatrix}$$

$$\therefore \quad \begin{bmatrix}0 & 0 & 0\\ -3 & -1 & 0\\ -1 & -2 & -2\end{bmatrix}\begin{bmatrix}a_1\\ a_2\\ a_3\end{bmatrix} = \begin{bmatrix}0\\ 0\\ 0\end{bmatrix}$$

$$-3a_1 - a_2 = 0$$

$$-a_1 + a_2 - 2a_3 = 0$$

$$\therefore \quad a_2 = -3a_1$$

Hence, $-a_1 + 6a_1 - 2a_3 = 0$. Therefore $a_3 = 5/2a_1$. Let $a_1 = t$. Therefore,

$$\begin{bmatrix}t\\ -3t\\ \frac{5}{2}t\end{bmatrix} = t\begin{bmatrix}1\\ -3\\ \frac{5}{2}\end{bmatrix} = \frac{t}{2}\begin{bmatrix}2\\ -6\\ 5\end{bmatrix}$$

Thus $\begin{bmatrix}2\\ -6\\ 5\end{bmatrix}$ is a basis for the eigenvectors associated with the eigenvalue 1 of matrix B.

$$(2I - B)\mathbf{a} = \begin{bmatrix}2 - 1 & 0 & 0\\ -3 & 2 - 2 & 0\\ -1 & -2 & 2 - 3\end{bmatrix}\begin{bmatrix}a_1\\ a_2\\ a_3\end{bmatrix} = \begin{bmatrix}0\\ 0\\ 0\end{bmatrix}$$

$$\therefore \quad \begin{bmatrix}1 & 0 & 0\\ -3 & 0 & 0\\ -1 & -2 & -1\end{bmatrix}\begin{bmatrix}a_1\\ a_2\\ a_3\end{bmatrix} = \begin{bmatrix}0\\ 0\\ 0\end{bmatrix}$$

The solution is $\begin{bmatrix} 0 \\ t \\ -2t \end{bmatrix} = t\begin{bmatrix} 0 \\ 1 \\ -2 \end{bmatrix}$. Thus, $\begin{bmatrix} 0 \\ 1 \\ -2 \end{bmatrix}$ is a basis for the eigenvectors associated with the eigenvalue 2 of matrix B.

$$(3I - B)\mathbf{a} = \begin{bmatrix} 3-1 & 0 & 0 \\ -3 & 3-2 & 0 \\ -1 & -2 & 3-3 \end{bmatrix}\begin{bmatrix} a_1 \\ a_2 \\ a_3 \end{bmatrix} = \begin{bmatrix} 0 \\ 0 \\ 0 \end{bmatrix}$$

$$\therefore \quad \begin{bmatrix} 2 & 0 & 0 \\ -3 & 1 & 0 \\ -1 & -2 & 0 \end{bmatrix}\begin{bmatrix} a_1 \\ a_2 \\ a_3 \end{bmatrix} = \begin{bmatrix} 0 \\ 0 \\ 0 \end{bmatrix}$$

The solution is $\begin{bmatrix} 0 \\ 0 \\ t \end{bmatrix} = t\begin{bmatrix} 0 \\ 0 \\ 1 \end{bmatrix}$. Thus, $\begin{bmatrix} 0 \\ 0 \\ 1 \end{bmatrix}$ is a basis for the eigenvectors associated with the eigenvalue 3 of matrix B.

15-2. The Linear Transformation Eigenproblem

A. Definition of eigenvalue and eigenvector of a linear transformation

If $L: V \to V$ is a linear transformation of an n-dimensional vector space V into itself, then a nonzero vector $\mathbf{a}$ in V is an eigenvector of L corresponding to the eigenvalue λ provided $L(\mathbf{a}) = \lambda\mathbf{a}$ and λ is a real number.

EXAMPLE 15-4: If $L: R^3 \to R^3$ is the linear transformation defined by $L\left(\begin{bmatrix} a_1 \\ a_2 \\ a_3 \end{bmatrix}\right) = \begin{bmatrix} a_2 \\ a_1 \\ 0 \end{bmatrix}$, verify that bases for eigenvectors corresponding to the eigenvalues 1, -1, and 0 are $\begin{bmatrix} 1 \\ 1 \\ 0 \end{bmatrix}$, $\begin{bmatrix} 1 \\ -1 \\ 0 \end{bmatrix}$, $\begin{bmatrix} 0 \\ 0 \\ 1 \end{bmatrix}$, respectively.

Solution

$$L\left(\begin{bmatrix} a \\ a \\ 0 \end{bmatrix}\right) = \begin{bmatrix} a \\ a \\ 0 \end{bmatrix} = 1\begin{bmatrix} a \\ a \\ 0 \end{bmatrix} = 1(a)\begin{bmatrix} 1 \\ 1 \\ 0 \end{bmatrix}$$

Therefore, $\begin{bmatrix} 1 \\ 1 \\ 0 \end{bmatrix}$ is a basis for the eigenvectors corresponding to the eigenvalue 1.

$$L\left(\begin{bmatrix} a \\ -a \\ 0 \end{bmatrix}\right) = \begin{bmatrix} -a \\ a \\ 0 \end{bmatrix} = -\begin{bmatrix} a \\ -a \\ 0 \end{bmatrix} = (-1)(a)\begin{bmatrix} 1 \\ -1 \\ 0 \end{bmatrix}$$

Therefore, $\begin{bmatrix} 1 \\ -1 \\ 0 \end{bmatrix}$ is a basis for the eigenvectors corresponding to the eigenvalue -1.

$$L\left(\begin{bmatrix} 0 \\ 0 \\ a \end{bmatrix}\right) = \begin{bmatrix} 0 \\ 0 \\ 0 \end{bmatrix} = 0\begin{bmatrix} 0 \\ 0 \\ a \end{bmatrix} = 0(a)\begin{bmatrix} 0 \\ 0 \\ 1 \end{bmatrix}$$

Therefore, $\begin{bmatrix} 0 \\ 0 \\ 1 \end{bmatrix}$ is a basis for the eigenvectors corresponding to the eigenvalue 0.

B. Equivalency of matrix and linear transformation eigenproblems

If $L\colon V \to V$ is a linear transformation of an n-dimensional vector space V into itself and A is the matrix representation of L with respect to the ordered basis B, then $L(\mathbf{a}) = \lambda\mathbf{a}$ can be written as $A[\mathbf{a}]_B = \lambda[\mathbf{a}]_B$ where $[\mathbf{a}]_B$ is the coordinate vector with respect to the basis B of the eigenvector $\mathbf{a}$.

EXAMPLE 15-5: Use the basis

$$B = \left\{\begin{bmatrix} 1 \\ 0 \\ 1 \end{bmatrix}, \begin{bmatrix} 0 \\ 1 \\ 1 \end{bmatrix}, \begin{bmatrix} 1 \\ 2 \\ 0 \end{bmatrix}\right\}$$

for R^3 to find a matrix representation of the linear transformation in Example 15-4 and then find the eigenvalues and corresponding eigenvectors.

Solution

$$L\left(\begin{bmatrix} 1 \\ 0 \\ 1 \end{bmatrix}\right) = \begin{bmatrix} 0 \\ 1 \\ 0 \end{bmatrix} = c_1\begin{bmatrix} 1 \\ 0 \\ 1 \end{bmatrix} + c_2\begin{bmatrix} 0 \\ 1 \\ 1 \end{bmatrix} + c_3\begin{bmatrix} 1 \\ 2 \\ 0 \end{bmatrix}$$

Thus, $c_1 = -1/3$, $c_2 = 1/3$, and $c_3 = 1/3$. Therefore, $\begin{bmatrix} 0 \\ 1 \\ 0 \end{bmatrix}_B = \begin{bmatrix} -\frac{1}{3} \\ \frac{1}{3} \\ \frac{1}{3} \end{bmatrix}$.

$$L\left(\begin{bmatrix} 0 \\ 1 \\ 1 \end{bmatrix}\right) = \begin{bmatrix} 1 \\ 0 \\ 0 \end{bmatrix} = c_1\begin{bmatrix} 1 \\ 0 \\ 1 \end{bmatrix} + c_2\begin{bmatrix} 0 \\ 1 \\ 1 \end{bmatrix} + c_3\begin{bmatrix} 1 \\ 2 \\ 0 \end{bmatrix}$$

Thus, $c_1 = 2/3$, $c_2 = -2/3$, $c_3 = 1/3$. Therefore, $\begin{bmatrix} 1 \\ 0 \\ 0 \end{bmatrix}_B = \begin{bmatrix} \frac{2}{3} \\ -\frac{2}{3} \\ \frac{1}{3} \end{bmatrix}$.

$$L\left(\begin{bmatrix} 1 \\ 2 \\ 0 \end{bmatrix}\right) = \begin{bmatrix} 2 \\ 1 \\ 0 \end{bmatrix} = c_1\begin{bmatrix} 1 \\ 0 \\ 1 \end{bmatrix} + c_2\begin{bmatrix} 0 \\ 1 \\ 1 \end{bmatrix} + c_3\begin{bmatrix} 1 \\ 2 \\ 0 \end{bmatrix}$$

Thus, $c_1 = 1$, $c_2 = -1$, and $c_3 = 1$. Therefore,

$$\begin{bmatrix} 2 \\ 1 \\ 0 \end{bmatrix}_B = \begin{bmatrix} 1 \\ -1 \\ 1 \end{bmatrix}$$

and thus

$$A = \begin{bmatrix} -\frac{1}{3} & \frac{2}{3} & 1 \\ \frac{1}{3} & -\frac{2}{3} & -1 \\ \frac{1}{3} & \frac{1}{3} & 1 \end{bmatrix}.$$

We will now find the eigenvalues and eigenvectors of matrix A.

$$\det(\lambda I - A) = \begin{vmatrix} \lambda + \frac{1}{3} & -\frac{2}{3} & -1 \\ -\frac{1}{3} & \lambda + \frac{2}{3} & 1 \\ -\frac{1}{3} & -\frac{1}{3} & \lambda - 1 \end{vmatrix} = 0$$

This reduces to $\lambda^3 - \lambda = 0$ or $\lambda(\lambda - 1)(\lambda + 1) = 0$. Therefore, the eigenvalues are 0, 1, and -1.

Observe that these are the same as the eigenvalues for the linear transformation in Example 15-4. This is not mere happenstance. The eigenvalues remain the same regardless of the basis used in the matrix representation of the linear transformation.

$$(0I - A)\mathbf{a} = \begin{bmatrix} \frac{1}{3} & -\frac{2}{3} & -1 \\ -\frac{1}{3} & \frac{2}{3} & 1 \\ -\frac{1}{3} & -\frac{1}{3} & -1 \end{bmatrix} \begin{bmatrix} a_1 \\ a_2 \\ a_3 \end{bmatrix} = \begin{bmatrix} 0 \\ 0 \\ 0 \end{bmatrix}$$

$$\therefore \quad \begin{bmatrix} a_1 \\ a_2 \\ a_3 \end{bmatrix} = \begin{bmatrix} -t \\ -2t \\ t \end{bmatrix} = -t \begin{bmatrix} 1 \\ 2 \\ -1 \end{bmatrix}$$

where t is arbitrary. Therefore, $\begin{bmatrix} 1 \\ 2 \\ -1 \end{bmatrix}$ is the coordinate vector representation with respect to basis B of a basis for the eigenvectors corresponding to the eigenvalue 0 of L.

$$1 \begin{bmatrix} 1 \\ 0 \\ 1 \end{bmatrix} + 2 \begin{bmatrix} 0 \\ 1 \\ 1 \end{bmatrix} - 1 \begin{bmatrix} 1 \\ 2 \\ 0 \end{bmatrix} = \begin{bmatrix} 0 \\ 0 \\ 1 \end{bmatrix}$$

Therefore, $\begin{bmatrix} 0 \\ 0 \\ 1 \end{bmatrix}$ is a basis for the eigenvectors corresponding to the eigenvalue 0 of L.

$$(1I - A)\mathbf{a} = \begin{bmatrix} \frac{4}{3} & -\frac{2}{3} & -1 \\ -\frac{1}{3} & \frac{5}{3} & 1 \\ -\frac{1}{3} & -\frac{1}{3} & 0 \end{bmatrix} \begin{bmatrix} a_1 \\ a_2 \\ a_3 \end{bmatrix} = \begin{bmatrix} 0 \\ 0 \\ 0 \end{bmatrix}$$

$$\therefore \quad \begin{bmatrix} a_1 \\ a_2 \\ a_3 \end{bmatrix} = \begin{bmatrix} t \\ -t \\ 2t \end{bmatrix} = t \begin{bmatrix} 1 \\ -1 \\ 2 \end{bmatrix}$$

where t is arbitrary.

Therefore, $\begin{bmatrix} 1 \\ -1 \\ 2 \end{bmatrix}$ is the coordinate vector representation with respect to basis B of a basis for the eigenvectors corresponding to the eigenvalue 1 of L.

$$1 \begin{bmatrix} 1 \\ 0 \\ 1 \end{bmatrix} - 1 \begin{bmatrix} 0 \\ 1 \\ 1 \end{bmatrix} + 2 \begin{bmatrix} 1 \\ 2 \\ 0 \end{bmatrix} = \begin{bmatrix} 3 \\ 3 \\ 0 \end{bmatrix} = 3 \begin{bmatrix} 1 \\ 1 \\ 0 \end{bmatrix}$$

Therefore, $\begin{bmatrix} 1 \\ 1 \\ 0 \end{bmatrix}$ is a basis for the eigenvectors corresponding to the eigenvalue 1 of L.

$$(-1I - A)\mathbf{a} = \begin{bmatrix} -\frac{2}{3} & -\frac{2}{3} & -1 \\ -\frac{1}{3} & -\frac{1}{3} & 1 \\ -\frac{1}{3} & -\frac{1}{3} & -2 \end{bmatrix} \begin{bmatrix} a_1 \\ a_2 \\ a_3 \end{bmatrix} = \begin{bmatrix} 0 \\ 0 \\ 0 \end{bmatrix}$$

$$\therefore \quad \begin{bmatrix} a_1 \\ a_2 \\ a_3 \end{bmatrix} = \begin{bmatrix} t \\ -t \\ 0 \end{bmatrix} = t \begin{bmatrix} 1 \\ -1 \\ 0 \end{bmatrix}$$

where t is arbitrary. Therefore, $\begin{bmatrix} 1 \\ -1 \\ 0 \end{bmatrix}$ is the coordinate vector representation with respect to basis B of a basis for the eigenvectors corresponding to the eigenvalue -1 of L.

$$1\begin{bmatrix} 1 \\ 0 \\ 1 \end{bmatrix} - 1\begin{bmatrix} 0 \\ 1 \\ 1 \end{bmatrix} + 0\begin{bmatrix} 1 \\ 2 \\ 0 \end{bmatrix} = \begin{bmatrix} 1 \\ -1 \\ 0 \end{bmatrix}$$

Therefore, $\begin{bmatrix} 1 \\ -1 \\ 0 \end{bmatrix}$ is a basis for the eigenvectors corresponding to the eigenvalue -1 of L.

Observe that, after changing from coordinate vectors, the bases for the eigenvectors in this example are the same as in Example 15-4.

C. Eigenspace

If $L: V \to V$ is a linear transformation of an n-dimensional vector space V into itself and λ is an eigenvalue of L, then the subset of V consisting of the zero vector and all eigenvectors of L corresponding to λ is a subspace of V called an **eigenspace**. Since every such linear transformation has a matrix representation with respect to any basis B of V, then we often refer to the eigenspace of matrix A corresponding to eigenvalue λ.

EXAMPLE 15-6: Find bases for the eigenspaces of

$$A = \begin{bmatrix} 2 & 1 & 1 \\ 0 & 3 & 1 \\ 0 & 0 & 1 \end{bmatrix}$$

Solution

$$\det(\lambda I - A) = \begin{bmatrix} \lambda - 2 & -1 & -1 \\ 0 & \lambda - 3 & -1 \\ 0 & 0 & \lambda - 1 \end{bmatrix} = (\lambda - 2)(\lambda - 3)(\lambda - 1) = 0$$

Therefore, the eigenvalues are 1, 2, and 3.

$$(1I - A)\mathbf{a} = \begin{bmatrix} -1 & -1 & -1 \\ 0 & -2 & -1 \\ 0 & 0 & 0 \end{bmatrix} \begin{bmatrix} a_1 \\ a_2 \\ a_3 \end{bmatrix} = \begin{bmatrix} 0 \\ 0 \\ 0 \end{bmatrix}$$

$$\therefore \quad \begin{bmatrix} a_1 \\ a_2 \\ a_3 \end{bmatrix} = \begin{bmatrix} t \\ t \\ -2t \end{bmatrix} = t \begin{bmatrix} 1 \\ 1 \\ -2 \end{bmatrix}$$

where t is arbitrary. Therefore, $\begin{bmatrix} 1 \\ 1 \\ -2 \end{bmatrix}$ is a basis for the eigenspace corresponding to eigenvalue 1.

$$(2I - A)\mathbf{a} = \begin{bmatrix} 0 & -1 & -1 \\ 0 & -1 & -1 \\ 0 & 0 & 2 \end{bmatrix} \begin{bmatrix} a_1 \\ a_2 \\ a_3 \end{bmatrix} = \begin{bmatrix} 0 \\ 0 \\ 0 \end{bmatrix}$$

$$\therefore \quad \begin{bmatrix} a_1 \\ a_2 \\ a_3 \end{bmatrix} = \begin{bmatrix} t \\ 0 \\ 0 \end{bmatrix} = t \begin{bmatrix} 1 \\ 0 \\ 0 \end{bmatrix}$$

where t is arbitrary. Therefore, $\begin{bmatrix} 1 \\ 0 \\ 0 \end{bmatrix}$ is a basis for the eigenspace corresponding to eigenvalue 2.

$$(3I - A)\mathbf{a} = \begin{bmatrix} 1 & -1 & -1 \\ 0 & 0 & -1 \\ 0 & 0 & 2 \end{bmatrix} \begin{bmatrix} a_1 \\ a_2 \\ a_3 \end{bmatrix} = \begin{bmatrix} 0 \\ 0 \\ 0 \end{bmatrix}$$

$$\therefore \quad \begin{bmatrix} a_1 \\ a_2 \\ a_3 \end{bmatrix} = \begin{bmatrix} t \\ t \\ 0 \end{bmatrix} = t \begin{bmatrix} 1 \\ 1 \\ 0 \end{bmatrix}$$

where t is arbitrary. Therefore, $\begin{bmatrix} 1 \\ 1 \\ 0 \end{bmatrix}$ is a basis for the eigenspace corresponding to the eigenvalue 3.

15-3. Properties of Eigenvalues and Eigenvectors

A. Independence of eigenvectors

If $\lambda_1, \lambda_2, \ldots, \lambda_n$ are distinct eigenvalues of an $n \times n$ matrix A and $B_1, B_2, \ldots, B_n$ are bases of eigenvectors associated with $\lambda_1, \lambda_2, \ldots, \lambda_n$, respectively, then the union $B_1 \cup B_2 \cup \cdots \cup B_n$ is a linearly independent set.

EXAMPLE 15-7: Verify that the set of basis vectors for the eigenspaces of matrix A in Example 15-6 is linearly independent.

Solution: All that we need to do is show that the determinant of the matrix having the basis vectors as columns is nonzero.

$$\begin{vmatrix} 1 & 1 & 1 \\ 1 & 0 & 1 \\ -2 & 0 & 0 \end{vmatrix} = -2\begin{vmatrix} 1 & 1 \\ 0 & 1 \end{vmatrix} = -2(1) = -2 \neq 0$$

Therefore, the three column vectors are linearly independent.

B. Properties of eigenvalues

If A is an $n \times n$ matrix with eigenvalue λ, then

(a) λ can be zero only if A is singular;
(b) $c\lambda$ is an eigenvalue of cA where c is a nonzero real number;
(c) λ is an eigenvalue of A^T;
(d) λ^k is an eigenvalue of A^k where k is a positive integer;
(e) where $\lambda \neq 0$ and A is invertible, $1/\lambda$ is an eigenvalue of A^{-1}; and
(f) where A and B are similar matrices, λ is an eigenvalue of B.

EXAMPLE 15-8: If $A = \begin{bmatrix} 2 & 0 & 1 \\ 1 & 2 & 1 \\ -2 & 0 & -1 \end{bmatrix}$, verify that **(a)** the eigenvalues of A are 0, 1, and 2 and A is singular; **(b)** the eigenvalues of $3A$ are 0, 3, and 6; **(c)** the eigenvalues of A^T are 0, 1, and 2; and **(d)** the eigenvalues of A^2 are 0, 1, and 4.

Solution

(a)

$$\det(\lambda I - A) = \begin{vmatrix} \lambda - 2 & 0 & -1 \\ -1 & \lambda - 2 & -1 \\ 2 & 0 & \lambda + 1 \end{vmatrix} = \lambda^3 - 3\lambda^2 + 2\lambda$$

$$\therefore \quad \lambda(\lambda - 1)(\lambda - 2) = 0$$

Hence, the eigenvalues of A are 0, 1, and 2. Since the first and third rows of A are negatives of each other, $\det A = 0$. Therefore, A is singular.

(b)

$$3A = \begin{bmatrix} 6 & 0 & 3 \\ 3 & 6 & 3 \\ -6 & 0 & -3 \end{bmatrix}$$

$$\det(\lambda I - A) = \begin{vmatrix} \lambda - 6 & 0 & -3 \\ -3 & \lambda - 6 & -3 \\ 6 & 0 & \lambda + 3 \end{vmatrix} = \lambda(\lambda - 6)(\lambda - 3)$$

$$\therefore \quad \lambda(\lambda - 6)(\lambda - 3) = 0$$

Hence, the eigenvalues of $3A$ are 0, 3, and 6.

(c)

$$A^T = \begin{bmatrix} 2 & 1 & -2 \\ 0 & 2 & 0 \\ 1 & 1 & -1 \end{bmatrix}$$

$$\det(\lambda I - A^T) = \begin{vmatrix} \lambda - 2 & -1 & 2 \\ 0 & \lambda - 2 & 0 \\ -1 & -1 & \lambda + 1 \end{vmatrix} = \lambda(\lambda - 1)(\lambda - 2)$$

$$\therefore \quad \lambda(\lambda - 1)(\lambda - 2) = 0$$

Hence, the eigenvalues of A^T are 0, 1, and 2.

(d)
$$A^2 = \begin{bmatrix} 2 & 0 & 1 \\ 1 & 2 & 1 \\ -2 & 0 & -1 \end{bmatrix}\begin{bmatrix} 2 & 0 & 1 \\ 1 & 2 & 1 \\ -2 & 0 & -1 \end{bmatrix} = \begin{bmatrix} 2 & 0 & 1 \\ 2 & 4 & 2 \\ -2 & 0 & -1 \end{bmatrix}$$

$$\det(\lambda I - A^2) = \begin{vmatrix} \lambda - 2 & 0 & -1 \\ -2 & \lambda - 4 & -2 \\ 2 & 0 & \lambda + 1 \end{vmatrix} = \lambda(\lambda - 1)(\lambda - 4)$$

$$\therefore \quad \lambda(\lambda - 1)(\lambda - 4) = 0$$

Hence, the eigenvalues of A^2 are 0, 1, and 4.

EXAMPLE 15-9: If $A = \begin{bmatrix} 1 & 0 & 0 \\ 3 & 2 & 0 \\ 1 & 2 & 3 \end{bmatrix}$ with eigenvalues 1, 2, and 3, find A^{-1} and verify Property **(e)**, Section 15-3.B which states that if $\lambda \neq 0$ and A is invertible, then $1/\lambda$ is an eigenvalue of A^{-1}.

Solution

$$\left[\begin{array}{ccc|ccc} 1 & 0 & 0 & 1 & 0 & 0 \\ 3 & 2 & 0 & 0 & 1 & 0 \\ 1 & 2 & 3 & 0 & 0 & 1 \end{array}\right] \xrightarrow[R_3 \to R_3 - R_1]{R_2 \to R_2 - 3R_1} \left[\begin{array}{ccc|ccc} 1 & 0 & 0 & 1 & 0 & 0 \\ 0 & 2 & 0 & -3 & 1 & 0 \\ 0 & 2 & 3 & -1 & 0 & 1 \end{array}\right] \xrightarrow{R_3 \to R_3 - R_2} \left[\begin{array}{ccc|ccc} 1 & 0 & 0 & 1 & 0 & 0 \\ 0 & 2 & 0 & -3 & 1 & 0 \\ 0 & 0 & 3 & 2 & -1 & 1 \end{array}\right]$$

$$\xrightarrow[R_3 \to 1/3 R_3]{R_2 \to 1/2 R_2} \left[\begin{array}{ccc|ccc} 1 & 0 & 0 & 1 & 0 & 0 \\ 0 & 1 & 0 & -\frac{3}{2} & \frac{1}{2} & 0 \\ 0 & 0 & 1 & \frac{2}{3} & -\frac{1}{3} & \frac{1}{3} \end{array}\right]$$

$$\therefore \quad A^{-1} = \begin{bmatrix} 1 & 0 & 0 \\ -\frac{3}{2} & \frac{1}{2} & 0 \\ \frac{2}{3} & -\frac{1}{3} & \frac{1}{3} \end{bmatrix}$$

$$\det(\lambda I - A^{-1}) = \begin{vmatrix} \lambda - 1 & 0 & 0 \\ \frac{3}{2} & \lambda - \frac{1}{2} & 0 \\ -\frac{2}{3} & \frac{1}{3} & \lambda - \frac{1}{3} \end{vmatrix} = (\lambda - 1)(\lambda - \tfrac{1}{2})(\lambda - \tfrac{1}{3})$$

$$\therefore \quad (\lambda - 1)(\lambda - \tfrac{1}{2})(\lambda - \tfrac{1}{3}) = 0$$

Hence, the eigenvalues of A^{-1} are 1, 1/2, and 1/3.

EXAMPLE 15-10: If

$$A = \begin{bmatrix} 2 & 0 & 1 \\ 1 & 2 & 1 \\ -2 & 0 & -1 \end{bmatrix}, \qquad P = \begin{bmatrix} 1 & 0 & 0 \\ 0 & 1 & 0 \\ -1 & 0 & 1 \end{bmatrix}, \qquad P^{-1} = \begin{bmatrix} 1 & 0 & 0 \\ 0 & 1 & 0 \\ 1 & 0 & 1 \end{bmatrix}$$

and the eigenvalues of A are 0, 1, and 2, find a matrix B which is similar to A and verify that the eigenvalues of A and B are identical.

Solution

$$B = P^{-1}AP = \begin{bmatrix} 1 & 0 & 0 \\ 0 & 1 & 0 \\ 1 & 0 & 1 \end{bmatrix}\begin{bmatrix} 2 & 0 & 1 \\ 1 & 2 & 1 \\ -2 & 0 & -1 \end{bmatrix}\begin{bmatrix} 1 & 0 & 0 \\ 0 & 1 & 0 \\ -1 & 0 & 1 \end{bmatrix} = \begin{bmatrix} 1 & 0 & 1 \\ 0 & 2 & 1 \\ 0 & 0 & 0 \end{bmatrix}$$

$$\det(\lambda I - B) = \begin{vmatrix} \lambda - 1 & 0 & -1 \\ 0 & \lambda - 2 & -1 \\ 0 & 0 & \lambda \end{vmatrix} = \lambda(\lambda - 1)(\lambda - 2)$$

Therefore, $\lambda(\lambda - 1)(\lambda - 2) = 0$ gives eigenvalues of B as 0, 1, and 2. These are identical to the eigenvalues of A.

C. Properties of eigenvectors

If A and B are $n \times n$ matrices with $\mathbf{a}$ an eigenvector of both A and B associated with the eigenvalues λ_1 of A and λ_2 of B, then

(a) $c\mathbf{a}$ is an eigenvector of both A and B when c is a nonzero real number;
(b) $A + B$ has eigenvector $\mathbf{a}$ with its associated eigenvalue being $\lambda_1 + \lambda_2$; and
(c) AB has eigenvector $\mathbf{a}$ with its associated eigenvalue being $\lambda_1 \lambda_2$.

EXAMPLE 15-11: The eigenvalues of $A = \begin{bmatrix} 2 & 1 & 3 \\ 0 & 3 & -1 \\ 0 & 0 & 5 \end{bmatrix}$ are 2, 3, and 5 and the corresponding basis eigenvectors are $\begin{bmatrix} 1 \\ 0 \\ 0 \end{bmatrix}$, $\begin{bmatrix} 1 \\ 1 \\ 0 \end{bmatrix}$, $\begin{bmatrix} 5 \\ -3 \\ 6 \end{bmatrix}$, respectively. The eigenvalues of $B = \begin{bmatrix} 4 & 1 & 3 \\ 0 & 5 & -1 \\ 0 & 0 & 7 \end{bmatrix}$ are 4, 5, and 7 and the corresponding basis eigenvectors are $\begin{bmatrix} 1 \\ 0 \\ 0 \end{bmatrix}$, $\begin{bmatrix} 1 \\ 1 \\ 0 \end{bmatrix}$, $\begin{bmatrix} 5 \\ -3 \\ 6 \end{bmatrix}$, respectively. **(a)** Verify that $c\begin{bmatrix} 5 \\ -3 \\ 6 \end{bmatrix}$ is also an eigenvector of A. **(b)** Verify Property **(b)** (Section 15-3.C) of eigenvectors by finding the eigenvalues and corresponding eigenvectors of $A + B$. **(c)** Verify Property **(c)** (Section 15-3.C) of eigenvectors by finding the eigenvalues and corresponding eigenvectors of AB.

Solution

(a) We must show that $A(c\mathbf{a}) = 5(c\mathbf{a})$ where $\mathbf{a} = \begin{bmatrix} 5 \\ -3 \\ 6 \end{bmatrix}$.

$$A(c\mathbf{a}) = \begin{bmatrix} 2 & 1 & 3 \\ 0 & 3 & -1 \\ 0 & 0 & 5 \end{bmatrix}\begin{bmatrix} 5c \\ -3c \\ 6c \end{bmatrix} = \begin{bmatrix} 25c \\ -15c \\ 30c \end{bmatrix} = 5\begin{bmatrix} 5c \\ -3c \\ 6c \end{bmatrix} = 5(c\mathbf{a})$$

Therefore, $c\begin{bmatrix} 5 \\ -3 \\ 6 \end{bmatrix}$ is an eigenvector of A.

(b)

$$A + B = \begin{bmatrix} 2 & 1 & 3 \\ 0 & 3 & -1 \\ 0 & 0 & 5 \end{bmatrix} + \begin{bmatrix} 4 & 1 & 3 \\ 0 & 5 & -1 \\ 0 & 0 & 7 \end{bmatrix} = \begin{bmatrix} 6 & 2 & 6 \\ 0 & 8 & -2 \\ 0 & 0 & 12 \end{bmatrix}$$

$$\det[\lambda I - (A + B)] = \begin{vmatrix} \lambda - 6 & -2 & -6 \\ 0 & \lambda - 8 & 2 \\ 0 & 0 & \lambda - 12 \end{vmatrix}$$

$$\therefore \quad (\lambda - 6)(\lambda - 8)(\lambda - 12) = 0$$

Hence, the eigenvalues of $A + B$ are 6, 8, and 12.

$$[6I - (A + B)]\mathbf{a} = \begin{bmatrix} 0 & -2 & -6 \\ 0 & -2 & 2 \\ 0 & 0 & -6 \end{bmatrix} \begin{bmatrix} a_1 \\ a_2 \\ a_3 \end{bmatrix} = \begin{bmatrix} 0 \\ 0 \\ 0 \end{bmatrix}$$

$$\therefore \quad \begin{bmatrix} a_1 \\ a_2 \\ a_3 \end{bmatrix} = \begin{bmatrix} t \\ 0 \\ 0 \end{bmatrix} = t \begin{bmatrix} 1 \\ 0 \\ 0 \end{bmatrix}$$

Therefore, $\begin{bmatrix} 1 \\ 0 \\ 0 \end{bmatrix}$ is a basis eigenvector corresponding to the eigenvalue 6 which is the sum of the eigenvalues associated with $\begin{bmatrix} 1 \\ 0 \\ 0 \end{bmatrix}$ in A and B: $(6 = 2 + 4)$.

$$[8I - (A + B)]\mathbf{a} = \begin{bmatrix} 2 & -2 & -6 \\ 0 & 0 & 2 \\ 0 & 0 & -4 \end{bmatrix} \begin{bmatrix} a_1 \\ a_2 \\ a_3 \end{bmatrix} = \begin{bmatrix} 0 \\ 0 \\ 0 \end{bmatrix}$$

$$\therefore \quad \begin{bmatrix} a_1 \\ a_2 \\ a_3 \end{bmatrix} = \begin{bmatrix} t \\ t \\ 0 \end{bmatrix} = \begin{bmatrix} 1 \\ 1 \\ 0 \end{bmatrix}$$

Therefore, $\begin{bmatrix} 1 \\ 1 \\ 0 \end{bmatrix}$ is a basis eigenvector corresponding to the eigenvalue 8 which is the sum of the eigenvalues associated with $\begin{bmatrix} 1 \\ 1 \\ 0 \end{bmatrix}$ in A and B: $(8 = 3 + 5)$.

$$[12I - (A + B)]\mathbf{a} = \begin{bmatrix} 6 & -2 & -6 \\ 0 & 4 & 2 \\ 0 & 0 & 0 \end{bmatrix} \begin{bmatrix} a_1 \\ a_2 \\ a_3 \end{bmatrix} = \begin{bmatrix} 0 \\ 0 \\ 0 \end{bmatrix}$$

$$\therefore \quad \begin{bmatrix} a_1 \\ a_2 \\ a_3 \end{bmatrix} = \begin{bmatrix} 5t \\ -3t \\ 6t \end{bmatrix} = t \begin{bmatrix} 5 \\ -3 \\ 6 \end{bmatrix}$$

Therefore, $\begin{bmatrix} 5 \\ -3 \\ 6 \end{bmatrix}$ is a basis eigenvector corresponding to the eigenvalue 12 which is the sum of the eigenvalues associated with $\begin{bmatrix} 5 \\ -3 \\ 6 \end{bmatrix}$ in A and B: $(12 = 5 + 7)$. Therefore, Property **(b)** (Section 15-3.C) holds for the given matrices A and B.

(c)

$$AB = \begin{bmatrix} 2 & 1 & 3 \\ 0 & 3 & -1 \\ 0 & 0 & 5 \end{bmatrix} \begin{bmatrix} 4 & 1 & 3 \\ 0 & 5 & -1 \\ 0 & 0 & 7 \end{bmatrix} = \begin{bmatrix} 8 & 7 & 26 \\ 0 & 15 & -10 \\ 0 & 0 & 35 \end{bmatrix}$$

$$\det(\lambda I - AB) = \begin{vmatrix} \lambda - 8 & -7 & -26 \\ 0 & \lambda - 15 & 10 \\ 0 & 0 & \lambda - 35 \end{vmatrix}$$

$$\therefore \quad (\lambda - 8)(\lambda - 15)(\lambda - 35) = 0$$

Hence, the eigenvalues of AB are 8, 15, and 35.

$$[8I - AB]\mathbf{a} = \begin{bmatrix} 0 & -7 & -26 \\ 0 & -7 & 10 \\ 0 & 0 & -27 \end{bmatrix} \begin{bmatrix} a_1 \\ a_2 \\ a_3 \end{bmatrix} = \begin{bmatrix} 0 \\ 0 \\ 0 \end{bmatrix}$$

$$\therefore \quad \begin{bmatrix} a_1 \\ a_2 \\ a_3 \end{bmatrix} = \begin{bmatrix} t \\ 0 \\ 0 \end{bmatrix} = t \begin{bmatrix} 1 \\ 0 \\ 0 \end{bmatrix}$$

Therefore, $\begin{bmatrix} 1 \\ 0 \\ 0 \end{bmatrix}$ is a basis eigenvector corresponding to the eigenvalue 8 which is the product of the eigenvalues associated with $\begin{bmatrix} 1 \\ 0 \\ 0 \end{bmatrix}$ in A and B: $(8 = 2 \cdot 4)$.

$$[15I - AB]\mathbf{a} = \begin{bmatrix} 7 & -7 & -26 \\ 0 & 0 & 0 \\ 0 & 0 & -20 \end{bmatrix} \begin{bmatrix} a_1 \\ a_2 \\ a_3 \end{bmatrix} = \begin{bmatrix} 0 \\ 0 \\ 0 \end{bmatrix}$$

$$\therefore \quad \begin{bmatrix} a_1 \\ a_2 \\ a_3 \end{bmatrix} = \begin{bmatrix} t \\ t \\ 0 \end{bmatrix} = t \begin{bmatrix} 1 \\ 1 \\ 0 \end{bmatrix}$$

Therefore, $\begin{bmatrix} 1 \\ 1 \\ 0 \end{bmatrix}$ is a basis eigenvector corresponding to the eigenvalue 15 which is the product of the eigenvalues associated with $\begin{bmatrix} 1 \\ 1 \\ 0 \end{bmatrix}$ in A and B: $(15 = 3 \cdot 5)$.

$$[35I - AB]\mathbf{a} = \begin{bmatrix} 27 & -7 & -26 \\ 0 & 20 & 10 \\ 0 & 0 & 0 \end{bmatrix} \begin{bmatrix} a_1 \\ a_2 \\ a_3 \end{bmatrix} = \begin{bmatrix} 0 \\ 0 \\ 0 \end{bmatrix}$$

$$\therefore \quad \begin{bmatrix} a_1 \\ a_2 \\ a_3 \end{bmatrix} = \begin{bmatrix} -\frac{5}{3}t \\ t \\ -2t \end{bmatrix} = -\frac{t}{3} \begin{bmatrix} 5 \\ -3 \\ 6 \end{bmatrix}$$

Therefore, $\begin{bmatrix} 5 \\ -3 \\ 6 \end{bmatrix}$ is a basis eigenvector corresponding to the eigenvalue 35 which is the product of the eigenvalues associated with $\begin{bmatrix} 5 \\ -3 \\ 6 \end{bmatrix}$ in A and B: ($35 = 5 \cdot 7$). Therefore, Property (c) (Section 15-3.C) holds for the given matrices A and B.

SUMMARY

1. If A is an $n \times n$ matrix, then a nonzero vector $\mathbf{a}$ in R^n is called an eigenvector corresponding to the eigenvalue λ provided $A\mathbf{a} = \lambda\mathbf{a}$ and λ is a real number.
2. Eigenvalues are also called characteristic values, proper values, or latent roots.
3. If A is an $n \times n$ matrix, then $\det(\lambda I - A)$ is called the characteristic determinant and, when expanded, the characteristic polynomial of A.
4. $\text{Det}(\lambda I - A) = 0$ is called the characteristic equation of A.
5. The real solutions of the characteristic equation are the eigenvalues of matrix A.
6. A basis of the eigenvectors corresponding to an eigenvalue λ can be found by solving the homogeneous equations $(\lambda I - A)\mathbf{a} = \mathbf{0}$.
7. If $L: V \to V$ is a linear transformation of an n-dimensional vector space V into itself, then a nonzero vector $\mathbf{a}$ is an eigenvector of L corresponding to the eigenvalue λ provided $L(\mathbf{a}) = \lambda\mathbf{a}$ and λ is a real number.
8. If $L: V \to V$ is a linear transformation of an n-dimensional vector space V into itself and A is the matrix representation of L with respect to the ordered basis B, then $L(\mathbf{a}) = \lambda\mathbf{a}$ can be written as $A[\mathbf{a}]_B = \lambda[\mathbf{a}]_B$ where $[\mathbf{a}]_B$ is the coordinate vector with respect to the basis B of the eigenvector $\mathbf{a}$.
9. The eigenvalues of a linear transformation remain the same regardless of the basis used in the matrix representation of the transformation.
10. If $L: V \to V$ is a linear transformation of an n-dimensional vector space V into itself and λ is an eigenvalue of L, then the subset of V consisting of the zero vector and all eigenvectors of L corresponding to λ is a subspace of V called an eigenspace.
11. If $\lambda_1, \lambda_2, \ldots, \lambda_n$ are distinct eigenvalues of an $n \times n$ matrix A and $B_1, B_2, \ldots, B_n$ are bases of eigenvectors associated with $\lambda_1, \lambda_2, \ldots, \lambda_n$, respectively, then the union $B_1 \cup B_2 \cup \cdots \cup B_n$ is a linearly independent set.
12. If A is an $n \times n$ matrix with eigenvalue λ, then
    - λ can be zero only if A is singular;
    - $c\lambda$ is an eigenvalue of cA where c is a nonzero real number;
    - λ is an eigenvalue of A^T;
    - λ^k is an eigenvalue of A^k where k is a positive integer;
    - where $\lambda \neq 0$ and A is invertible, $1/\lambda$ is an eigenvalue of A^{-1}; and
    - where A and B are similar matrices, λ is an eigenvalue of B.
13. If A and B are $n \times n$ matrices with $\mathbf{a}$ an eigenvector of both A and B associated with eigenvalues λ_1 of A and λ_2 of B, then
    - $c\mathbf{a}$ is an eigenvector of A where c is a nonzero real number;
    - $A + B$ has eigenvector $\mathbf{a}$ with its associated eigenvalue being $\lambda_1 + \lambda_2$; and
    - AB has eigenvector $\mathbf{a}$ with its associated eigenvalue being $\lambda_1\lambda_2$.

RAISE YOUR GRADES

Can you ...?

☑ determine if a given vector is an eigenvector of a matrix
☑ find the characteristic polynomial and equation of a matrix
☑ find the eigenvalues of a matrix
☑ find a basis for the eigenvectors corresponding to a given eigenvalue of a matrix
☑ change a linear transformation eigenproblem to its equivalent matrix eigenproblem

☑ use the properties of eigenvalues to find eigenvalues of cA, A^T, A^k, A^{-1}, and any matrix similar to A when the eigenvalues of A are known

☑ use the properties of eigenvectors to find the eigenvectors and eigenvalues of $A + B$ and AB when the eigenvectors of A and B are equal

SOLVED PROBLEMS

PROBLEM 15-1 Verify that $\begin{bmatrix} 1 \\ 1 \end{bmatrix}$ and $\begin{bmatrix} -2 \\ 3 \end{bmatrix}$ are eigenvectors of $A = \begin{bmatrix} 1 & -2 \\ -3 & 2 \end{bmatrix}$.

Solution

$$\begin{bmatrix} 1 & -2 \\ -3 & 2 \end{bmatrix}\begin{bmatrix} 1 \\ 1 \end{bmatrix} = \begin{bmatrix} -1 \\ -1 \end{bmatrix} = -1\begin{bmatrix} 1 \\ 1 \end{bmatrix}$$

Therefore, $\begin{bmatrix} 1 \\ 1 \end{bmatrix}$ is an eigenvector corresponding to eigenvalue -1.

$$\begin{bmatrix} 1 & -2 \\ -3 & 2 \end{bmatrix}\begin{bmatrix} -2 \\ 3 \end{bmatrix} = \begin{bmatrix} -8 \\ 12 \end{bmatrix} = 4\begin{bmatrix} -2 \\ 3 \end{bmatrix}$$

Therefore, $\begin{bmatrix} -2 \\ 3 \end{bmatrix}$ is an eigenvector corresponding to eigenvalue 4.

PROBLEM 15-2 If $A = \begin{bmatrix} 1 & 0 & -1 \\ 2 & 0 & 1 \\ 3 & 1 & 2 \end{bmatrix}$, find **(a)** the characteristic determinant, **(b)** the characteristic polynomial, and **(c)** the characteristic equation.

Solution

(a)
$$\det(\lambda I - A) = \left| \begin{bmatrix} \lambda & 0 & 0 \\ 0 & \lambda & 0 \\ 0 & 0 & \lambda \end{bmatrix} - \begin{bmatrix} 1 & 0 & -1 \\ 2 & 0 & 1 \\ 3 & 1 & 2 \end{bmatrix} \right|$$
$$= \begin{vmatrix} \lambda - 1 & 0 & 1 \\ -2 & \lambda & -1 \\ -3 & -1 & \lambda - 2 \end{vmatrix}$$

which is the characteristic determinant.

(b)
$$\begin{vmatrix} \lambda - 1 & 0 & 1 \\ -2 & \lambda & -1 \\ -3 & -1 & \lambda - 2 \end{vmatrix} = (\lambda - 1)(\lambda)(\lambda - 2) + 2 + 3\lambda - (\lambda - 1)$$
$$= \lambda^3 - 3\lambda^2 + 4\lambda + 3$$

which is the characteristic polynomial.

(c) $\lambda^3 - 3\lambda^2 + 4\lambda + 3 = 0$ is the characteristic equation.

PROBLEM 15-3 Find the characteristic equation and the eigenvalues of each matrix.

(a) $A = \begin{bmatrix} 1 & -2 \\ -3 & 2 \end{bmatrix}$ **(b)** $B = \begin{bmatrix} 2 & 0 & 0 \\ 1 & 3 & 0 \\ 3 & -1 & 4 \end{bmatrix}$

Solution

(a)

$$\det(\lambda I - A) = \begin{bmatrix} \lambda - 1 & 2 \\ 3 & \lambda - 2 \end{bmatrix} = (\lambda - 1)(\lambda - 2) - 6 = 0$$

$$\therefore \quad \lambda^2 - 3\lambda - 4 = 0$$

$$(\lambda + 1)(\lambda - 4) = 0$$

Therefore, $\lambda = -1$ and $\lambda = 4$ are eigenvalues.

(b)

$$\det(\lambda I - B) = \begin{vmatrix} \lambda - 2 & 0 & 0 \\ -1 & \lambda - 3 & 0 \\ -3 & 1 & \lambda - 4 \end{vmatrix} = (\lambda - 2)(\lambda - 3)(\lambda - 4) = 0$$

$$\therefore \quad (\lambda - 2)(\lambda - 3)(\lambda - 4) = 0$$

Therefore, $\lambda = 2$, $\lambda = 3$, and $\lambda = 4$ are eigenvalues.

PROBLEM 15-4 Find the eigenvectors associated with the eigenvalues in Problem 15-3.

Solution

(a) For $\lambda = -1$,

$$\left(-1\begin{bmatrix} 1 & 0 \\ 0 & 1 \end{bmatrix} - \begin{bmatrix} 1 & -2 \\ -3 & 2 \end{bmatrix}\right)\begin{bmatrix} a_1 \\ a_2 \end{bmatrix} = \begin{bmatrix} 0 \\ 0 \end{bmatrix}$$

$$\begin{bmatrix} -2 & 2 \\ 3 & -3 \end{bmatrix}\begin{bmatrix} a_1 \\ a_2 \end{bmatrix} = \begin{bmatrix} 0 \\ 0 \end{bmatrix}$$

$$-2a_1 + 2a_2 = 0$$

$$3a_1 - 3a_2 = 0$$

Hence, $a_1 = a_2$. Let $a_2 = t$. Therefore,

$$\begin{bmatrix} t \\ t \end{bmatrix} = t\begin{bmatrix} 1 \\ 1 \end{bmatrix}$$

Thus, $\begin{bmatrix} 1 \\ 1 \end{bmatrix}$ is a basis for the eigenvectors associated with eigenvalue -1 of matrix $\begin{bmatrix} 1 & -2 \\ -3 & 2 \end{bmatrix}$.

For $\lambda = 4$,

$$\left(4\begin{bmatrix} 1 & 0 \\ 0 & 1 \end{bmatrix} - \begin{bmatrix} 1 & -2 \\ -3 & 2 \end{bmatrix}\right)\begin{bmatrix} a_1 \\ a_2 \end{bmatrix} = \begin{bmatrix} 0 \\ 0 \end{bmatrix}$$

$$\begin{bmatrix} 3 & 2 \\ 3 & 2 \end{bmatrix}\begin{bmatrix} a_1 \\ a_2 \end{bmatrix} = \begin{bmatrix} 0 \\ 0 \end{bmatrix}$$

$$3a_1 + 2a_2 = 0$$

$$3a_1 + 2a_2 = 0$$

Hence, $a_1 = -(2/3)a_2$. Let $a_2 = t$. Then

$$\begin{bmatrix} -\frac{2}{3}t \\ t \end{bmatrix} = \frac{t}{3}\begin{bmatrix} -2 \\ 3 \end{bmatrix}$$

Thus, $\begin{bmatrix} -2 \\ 3 \end{bmatrix}$ is a basis for the eigenvectors associated with eigenvalue 4 of $\begin{bmatrix} 1 & -2 \\ -3 & 2 \end{bmatrix}$.

(b) For $\lambda = 2$,

$$\left(2\begin{bmatrix} 1 & 0 & 0 \\ 0 & 1 & 0 \\ 0 & 0 & 1 \end{bmatrix} - \begin{bmatrix} 2 & 0 & 0 \\ 1 & 3 & 0 \\ 3 & -1 & 4 \end{bmatrix}\right)\begin{bmatrix} a_1 \\ a_2 \\ a_3 \end{bmatrix} = \begin{bmatrix} 0 \\ 0 \\ 0 \end{bmatrix}$$

$$\begin{bmatrix} 0 & 0 & 0 \\ -1 & -1 & 0 \\ -3 & 1 & -2 \end{bmatrix}\begin{bmatrix} a_1 \\ a_2 \\ a_3 \end{bmatrix} = \begin{bmatrix} 0 \\ 0 \\ 0 \end{bmatrix}$$

$$\begin{aligned} -a_1 - a_2 \qquad &= 0 \\ -3a_1 + a_2 - 2a_3 &= 0 \end{aligned}$$

Therefore, $a_1 = -a_2$ and $a_3 = 2a_2$. Let $a_2 = t$. Thus,

$$\begin{bmatrix} -t \\ t \\ 2t \end{bmatrix} = t\begin{bmatrix} -1 \\ 1 \\ 2 \end{bmatrix}$$

Therefore, $\begin{bmatrix} -1 \\ 1 \\ 2 \end{bmatrix}$ is a basis for eigenvectors associated with eigenvalue 2 of matrix $\begin{bmatrix} 2 & 0 & 0 \\ 1 & 3 & 0 \\ 3 & -1 & 4 \end{bmatrix}$.

For $\lambda = 3$,

$$\left(3\begin{bmatrix} 1 & 0 & 0 \\ 0 & 1 & 0 \\ 0 & 0 & 1 \end{bmatrix} - \begin{bmatrix} 2 & 0 & 0 \\ 1 & 3 & 0 \\ 3 & -1 & 4 \end{bmatrix}\right)\begin{bmatrix} a_1 \\ a_2 \\ a_3 \end{bmatrix} = \begin{bmatrix} 0 \\ 0 \\ 0 \end{bmatrix}$$

$$\begin{bmatrix} 1 & 0 & 0 \\ -1 & 0 & 0 \\ -3 & 1 & -1 \end{bmatrix}\begin{bmatrix} a_1 \\ a_2 \\ a_3 \end{bmatrix} = \begin{bmatrix} 0 \\ 0 \\ 0 \end{bmatrix}$$

$$\begin{aligned} a_1 \qquad\qquad &= 0 \\ a_1 \qquad\qquad &\ 0 \\ -3a_1 + a_2 - a_3 &= 0 \end{aligned}$$

Therefore, $a_1 = 0$ and $a_2 = a_3$. Let $a_3 = t$. Thus,

$$\begin{bmatrix} 0 \\ t \\ t \end{bmatrix} = t\begin{bmatrix} 0 \\ 1 \\ 1 \end{bmatrix}$$

Therefore, $\begin{bmatrix} 0 \\ 1 \\ 1 \end{bmatrix}$ is a basis for eigenvectors associated with eigenvalue 3 of matrix B.

For $\lambda = 4$,

$$\left(4\begin{bmatrix}1 & 0 & 0\\ 0 & 1 & 0\\ 0 & 0 & 1\end{bmatrix} - \begin{bmatrix}2 & 0 & 0\\ 1 & 3 & 0\\ 3 & -1 & 4\end{bmatrix}\right)\begin{bmatrix}a_1\\ a_2\\ a_3\end{bmatrix} = \begin{bmatrix}0\\ 0\\ 0\end{bmatrix}$$

$$\begin{bmatrix}2 & 0 & 0\\ -1 & 1 & 0\\ -3 & 1 & 0\end{bmatrix}\begin{bmatrix}a_1\\ a_2\\ a_3\end{bmatrix} = \begin{bmatrix}0\\ 0\\ 0\end{bmatrix}$$

$$\begin{aligned} 2a_1 \qquad &= 0 \\ -a_1 + a_2 &= 0 \\ -3a_1 + a_2 &= 0 \end{aligned}$$

The solution to this system is

$$\begin{bmatrix}0\\ 0\\ t\end{bmatrix} = t\begin{bmatrix}0\\ 0\\ 1\end{bmatrix}$$

Therefore, $\begin{bmatrix}0\\ 0\\ 1\end{bmatrix}$ is a basis for the eigenvectors associated with eigenvalue 4 of matrix B.

PROBLEM 15-5 Find the eigenvalues and their corresponding eigenvectors for

$$A = \begin{bmatrix}1 & 2 & 2\\ 0 & 4 & 3\\ 0 & -3 & -2\end{bmatrix}$$

Solution

$$\det(\lambda I - A) = \begin{vmatrix}\lambda - 1 & -2 & -2\\ 0 & \lambda - 4 & -3\\ 0 & 3 & \lambda + 2\end{vmatrix} = (\lambda - 1)(\lambda - 4)(\lambda + 2) + 9(\lambda - 1)$$

$$\therefore \quad (\lambda - 1)[(\lambda - 4)(\lambda + 2) + 9] = 0$$

$$(\lambda - 1)(\lambda^2 - 2\lambda - 8 + 9) = 0$$

$$(\lambda - 1)(\lambda^2 - 2\lambda + 1) = 0$$

$$(\lambda - 1)^3 = 0$$

Thus, $\lambda = 1$ is an eigenvalue of multiplicity 3.

For $\lambda = 1$,

$$\left(1\begin{bmatrix}0 & 1 & 0\\ 0 & 1 & 0\\ 0 & 0 & 1\end{bmatrix} - \begin{bmatrix}1 & 2 & 2\\ 0 & 4 & 3\\ 0 & -3 & -2\end{bmatrix}\right)\begin{bmatrix}a_1\\ a_2\\ a_3\end{bmatrix} = \begin{bmatrix}0\\ 0\\ 0\end{bmatrix}$$

$$\begin{bmatrix}0 & -2 & -2\\ 0 & -3 & -3\\ 0 & 3 & 3\end{bmatrix}\begin{bmatrix}a_1\\ a_2\\ a_3\end{bmatrix} = \begin{bmatrix}0\\ 0\\ 0\end{bmatrix}$$

$$-2a_2 - 2a_3 = 0$$
$$-3a_2 - 3a_3 = 0$$
$$3a_2 + 3a_3 = 0$$

This gives $a_2 = a_3$, and a_1 is arbitrary. Let $a_1 = s$ and $a_3 = t$. Then

$$\begin{bmatrix} s \\ -t \\ t \end{bmatrix} = s\begin{bmatrix} 1 \\ 0 \\ 0 \end{bmatrix} + t\begin{bmatrix} 0 \\ -1 \\ 1 \end{bmatrix}$$

Thus $\left\{\begin{bmatrix} 1 \\ 0 \\ 0 \end{bmatrix}, \begin{bmatrix} 0 \\ -1 \\ 1 \end{bmatrix}\right\}$ is a basis for the eigenvectors associated with $\lambda = 1$.

PROBLEM 15-6 If $L: R^3 \to R^3$ is the linear transformation defined by $L\left(\begin{bmatrix} a_1 \\ a_2 \\ a_3 \end{bmatrix}\right) = \begin{bmatrix} 0 \\ a_3 \\ a_2 \end{bmatrix}$, verify that bases for eigenvectors corresponding to the eigenvalues 1, -1, and 0 are $\begin{bmatrix} 0 \\ 1 \\ 1 \end{bmatrix}$, $\begin{bmatrix} 0 \\ -1 \\ 1 \end{bmatrix}$, $\begin{bmatrix} 1 \\ 0 \\ 0 \end{bmatrix}$, respectively.

Solution

$$L\left(\begin{bmatrix} 0 \\ a \\ a \end{bmatrix}\right) = \begin{bmatrix} 0 \\ a \\ a \end{bmatrix} = 1\begin{bmatrix} 0 \\ a \\ a \end{bmatrix} = 1(a)\begin{bmatrix} 0 \\ 1 \\ 1 \end{bmatrix}$$

Therefore, $\begin{bmatrix} 0 \\ 1 \\ 1 \end{bmatrix}$ is a basis for the eigenvectors corresponding to 1.

$$L\left(\begin{bmatrix} 0 \\ -a \\ a \end{bmatrix}\right) = \begin{bmatrix} 0 \\ a \\ -a \end{bmatrix} = -1\begin{bmatrix} 0 \\ -a \\ a \end{bmatrix} = (-1)(a)\begin{bmatrix} 0 \\ -1 \\ 1 \end{bmatrix}$$

Therefore, $\begin{bmatrix} 1 \\ -1 \\ 0 \end{bmatrix}$ is a basis for the eigenvectors corresponding to -1.

$$L\left(\begin{bmatrix} a \\ 0 \\ 0 \end{bmatrix}\right) = \begin{bmatrix} 0 \\ 0 \\ 0 \end{bmatrix} = 0\begin{bmatrix} a \\ 0 \\ 0 \end{bmatrix} = 0(a)\begin{bmatrix} 1 \\ 0 \\ 0 \end{bmatrix}$$

Therefore, $\begin{bmatrix} 1 \\ 0 \\ 0 \end{bmatrix}$ is a basis for the eigenvectors corresponding to 0.

PROBLEM 15-7 Use the basis $B = \left\{\begin{bmatrix} 1 \\ 1 \\ 0 \end{bmatrix}, \begin{bmatrix} 0 \\ -2 \\ 1 \end{bmatrix}, \begin{bmatrix} 1 \\ 0 \\ -1 \end{bmatrix}\right\}$ for R^3 to find the matrix representation of the linear transformation defined by

$$L\left(\begin{bmatrix} a_1 \\ a_2 \\ a_3 \end{bmatrix}\right) = \begin{bmatrix} 0 \\ a_3 \\ a_2 \end{bmatrix}$$

Solution

$$L\left(\begin{bmatrix}1\\1\\0\end{bmatrix}\right)=\begin{bmatrix}0\\0\\1\end{bmatrix}=c_1\begin{bmatrix}1\\1\\0\end{bmatrix}+c_2\begin{bmatrix}0\\-2\\1\end{bmatrix}+c_3\begin{bmatrix}1\\0\\-1\end{bmatrix}$$

This has solutions of $c_1 = 2/3, c_2 = 1/3, c_3 = -2/3$. Therefore, $\begin{bmatrix}0\\1\\1\end{bmatrix}_B = \begin{bmatrix}2/3\\1/3\\-2/3\end{bmatrix}$.

$$L\left(\begin{bmatrix}0\\-2\\1\end{bmatrix}\right)=\begin{bmatrix}0\\1\\-2\end{bmatrix}=c_1\begin{bmatrix}1\\1\\0\end{bmatrix}+c_2\begin{bmatrix}0\\-2\\1\end{bmatrix}+c_3\begin{bmatrix}1\\0\\-1\end{bmatrix}$$

This has solution of $c_1 = -1, c_2 = -1, c_3 = 1$. Therefore, $\begin{bmatrix}1\\0\\-2\end{bmatrix}_B = \begin{bmatrix}-1\\-1\\1\end{bmatrix}$.

$$L\left(\begin{bmatrix}1\\0\\-1\end{bmatrix}\right)=\begin{bmatrix}0\\-1\\0\end{bmatrix}=c_1\begin{bmatrix}1\\1\\0\end{bmatrix}+c_2\begin{bmatrix}0\\-2\\1\end{bmatrix}+c_3\begin{bmatrix}1\\0\\-1\end{bmatrix}$$

This has solution of $c_1 = -1/3,\ c_2 = 1/3,\ c_3 = 1/3$. Therefore, $\begin{bmatrix}-1\\1\\0\end{bmatrix}_B = \begin{bmatrix}-1/3\\1/3\\1/3\end{bmatrix}$. Therefore, the matrix representation of the linear transformation as defined is

$$A=\begin{bmatrix}\frac{2}{3} & -1 & -\frac{1}{3}\\ \frac{1}{3} & -1 & \frac{1}{3}\\ -\frac{2}{3} & 1 & \frac{1}{3}\end{bmatrix}$$

PROBLEM 15-8 Find the eigenvalues and the corresponding eigenvectors of matrix A in Problem 15-7.

Solution

$$\det(\lambda I - A)=\begin{vmatrix}\lambda-\frac{2}{3} & 1 & \frac{1}{3}\\ -\frac{1}{3} & \lambda+1 & -\frac{1}{3}\\ \frac{2}{3} & -1 & \lambda-\frac{1}{3}\end{vmatrix}=0$$

$$\therefore\quad \lambda^3-\lambda=0$$

$$\lambda(\lambda^2-1)=0$$

Thus, $\lambda = 0$, $\lambda = 1$, and $\lambda = -1$ are the eigenvalues.

For $\lambda = 0$,

$$\left(0\begin{bmatrix}1&0&0\\0&1&0\\0&0&1\end{bmatrix}-\begin{bmatrix}\frac{2}{3} & -1 & -\frac{1}{3}\\ \frac{1}{3} & -1 & \frac{1}{3}\\ -\frac{2}{3} & 1 & \frac{1}{3}\end{bmatrix}\right)\begin{bmatrix}a_1\\a_2\\a_3\end{bmatrix}=\begin{bmatrix}0\\0\\0\end{bmatrix}$$

$$\begin{bmatrix}-\frac{2}{3} & 1 & \frac{1}{3}\\ -\frac{1}{3} & 1 & -\frac{1}{3}\\ \frac{2}{3} & -1 & -\frac{1}{3}\end{bmatrix}\begin{bmatrix}a_1\\a_2\\a_3\end{bmatrix}=\begin{bmatrix}0\\0\\0\end{bmatrix}$$

$$\therefore\quad \begin{bmatrix}a_1\\a_2\\a_3\end{bmatrix}=\begin{bmatrix}2t\\t\\t\end{bmatrix}=t\begin{bmatrix}2\\1\\1\end{bmatrix}$$

where t is arbitrary. Thus, $\begin{bmatrix} 2 \\ 1 \\ 1 \end{bmatrix}$ is the coordinate vector of $\lambda = 0$. Therefore,

$$2\begin{bmatrix} 1 \\ 1 \\ 0 \end{bmatrix} + 1\begin{bmatrix} 0 \\ -2 \\ 1 \end{bmatrix} + 1\begin{bmatrix} 1 \\ 0 \\ -1 \end{bmatrix} = \begin{bmatrix} 1 \\ 0 \\ 0 \end{bmatrix}$$

and $\begin{bmatrix} 1 \\ 0 \\ 0 \end{bmatrix}$ is a basis of the eigenvectors corresponding to eigenvalue 0.

For $\lambda = 1$,

$$\left(1\begin{bmatrix} 1 & 0 & 0 \\ 0 & 1 & 0 \\ 0 & 0 & 1 \end{bmatrix} - \begin{bmatrix} \frac{2}{3} & -1 & -\frac{1}{3} \\ \frac{1}{3} & -1 & \frac{1}{3} \\ -\frac{2}{3} & 1 & \frac{1}{3} \end{bmatrix}\right)\begin{bmatrix} a_1 \\ a_2 \\ a_3 \end{bmatrix} = \begin{bmatrix} 0 \\ 0 \\ 0 \end{bmatrix}$$

$$\begin{bmatrix} \frac{1}{3} & 1 & \frac{1}{3} \\ -\frac{1}{3} & 2 & -\frac{1}{3} \\ \frac{2}{3} & -1 & \frac{2}{3} \end{bmatrix}\begin{bmatrix} a_1 \\ a_2 \\ a_3 \end{bmatrix} = \begin{bmatrix} 0 \\ 0 \\ 0 \end{bmatrix}$$

$$\therefore \quad \begin{bmatrix} a_1 \\ a_2 \\ a_3 \end{bmatrix} = \begin{bmatrix} t \\ 0 \\ -t \end{bmatrix} = t\begin{bmatrix} 1 \\ 0 \\ -1 \end{bmatrix}$$

where t is arbitrary. Thus, $\begin{bmatrix} 1 \\ 0 \\ -1 \end{bmatrix}$ is the coordinate vector of $\lambda = 1$. Therefore,

$$1\begin{bmatrix} 1 \\ 1 \\ 0 \end{bmatrix} + 0\begin{bmatrix} 0 \\ -2 \\ 1 \end{bmatrix} - 1\begin{bmatrix} 1 \\ 0 \\ -1 \end{bmatrix} = \begin{bmatrix} 0 \\ 1 \\ 1 \end{bmatrix}$$

Hence, $\begin{bmatrix} 0 \\ 1 \\ 1 \end{bmatrix}$ is a basis of the eigenvectors corresponding to eigenvalue 1.

For $\lambda = -1$,

$$\left(-1\begin{bmatrix} 1 & 0 & 0 \\ 0 & 1 & 0 \\ 0 & 0 & 1 \end{bmatrix} - \begin{bmatrix} \frac{2}{3} & -1 & -\frac{1}{3} \\ \frac{1}{3} & -1 & \frac{1}{3} \\ -\frac{2}{3} & 1 & \frac{1}{3} \end{bmatrix}\right)\begin{bmatrix} a_1 \\ a_2 \\ a_3 \end{bmatrix} = \begin{bmatrix} 0 \\ 0 \\ 0 \end{bmatrix}$$

$$\begin{bmatrix} -\frac{5}{3} & 1 & \frac{1}{3} \\ -\frac{1}{3} & 0 & -\frac{1}{3} \\ \frac{2}{3} & -1 & -\frac{4}{3} \end{bmatrix}\begin{bmatrix} a_1 \\ a_2 \\ a_3 \end{bmatrix} = \begin{bmatrix} 0 \\ 0 \\ 0 \end{bmatrix}$$

$$\therefore \quad \begin{bmatrix} a_1 \\ a_2 \\ a_3 \end{bmatrix} = \begin{bmatrix} -t \\ -2t \\ t \end{bmatrix} = t\begin{bmatrix} -1 \\ -2 \\ 1 \end{bmatrix}$$

where t is arbitrary. Thus, $\begin{bmatrix} -1 \\ -2 \\ 1 \end{bmatrix}$ is the coordinate vector of $\lambda = -1$. Therefore,

$$-1\begin{bmatrix}1\\1\\0\end{bmatrix}-2\begin{bmatrix}0\\-2\\1\end{bmatrix}+1\begin{bmatrix}1\\0\\-1\end{bmatrix}=\begin{bmatrix}0\\3\\-3\end{bmatrix}=-3\begin{bmatrix}0\\-1\\1\end{bmatrix}$$

Hence, $\begin{bmatrix}0\\-1\\1\end{bmatrix}$ is a basis of the eigenvectors corresponding to eigenvalue -1.

Observe that the bases for the eigenvectors in this problem are the same as those given in Problem 15-6.

PROBLEM 15-9 Find bases for the eigenspaces of

$$A=\begin{bmatrix}-1&3&3\\0&2&3\\0&0&1\end{bmatrix}$$

Solution

$$\det(\lambda I-A)=\begin{vmatrix}\lambda+1&-3&-3\\0&\lambda-2&-3\\0&0&\lambda-1\end{vmatrix}=(\lambda+1)(\lambda-2)(\lambda-1)=0$$

Therefore, the eigenvalues are $\lambda=-1$, $\lambda=1$, and $\lambda=2$.

For $\lambda=-1$,

$$\begin{bmatrix}0&-3&-3\\0&-3&-3\\0&0&-2\end{bmatrix}\begin{bmatrix}a_1\\a_2\\a_3\end{bmatrix}=\begin{bmatrix}0\\0\\0\end{bmatrix}$$

$$\therefore\quad\begin{bmatrix}a_1\\a_2\\a_3\end{bmatrix}=\begin{bmatrix}t\\0\\0\end{bmatrix}=t\begin{bmatrix}1\\0\\0\end{bmatrix}$$

where t is arbitrary.

For $\lambda=1$,

$$\begin{bmatrix}2&-3&-3\\0&-1&-3\\0&0&0\end{bmatrix}\begin{bmatrix}a_1\\a_2\\a_3\end{bmatrix}=\begin{bmatrix}0\\0\\0\end{bmatrix}$$

$$\therefore\quad\begin{bmatrix}a_1\\a_2\\a_3\end{bmatrix}=\begin{bmatrix}-3t\\-3t\\t\end{bmatrix}=t\begin{bmatrix}-3\\-3\\1\end{bmatrix}$$

where t is arbitrary.

For $\lambda=2$,

$$\begin{bmatrix}3&-3&-3\\0&0&-3\\0&0&1\end{bmatrix}\begin{bmatrix}a_1\\a_2\\a_3\end{bmatrix}=\begin{bmatrix}0\\0\\0\end{bmatrix}$$

$$\therefore\quad\begin{bmatrix}a_1\\a_2\\a_3\end{bmatrix}=\begin{bmatrix}t\\t\\0\end{bmatrix}=t\begin{bmatrix}1\\1\\0\end{bmatrix}$$

where t is arbitrary. Therefore, $\begin{bmatrix} 1 \\ 0 \\ 0 \end{bmatrix}$ is a basis for the eigenspace corresponding to the eigenvalue -1; $\begin{bmatrix} -3 \\ -3 \\ 1 \end{bmatrix}$ is a basis for the eigenspace corresponding to the eigenvalue 1; and $\begin{bmatrix} 1 \\ 1 \\ 0 \end{bmatrix}$ is a basis for the eigenspace corresponding to the eigenvalue 2.

PROBLEM 15-10 Verify that the set of basis vectors for the eigenspaces of matrix A in Problem 15-9 is linearly independent.

Solution If the matrix $\begin{bmatrix} 1 & -3 & 1 \\ 0 & -3 & 1 \\ 0 & 1 & 0 \end{bmatrix}$ has a nonzero determinant value, the set is linearly independent:

$$\begin{vmatrix} 1 & -3 & 1 \\ 0 & -3 & 1 \\ 0 & 1 & 0 \end{vmatrix} = -1 \neq 0$$

Therefore, the three column vectors are linearly independent.

PROBLEM 15-11 If $A = \begin{bmatrix} 1 & 0 & -2 \\ 1 & 2 & 1 \\ -1 & 0 & 2 \end{bmatrix}$, verify that **(a)** the eigenvalues of A are 0, 2, and 3, and A is singular; **(b)** the eigenvalues of $2A$ are 0, 4, and 6; **(c)** the eigenvalues of A^T are 0, 2, and 3; and **(d)** the eigenvalues of A^2 are 0, 4, and 9.

Solution

(a)
$$\det(\lambda I - A) = \begin{vmatrix} \lambda - 1 & 0 & 2 \\ -1 & \lambda - 2 & -1 \\ 1 & 0 & \lambda - 2 \end{vmatrix} = (\lambda - 1)(\lambda - 2)^2 - 2(\lambda - 2) = 0$$

$$\therefore \quad (\lambda - 2)[(\lambda - 1)(\lambda - 2) - 2] = (\lambda - 2)(\lambda^2 - 3\lambda) = 0$$

or

$$\lambda(\lambda - 2)(\lambda - 3) = 0$$

Hence, the eigenvalues of A are 0, 2, and 3.

Since $\begin{vmatrix} 1 & 0 & -2 \\ 1 & 2 & 1 \\ -1 & 0 & 2 \end{vmatrix} = 4 - 4 = 0$, A is singular.

(b)
$$2A = \begin{bmatrix} 2 & 0 & -4 \\ 2 & 4 & 2 \\ -2 & 0 & 4 \end{bmatrix}$$

$$\det(\lambda I - A) = \begin{vmatrix} \lambda - 2 & 0 & 4 \\ -2 & \lambda - 4 & -2 \\ 2 & 0 & \lambda - 4 \end{vmatrix} = 0$$

$$\therefore \quad \lambda(\lambda - 4)(\lambda - 6) = 0$$

Hence, the eigenvalues of $2A$ are 0, 4, and 6.

(c)

$$A^T = \begin{bmatrix} 1 & 1 & -1 \\ 0 & 2 & 0 \\ -2 & 1 & 2 \end{bmatrix}$$

$$\det(\lambda I - A^T) = \begin{vmatrix} \lambda - 1 & -1 & 1 \\ 0 & \lambda - 2 & 0 \\ 2 & -1 & \lambda - 2 \end{vmatrix} = 0$$

$$\therefore \quad \lambda(\lambda - 2)(\lambda - 3) = 0$$

Hence, the eigenvalues of A^T are 0, 2, and 3.

(d)

$$A^2 = \begin{bmatrix} 1 & 0 & -2 \\ 1 & 2 & 1 \\ -1 & 0 & 2 \end{bmatrix}\begin{bmatrix} 1 & 0 & -2 \\ 1 & 2 & 1 \\ -1 & 0 & 2 \end{bmatrix} = \begin{bmatrix} 3 & 0 & -6 \\ 2 & 4 & 2 \\ -3 & 0 & 6 \end{bmatrix}$$

$$\det(\lambda I - A^2) = \begin{vmatrix} \lambda - 3 & 0 & 6 \\ -2 & \lambda - 4 & -2 \\ 3 & 0 & \lambda - 6 \end{vmatrix} = 0$$

$$\therefore \quad \lambda(\lambda - 4)(\lambda - 9) = 0$$

Hence, the eigenvalues of A^2 are 0, 4, and 9.

PROBLEM 15-12 Assume $A = \begin{bmatrix} -1 & 0 & 0 \\ 3 & 2 & 0 \\ 3 & 3 & 1 \end{bmatrix}$ with eigenvalues of -1, 1, and 2. **(a)** Find A^{-1} and **(b)** show that eigenvalues of A^{-1} are $1/-1$, $1/1$, and $1/2$ or $\lambda_1 = -1$, $\lambda_2 = 1$, and $\lambda_3 = 1/2$.

Solution

(a)

$$\left[\begin{array}{ccc|ccc} -1 & 0 & 0 & 1 & 0 & 0 \\ 3 & 2 & 0 & 0 & 1 & 0 \\ 3 & 3 & 1 & 0 & 0 & 1 \end{array}\right] \xrightarrow[R_3 \to R_3 + 3R_1]{R_2 \to R_2 + 3R_1} \left[\begin{array}{ccc|ccc} -1 & 0 & 0 & 1 & 0 & 0 \\ 0 & 2 & 0 & 3 & 1 & 0 \\ 0 & 3 & 1 & 3 & 0 & 1 \end{array}\right]$$

$$\xrightarrow[R_3 \to R_3 - R_2]{R_1 \to -R_1} \left[\begin{array}{ccc|ccc} 1 & 0 & 0 & -1 & 0 & 0 \\ 0 & 2 & 0 & 3 & 1 & 0 \\ 0 & 1 & 1 & 0 & -1 & 1 \end{array}\right] \xrightarrow{R_2 \to R_2 - R_3} \left[\begin{array}{ccc|ccc} 1 & 0 & 0 & -1 & 0 & 0 \\ 0 & 1 & -1 & 3 & 2 & -1 \\ 0 & 1 & 1 & 0 & -1 & 1 \end{array}\right]$$

$$\xrightarrow[R_3 \to R_3 - R_2]{} \left[\begin{array}{ccc|ccc} 1 & 0 & 0 & -1 & 0 & 0 \\ 0 & 1 & -1 & 3 & 2 & -1 \\ 0 & 0 & 2 & -3 & -3 & 2 \end{array}\right] \xrightarrow[R_3 \to 1/2 R_3]{} \left[\begin{array}{ccc|ccc} 1 & 0 & 0 & -1 & 0 & 0 \\ 0 & 1 & -1 & 3 & 2 & -1 \\ 0 & 0 & 1 & -\frac{3}{2} & -\frac{3}{2} & 1 \end{array}\right]$$

$$\xrightarrow{R_2 \to R_2 + R_3} \left[\begin{array}{ccc|ccc} 1 & 0 & 0 & -1 & 0 & 0 \\ 0 & 1 & 0 & \frac{3}{2} & \frac{1}{2} & 0 \\ 0 & 0 & 1 & -\frac{3}{2} & -\frac{3}{2} & 1 \end{array}\right] \quad \therefore \quad A^{-1} = \begin{bmatrix} -1 & 0 & 0 \\ \frac{3}{2} & \frac{1}{2} & 0 \\ -\frac{3}{2} & -\frac{3}{2} & 1 \end{bmatrix}$$

(b)

$$\det(\lambda I - A^{-1}) = \begin{vmatrix} \lambda + 1 & 0 & 0 \\ -\frac{3}{2} & \lambda - \frac{1}{2} & 0 \\ \frac{3}{2} & \frac{3}{2} & \lambda - 1 \end{vmatrix} = (\lambda + 1)(\lambda - 1/2)(\lambda - 1) = 0$$

Therefore, $\lambda_1 = -1$, $\lambda_2 = 1$, and $\lambda_3 = 1/2$ are eigenvalues of A^{-1}.

PROBLEM 15-13 Assume

$$A = \begin{bmatrix} 1 & 0 & -2 \\ 1 & 2 & 1 \\ -1 & 0 & 2 \end{bmatrix}, \qquad P = \begin{bmatrix} 1 & 0 & -1 \\ 0 & 1 & 0 \\ 0 & 0 & 1 \end{bmatrix}, \qquad P^{-1} = \begin{bmatrix} 1 & 0 & 1 \\ 0 & 1 & 0 \\ 0 & 0 & 1 \end{bmatrix}$$

and the eigenvalues of A are 0, 2, and 3. **(a)** Find a matrix B similar to A and **(b)** verify that eigenvalues of A and B are identical.

Solution

(a)
$$B = P^{-1}AP = \begin{bmatrix} 1 & 0 & 1 \\ 0 & 1 & 0 \\ 0 & 0 & 1 \end{bmatrix}\begin{bmatrix} 1 & 0 & -2 \\ 1 & 2 & 1 \\ -1 & 0 & 2 \end{bmatrix}\begin{bmatrix} 1 & 0 & -1 \\ 0 & 1 & 0 \\ 0 & 0 & 1 \end{bmatrix}$$
$$= \begin{bmatrix} 0 & 0 & 0 \\ 1 & 2 & 1 \\ -1 & 0 & 2 \end{bmatrix}\begin{bmatrix} 1 & 0 & -1 \\ 0 & 1 & 0 \\ 0 & 0 & 1 \end{bmatrix} = \begin{bmatrix} 0 & 0 & 0 \\ 1 & 2 & 0 \\ -1 & 0 & 3 \end{bmatrix}$$

(b)
$$\det(\lambda I - B) = \begin{vmatrix} \lambda & 0 & 0 \\ -1 & \lambda - 2 & 0 \\ 1 & 0 & \lambda - 3 \end{vmatrix} = \lambda(\lambda - 2)(\lambda - 3) = 0$$

Therefore, $\lambda = 0$, $\lambda = 2$, and $\lambda = 3$ are the eigenvalues of B which are the same as the eigenvalues of A.

PROBLEM 15-14 The eigenvalues of $A = \begin{bmatrix} 2 & 1 & 1 \\ 0 & 1 & 2 \\ 0 & 0 & -1 \end{bmatrix}$ are -1, 1, and 2. The eigenvalues of $B = \begin{bmatrix} 3 & 1 & 1 \\ 0 & 2 & 1 \\ 0 & 0 & 1 \end{bmatrix}$ are 1, 2, and 3. Find **(a)** the eigenvalues of $A + B$ and **(b)** the eigenvalues of AB.

Solution

(a)
$$\det[\lambda I - (A + B)] = \begin{vmatrix} \lambda - 5 & -2 & -2 \\ 0 & \lambda - 3 & -3 \\ 0 & 0 & \lambda \end{vmatrix} = 0$$

Therefore, $\lambda(\lambda - 5)(\lambda - 3) = 0$. The eigenvalues of $A + B$ are 0, 3, and 5.

(b)
$$\det(\lambda I - AB) = \begin{vmatrix} \lambda - 6 & -4 & -4 \\ 0 & \lambda - 2 & -3 \\ 0 & 0 & \lambda + 1 \end{vmatrix} = 0$$

Therefore, $(\lambda - 6)(\lambda - 2)(\lambda + 1) = 0$. The eigenvalues of AB are -1, 2, and 6.

PROBLEM 15-15 In Problem 15-14 the eigenvectors corresponding to eigenvalues -1, 1, and 2 of A, and the eigenvectors corresponding to eigenvalues 1, 2, and 3 of B are $\begin{bmatrix} 0 \\ 1 \\ -1 \end{bmatrix}$, $\begin{bmatrix} -1 \\ 1 \\ 0 \end{bmatrix}$, and $\begin{bmatrix} 1 \\ 0 \\ 0 \end{bmatrix}$, respectively. Verify that **(a)** $c\begin{bmatrix} 0 \\ 1 \\ -1 \end{bmatrix}$ is also an eigenvector of A and **(b)** the eigenvectors of the eigenvalues of $A + B$ are the same as those for A and B.

Solution

(a) We must show that $A(c\mathbf{a}) = -1(c\mathbf{a})$ where $\mathbf{a} = \begin{bmatrix} 0 \\ 1 \\ -1 \end{bmatrix}$.

$$\begin{bmatrix} 2 & 1 & 1 \\ 0 & 1 & 2 \\ 0 & 0 & -1 \end{bmatrix} \begin{bmatrix} 0 \\ c \\ -c \end{bmatrix} = \begin{bmatrix} 0 \\ -c \\ c \end{bmatrix} = -1 \begin{bmatrix} 0 \\ c \\ -c \end{bmatrix} = -1(c) \begin{bmatrix} 0 \\ 1 \\ -1 \end{bmatrix}$$

Therefore, $c\begin{bmatrix} 0 \\ 1 \\ -1 \end{bmatrix}$ is an eigenvector of A.

(b) The eigenvalues of $A + B$ are 0, 3, and 5 from Problem 15-14. We must find the corresponding eigenvectors for these eigenvalues.

For $\lambda = 0$,

$$\begin{bmatrix} -5 & -2 & -2 \\ 0 & -3 & -3 \\ 0 & 0 & 0 \end{bmatrix} \begin{bmatrix} a_1 \\ a_2 \\ a_3 \end{bmatrix} = \begin{bmatrix} 0 \\ 0 \\ 0 \end{bmatrix}$$

$$\begin{bmatrix} a_1 \\ a_2 \\ a_3 \end{bmatrix} = \begin{bmatrix} 0 \\ t \\ -t \end{bmatrix} = t \begin{bmatrix} 0 \\ 1 \\ -1 \end{bmatrix}$$

Therefore, $\begin{bmatrix} 0 \\ 1 \\ -1 \end{bmatrix}$ is the eigenvector.

For $\lambda = 3$,

$$\begin{bmatrix} -2 & -2 & -2 \\ 0 & 0 & -3 \\ 0 & 0 & 3 \end{bmatrix} \begin{bmatrix} a_1 \\ a_2 \\ a_3 \end{bmatrix} = \begin{bmatrix} 0 \\ 0 \\ 0 \end{bmatrix}$$

$$\begin{bmatrix} a_1 \\ a_2 \\ a_3 \end{bmatrix} = \begin{bmatrix} -t \\ t \\ 0 \end{bmatrix} = t \begin{bmatrix} -1 \\ 1 \\ 0 \end{bmatrix}$$

Therefore, $\begin{bmatrix} -1 \\ 1 \\ 0 \end{bmatrix}$ is the eigenvector.

For $\lambda = 5$,

$$\begin{bmatrix} 0 & -2 & -2 \\ 0 & 2 & -3 \\ 0 & 0 & 5 \end{bmatrix} \begin{bmatrix} a_1 \\ a_2 \\ a_3 \end{bmatrix}$$

$$\begin{bmatrix} a_1 \\ a_2 \\ a_3 \end{bmatrix} = \begin{bmatrix} t \\ 0 \\ 0 \end{bmatrix} = t \begin{bmatrix} 1 \\ 0 \\ 0 \end{bmatrix}$$

Therefore, $\begin{bmatrix} 1 \\ 0 \\ 0 \end{bmatrix}$ is the eigenvector. Therefore, the eigenvectors of $A + B$ are the same as those for A and B.

$\lambda_A = -1$ and $\lambda_B = 1$ have the same eigenvector $\begin{bmatrix} 0 \\ 1 \\ -1 \end{bmatrix}$. Therefore, $\lambda_{A+B} = -1 + 1 = 0$ has the eigenvector $\begin{bmatrix} 0 \\ 1 \\ -1 \end{bmatrix}$.

$\lambda_A = 1$ and $\lambda_B = 2$ have same eigenvector $\begin{bmatrix} -1 \\ 1 \\ 0 \end{bmatrix}$. Therefore, $\lambda_{A+B} = 1 + 2 = 3$ has the eigenvector $\begin{bmatrix} -1 \\ 1 \\ 0 \end{bmatrix}$.

$\lambda_A = 2$ and $\lambda_B = 3$ have same eigenvector $\begin{bmatrix} 1 \\ 0 \\ 0 \end{bmatrix}$. Therefore, $\lambda_{A+B} = 2 + 3 = 5$ has the eigenvector $\begin{bmatrix} 1 \\ 0 \\ 0 \end{bmatrix}$. This is always the case if A and B are both $n \times n$ with the same eigenvectors. If A and B are both $n \times n$ but do not have the same eigenvectors, this is not usually the case.

PROBLEM 15-16 Using $AB = \begin{bmatrix} 6 & 4 & 4 \\ 0 & 2 & 3 \\ 0 & 0 & -1 \end{bmatrix}$ from Problem 15-14, verify that the eigenvectors of AB are the same as those for A and B.

Solution Using the eigenvalues of AB found in Problem 15-14, we will find the corresponding eigenvectors.

For $\lambda = -1$,

$$\begin{bmatrix} -7 & -4 & -4 \\ 0 & -3 & -3 \\ 0 & 0 & 0 \end{bmatrix} \begin{bmatrix} a_1 \\ a_2 \\ a_3 \end{bmatrix} = \begin{bmatrix} 0 \\ 0 \\ 0 \end{bmatrix}$$

$$\begin{bmatrix} a_1 \\ a_2 \\ a_3 \end{bmatrix} = \begin{bmatrix} 0 \\ t \\ -t \end{bmatrix} = t \begin{bmatrix} 0 \\ 1 \\ -1 \end{bmatrix}$$

Therefore, $\begin{bmatrix} 0 \\ 1 \\ -1 \end{bmatrix}$ is the eigenvector.

For $\lambda = 2$,

$$\begin{bmatrix} -4 & -4 & -4 \\ 0 & 0 & -3 \\ 0 & 0 & 3 \end{bmatrix} \begin{bmatrix} a_1 \\ a_2 \\ a_3 \end{bmatrix} = \begin{bmatrix} 0 \\ 0 \\ 0 \end{bmatrix}$$

$$\begin{bmatrix} a_1 \\ a_2 \\ a_3 \end{bmatrix} = \begin{bmatrix} -t \\ t \\ 0 \end{bmatrix} = t \begin{bmatrix} -1 \\ 1 \\ 0 \end{bmatrix}$$

Therefore, $\begin{bmatrix} -1 \\ 1 \\ 0 \end{bmatrix}$ is the eigenvector.

For $\lambda = 6$,

$$\begin{bmatrix} 0 & -4 & -4 \\ 0 & 4 & -3 \\ 0 & 0 & 7 \end{bmatrix} \begin{bmatrix} a_1 \\ a_2 \\ a_3 \end{bmatrix} = \begin{bmatrix} 0 \\ 0 \\ 0 \end{bmatrix}$$

$$\begin{bmatrix} a_1 \\ a_2 \\ a_3 \end{bmatrix} = \begin{bmatrix} t \\ 0 \\ 0 \end{bmatrix} = t \begin{bmatrix} 1 \\ 0 \\ 0 \end{bmatrix}$$

Therefore, $\begin{bmatrix} 1 \\ 0 \\ 0 \end{bmatrix}$ is the eigenvector. These are the same eigenvectors as those for A and B. Also $-1 = (-1)(1)$, $2 = 1(1)$, and $6 = 2(3)$, products of the corresponding eigenvalues of A and B.

PROBLEM 15-17 Find the eigenvalues of

$$A = \begin{bmatrix} 1 & 2 & 0 & 0 & 0 \\ 0 & 2 & 0 & 0 & 0 \\ 0 & 0 & 3 & 2 & 0 \\ 0 & 0 & 0 & 1 & 0 \\ 0 & 0 & 0 & 0 & 5 \end{bmatrix}$$

Solution

$$\begin{vmatrix} \lambda-1 & -2 & 0 & 0 & 0 \\ 0 & \lambda-2 & 0 & 0 & 0 \\ 0 & 0 & \lambda-3 & -2 & 0 \\ 0 & 0 & 0 & \lambda-1 & 0 \\ 0 & 0 & 0 & 0 & \lambda-5 \end{vmatrix} = 0$$

This matrix is upper triangular, so the determinant is simply the product of all terms on the main diagonal. Therefore, $(\lambda - 1)(\lambda - 2)(\lambda - 3)(\lambda - 1)(\lambda - 5) = 0$. Thus, the eigenvalues are 1, 1, 2, 3, 5.

PROBLEM 15-18 Find the eigenspace of the eigenvalue $\lambda = 1$ in Problem 15-17.

Solution

$$\begin{bmatrix} 0 & -2 & 0 & 0 & 0 \\ 0 & -1 & 0 & 0 & 0 \\ 0 & 0 & -2 & -2 & 0 \\ 0 & 0 & 0 & 0 & 0 \\ 0 & 0 & 0 & 0 & -4 \end{bmatrix} \begin{bmatrix} a_1 \\ a_2 \\ a_3 \\ a_4 \\ a_5 \end{bmatrix} = \begin{bmatrix} 0 \\ 0 \\ 0 \\ 0 \\ 0 \end{bmatrix}$$

$$\begin{bmatrix} a_1 \\ a_2 \\ a_3 \\ a_4 \\ a_5 \end{bmatrix} = \begin{bmatrix} s \\ 0 \\ -t \\ t \\ 0 \end{bmatrix} = s \begin{bmatrix} 1 \\ 0 \\ 0 \\ 0 \\ 0 \end{bmatrix} + t \begin{bmatrix} 0 \\ 0 \\ -1 \\ 1 \\ 0 \end{bmatrix}$$

Therefore, the eigenspace for $\lambda = 1$ has two eigenvectors in its basis.

$$\begin{bmatrix} 1 \\ 0 \\ 0 \\ 0 \\ 0 \end{bmatrix} \quad \text{and} \quad \begin{bmatrix} 0 \\ 0 \\ -1 \\ 1 \\ 0 \end{bmatrix}$$

PROBLEM 15-19 Show that $A = \begin{bmatrix} 1 & 2 \\ 3 & 2 \end{bmatrix}$ is a solution of $\lambda^2 - 3\lambda - 4 = 0$.

Solution Show

$$\begin{bmatrix} 1 & 2 \\ 3 & 2 \end{bmatrix}^2 - 3\begin{bmatrix} 1 & 2 \\ 3 & 2 \end{bmatrix} - 4\begin{bmatrix} 1 & 0 \\ 0 & 1 \end{bmatrix} = \begin{bmatrix} 0 & 0 \\ 0 & 0 \end{bmatrix}$$

$$\begin{bmatrix} 7 & 6 \\ 9 & 10 \end{bmatrix} - \begin{bmatrix} 3 & 6 \\ 9 & 6 \end{bmatrix} - \begin{bmatrix} 4 & 0 \\ 0 & 4 \end{bmatrix} = \begin{bmatrix} 0 & 0 \\ 0 & 0 \end{bmatrix}$$

Therefore, A satisfies its characteristic equation. This is true in general by the **Cayley–Hamilton Theorem** which says that every $n \times n$ matrix satisfies its characteristic equation.

PROBLEM 15-20 Consider the matrix

$$A = \begin{bmatrix} 1 & 3 & 2 & -2 \\ 0 & 2 & 3 & -4 \\ 0 & 0 & 3 & -1 \\ 0 & 0 & 0 & 4 \end{bmatrix}$$

Let $A_1 = \begin{bmatrix} 1 & 3 \\ 0 & 2 \end{bmatrix}$ and $A_2 = \begin{bmatrix} 3 & -1 \\ 0 & 4 \end{bmatrix}$ and show that the characteristic equation of A is equal to the product of the characteristic equation of A_1 and A_2.

Solution For A,

$$\begin{vmatrix} \lambda - 1 & -3 & -2 & 2 \\ 0 & \lambda - 2 & -3 & 4 \\ 0 & 0 & \lambda - 3 & 1 \\ 0 & 0 & 0 & \lambda - 4 \end{vmatrix} = (\lambda - 1)(\lambda - 2)(\lambda - 3)(\lambda - 4) = 0$$

For A_1,

$$\begin{vmatrix} \lambda - 1 & -3 \\ 0 & \lambda - 2 \end{vmatrix} = (\lambda - 1)(\lambda - 2) = 0$$

For A_2,

$$\begin{vmatrix} \lambda - 3 & 1 \\ 0 & \lambda - 4 \end{vmatrix} = (\lambda - 3)(\lambda - 4) = 0$$

Therefore, $(\lambda - 1)(\lambda - 2)(\lambda - 3)(\lambda - 4) = [(\lambda - 1) \quad (\lambda - 2)][(\lambda - 3) \quad (\lambda - 4)]$.

Supplementary Problems

PROBLEM 15-21 Find the eigenvalues of $A = \begin{bmatrix} 2 & 3 \\ 1 & 4 \end{bmatrix}$.

PROBLEM 15-22 Find the characteristic equation of $B = \begin{bmatrix} 1 & 2 & 1 \\ 2 & 2 & -2 \\ 3 & -1 & 3 \end{bmatrix}$.

PROBLEM 15-23 Find the eigenvalues and the corresponding eigenvectors for $A = \begin{bmatrix} 2 & 1 \\ 1 & 3 \end{bmatrix}$.

PROBLEM 15-24 Find all the roots of $\lambda^4 - 2\lambda^3 - 4\lambda^2 + 2\lambda + 3 = 0$.

PROBLEM 15-25 Find the eigenvalues and corresponding eigenvectors of $A = \begin{bmatrix} 0 & 0 & 0 \\ 0 & 1 & 0 \\ 2 & 0 & 1 \end{bmatrix}$.

PROBLEM 15-26 If $A^2 = A$, find all the possible eigenvalues of A.

PROBLEM 15-27 If $A = \begin{bmatrix} a_1 & b_1 \\ a_2 & b_2 \end{bmatrix}$, find the condition that A has equal eigenvalues.

PROBLEM 15-28 Given $A = \begin{bmatrix} 1 & 0 & -2 \\ 0 & -1 & -2 \\ -2 & -2 & 0 \end{bmatrix}$, find the eigenvalues and the corresponding eigenvectors.

PROBLEM 15-29 Show that the eigenvectors of $A = \begin{bmatrix} 1 & 0 & -2 \\ 0 & -1 & -2 \\ -2 & -2 & 0 \end{bmatrix}$ are orthogonal.

PROBLEM 15-30 Normalize the eigenvectors of Problem 15-29 to find an orthogonal matrix P.

PROBLEM 15-31 Find a 3×3 matrix with eigenvalues of $\lambda_1 = -1$, $\lambda_2 = 1$, and $\lambda_2 = 2$ and associated eigenvectors $\begin{bmatrix} 1 \\ -1 \\ 0 \end{bmatrix}$, $\begin{bmatrix} 0 \\ 1 \\ 0 \end{bmatrix}$, and $\begin{bmatrix} 1 \\ 2 \\ 1 \end{bmatrix}$, respectively.

PROBLEM 15-32 Given $A = \begin{bmatrix} 2 & 0 \\ 0 & 3 \end{bmatrix}$ and $P = \begin{bmatrix} 1 & -1 \\ 2 & 1 \end{bmatrix}$, find $B = P^{-1}AP$ and the eigenvalues for B.

Answers to Supplementary Problems

15-21 $\lambda_1 = 1, \lambda_2 = 5$

15-22 $\lambda^3 - 6\lambda^2 + 2\lambda + 28 = 0$

15-23 $\lambda_1 = (5 + \sqrt{5})/2$; eigenvector of $\begin{bmatrix} (-1 + \sqrt{5})/2 \\ 1 \end{bmatrix}$. $\lambda_2 = (5 - \sqrt{5})/2$; eigenvector of $\begin{bmatrix} (-1 - \sqrt{5})/2 \\ 1 \end{bmatrix}$.

15-24 $\lambda_1 = -1, \lambda_2 = -1, \lambda_3 = 1, \lambda_4 = 3$

15-25 $\lambda = 0$; eigenvector of $\begin{bmatrix} 1 \\ 0 \\ -2 \end{bmatrix}$. $\lambda = 1$; eigenvectors of $\begin{bmatrix} 0 \\ 1 \\ 0 \end{bmatrix}$ and $\begin{bmatrix} 0 \\ 0 \\ 1 \end{bmatrix}$.

15-26 $\lambda = 0$ and $\lambda = 1$

15-27 $(b_2 + a_1)^2 - 4(a_1 b_2 - a_2 b_1) = 0$

15-28 $\lambda_1 = -3$; eigenvector is $\begin{bmatrix} 1 \\ 2 \\ 2 \end{bmatrix} = \mathbf{u}_1$. $\lambda_2 = 0$; eigenvector is $\begin{bmatrix} -2 \\ 2 \\ -1 \end{bmatrix} = \mathbf{u}_2$. $\lambda_3 = 3$; eigenvector is $\begin{bmatrix} -2 \\ -1 \\ 2 \end{bmatrix} = \mathbf{u}_3$.

15-29 $\langle \mathbf{u}_1, \mathbf{u}_2 \rangle = -2 + 4 - 2 = 0$;
$\langle \mathbf{u}_1, \mathbf{u}_3 \rangle = -2 - 2 + 4 = 0$;
$\langle \mathbf{u}_2, \mathbf{u}_3 \rangle = 4 - 2 - 2 = 0$

15-30 $\dfrac{\mathbf{u}_1}{\|\mathbf{u}_1\|} = \begin{bmatrix} \frac{1}{3} \\ \frac{2}{3} \\ \frac{2}{3} \end{bmatrix}, \quad \dfrac{\mathbf{u}_2}{\|\mathbf{u}_2\|} = \begin{bmatrix} -\frac{2}{3} \\ \frac{2}{3} \\ -\frac{1}{3} \end{bmatrix},$

$$\frac{\mathbf{u}_3}{\|\mathbf{u}_3\|} = \begin{bmatrix} -\frac{2}{3} \\ -\frac{1}{3} \\ \frac{2}{3} \end{bmatrix}$$

$$P = \begin{bmatrix} \frac{1}{3} & -\frac{2}{3} & -\frac{2}{3} \\ \frac{2}{3} & \frac{2}{3} & -\frac{1}{3} \\ \frac{2}{3} & -\frac{1}{3} & \frac{2}{3} \end{bmatrix}$$

15-31 $A = \begin{bmatrix} -1 & 0 & 3 \\ 2 & 1 & 0 \\ 0 & 0 & 2 \end{bmatrix}$

15-32 $P^{-1} = \begin{bmatrix} \frac{1}{3} & \frac{1}{3} \\ -\frac{2}{3} & \frac{1}{3} \end{bmatrix} \qquad B = \begin{bmatrix} \frac{8}{3} & \frac{1}{3} \\ \frac{2}{3} & \frac{7}{3} \end{bmatrix}$

$\lambda_1 = 2 \qquad \lambda_2 = 3$

16 DIAGONALIZATION

THIS CHAPTER IS ABOUT

☑ **The Meaning of Diagonalization**
☑ **Diagonalization of Matrices**
☑ **Diagonalization of Symmetric Matrices**

16-1. The Meaning of Diagonalization

Remember that a diagonal matrix is a square matrix in which all entries off the main diagonal are zeros.

A. Diagonalizable matrices

An $n \times n$ matrix A is **diagonalizable** if it is similar to a diagonal matrix D.

EXAMPLE 16-1: Show that $A = \begin{bmatrix} 3 & 1 \\ -2 & 0 \end{bmatrix}$ is similar to $D = \begin{bmatrix} 1 & 0 \\ 0 & 2 \end{bmatrix}$ and is, therefore, diagonalizable.

Solution: We must find an invertible matrix P such that $P^{-1}AP = D$. If we multiply on the left by P, we get $AP = PD$. Therefore,

$$\begin{bmatrix} 3 & 1 \\ -2 & 0 \end{bmatrix}\begin{bmatrix} a & b \\ c & d \end{bmatrix} = \begin{bmatrix} a & b \\ c & d \end{bmatrix}\begin{bmatrix} 1 & 0 \\ 0 & 2 \end{bmatrix}$$

Expanding both sides and setting corresponding entries equal, we get

$$\begin{aligned} 3a \quad + c \quad &= a \\ 3b \quad + d &= 2b \\ -2a \quad \quad &= c \\ -2b \quad \quad &= 2d \end{aligned}$$

A solution to this system is $a = 1$, $b = -1$, $c = -2$, and $d = 1$. Therefore,

$$P = \begin{bmatrix} 1 & -1 \\ -2 & 1 \end{bmatrix} \quad \text{and} \quad P^{-1}\begin{bmatrix} -1 & -1 \\ -2 & -1 \end{bmatrix}$$

To check,

$$P^{-1}AP = \begin{bmatrix} -1 & -1 \\ -2 & -1 \end{bmatrix}\begin{bmatrix} 3 & 1 \\ -2 & 0 \end{bmatrix}\begin{bmatrix} 1 & -1 \\ -2 & 1 \end{bmatrix} = \begin{bmatrix} 1 & 0 \\ 0 & 2 \end{bmatrix} = D$$

Therefore, matrix A is diagonalizable because it is similar to a diagonalizable matrix D.

B. Diagonalizable linear transformations

If $L: V \to V$ is a linear transformation of an n-dimensional vector space V into itself, then L is diagonalizable provided an ordered basis S for V exists such that the matrix representation of L with respect to S is diagonalizable.

EXAMPLE 16-2: Show that the linear transformation $L: R^2 \to R^2$ defined by

$$L\left(\begin{bmatrix} a_1 \\ a_2 \end{bmatrix}\right) = \begin{bmatrix} 3a_1 \\ -2a_2 \end{bmatrix}$$

is diagonalizable by finding A, the matrix representation of L with respect to the natural ordered basis $B = \left\{\begin{bmatrix} 1 \\ 0 \end{bmatrix}, \begin{bmatrix} 0 \\ 1 \end{bmatrix}\right\}$, and showing that A is diagonalizable.

Solution

$$L\left(\begin{bmatrix} 1 \\ 0 \end{bmatrix}\right) = \begin{bmatrix} 3 \\ 0 \end{bmatrix} = c_1\begin{bmatrix} 1 \\ 0 \end{bmatrix} + c_2\begin{bmatrix} 0 \\ 1 \end{bmatrix}$$

which has the solution $c_1 = 3$ and $c_2 = 0$. Therefore, $\begin{bmatrix} 3 \\ 0 \end{bmatrix}_B = \begin{bmatrix} 3 \\ 0 \end{bmatrix}$.

$$L\left(\begin{bmatrix} 0 \\ 1 \end{bmatrix}\right) = \begin{bmatrix} 0 \\ -2 \end{bmatrix} = c_1\begin{bmatrix} 1 \\ 0 \end{bmatrix} + c_2\begin{bmatrix} 0 \\ 1 \end{bmatrix}$$

which has the solution $c_1 = 0$ and $c_2 = -2$. Therefore,

$$\begin{bmatrix} 0 \\ -2 \end{bmatrix}_B = \begin{bmatrix} 0 \\ -2 \end{bmatrix}$$

Therefore, $A = \begin{bmatrix} 3 & 0 \\ 0 & -2 \end{bmatrix}$. Since this matrix representation of L is already diagonal, then our problem is completed.

When A is not diagonal, we must find a diagonal matrix D which is similar to A to show that L is diagonalizable. Since the diagonalization of a linear transformation first involves finding the matrix representation of the linear transformation, the remainder of this chapter concentrates on matrices.

16-2. Diagonalization of Matrices

A. Determining if square matrices are diagonalizable

An $n \times n$ matrix A is diagonalizable if and only if A has n linearly independent eigenvectors. Since eigenvectors associated with distinct eigenvalues are linearly independent, then the $n \times n$ matrix A is diagonalizable if it has n distinct eigenvalues. Matrix A may still have n linearly independent eigenvectors, however, if A has less than n distinct eigenvalues.

note: Although in some areas of mathematics complex eigenvalues are permitted, we shall continue to restrict our eigenvalues to real numbers.

EXAMPLE 16-3: Determine if each of the following matrices is diagonalizable.

(a) $A = \begin{bmatrix} 2 & 1 \\ 2 & 3 \end{bmatrix}$ **(b)** $B = \begin{bmatrix} 1 & 0 & -1 \\ 0 & 2 & 0 \\ -8 & 0 & 3 \end{bmatrix}$

(c) $C = \begin{bmatrix} 2 & 0 & 3 \\ 0 & 2 & 1 \\ 0 & 0 & 1 \end{bmatrix}$ **(d)** $D = \begin{bmatrix} 0 & -1 & -1 \\ 1 & 2 & 1 \\ 0 & 0 & 1 \end{bmatrix}$

Solution

(a)

$$\det(\lambda I - A) = \begin{vmatrix} \lambda - 2 & -1 \\ -2 & \lambda - 3 \end{vmatrix} = \lambda^2 - 5\lambda + 4 = 0$$

or $(\lambda - 4)(\lambda - 1) = 0$. Therefore, the eigenvalues are 1 and 4. Hence, since we have two distinct eigenvalues, the 2×2 matrix A is diagonalizable.

(b)

$$\det(\lambda I - B) = \begin{vmatrix} \lambda - 1 & 0 & 1 \\ 0 & \lambda - 2 & 0 \\ 8 & 0 & \lambda - 3 \end{vmatrix} = \lambda^3 - 6\lambda^2 + 3\lambda + 10$$

$$\therefore \quad (\lambda + 1)(\lambda - 2)(\lambda - 5) = 0$$

and the eigenvalues are -1, 2, and 5. Hence, the 3×3 matrix B is diagonalizable because we have three distinct eigenvalues.

(c) Since C is triangular, its eigenvalues are its diagonal entries of 2 and 1. Although there are only two distinct eigenvalues, there may still be three linearly independent eigenvectors:

$$\lambda I - C = \begin{bmatrix} \lambda - 2 & 0 & -3 \\ 0 & \lambda - 2 & -1 \\ 0 & 0 & \lambda - 1 \end{bmatrix}$$

For $\lambda = 1$, we have

$$\begin{bmatrix} -1 & 0 & -3 \\ 0 & -1 & -1 \\ 0 & 0 & 0 \end{bmatrix} \begin{bmatrix} a_1 \\ a_2 \\ a_3 \end{bmatrix} = \begin{bmatrix} 0 \\ 0 \\ 0 \end{bmatrix}$$

$$\begin{aligned} -a_1 \qquad - 3a_3 &= 0 \qquad \text{or} \qquad a_1 = -3a_3 \\ -a_2 - a_3 &= 0 \qquad\qquad\qquad a_2 = -a_3 \end{aligned}$$

Thus, the solution is

$$\begin{bmatrix} -3t \\ -t \\ t \end{bmatrix} = -t \begin{bmatrix} 3 \\ 1 \\ -1 \end{bmatrix}$$

Hence, the dimension of the eigenspace associated with $\lambda = 1$ is one since a basis contains one vector $\begin{bmatrix} 3 \\ 1 \\ -1 \end{bmatrix}$.

For $\lambda = 2$, we have

$$\begin{bmatrix} 0 & 0 & -3 \\ 0 & 0 & -1 \\ 0 & 0 & 1 \end{bmatrix} \begin{bmatrix} a_1 \\ a_2 \\ a_3 \end{bmatrix} = \begin{bmatrix} 0 \\ 0 \\ 0 \end{bmatrix}$$

Therefore, the solution is

$$\begin{bmatrix} s \\ t \\ 0 \end{bmatrix} = s \begin{bmatrix} 1 \\ 0 \\ 0 \end{bmatrix} + t \begin{bmatrix} 0 \\ 1 \\ 0 \end{bmatrix}$$

Therefore, the dimension of the eigenspace associated with $\lambda = 2$ is two of which basis vectors are

$\begin{bmatrix}1\\0\\0\end{bmatrix}$ and $\begin{bmatrix}0\\1\\0\end{bmatrix}$. Therefore, the 3×3 matrix C is diagonalizable since it has three linearly independent eigenvectors: $\begin{bmatrix}3\\1\\-1\end{bmatrix}$, and $\begin{bmatrix}1\\0\\0\end{bmatrix}$, and $\begin{bmatrix}0\\1\\0\end{bmatrix}$.

(d)
$$\det(\lambda I - D) = \begin{vmatrix}\lambda & 1 & 1\\ -1 & \lambda-2 & -1\\ 0 & 0 & \lambda-1\end{vmatrix} = \lambda^3 - 3\lambda^2 + 3\lambda - 1 = 0$$

$$\therefore \quad (\lambda - 1)^3 = 0$$

Hence, 1 is the only eigenvalue. We still must check to see if there are three linearly independent eigenvectors.

$$\begin{bmatrix}1 & 1 & 1\\ -1 & -1 & -1\\ 0 & 0 & 0\end{bmatrix}\begin{bmatrix}a_1\\a_2\\a_3\end{bmatrix} = \begin{bmatrix}0\\0\\0\end{bmatrix}$$

$$a_1 + a_2 + a_3 = 0$$

$$-a_1 - a_2 - a_3 = 0$$

Let $a_1 = s$ and $a_2 = t$. Therefore,

$$\begin{bmatrix}a_1\\a_2\\a_3\end{bmatrix} = \begin{bmatrix}s\\t\\-s-t\end{bmatrix} = s\begin{bmatrix}1\\0\\-1\end{bmatrix} + t\begin{bmatrix}0\\1\\-1\end{bmatrix}$$

Since the eigenspace associated with the eigenvalue 1 has dimension 2, then there are only two linearly independent vectors. Hence, D is not diagonalizable.

B. Procedure for diagonalizing a diagonalizable $n \times n$ matrix A

Step 1: Find the eigenvalues of A.
Step 2: Find n linearly independent eigenvectors associated with the eigenvalues of A. Call them $\mathbf{a}_1, \mathbf{a}_2, \ldots, \mathbf{a}_n$.
Step 3: Form the matrix P having $\mathbf{a}_1, \mathbf{a}_2, \ldots, \mathbf{a}_n$ as columns; that is, $P = [\mathbf{a}_1, \mathbf{a}_2, \ldots, \mathbf{a}_n]$.
Step 4: Find P^{-1}.
Step 5: Find $D = P^{-1}AP$ which will be diagonal, with the diagonal entry in the ith column being the eigenvalue corresponding to the eigenvector $\mathbf{a}_i$.

note: If you interchange two columns in P to get a matrix Q, then Q also diagonalizes A. The resulting matrix $D_2 = Q^{-1}AQ$ is the same as interchanging the same two columns in $P^{-1}AP$ as were interchanged in P to get Q.

EXAMPLE 16-4: For each of the following matrices, find a nonsingular matrix P that diagonalizes it and also find the similar diagonal matrix.

(a) $A = \begin{bmatrix}2 & 1\\ 2 & 3\end{bmatrix}$ **(b)** $B = \begin{bmatrix}1 & 0 & -1\\ 0 & 2 & 0\\ -8 & 0 & 3\end{bmatrix}$ **(c)** $C = \begin{bmatrix}2 & 0 & 3\\ 0 & 2 & 1\\ 0 & 0 & 1\end{bmatrix}$

Solution

(a) In Example 16-3, we found that A is diagonalizable but did not find eigenvectors. For the eigenvalue 1, we get

$$\begin{bmatrix} -1 & -1 \\ -2 & -2 \end{bmatrix}\begin{bmatrix} a_1 \\ a_2 \end{bmatrix} = \begin{bmatrix} 0 \\ 0 \end{bmatrix}$$

which has the solution

$$\begin{bmatrix} a_1 \\ a_2 \end{bmatrix} = \begin{bmatrix} t \\ -t \end{bmatrix} = t\begin{bmatrix} 1 \\ -1 \end{bmatrix}$$

Therefore, an eigenvector corresponding to 1 is $\begin{bmatrix} 1 \\ -1 \end{bmatrix}$.

For the eigenvalue 4, we get

$$\begin{bmatrix} 2 & -1 \\ -2 & 1 \end{bmatrix}\begin{bmatrix} a_1 \\ a_2 \end{bmatrix} = \begin{bmatrix} 0 \\ 0 \end{bmatrix}$$

which has the solution

$$\begin{bmatrix} a_1 \\ a_2 \end{bmatrix} = \begin{bmatrix} t \\ 2t \end{bmatrix} = t\begin{bmatrix} 1 \\ 2 \end{bmatrix}$$

Therefore, an eigenvector corresponding to 4 is $\begin{bmatrix} 1 \\ 2 \end{bmatrix}$. Therefore, $P = \begin{bmatrix} 1 & 1 \\ -1 & 2 \end{bmatrix}$ and $P^{-1} = \begin{bmatrix} 2/3 & -1/3 \\ 1/3 & 1/3 \end{bmatrix}$.

$$D = P^{-1}AP = \begin{bmatrix} \frac{2}{3} & -\frac{1}{3} \\ \frac{1}{3} & \frac{1}{3} \end{bmatrix}\begin{bmatrix} 2 & 1 \\ 2 & 3 \end{bmatrix}\begin{bmatrix} 1 & 1 \\ -1 & 2 \end{bmatrix} = \begin{bmatrix} 1 & 0 \\ 0 & 4 \end{bmatrix}$$

Observe that not only is this diagonal, but that each diagonal entry is the eigenvalue corresponding to the eigenvector for that same column of P.

(b) In Example 16-3, we found that B is diagonalizable but did not find eigenvectors. For $\lambda = -1$, we get

$$\begin{bmatrix} -2 & 0 & 1 \\ 0 & -3 & 0 \\ 8 & 0 & -4 \end{bmatrix}\begin{bmatrix} a_1 \\ a_2 \\ a_3 \end{bmatrix} = \begin{bmatrix} 0 \\ 0 \\ 0 \end{bmatrix}$$

which has the solution

$$\begin{bmatrix} a_1 \\ a_2 \\ a_3 \end{bmatrix} = \begin{bmatrix} t \\ 0 \\ 2t \end{bmatrix} = t\begin{bmatrix} 1 \\ 0 \\ 2 \end{bmatrix}$$

Therefore, an eigenvector corresponding to -1 is $\begin{bmatrix} 1 \\ 0 \\ 2 \end{bmatrix}$.

For $\lambda = 2$, we get

$$\begin{bmatrix} 1 & 0 & 1 \\ 0 & 0 & 0 \\ 8 & 0 & -1 \end{bmatrix}\begin{bmatrix} a_1 \\ a_2 \\ a_3 \end{bmatrix} = \begin{bmatrix} 0 \\ 0 \\ 0 \end{bmatrix}$$

which has the solution

$$\begin{bmatrix} a_1 \\ a_2 \\ a_3 \end{bmatrix} = \begin{bmatrix} 0 \\ t \\ 0 \end{bmatrix} = t\begin{bmatrix} 0 \\ 1 \\ 0 \end{bmatrix}$$

Therefore, an eigenvector corresponding to 2 is $\begin{bmatrix} 0 \\ 1 \\ 0 \end{bmatrix}$.

For $\lambda = 5$, we get

$$\begin{bmatrix} 4 & 0 & 1 \\ 0 & 3 & 0 \\ 8 & 0 & 2 \end{bmatrix}\begin{bmatrix} a_1 \\ a_2 \\ a_3 \end{bmatrix} = \begin{bmatrix} 0 \\ 0 \\ 0 \end{bmatrix}$$

which has the solution

$$\begin{bmatrix} a_1 \\ a_2 \\ a_3 \end{bmatrix} = \begin{bmatrix} t \\ 0 \\ -4t \end{bmatrix} = t\begin{bmatrix} 1 \\ 0 \\ -4 \end{bmatrix}$$

Therefore, an eigenvector corresponding to 5 is $\begin{bmatrix} 1 \\ 0 \\ -4 \end{bmatrix}$. Therefore,

$$P = \begin{bmatrix} 1 & 0 & 1 \\ 0 & 1 & 0 \\ 2 & 0 & -4 \end{bmatrix} \quad \text{and} \quad P^{-1} = \begin{bmatrix} \frac{2}{3} & 0 & \frac{1}{6} \\ 0 & 1 & 0 \\ \frac{1}{3} & 0 & -\frac{1}{6} \end{bmatrix}$$

$$D = P^{-1}BP = \begin{bmatrix} \frac{2}{3} & 0 & \frac{1}{6} \\ 0 & 1 & 0 \\ \frac{1}{3} & 0 & -\frac{1}{6} \end{bmatrix}\begin{bmatrix} 1 & 0 & -1 \\ 0 & 2 & 0 \\ -8 & 0 & 3 \end{bmatrix}\begin{bmatrix} 1 & 0 & 1 \\ 0 & 1 & 0 \\ 2 & 0 & -4 \end{bmatrix} = \begin{bmatrix} -1 & 0 & 0 \\ 0 & 2 & 0 \\ 0 & 0 & 5 \end{bmatrix}$$

Once more, notice that the eigenvalues are the diagonal entries.

(c) In Example 16-3, we found that C is diagonalizable with eigenvalues 2, 2, 1 and eigenvectors $\begin{bmatrix} 1 \\ 0 \\ 0 \end{bmatrix}, \begin{bmatrix} 0 \\ 1 \\ 0 \end{bmatrix}, \begin{bmatrix} 3 \\ 1 \\ -1 \end{bmatrix}$, respectively. Therefore,

$$P = \begin{bmatrix} 1 & 0 & 3 \\ 0 & 1 & 1 \\ 0 & 0 & -1 \end{bmatrix} \quad \text{and} \quad P^{-1} = \begin{bmatrix} 1 & 0 & 3 \\ 0 & 1 & 1 \\ 0 & 0 & -1 \end{bmatrix}$$

$$D = P^{-1}CP = \begin{bmatrix} 1 & 0 & 3 \\ 0 & 1 & 1 \\ 0 & 0 & -1 \end{bmatrix}\begin{bmatrix} 2 & 0 & 3 \\ 0 & 2 & 1 \\ 0 & 0 & 1 \end{bmatrix}\begin{bmatrix} 1 & 0 & 3 \\ 0 & 1 & 1 \\ 0 & 0 & -1 \end{bmatrix} = \begin{bmatrix} 2 & 0 & 0 \\ 0 & 2 & 0 \\ 0 & 0 & 1 \end{bmatrix}$$

EXAMPLE 16-5: For matrix B in Example 16-4, show that $Q = \begin{bmatrix} 1 & 0 & 1 \\ 0 & 1 & 0 \\ -4 & 0 & 2 \end{bmatrix}$ obtained by interchanging the eigenvectors in the first and third columns of P also diagonalizes B.

Solution: We find

$$Q^{-1} = \begin{bmatrix} \frac{1}{3} & 0 & -\frac{1}{6} \\ 0 & 1 & 0 \\ \frac{2}{3} & 0 & \frac{1}{6} \end{bmatrix}$$

$$\therefore \quad D = Q^{-1}BQ = \begin{bmatrix} \frac{1}{3} & 0 & -\frac{1}{6} \\ 0 & 1 & 0 \\ \frac{2}{3} & 0 & \frac{1}{6} \end{bmatrix}\begin{bmatrix} 1 & 0 & -1 \\ 0 & 2 & 0 \\ -8 & 0 & 3 \end{bmatrix}\begin{bmatrix} 1 & 0 & 1 \\ 0 & 1 & 0 \\ -4 & 0 & 2 \end{bmatrix} = \begin{bmatrix} 5 & 0 & 0 \\ 0 & 2 & 0 \\ 0 & 0 & -1 \end{bmatrix}$$

Observe that the diagonal elements of D are eigenvalues, but the order has been changed to correspond to the order in which the eigenvectors of Q were chosen. In general, any permutation of the linearly independent eigenvectors corresponding to the eigenvalues of a matrix A gives a matrix P that diagonalizes A.

16-3. Diagonalization of Symmetric Matrices

A. Definitions of symmetric and orthogonal matrices

An $n \times n$ matrix A is symmetric if $A = A^T$ where A^T is the transpose of A. It is orthogonal if and only if its columns form an orthonormal set. (*Note:* We shall use only the standard inner product.) Furthermore, a nonsingular matrix A is orthogonal if and only if $A^{-1} = A^T$.

EXAMPLE 16-6: Which of the following matrices are symmetric?

(a) $A = \begin{bmatrix} 1 & 2 & 3 \\ 2 & 0 & -2 \\ 3 & -2 & 4 \end{bmatrix}$ **(b)** $B = \begin{bmatrix} 1 & 2 & 3 \\ 3 & 1 & 4 \\ 2 & 4 & 5 \end{bmatrix}$

Solution

(a) $$A^T = \begin{bmatrix} 1 & 2 & 3 \\ 2 & 0 & -2 \\ 3 & -2 & 4 \end{bmatrix} = A, \qquad \therefore \quad A \text{ is symmetric.}$$

(b) $$B^T = \begin{bmatrix} 1 & 3 & 2 \\ 2 & 1 & 4 \\ 3 & 4 & 5 \end{bmatrix} \neq B, \qquad \therefore \quad B \text{ is not symmetric.}$$

EXAMPLE 16-7: Which of the following matrices are orthogonal?

(a) $A = \begin{bmatrix} \frac{1}{\sqrt{5}} & -\frac{2}{\sqrt{5}} \\ \frac{2}{\sqrt{5}} & \frac{1}{\sqrt{5}} \end{bmatrix}$ **(b)** $B = \begin{bmatrix} \frac{1}{3} & \frac{2}{3} & \frac{2}{3} \\ -\frac{2}{3} & -\frac{1}{3} & \frac{2}{3} \\ -\frac{2}{3} & \frac{2}{3} & -\frac{1}{3} \end{bmatrix}$

(c) $C = \begin{bmatrix} \frac{1}{\sqrt{2}} & \frac{1}{\sqrt{2}} & 0 \\ \frac{1}{\sqrt{2}} & -\frac{1}{\sqrt{2}} & 0 \\ 0 & 0 & 1 \end{bmatrix}$ **(d)** $D = \begin{bmatrix} \frac{1}{\sqrt{2}} & 0 & \frac{1}{\sqrt{2}} \\ \frac{1}{\sqrt{2}} & \frac{1}{\sqrt{2}} & 0 \\ 0 & \frac{1}{\sqrt{2}} & \frac{1}{\sqrt{2}} \end{bmatrix}$

Solution: Usually the shortest way to determine if a matrix A is orthogonal is to compute A^TA. If this product is the identity I, then A is orthogonal. If it is not I, then A is not orthogonal.

(a) $$A^TA = \begin{bmatrix} \frac{1}{\sqrt{5}} & \frac{2}{\sqrt{5}} \\ -\frac{2}{\sqrt{5}} & \frac{1}{\sqrt{5}} \end{bmatrix} \begin{bmatrix} \frac{1}{\sqrt{5}} & -\frac{2}{\sqrt{5}} \\ \frac{2}{\sqrt{5}} & \frac{1}{\sqrt{5}} \end{bmatrix} = \begin{bmatrix} 1 & 0 \\ 0 & 1 \end{bmatrix} = I$$

$\therefore$ A is orthogonal, $A^{-1} = A^T$

(b)

$$B^TB = \begin{bmatrix} \frac{1}{3} & -\frac{2}{3} & -\frac{2}{3} \\ \frac{2}{3} & -\frac{1}{3} & \frac{2}{3} \\ \frac{2}{3} & \frac{2}{3} & -\frac{1}{3} \end{bmatrix} \begin{bmatrix} \frac{1}{3} & \frac{2}{3} & \frac{2}{3} \\ -\frac{2}{3} & -\frac{1}{3} & \frac{2}{3} \\ -\frac{2}{3} & \frac{2}{3} & -\frac{1}{3} \end{bmatrix} = \begin{bmatrix} 1 & 0 & 0 \\ 0 & 1 & 0 \\ 0 & 0 & 1 \end{bmatrix} = I$$

$\therefore$ B is orthogonal, $B^{-1} = B^T$

(c)

$$C^TC = \begin{bmatrix} \frac{1}{\sqrt{2}} & \frac{1}{\sqrt{2}} & 0 \\ \frac{1}{\sqrt{2}} & -\frac{1}{\sqrt{2}} & 0 \\ 0 & 0 & 1 \end{bmatrix} \begin{bmatrix} \frac{1}{\sqrt{2}} & \frac{1}{\sqrt{2}} & 0 \\ \frac{1}{\sqrt{2}} & -\frac{1}{\sqrt{2}} & 0 \\ 0 & 0 & 1 \end{bmatrix} = \begin{bmatrix} 1 & 0 & 0 \\ 0 & 1 & 0 \\ 0 & 0 & 1 \end{bmatrix} = I$$

$\therefore$ C is orthogonal, $C^{-1} = C^T$

(d)

$$D^TD = \begin{bmatrix} \frac{1}{\sqrt{2}} & \frac{1}{\sqrt{2}} & 0 \\ 0 & \frac{1}{\sqrt{2}} & \frac{1}{\sqrt{2}} \\ \frac{1}{\sqrt{2}} & 0 & \frac{1}{\sqrt{2}} \end{bmatrix} \begin{bmatrix} \frac{1}{\sqrt{2}} & 0 & \frac{1}{\sqrt{2}} \\ \frac{1}{\sqrt{2}} & \frac{1}{\sqrt{2}} & 0 \\ 0 & \frac{1}{\sqrt{2}} & \frac{1}{\sqrt{2}} \end{bmatrix} = \begin{bmatrix} 1 & \frac{1}{2} & \frac{1}{2} \\ \frac{1}{2} & 1 & \frac{1}{2} \\ \frac{1}{2} & \frac{1}{2} & 1 \end{bmatrix} \neq I$$

Since $D^TD \neq I$, D is not orthogonal.

B. Properties of symmetric matrices

If A is an $n \times n$ symmetric matrix with real entries, then

(a) the characteristic equation has real roots;
(b) eigenvectors associated with distinct eigenvalues of A are orthogonal;
(c) the eigenspace of each eigenvalue has an orthogonal basis and the dimension of the eigenspace is the multiplicity of the eigenvalue;
(d) A has an orthonormal set of n eigenvectors which is a basis for R^n; and
(e) an orthogonal matrix P and a diagonal matrix D exist such that $D = P^TAP$, in which case we say A is orthogonally diagonalizable and $P^{-1} = P^T$. The diagonal entries of D are the eigenvalues of A.

EXAMPLE 16-8: Verify that the symmetric matrix $A = \begin{bmatrix} 1 & 0 & 0 \\ 0 & 0 & 1 \\ 0 & 1 & 0 \end{bmatrix}$ satisfies all five of the foregoing properties.

Solution

(a)

$$\det(\lambda I - A) = \begin{vmatrix} \lambda - 1 & 0 & 0 \\ 0 & \lambda & -1 \\ 0 & -1 & \lambda \end{vmatrix} = \lambda^3 - \lambda^2 - \lambda + 1 = 0$$

or

$$(\lambda - 1)^2(\lambda + 1) = 0$$

Therefore, the eigenvalues that are the roots of the characteristic equation are 1, with multiplicity 2, and -1 with multiplicity 1. Thus, property (**a**) is satisfied.

(**b**) For $\lambda = -1$, we get

$$\begin{bmatrix} -2 & 0 & 0 \\ 0 & -1 & -1 \\ 0 & -1 & -1 \end{bmatrix} \begin{bmatrix} a_1 \\ a_2 \\ a_3 \end{bmatrix} = \begin{bmatrix} 0 \\ 0 \\ 0 \end{bmatrix}$$

which has the solution

$$\begin{bmatrix} a_1 \\ a_2 \\ a_3 \end{bmatrix} = \begin{bmatrix} 0 \\ t \\ -t \end{bmatrix} = t \begin{bmatrix} 0 \\ 1 \\ -1 \end{bmatrix}$$

Therefore, an eigenvector associated with the eigenvalue -1 is $\begin{bmatrix} 0 \\ 1 \\ -1 \end{bmatrix}$.

For $\lambda = 1$, we get

$$\begin{bmatrix} 0 & 0 & 0 \\ 0 & 1 & -1 \\ 0 & -1 & 1 \end{bmatrix} \begin{bmatrix} a_1 \\ a_2 \\ a_3 \end{bmatrix} = \begin{bmatrix} 0 \\ 0 \\ 0 \end{bmatrix}$$

which has the solution

$$\begin{bmatrix} a_1 \\ a_2 \\ a_3 \end{bmatrix} = \begin{bmatrix} s \\ t \\ t \end{bmatrix} = s \begin{bmatrix} 1 \\ 0 \\ 0 \end{bmatrix} + t \begin{bmatrix} 0 \\ 1 \\ 1 \end{bmatrix}$$

Therefore, the eigenvectors associated with eigenvalue 1 are $\begin{bmatrix} 1 \\ 0 \\ 0 \end{bmatrix}$ and $\begin{bmatrix} 0 \\ 1 \\ 1 \end{bmatrix}$.

The inner products

$$\left\langle \begin{bmatrix} 1 \\ 0 \\ 0 \end{bmatrix}, \begin{bmatrix} 0 \\ 1 \\ -1 \end{bmatrix} \right\rangle = 1(0) + 0(1) + (0)(-1) = 0$$

and

$$\left\langle \begin{bmatrix} 0 \\ 1 \\ 1 \end{bmatrix}, \begin{bmatrix} 0 \\ 1 \\ -1 \end{bmatrix} \right\rangle = 0(0) + 1(1) + 1(-1) = 0$$

Therefore, $\begin{bmatrix} 1 \\ 0 \\ 0 \end{bmatrix}$ and $\begin{bmatrix} 0 \\ 1 \\ 1 \end{bmatrix}$ are each orthogonal to $\begin{bmatrix} 0 \\ 1 \\ -1 \end{bmatrix}$, and hence, property (**b**) is satisfied.

(**c**) Since $\begin{bmatrix} 1 \\ 0 \\ 0 \end{bmatrix}$ and $\begin{bmatrix} 0 \\ 1 \\ 1 \end{bmatrix}$ are already orthogonal, we see that property (**c**) is satisfied.

(d) Since $\begin{bmatrix}1\\0\\0\end{bmatrix}$ is already a unit vector, we need to normalize only $\begin{bmatrix}0\\1\\1\end{bmatrix}$ and $\begin{bmatrix}0\\1\\-1\end{bmatrix}$. When we divide each by its length, we get $\begin{bmatrix}0\\1/\sqrt{2}\\1/\sqrt{2}\end{bmatrix}$ and $\begin{bmatrix}0\\1/\sqrt{2}\\-1/\sqrt{2}\end{bmatrix}$. The three vectors $\begin{bmatrix}1\\0\\0\end{bmatrix}$, $\begin{bmatrix}0\\1/\sqrt{2}\\1/\sqrt{2}\end{bmatrix}$, and $\begin{bmatrix}0\\1/\sqrt{2}\\-1/\sqrt{2}\end{bmatrix}$ form an orthonormal basis for R^3 because each is a unit vector and

$$\begin{vmatrix}1 & 0 & 0\\ 0 & \dfrac{1}{\sqrt{2}} & \dfrac{1}{\sqrt{2}}\\ 0 & \dfrac{1}{\sqrt{2}} & -\dfrac{1}{\sqrt{2}}\end{vmatrix} = -\frac{1}{2} - \frac{1}{2} = -1 \neq 0$$

(e) Let

$$P = \begin{bmatrix}1 & 0 & 0\\ 0 & \dfrac{1}{\sqrt{2}} & \dfrac{1}{\sqrt{2}}\\ 0 & \dfrac{1}{\sqrt{2}} & -\dfrac{1}{\sqrt{2}}\end{bmatrix}$$

It is orthogonal because

$$P^TP = \begin{bmatrix}1 & 0 & 0\\ 0 & \dfrac{1}{\sqrt{2}} & \dfrac{1}{\sqrt{2}}\\ 0 & \dfrac{1}{\sqrt{2}} & -\dfrac{1}{\sqrt{2}}\end{bmatrix}\begin{bmatrix}1 & 0 & 0\\ 0 & \dfrac{1}{\sqrt{2}} & \dfrac{1}{\sqrt{2}}\\ 0 & \dfrac{1}{\sqrt{2}} & -\dfrac{1}{\sqrt{2}}\end{bmatrix} = \begin{bmatrix}1 & 0 & 0\\ 0 & 1 & 0\\ 0 & 0 & 1\end{bmatrix}$$

Thus,

$$P^{-1} = P^T$$

$$\therefore \quad D = P^TAP = \begin{bmatrix}1 & 0 & 0\\ 0 & \dfrac{1}{\sqrt{2}} & \dfrac{1}{\sqrt{2}}\\ 0 & \dfrac{1}{\sqrt{2}} & -\dfrac{1}{\sqrt{2}}\end{bmatrix}\begin{bmatrix}1 & 0 & 0\\ 0 & 0 & 1\\ 0 & 1 & 0\end{bmatrix}\begin{bmatrix}1 & 0 & 0\\ 0 & \dfrac{1}{\sqrt{2}} & \dfrac{1}{\sqrt{2}}\\ 0 & \dfrac{1}{\sqrt{2}} & -\dfrac{1}{\sqrt{2}}\end{bmatrix} = \begin{bmatrix}1 & 0 & 0\\ 0 & 1 & 0\\ 0 & 0 & -1\end{bmatrix}$$

The diagonal entries of D are the eigenvalues of A.

C. Procedure for orthogonally diagonalizing an $n \times n$ symmetric matrix A

Step 1: Find the eigenvalues of A.

Step 2: Find basis eigenvectors associated with each eigenvalue.

Step 3: If a basis for an eigenspace contains more than one eigenvector, use the Gram–Schmidt process to obtain an orthonormal basis for that eigenspace. If a basis for an eigenspace contains only one eigenvector, change it to a unit vector.

Step 4: Form the matrix P having the resulting eigenvectors $\mathbf{a}_1, \mathbf{a}_2, \ldots, \mathbf{a}_n$ from Step 3 as columns.

Step 5: Find P^T. (This is also P^{-1}.)

Step 6: Find $D = P^TAP$ which will be diagonal with the diagonal entry in the ith column being the eigenvalue corresponding to the eigenvector $\mathbf{a}_i$.

EXAMPLE 16-9: Orthogonally diagonalize the matrix $A = \begin{bmatrix} 0 & 3 & 3 \\ 3 & 0 & 3 \\ 3 & 3 & 0 \end{bmatrix}$.

Solution

Step 1:

$$\det(\lambda I - A) = \begin{vmatrix} \lambda & -3 & -3 \\ -3 & \lambda & -3 \\ -3 & -3 & \lambda \end{vmatrix} = \lambda^3 - 27\lambda - 54$$

$$\therefore \quad (\lambda + 3)^2(\lambda - 6) = 0$$

Thus, the eigenvalues are -3 and 6 with multiplicities 2 and 1, respectively.

Step 2: For $\lambda = -3$, we get

$$\begin{bmatrix} -3 & -3 & -3 \\ -3 & -3 & -3 \\ -3 & -3 & -3 \end{bmatrix} \begin{bmatrix} a_1 \\ a_2 \\ a_3 \end{bmatrix} = \begin{bmatrix} 0 \\ 0 \\ 0 \end{bmatrix}$$

which has the solution

$$\begin{bmatrix} a_1 \\ a_2 \\ a_3 \end{bmatrix} = \begin{bmatrix} s \\ t \\ -(s+t) \end{bmatrix} = s \begin{bmatrix} 1 \\ 0 \\ -1 \end{bmatrix} + t \begin{bmatrix} 0 \\ 1 \\ -1 \end{bmatrix}$$

For $\lambda = 6$, we get

$$\begin{bmatrix} 6 & -3 & -3 \\ -3 & 6 & -3 \\ -3 & -3 & 6 \end{bmatrix} \begin{bmatrix} a_1 \\ a_2 \\ a_3 \end{bmatrix} = \begin{bmatrix} 0 \\ 0 \\ 0 \end{bmatrix}$$

which has the solution

$$\begin{bmatrix} a_1 \\ a_2 \\ a_3 \end{bmatrix} = \begin{bmatrix} t \\ t \\ t \end{bmatrix} = t \begin{bmatrix} 1 \\ 1 \\ 1 \end{bmatrix}$$

Step 3: Since a basis for the eigenspace associated with the eigenvalue -3 contains two eigenvectors, we must apply Gram–Schmidt to obtain an orthonormal basis.

Let $\mathbf{b}_1 = \begin{bmatrix} 1 \\ 0 \\ -1 \end{bmatrix}$ and $\mathbf{b}_2 = \begin{bmatrix} 0 \\ 1 \\ -1 \end{bmatrix}$. Therefore,

$$\mathbf{a}_1 = \frac{\mathbf{b}_1}{\|\mathbf{b}_1\|} = \begin{bmatrix} \frac{1}{\sqrt{2}} \\ 0 \\ -\frac{1}{\sqrt{2}} \end{bmatrix}$$

$$\mathbf{c}_2 = \mathbf{b}_2 - \frac{\langle \mathbf{b}_2, \mathbf{a}_1 \rangle}{\langle \mathbf{a}_1, \mathbf{a}_1 \rangle} \mathbf{a}_1$$

$$= \begin{bmatrix} 0 \\ 1 \\ -1 \end{bmatrix} - \frac{\left\langle \begin{bmatrix} 0 \\ 1 \\ -1 \end{bmatrix}, \begin{bmatrix} 1/\sqrt{2} \\ 0 \\ -1/\sqrt{2} \end{bmatrix} \right\rangle}{\left\langle \begin{bmatrix} 1/\sqrt{2} \\ 0 \\ -1/\sqrt{2} \end{bmatrix}, \begin{bmatrix} 1/\sqrt{2} \\ 0 \\ -1/\sqrt{2} \end{bmatrix} \right\rangle} \begin{bmatrix} \frac{1}{\sqrt{2}} \\ 0 \\ -\frac{1}{\sqrt{2}} \end{bmatrix}$$

$$= \begin{bmatrix} 0 \\ 1 \\ -1 \end{bmatrix} - \frac{1/\sqrt{2}}{1} \begin{bmatrix} \frac{1}{\sqrt{2}} \\ 0 \\ -\frac{1}{\sqrt{2}} \end{bmatrix} = \begin{bmatrix} 0 \\ 1 \\ -1 \end{bmatrix} - \begin{bmatrix} \frac{1}{2} \\ 0 \\ -\frac{1}{2} \end{bmatrix} = \begin{bmatrix} -\frac{1}{2} \\ 1 \\ -\frac{1}{2} \end{bmatrix}$$

To get $\mathbf{a}_2$ we must change $\mathbf{c}_2$ to a unit vector. Therefore,

$$\mathbf{a}_2 = \frac{\mathbf{c}_2}{\|\mathbf{c}_2\|} = \begin{bmatrix} -\frac{1}{\sqrt{6}} \\ \frac{2}{\sqrt{6}} \\ -\frac{1}{\sqrt{6}} \end{bmatrix}$$

Changing $\begin{bmatrix} 1 \\ 1 \\ 1 \end{bmatrix}$ to a unit vector, we get $\begin{bmatrix} 1/\sqrt{3} \\ 1/\sqrt{3} \\ 1/\sqrt{3} \end{bmatrix}$.

Step 4:

$$P = \begin{bmatrix} \frac{1}{\sqrt{2}} & -\frac{1}{\sqrt{6}} & \frac{1}{\sqrt{3}} \\ 0 & \frac{2}{\sqrt{6}} & \frac{1}{\sqrt{3}} \\ -\frac{1}{\sqrt{2}} & -\frac{1}{\sqrt{6}} & \frac{1}{\sqrt{3}} \end{bmatrix}$$

Step 5:

$$P^T = \begin{bmatrix} \frac{1}{\sqrt{2}} & 0 & -\frac{1}{\sqrt{2}} \\ -\frac{1}{\sqrt{6}} & \frac{2}{\sqrt{6}} & -\frac{1}{\sqrt{6}} \\ \frac{1}{\sqrt{3}} & \frac{1}{\sqrt{3}} & \frac{1}{\sqrt{3}} \end{bmatrix}$$

Step 6:

$$D = P^TAP = \begin{bmatrix} \frac{1}{\sqrt{2}} & 0 & -\frac{1}{\sqrt{2}} \\ -\frac{1}{\sqrt{6}} & \frac{2}{\sqrt{6}} & -\frac{1}{\sqrt{6}} \\ \frac{1}{\sqrt{3}} & \frac{1}{\sqrt{3}} & \frac{1}{\sqrt{3}} \end{bmatrix} \begin{bmatrix} 0 & 3 & 3 \\ 3 & 0 & 3 \\ 3 & 3 & 0 \end{bmatrix} \begin{bmatrix} \frac{1}{\sqrt{2}} & -\frac{1}{\sqrt{6}} & -\frac{1}{\sqrt{3}} \\ 0 & \frac{2}{\sqrt{6}} & \frac{1}{\sqrt{3}} \\ -\frac{1}{\sqrt{2}} & -\frac{1}{\sqrt{6}} & \frac{1}{\sqrt{3}} \end{bmatrix}$$

$$= \begin{bmatrix} -3 & 0 & 0 \\ 0 & -3 & 0 \\ 0 & 0 & 6 \end{bmatrix}$$

SUMMARY

1. A diagonal matrix is a square matrix in which all entries off the main diagonal are zeros.
2. An $n \times n$ matrix A is diagonalizable if it is similar to a diagonal matrix D.
3. If $L: V \to V$ is a linear transformation of an n-dimensional vector space V into itself, then L is diagonalizable provided an ordered basis S for V exists such that the matrix representation of L with respect to S is diagonalizable.
4. An $n \times n$ matrix A is diagonalizable if and only if it has n linearly independent eigenvectors.
5. An $n \times n$ matrix A is diagonalizable if it has n distinct eigenvalues.
6. The procedure for diagonalizing a diagonalizable $n \times n$ matrix A is as follows.

   Step 1: Find the eigenvalues of A.
   Step 2: Find n linearly independent eigenvectors associated with the eigenvalues of A. Call them $\mathbf{a}_1, \mathbf{a}_2, \ldots, \mathbf{a}_n$.
   Step 3: Form the matrix P having $\mathbf{a}_1, \mathbf{a}_2, \ldots, \mathbf{a}_n$ as columns.
   Step 4: Find P^{-1}.
   Step 5: Find $D = P^{-1}AP$.

7. To diagonalize orthogonally an $n \times n$ symmetric matrix, follow the foregoing steps with the following changes. Replace Step 2 with, "Find basis eigenvectors associated with each eigenvalue. If a basis for an eigenspace contains more than one eigenvector, use the Gram–Schmidt process to obtain an orthonormal basis for that eigenspace. If a basis for an eigenspace contains only one eigenvector, change it to a unit eigenvector." In Steps 4 and 5, replace P^{-1} with P^T.
8. In $D = P^{-1}AP$ (or P^TAP) the diagonal entry in the ith column is the eigenvalue corresponding to the eigenvector $\mathbf{a}_i$.
9. In general, any permutation of the linearly independent eigenvectors corresponding to the eigenvalues of a matrix A gives a matrix P that diagonalizes A.
10. An $n \times n$ matrix A is symmetric if $A = A^T$ where A^T is the transpose of A.
11. A matrix is orthogonal if and only if its columns form an orthonormal set.
12. A nonsingular matrix A is orthogonal if and only if $A^{-1} = A^T$.
13. Usually, the shortest way to determine whether a matrix A is orthogonal is to compute A^TA. If this is I, then A is orthogonal; otherwise, it is not.
14. If A is an $n \times n$ symmetric matrix with real entries, then

    - the characteristic equation has real roots;
    - eigenvectors associated with distinct eigenvalues of A are orthogonal;

- the eigenspace of each eigenvalue has an orthogonal basis and the dimension of the eigenspace is the multiplicity of the eigenvalue;
- A has an orthonormal set of n eigenvectors that is a basis for R^n; and
- there exist an orthogonal matrix P and a diagonal matrix D such that $D = P^TAP$, in which case A is orthogonally diagonalizable and $P^{-1} = P^T$.

RAISE YOUR GRADES

Can you . . . ?

☑ determine if an $n \times n$ matrix A is diagonalizable
☑ diagonalize an $n \times n$ matrix
☑ orthogonally diagonalize an $n \times n$ symmetric matrix

SOLVED PROBLEMS

PROBLEM 16-1 Show that $A = \begin{bmatrix} 1 & 4 \\ 2 & -1 \end{bmatrix}$ is similar to $D = \begin{bmatrix} -3 & 0 \\ 0 & 3 \end{bmatrix}$ and is, therefore, diagonalizable.

Solution We must find an invertible matrix P such that $P^{-1}AP = D$, or $AP = PD$.

$$\begin{bmatrix} 1 & 4 \\ 2 & -1 \end{bmatrix}\begin{bmatrix} a & b \\ c & d \end{bmatrix} = \begin{bmatrix} a & b \\ c & d \end{bmatrix}\begin{bmatrix} -3 & 0 \\ 0 & 3 \end{bmatrix}$$

Expanding this system, we get

$$\begin{aligned} a \quad\quad + 4c \quad\quad &= -3a \\ b \quad\quad + 4d &= 3b \\ 2a \quad\quad - c \quad\quad &= -3c \\ 2b \quad\quad - d &= 3d \end{aligned}$$

This reduces to

$$\begin{aligned} a \quad + c \quad &= 0 \\ b \quad - 2d &= 0 \end{aligned}$$

A solution to this system is

$$a = 1, \qquad b = 2, \qquad c = -1, \qquad d = 1$$

$$\therefore \quad P = \begin{bmatrix} 1 & 2 \\ -1 & 1 \end{bmatrix} \quad \text{and} \quad P^{-1} = \begin{bmatrix} \frac{1}{3} & -\frac{2}{3} \\ \frac{1}{3} & \frac{1}{3} \end{bmatrix}.$$

Thus, we check

$$P^{-1}AP = \begin{bmatrix} \frac{1}{3} & -\frac{2}{3} \\ \frac{1}{3} & \frac{1}{3} \end{bmatrix}\begin{bmatrix} 1 & 4 \\ 2 & -1 \end{bmatrix}\begin{bmatrix} 1 & 2 \\ -1 & 1 \end{bmatrix} = \begin{bmatrix} -3 & 0 \\ 0 & 3 \end{bmatrix} = D$$

PROBLEM 16-2 Show that $A = \begin{bmatrix} 0 & 2 \\ -1 & 3 \end{bmatrix}$ is diagonalizable.

Solution A will be diagonalizable if it has two distinct eigenvalues.

$$\det(\lambda I - A) = \begin{vmatrix} \lambda & -2 \\ 1 & \lambda - 3 \end{vmatrix} = \lambda^2 - 3\lambda + 2 = (\lambda - 2)(\lambda - 1)$$

Therefore, the eigenvalues are 1 and 2, and so A is diagonalizable.

PROBLEM 16-3 Show that the linear transformation $L: R^2 \to R^2$ defined by $L\left(\begin{bmatrix} a_1 \\ a_2 \end{bmatrix}\right) = \begin{bmatrix} -a_1 + a_2 \\ -3a_2 \end{bmatrix}$ is diagonalizable by finding A, the matrix representation of L with respect to the natural ordered basis $B = \left\{\begin{bmatrix} 1 \\ 0 \end{bmatrix}, \begin{bmatrix} 0 \\ 1 \end{bmatrix}\right\}$, and showing that A is diagonalizable by finding D.

Solution Since we are using the natural ordered basis,

$$L\left(\begin{bmatrix} a_1 \\ a_2 \end{bmatrix}\right) = a_1 \begin{bmatrix} -2 \\ 0 \end{bmatrix} + a_2 \begin{bmatrix} 1 \\ -3 \end{bmatrix}$$

From this we read off $A = \begin{bmatrix} -2 & 1 \\ 0 & -3 \end{bmatrix}$ as the matrix representation of L.

Next we must diagonalize A by finding the eigenvalues and the corresponding eigenvectors that will give us the matrix P.

$$\det(\lambda I - A) = \begin{vmatrix} \lambda + 2 & -1 \\ 0 & \lambda + 3 \end{vmatrix} = (\lambda + 2)(\lambda + 3) = 0$$

There are two distinct eigenvalues, which tells us A is diagonalizable.

For $\lambda = -2$,

$$\begin{bmatrix} 0 & -1 \\ 0 & 1 \end{bmatrix} \begin{bmatrix} a_1 \\ a_2 \end{bmatrix} = \begin{bmatrix} 0 \\ 0 \end{bmatrix}$$

which has the solution

$$\begin{bmatrix} a_1 \\ a_2 \end{bmatrix} = \begin{bmatrix} t \\ 0 \end{bmatrix} = t\begin{bmatrix} 1 \\ 0 \end{bmatrix}$$

For $\lambda = -3$,

$$\begin{bmatrix} -1 & -1 \\ 0 & 0 \end{bmatrix} \begin{bmatrix} a_1 \\ a_2 \end{bmatrix} = \begin{bmatrix} 0 \\ 0 \end{bmatrix}$$

which has the solution

$$\begin{bmatrix} a_1 \\ a_2 \end{bmatrix} = \begin{bmatrix} t \\ -t \end{bmatrix} = t\begin{bmatrix} 1 \\ -1 \end{bmatrix}$$

$$\therefore \quad P = \begin{bmatrix} 1 & 1 \\ 0 & -1 \end{bmatrix} \quad \text{and} \quad P^{-1} = \begin{bmatrix} 1 & 1 \\ 0 & -1 \end{bmatrix}$$

Thus,

$$P = P^{-1}AP = \begin{bmatrix} 1 & 1 \\ 0 & -1 \end{bmatrix} \begin{bmatrix} -2 & 1 \\ 0 & -3 \end{bmatrix} \begin{bmatrix} 1 & 1 \\ 0 & -1 \end{bmatrix} = \begin{bmatrix} -2 & 0 \\ 0 & -3 \end{bmatrix} = D$$

PROBLEM 16-4 Determine if each of the following matrices is diagonalizable.

(a) $A = \begin{bmatrix} 2 & 0 & 1 \\ 0 & 2 & -1 \\ 3 & -1 & 2 \end{bmatrix}$ **(b)** $B = \begin{bmatrix} -3 & 0 & 1 \\ 0 & -3 & 0 \\ 0 & 0 & 2 \end{bmatrix}$ **(c)** $C = \begin{bmatrix} 2 & 1 \\ 0 & 2 \end{bmatrix}$

Solution

(a) $$\det(\lambda I - A) = \begin{vmatrix} \lambda - 2 & 0 & -1 \\ 0 & \lambda - 2 & 1 \\ -3 & 1 & \lambda - 2 \end{vmatrix} = (\lambda - 2)^3 - 3(\lambda - 2) - (\lambda - 2) = \lambda(\lambda - 2)(\lambda - 4)$$

There are three eigenvalues 0, 2, and 4. Thus, A is diagonalizable.

(b) $$\det(\lambda I - B) = \begin{vmatrix} \lambda + 3 & 0 & -1 \\ 0 & \lambda + 3 & 0 \\ 0 & 0 & \lambda - 2 \end{vmatrix} = (\lambda + 3)^2(\lambda - 2)$$

There are two eigenvalues -3, with multiplicity 2, and 2. We must check to see if there are two linearly independent eigenvectors for $\lambda = -3$.

For $\lambda = -3$,

$$\begin{bmatrix} 0 & 0 & -1 \\ 0 & 0 & 0 \\ 0 & 0 & -5 \end{bmatrix} \begin{bmatrix} a_1 \\ a_2 \\ a_3 \end{bmatrix} = \begin{bmatrix} 0 \\ 0 \\ 0 \end{bmatrix}$$

which has the solution

$$\begin{bmatrix} a_1 \\ a_2 \\ a_3 \end{bmatrix} = \begin{bmatrix} s \\ t \\ 0 \end{bmatrix} = s \begin{bmatrix} 1 \\ 0 \\ 0 \end{bmatrix} + t \begin{bmatrix} 0 \\ 1 \\ 0 \end{bmatrix}$$

Therefore, the 3×3 matrix B is diagonalizable because it has three linearly independent eigenvectors $\begin{bmatrix} 1 \\ 0 \\ 0 \end{bmatrix}$, $\begin{bmatrix} 0 \\ 1 \\ 0 \end{bmatrix}$, and $\begin{bmatrix} 1 \\ 0 \\ 5 \end{bmatrix}$. This last vector is the eigenvector for $\lambda = 2$.

(c) $$\det(\lambda I - C) = \begin{vmatrix} \lambda - 2 & -1 \\ 0 & \lambda - 2 \end{vmatrix} = (\lambda - 2)^2$$

There is only one eigenvalue $\lambda = 2$ of multiplicity 2. If there are two linearly independent eigenvectors corresponding to $\lambda = 2$, then C is diagonalizable.

For $\lambda = 2$,

$$\begin{bmatrix} 0 & -1 \\ 0 & 0 \end{bmatrix} \begin{bmatrix} a_1 \\ a_2 \end{bmatrix} = \begin{bmatrix} 0 \\ 0 \end{bmatrix}$$

which has the solution

$$\begin{bmatrix} a_1 \\ a_2 \end{bmatrix} = \begin{bmatrix} t \\ 0 \end{bmatrix} = t \begin{bmatrix} 1 \\ 0 \end{bmatrix}$$

Therefore, there is only one eigenvector associated with $\lambda = 2$, and C is not diagonalizable.

PROBLEM 16-5 For each of the following matrices, find a nonsingular matrix P that diagonalizes it and find the similar diagonal matrix D.

(a) $A = \begin{bmatrix} 2 & 0 & 1 \\ 0 & 2 & -1 \\ 3 & -1 & 2 \end{bmatrix}$ **(b)** $B = \begin{bmatrix} -3 & 0 & 1 \\ 0 & -3 & 0 \\ 0 & 0 & 2 \end{bmatrix}$

Solution

(a) In Problem 16-4, we found that A is diagonalizable with eigenvalues 0, 2, and 4. We must now find the corresponding eigenvectors.

For $\lambda = 0$,

$$\begin{bmatrix} -2 & 0 & -1 \\ 0 & -2 & 1 \\ -3 & 1 & -2 \end{bmatrix} \begin{bmatrix} a_1 \\ a_2 \\ a_3 \end{bmatrix} = \begin{bmatrix} 0 \\ 0 \\ 0 \end{bmatrix}$$

$$\begin{aligned} -2a_1 \qquad\quad - \; a_3 &= 0 \\ -2a_2 + \; a_3 &= 0 \\ -3a_1 + \; a_2 - 2a_3 &= 0 \end{aligned}$$

which has the solution

$$\begin{bmatrix} a_1 \\ a_2 \\ a_3 \end{bmatrix} = \begin{bmatrix} t \\ -t \\ -2t \end{bmatrix} = t\begin{bmatrix} 1 \\ -1 \\ -2 \end{bmatrix}$$

For $\lambda = 2$,

$$\begin{bmatrix} 0 & 0 & -1 \\ 0 & 0 & 1 \\ -3 & 1 & 0 \end{bmatrix} \begin{bmatrix} a_1 \\ a_2 \\ a_3 \end{bmatrix} = \begin{bmatrix} 0 \\ 0 \\ 0 \end{bmatrix}$$

which has the solution

$$\begin{bmatrix} a_1 \\ a_2 \\ a_3 \end{bmatrix} = \begin{bmatrix} t \\ 3t \\ 0 \end{bmatrix} = t\begin{bmatrix} 1 \\ 3 \\ 0 \end{bmatrix}$$

For $\lambda = 4$,

$$\begin{bmatrix} 2 & 0 & -1 \\ 0 & 2 & 1 \\ -3 & 1 & 2 \end{bmatrix} \begin{bmatrix} a_1 \\ a_2 \\ a_3 \end{bmatrix} = \begin{bmatrix} 0 \\ 0 \\ 0 \end{bmatrix}$$

which has the solution

$$\begin{bmatrix} a_1 \\ a_2 \\ a_3 \end{bmatrix} = \begin{bmatrix} t \\ -t \\ 2t \end{bmatrix} = t\begin{bmatrix} 1 \\ -1 \\ 2 \end{bmatrix}$$

Therefore,

$$P = \begin{bmatrix} 1 & 1 & 1 \\ -1 & 3 & -1 \\ -2 & 0 & 2 \end{bmatrix} \quad \text{and} \quad P^{-1} = \begin{bmatrix} \frac{3}{8} & -\frac{1}{8} & -\frac{1}{4} \\ \frac{1}{4} & \frac{1}{4} & 0 \\ \frac{3}{8} & -\frac{1}{8} & \frac{1}{4} \end{bmatrix}$$

Thus,

$$P^{-1}AP = \begin{bmatrix} \frac{3}{8} & -\frac{1}{8} & -\frac{1}{4} \\ \frac{1}{4} & \frac{1}{4} & 0 \\ \frac{3}{8} & -\frac{1}{8} & \frac{1}{4} \end{bmatrix} \begin{bmatrix} 2 & 0 & 1 \\ 0 & 2 & -1 \\ 3 & -1 & 2 \end{bmatrix} \begin{bmatrix} 1 & 1 & 1 \\ -1 & 3 & -1 \\ -2 & 0 & 2 \end{bmatrix}$$

$$= \begin{bmatrix} 0 & 0 & 0 \\ \frac{1}{2} & \frac{1}{2} & 0 \\ \frac{3}{2} & -\frac{1}{2} & 1 \end{bmatrix} \begin{bmatrix} 1 & 1 & 1 \\ -1 & 3 & -1 \\ -2 & 0 & 2 \end{bmatrix} = \begin{bmatrix} 0 & 0 & 0 \\ 0 & 2 & 0 \\ 0 & 0 & 4 \end{bmatrix}$$

$$\therefore \quad D = \begin{bmatrix} 0 & 0 & 0 \\ 0 & 2 & 0 \\ 0 & 0 & 4 \end{bmatrix}$$

(b) In Problem 16-4 we found the eigenvectors associated with eigenvalues -3 and 2 to be $\begin{bmatrix} 1 \\ 0 \\ 0 \end{bmatrix}$, $\begin{bmatrix} 0 \\ 1 \\ 0 \end{bmatrix}$, and $\begin{bmatrix} 1 \\ 0 \\ 5 \end{bmatrix}$. Thus, $P = \begin{bmatrix} 1 & 0 & 1 \\ 0 & 1 & 0 \\ 0 & 0 & 5 \end{bmatrix}$. We can find P^{-1} to be $\begin{bmatrix} 1 & 0 & -\frac{1}{5} \\ 0 & 1 & 0 \\ 0 & 0 & \frac{1}{5} \end{bmatrix}$. Therefore,

$$P^{-1}BP = \begin{bmatrix} 1 & 0 & -\frac{1}{5} \\ 0 & 1 & 0 \\ 0 & 0 & \frac{1}{5} \end{bmatrix} \begin{bmatrix} -3 & 0 & 1 \\ 0 & -3 & 0 \\ 0 & 0 & 2 \end{bmatrix} \begin{bmatrix} 1 & 0 & 1 \\ 0 & 1 & 0 \\ 0 & 0 & 5 \end{bmatrix}$$

$$= \begin{bmatrix} -3 & 0 & 0 \\ 0 & -3 & 0 \\ 0 & 0 & 2 \end{bmatrix} = D$$

PROBLEM 16-6 If we interchange the first two columns of matrix P found in Problem 16-4 **(b)** we get $Q = \begin{bmatrix} 0 & 1 & 1 \\ 1 & 0 & 0 \\ 0 & 0 & 5 \end{bmatrix}$. Show that this matrix will diagonalize $B = \begin{bmatrix} -3 & 0 & 1 \\ 0 & -3 & 0 \\ 0 & 0 & 2 \end{bmatrix}$.

Solution We find

$$Q^{-1} = \begin{bmatrix} 0 & 1 & 0 \\ 1 & 0 & -\frac{1}{5} \\ 0 & 0 & \frac{1}{5} \end{bmatrix}$$

Then,

$$Q^{-1}BQ = \begin{bmatrix} 0 & 1 & 0 \\ 1 & 0 & -\frac{1}{5} \\ 0 & 0 & \frac{1}{5} \end{bmatrix} \begin{bmatrix} -3 & 0 & 1 \\ 0 & -3 & 0 \\ 0 & 0 & 2 \end{bmatrix} \begin{bmatrix} 0 & 1 & 1 \\ 1 & 0 & 0 \\ 0 & 0 & 5 \end{bmatrix}$$

$$= \begin{bmatrix} -3 & 0 & 0 \\ 0 & -3 & 0 \\ 0 & 0 & 2 \end{bmatrix}$$

We cannot tell that the eigenvalues have been rearranged because the first two columns contained the same eigenvalue -3.

PROBLEM 16-7 Which of the following matrices are symmetric?

(a) $A = \begin{bmatrix} 2 & 1 & -4 \\ 1 & 1 & 5 \\ -4 & 5 & 3 \end{bmatrix}$ **(b)** $B = \begin{bmatrix} 3 & 2 & -1 \\ -1 & 1 & 1 \\ 2 & 1 & 4 \end{bmatrix}$

Solution

(a)
$$A^T = \begin{bmatrix} 2 & 1 & -4 \\ 1 & 1 & 5 \\ -4 & 5 & 3 \end{bmatrix} = A, \qquad \therefore \quad A \text{ is symmetric}$$

(b)
$$B^T = \begin{bmatrix} 3 & -1 & 2 \\ 2 & 1 & 1 \\ -1 & 1 & 4 \end{bmatrix} \neq B, \qquad \therefore \quad B \text{ is not symmetric}$$

PROBLEM 16-8 Which of the following matrices are orthogonal?

(a) $A = \begin{bmatrix} \frac{1}{\sqrt{10}} & \frac{3}{\sqrt{10}} \\ -\frac{3}{\sqrt{10}} & \frac{1}{\sqrt{10}} \end{bmatrix}$ **(b)** $B = \begin{bmatrix} \frac{1}{\sqrt{2}} & \frac{1}{\sqrt{6}} & -\frac{1}{\sqrt{3}} \\ -\frac{1}{\sqrt{2}} & \frac{1}{\sqrt{6}} & -\frac{1}{\sqrt{3}} \\ 0 & \frac{2}{\sqrt{6}} & \frac{1}{\sqrt{3}} \end{bmatrix}$

(c) $C = \begin{bmatrix} \frac{1}{\sqrt{10}} & \frac{3}{\sqrt{10}} & 0 \\ \frac{3}{\sqrt{10}} & -\frac{1}{\sqrt{10}} & 0 \\ 0 & 0 & 1 \end{bmatrix}$ **(d)** $D = \begin{bmatrix} \frac{1}{\sqrt{10}} & \frac{3}{\sqrt{10}} & 0 \\ 0 & \frac{1}{\sqrt{10}} & -\frac{3}{\sqrt{10}} \\ \frac{3}{\sqrt{10}} & 0 & \frac{1}{\sqrt{10}} \end{bmatrix}$

Solution

(a)
$$A^TA = \begin{bmatrix} \frac{1}{\sqrt{10}} & -\frac{3}{\sqrt{10}} \\ \frac{3}{\sqrt{10}} & \frac{1}{\sqrt{10}} \end{bmatrix} \begin{bmatrix} \frac{1}{\sqrt{10}} & \frac{3}{\sqrt{10}} \\ -\frac{3}{\sqrt{10}} & \frac{1}{\sqrt{10}} \end{bmatrix} = \begin{bmatrix} 1 & 0 \\ 0 & 1 \end{bmatrix} = I$$

$$\therefore \quad A \text{ is orthogonal, } A^{-1} = A^T$$

(b)
$$B^TB = \begin{bmatrix} \frac{1}{\sqrt{2}} & -\frac{1}{\sqrt{2}} & 0 \\ \frac{1}{\sqrt{6}} & \frac{1}{\sqrt{6}} & \frac{2}{\sqrt{6}} \\ -\frac{1}{\sqrt{3}} & -\frac{1}{\sqrt{3}} & \frac{1}{\sqrt{3}} \end{bmatrix} \begin{bmatrix} \frac{1}{\sqrt{2}} & \frac{1}{\sqrt{6}} & -\frac{1}{\sqrt{3}} \\ -\frac{1}{\sqrt{2}} & \frac{1}{\sqrt{6}} & -\frac{1}{\sqrt{3}} \\ 0 & \frac{2}{\sqrt{6}} & \frac{1}{\sqrt{3}} \end{bmatrix} = \begin{bmatrix} 1 & 0 & 0 \\ 0 & 1 & 0 \\ 0 & 0 & 1 \end{bmatrix} = I$$

$$\therefore \quad B \text{ is orthogonal, } B^{-1} = B^T$$

(c)

$$C^TC = \begin{bmatrix} \frac{1}{\sqrt{10}} & \frac{3}{\sqrt{10}} & 0 \\ \frac{3}{\sqrt{10}} & -\frac{1}{\sqrt{10}} & 0 \\ 0 & 0 & 1 \end{bmatrix} \begin{bmatrix} \frac{1}{\sqrt{10}} & \frac{3}{\sqrt{10}} & 0 \\ \frac{3}{\sqrt{10}} & -\frac{1}{\sqrt{10}} & 0 \\ 0 & 0 & 1 \end{bmatrix} = \begin{bmatrix} 1 & 0 & 0 \\ 0 & 1 & 0 \\ 0 & 0 & 1 \end{bmatrix} = I$$

$$\therefore \quad C \text{ is orthogonal, } C = C^T = C^{-1}$$

(d)

$$D^TD = \begin{bmatrix} \frac{1}{\sqrt{10}} & 0 & \frac{3}{\sqrt{10}} \\ \frac{3}{\sqrt{10}} & \frac{1}{\sqrt{10}} & 0 \\ 0 & -\frac{3}{\sqrt{10}} & \frac{1}{\sqrt{10}} \end{bmatrix} \begin{bmatrix} \frac{1}{\sqrt{10}} & \frac{3}{\sqrt{10}} & 0 \\ 0 & \frac{1}{\sqrt{10}} & -\frac{3}{\sqrt{10}} \\ \frac{3}{\sqrt{10}} & 0 & \frac{1}{\sqrt{10}} \end{bmatrix}$$

$$= \begin{bmatrix} 1 & \frac{3}{10} & \frac{3}{10} \\ \frac{3}{10} & 1 & -\frac{3}{10} \\ \frac{3}{10} & -\frac{3}{10} & 1 \end{bmatrix} \neq I$$

$$\therefore \quad D \text{ is not orthogonal}$$

PROBLEM 16-9 Verify the five properties for symmetric matrices given in Section 16-3.B for the symmetric matrix

$$A = \begin{bmatrix} 1 & 0 & 0 \\ 0 & 2 & \sqrt{2} \\ 0 & \sqrt{2} & 3 \end{bmatrix}$$

Solution

(a)

$$\det(\lambda I - A) = \begin{vmatrix} \lambda - 1 & 0 & 0 \\ 0 & \lambda - 2 & -\sqrt{2} \\ 0 & -\sqrt{2} & \lambda - 3 \end{vmatrix} = (\lambda - 1)(\lambda - 2)(\lambda - 3) - 2(\lambda - 1)$$

$$= (\lambda - 1)^2(\lambda - 4)$$

the eigenvalues are 1, with multiplicity 2, and 4. Therefore, property **(a)** of Section 16-3.B is satisfied.

(b) For $\lambda = 1$, we get

$$\begin{bmatrix} 0 & 0 & 0 \\ 0 & -1 & -\sqrt{2} \\ 0 & -\sqrt{2} & -2 \end{bmatrix} \begin{bmatrix} a_1 \\ a_2 \\ a_3 \end{bmatrix} = \begin{bmatrix} 0 \\ 0 \\ 0 \end{bmatrix}$$

which has the solution

$$\begin{bmatrix} a_1 \\ a_2 \\ a_3 \end{bmatrix} = s \begin{bmatrix} 1 \\ 0 \\ 0 \end{bmatrix} + t \begin{bmatrix} 0 \\ -\sqrt{2} \\ 1 \end{bmatrix}$$

Therefore, the eigenvectors associated with $\lambda = 1$ are $\begin{bmatrix} 1 \\ 0 \\ 0 \end{bmatrix}$ and $\begin{bmatrix} 0 \\ -\sqrt{2} \\ 1 \end{bmatrix}$

For $\lambda = 4$, we get

$$\begin{bmatrix} 3 & 0 & 0 \\ 0 & 2 & -\sqrt{2} \\ 0 & -\sqrt{2} & 1 \end{bmatrix} \begin{bmatrix} a_1 \\ a_2 \\ a_3 \end{bmatrix} = \begin{bmatrix} 0 \\ 0 \\ 0 \end{bmatrix}$$

which has the solution

$$\begin{bmatrix} a_1 \\ a_2 \\ a_3 \end{bmatrix} = \begin{bmatrix} 0 \\ t \\ \sqrt{2}t \end{bmatrix} = t \begin{bmatrix} 0 \\ 1 \\ \sqrt{2} \end{bmatrix}$$

Therefore, the eigenvector associated with $\lambda = 4$ is $\begin{bmatrix} 0 \\ 1 \\ \sqrt{2} \end{bmatrix}$. The inner products

$$\left\langle \begin{bmatrix} 1 \\ 0 \\ 0 \end{bmatrix}, \begin{bmatrix} 0 \\ 1 \\ \sqrt{2} \end{bmatrix} \right\rangle = 1(0) + 0(1) + 0(\sqrt{2}) = 0$$

and

$$\left\langle \begin{bmatrix} 0 \\ -\sqrt{2} \\ 1 \end{bmatrix}, \begin{bmatrix} 0 \\ 1 \\ \sqrt{2} \end{bmatrix} \right\rangle = 0(0) + (-\sqrt{2})(1) + 1(\sqrt{2}) = 0$$

Therefore, $\begin{bmatrix} 1 \\ 0 \\ 0 \end{bmatrix}$ and $\begin{bmatrix} 0 \\ -\sqrt{2} \\ 1 \end{bmatrix}$ are each orthogonal to $\begin{bmatrix} 0 \\ 1 \\ \sqrt{2} \end{bmatrix}$, and hence property **(b)** of Section 16-3B is satisfied.

(c) Since $\begin{bmatrix} 1 \\ 0 \\ 0 \end{bmatrix}$ and $\begin{bmatrix} 0 \\ -\sqrt{2} \\ 1 \end{bmatrix}$ are also orthogonal, the dimension of the eigenspace is the multiplicity of the eigenvalue. Thus, property **(c)** of Section 16-3B is satisfied.

(d) Since $\begin{bmatrix} 1 \\ 0 \\ 0 \end{bmatrix}$ is a unit vector, we only need to normalize $\begin{bmatrix} 0 \\ -\sqrt{2} \\ 1 \end{bmatrix}$ and $\begin{bmatrix} 0 \\ 1 \\ \sqrt{2} \end{bmatrix}$. When we divide each of these by its length, we get $\begin{bmatrix} 0 \\ -\sqrt{2}/\sqrt{3} \\ 1/\sqrt{3} \end{bmatrix}$ and $\begin{bmatrix} 0 \\ 1/\sqrt{3} \\ \sqrt{2}/\sqrt{3} \end{bmatrix}$. The three vectors $\begin{bmatrix} 1 \\ 0 \\ 0 \end{bmatrix}$, $\begin{bmatrix} 0 \\ -\sqrt{2}/\sqrt{3} \\ 1/\sqrt{3} \end{bmatrix}$, and $\begin{bmatrix} 0 \\ 1/\sqrt{3} \\ \sqrt{2}/\sqrt{3} \end{bmatrix}$ form an orthonormal basis for R^3 because each is a unit vector and

$$\begin{vmatrix} 1 & 0 & 0 \\ 0 & -\dfrac{\sqrt{2}}{\sqrt{3}} & \dfrac{1}{\sqrt{3}} \\ 0 & \dfrac{1}{\sqrt{3}} & \dfrac{\sqrt{2}}{\sqrt{3}} \end{vmatrix} = -\frac{2}{3} - \frac{1}{3} = -1 \neq 0$$

(e) Let

$$P = \begin{bmatrix} 1 & 0 & 0 \\ 0 & -\frac{\sqrt{2}}{\sqrt{3}} & \frac{1}{\sqrt{3}} \\ 0 & \frac{1}{\sqrt{3}} & \frac{\sqrt{2}}{\sqrt{3}} \end{bmatrix}$$

Since

$$P^T P = \begin{bmatrix} 1 & 0 & 0 \\ 0 & -\frac{\sqrt{2}}{\sqrt{3}} & \frac{1}{\sqrt{3}} \\ 0 & \frac{1}{\sqrt{3}} & \frac{\sqrt{2}}{\sqrt{3}} \end{bmatrix} \begin{bmatrix} 1 & 0 & 0 \\ 0 & -\frac{\sqrt{2}}{\sqrt{3}} & \frac{1}{\sqrt{3}} \\ 0 & \frac{1}{\sqrt{3}} & \frac{\sqrt{2}}{\sqrt{3}} \end{bmatrix} = \begin{bmatrix} 1 & 0 & 0 \\ 0 & 1 & 0 \\ 0 & 0 & 1 \end{bmatrix} = I$$

it is orthogonal, and $P^{-1} = P^T$. Therefore,

$$D = P^T A P = \begin{bmatrix} 1 & 0 & 0 \\ 0 & -\frac{\sqrt{2}}{\sqrt{3}} & \frac{1}{\sqrt{3}} \\ 0 & \frac{1}{\sqrt{3}} & \frac{\sqrt{2}}{\sqrt{3}} \end{bmatrix} \begin{bmatrix} 1 & 0 & 0 \\ 0 & 2 & \sqrt{2} \\ 0 & \sqrt{2} & 3 \end{bmatrix} \begin{bmatrix} 1 & 0 & 0 \\ 0 & -\frac{\sqrt{2}}{\sqrt{3}} & \frac{1}{\sqrt{3}} \\ 0 & \frac{1}{\sqrt{3}} & \frac{\sqrt{2}}{\sqrt{3}} \end{bmatrix}$$

$$= \begin{bmatrix} 1 & 0 & 0 \\ 0 & 1 & 0 \\ 0 & 0 & 4 \end{bmatrix}$$

Therefore, the diagonal entries of D are the eigenvalues of A.

PROBLEM 16-10 Orthogonally diagonalize the matrix $A = \begin{bmatrix} 2 & 1 & 1 \\ 1 & 2 & 1 \\ 1 & 1 & 2 \end{bmatrix}$.

Solution

Step 1: $$\det(\lambda I - A) = \begin{vmatrix} \lambda - 2 & -1 & -1 \\ -1 & \lambda - 2 & -1 \\ -1 & -1 & \lambda - 2 \end{vmatrix} = (\lambda - 1)^2(\lambda - 4)$$

Therefore, the eigenvalues are 1, with multiplicity 2, and 4.

Step 2: For $\lambda = 1$, we get

$$\begin{bmatrix} -1 & -1 & -1 \\ -1 & -1 & -1 \\ -1 & -1 & -1 \end{bmatrix} \begin{bmatrix} a_1 \\ a_2 \\ a_3 \end{bmatrix} = \begin{bmatrix} 0 \\ 0 \\ 0 \end{bmatrix}$$

which has the solution

$$\begin{bmatrix} a_1 \\ a_2 \\ a_3 \end{bmatrix} = \begin{bmatrix} s \\ t \\ -(s+t) \end{bmatrix} = s\begin{bmatrix} 1 \\ 0 \\ -1 \end{bmatrix} + t\begin{bmatrix} 0 \\ 1 \\ -1 \end{bmatrix}$$

For $\lambda = 4$, we get

$$\begin{bmatrix} 2 & -1 & -1 \\ -1 & 2 & -1 \\ -1 & -1 & 2 \end{bmatrix}\begin{bmatrix} a_1 \\ a_2 \\ a_3 \end{bmatrix} = \begin{bmatrix} 0 \\ 0 \\ 0 \end{bmatrix}$$

which has the solution

$$\begin{bmatrix} a_1 \\ a_2 \\ a_3 \end{bmatrix} = \begin{bmatrix} t \\ t \\ t \end{bmatrix} = t\begin{bmatrix} 1 \\ 1 \\ 1 \end{bmatrix}$$

Step 3: We must apply the Gram–Schmidt process to $\begin{bmatrix} 1 \\ 0 \\ -1 \end{bmatrix}$ and $\begin{bmatrix} 0 \\ 1 \\ -1 \end{bmatrix}$. Let

$$\mathbf{b}_1 = \begin{bmatrix} 1 \\ 0 \\ -1 \end{bmatrix} \quad \text{and} \quad \mathbf{b}_2 = \begin{bmatrix} 0 \\ 1 \\ -1 \end{bmatrix}$$

$$\therefore \quad \mathbf{a}_1 = \frac{\mathbf{b}_1}{\|\mathbf{b}_1\|} = \begin{bmatrix} \frac{1}{\sqrt{2}} \\ 0 \\ -\frac{1}{\sqrt{2}} \end{bmatrix}$$

$$\mathbf{c}_2 = \mathbf{b}_2 - \frac{\langle \mathbf{b}_2, \mathbf{a}_1 \rangle}{\langle \mathbf{a}_1, \mathbf{a}_1 \rangle}\mathbf{a}_1$$

$$= \begin{bmatrix} 0 \\ 1 \\ -1 \end{bmatrix} - \frac{\left\langle \begin{bmatrix} 0 \\ 1 \\ -1 \end{bmatrix}, \begin{bmatrix} 1/\sqrt{2} \\ 0 \\ -1/\sqrt{2} \end{bmatrix} \right\rangle}{\left\langle \begin{bmatrix} 1/\sqrt{2} \\ 0 \\ -1/\sqrt{2} \end{bmatrix}, \begin{bmatrix} 1/\sqrt{2} \\ 0 \\ -1/\sqrt{2} \end{bmatrix} \right\rangle} \begin{bmatrix} \frac{1}{\sqrt{2}} \\ 0 \\ -\frac{1}{\sqrt{2}} \end{bmatrix}$$

$$= \begin{bmatrix} 0 \\ 1 \\ -1 \end{bmatrix} - \frac{1/\sqrt{2}}{1}\begin{bmatrix} \frac{1}{\sqrt{2}} \\ 0 \\ -\frac{1}{\sqrt{2}} \end{bmatrix} = \begin{bmatrix} -\frac{1}{2} \\ 1 \\ -\frac{1}{2} \end{bmatrix}$$

To get $\mathbf{a}_2$, we must normalize $\mathbf{c}_2$. Therefore,

$$\mathbf{a}_2 = \frac{\mathbf{c}_2}{\|\mathbf{c}_2\|} = \begin{bmatrix} -\dfrac{1}{\sqrt{6}} \\ \dfrac{2}{\sqrt{6}} \\ -\dfrac{1}{\sqrt{6}} \end{bmatrix}$$

Changing $\begin{bmatrix} 1 \\ 1 \\ 1 \end{bmatrix}$ to a unit vector, we get $\begin{bmatrix} 1/\sqrt{3} \\ 1/\sqrt{3} \\ 1/\sqrt{3} \end{bmatrix}$.

Step 4:

$$P = \begin{bmatrix} \dfrac{1}{\sqrt{2}} & -\dfrac{1}{\sqrt{6}} & \dfrac{1}{\sqrt{3}} \\ 0 & \dfrac{2}{\sqrt{6}} & \dfrac{1}{\sqrt{3}} \\ -\dfrac{1}{\sqrt{2}} & -\dfrac{1}{\sqrt{6}} & \dfrac{1}{\sqrt{3}} \end{bmatrix}$$

Step 5:

$$P^T = \begin{bmatrix} \dfrac{1}{\sqrt{2}} & 0 & -\dfrac{1}{\sqrt{2}} \\ -\dfrac{1}{\sqrt{6}} & \dfrac{2}{\sqrt{6}} & -\dfrac{1}{\sqrt{6}} \\ \dfrac{1}{\sqrt{3}} & \dfrac{1}{\sqrt{3}} & \dfrac{1}{\sqrt{3}} \end{bmatrix}$$

Step 6:

$$D = P^TAP = \begin{bmatrix} \dfrac{1}{\sqrt{2}} & 0 & -\dfrac{1}{\sqrt{2}} \\ -\dfrac{1}{\sqrt{6}} & \dfrac{2}{\sqrt{6}} & -\dfrac{1}{\sqrt{6}} \\ \dfrac{1}{\sqrt{3}} & \dfrac{1}{\sqrt{3}} & \dfrac{1}{\sqrt{3}} \end{bmatrix} \begin{bmatrix} 2 & 1 & 1 \\ 1 & 2 & 1 \\ 1 & 1 & 2 \end{bmatrix} \begin{bmatrix} \dfrac{1}{\sqrt{2}} & -\dfrac{1}{\sqrt{6}} & \dfrac{1}{\sqrt{3}} \\ 0 & \dfrac{2}{\sqrt{6}} & \dfrac{1}{\sqrt{3}} \\ -\dfrac{1}{\sqrt{2}} & -\dfrac{1}{\sqrt{6}} & \dfrac{1}{\sqrt{3}} \end{bmatrix}$$

$$= \begin{bmatrix} 1 & 0 & 0 \\ 0 & 1 & 0 \\ 0 & 0 & 4 \end{bmatrix}$$

PROBLEM 16-11 Finding a matrix P that will diagonalize a matrix A that is diagonalizable is a powerful tool for solving a **system of differential equations** defined by

$$\begin{aligned} y_1' &= a_{11}y_1 + a_{12}y_2 + \cdots + a_{1n}y_n \\ y_2' &= a_{21}y_1 + a_{22}y_2 + \cdots + a_{2n}y_n \\ &\vdots \qquad \cdots \qquad\qquad \cdots \\ y_n' &= a_{n1}y_1 + a_{n2}y_2 + \cdots + a_{nn}y_n \end{aligned}$$

or, in matrix form,

$$\frac{d\mathbf{Y}}{dt} = \begin{bmatrix} a_{11} & a_{12} & \cdots & a_{1n} \\ a_{21} & a_{22} & \cdots & a_{2n} \\ \vdots & \cdots & & \\ a_{n1} & a_{n2} & \cdots & a_{nn} \end{bmatrix} \mathbf{Y} \quad \text{where } \mathbf{Y} = \begin{bmatrix} y_1 \\ y_2 \\ \vdots \\ y_n \end{bmatrix}, \qquad y_i = f_i(t)$$

and

$$\frac{d\mathbf{Y}}{dt} = \begin{bmatrix} y_1' \\ y_2' \\ \vdots \\ y_n' \end{bmatrix}$$

which is often written as $d\mathbf{Y}/dt = A\mathbf{Y}$. Write the system

$$\begin{aligned} y_1' &= -3y_1 + 4y_2 \\ y_2' &= 2y_1 - y_2 \end{aligned}$$

in matrix form.

Solution

$$\begin{bmatrix} y_1' \\ y_2' \end{bmatrix} = \begin{bmatrix} -3 & 4 \\ 2 & -1 \end{bmatrix} \begin{bmatrix} y_1 \\ y_2 \end{bmatrix} \quad \text{or} \quad \frac{d\mathbf{Y}}{dt} = \begin{bmatrix} -3 & 4 \\ 2 & -1 \end{bmatrix} \mathbf{Y}$$

PROBLEM 16-12 Find a matrix P that diagonalizes the matrix $A = \begin{bmatrix} -3 & 4 \\ 2 & -1 \end{bmatrix}$.

Solution

$$\det(\lambda I - A) = \begin{vmatrix} \lambda + 3 & -4 \\ -2 & \lambda + 1 \end{vmatrix} = \lambda^2 + 4\lambda - 5 = (\lambda + 5)(\lambda - 1)$$

For $\lambda = 1$, we get

$$\begin{bmatrix} 4 & -4 \\ -2 & 2 \end{bmatrix} \begin{bmatrix} a_1 \\ a_2 \end{bmatrix} = \begin{bmatrix} 0 \\ 0 \end{bmatrix}$$

which has the solution

$$\begin{bmatrix} a_1 \\ a_2 \end{bmatrix} = \begin{bmatrix} k \\ k \end{bmatrix} = k \begin{bmatrix} 1 \\ 1 \end{bmatrix}, \qquad k \text{ arbitrary}$$

For $\lambda = -5$, we get

$$\begin{bmatrix} -2 & -4 \\ -2 & -4 \end{bmatrix} \begin{bmatrix} a_1 \\ a_2 \end{bmatrix} = \begin{bmatrix} 0 \\ 0 \end{bmatrix}$$

which has the solution

$$\begin{bmatrix} a_1 \\ a_2 \end{bmatrix} = \begin{bmatrix} -2k \\ k \end{bmatrix} = k \begin{bmatrix} -2 \\ 1 \end{bmatrix}, \qquad k \text{ arbitrary}$$

$$\therefore \quad P = \begin{bmatrix} 1 & -2 \\ 1 & 1 \end{bmatrix} \qquad P^{-1} = \begin{bmatrix} \frac{1}{3} & \frac{2}{3} \\ -\frac{1}{3} & \frac{1}{3} \end{bmatrix}$$

Therefore,

$$D = P^{-1}AP = \begin{bmatrix} \frac{1}{3} & \frac{2}{3} \\ -\frac{1}{3} & \frac{1}{3} \end{bmatrix} \begin{bmatrix} -3 & 4 \\ 2 & -1 \end{bmatrix} \begin{bmatrix} 1 & -2 \\ 1 & 1 \end{bmatrix} = \begin{bmatrix} 1 & 0 \\ 0 & -5 \end{bmatrix}$$

PROBLEM 16-13 If we make the substitutions $\mathbf{Y} = P\mathbf{U}$ and $d\mathbf{Y} = P\mathbf{U}'$ in Problem 16-11, we will get a new diagonal system $U' = DU$ which has solution of

$$\mathbf{U} = \begin{bmatrix} c_1 e^{\lambda_1 t} \\ c_2 e^{\lambda_2 t} \end{bmatrix}$$

Use $D = \begin{bmatrix} 1 & 0 \\ 0 & -5 \end{bmatrix}$ and solve the system of equations in Problem 16-11.

Solution

$$\mathbf{U}' = \begin{bmatrix} 1 & 0 \\ 0 & -5 \end{bmatrix}\mathbf{U} \quad \text{or} \quad \begin{aligned} u_1' &= u_1 \\ u_2' &= -5u_2 \end{aligned}$$

$$\therefore \quad \mathbf{U} = \begin{bmatrix} c_1 e^{t} \\ c_2 e^{-5t} \end{bmatrix}$$

Thus, the equation $\mathbf{Y} = P\mathbf{U}$ gives

$$\begin{bmatrix} y_1 \\ y_2 \end{bmatrix} = \begin{bmatrix} 1 & -2 \\ 1 & 1 \end{bmatrix}\begin{bmatrix} c_1 e^{t} \\ c_2 e^{-5t} \end{bmatrix} = \begin{bmatrix} c_1 e^{t} - 2c_2 e^{-5t} \\ c_1 e^{t} + c_2 e^{-5t} \end{bmatrix}$$

or

$$y_1 = c_1 e^{t} - 2c_2 e^{-5t} \qquad y_2 = c_1 e^{t} + c_2 e^{-5t}$$

Supplementary Problems

PROBLEM 16-14 Find a matrix that will diagonalize $A = \begin{bmatrix} 3 & -2 \\ -4 & 5 \end{bmatrix}$.

PROBLEM 16-15 Show that $A = \begin{bmatrix} 1 & 2 & -1 \\ 0 & 1 & -2 \\ 0 & 1 & 4 \end{bmatrix}$ is diagonalizable.

PROBLEM 16-16 Determine if each of the following matrices is diagonalizable.

(a) $A = \begin{bmatrix} 1 & 1 \\ 0 & 1 \end{bmatrix}$ **(b)** $B = \begin{bmatrix} 2 & 0 & 0 \\ 0 & 3 & 2 \\ 1 & 0 & 1 \end{bmatrix}$ **(c)** $C = \begin{bmatrix} -3 & 0 & 1 \\ 0 & 2 & -1 \\ 0 & 0 & 2 \end{bmatrix}$

PROBLEM 16-17 Find the matrix P that will diagonalize $A = \begin{bmatrix} 2 & 0 & 0 \\ 0 & 3 & 2 \\ 1 & 0 & 1 \end{bmatrix}$.

PROBLEM 16-18 Find the matrix P that will diagonalize $A = \begin{bmatrix} 2 & 1 \\ 2 & 3 \end{bmatrix}$.

PROBLEM 16-19 Find the diagonal matrix D that is similar to $A = \begin{bmatrix} 2 & 1 \\ 2 & 3 \end{bmatrix}$.

PROBLEM 16-20 Find the eigenvectors of $A = \begin{bmatrix} 1 & 0 & 0 \\ 0 & 0 & -2 \\ 0 & -2 & 0 \end{bmatrix}$.

PROBLEM 16-21 Find the matrix P and P^T that will orthogonally diagonalize $A = \begin{bmatrix} 1 & 0 & 0 \\ 0 & 0 & -2 \\ 0 & -2 & 0 \end{bmatrix}$.

PROBLEM 16-22 Find the diagonal matrix D of Problem 16-21.

PROBLEM 16-23 Write the system of differential equations

$$y_1' = -5y_1 + 3y_2$$
$$y_2' = 4y_1 - y_2$$

as a matrix equation.

PROBLEM 16-24 Find the matrix P that will diagonalize $A = \begin{bmatrix} -5 & 3 \\ 4 & -1 \end{bmatrix}$.

PROBLEM 16-25 Find the diagonal matrix D and solve the system in Problem 16-23.

Answers to Supplementary Problems

16-14 $P = \begin{bmatrix} 1 & 1 \\ 1 & -2 \end{bmatrix}$.

16-15 There are three distinct eigenvalues 2, 3, 1.

16-16 **(a)** Not diagonalizable, only one eigenvector
(b) Diagonalizable since there are three distinct eigenvalues
(c) Not diagonalizable

16-17 $P = \begin{bmatrix} 0 & 1 & 0 \\ 1 & -2 & 1 \\ -1 & 1 & 0 \end{bmatrix}$

16-18 $P = \begin{bmatrix} 1 & 1 \\ -1 & 2 \end{bmatrix}$

16-19 $D = \begin{bmatrix} 1 & 0 \\ 0 & 4 \end{bmatrix}$

16-20 $\begin{bmatrix} 1 \\ 0 \\ 0 \end{bmatrix}, \begin{bmatrix} 0 \\ 1 \\ -1 \end{bmatrix}, \begin{bmatrix} 0 \\ 1 \\ 1 \end{bmatrix}$

16-21

$$P = \begin{bmatrix} 1 & 0 & 0 \\ 0 & \frac{1}{\sqrt{2}} & \frac{1}{\sqrt{2}} \\ 0 & -\frac{1}{\sqrt{2}} & \frac{1}{\sqrt{2}} \end{bmatrix} \qquad P^T = \begin{bmatrix} 1 & 0 & 0 \\ 0 & \frac{1}{\sqrt{2}} & -\frac{1}{\sqrt{2}} \\ 0 & \frac{1}{\sqrt{2}} & \frac{1}{\sqrt{2}} \end{bmatrix}$$

16-22 $D = P^TAP = \begin{bmatrix} 1 & 0 & 0 \\ 0 & 2 & 0 \\ 0 & 0 & -2 \end{bmatrix}$

16-23 $\begin{bmatrix} y_1' \\ y_2' \end{bmatrix} = \begin{bmatrix} -5 & 3 \\ 4 & -1 \end{bmatrix} \begin{bmatrix} y_1 \\ y_2 \end{bmatrix}$

16-24 $P = \begin{bmatrix} 1 & -3 \\ 2 & 2 \end{bmatrix} \qquad P^{-1} = \begin{bmatrix} \frac{2}{8} & \frac{3}{8} \\ -\frac{2}{8} & \frac{1}{8} \end{bmatrix}$

16-25 $D = \begin{bmatrix} 1 & 0 \\ 0 & -7 \end{bmatrix}$

$$\begin{bmatrix} y_1 \\ y_2 \end{bmatrix} = \begin{bmatrix} 1 & -3 \\ 2 & 2 \end{bmatrix} \begin{bmatrix} c_1e^t \\ c_2e^{-7t} \end{bmatrix} = \begin{bmatrix} c_1e^t - 3c_2e^{-7t} \\ 2c_1e^t + 2c_2e^{-7t} \end{bmatrix}$$

17 APPLICATIONS TO QUADRATIC FORMS

THIS CHAPTER IS ABOUT

☑ **Quadratic Forms**
☑ **Principal Axes Property for Quadratic Forms**
☑ **Conics**
☑ **Quadric Surfaces**
☑ **Definiteness of Quadratic Forms**

17-1. Quadratic Forms

A. Quadratic equations with real coefficients

An equation of the form

$$Ax^2 + By^2 + Cz^2 + Dxy + Exz + Fyz + Gx + Hy + Iz = J$$

with not all of A, B, C, D, E, F equal to zero is a **quadratic equation** in the three variables x, y, and z.

B. Quadratic forms associated with equations

The part of a quadratic equation consisting of

$$Ax^2 + By^2 + Cxy \qquad \text{or} \qquad Ax^2 + By^2 + Cz^2 + Dxy + Exz + Fyz$$

are called **quadratic forms**.

EXAMPLE 17-1: In each of the following quadratic equations, find the associated quadratic form: **(a)** $5x^2 + 3y = 4$; **(b)** $4x^2 + 3y^2 + 2xy + x - y = 2$; **(c)** $4x^2 + 3y^2 + z^2 + yz + x + z = 0$.

Solution: The quadratic forms are the second-degree terms.

(a) $5x^2$
(b) $4x^2 + 3y^2 + 2xy$
(c) $4x^2 + 3y^2 + z^2 + yz$

C. Matrix of a quadratic form

A symmetric matrix A is called the **matrix of a quadratic form** if X^TAX equals the quadratic form where

$$X = \begin{bmatrix} x \\ y \end{bmatrix}, \begin{bmatrix} x \\ y \\ z \end{bmatrix}, \quad \text{or} \quad \begin{bmatrix} x_1 \\ x_2 \\ \vdots \\ x_n \end{bmatrix}$$

For the quadratic equation $ax^2 + by^2 + cxy + dx + ey = f$ in two variables, $A = \begin{bmatrix} a & c/2 \\ c/2 & b \end{bmatrix}$.

For the quadratic equation $ax^2 + by^2 + cz^2 + dxy + exz + fyz + gx + hy + iz = j$ in three variables, $A = \begin{bmatrix} a & d/2 & e/2 \\ d/2 & b & f/2 \\ e/2 & f/2 & c \end{bmatrix}$.

A good way to remember this is to think in terms of the subscripts of a_{ij} in the matrix $\begin{bmatrix} a_{11} & a_{12} & a_{13} \\ a_{21} & a_{22} & a_{23} \\ a_{31} & a_{32} & a_{33} \end{bmatrix}$ where 1, 2, and 3 denote the variables x, y, and z, respectively. For example, a_{12} refers to $d/2$ which comes from the xy terms.

EXAMPLE 17-2: In Example 17-1, find A in each part and verify that X^TAX gives the associated quadratic form.

(a)

$$A = \begin{bmatrix} 5 & 0 \\ 0 & 0 \end{bmatrix}$$

$$X^TAX = [x \quad y]\begin{bmatrix} 5 & 0 \\ 0 & 0 \end{bmatrix}\begin{bmatrix} x \\ y \end{bmatrix} = 5x^2$$

(b)

$$A = \begin{bmatrix} 4 & 1 \\ 1 & 3 \end{bmatrix}$$

$$X^TAX = [x \quad y]\begin{bmatrix} 4 & 1 \\ 1 & 3 \end{bmatrix}\begin{bmatrix} x \\ y \end{bmatrix} = 4x^2 + 2xy + 3y^2$$

(c)

$$A = \begin{bmatrix} 4 & 0 & 0 \\ 0 & 3 & \frac{1}{2} \\ 0 & \frac{1}{2} & 1 \end{bmatrix}$$

$$X^TAX = [x \quad y \quad z]\begin{bmatrix} 4 & 0 & 0 \\ 0 & 3 & \frac{1}{2} \\ 0 & \frac{1}{2} & 1 \end{bmatrix}\begin{bmatrix} x \\ y \\ z \end{bmatrix} = 4x^2 + 3y^2 + yz + z^2$$

D. Matrix form of a quadratic equation

The **matrix form of a quadratic equation** in two variables is $X^TAX + BX = f$ where $X = \begin{bmatrix} x \\ y \end{bmatrix}$, A is the matrix of the quadratic form, and $B = [d \quad e]$. The matrix form of a quadratic equation in three variables is $X^TAX + BX = j$ where $X = \begin{bmatrix} x \\ y \\ z \end{bmatrix}$, A is the matrix of the quadratic form, and $B = [g \quad h \quad i]$.

17-2. Principal Axes Property for Quadratic Forms

If A is the matrix of the quadratic form X^TAX where

$$X = \begin{bmatrix} x_1 \\ x_2 \\ \vdots \\ x_n \end{bmatrix}$$

then a matrix P exists that orthogonally diagonalizes A into the similar matrix $D = P^TAP$ and transforms the quadratic equation $X^TAX + BX = f$ into the equivalent quadratic equation

$$Y^TDY + (BP)Y = f, \qquad \text{where } Y = \begin{bmatrix} x_1' \\ x_2' \\ \vdots \\ x_n' \end{bmatrix}$$

The quadratic form Y^TDY has no mixed variable terms, and the coefficients of $(x_1')^2, (x_2')^2, \ldots, (x_n')^2$ are the eigenvalues of A. Also, $\det(P) = 1$. (This is also referred to as **rotation of axes**.)

EXAMPLE 17-3: Transform the quadratic form $4x^2 + 4xy + 4y^2$ into an equivalent quadratic form having no xy term.

Solution

$$4x^2 + 4xy + 4y^2 = [x \quad y]\begin{bmatrix} 4 & 2 \\ 2 & 4 \end{bmatrix}\begin{bmatrix} x \\ y \end{bmatrix} = X^TAX$$

To diagonalize the quadratic form, we need the eigenvalues of A, but we do not need the eigenvectors nor do we need the matrix P.

$$\det(\lambda I - A) = \begin{vmatrix} \lambda - 4 & -2 \\ -2 & \lambda - 4 \end{vmatrix} = \lambda^2 - 8\lambda + 12 = (\lambda - 2)(\lambda - 6) = 0$$

Therefore, the eigenvalues are 2 and 6. Hence, the equivalent quadratic form is

$$Y^TDY = [x' \quad y']\begin{bmatrix} 2 & 0 \\ 0 & 6 \end{bmatrix}\begin{bmatrix} x' \\ y' \end{bmatrix} = 2(x')^2 + 6(y')^2$$

17-3. Conics

The graphs of quadratic equations in two variables are called **conics** or **conic sections**.

A. The three types of nondegenerate conics

Standard form of equation	Name
(1) $x^2/a^2 + y^2/b^2 = 1;\ a, b > 0$	ellipse or a circle for special case $a = b$.
(2) $x^2/a^2 - y^2/b^2 = 1$ or $y^2/a^2 - x^2/b^2 = 1;\ a, b > 0$	hyperbola
(3) $x^2 = ky$ or $y^2 = kx;\ k \neq 0$	parabola

In its degenerate form a conic may be the empty set, a point, a line, two parallel lines, or two intersecting lines.

B. Finding standard forms of conics

Step 1: Express the quadratic equation $ax^2 + by^2 + cxy + dx + ey = f$ in the matrix form $X^TAX + BX = f$.

Step 2: Find a matrix P_1 which orthogonally diagonalizes A.

Step 3: Interchange two columns of P_1, if necessary, to get a matrix P such that $\det(P) = 1$.

Step 4: Replace $X = \begin{bmatrix} x \\ y \end{bmatrix}$ in the matrix form of the equation by $PY = P\begin{bmatrix} x' \\ y' \end{bmatrix}$ to obtain $Y^TDY + (BP)Y = f$ where $D = P^TAP$.

Step 5: If necessary, complete the squares on the variables in the equation to obtain the standard form of the conic. (This is equivalent to a translation of axes.)

EXAMPLE 17-4: Find the standard form and name the conic represented by the equation

$$2y^2 - x^2 + 4xy - 2y + 3x = 6$$

Solution

Step 1:

$$[x \quad y]\begin{bmatrix} -1 & 2 \\ 2 & 2 \end{bmatrix}\begin{bmatrix} x \\ y \end{bmatrix} + [3 \quad -2]\begin{bmatrix} x \\ y \end{bmatrix} = 6$$

where

$$\begin{bmatrix} -1 & 2 \\ 2 & 2 \end{bmatrix} = A, \qquad [3 \quad -2] = B$$

Step 2:

$$\det(\lambda I - A) = \begin{vmatrix} \lambda + 1 & -2 \\ -2 & \lambda - 2 \end{vmatrix} = \lambda^2 - \lambda - 6 = (\lambda - 3)(\lambda + 2)$$

Therefore, the eigenvalues are -2 and 3. For $\lambda = -2$, we get

$$\begin{bmatrix} -1 & -2 \\ -2 & -4 \end{bmatrix}\begin{bmatrix} a_1 \\ a_2 \end{bmatrix} = \begin{bmatrix} 0 \\ 0 \end{bmatrix}$$

which has the solution

$$\begin{bmatrix} a_1 \\ a_2 \end{bmatrix} = \begin{bmatrix} -2t \\ t \end{bmatrix} = t\begin{bmatrix} -2 \\ 1 \end{bmatrix}$$

For $\lambda = 3$, we get

$$\begin{bmatrix} 4 & -2 \\ -2 & 1 \end{bmatrix}\begin{bmatrix} a_1 \\ a_2 \end{bmatrix} = \begin{bmatrix} 0 \\ 0 \end{bmatrix}$$

which has the solution

$$\begin{bmatrix} a_1 \\ a_2 \end{bmatrix} = \begin{bmatrix} t \\ 2t \end{bmatrix} = t\begin{bmatrix} 1 \\ 2 \end{bmatrix}$$

$\begin{bmatrix} -2 \\ 1 \end{bmatrix}$ and $\begin{bmatrix} 1 \\ 2 \end{bmatrix}$ are orthogonal, but we must change these to unit vectors. Therefore, the matrix P_1 having these unit eigenvectors as columns is $\begin{bmatrix} -2/\sqrt{5} & 1/\sqrt{5} \\ 1/\sqrt{5} & 2/\sqrt{5} \end{bmatrix}$.

Step 3: Since $\det(P_1) = -1$, we interchange the columns to get $P = \begin{bmatrix} 1/\sqrt{5} & -2/\sqrt{5} \\ 2/\sqrt{5} & 1/\sqrt{5} \end{bmatrix}$ for which $\det(P) = 1$.

Step 4: Since the columns of P are eigenvectors for the eigenvalues 3 and -2, respectively, then $D = \begin{bmatrix} 3 & 0 \\ 0 & -2 \end{bmatrix}$. Therefore, $Y^TDY + (BP)Y = f$ becomes

$$[x' \quad y']\begin{bmatrix} 3 & 0 \\ 0 & -2 \end{bmatrix}\begin{bmatrix} x' \\ y' \end{bmatrix} + [3 \quad -2]\begin{bmatrix} \frac{1}{\sqrt{5}} & -\frac{2}{\sqrt{5}} \\ \frac{2}{\sqrt{5}} & \frac{1}{\sqrt{5}} \end{bmatrix}\begin{bmatrix} x' \\ y' \end{bmatrix} = 6$$

This simplifies to

$$3(x')^2 - 2(y')^2 - (1/\sqrt{5})x' - (8/\sqrt{5})y' = 6$$

Step 5: After completing the square, this becomes $3(x' - 1/(6\sqrt{5}))^2 - 2(y' + 2/\sqrt{5})^2 = 289/60$. Dividing by 289/60, we get

$$\frac{(x' - 1/(6\sqrt{5}))^2}{289/180} - \frac{(y' + 2/\sqrt{5})^2}{289/120} = 1$$

which is the standard form of a hyperbola.

EXAMPLE 17-5: Find the standard form and name the conic represented by $x^2 + xy + y^2 = 1$.

Solution

Step 1:
$$[x \quad y]\begin{bmatrix} 1 & \frac{1}{2} \\ \frac{1}{2} & 1 \end{bmatrix}\begin{bmatrix} x \\ y \end{bmatrix} + [0 \quad 0]\begin{bmatrix} x \\ y \end{bmatrix} = 1$$

where

$$\begin{bmatrix} 1 & \frac{1}{2} \\ \frac{1}{2} & 1 \end{bmatrix} = A, \qquad [0 \quad 0] = B$$

Step 2:
$$\det(\lambda I - A) = \begin{vmatrix} \lambda - 1 & -\frac{1}{2} \\ -\frac{1}{2} & \lambda - 1 \end{vmatrix} = \lambda^2 - 2\lambda + \frac{3}{4} = \left(\lambda - \frac{1}{2}\right)\left(\lambda - \frac{3}{2}\right)$$

Therefore, the eigenvalues are 1/2 and 3/2. For $\lambda = 1/2$, we get

$$\begin{bmatrix} -\frac{1}{2} & -\frac{1}{2} \\ -\frac{1}{2} & -\frac{1}{2} \end{bmatrix}\begin{bmatrix} a_1 \\ a_2 \end{bmatrix} = \begin{bmatrix} 0 \\ 0 \end{bmatrix}$$

which has the solution

$$\begin{bmatrix} a_1 \\ a_2 \end{bmatrix} = \begin{bmatrix} t \\ -t \end{bmatrix} = t\begin{bmatrix} 1 \\ -1 \end{bmatrix}$$

For $\lambda = 3/2$, we get

$$\begin{bmatrix} \frac{1}{2} & -\frac{1}{2} \\ -\frac{1}{2} & \frac{1}{2} \end{bmatrix}\begin{bmatrix} a_1 \\ a_2 \end{bmatrix} = \begin{bmatrix} 0 \\ 0 \end{bmatrix}$$

which has the solution

$$\begin{bmatrix} a_1 \\ a_2 \end{bmatrix} = \begin{bmatrix} t \\ t \end{bmatrix} = t\begin{bmatrix} 1 \\ 1 \end{bmatrix}$$

$\begin{bmatrix} 1 \\ -1 \end{bmatrix}$ and $\begin{bmatrix} 1 \\ 1 \end{bmatrix}$ are orthogonal, but we must change each to unit vectors by dividing each by its length. Therefore, the matrix P_1 having these unit eigenvectors as columns is $\begin{bmatrix} 1/\sqrt{2} & 1/\sqrt{2} \\ -1/\sqrt{2} & 1/\sqrt{2} \end{bmatrix}$.

Step 3: Since $\det(P_1) = 1$, $P = P_1$.

Step 4: Since the columns of P are eigenvectors for the eigenvalues 1/2 and 3/2, respectively, then

$$D = \begin{bmatrix} \frac{1}{2} & 0 \\ 0 & \frac{1}{2} \end{bmatrix}$$

Therefore, $Y^TDY + (BP)Y = f$ becomes

$$[x' \quad y']\begin{bmatrix} \frac{1}{2} & 0 \\ 0 & \frac{3}{2} \end{bmatrix}\begin{bmatrix} x' \\ y' \end{bmatrix} + [0 \quad 0]\begin{bmatrix} \frac{1}{\sqrt{2}} & \frac{1}{\sqrt{2}} \\ -\frac{1}{\sqrt{2}} & \frac{1}{\sqrt{2}} \end{bmatrix}\begin{bmatrix} x' \\ y' \end{bmatrix} = 1$$

This simplifies to

$$\frac{1}{2}(x')^2 + \frac{3}{2}(y')^2 = 1 \qquad \text{or} \qquad \frac{(x')^2}{2} + \frac{(y')^2}{2/3} = 1$$

which is the standard form of an ellipse.

EXAMPLE 17-6: Find the standard form and name the conic represented by the equation $4x^2 + 4xy + y^2 - 3y = 6$.

Solution

Step 1:

$$[x \quad y]\begin{bmatrix} 4 & 2 \\ 2 & 1 \end{bmatrix}\begin{bmatrix} x \\ y \end{bmatrix} + [0 \quad -3]\begin{bmatrix} x \\ y \end{bmatrix} = 6$$

where

$$\begin{bmatrix} 4 & 2 \\ 2 & 1 \end{bmatrix} = A, \qquad [0 \quad -3] = B$$

Step 2:

$$\det(\lambda I - A) = \begin{vmatrix} \lambda - 4 & -2 \\ -2 & \lambda - 1 \end{vmatrix} = \lambda(\lambda - 5)$$

Therefore, the eigenvalues are 0 and 5. For $\lambda = 0$, we get

$$\begin{bmatrix} -4 & -2 \\ -2 & -1 \end{bmatrix}\begin{bmatrix} a_1 \\ a_2 \end{bmatrix} = \begin{bmatrix} 0 \\ 0 \end{bmatrix}$$

which has the solution

$$\begin{bmatrix} a_1 \\ a_2 \end{bmatrix} = \begin{bmatrix} t \\ -2t \end{bmatrix} = t\begin{bmatrix} 1 \\ -2 \end{bmatrix}$$

For $\lambda = 5$, we get

$$\begin{bmatrix} 1 & -2 \\ -2 & 4 \end{bmatrix}\begin{bmatrix} a_1 \\ a_2 \end{bmatrix} = \begin{bmatrix} 0 \\ 0 \end{bmatrix}$$

which has the solution

$$\begin{bmatrix} a_1 \\ a_2 \end{bmatrix} = \begin{bmatrix} 2t \\ t \end{bmatrix} = t\begin{bmatrix} 2 \\ 1 \end{bmatrix}$$

$\begin{bmatrix} 1 \\ -2 \end{bmatrix}$ and $\begin{bmatrix} 1 \\ 2 \end{bmatrix}$ are orthogonal, but we must change each to unit vectors by dividing each by its length. Therefore, the matrix having these unit eigenvectors as columns is $P_1 = \begin{bmatrix} 1/\sqrt{5} & 2/\sqrt{5} \\ -2/\sqrt{5} & 1/\sqrt{5} \end{bmatrix}$.

Step 3: Since $\det(P_1) = 1$, then $P = P_1$.

Step 4: Since the columns of P are eigenvectors for the eigenvalues 0 and 5, respectively, then $D = \begin{bmatrix} 0 & 0 \\ 0 & 5 \end{bmatrix}$. Therefore, $Y^T D Y + (BP)Y = f$ becomes

$$[x' \quad y']\begin{bmatrix} 0 & 0 \\ 0 & 5 \end{bmatrix}\begin{bmatrix} x' \\ y' \end{bmatrix} + [0 \quad -3]\begin{bmatrix} \frac{1}{\sqrt{5}} & \frac{2}{\sqrt{5}} \\ -\frac{2}{\sqrt{5}} & \frac{1}{\sqrt{5}} \end{bmatrix}\begin{bmatrix} x' \\ y' \end{bmatrix} = 6$$

This simplifies to $5(y')^2 + (6/\sqrt{5})x' - (3/\sqrt{5})y' = 6$ which becomes $(y - 3/(10\sqrt{5}))^2 = -6/(5\sqrt{5})(x' - 203\sqrt{5}/200)$. This is the standard form of a parabola.

17-4. Quadric Surfaces

The graphs of quadratic equations in three variables are called quadrics or quadric surfaces.

A. The nine types of nondegenerate quadrics

Standard form of equation	Name
(1) $x^2/a^2 + y^2/b^2 + z^2/c^2 = 1$; $a, b, c > 0$	ellipsoid or sphere for special case $a = b = c$.
(2) $x^2/a^2 + y^2/b^2 = \pm z$; $a, b > 0$ or any permutation of the variables x, y, and z	elliptic paraboloid
(3) $x^2/a^2 - y^2/b^2 = \pm z$; $a, b > 0$ or any permutation of the variables x, y, and z	hyperbolic paraboloid
(4) $x^2/a^2 + y^2/b^2 - z^2/c^2 = 1$; $a, b, c > 0$ or any permutation of the variables x, y, and z	hyperboloid of one sheet
(5) $x^2/a^2 - y^2/b^2 - z^2/c^2 = 1$; $a, b, c > 0$ or any permutation of the variables x, y, and z	hyperboloid of two sheets
(6) $x^2/a^2 + y^2/b^2 - z^2/c^2 = 0$; $a, b, c > 0$ or any permutation of the variables x, y, and z	cone
(7) $x^2/a^2 + y^2/b^2 = 1$; $a, b > 0$ or either x or y replaced by z	elliptic cylinder
(8) $y^2 = ax + bz$; not both $a, b = 0$ or any permutation of the variables x, y, and z	parabolic cylinder
(9) $x^2/a^2 - y^2/b^2 = \pm 1$; $a, b > 0$ or either x or y replaced by z	hyperbolic cylinder

In its degenerate form a quadric may be the empty set, a point, a plane, two intersecting planes, or two parallel planes.

B. Finding standard forms of quadric surfaces

Step 1: Express the quadratic equation

$$ax^2 + by^2 + cz^2 + dxy + exz + fyz + gx + hy + iz = j$$

in the matrix form $X^TAX + BX = j$.

Step 2: Find a matrix P_1 which orthogonally diagonalizes A.

Step 3: Interchange the columns of P_1, if necessary, to get a matrix P such that $\det(P) = 1$.

Step 4: Replace $X = \begin{bmatrix} x \\ y \\ z \end{bmatrix}$ in the matrix form of the equation by $PY = P\begin{bmatrix} x' \\ y' \\ z' \end{bmatrix}$ to obtain $Y^TDY + (BP)Y = j$ where $D = P^TAP$.

Step 5: If necessary, complete the squares on the variables in the equation to obtain the standard form of the quadric surface. (This is equivalent to a translation of axes.)

EXAMPLE 17-7: Find the standard form and name the quadric surface represented by the equation $6xy + 6xz + 6yz + x + y + z = 4$.

Solution

Step 1:

$$[x \quad y \quad z]\begin{bmatrix} 0 & 3 & 3 \\ 3 & 0 & 3 \\ 3 & 3 & 0 \end{bmatrix}\begin{bmatrix} x \\ y \\ z \end{bmatrix} + [1 \quad 1 \quad 1]\begin{bmatrix} x \\ y \\ z \end{bmatrix} = 4$$

where

$$\begin{bmatrix} 0 & 3 & 3 \\ 3 & 0 & 3 \\ 3 & 3 & 0 \end{bmatrix} = A \quad \text{and} \quad [1 \quad 1 \quad 1] = B$$

Step 2: In Example 16-8, we found the eigenvalues of A to be -3 and 6 with multiplicities 2 and 1, respectively. We also found that

$$P_1 = \begin{bmatrix} \frac{1}{\sqrt{2}} & -\frac{1}{\sqrt{6}} & \frac{1}{\sqrt{3}} \\ 0 & \frac{2}{\sqrt{6}} & \frac{1}{\sqrt{3}} \\ -\frac{1}{\sqrt{2}} & -\frac{1}{\sqrt{6}} & \frac{1}{\sqrt{3}} \end{bmatrix}$$

orthogonally diagonalized A into

$$D = \begin{bmatrix} -3 & 0 & 0 \\ 0 & -3 & 0 \\ 0 & 0 & 6 \end{bmatrix}$$

Step 3: Since $\det(P_1) = 1$, then $P = P_1$.
Step 4: Therefore, $Y^T DY + (BP)Y = j$ becomes

$$[x' \quad y' \quad z'] \begin{bmatrix} -3 & 0 & 0 \\ 0 & -3 & 0 \\ 0 & 0 & 6 \end{bmatrix} \begin{bmatrix} x' \\ y' \\ z' \end{bmatrix} + [1 \quad 1 \quad 1] \begin{bmatrix} \frac{1}{\sqrt{2}} & -\frac{1}{\sqrt{6}} & \frac{1}{\sqrt{3}} \\ 0 & \frac{2}{\sqrt{6}} & \frac{1}{\sqrt{3}} \\ -\frac{1}{\sqrt{2}} & -\frac{1}{\sqrt{6}} & \frac{1}{\sqrt{3}} \end{bmatrix} \begin{bmatrix} x' \\ y' \\ z' \end{bmatrix} = 4$$

$= 4$

This simplifies to $-3(x')^2 - 3(y')^2 + 6(z')^2 + \sqrt{3}z' = 4$ which becomes

$$\frac{(z' + \sqrt{3}/12)^2}{11/16} - \frac{(x')^2}{11/8} - \frac{(y')^2}{11/8} = 1$$

The final equation is the standard form of a hyperboloid of two sheets.

17-5. Definiteness of Quadratic Forms

A. Definition of definiteness

If A is an $n \times n$ symmetric matrix, then the quadratic form $X^T AX$ where $X = \begin{bmatrix} x_1 \\ x_2 \\ \vdots \\ x_n \end{bmatrix}$ is

(1) positive definite if $X^T AX > 0$ for all $X \neq \mathbf{0}$;
(2) negative definite if $X^T AX < 0$ for all $X \neq \mathbf{0}$;

(3) indefinite if X^TAX is positive for some X and negative for some X (it may also be 0 for some nonzero X);

(4) neither positive definite, negative definite, nor indefinite, otherwise.

EXAMPLE 17-8: Determine whether each of the following quadratic forms is positive definite, negative definite, indefinite, or neither: **(a)** $7x^2 + 8y^2 + 5z^2$ in x, y, and z; **(b)** $-5x^2 - 4y^2$ in x and y; **(c)** $5x^2 - 4y^2$ in x and y; **(d)** $-5x^2 - 4y^2$ in x, y, and z.

Solution

(a) It is positive definite since it is zero only if $X = \begin{bmatrix} 0 \\ 0 \\ 0 \end{bmatrix}$ and is positive otherwise.

(b) It is negative definite since it is zero only if $X = \begin{bmatrix} 0 \\ 0 \end{bmatrix}$ and is negative otherwise.

(c) It is indefinite because $5(3)^2 - 4(2)^2 = 29$ and $5(2)^2 - 4(3)^2 = -16$. The fact that it is zero for $\begin{bmatrix} 2 \\ \sqrt{5} \end{bmatrix}$ has no bearing on the answer.

(d) It is neither because it is never positive but is zero for $X = \begin{bmatrix} 0 \\ 0 \\ 5 \end{bmatrix}$.

B. Eigenvalues and definiteness of quadratic forms

If A is an $n \times n$ symmetric matrix, then the quadratic form X^TAX where $X = \begin{bmatrix} x_1 \\ x_2 \\ \vdots \\ x_n \end{bmatrix}$ is

(1) positive definite if all the eigenvalues of A are positive;

(2) negative definite if all the eigenvalues of A are negative;

(3) indefinite if at least one of the eigenvalues of A is positive and all the others are negative;

(4) neither positive definite, negative definite, nor indefinite if zero is an eigenvalue of A.

EXAMPLE 17-9: Determine whether each of the following quadratic forms is positive definite, negative definite, indefinite, or neither: **(a)** $6xy + 6xz + 6yz$; **(b)** $4x^2 + 4xy + 4y^2$; **(c)** $4x^2 + 4xy + y^2$; **(d)** $-3x^2 + 2xy - 3y^2$.

Solution

(a)

$$A = \begin{bmatrix} 0 & 3 & 3 \\ 3 & 0 & 3 \\ 3 & 3 & 0 \end{bmatrix}$$

$$\det(\lambda I - A) = \begin{vmatrix} \lambda & -3 & -3 \\ -3 & \lambda & -3 \\ -3 & -3 & \lambda \end{vmatrix} = \lambda^3 - 27\lambda - 54 = (\lambda + 3)^2(\lambda - 6)$$

Therefore, the eigenvalues are -3 and 6 with multiplicities 2 and 1, respectively. Thus, **(a)** is indefinite.

(b)

$$A = \begin{bmatrix} 4 & 2 \\ 2 & 4 \end{bmatrix}$$

$$\det(\lambda I - A) = \begin{vmatrix} \lambda - 4 & -2 \\ -2 & \lambda - 4 \end{vmatrix} = \lambda^2 - 8\lambda + 12 = (\lambda - 2)(\lambda - 6)$$

Therefore, the eigenvalues are 2 and 6. Thus, **(b)** is positive definite.

(c)

$$A = \begin{bmatrix} 4 & 2 \\ 2 & 1 \end{bmatrix}$$

$$\det(\lambda I - A) = \begin{vmatrix} \lambda - 4 & -2 \\ -2 & \lambda - 1 \end{vmatrix} = \lambda^2 - 5\lambda = \lambda(\lambda - 5)$$

Therefore, the eigenvalues are 0 and 5. Thus, **(c)** is neither.

(d)

$$A = \begin{bmatrix} -3 & 1 \\ 1 & -3 \end{bmatrix}$$

$$\det(\lambda I - A) = \begin{vmatrix} \lambda + 3 & -1 \\ -1 & \lambda + 3 \end{vmatrix} = \lambda^2 + 6\lambda + 8 = (\lambda + 2)(\lambda + 4)$$

Therefore, the eigenvalues are -2 and -4. Thus, **(d)** is negative definite.

SUMMARY

1. The part of a quadratic equation consisting of the second-degree terms such as $3x^2 + 5y^2 - 4xy$ is called a quadratic form.
2. The matrix of the quadratic form $X^TAX = ax^2 + by^2 + cxy$ in two variables is $A = \begin{bmatrix} a & c/2 \\ c/2 & b \end{bmatrix}$ where $x = \begin{bmatrix} x \\ y \end{bmatrix}$.
3. The matrix of the quadratic form $X^TAX = ax^2 + by^2 + cz^2 + dxy + exz + fyz$ is $A = \begin{bmatrix} a & d/2 & e/2 \\ d/2 & b & f/2 \\ e/2 & f/2 & c \end{bmatrix}$ where $X = \begin{bmatrix} x \\ y \\ z \end{bmatrix}$.
4. A good way to remember the matrix of a quadratic form is to think in terms of the subscripts of a_{ij} in the matrix $\begin{bmatrix} a_{11} & a_{12} & a_{13} \\ a_{21} & a_{22} & a_{23} \\ a_{31} & a_{32} & a_{33} \end{bmatrix}$ where 1, 2, and 3 denote the variables x, y, and z, respectively. a_{23} refers to the coefficient of yz term.
5. The matrix form of the two-variable equation $ax^2 + by^2 + cxy + dx + ey = f$ is $X^TAX + BX = f$ where $X = \begin{bmatrix} x \\ y \end{bmatrix}$, A is the matrix of the quadratic form, and $B = [d \quad e]$.
6. The matrix form of the three-variable equation $ax^2 + by^2 + cz^2 + dxy + exz + fyz + gx + hy + iz = j$ is $X^TAX + BX = j$ where $X = \begin{bmatrix} x \\ y \\ z \end{bmatrix}$, A is the matrix of the quadratic form, and $B = [g \quad h \quad i]$.
7. If A is the matrix of the quadratic form X^TAX, then
   - a matrix P exists that orthogonally diagonalizes A into the similar matrix $D = P^TAP$,
   - P transforms the equation $X^TAX + BX = f$ into the equivalent quadratic equation $Y^TDY +$

$(BP)Y = f$ where

$$X = \begin{bmatrix} x_1 \\ x_2 \\ \vdots \\ x_n \end{bmatrix} \quad \text{and} \quad Y = \begin{bmatrix} x'_1 \\ x'_2 \\ \vdots \\ x'_n \end{bmatrix}$$

- the quadratic form Y^TDY has no mixed variable terms,
- the coefficients of $(x'_1)^2, (x'_2)^2, \ldots, (x'_n)^2$ are the eigenvalues of A.

8. To diagonalize a quadratic form, we need the eigenvalues of A, but we do not need the eigenvectors of A nor the matrix P.

9. The graphs of quadratic equations in two variables are called conics or conic sections, and there are three types that are nondegenerate.

10. The graphs of quadratic equations in three variables are called quadrics or quadric surfaces, and there are nine types that are nondegenerate.

11. The five steps in finding the standard form of a conic or quadric are as follows:

Step 1: Express the quadratic equation in the matrix form $X^TAX + BX = f$ or j.
Step 2: Find a matrix P_1 that orthogonally diagonalizes A.
Step 3: Interchange the columns of P_1, if necessary, to get a matrix P such that $\det(P) = 1$.
Step 4: Replace $X = \begin{bmatrix} x_1 \\ x_2 \\ \vdots \\ x_n \end{bmatrix}$ in the matrix form of the equation by $PY = P\begin{bmatrix} x'_1 \\ x'_2 \\ \vdots \\ x'_n \end{bmatrix}$ to obtain $Y^TDY + (BP)Y = f$ or j where $D = P^TAP$.
Step 5: If necessary, complete the squares on the variables in the equation to obtain the standard form of the conic or quadric.

12. If A is an $n \times n$ symmetric matrix, then the two statements in each of the following are equivalent for the quadratic form X^TAX where $X = \begin{bmatrix} x_1 \\ x_2 \\ \vdots \\ x_n \end{bmatrix}$.

(1) X^TAX is positive definite if
  (a) $X^TAX > 0$ for all $X \neq 0$,
  (b) all the eigenvalues of A are positive.
(2) X^TAX is negative definite if
  (a) $X^TAX < 0$ for all $X \neq 0$,
  (b) all the eigenvalues of A are negative.
(3) X^TAX is indefinite if
  (a) X^TAX is positive for some X and negative for some X,
  (b) some of the eigenvalues of A are positive and all the others are negative.
(4) X^TAX is neither positive definite, negative definite, nor indefinite if
  (a) one of the previous statements fails to be true,
  (b) zero is an eigenvalue of A.

RAISE YOUR GRADES

Can you ...?

☑ find the matrix of a quadratic form
☑ express a quadratic equation in matrix form

☑ transform a quadratic form X^TAX into a similar quadratic form Y^TDY that has no mixed variable terms
☑ identify the three types of nondegenerate conics
☑ transform a quadratic equation in two variables to the standard form of a conic
☑ identify the nine types of nondegenerate quadrics
☑ transform a quadratic equation in three variables to the standard form of a quadric
☑ determine whether a quadratic form is positive definite, negative definite, indefinite, or neither

SOLVED PROBLEMS

PROBLEM 17-1 In each of the following quadratic equations, find the associated quadratic form: **(a)** $x^2 - 2xy + y^2 + x - 2y = 1$; **(b)** $y^2 + 4x - y = 4$; **(c)** $3x^2 + y^2 - z^2 + 2xz + z = 2$; **(d)** $xy + 4y = 0$.

Solution

(a) $x^2 - 2xy + y^2$
(b) y^2
(c) $3x^2 + y^2 - z^2 + 2xz$
(d) xy

PROBLEM 17-2 In Problem 17-1, find A in each part and verify that X^TAX gives the associated quadratic form.

Solution

(a)
$$A = \begin{bmatrix} 1 & -1 \\ -1 & 1 \end{bmatrix}$$

$$X^TAX = [x \quad y]\begin{bmatrix} 1 & -1 \\ -1 & 1 \end{bmatrix}\begin{bmatrix} x \\ y \end{bmatrix} = X^2 - 2xy + y^2$$

(b)
$$A = \begin{bmatrix} 0 & 0 \\ 0 & 1 \end{bmatrix}$$

$$X^TAX = [x \quad y]\begin{bmatrix} 0 & 0 \\ 0 & 1 \end{bmatrix}\begin{bmatrix} x \\ y \end{bmatrix} = y^2$$

(c)
$$A = \begin{bmatrix} 3 & 0 & 1 \\ 0 & 1 & 0 \\ 1 & 0 & -1 \end{bmatrix}$$

$$X^TAX = [x \quad y \quad z]\begin{bmatrix} 3 & 0 & 1 \\ 0 & 1 & 0 \\ 1 & 0 & -1 \end{bmatrix}\begin{bmatrix} x \\ y \\ z \end{bmatrix} = 3x^2 + y^2 - z^2 + 2xz$$

(d)
$$A = \begin{bmatrix} 0 & \frac{1}{2} \\ \frac{1}{2} & 0 \end{bmatrix}$$

$$X^TAX = [x \quad y]\begin{bmatrix} 0 & \frac{1}{2} \\ \frac{1}{2} & 0 \end{bmatrix}\begin{bmatrix} x \\ y \end{bmatrix} = \frac{1}{2}xy + \frac{1}{2}xy = xy$$

PROBLEM 17-3 Transform the quadratic form $x^2 - 2xy + y^2$ into an equivalent quadratic form with no xy term.

Solution

$$x^2 - 2xy + y^2 = [x \quad y]\begin{bmatrix} 1 & -1 \\ -1 & 1 \end{bmatrix}\begin{bmatrix} x \\ y \end{bmatrix} = X^TAX$$

To diagonalize the quadratic form, we need the eigenvalues of A.

$$\det(\lambda I - A) = \begin{vmatrix} \lambda - 1 & 1 \\ 1 & \lambda - 1 \end{vmatrix} = (\lambda - 1)^2 - 1 = \lambda(\lambda - 2) = 0$$

Therefore, the eigenvalues of A are 0 and 2. Thus, the equivalent quadratic form is

$$Y^TDY = [x' \quad y']\begin{bmatrix} 0 & 0 \\ 0 & 2 \end{bmatrix}\begin{bmatrix} x' \\ y' \end{bmatrix} = 2(y')^2$$

PROBLEM 17-4 Transform the quadratic form xy into an equivalent quadratic form with no xy term.

Solution

$$xy = [x \quad y]\begin{bmatrix} 0 & \frac{1}{2} \\ \frac{1}{2} & 0 \end{bmatrix}\begin{bmatrix} x \\ y \end{bmatrix} = X^TAX$$

$$\det(\lambda I - A) = \begin{vmatrix} \lambda & -\frac{1}{2} \\ -\frac{1}{2} & \lambda \end{vmatrix} = \lambda^2 - \frac{1}{4} = \left(\lambda - \frac{1}{2}\right)\left(\lambda + \frac{1}{2}\right) = 0$$

Thus,

$$Y^TDY = [x' \quad y']\begin{bmatrix} \frac{1}{2} & 0 \\ 0 & -\frac{1}{2} \end{bmatrix}\begin{bmatrix} x' \\ y' \end{bmatrix} = \frac{1}{2}(x')^2 - \frac{1}{2}(y')^2$$

PROBLEM 17-5 Find the standard form and name the conic represented by $x^2 + 4y^2 - 4xy + y - 2x = 4$

Solution

Step 1: $[x \quad y]\begin{bmatrix} 1 & -2 \\ -2 & 4 \end{bmatrix}\begin{bmatrix} x \\ y \end{bmatrix} + [-2 \quad 1]\begin{bmatrix} x \\ y \end{bmatrix} = 4$ where $A = \begin{bmatrix} 1 & -2 \\ -2 & 4 \end{bmatrix}$ and $[-2 \quad 1] = B$.

Step 2:

$$\det(\lambda I - A) = \begin{vmatrix} \lambda - 1 & 2 \\ 2 & \lambda - 4 \end{vmatrix} = \lambda(\lambda - 5) = 0$$

Therefore, the eigenvalues are 0 and 5. For $\lambda = 0$, we get

$$\begin{bmatrix} -1 & 2 \\ 2 & -4 \end{bmatrix}\begin{bmatrix} a_1 \\ a_2 \end{bmatrix} = \begin{bmatrix} 0 \\ 0 \end{bmatrix}$$

which has the solution

$$\begin{bmatrix} a_1 \\ a_2 \end{bmatrix} = \begin{bmatrix} 2t \\ t \end{bmatrix} = t\begin{bmatrix} 2 \\ 1 \end{bmatrix}$$

For $\lambda = 5$, we get

$$\begin{bmatrix} 4 & 2 \\ 2 & 1 \end{bmatrix}\begin{bmatrix} a_1 \\ a_2 \end{bmatrix} = \begin{bmatrix} 0 \\ 0 \end{bmatrix}$$

which has the solution

$$\begin{bmatrix} a_1 \\ a_2 \end{bmatrix} = \begin{bmatrix} t \\ -2t \end{bmatrix} = t\begin{bmatrix} 1 \\ -2 \end{bmatrix}$$

$\begin{bmatrix} 2 \\ 1 \end{bmatrix}$ and $\begin{bmatrix} 1 \\ -2 \end{bmatrix}$ are orthogonal, but we must change these to unit vectors.

Therefore, $P_1 = \begin{bmatrix} 2/\sqrt{5} & 1/\sqrt{5} \\ 1/\sqrt{5} & -2/\sqrt{5} \end{bmatrix}$.

Step 3: Since $\det(P_1) = -1$, we interchange the columns to get $P = \begin{bmatrix} 1/\sqrt{5} & 2/\sqrt{5} \\ -2/\sqrt{5} & 1/\sqrt{5} \end{bmatrix}$ for which $\det(P) = 1$.

Step 4: $D = \begin{bmatrix} 5 & 0 \\ 0 & 0 \end{bmatrix}$ since $\lambda = 5$ corresponds to $\begin{bmatrix} 1/\sqrt{5} \\ -2/\sqrt{5} \end{bmatrix}$ and $\lambda = 0$ to $\begin{bmatrix} 2/\sqrt{5} \\ 1/\sqrt{5} \end{bmatrix}$.

Therefore, $Y^T DY + (BP)Y = f$ becomes

$$[x' \quad y'] \begin{bmatrix} 5 & 0 \\ 0 & 0 \end{bmatrix} \begin{bmatrix} x' \\ y' \end{bmatrix} + [-2 \quad 1] \begin{bmatrix} 1/\sqrt{5} & 2/\sqrt{5} \\ -2/\sqrt{5} & 1/\sqrt{5} \end{bmatrix} \begin{bmatrix} x' \\ y' \end{bmatrix} = 4$$

This simplifies to $5(x')^2 - 4/\sqrt{5}x' - 3/\sqrt{5}y' = 4$ which becomes $(x' - 2/(5\sqrt{5}))^2 = 3/(5\sqrt{5})(y' + 104\sqrt{5}/75)$ on completing the square. This is the standard form of a parabola.

PROBLEM 17-6 Find the standard form and name the conic represented by $4x^2 + 4xy + 4y^2 + \sqrt{2}x - \sqrt{2}y = 7$

Solution

Step 1:

$$[x \quad y] \begin{bmatrix} 4 & 2 \\ 2 & 4 \end{bmatrix} \begin{bmatrix} x \\ y \end{bmatrix} + [\sqrt{2} \quad -\sqrt{2}] \begin{bmatrix} x \\ y \end{bmatrix} = 7$$

Step 2:

$$\det(\lambda I - A) = \begin{vmatrix} \lambda - 4 & -2 \\ -2 & \lambda - 4 \end{vmatrix} = (\lambda - 4)^2 - 4 = \lambda^2 - 8\lambda + 12$$

Therefore, the eigenvalues are 2 and 6. For $\lambda = 6$, we get

$$\begin{bmatrix} 2 & -2 \\ -2 & 2 \end{bmatrix} \begin{bmatrix} a_1 \\ a_2 \end{bmatrix} = \begin{bmatrix} 0 \\ 0 \end{bmatrix}$$

which has the solution

$$\begin{bmatrix} a_1 \\ a_2 \end{bmatrix} = \begin{bmatrix} t \\ t \end{bmatrix} = t \begin{bmatrix} 1 \\ 1 \end{bmatrix}$$

For $\lambda = 2$, we get

$$\begin{bmatrix} -2 & -2 \\ -2 & -2 \end{bmatrix} \begin{bmatrix} a_1 \\ a_2 \end{bmatrix} = \begin{bmatrix} 0 \\ 0 \end{bmatrix}$$

which has the solution

$$\begin{bmatrix} a_1 \\ a_2 \end{bmatrix} = \begin{bmatrix} -t \\ t \end{bmatrix} = t \begin{bmatrix} -1 \\ 1 \end{bmatrix}$$

$$\therefore \quad P_1 = \begin{bmatrix} \frac{1}{\sqrt{2}} & -\frac{1}{\sqrt{2}} \\ \frac{1}{\sqrt{2}} & \frac{1}{\sqrt{2}} \end{bmatrix}$$

Step 3: Since $\det(P_1) = 1$, let $P = \begin{bmatrix} 1/\sqrt{2} & -1/\sqrt{2} \\ 1/\sqrt{2} & 1/\sqrt{2} \end{bmatrix}$.

Step 4: $D = \begin{bmatrix} 6 & 0 \\ 0 & 2 \end{bmatrix}$, $B = [\sqrt{2} \quad -\sqrt{2}]$. Therefore, $Y^T DY + (BP)Y = f$ becomes

$$[x' \quad y']\begin{bmatrix} 6 & 0 \\ 0 & 2 \end{bmatrix}\begin{bmatrix} x' \\ y' \end{bmatrix} + [\sqrt{2} \quad -\sqrt{2}]\begin{bmatrix} \frac{1}{\sqrt{2}} & -\frac{1}{\sqrt{2}} \\ \frac{1}{\sqrt{2}} & \frac{1}{\sqrt{2}} \end{bmatrix}\begin{bmatrix} x' \\ y' \end{bmatrix} = 7$$

This simplifies to $6(x')^2 + 2(y')^2 - 2y' = 7$, which reduces to $3(x')^2 + (y' - 1/2)^2 = 15/4$ or

$$\frac{(x')^2}{\frac{5}{4}} + \frac{(y' - 1/2)^2}{\frac{15}{4}} = 1$$

which is the standard form of an ellipse.

PROBLEM 17-7 Find the standard form and name the conic represented by $2xy - \sqrt{2}x = 1$.

solution

Step 1:

$$[x \quad y]\begin{bmatrix} 0 & 1 \\ 1 & 0 \end{bmatrix}\begin{bmatrix} x \\ y \end{bmatrix} + [-\sqrt{2} \quad 0]\begin{bmatrix} x \\ y \end{bmatrix} = 1$$

Step 2:

$$\det(\lambda I - A) = \begin{vmatrix} \lambda & -1 \\ -1 & \lambda \end{vmatrix} = \lambda^2 - 1 = (\lambda + 1)(\lambda - 1)$$

Therefore, the eigenvalues are -1 and 1. For $\lambda = -1$, we get

$$\begin{bmatrix} -1 & -1 \\ -1 & -1 \end{bmatrix}\begin{bmatrix} a_1 \\ a_2 \end{bmatrix} = \begin{bmatrix} 0 \\ 0 \end{bmatrix}$$

which has the solution

$$\begin{bmatrix} a_1 \\ a_2 \end{bmatrix} = \begin{bmatrix} t \\ -t \end{bmatrix} = t\begin{bmatrix} 1 \\ -1 \end{bmatrix}$$

For $\lambda = 1$, we get

$$\begin{bmatrix} 1 & -1 \\ -1 & 1 \end{bmatrix}\begin{bmatrix} a_1 \\ a_2 \end{bmatrix} = \begin{bmatrix} 0 \\ 0 \end{bmatrix}$$

which has the solution

$$\begin{bmatrix} a_1 \\ a_2 \end{bmatrix} = \begin{bmatrix} t \\ t \end{bmatrix} = t\begin{bmatrix} 1 \\ 1 \end{bmatrix}$$

$$\therefore \quad P_1 = \begin{bmatrix} \frac{1}{\sqrt{2}} & \frac{1}{\sqrt{2}} \\ -\frac{1}{\sqrt{2}} & \frac{1}{\sqrt{2}} \end{bmatrix}$$

Since $\det(P_1) = 1$, $P = \begin{bmatrix} 1/\sqrt{2} & 1/\sqrt{2} \\ -1/\sqrt{2} & 1/\sqrt{2} \end{bmatrix}$.

Step 3: $P = \begin{bmatrix} 1/\sqrt{2} & 1/\sqrt{2} \\ -1/\sqrt{2} & 1/\sqrt{2} \end{bmatrix}$

Step 4: $D = \begin{bmatrix} -1 & 0 \\ 0 & 1 \end{bmatrix}$, $B = [-\sqrt{2} \quad 0]$. Therefore, $Y^TDY + (BP)Y = f$ becomes

$$[x' \quad y'] \begin{bmatrix} -1 & 0 \\ 0 & 1 \end{bmatrix} \begin{bmatrix} x' \\ y' \end{bmatrix} + [-\sqrt{2} \quad 0] \begin{bmatrix} \frac{1}{\sqrt{2}} & \frac{1}{\sqrt{2}} \\ -\frac{1}{\sqrt{2}} & \frac{1}{\sqrt{2}} \end{bmatrix} \begin{bmatrix} x' \\ y' \end{bmatrix} = 1$$

This simplifies to $(y' - 1/2)^2 - (x' + 1/2)^2 = 1$, which is the standard form of a hyperbola.

PROBLEM 17-8 Find the standard form and name the quadric surface represented by $4x^2 + 4y^2 + 4z^2 + 2yz + x + \sqrt{2}y + \sqrt{2}z = \frac{1}{5}$

Solution

Step 1:

$$[x \quad y \quad z] \begin{bmatrix} 4 & 0 & 0 \\ 0 & 4 & 1 \\ 0 & 1 & 4 \end{bmatrix} \begin{bmatrix} x \\ y \\ z \end{bmatrix} + [1 \quad \sqrt{2} \quad \sqrt{2}] \begin{bmatrix} x \\ y \\ z \end{bmatrix} = \frac{1}{5}$$

Step 2:

$$\det(\lambda I - A) = \begin{vmatrix} \lambda - 4 & 0 & 0 \\ 0 & \lambda - 4 & -1 \\ 0 & -1 & \lambda - 4 \end{vmatrix} = (\lambda - 3)(\lambda - 4)(\lambda - 5)$$

Therefore, the eigenvalues are 3, 4, and 5. For $\lambda = 3$, we get

$$\begin{bmatrix} -1 & 0 & 0 \\ 0 & -1 & -1 \\ 0 & -1 & -1 \end{bmatrix} \begin{bmatrix} a_1 \\ a_2 \\ a_3 \end{bmatrix} = \begin{bmatrix} 0 \\ 0 \\ 0 \end{bmatrix}$$

which has the solution

$$\begin{bmatrix} a_1 \\ a_2 \\ a_3 \end{bmatrix} = \begin{bmatrix} 0 \\ t \\ -t \end{bmatrix} = t \begin{bmatrix} 0 \\ 1 \\ -1 \end{bmatrix}$$

For $\lambda = 4$, we get

$$\begin{bmatrix} 0 & 0 & 0 \\ 0 & 0 & -1 \\ 0 & -1 & 0 \end{bmatrix} \begin{bmatrix} a_1 \\ a_2 \\ a_3 \end{bmatrix} = \begin{bmatrix} 0 \\ 0 \\ 0 \end{bmatrix}$$

which has the solution

$$\begin{bmatrix} a_1 \\ a_2 \\ a_3 \end{bmatrix} = \begin{bmatrix} t \\ 0 \\ 0 \end{bmatrix} = t \begin{bmatrix} 1 \\ 0 \\ 0 \end{bmatrix}$$

For $\lambda = 5$, we get

$$\begin{bmatrix} 1 & 0 & 0 \\ 0 & 1 & -1 \\ 0 & -1 & 1 \end{bmatrix} \begin{bmatrix} a_1 \\ a_2 \\ a_3 \end{bmatrix} = \begin{bmatrix} 0 \\ 0 \\ 0 \end{bmatrix}$$

$$\begin{bmatrix} a_1 \\ a_2 \\ a_3 \end{bmatrix} = \begin{bmatrix} 0 \\ t \\ t \end{bmatrix} = t\begin{bmatrix} 0 \\ 1 \\ 1 \end{bmatrix}$$

$$\therefore \quad P_1 = \begin{bmatrix} 0 & 1 & 0 \\ \frac{1}{\sqrt{2}} & 0 & \frac{1}{\sqrt{2}} \\ -\frac{1}{\sqrt{2}} & 0 & \frac{1}{\sqrt{2}} \end{bmatrix}$$

Step 3: Since $\det(P_1) = -1/2 - 1/2 = -1$, choose $P = \begin{bmatrix} 1 & 0 & 0 \\ 0 & 1/\sqrt{2} & 1/\sqrt{2} \\ 0 & -1/\sqrt{2} & 1/\sqrt{2} \end{bmatrix}$.

Step 4: Therefore, $Y^TDY + (BP)Y = j$ becomes

$$[x' \quad y' \quad z']\begin{bmatrix} 4 & 0 & 0 \\ 0 & 3 & 0 \\ 0 & 0 & 5 \end{bmatrix}\begin{bmatrix} x' \\ y' \\ z' \end{bmatrix} + [1 \quad \sqrt{2} \quad \sqrt{2}]\begin{bmatrix} 1 & 0 & 0 \\ 0 & \frac{1}{\sqrt{2}} & \frac{1}{\sqrt{2}} \\ 0 & -\frac{1}{\sqrt{2}} & \frac{1}{\sqrt{2}} \end{bmatrix}\begin{bmatrix} x' \\ y' \\ z' \end{bmatrix}$$

$$= \frac{1}{5}$$

This expands into

$$4(x')^2 + 3(y')^2 + 5(z')^2 + x' + z' = \frac{1}{5}$$

$$4(x')^2 + x' + 3(y')^2 + 5(z')^2 + z' = \frac{1}{5}$$

$$4\left[(x')^2 + \frac{1}{4}x' + \frac{1}{64}\right] + 3(y')^2 + 5\left[(z')^2 + \frac{1}{5}z' + \frac{1}{100}\right] = \frac{1}{5} + \frac{1}{16} + \frac{1}{20}$$

$$4\left(x' + \frac{1}{8}\right)^2 + 3(y')^2 + 5\left(z' + \frac{1}{10}\right)^2 = \frac{25}{80}$$

$$\therefore \quad \frac{(x' + 1/8)^2}{\frac{25}{320}} + \frac{(y')^2}{\frac{25}{240}} + \frac{(z' + 1/10)^2}{\frac{5}{80}} = 1$$

This is the standard form of an ellipsoid.

PROBLEM 17-9 Determine whether each of the following quadratic forms is positive definite, negative definite, indefinite, or neither: **(a)** $5x^2 - 4y^2 + z^2$ in x, y, and z; **(b)** $3x^2 + y^2$ in x and y; **(c)** $-3x^2 - y^2$ in x and y; **(d)** $3x^2 - y^2 + 2z^2 + 2xz$ in x, y, and z.

Solution

(a) Indefinite, since $5(1)^2 - 4(3)^2 + (2)^2 = -27$ and $5(1)^2 - 4(2)^2 + (4)^2 = 5$.

(b) Positive definite, since it is zero only if $X = \begin{bmatrix} 0 \\ 0 \end{bmatrix}$ and is positive otherwise.

(c) Negative definite, since it is zero only if $X = \begin{bmatrix} 0 \\ 0 \end{bmatrix}$ and is negative otherwise.

(d) Indefinite, since $3(1)^2 - (2)^2 + 2(3)^2 + 2(1)(3) = 29$ and $3(0)^2 - (5)^2 + 2(1)^2 + 2(0)(3) = -23$.

PROBLEM 17-10 Determine whether each of the following quadratic forms is positive definite, negative definite, indefinite, or neither: **(a)** $2y^2 - x^2 + 4xy$, **(b)** $4x^2 + 4xy + 4y^2$, **(c)** $x^2 + 4xy + y^2$, **(d)** $5x^2 - 4xy + 5y^2$.

Solution

(a)
$$A = \begin{bmatrix} -1 & 2 \\ 2 & 2 \end{bmatrix}, \qquad \det(\lambda I - A) = \begin{vmatrix} \lambda + 1 & -2 \\ -2 & \lambda - 2 \end{vmatrix} = (\lambda - 3)(\lambda + 2)$$

Therefore, the eigenvalues are -2 and 3. Hence, **(a)** is indefinite, since one root is positive and the other is negative.

(b)
$$A = \begin{bmatrix} 4 & 2 \\ 2 & 4 \end{bmatrix}, \qquad \det(\lambda I - A) = \begin{vmatrix} \lambda - 4 & -2 \\ -2 & \lambda - 4 \end{vmatrix} = (\lambda - 2)(\lambda - 6)$$

Therefore, the eigenvalues are 2 and 6. Hence, **(b)** is positive definite.

(c)
$$A = \begin{bmatrix} 4 & 2 \\ 2 & 1 \end{bmatrix}, \qquad \det(\lambda I - A) = \begin{vmatrix} \lambda - 4 & -2 \\ -2 & \lambda - 1 \end{vmatrix} = \lambda(\lambda - 5)$$

Therefore, the eigenvalues are 0 and 5. Hence, **(c)** is neither.

(d)
$$A = \begin{bmatrix} 5 & -2 \\ -2 & 5 \end{bmatrix}, \qquad \det(\lambda I - A) = \begin{vmatrix} \lambda - 5 & 2 \\ 2 & \lambda - 5 \end{vmatrix} = \lambda^2 + 10\lambda + 21 = (\lambda + 7)(\lambda + 3)$$

Therefore, the eigenvalues are -3 and -7. Hence, **(d)** is negative definite.

PROBLEM 17-11 Determine whether the quadratic form $x^2 + y^2 + 2z^2 + 4xz$ is positive definite, negative definite, indefinite, or neither.

Solution

$$A = \begin{bmatrix} 1 & 0 & 2 \\ 0 & 1 & 0 \\ 2 & 0 & 2 \end{bmatrix}$$

$$\det(\lambda I - A) = \begin{vmatrix} \lambda - 1 & 0 & -2 \\ 0 & \lambda - 1 & 0 \\ -2 & 0 & \lambda - 2 \end{vmatrix} = (\lambda - 1)^2(\lambda - 2) - 4(\lambda - 1) = (\lambda - 1)(\lambda - 2)(\lambda + 1)$$

Therefore, the eigenvalues are -1, 1, and 2. Thus, the quadratic form is indefinite.

Supplementary Problems

PROBLEM 17-12 In each of the following quadratic equations find the associated quadratic form: **(a)** $x^2 - xy + y^2 - 2x = 4$, **(b)** $3x^2 + y^2 - z^2 + xz - 2yz + y = 2$, **(c)** $4x^2 + y^2 + 5z^2 + 3xz + x - 3z = 0$.

PROBLEM 17-13 Find the matrix A of each quadratic form in Problem 17-12.

PROBLEM 17-14 Transform $3x^2 - 2xy + 3y^2$ into quadratic form with no xy term.

PROBLEM 17-15 Find the standard form and identify the conic $x^2 - 4xy + y^2 = 3$.

PROBLEM 17-16 Find the standard form and name the quadric surface $3x^2 + 3y^2 + 3z^2 + 4xz = 15$.

PROBLEM 17-17 Determine which of the following quadratic forms are positive definite, negative definite, indefinite, or neither: **(a)** $x^2 + 5y^2 + 3z^2$, **(b)** $-2x^2 + 3y^2$, **(c)** $-2x^2 - 3y^2$, **(d)** $x^2 - 2xy + y^2$, **(e)** $y^2 + 2xz + z^2$.

PROBLEM 17-18 A well-known theorem from analytic geometry states that, if $ax^2 + bxy + cy^2 + dx + ey = f$,

(1) $\begin{vmatrix} a & b/2 \\ b/2 & c \end{vmatrix} = 0$, the quadratic equation is a parabola, or two parallel lines.

(2) $\begin{vmatrix} a & b/2 \\ b/2 & c \end{vmatrix} < 0$, the quadratic equation is an hyperbola, or two intersecting lines.

(3) $\begin{vmatrix} a & b/2 \\ b/2 & c \end{vmatrix} > 0$, the quadratic equation is an ellipse, a circle, or a point.

Classify each of the conics represented by the following equations: **(a)** $x^2 - 2xy + 3y^2 + 3x = 4$, **(b)** $y^2 + 2xy = 3$, **(c)** $x^2 + 2xy + y^2 + x - y = 1$.

PROBLEM 17-19 If $\begin{vmatrix} a & b/2 \\ b/2 & c \end{vmatrix} > 0$, the quadratic form is positive definite. Which of the following quadratic forms are positive definite? **(a)** $3x^2 + 2xy + y^2$; **(b)** $x^2 - 3xy - y^2$; **(c)** $y^2 + 2xy$.

Answers to Supplementary Problems

17-12 **(a)** $x^2 - xy + y^2$
(b) $3x^2 + y^2 - z^2 + xz - 2yz$
(c) $4x^2 + y^2 + 5z^2 + 3xz$

17-13 **(a)** $A = \begin{bmatrix} 1 & -\frac{1}{2} \\ -\frac{1}{2} & 1 \end{bmatrix}$

(b) $A = \begin{bmatrix} 3 & 0 & \frac{1}{2} \\ 0 & 1 & -1 \\ \frac{1}{2} & -1 & -1 \end{bmatrix}$

(c) $A = \begin{bmatrix} 4 & 0 & \frac{3}{2} \\ 0 & 1 & 0 \\ \frac{3}{2} & 0 & 5 \end{bmatrix}$

17-14 $4(x')^2 + 2(y')^2$

17-15 $[(x')^2/1] - [(y')^2/3] = 1$: hyperbola

17-16 $[(x')^2/15] + [(y')^2/5] + [(z')^2/3] = 1$; ellipsoid

17-17 **(a)** Positive definite
(b) Indefinite
(c) Negative definite
(d) Neither
(e) Indefinite

17-18 **(a)** Ellipse
(b) Hyperbola
(c) Parabola

17-19 **(a)** Positive definite
(b) Not positive definite
(c) Not positive definite

18 APPLICATIONS TO GEOMETRIC TRANSFORMATIONS

THIS CHAPTER IS ABOUT

- ☑ **Transformations in Euclidean Space R_2**
- ☑ **Properties of R_2 Transformations**
- ☑ **Transformations in Euclidean Space R_3**
- ☑ **Properties of R_3 Transformations**
- ☑ **Isometries**

18-1. Transformations in Euclidean Space R_2

Although all of the following are transformations of coordinates of individual points; in most cases, we must transform the coordinates of each point in a set of points. Except for translation, all of the following are linear transformations. Each illustration represents the images of all points of the rectangle having vertices at $A(2,2)$, $B(-1,2)$, $C(-1,-1)$, and $D(2,-1)$. In each case, the points A', B', C', and D' are the images of the vertices A, B, C, and D, respectively. Furthermore, (x, y) and $\begin{bmatrix} x \\ y \end{bmatrix}$ should be thought of as coordinates of the same point. The rectangle in Figure 18-1 is referred to as $ABCD$ in this chapter.

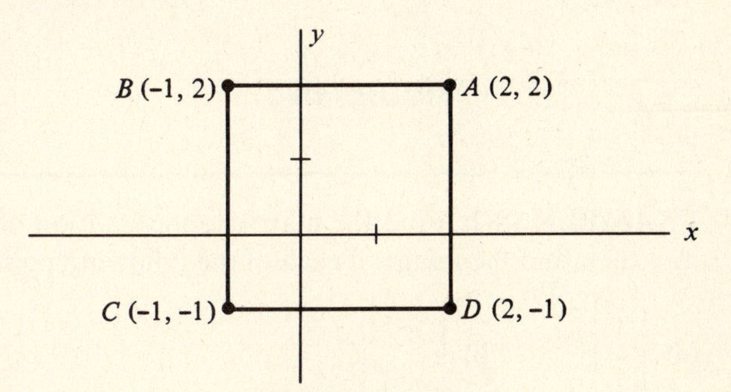

FIGURE 18-1

A. Translation

Translation by **a** units in the x-direction and **b** units in the y-direction is illustrated in Figure 18-2. This translation is represented by

$$\begin{bmatrix} x' \\ y' \end{bmatrix} = \begin{bmatrix} x \\ y \end{bmatrix} + \begin{bmatrix} a \\ b \end{bmatrix}$$

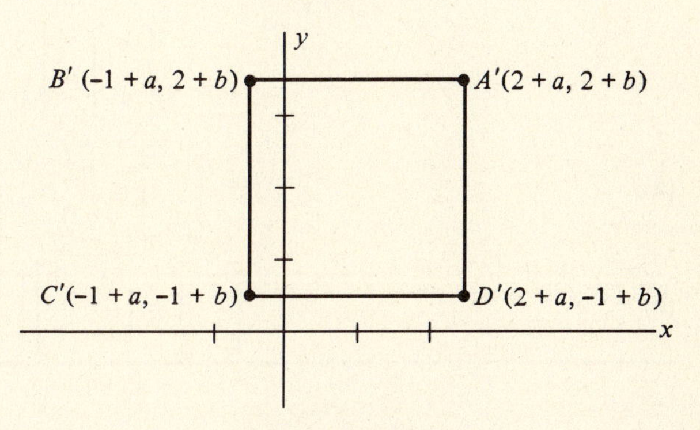

FIGURE 18-2

EXAMPLE 18-1: Find the matrix representation of a translation by 3 units in the x-direction and -2 units in the y-direction, and then find the image of each of the following points: **(a)** $\begin{bmatrix} 1 \\ 1 \end{bmatrix}$, **(b)** $\begin{bmatrix} -5 \\ 4 \end{bmatrix}$.

Solution

$$\begin{bmatrix} x' \\ y' \end{bmatrix} = \begin{bmatrix} x \\ y \end{bmatrix} + \begin{bmatrix} 3 \\ -2 \end{bmatrix} = \begin{bmatrix} x + 3 \\ y - 2 \end{bmatrix}$$

(a) Therefore, for

$$\begin{bmatrix} 1 \\ 1 \end{bmatrix}, \begin{bmatrix} x' \\ y' \end{bmatrix} = \begin{bmatrix} 1+3 \\ 1-2 \end{bmatrix} = \begin{bmatrix} 4 \\ -1 \end{bmatrix}$$

(b) For

$$\begin{bmatrix} -5 \\ 4 \end{bmatrix}, \begin{bmatrix} x' \\ y' \end{bmatrix} = \begin{bmatrix} -5+3 \\ 4-2 \end{bmatrix} = \begin{bmatrix} -2 \\ 2 \end{bmatrix}$$

B. Expansions or dilations

Expansion or **dilation** in the x-direction by a factor of k where $k > 1$ is illustrated in Figure 18-3 and is represented by

$$\begin{bmatrix} x' \\ y' \end{bmatrix} = \begin{bmatrix} k & 0 \\ 0 & 1 \end{bmatrix} \begin{bmatrix} x \\ y \end{bmatrix}, \qquad k > 1$$

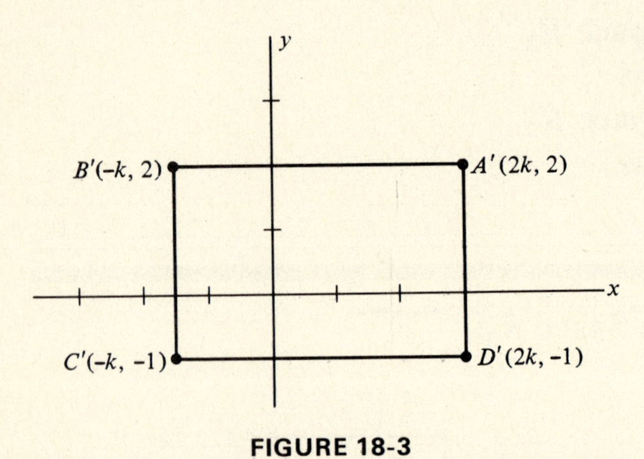

FIGURE 18-3

EXAMPLE 18-2: Find the matrix representation of an expansion in the x-direction by a factor of 3, and then find the image of each of the following points.

(a) $\begin{bmatrix} 1 \\ 1 \end{bmatrix}$ **(b)** $\begin{bmatrix} -5 \\ 4 \end{bmatrix}$

Solution

$$\begin{bmatrix} x' \\ y' \end{bmatrix} = \begin{bmatrix} 3 & 0 \\ 0 & 1 \end{bmatrix} \begin{bmatrix} x \\ y \end{bmatrix}$$

(a)
$$\begin{bmatrix} x' \\ y' \end{bmatrix} = \begin{bmatrix} 3 & 0 \\ 0 & 1 \end{bmatrix} \begin{bmatrix} 1 \\ 1 \end{bmatrix} = \begin{bmatrix} 3 \\ 1 \end{bmatrix}$$

(b)
$$\begin{bmatrix} x' \\ y' \end{bmatrix} = \begin{bmatrix} 3 & 0 \\ 0 & 1 \end{bmatrix} \begin{bmatrix} -5 \\ 4 \end{bmatrix} = \begin{bmatrix} -15 \\ 4 \end{bmatrix}$$

Expansion or dilation in the y-direction by a factor of k where $k > 1$ is illustrated in Figure 18-4 and is represented by

$$\begin{bmatrix} x' \\ y' \end{bmatrix} = \begin{bmatrix} 1 & 0 \\ 0 & k \end{bmatrix} \begin{bmatrix} x \\ y \end{bmatrix}, \qquad k > 1$$

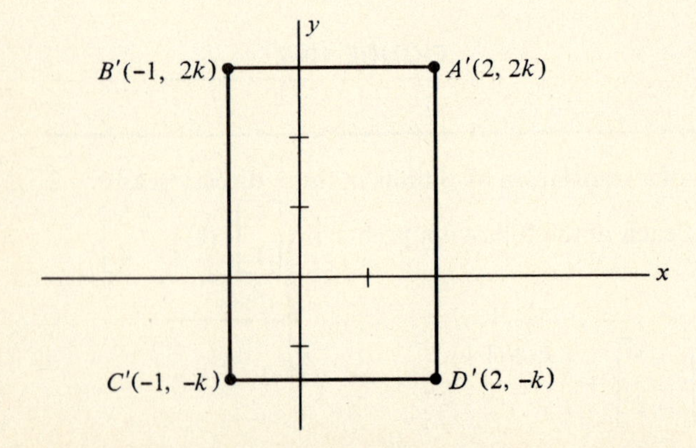

FIGURE 18-4

EXAMPLE 18-3: Find the matrix representation of an expansion in the y-direction by a factor of 3, and then find the image of each of the following points.

(a) $\begin{bmatrix} 1 \\ 1 \end{bmatrix}$ **(b)** $\begin{bmatrix} -5 \\ 4 \end{bmatrix}$

Solution

$$\begin{bmatrix} x' \\ y' \end{bmatrix} = \begin{bmatrix} 1 & 0 \\ 0 & 3 \end{bmatrix} \begin{bmatrix} x \\ y \end{bmatrix}$$

(a) $$\begin{bmatrix} x' \\ y' \end{bmatrix} = \begin{bmatrix} 1 & 0 \\ 0 & 3 \end{bmatrix} \begin{bmatrix} 1 \\ 1 \end{bmatrix} = \begin{bmatrix} 1 \\ 3 \end{bmatrix}$$

(b) $$\begin{bmatrix} x' \\ y' \end{bmatrix} = \begin{bmatrix} 1 & 0 \\ 0 & 3 \end{bmatrix} \begin{bmatrix} -5 \\ 4 \end{bmatrix} = \begin{bmatrix} -5 \\ 12 \end{bmatrix}$$

C. Compressions or contractions

Compression or **contraction** in the x-direction by a factor of k where $0 < k < 1$ is illustrated in Figure 18-5 and is represented by

$$\begin{bmatrix} x' \\ y' \end{bmatrix} = \begin{bmatrix} k & 0 \\ 0 & 1 \end{bmatrix} \begin{bmatrix} x \\ y \end{bmatrix}, \qquad 0 < k < 1$$

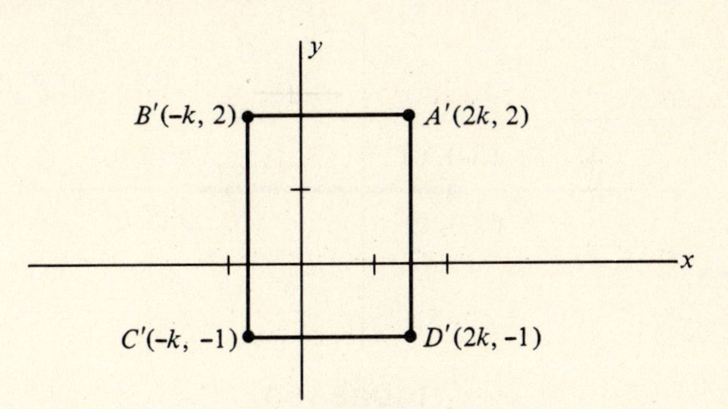

FIGURE 18-5

EXAMPLE 18-4: Find the matrix representation of a compression in the x-direction by a factor of 1/2, and then find the image of each of the following points.

(a) $\begin{bmatrix} 2 \\ 3 \end{bmatrix}$ **(b)** $\begin{bmatrix} 4 \\ -8 \end{bmatrix}$

Solution

$$\begin{bmatrix} x' \\ y' \end{bmatrix} = \begin{bmatrix} \frac{1}{2} & 0 \\ 0 & 1 \end{bmatrix} \begin{bmatrix} x \\ y \end{bmatrix}$$

(a) $$\begin{bmatrix} x' \\ y' \end{bmatrix} = \begin{bmatrix} \frac{1}{2} & 0 \\ 0 & 1 \end{bmatrix} \begin{bmatrix} 2 \\ 3 \end{bmatrix} = \begin{bmatrix} 1 \\ 3 \end{bmatrix}$$

$$\begin{bmatrix} x' \\ y' \end{bmatrix} = \begin{bmatrix} \frac{1}{2} & 0 \\ 0 & 1 \end{bmatrix} \begin{bmatrix} 4 \\ -8 \end{bmatrix} = \begin{bmatrix} 2 \\ -8 \end{bmatrix}$$

Compression or contraction in the y-direction by a factor of k where $0 < k < 1$ is illustrated in Figure 18-6 and is represented by

$$\begin{bmatrix} x' \\ y' \end{bmatrix} = \begin{bmatrix} 1 & 0 \\ 0 & k \end{bmatrix} \begin{bmatrix} x \\ y \end{bmatrix}, \qquad 0 < k < 1$$

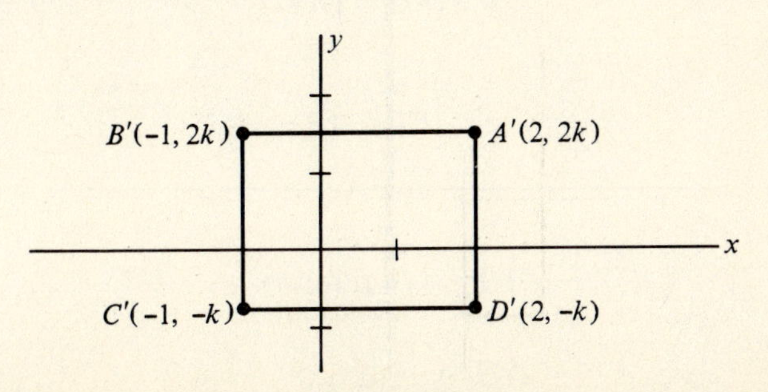

FIGURE 18-6

EXAMPLE 18-5: Find the matrix representation of a compression in the y-direction by a factor of 1/2, and then find the image of the following points.

(a) $\begin{bmatrix} 2 \\ 3 \end{bmatrix}$ **(b)** $\begin{bmatrix} 4 \\ -8 \end{bmatrix}$

Solution

$$\begin{bmatrix} x' \\ y' \end{bmatrix} = \begin{bmatrix} 1 & 0 \\ 0 & \frac{1}{2} \end{bmatrix} \begin{bmatrix} x \\ y \end{bmatrix}$$

(a)
$$\begin{bmatrix} x' \\ y' \end{bmatrix} = \begin{bmatrix} 1 & 0 \\ 0 & \frac{1}{2} \end{bmatrix} \begin{bmatrix} 2 \\ 3 \end{bmatrix} = \begin{bmatrix} 2 \\ \frac{3}{2} \end{bmatrix}$$

(b)
$$\begin{bmatrix} x' \\ y' \end{bmatrix} = \begin{bmatrix} 1 & 0 \\ 0 & \frac{1}{2} \end{bmatrix} \begin{bmatrix} 4 \\ -8 \end{bmatrix} = \begin{bmatrix} 4 \\ -4 \end{bmatrix}$$

D. Projections

Projection onto the x-axis is illustrated in Figure 18-7 and is represented by

$$\begin{bmatrix} x' \\ y' \end{bmatrix} = \begin{bmatrix} 1 & 0 \\ 0 & 0 \end{bmatrix} \begin{bmatrix} x \\ y \end{bmatrix}$$

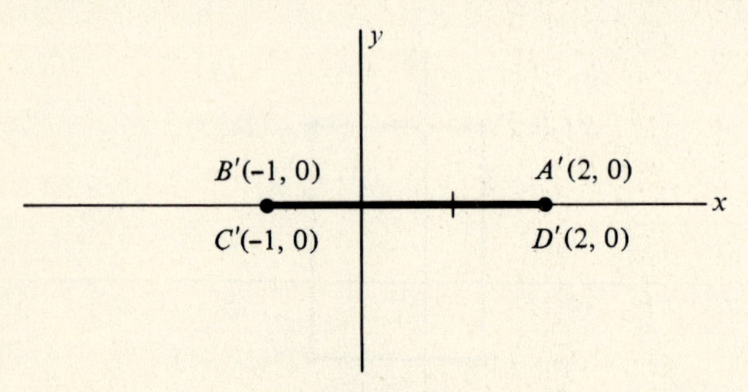

FIGURE 18-7

EXAMPLE 18-6: Find the matrix representation for a projection onto the x-axis, and then find the image of each of the following points.

(a) $\begin{bmatrix} 5 \\ 7 \end{bmatrix}$ **(b)** $\begin{bmatrix} -6 \\ 4 \end{bmatrix}$

Solution

$$\begin{bmatrix} x' \\ y' \end{bmatrix} = \begin{bmatrix} 1 & 0 \\ 0 & 0 \end{bmatrix} \begin{bmatrix} x \\ y \end{bmatrix}$$

(a)
$$\begin{bmatrix} x' \\ y' \end{bmatrix} = \begin{bmatrix} 1 & 0 \\ 0 & 0 \end{bmatrix} \begin{bmatrix} 5 \\ 7 \end{bmatrix} = \begin{bmatrix} 5 \\ 0 \end{bmatrix}$$

(b)
$$\begin{bmatrix} x' \\ y' \end{bmatrix} = \begin{bmatrix} 1 & 0 \\ 0 & 0 \end{bmatrix} \begin{bmatrix} -6 \\ 4 \end{bmatrix} = \begin{bmatrix} -6 \\ 0 \end{bmatrix}$$

Projection onto the y-axis is illustrated in Figure 18-8 and is represented by

$$\begin{bmatrix} x' \\ y' \end{bmatrix} = \begin{bmatrix} 0 & 0 \\ 0 & 1 \end{bmatrix} \begin{bmatrix} x \\ y \end{bmatrix}$$

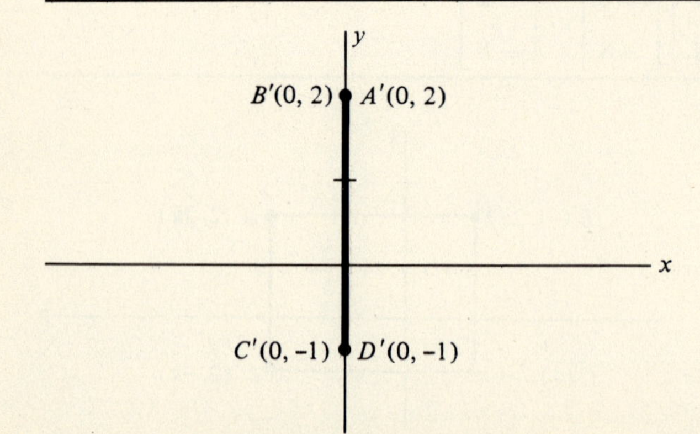

FIGURE 18-8

EXAMPLE 18-7: Find the matrix representation for a projection onto the y-axis, and then find the image of each of the following points.

(a) $\begin{bmatrix} 5 \\ 7 \end{bmatrix}$ **(b)** $\begin{bmatrix} -6 \\ 4 \end{bmatrix}$

Solution

$$\begin{bmatrix} x' \\ y' \end{bmatrix} = \begin{bmatrix} 0 & 0 \\ 0 & 1 \end{bmatrix} \begin{bmatrix} x \\ y \end{bmatrix}$$

(a) $$\begin{bmatrix} x' \\ y' \end{bmatrix} = \begin{bmatrix} 0 & 0 \\ 0 & 1 \end{bmatrix} \begin{bmatrix} 5 \\ 7 \end{bmatrix} = \begin{bmatrix} 0 \\ 7 \end{bmatrix}$$

(b) $$\begin{bmatrix} x' \\ y' \end{bmatrix} = \begin{bmatrix} 0 & 0 \\ 0 & 1 \end{bmatrix} \begin{bmatrix} -6 \\ 4 \end{bmatrix} = \begin{bmatrix} 0 \\ 4 \end{bmatrix}$$

E. Shears

Shear in the x-direction is illustrated in Figure 18-9 and is represented by

$$\begin{bmatrix} x' \\ y' \end{bmatrix} = \begin{bmatrix} 1 & c \\ 0 & 1 \end{bmatrix} \begin{bmatrix} x \\ y \end{bmatrix}, \qquad c \neq 0$$

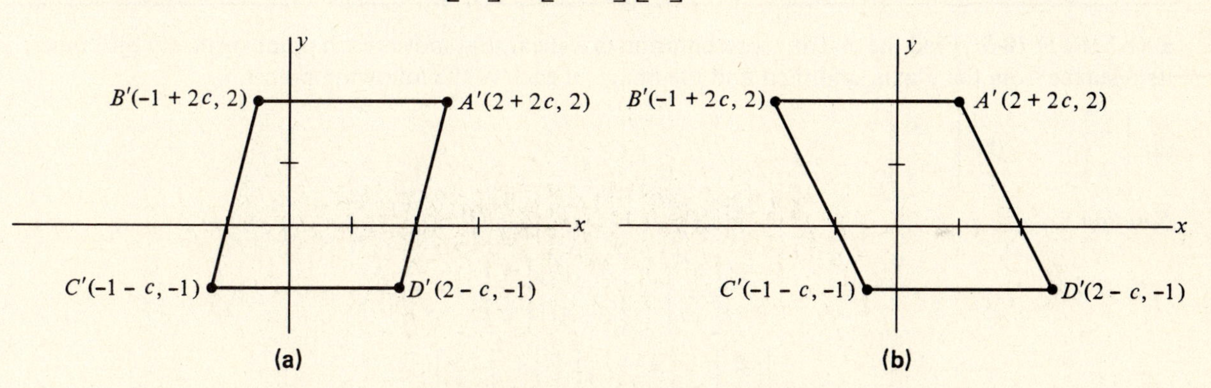

FIGURE 18-9. (a) $c > 0$. (b) $c < 0$.

A shear in the x-direction leaves the points on the x-axis fixed and moves each point off the x-axis parallel to its initial position by an amount proportional to its distance from the x-axis. The number c is the constant of proportionality.

EXAMPLE 18-8: Find the matrix representation of a shear that moves each point horizontally by 2 times its distance from the x-axis, and then find the image of the following points.

(a) $\begin{bmatrix} 5 \\ 7 \end{bmatrix}$ **(b)** $\begin{bmatrix} -6 \\ 4 \end{bmatrix}$

Solution

$$\begin{bmatrix} x' \\ y' \end{bmatrix} = \begin{bmatrix} 1 & 2 \\ 0 & 1 \end{bmatrix} \begin{bmatrix} x \\ y \end{bmatrix}$$

(a) $$\begin{bmatrix} x' \\ y' \end{bmatrix} = \begin{bmatrix} 1 & 2 \\ 0 & 1 \end{bmatrix} \begin{bmatrix} 5 \\ 7 \end{bmatrix} = \begin{bmatrix} 19 \\ 7 \end{bmatrix}$$

(b) $$\begin{bmatrix} x' \\ y' \end{bmatrix} = \begin{bmatrix} 1 & 2 \\ 0 & 1 \end{bmatrix} \begin{bmatrix} -6 \\ 4 \end{bmatrix} = \begin{bmatrix} 2 \\ 4 \end{bmatrix}$$

Shear in the y-direction is illustrated in Figure 18-10 and is represented by

$$\begin{bmatrix} x' \\ y' \end{bmatrix} = \begin{bmatrix} 1 & 0 \\ c & 1 \end{bmatrix} \begin{bmatrix} x \\ y \end{bmatrix}, \qquad c \neq 0$$

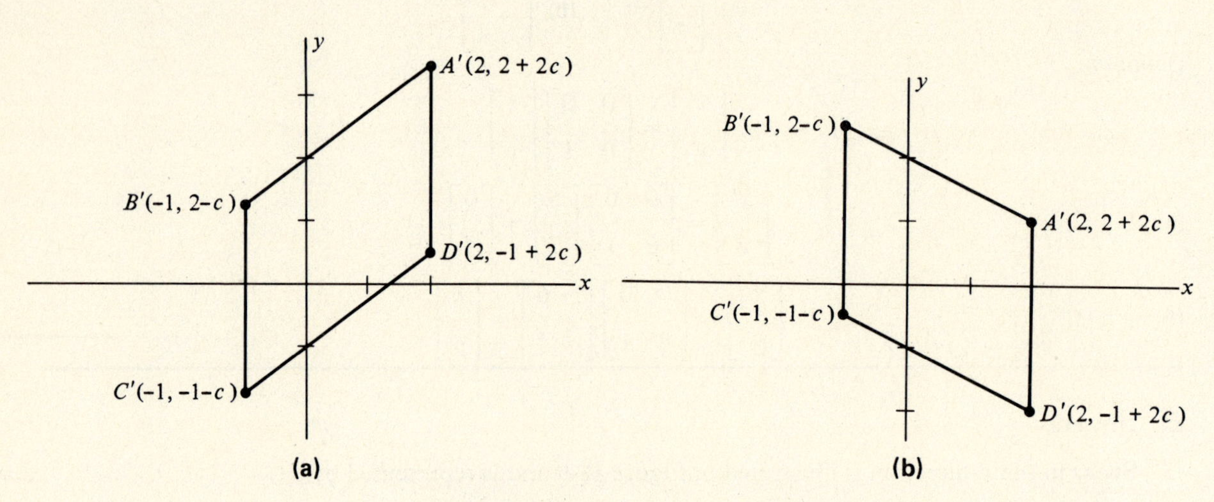

FIGURE 18-10. (a) $c > 0$. (b) $c < 0$.

EXAMPLE 18-9: Find the matrix representation of a shear that moves each point vertically by 3 times its distance from the y-axis, and then find the image of each of the following points.

(a) $\begin{bmatrix} 5 \\ 7 \end{bmatrix}$ **(b)** $\begin{bmatrix} -6 \\ 4 \end{bmatrix}$

Solution

$$\begin{bmatrix} x' \\ y' \end{bmatrix} = \begin{bmatrix} 1 & 0 \\ 3 & 1 \end{bmatrix} \begin{bmatrix} x \\ y \end{bmatrix}$$

(a)

$$\begin{bmatrix} x' \\ y' \end{bmatrix} = \begin{bmatrix} 1 & 0 \\ 3 & 1 \end{bmatrix} \begin{bmatrix} 5 \\ 7 \end{bmatrix} = \begin{bmatrix} 5 \\ 22 \end{bmatrix}$$

(b)

$$\begin{bmatrix} x' \\ y' \end{bmatrix} = \begin{bmatrix} 1 & 0 \\ 3 & 1 \end{bmatrix} \begin{bmatrix} -6 \\ 4 \end{bmatrix} = \begin{bmatrix} -6 \\ -14 \end{bmatrix}$$

F. Reflections

Reflection about the x-axis is illustrated in Figure 18-11 and is represented by

$$\begin{bmatrix} x' \\ y' \end{bmatrix} = \begin{bmatrix} 1 & 0 \\ 0 & -1 \end{bmatrix} \begin{bmatrix} x \\ y \end{bmatrix}$$

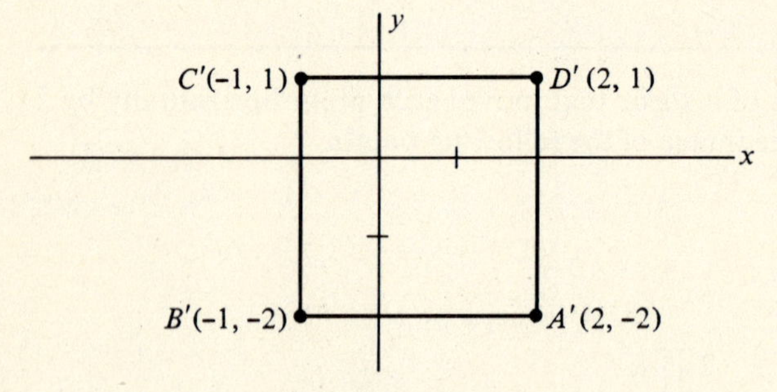

FIGURE 18-11

EXAMPLE 18-10: Find the image if each of the following points is reflected about the x-axis.

(a) $\begin{bmatrix} 5 \\ 7 \end{bmatrix}$ **(b)** $\begin{bmatrix} -3 \\ 4 \end{bmatrix}$

Solution

$$\begin{bmatrix} x' \\ y' \end{bmatrix} = \begin{bmatrix} 1 & 0 \\ 0 & -1 \end{bmatrix} \begin{bmatrix} x \\ y \end{bmatrix}$$

(a) $$\begin{bmatrix} x' \\ y' \end{bmatrix} = \begin{bmatrix} 1 & 0 \\ 0 & -1 \end{bmatrix} \begin{bmatrix} 5 \\ 7 \end{bmatrix} = \begin{bmatrix} 5 \\ -7 \end{bmatrix}$$

(b) $$\begin{bmatrix} x' \\ y' \end{bmatrix} = \begin{bmatrix} 1 & 0 \\ 0 & -1 \end{bmatrix} \begin{bmatrix} -3 \\ 4 \end{bmatrix} = \begin{bmatrix} -3 \\ -4 \end{bmatrix}$$

Reflection about the y-axis is illustrated in Figure 18-12 and is represented by

$$\begin{bmatrix} x' \\ y' \end{bmatrix} = \begin{bmatrix} -1 & 0 \\ 0 & 1 \end{bmatrix} \begin{bmatrix} x \\ y \end{bmatrix}$$

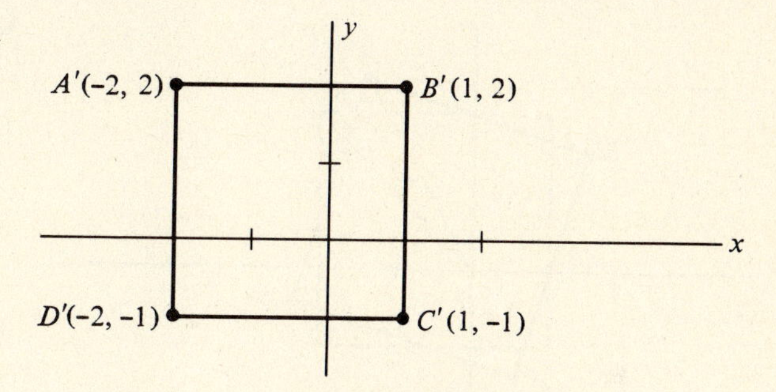

FIGURE 18-12

EXAMPLE 18-11: Find the image if each of the following points is reflected about the y-axis.

(a) $\begin{bmatrix} 5 \\ 7 \end{bmatrix}$ **(b)** $\begin{bmatrix} -3 \\ 4 \end{bmatrix}$

Solution

$$\begin{bmatrix} x' \\ y' \end{bmatrix} = \begin{bmatrix} -1 & 0 \\ 0 & 1 \end{bmatrix} \begin{bmatrix} x \\ y \end{bmatrix}$$

(a) $$\begin{bmatrix} x' \\ y' \end{bmatrix} = \begin{bmatrix} -1 & 0 \\ 0 & 1 \end{bmatrix} \begin{bmatrix} 5 \\ 7 \end{bmatrix} = \begin{bmatrix} -5 \\ 7 \end{bmatrix}$$

(b) $$\begin{bmatrix} x' \\ y' \end{bmatrix} = \begin{bmatrix} -1 & 0 \\ 0 & 1 \end{bmatrix} \begin{bmatrix} -3 \\ 4 \end{bmatrix} = \begin{bmatrix} 3 \\ 4 \end{bmatrix}$$

Reflection about the line $y = x$ is illustrated in Figure 18-13 and is represented by

$$\begin{bmatrix} x' \\ y' \end{bmatrix} = \begin{bmatrix} 0 & 1 \\ 1 & 0 \end{bmatrix} \begin{bmatrix} x \\ y \end{bmatrix}$$

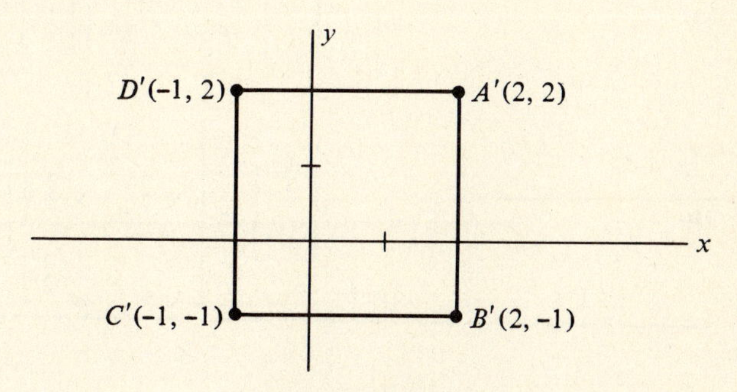

FIGURE 18-13

EXAMPLE 18-12: Find the image if each of the following points is reflected about the line $y = x$.

(a) $\begin{bmatrix} 5 \\ 7 \end{bmatrix}$ **(b)** $\begin{bmatrix} -3 \\ 4 \end{bmatrix}$

Solution

$$\begin{bmatrix} x' \\ y' \end{bmatrix} = \begin{bmatrix} 0 & 1 \\ 1 & 0 \end{bmatrix} \begin{bmatrix} x \\ y \end{bmatrix}$$

(a) $$\begin{bmatrix} x' \\ y' \end{bmatrix} = \begin{bmatrix} 0 & 1 \\ 1 & 0 \end{bmatrix} \begin{bmatrix} 5 \\ 7 \end{bmatrix} = \begin{bmatrix} 7 \\ 5 \end{bmatrix}$$

(b) $$\begin{bmatrix} x' \\ y' \end{bmatrix} = \begin{bmatrix} 0 & 1 \\ 1 & 0 \end{bmatrix} \begin{bmatrix} -3 \\ 4 \end{bmatrix} = \begin{bmatrix} 4 \\ -3 \end{bmatrix}$$

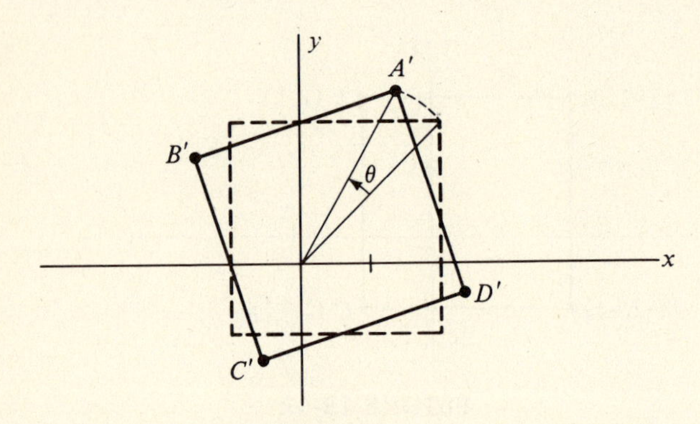

FIGURE 18-14

G. Rotation through θ

Rotation through the angle θ is illustrated in Figure 18-14. (If $\theta > 0$, the rotation is counterclockwise. If $\theta < 0$, the rotation is clockwise.) This is represented by

$$\begin{bmatrix} x' \\ y' \end{bmatrix} = \begin{bmatrix} \cos\theta & -\sin\theta \\ \sin\theta & \cos\theta \end{bmatrix} \begin{bmatrix} x \\ y \end{bmatrix}$$

EXAMPLE 18-13: Find the image if each of the following points is rotated counterclockwise through 45°.

(a) $\begin{bmatrix} 5 \\ 7 \end{bmatrix}$ **(b)** $\begin{bmatrix} -3 \\ 4 \end{bmatrix}$

Solution

$$\begin{bmatrix} x' \\ y' \end{bmatrix} = \begin{bmatrix} \cos 45^\circ & -\sin 45^\circ \\ \sin 45^\circ & \cos 45^\circ \end{bmatrix} \begin{bmatrix} x \\ y \end{bmatrix} = \begin{bmatrix} \frac{1}{\sqrt{2}} & -\frac{1}{\sqrt{2}} \\ \frac{1}{\sqrt{2}} & \frac{1}{\sqrt{2}} \end{bmatrix} \begin{bmatrix} x \\ y \end{bmatrix}$$

(a) $$\begin{bmatrix} x' \\ y' \end{bmatrix} = \begin{bmatrix} \frac{1}{\sqrt{2}} & -\frac{1}{\sqrt{2}} \\ \frac{1}{\sqrt{2}} & \frac{1}{\sqrt{2}} \end{bmatrix} \begin{bmatrix} 5 \\ 7 \end{bmatrix} = \begin{bmatrix} -\frac{2}{\sqrt{2}} \\ \frac{12}{\sqrt{2}} \end{bmatrix} = \begin{bmatrix} -\sqrt{2} \\ 6\sqrt{2} \end{bmatrix}$$

(b) $$\begin{bmatrix} x' \\ y' \end{bmatrix} = \begin{bmatrix} \frac{1}{\sqrt{2}} & -\frac{1}{\sqrt{2}} \\ \frac{1}{\sqrt{2}} & \frac{1}{\sqrt{2}} \end{bmatrix} \begin{bmatrix} -3 \\ 4 \end{bmatrix} = \begin{bmatrix} -\frac{7\sqrt{2}}{2} \\ \frac{\sqrt{2}}{2} \end{bmatrix}$$

18-2. Properties of R_2 Transformations

Property A Every 2×2 elementary matrix used as a premultiplier accomplishes exactly one of the following geometric transformations in R_2:

(1) an expansion or compression in the x- or y-direction;
(2) an expansion or compression in the x- or y-direction followed by a reflection about a coordinate axis;
(3) a shear in the x- or y-direction; or
(4) a reflection about $y = x$ or a coordinate axis.

EXAMPLE 18-14: Determine the geometric transformation accomplished in R_2 by using each of the following elementary matrices as a premultiplier.

(a) $\begin{bmatrix} \frac{1}{2} & 0 \\ 0 & 1 \end{bmatrix}$ **(b)** $\begin{bmatrix} 1 & 0 \\ 0 & 5 \end{bmatrix}$ **(c)** $\begin{bmatrix} 1 & 0 \\ 0 & -5 \end{bmatrix}$ **(d)** $\begin{bmatrix} 1 & 0 \\ 2 & 1 \end{bmatrix}$ **(e)** $\begin{bmatrix} 1 & 0 \\ 0 & -1 \end{bmatrix}$

(f) $\begin{bmatrix} 0 & 1 \\ 1 & 0 \end{bmatrix}$

Solution

(a) A compression by a factor of 1/2 in the x-direction.
(b) An expansion by a factor of 5 in the y-direction.
(c) An expansion by a factor of 5 in the y-direction followed by a reflection about the x-axis.
(d) A shear in the y-direction where $c = 2$.
(e) A reflection about the x-axis.
(f) A reflection about the line $y = x$.

Property B Every 2×2 orthogonal matrix A such that $\det(A) = 1$ is a rotation matrix in R_2. If A is orthogonal and $\det(A) = -1$, then A represents a reflection followed by a rotation.

EXAMPLE 18-15: Determine whether each of the following matrices is a rotation, a reflection followed by a rotation, or neither.

(a) $A = \begin{bmatrix} -\dfrac{2}{\sqrt{5}} & \dfrac{1}{\sqrt{5}} \\ \dfrac{1}{\sqrt{5}} & \dfrac{2}{\sqrt{5}} \end{bmatrix}$ **(b)** $B = \begin{bmatrix} \dfrac{1}{\sqrt{5}} & -\dfrac{2}{\sqrt{5}} \\ \dfrac{2}{\sqrt{5}} & \dfrac{1}{\sqrt{5}} \end{bmatrix}$ **(c)** $C = \begin{bmatrix} \dfrac{1}{\sqrt{2}} & -\dfrac{1}{\sqrt{2}} \\ -\dfrac{1}{\sqrt{2}} & \dfrac{1}{\sqrt{2}} \end{bmatrix}$

Solution

(a) $(-2/\sqrt{5})(1/\sqrt{5}) + (1/\sqrt{5})(2/\sqrt{5}) = 0$ and $\det(A) = -1$. Therefore, since A is orthogonal and $\det(A) = -1$, the matrix A represents a reflection followed by a rotation.

(b) $(1/\sqrt{5})(-2/\sqrt{5}) + (2/\sqrt{5})(1/\sqrt{5}) = 0$ and $\det(B) = 1$. Therefore, since B is orthogonal and $\det(B) = 1$, the matrix B represents a rotation. The a_{11} position element is $1/\sqrt{5} = \cos\theta$; therefore, $\theta = \arccos 1/\sqrt{5}$.

(c) $(1/\sqrt{2})(-1/\sqrt{2}) + (-1/\sqrt{2})(1/\sqrt{2}) = -1/2 - 1/2 = -1$ and $\det(C) = 0$. Therefore, C is not orthogonal and is neither a rotation nor a reflection followed by a rotation.

Property C Every 2×2 invertible matrix is equivalent to a sequence of linear transformations representing expansions, compressions, reflections, and shears in R_2.

EXAMPLE 18-16: Express the rotation matrix $A = \begin{bmatrix} 1/\sqrt{5} & -2/\sqrt{5} \\ 2/\sqrt{5} & 1/\sqrt{5} \end{bmatrix}$ as a sequence of expansions, compressions, reflections, and shears.

Solution: Use the method of Example 4-11 to express A as a product of elementary matrices. Therefore,

$$A = \begin{bmatrix} \frac{1}{\sqrt{5}} & -\frac{2}{\sqrt{5}} \\ \frac{2}{\sqrt{5}} & \frac{1}{\sqrt{5}} \end{bmatrix} = \begin{bmatrix} 1 & 0 \\ 0 & \frac{1}{\sqrt{5}} \end{bmatrix} \begin{bmatrix} \frac{1}{\sqrt{5}} & 0 \\ 0 & 1 \end{bmatrix} \begin{bmatrix} 1 & 0 \\ 2 & 1 \end{bmatrix} \begin{bmatrix} 1 & 0 \\ 0 & 5 \end{bmatrix} \begin{bmatrix} 1 & -2 \\ 0 & 1 \end{bmatrix}$$
$$= E_5E_4E_3E_2E_1$$

Now read from right to left. Therefore, A is equivalent to

(1) a shear in the x-direction with $c = -2$;
(2) an expansion in y-direction by a factor of 5;
(3) a shear in y-direction with $c = 2$;
(4) a compression in the x-direction by a factor of $1/\sqrt{5}$;
(5) a compression in the y-direction by a factor of $1/\sqrt{5}$.

note 1: Observe that the rotation was accomplished by a sequence of shears, expansions, and compressions.
note 2: A sequence of shears, compressions, expansions, and reflections is not necessarily unique.

Property D If the invertible matrix A is the matrix representation of a linear transformation $L: R_2 \to R_2$, then

(1) the images of distinct points are distinct;
(2) the images of three points are collinear if and only if the three points are collinear;
(3) the image of the line segment joining points P and Q is the line segment joining the images of P and Q;
(4) the image of a line is a line;
(5) the image of a line through the origin is a line through the origin; and
(6) the images of parallel lines are parallel lines.

EXAMPLE 18-17: Find the image of the square having vertices $A(0,0)$, $B(1,0)$, $C(1,1)$, and $D(0,1)$ under the transformation represented by the matrix $\begin{bmatrix} 2 & 5 \\ 1 & 1 \end{bmatrix}$.

Solution: The image of A is

$$A' = \begin{bmatrix} 2 & 5 \\ 1 & 1 \end{bmatrix} \begin{bmatrix} 0 \\ 0 \end{bmatrix} = \begin{bmatrix} 0 \\ 0 \end{bmatrix}$$

The image of B is

$$B' = \begin{bmatrix} 2 & 5 \\ 1 & 1 \end{bmatrix} \begin{bmatrix} 1 \\ 0 \end{bmatrix} = \begin{bmatrix} 2 \\ 1 \end{bmatrix}$$

The image of C is

$$C' = \begin{bmatrix} 2 & 5 \\ 1 & 1 \end{bmatrix} \begin{bmatrix} 1 \\ 1 \end{bmatrix} = \begin{bmatrix} 7 \\ 2 \end{bmatrix}$$

The image of D is

$$D' = \begin{bmatrix} 2 & 5 \\ 1 & 1 \end{bmatrix} \begin{bmatrix} 0 \\ 1 \end{bmatrix} = \begin{bmatrix} 5 \\ 1 \end{bmatrix}$$

Therefore, to find the entire image, plot these four points and find the parallelogram having line segments $\overline{A'B'}$, $\overline{B'C'}$, $\overline{C'D'}$, and $\overline{D'A'}$ as sides (as in Figure 18-15).

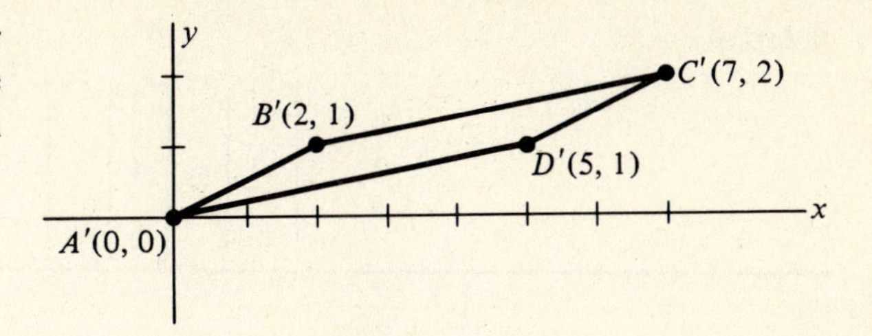

FIGURE 18-15

EXAMPLE 18-18: Find the image of the triangle having vertices $A(0,0)$, $B(3,0)$, and $C(2,4)$ under the transformation represented by the matrix $\begin{bmatrix} 3 & 4 \\ 1 & 2 \end{bmatrix}$.

Solution: The image of A is

$$A' = \begin{bmatrix} 3 & 4 \\ 1 & 2 \end{bmatrix}\begin{bmatrix} 0 \\ 0 \end{bmatrix} = \begin{bmatrix} 0 \\ 0 \end{bmatrix}$$

The image of B is

$$B' = \begin{bmatrix} 3 & 4 \\ 1 & 2 \end{bmatrix}\begin{bmatrix} 3 \\ 0 \end{bmatrix} = \begin{bmatrix} 9 \\ 3 \end{bmatrix}$$

The image of C is

$$C' = \begin{bmatrix} 3 & 4 \\ 1 & 2 \end{bmatrix}\begin{bmatrix} 2 \\ 4 \end{bmatrix} = \begin{bmatrix} 22 \\ 10 \end{bmatrix}$$

Therefore, the image of triangle ABC is triangle $A'B'C'$ which has vertices at $A'(0,0)$, $B'(9,3)$, and $C'(22,10)$.

18-3. Transformations in Euclidean Space R_3

As in Section 18-1, (x, y, z) and $\begin{bmatrix} x \\ y \\ z \end{bmatrix}$ should be thought of as coordinates of the same point.

A. **Translation** by **a** units in the x-direction, **b** units in the y-direction, and **c** units in the z-direction is represented by

$$\begin{bmatrix} x' \\ y' \\ z' \end{bmatrix} = \begin{bmatrix} x \\ y \\ z \end{bmatrix} + \begin{bmatrix} a \\ b \\ c \end{bmatrix}$$

This is the only nonlinear transformation in this section.

EXAMPLE 18-19: Find the image of $(3, 4, 5)$ when translated by 2 units in the x-direction, -5 units in the y-direction, and 3 units in the z-direction.

Solution

$$\begin{bmatrix} x' \\ y' \\ z' \end{bmatrix} = \begin{bmatrix} 3 \\ 4 \\ 5 \end{bmatrix} + \begin{bmatrix} 2 \\ -5 \\ 3 \end{bmatrix} = \begin{bmatrix} 5 \\ -1 \\ 8 \end{bmatrix}$$

B. An expansion ($k > 1$) or compression ($0 < k < 1$) is represented by

$$\begin{bmatrix} x' \\ y' \\ z' \end{bmatrix} = \begin{bmatrix} k & 0 & 0 \\ 0 & 1 & 0 \\ 0 & 0 & 1 \end{bmatrix} \begin{bmatrix} x \\ y \\ z \end{bmatrix} \quad \text{in the } x\text{-direction}$$

$$\begin{bmatrix} x' \\ y' \\ z' \end{bmatrix} = \begin{bmatrix} 1 & 0 & 0 \\ 0 & k & 0 \\ 0 & 0 & 1 \end{bmatrix} \begin{bmatrix} x \\ y \\ z \end{bmatrix} \quad \text{in the } y\text{-direction}$$

$$\begin{bmatrix} x' \\ y' \\ z' \end{bmatrix} = \begin{bmatrix} 1 & 0 & 0 \\ 0 & 1 & 0 \\ 0 & 0 & k \end{bmatrix} \begin{bmatrix} x \\ y \\ z \end{bmatrix} \quad \text{in the } z\text{-direction}$$

EXAMPLE 18-20: Find the matrix equation for each of the following transformations: **(a)** expansion by a factor of 3 in the y-direction; **(b)** expansion by a factor of 3 in the x-direction and a compression by a factor of 1/2 in the z-direction.

Solution

(a)

$$\begin{bmatrix} x' \\ y' \\ z' \end{bmatrix} = \begin{bmatrix} 1 & 0 & 0 \\ 0 & 3 & 0 \\ 0 & 0 & 1 \end{bmatrix} \begin{bmatrix} x \\ y \\ z \end{bmatrix}$$

(b)

$$\begin{bmatrix} x' \\ y' \\ z' \end{bmatrix} = \begin{bmatrix} 3 & 0 & 0 \\ 0 & 1 & 0 \\ 0 & 0 & \frac{1}{2} \end{bmatrix} \begin{bmatrix} x \\ y \\ z \end{bmatrix}$$

C. A projection is represented by

$$\begin{bmatrix} x' \\ y' \\ z' \end{bmatrix} = \begin{bmatrix} 1 & 0 & 0 \\ 0 & 0 & 0 \\ 0 & 0 & 0 \end{bmatrix} \begin{bmatrix} x \\ y \\ z \end{bmatrix}, \quad \text{if onto the } x\text{-axis}$$

$$\begin{bmatrix} x' \\ y' \\ z' \end{bmatrix} = \begin{bmatrix} 0 & 0 & 0 \\ 0 & 1 & 0 \\ 0 & 0 & 0 \end{bmatrix} \begin{bmatrix} x \\ y \\ z \end{bmatrix}, \quad \text{if onto the } y\text{-axis}$$

$$\begin{bmatrix} x' \\ y' \\ z' \end{bmatrix} = \begin{bmatrix} 0 & 0 & 0 \\ 0 & 0 & 0 \\ 0 & 0 & 1 \end{bmatrix} \begin{bmatrix} x \\ y \\ z \end{bmatrix}, \quad \text{if onto the } z\text{-axis}$$

$$\begin{bmatrix} x' \\ y' \\ z' \end{bmatrix} = \begin{bmatrix} 1 & 0 & 0 \\ 0 & 1 & 0 \\ 0 & 0 & 0 \end{bmatrix} \begin{bmatrix} x \\ y \\ z \end{bmatrix}, \quad \text{if onto the } xy\text{-plane}$$

$$\begin{bmatrix} x' \\ y' \\ z' \end{bmatrix} = \begin{bmatrix} 1 & 0 & 0 \\ 0 & 0 & 0 \\ 0 & 0 & 1 \end{bmatrix} \begin{bmatrix} x \\ y \\ z \end{bmatrix}, \quad \text{if onto the } xz\text{-plane}$$

$$\begin{bmatrix} x' \\ y' \\ z' \end{bmatrix} = \begin{bmatrix} 0 & 0 & 0 \\ 0 & 1 & 0 \\ 0 & 0 & 1 \end{bmatrix} \begin{bmatrix} x \\ y \\ z \end{bmatrix}, \quad \text{if onto the } yz\text{-plane}$$

EXAMPLE 18-21: Find the image of $(5, 3, 7)$ if projected onto **(a)** the y-axis and **(b)** the xz-plane.

Solution

(a)
$$\begin{bmatrix} x' \\ y' \\ z' \end{bmatrix} = \begin{bmatrix} 0 & 0 & 0 \\ 0 & 1 & 0 \\ 0 & 0 & 0 \end{bmatrix} \begin{bmatrix} 5 \\ 3 \\ 7 \end{bmatrix} = \begin{bmatrix} 0 \\ 3 \\ 0 \end{bmatrix}$$

(b)
$$\begin{bmatrix} x' \\ y' \\ z' \end{bmatrix} = \begin{bmatrix} 1 & 0 & 0 \\ 0 & 0 & 0 \\ 0 & 0 & 1 \end{bmatrix} \begin{bmatrix} 5 \\ 3 \\ 7 \end{bmatrix} = \begin{bmatrix} 5 \\ 0 \\ 7 \end{bmatrix}$$

D. A shear is represented by

$$\begin{bmatrix} x' \\ y' \\ z' \end{bmatrix} = \begin{bmatrix} 1 & 0 & c \\ 0 & 1 & 0 \\ 0 & 0 & 1 \end{bmatrix} \begin{bmatrix} x \\ y \\ z \end{bmatrix}$$

if a point is moved parallel to the x-axis by an amount proportional to its distance from the xy-plane;

$$\begin{bmatrix} x' \\ y' \\ z' \end{bmatrix} = \begin{bmatrix} 1 & c & 0 \\ 0 & 1 & 0 \\ 0 & 0 & 1 \end{bmatrix} \begin{bmatrix} x \\ y \\ z \end{bmatrix}$$

if a point is moved parallel to the x-axis by an amount proportional to its distance from the xz-plane;

$$\begin{bmatrix} x' \\ y' \\ z' \end{bmatrix} = \begin{bmatrix} 1 & 0 & 0 \\ 0 & 1 & c \\ 0 & 0 & 1 \end{bmatrix} \begin{bmatrix} x \\ y \\ z \end{bmatrix}$$

if a point is moved parallel to the y-axis by an amount proportional to its distance from the xy-plane;

$$\begin{bmatrix} x' \\ y' \\ z' \end{bmatrix} = \begin{bmatrix} 1 & 0 & 0 \\ c & 1 & 0 \\ 0 & 0 & 1 \end{bmatrix} \begin{bmatrix} x \\ y \\ z \end{bmatrix}$$

if a point is moved parallel to the y-axis by an amount proportional to its distance from the yz-plane;

$$\begin{bmatrix} x' \\ y' \\ z' \end{bmatrix} = \begin{bmatrix} 1 & 0 & 0 \\ 0 & 1 & 0 \\ 0 & c & 1 \end{bmatrix} \begin{bmatrix} x \\ y \\ z \end{bmatrix}$$

if a point is moved parallel to the z-axis by an amount proportional to its distance from the xz-plane; and

$$\begin{bmatrix} x' \\ y' \\ z' \end{bmatrix} = \begin{bmatrix} 1 & 0 & 0 \\ 0 & 1 & 0 \\ c & 0 & 1 \end{bmatrix} \begin{bmatrix} x \\ y \\ z \end{bmatrix}$$

if a point is moved parallel to the z-axis by an amount proportional to its distance from the yz-plane.

EXAMPLE 18-22: Find the image of $(3, 5, -7)$ if **(a)** it is moved parallel to the x-axis by 3 times its distance from the xz-plane and **(b)** it is moved parallel to the x-axis by 2 times its distance from the xy-plane, then moved parallel to the z-axis by -3 times its distance from the yz-plane.

Solution

(a)

$$\begin{bmatrix} x' \\ y' \\ z' \end{bmatrix} = \begin{bmatrix} 1 & 3 & 0 \\ 0 & 1 & 0 \\ 0 & 0 & 1 \end{bmatrix} \begin{bmatrix} 3 \\ 5 \\ -7 \end{bmatrix} = \begin{bmatrix} 18 \\ 5 \\ -7 \end{bmatrix}$$

(b)

$$\begin{bmatrix} x' \\ y' \\ z' \end{bmatrix} = \begin{bmatrix} 1 & 0 & 0 \\ 0 & 1 & 0 \\ -3 & 0 & 1 \end{bmatrix} \begin{bmatrix} 1 & 0 & 2 \\ 0 & 1 & 0 \\ 0 & 0 & 1 \end{bmatrix} \begin{bmatrix} 3 \\ 5 \\ -7 \end{bmatrix} = \begin{bmatrix} 1 & 0 & 0 \\ 0 & 1 & 0 \\ -3 & 0 & 1 \end{bmatrix} \begin{bmatrix} -11 \\ 5 \\ -7 \end{bmatrix} = \begin{bmatrix} -11 \\ 5 \\ 26 \end{bmatrix}$$

E. A reflection is represented by

$$\begin{bmatrix} x' \\ y' \\ z' \end{bmatrix} = \begin{bmatrix} 1 & 0 & 0 \\ 0 & -1 & 0 \\ 0 & 0 & -1 \end{bmatrix} \begin{bmatrix} x \\ y \\ z \end{bmatrix}, \quad \text{if it is about the } x\text{-axis}$$

$$\begin{bmatrix} x' \\ y' \\ z' \end{bmatrix} = \begin{bmatrix} -1 & 0 & 0 \\ 0 & 1 & 0 \\ 0 & 0 & -1 \end{bmatrix} \begin{bmatrix} x \\ y \\ z \end{bmatrix}, \quad \text{if it is about the } y\text{-axis}$$

$$\begin{bmatrix} x' \\ y' \\ z' \end{bmatrix} = \begin{bmatrix} -1 & 0 & 0 \\ 0 & -1 & 0 \\ 0 & 0 & 1 \end{bmatrix} \begin{bmatrix} x \\ y \\ z \end{bmatrix}, \quad \text{if it is about the } z\text{-axis}$$

$$\begin{bmatrix} x' \\ y' \\ z' \end{bmatrix} = \begin{bmatrix} 1 & 0 & 0 \\ 0 & -1 & 0 \\ 0 & 0 & 1 \end{bmatrix} \begin{bmatrix} x \\ y \\ z \end{bmatrix}, \quad \text{if it is about the } xz\text{-plane}$$

$$\begin{bmatrix} x' \\ y' \\ z' \end{bmatrix} = \begin{bmatrix} -1 & 0 & 0 \\ 0 & 1 & 0 \\ 0 & 0 & 1 \end{bmatrix} \begin{bmatrix} x \\ y \\ z \end{bmatrix}, \quad \text{if it is about the } yz\text{-plane}$$

$$\begin{bmatrix} x' \\ y' \\ z' \end{bmatrix} = \begin{bmatrix} 1 & 0 & 0 \\ 0 & 1 & 0 \\ 0 & 0 & -1 \end{bmatrix} \begin{bmatrix} x \\ y \\ z \end{bmatrix}, \quad \text{if it is about the } xy\text{-plane}$$

$$\begin{bmatrix} x' \\ y' \\ z' \end{bmatrix} = \begin{bmatrix} 0 & 0 & 1 \\ 0 & 1 & 0 \\ 1 & 0 & 0 \end{bmatrix} \begin{bmatrix} x \\ y \\ z \end{bmatrix}, \quad \text{if it is about the plane } x = z$$

$$\begin{bmatrix} x' \\ y' \\ z' \end{bmatrix} = \begin{bmatrix} 0 & 1 & 0 \\ 1 & 0 & 0 \\ 0 & 0 & 1 \end{bmatrix} \begin{bmatrix} x \\ y \\ z \end{bmatrix}, \quad \text{if it is about the plane } y = x$$

$$\begin{bmatrix} x' \\ y' \\ z' \end{bmatrix} = \begin{bmatrix} 1 & 0 & 0 \\ 0 & 0 & 1 \\ 0 & 1 & 0 \end{bmatrix} \begin{bmatrix} x \\ y \\ z \end{bmatrix}, \quad \text{if it is about the plane } y = z$$

$$\begin{bmatrix} x' \\ y' \\ z' \end{bmatrix} = \begin{bmatrix} -1 & 0 & 0 \\ 0 & -1 & 0 \\ 0 & 0 & -1 \end{bmatrix} \begin{bmatrix} x \\ y \\ z \end{bmatrix}, \quad \text{if it is about the origin}$$

EXAMPLE 18-23: Find the image of $(-4, 5, 6)$ if **(a)** it is reflected about the x-axis, **(b)** it is reflected about the yz-plane, and **(c)** it is reflected about the y-axis, then reflected about the xz-plane.

Solution

(a)
$$\begin{bmatrix} x' \\ y' \\ z' \end{bmatrix} = \begin{bmatrix} 1 & 0 & 0 \\ 0 & -1 & 0 \\ 0 & 0 & -1 \end{bmatrix} \begin{bmatrix} -4 \\ 5 \\ 6 \end{bmatrix} = \begin{bmatrix} -4 \\ -5 \\ -6 \end{bmatrix}$$

(b)
$$\begin{bmatrix} x' \\ y' \\ z' \end{bmatrix} = \begin{bmatrix} -1 & 0 & 0 \\ 0 & 1 & 0 \\ 0 & 0 & 1 \end{bmatrix} \begin{bmatrix} -4 \\ 5 \\ 6 \end{bmatrix} = \begin{bmatrix} 4 \\ 5 \\ 6 \end{bmatrix}$$

(c)
$$\begin{bmatrix} x' \\ y' \\ z' \end{bmatrix} = \begin{bmatrix} 1 & 0 & 0 \\ 0 & -1 & 0 \\ 0 & 0 & 1 \end{bmatrix} \begin{bmatrix} -1 & 0 & 0 \\ 0 & 1 & 0 \\ 0 & 0 & -1 \end{bmatrix} \begin{bmatrix} -4 \\ 5 \\ 6 \end{bmatrix} = \begin{bmatrix} 1 & 0 & 0 \\ 0 & -1 & 0 \\ 0 & 0 & 1 \end{bmatrix} \begin{bmatrix} 4 \\ 5 \\ -6 \end{bmatrix} = \begin{bmatrix} 4 \\ -5 \\ -6 \end{bmatrix}$$

note: The product of the two reflections in part **(c)** is equivalent to a single reflection about the origin.

F. A rotation is represented by

$$\begin{bmatrix} x' \\ y' \\ z' \end{bmatrix} = \begin{bmatrix} \cos\theta & -\sin\theta & 0 \\ \sin\theta & \cos\theta & 0 \\ 0 & 0 & 1 \end{bmatrix} \begin{bmatrix} x \\ y \\ z \end{bmatrix}$$

if the axis of rotation is the z-axis and θ is measured from the positive x-axis in a counterclockwise direction toward the y-axis,

$$\begin{bmatrix} x' \\ y' \\ z' \end{bmatrix} = \begin{bmatrix} \cos\theta & 0 & -\sin\theta \\ 0 & 1 & 0 \\ \sin\theta & 0 & \cos\theta \end{bmatrix} \begin{bmatrix} x \\ y \\ z \end{bmatrix}$$

if the axis of rotation is the y-axis and θ is measured from the positive x-axis in a counterclockwise direction toward the z-axis, and

$$\begin{bmatrix} x' \\ y' \\ z' \end{bmatrix} = \begin{bmatrix} 1 & 0 & 0 \\ 0 & \cos\theta & -\sin\theta \\ 0 & \sin\theta & \cos\theta \end{bmatrix} \begin{bmatrix} x \\ y \\ z \end{bmatrix}$$

if the axis of rotation is the x-axis and θ is measured from the positive y-axis in a counterclockwise direction toward the z-axis.

EXAMPLE 18-24: Find the image of $(3, \sqrt{2}, -4)$ under the following rotations: **(a)** a rotation of 45° about the y-axis and **(b)** a rotation of 45° about the y-axis followed by a rotation of 45° about the z-axis.

Solution

(a)
$$\begin{bmatrix} x' \\ y' \\ z' \end{bmatrix} = \begin{bmatrix} \cos 45^\circ & 0 & -\sin 45^\circ \\ 0 & 1 & 0 \\ \sin 45^\circ & 0 & \cos 45^\circ \end{bmatrix} \begin{bmatrix} 3 \\ \sqrt{2} \\ -4 \end{bmatrix}$$

$$= \begin{bmatrix} \frac{1}{\sqrt{2}} & 0 & -\frac{1}{\sqrt{2}} \\ 0 & 1 & 0 \\ \frac{1}{\sqrt{2}} & 0 & \frac{1}{\sqrt{2}} \end{bmatrix} \begin{bmatrix} 3 \\ \sqrt{2} \\ -4 \end{bmatrix} = \begin{bmatrix} \frac{7}{\sqrt{2}} \\ \sqrt{2} \\ -\frac{1}{\sqrt{2}} \end{bmatrix}$$

(b)
$$\begin{bmatrix} x' \\ y' \\ z' \end{bmatrix} = \begin{bmatrix} \frac{1}{\sqrt{2}} & -\frac{1}{\sqrt{2}} & 0 \\ \frac{1}{\sqrt{2}} & \frac{1}{\sqrt{2}} & 0 \\ 0 & 0 & 1 \end{bmatrix} \begin{bmatrix} \frac{1}{\sqrt{2}} & 0 & -\frac{1}{\sqrt{2}} \\ 0 & 1 & 0 \\ \frac{1}{\sqrt{2}} & 0 & \frac{1}{\sqrt{2}} \end{bmatrix} \begin{bmatrix} 3 \\ \sqrt{2} \\ -4 \end{bmatrix} = \begin{bmatrix} \frac{5}{2} \\ \frac{9}{2} \\ -\frac{1}{\sqrt{2}} \end{bmatrix}$$

18-4. Properties of R_3 Transformations

Property A Every 3×3 elementary matrix as a premultiplier accomplishes exactly one of the following geometric transformations in R_3:

(1) an expansion or compression in the x, y, or z-direction;
(2) an expansion or compression in the x, y, or z-direction followed by a reflection about a coordinate plane;
(3) a shear parallel to a coordinate axis by an amount proportional to the distance from a coordinate plane; or
(4) a reflection about $y = x$, $x = z$, $y = z$, or a coordinate plane.

EXAMPLE 18-25: Determine the geometric transformation accomplished in R_3 by using each of the following elementary matrices as a premultiplier.

(a) $\begin{bmatrix} 1 & 0 & 0 \\ 0 & \frac{1}{2} & 0 \\ 0 & 0 & 1 \end{bmatrix}$ **(b)** $\begin{bmatrix} 1 & 0 & 0 \\ 0 & 1 & 0 \\ 0 & 0 & -2 \end{bmatrix}$ **(c)** $\begin{bmatrix} 1 & 3 & 0 \\ 0 & 1 & 0 \\ 0 & 0 & 1 \end{bmatrix}$

Solution

(a) This matrix gives a compression by a factor of 1/2 in the y-direction.
(b) An expansion by a factor of 2 in the z-direction followed by a reflection about the xy-plane.
(c) A shear parallel to the x-axis by three times the distance of the point from the xz-plane.

Property B Every 3×3 orthogonal matrix A such that $\det(A) = 1$ is a rotation matrix in R_3. If A is orthogonal and $\det(A) = -1$, then A represents a rotation along with a reflection about one of the coordinate planes.

EXAMPLE 18-26: Determine whether each of the following matrices is a rotation, a combination of a rotation and a reflection, or neither.

$$\text{(a) } A = \begin{bmatrix} \frac{1}{3} & \frac{2}{3} & \frac{2}{3} \\ -\frac{2}{3} & -\frac{1}{3} & \frac{2}{3} \\ -\frac{2}{3} & \frac{2}{3} & -\frac{1}{3} \end{bmatrix} \qquad \text{(b) } B = \begin{bmatrix} \frac{1}{3} & \frac{2}{3} & \frac{2}{3} \\ -\frac{2}{3} & \frac{2}{3} & -\frac{1}{3} \\ -\frac{2}{3} & -\frac{1}{3} & \frac{2}{3} \end{bmatrix} \qquad \text{(c) } C = \begin{bmatrix} \frac{1}{\sqrt{2}} & 0 & \frac{1}{\sqrt{2}} \\ \frac{1}{\sqrt{2}} & \frac{1}{\sqrt{2}} & 0 \\ 0 & \frac{1}{\sqrt{2}} & \frac{1}{\sqrt{2}} \end{bmatrix}$$

Solution

(a) The column vectors are orthonormal, but $\det(A) = -1$. Therefore, A represents a combination of a rotation and a reflection.
(b) The column vectors are orthonormal, and $\det(B) = 1$. Therefore, B represents a rotation.
(c) Since $1/\sqrt{2}(0) + 1/\sqrt{2}(1/\sqrt{2}) + 0(1/\sqrt{2}) = 1/2 \neq 0$, then C is not orthogonal. Therefore, C represents neither.

Property C Every 3×3 invertible matrix is equivalent to a sequence of linear transformations representing expansions, compressions, reflections, and shears in R_3.

EXAMPLE 18-27: Express the invertible matrix $A = \begin{bmatrix} 2 & 1 & 1 \\ 1 & 2 & 3 \\ -1 & 0 & 2 \end{bmatrix}$ as a sequence of expansions, compressions, reflections, and shears.

Solution: Use the method of Example 4-11 to express A as a product of elementary matrices. Therefore, $A = E_8E_7E_6E_5E_4E_3E_2E_1$ where

$$E_1 = \begin{bmatrix} 1 & 0 & -2 \\ 0 & 1 & 0 \\ 0 & 0 & 1 \end{bmatrix} \qquad E_2 = \begin{bmatrix} 1 & 0 & 0 \\ 0 & 1 & 0 \\ 0 & 0 & 5 \end{bmatrix}$$

$$E_3 = \begin{bmatrix} 1 & 0 & 0 \\ 0 & 1 & 0 \\ 0 & 1 & 1 \end{bmatrix} \qquad E_4 = \begin{bmatrix} 1 & 0 & 0 \\ 0 & 1 & 1 \\ 0 & 0 & 1 \end{bmatrix}$$

$$E_5 = \begin{bmatrix} -1 & 0 & 0 \\ 0 & 1 & 0 \\ 0 & 0 & 1 \end{bmatrix} \qquad E_6 = \begin{bmatrix} 1 & 0 & 0 \\ 0 & 1 & 0 \\ -2 & 0 & 1 \end{bmatrix}$$

$$E_7 = \begin{bmatrix} 1 & 0 & 0 \\ -1 & 1 & 0 \\ 0 & 0 & 1 \end{bmatrix} \qquad E_8 = \begin{bmatrix} 0 & 0 & 1 \\ 0 & 1 & 0 \\ 1 & 0 & 0 \end{bmatrix}$$

Now read the geometric effect of $E_8E_7E_6E_5E_4E_3E_2E_1$ from right to left. Therefore, A is equivalent to

(1) a shear parallel to the x-axis by -2 times its distance from the xy-plane;
(2) an expansion by a factor of 5 in the z-direction;
(3) a shear parallel to the z-axis by an amount equal to its distance from the xz-plane;
(4) a shear parallel to the y-axis by an amount equal to its distance from the xy-plane;
(5) a reflection about the yz-plane;
(6) a shear parallel to the z-axis by -2 times its distance from the yz-plane;
(7) a shear parallel to the y-axis by an amount equal to the negative of its distance from the yz-plane;
(8) a reflection about the plane $x = z$.

Property D If the invertible matrix A is the matrix representation of a linear transformation $L: R_3 \to R_3$, then

(1) the images of distinct points are distinct;
(2) the images of three points are collinear if and only if the three points are collinear;
(3) the image of a line segment joining points P and Q is the line segment joining the images of P and Q;
(4) the image of a line is a line;
(5) the image of a line through the origin is a line through the origin;
(6) the images of parallel lines are parallel lines;
(7) the images of four points are coplanar if and only if the four points are coplanar;
(8) the image of a plane through the origin is a plane through the origin;
(9) the images of parallel planes are parallel planes.

EXAMPLE 18-28: Find the image of the rectangular parallelepiped having vertices $A(0, 0, 0)$, $B(1, 0, 0)$, $C(1, 1, 0)$, $D(0, 1, 0)$, $E(0, 1, 1)$, $F(1, 1, 1)$, $G(1, 0, 1)$, and $H(0, 0, 1)$ under the transformation represented by the matrix $\begin{bmatrix} 2 & -1 & 3 \\ 0 & 1 & -2 \\ 3 & -1 & 2 \end{bmatrix}$.

Solution: The image of A is

$$A' = \begin{bmatrix} 2 & -1 & 3 \\ 0 & 1 & -2 \\ 3 & -1 & 2 \end{bmatrix} \begin{bmatrix} 0 \\ 0 \\ 0 \end{bmatrix} = \begin{bmatrix} 0 \\ 0 \\ 0 \end{bmatrix}$$

The image of B is

$$B' = \begin{bmatrix} 2 & -1 & 3 \\ 0 & 1 & -2 \\ 3 & -1 & 2 \end{bmatrix} \begin{bmatrix} 1 \\ 0 \\ 0 \end{bmatrix} = \begin{bmatrix} 2 \\ 0 \\ 3 \end{bmatrix}$$

The image of C is

$$C' = \begin{bmatrix} 2 & -1 & 3 \\ 0 & 1 & -2 \\ 3 & -1 & 2 \end{bmatrix} \begin{bmatrix} 1 \\ 1 \\ 0 \end{bmatrix} = \begin{bmatrix} 1 \\ 1 \\ 2 \end{bmatrix}$$

The image of D is

$$D' = \begin{bmatrix} 2 & -1 & 3 \\ 0 & 1 & -2 \\ 3 & -1 & 2 \end{bmatrix}\begin{bmatrix} 0 \\ 1 \\ 0 \end{bmatrix} = \begin{bmatrix} -1 \\ 1 \\ -1 \end{bmatrix}$$

The image of E is

$$E' = \begin{bmatrix} 2 & -1 & 3 \\ 0 & 1 & -2 \\ 3 & -1 & 2 \end{bmatrix}\begin{bmatrix} 0 \\ 1 \\ 1 \end{bmatrix} = \begin{bmatrix} 2 \\ -1 \\ 1 \end{bmatrix}$$

The image of F is

$$F' = \begin{bmatrix} 2 & -1 & 3 \\ 0 & 1 & -2 \\ 3 & -1 & 2 \end{bmatrix}\begin{bmatrix} 1 \\ 1 \\ 1 \end{bmatrix} = \begin{bmatrix} 4 \\ -1 \\ 4 \end{bmatrix}$$

The image of G is

$$G' = \begin{bmatrix} 2 & -1 & 3 \\ 0 & 1 & -2 \\ 3 & -1 & 2 \end{bmatrix}\begin{bmatrix} 1 \\ 0 \\ 1 \end{bmatrix} = \begin{bmatrix} 5 \\ -2 \\ 5 \end{bmatrix}$$

The image of H is

$$H' = \begin{bmatrix} 2 & -1 & 3 \\ 0 & 1 & -2 \\ 3 & -1 & 2 \end{bmatrix}\begin{bmatrix} 0 \\ 0 \\ 1 \end{bmatrix} = \begin{bmatrix} 3 \\ -2 \\ 2 \end{bmatrix}$$

Therefore, the image of the given figure is the parallelepiped having vertices $A', B', C', D', E', F', G'$, and H'.

18-5. Isometries

An **isometry** is a one-to-one, onto transformation from a set of points onto itself which preserves distance. A linear transformation $L: R_2 \to R_2$ or $L: R_3 \to R_3$ is an isometry if and only if the matrix of the transformation is orthogonal. Consequently, the linear transformations of R_2 and R_3 that are isometries are rotations and reflections. Although not linear, translations are isometries.

EXAMPLE 18-29: Which of the following matrices represent linear transformations that are isometries in R_2?

(a) $A = \begin{bmatrix} 2 & 0 \\ 0 & 3 \end{bmatrix}$ **(b)** $B = \begin{bmatrix} \frac{3}{\sqrt{10}} & \frac{1}{\sqrt{10}} \\ -\frac{1}{\sqrt{10}} & \frac{3}{\sqrt{10}} \end{bmatrix}$ **(c)** $C = \begin{bmatrix} \frac{5}{\sqrt{41}} & -\frac{4}{\sqrt{41}} \\ \frac{4}{\sqrt{41}} & \frac{5}{\sqrt{41}} \end{bmatrix}$

Solution

(a) Although the column vectors of A are orthogonal, they are not orthonormal. Therefore, matrix A is not orthogonal, and thus A does not represent an isometry. $\text{Det}(A) \neq 1$ or -1.

(b) The column vectors of B are orthonormal; thus, B is orthogonal. Hence, B represents an isometry.

(c) The column vectors of C are orthonormal; thus, C is orthogonal. Hence, C represents an isometry.

EXAMPLE 18-30: Which of the following matrices represent linear transformations that are isometries in R_3?

(a) $A = \begin{bmatrix} 1 & 1 & 1 \\ 2 & -1 & 0 \\ 1 & 1 & -1 \end{bmatrix}$ (b) $B = \begin{bmatrix} \frac{1}{\sqrt{6}} & \frac{1}{\sqrt{3}} & \frac{1}{\sqrt{2}} \\ \frac{2}{\sqrt{6}} & -\frac{1}{\sqrt{3}} & 0 \\ \frac{1}{\sqrt{6}} & \frac{1}{\sqrt{3}} & -\frac{1}{\sqrt{2}} \end{bmatrix}$

Solution

(a) Although the column vectors of A are orthogonal, they are not orthonormal. Therefore, matrix A is not orthogonal; thus, A does not represent an isometry.

(b) The column vectors of B are orthonormal; thus, B is orthogonal. Hence, B represents an isometry.

SUMMARY

1. R_2 linear transformations

Name of transformation	Matrix representation		
Expansion ($k > 1$) Compression ($0 < k < 1$)	$\begin{bmatrix} k & 0 \\ 0 & 1 \end{bmatrix}$ x-direction	$\begin{bmatrix} 1 & 0 \\ 0 & k \end{bmatrix}$ y-direction	
Projection	$\begin{bmatrix} 1 & 0 \\ 0 & 0 \end{bmatrix}$ onto x-axis	$\begin{bmatrix} 0 & 0 \\ 0 & 1 \end{bmatrix}$ onto y-axis	
Shear ($c \neq 0$)	$\begin{bmatrix} 1 & c \\ 0 & 1 \end{bmatrix}$ x-direction	$\begin{bmatrix} 1 & 0 \\ c & 1 \end{bmatrix}$ y-direction	
Reflection	$\begin{bmatrix} 1 & 0 \\ 0 & -1 \end{bmatrix}$ about x-axis	$\begin{bmatrix} -1 & 0 \\ 0 & 1 \end{bmatrix}$ about y-axis	$\begin{bmatrix} 0 & 1 \\ 1 & 0 \end{bmatrix}$ about $y = x$
Rotation (counterclockwise for $\theta > 0$)	$\begin{bmatrix} \cos\theta & -\sin\theta \\ \sin\theta & \cos\theta \end{bmatrix}$		

2. R_3 linear transformations

Name of transformation	Matrix representation		
Expansion ($k > 1$) Compression ($0 < k < 1$)	$\begin{bmatrix} k & 0 & 0 \\ 0 & 1 & 0 \\ 0 & 0 & 1 \end{bmatrix}$ x-direction	$\begin{bmatrix} 1 & 0 & 0 \\ 0 & k & 0 \\ 0 & 0 & 1 \end{bmatrix}$ y-direction	$\begin{bmatrix} 1 & 0 & 0 \\ 0 & 1 & 0 \\ 0 & 0 & k \end{bmatrix}$ z-direction

Name of transformation	Matrix representation		
Projection	$\begin{bmatrix} 1 & 0 & 0 \\ 0 & 1 & 0 \\ 0 & 0 & 0 \end{bmatrix}$ onto xy-plane	$\begin{bmatrix} 1 & 0 & 0 \\ 0 & 0 & 0 \\ 0 & 0 & 1 \end{bmatrix}$ onto xz-plane	$\begin{bmatrix} 0 & 0 & 0 \\ 0 & 1 & 0 \\ 0 & 0 & 1 \end{bmatrix}$ onto yz-plane
	$\begin{bmatrix} 1 & 0 & 0 \\ 0 & 0 & 0 \\ 0 & 0 & 0 \end{bmatrix}$ onto x-axis	$\begin{bmatrix} 0 & 0 & 0 \\ 0 & 1 & 0 \\ 0 & 0 & 0 \end{bmatrix}$ onto y-axis	$\begin{bmatrix} 0 & 0 & 0 \\ 0 & 0 & 0 \\ 0 & 0 & 1 \end{bmatrix}$ onto z-axis
Shear ($c \neq 0$)	$\begin{bmatrix} 1 & 0 & c \\ 0 & 1 & 0 \\ 0 & 0 & 1 \end{bmatrix}$ parallel to x-axis by distance from xy-plane	$\begin{bmatrix} 1 & c & 0 \\ 0 & 1 & 0 \\ 0 & 0 & 1 \end{bmatrix}$ parallel to x-axis by distance from xz-plane	$\begin{bmatrix} 1 & 0 & 0 \\ 0 & 1 & c \\ 0 & 0 & 1 \end{bmatrix}$ parallel to y-axis by distance from xy-plane
	$\begin{bmatrix} 1 & 0 & 0 \\ c & 1 & 0 \\ 0 & 0 & 1 \end{bmatrix}$ parallel to y-axis by distance from yz-plane	$\begin{bmatrix} 1 & 0 & 0 \\ 0 & 1 & 0 \\ 0 & c & 1 \end{bmatrix}$ parallel to z-axis by distance from xz-plane	$\begin{bmatrix} 1 & 0 & 0 \\ 0 & 1 & 0 \\ c & 0 & 1 \end{bmatrix}$ parallel to z-axis by distance from yz-plane
Reflection	$\begin{bmatrix} 1 & 0 & 0 \\ 0 & -1 & 0 \\ 0 & 0 & -1 \end{bmatrix}$ about x-axis	$\begin{bmatrix} -1 & 0 & 0 \\ 0 & 1 & 0 \\ 0 & 0 & -1 \end{bmatrix}$ about y-axis	$\begin{bmatrix} -1 & 0 & 0 \\ 0 & -1 & 0 \\ 0 & 0 & 1 \end{bmatrix}$ about z-axis
	$\begin{bmatrix} 1 & 0 & 0 \\ 0 & -1 & 0 \\ 0 & 0 & 1 \end{bmatrix}$ about xz-plane	$\begin{bmatrix} -1 & 0 & 0 \\ 0 & 1 & 0 \\ 0 & 0 & 1 \end{bmatrix}$ about yz-plane	$\begin{bmatrix} 1 & 0 & 0 \\ 0 & 1 & 0 \\ 0 & 0 & -1 \end{bmatrix}$ about xy-plane
	$\begin{bmatrix} 0 & 0 & 1 \\ 0 & 1 & 0 \\ 1 & 0 & 0 \end{bmatrix}$ about $x = z$	$\begin{bmatrix} 0 & 1 & 0 \\ 1 & 0 & 0 \\ 0 & 0 & 1 \end{bmatrix}$ about $y = x$	$\begin{bmatrix} 1 & 0 & 0 \\ 0 & 0 & 1 \\ 0 & 1 & 0 \end{bmatrix}$ about $y = z$
Rotation	$\begin{bmatrix} \cos\theta & -\sin\theta & 0 \\ \sin\theta & \cos\theta & 0 \\ 0 & 0 & 1 \end{bmatrix}$ axis of rotation is z-axis, $\theta > 0$ is from positive x-axis toward positive y-axis		

<table>
<tr><th>Name of transformation</th><th>Matrix representation</th></tr>
<tr><td rowspan="2">Rotation</td><td>$$\begin{bmatrix} \cos\theta & 0 & -\sin\theta \\ 0 & 1 & 0 \\ \sin\theta & 0 & \cos\theta \end{bmatrix}$$ axis of rotation is y-axis, $\theta > 0$ is from positive x-axis toward positive z-axis</td></tr>
<tr><td>$$\begin{bmatrix} 1 & 0 & 0 \\ 0 & \cos\theta & -\sin\theta \\ 0 & \sin\theta & \cos\theta \end{bmatrix}$$ axis of rotation is x-axis, $\theta > 0$ is from positive y-axis toward positive z-axis</td></tr>
</table>

3. Every 2×2 or 3×3 elementary matrix is equivalent to exactly one of
   - an expansion or compression in the direction of an axis;
   - an expansion or compression in the direction of an axis followed by a reflection about a coordinate axis if in R_2 or a coordinate plane if in R_3;
   - a shear in the direction of an axis; or
   - a reflection.
4. Every 2×2 or 3×3 orthogonal matrix A is a rotation matrix if $\det(A) = 1$ and a rotation along with a reflection if $\det(A) = -1$.
5. Every 2×2 or 3×3 invertible matrix is equivalent to a sequence of expansions, compressions, reflections, and shears.
6. If the invertible matrix A represents a linear transformation $L: R_2 \to R_2$ or $L: R_3 \to R_3$, then
   - the images of distinct points are distinct;
   - the images of three points are collinear if and only if the three points are collinear;
   - the image of the line segment joining points P and Q is the line segment joining the images of P and Q;
   - the image of a line is a line;
   - the image of a line through the origin is a line through the origin;
   - the images of parallel lines are parallel lines;
   - the images of four points in R_3 are coplanar if and only if the four points are coplanar;
   - the image of a plane through the origin is a plane through the origin; and
   - the images of parallel planes are parallel planes.
7. An isometry is a one-to-one, onto transformation from a set of points onto itself that preserves distance.
8. A linear transformation $L: R_2 \to R_2$ or $L: R_3 \to R_3$ is an isometry if and only if the matrix of the transformation is orthogonal.
9. The linear transformations of R_2 or R_3 that are isometries are rotations and reflections.
10. Translations are isometries but are not linear.

RAISE YOUR GRADES

Can you ...?

☑ represent expansions, compressions, projections, shears, reflections, and rotations in R_2 and R_3 as matrices

☑ when given a 2×2 or 3×3 elementary matrix, determine whether it represents an expansion, compression, shear, or reflection or two of these

☑ determine if a 2×2 or 3×3 matrix is a rotation

☑ express a 2×2 or 3×3 invertible matrix as a sequence of expansions, compressions, reflections, and shears

☑ determine the image of a set of points under a linear transformation having an invertible matrix representation
☑ determine if a linear transformation in R_2 or R_3 is an isometry

SOLVED PROBLEMS

PROBLEM 18-1 Find the matrix representation of a translation by -4 units in the x-direction and 3 units in the y-direction and then the image of each of the following points.

(a) $\begin{bmatrix} -2 \\ -3 \end{bmatrix}$ **(b)** $\begin{bmatrix} 3 \\ -1 \end{bmatrix}$

Solution

$$\begin{bmatrix} x' \\ y' \end{bmatrix} = \begin{bmatrix} x \\ y \end{bmatrix} + \begin{bmatrix} -4 \\ 3 \end{bmatrix} = \begin{bmatrix} x-4 \\ y+3 \end{bmatrix}$$

(a) Therefore, for $\begin{bmatrix} -2 \\ -3 \end{bmatrix}$, $\begin{bmatrix} x' \\ y' \end{bmatrix} = \begin{bmatrix} -2-4 \\ -3+3 \end{bmatrix} = \begin{bmatrix} -6 \\ 0 \end{bmatrix}$.

(b) For $\begin{bmatrix} 3 \\ -1 \end{bmatrix}$, $\begin{bmatrix} x' \\ y' \end{bmatrix} = \begin{bmatrix} 3-4 \\ -1+3 \end{bmatrix} = \begin{bmatrix} -1 \\ 2 \end{bmatrix}$.

PROBLEM 18-2 Find the matrix representation of an expansion in the x-direction by a factor of 2, and then find the image of each of the following points.

(a) $\begin{bmatrix} -2 \\ -3 \end{bmatrix}$ **(b)** $\begin{bmatrix} 3 \\ -1 \end{bmatrix}$

Solution

$$\begin{bmatrix} x' \\ y' \end{bmatrix} = \begin{bmatrix} 2 & 0 \\ 0 & 1 \end{bmatrix}\begin{bmatrix} x \\ y \end{bmatrix}$$

(a) $$\begin{bmatrix} x' \\ y' \end{bmatrix} = \begin{bmatrix} 2 & 0 \\ 0 & 1 \end{bmatrix}\begin{bmatrix} -2 \\ -3 \end{bmatrix} = \begin{bmatrix} -4 \\ -3 \end{bmatrix}$$

(b) $$\begin{bmatrix} x' \\ y' \end{bmatrix} = \begin{bmatrix} 2 & 0 \\ 0 & 1 \end{bmatrix}\begin{bmatrix} 3 \\ -1 \end{bmatrix} = \begin{bmatrix} 6 \\ -1 \end{bmatrix}$$

PROBLEM 18-3 Find the matrix representation of an expansion in the y-direction by a factor of 2, and then find the image of each of the following points.

(a) $\begin{bmatrix} -2 \\ -3 \end{bmatrix}$ **(b)** $\begin{bmatrix} 3 \\ -1 \end{bmatrix}$

Solution

$$\begin{bmatrix} x' \\ y' \end{bmatrix} = \begin{bmatrix} 1 & 0 \\ 0 & 2 \end{bmatrix}\begin{bmatrix} x \\ y \end{bmatrix}$$

(a) $$\begin{bmatrix} x' \\ y' \end{bmatrix} = \begin{bmatrix} 1 & 0 \\ 0 & 2 \end{bmatrix}\begin{bmatrix} -2 \\ -3 \end{bmatrix} = \begin{bmatrix} -2 \\ -6 \end{bmatrix}$$

(b) $$\begin{bmatrix} x' \\ y' \end{bmatrix} = \begin{bmatrix} 1 & 0 \\ 0 & 2 \end{bmatrix}\begin{bmatrix} 3 \\ -1 \end{bmatrix} = \begin{bmatrix} 3 \\ -2 \end{bmatrix}$$

PROBLEM 18-4 Find the matrix representation of a compression in the x-direction by a factor of 1/3, and then find the image of each of the following points.

(a) $\begin{bmatrix} 3 \\ -4 \end{bmatrix}$ **(b)** $\begin{bmatrix} 6 \\ 3 \end{bmatrix}$

Solution

$$\begin{bmatrix} x' \\ y' \end{bmatrix} = \begin{bmatrix} \frac{1}{3} & 0 \\ 0 & 1 \end{bmatrix} \begin{bmatrix} x \\ y \end{bmatrix}$$

(a) $$\begin{bmatrix} x' \\ y' \end{bmatrix} = \begin{bmatrix} \frac{1}{3} & 0 \\ 0 & 1 \end{bmatrix} \begin{bmatrix} 3 \\ -4 \end{bmatrix} = \begin{bmatrix} 1 \\ -4 \end{bmatrix}$$

(b) $$\begin{bmatrix} x' \\ y' \end{bmatrix} = \begin{bmatrix} \frac{1}{3} & 0 \\ 0 & 1 \end{bmatrix} \begin{bmatrix} 6 \\ 3 \end{bmatrix} = \begin{bmatrix} 2 \\ 3 \end{bmatrix}$$

PROBLEM 18-5 Find the matrix representation of a compression in the y-direction by a factor of 1/3, and then find the image of each of the following points.

(a) $\begin{bmatrix} 3 \\ -4 \end{bmatrix}$ **(b)** $\begin{bmatrix} 6 \\ 3 \end{bmatrix}$

Solution

$$\begin{bmatrix} x' \\ y' \end{bmatrix} = \begin{bmatrix} 1 & 0 \\ 0 & \frac{1}{3} \end{bmatrix} \begin{bmatrix} x \\ y \end{bmatrix}$$

(a) $$\begin{bmatrix} x' \\ y' \end{bmatrix} = \begin{bmatrix} 1 & 0 \\ 0 & \frac{1}{3} \end{bmatrix} \begin{bmatrix} 3 \\ -4 \end{bmatrix} = \begin{bmatrix} 3 \\ -\frac{4}{3} \end{bmatrix}$$

(b) $$\begin{bmatrix} x' \\ y' \end{bmatrix} = \begin{bmatrix} 1 & 0 \\ 0 & \frac{1}{3} \end{bmatrix} \begin{bmatrix} 6 \\ 3 \end{bmatrix} = \begin{bmatrix} 6 \\ 1 \end{bmatrix}$$

PROBLEM 18-6 Find the matrix representation for a projection onto the x-axis, and then find the image of each of the following points.

(a) $\begin{bmatrix} 3 \\ 5 \end{bmatrix}$ **(b)** $\begin{bmatrix} -7 \\ -3 \end{bmatrix}$

Solution

$$\begin{bmatrix} x' \\ y' \end{bmatrix} = \begin{bmatrix} 1 & 0 \\ 0 & 0 \end{bmatrix} \begin{bmatrix} x \\ y \end{bmatrix}$$

(a) $$\begin{bmatrix} x' \\ y' \end{bmatrix} = \begin{bmatrix} 1 & 0 \\ 0 & 0 \end{bmatrix} \begin{bmatrix} 3 \\ 5 \end{bmatrix} = \begin{bmatrix} 3 \\ 0 \end{bmatrix}$$

(b) $$\begin{bmatrix} x' \\ y' \end{bmatrix} = \begin{bmatrix} 1 & 0 \\ 0 & 0 \end{bmatrix} \begin{bmatrix} -7 \\ -3 \end{bmatrix} = \begin{bmatrix} -7 \\ 0 \end{bmatrix}$$

PROBLEM 18-7 Find the matrix representation for a projection onto the y-axis, and then find the image of each of the following points.

(a) $\begin{bmatrix} 3 \\ 5 \end{bmatrix}$ **(b)** $\begin{bmatrix} -7 \\ -3 \end{bmatrix}$

Solution

$$\begin{bmatrix} x' \\ y' \end{bmatrix} = \begin{bmatrix} 0 & 0 \\ 0 & 1 \end{bmatrix} \begin{bmatrix} x \\ y \end{bmatrix}$$

(a)
$$\begin{bmatrix} x' \\ y' \end{bmatrix} = \begin{bmatrix} 0 & 0 \\ 0 & 1 \end{bmatrix} \begin{bmatrix} 3 \\ 5 \end{bmatrix} = \begin{bmatrix} 0 \\ 5 \end{bmatrix}$$

(b)
$$\begin{bmatrix} x' \\ y' \end{bmatrix} = \begin{bmatrix} 0 & 0 \\ 0 & 1 \end{bmatrix} \begin{bmatrix} -7 \\ -3 \end{bmatrix} = \begin{bmatrix} 0 \\ -3 \end{bmatrix}$$

PROBLEM 18-8 Find the matrix representation of a shear that moves each point horizontally by 3 times its distance from the x-axis, and then find the image of each of the following points.

(a) $\begin{bmatrix} 4 \\ 2 \end{bmatrix}$ **(b)** $\begin{bmatrix} 5 \\ -3 \end{bmatrix}$

Solution

$$\begin{bmatrix} x' \\ y' \end{bmatrix} = \begin{bmatrix} 1 & 3 \\ 0 & 1 \end{bmatrix} \begin{bmatrix} x \\ y \end{bmatrix}$$

(a)
$$\begin{bmatrix} x' \\ y' \end{bmatrix} = \begin{bmatrix} 1 & 3 \\ 0 & 1 \end{bmatrix} \begin{bmatrix} 4 \\ 2 \end{bmatrix} = \begin{bmatrix} 10 \\ 2 \end{bmatrix}$$

(b)
$$\begin{bmatrix} x' \\ y' \end{bmatrix} = \begin{bmatrix} 1 & 3 \\ 0 & 1 \end{bmatrix} \begin{bmatrix} 5 \\ -3 \end{bmatrix} = \begin{bmatrix} -4 \\ -3 \end{bmatrix}$$

PROBLEM 18-9 Find the matrix representation of a shear that moves each point vertically by -3 times its distance from the y-axis, and then find the image of each of the following points.

(a) $\begin{bmatrix} 4 \\ 2 \end{bmatrix}$ **(b)** $\begin{bmatrix} 5 \\ -3 \end{bmatrix}$

Solution

$$\begin{bmatrix} x' \\ y' \end{bmatrix} = \begin{bmatrix} 1 & 0 \\ -3 & 1 \end{bmatrix} \begin{bmatrix} x \\ y \end{bmatrix}$$

(a)
$$\begin{bmatrix} x' \\ y' \end{bmatrix} = \begin{bmatrix} 1 & 0 \\ -3 & 1 \end{bmatrix} \begin{bmatrix} 4 \\ 2 \end{bmatrix} = \begin{bmatrix} 4 \\ -10 \end{bmatrix}$$

(b)
$$\begin{bmatrix} x' \\ y' \end{bmatrix} = \begin{bmatrix} 1 & 0 \\ -3 & 1 \end{bmatrix} \begin{bmatrix} 5 \\ -3 \end{bmatrix} = \begin{bmatrix} 5 \\ -18 \end{bmatrix}$$

PROBLEM 18-10 Find the image if each of the following points is reflected about the x-axis.

(a) $\begin{bmatrix} 4 \\ 2 \end{bmatrix}$ **(b)** $\begin{bmatrix} 5 \\ -3 \end{bmatrix}$

Solution

$$\begin{bmatrix} x' \\ y' \end{bmatrix} = \begin{bmatrix} 1 & 0 \\ 0 & -1 \end{bmatrix} \begin{bmatrix} x \\ y \end{bmatrix}$$

(a)
$$\begin{bmatrix} x' \\ y' \end{bmatrix} = \begin{bmatrix} 1 & 0 \\ 0 & -1 \end{bmatrix} \begin{bmatrix} 4 \\ 2 \end{bmatrix} = \begin{bmatrix} 4 \\ -2 \end{bmatrix}$$

(b)
$$\begin{bmatrix} x' \\ y' \end{bmatrix} = \begin{bmatrix} 1 & 0 \\ 0 & -1 \end{bmatrix} \begin{bmatrix} 5 \\ -3 \end{bmatrix} = \begin{bmatrix} 5 \\ 3 \end{bmatrix}$$

PROBLEM 18-11 Find the image if each of the following points is reflected about the y-axis.

(a) $\begin{bmatrix} 4 \\ 2 \end{bmatrix}$ **(b)** $\begin{bmatrix} 5 \\ -3 \end{bmatrix}$

Solution

$$\begin{bmatrix} x' \\ y' \end{bmatrix} = \begin{bmatrix} -1 & 0 \\ 0 & 1 \end{bmatrix} \begin{bmatrix} x \\ y \end{bmatrix}$$

(a) $$\begin{bmatrix} x' \\ y' \end{bmatrix} = \begin{bmatrix} -1 & 0 \\ 0 & 1 \end{bmatrix} \begin{bmatrix} 4 \\ 2 \end{bmatrix} = \begin{bmatrix} -4 \\ 2 \end{bmatrix}$$

(b) $$\begin{bmatrix} x' \\ y' \end{bmatrix} = \begin{bmatrix} -1 & 0 \\ 0 & 1 \end{bmatrix} \begin{bmatrix} 5 \\ -3 \end{bmatrix} = \begin{bmatrix} -5 \\ -3 \end{bmatrix}$$

PROBLEM 18-12 Find the image if each of the following points is reflected about the line $y = x$.

(a) $\begin{bmatrix} 4 \\ 2 \end{bmatrix}$ **(b)** $\begin{bmatrix} 5 \\ -3 \end{bmatrix}$

Solution

$$\begin{bmatrix} x' \\ y' \end{bmatrix} = \begin{bmatrix} 0 & 1 \\ 1 & 0 \end{bmatrix} \begin{bmatrix} x \\ y \end{bmatrix}$$

(a) $$\begin{bmatrix} x' \\ y' \end{bmatrix} = \begin{bmatrix} 0 & 1 \\ 1 & 0 \end{bmatrix} \begin{bmatrix} 4 \\ 2 \end{bmatrix} = \begin{bmatrix} 2 \\ 4 \end{bmatrix}$$

(b) $$\begin{bmatrix} x' \\ y' \end{bmatrix} = \begin{bmatrix} 0 & 1 \\ 1 & 0 \end{bmatrix} \begin{bmatrix} 5 \\ -3 \end{bmatrix} = \begin{bmatrix} -3 \\ 5 \end{bmatrix}$$

PROBLEM 18-13 Find the image if each of the following points is rotated counterclockwise through 45°.

(a) $\begin{bmatrix} 4 \\ 2 \end{bmatrix}$ **(b)** $\begin{bmatrix} 5 \\ -3 \end{bmatrix}$

Solution

$$\begin{bmatrix} x' \\ y' \end{bmatrix} = \begin{bmatrix} \cos 45^\circ & -\sin 45^\circ \\ \sin 45^\circ & \cos 45^\circ \end{bmatrix} \begin{bmatrix} x \\ y \end{bmatrix} = \begin{bmatrix} \frac{1}{\sqrt{2}} & -\frac{1}{\sqrt{2}} \\ \frac{1}{\sqrt{2}} & \frac{1}{\sqrt{2}} \end{bmatrix} \begin{bmatrix} x \\ y \end{bmatrix}$$

(a) $$\begin{bmatrix} x' \\ y' \end{bmatrix} = \begin{bmatrix} \frac{1}{\sqrt{2}} & -\frac{1}{\sqrt{2}} \\ \frac{1}{\sqrt{2}} & \frac{1}{\sqrt{2}} \end{bmatrix} \begin{bmatrix} 4 \\ 2 \end{bmatrix} = \begin{bmatrix} \frac{2}{\sqrt{2}} \\ \frac{6}{\sqrt{2}} \end{bmatrix} = \begin{bmatrix} \sqrt{2} \\ 3\sqrt{2} \end{bmatrix}$$

(b) $$\begin{bmatrix} x' \\ y' \end{bmatrix} = \begin{bmatrix} \frac{1}{\sqrt{2}} & -\frac{1}{\sqrt{2}} \\ \frac{1}{\sqrt{2}} & \frac{1}{\sqrt{2}} \end{bmatrix} \begin{bmatrix} 5 \\ -3 \end{bmatrix} = \begin{bmatrix} \frac{8}{\sqrt{2}} \\ \frac{2}{\sqrt{2}} \end{bmatrix} = \begin{bmatrix} 4\sqrt{2} \\ \sqrt{2} \end{bmatrix}$$

PROBLEM 18-14 Find the image if each of the following points is rotated clockwise through 30°.

(a) $\begin{bmatrix} 4 \\ 2 \end{bmatrix}$ **(b)** $\begin{bmatrix} 5 \\ -3 \end{bmatrix}$

Solution

$$\begin{bmatrix} x' \\ y' \end{bmatrix} = \begin{bmatrix} \cos(-30°) & -\sin(-30°) \\ \sin(-30°) & \cos(-30°) \end{bmatrix} \begin{bmatrix} x \\ y \end{bmatrix} = \begin{bmatrix} \frac{\sqrt{3}}{2} & \frac{1}{2} \\ -\frac{1}{2} & \frac{\sqrt{3}}{2} \end{bmatrix} \begin{bmatrix} x \\ y \end{bmatrix}$$

(a)
$$\begin{bmatrix} x' \\ y' \end{bmatrix} = \begin{bmatrix} \frac{\sqrt{3}}{2} & \frac{1}{2} \\ -\frac{1}{2} & \frac{\sqrt{3}}{2} \end{bmatrix} \begin{bmatrix} 4 \\ 2 \end{bmatrix} = \begin{bmatrix} 2\sqrt{3} + 1 \\ -2 + \sqrt{3} \end{bmatrix}$$

(b)
$$\begin{bmatrix} x' \\ y' \end{bmatrix} = \begin{bmatrix} \frac{\sqrt{3}}{2} & \frac{1}{2} \\ -\frac{1}{2} & \frac{\sqrt{3}}{2} \end{bmatrix} \begin{bmatrix} 5 \\ -3 \end{bmatrix} = \begin{bmatrix} \frac{5\sqrt{3} - 3}{2} \\ \frac{-5 - 3\sqrt{3}}{2} \end{bmatrix}$$

PROBLEM 18-15 Determine the geometric transformation accomplished in R_2 by using each of the following elementary matrices as a premultiplier.

(a) $\begin{bmatrix} 1 & 0 \\ 0 & \frac{1}{2} \end{bmatrix}$ **(b)** $\begin{bmatrix} 3 & 0 \\ 0 & 1 \end{bmatrix}$ **(c)** $\begin{bmatrix} -3 & 0 \\ 0 & 1 \end{bmatrix}$ **(d)** $\begin{bmatrix} 1 & \frac{1}{2} \\ 0 & 1 \end{bmatrix}$

(e) $\begin{bmatrix} -1 & 0 \\ 0 & 1 \end{bmatrix}$ **(f)** $\begin{bmatrix} 0 & 1 \\ -1 & 0 \end{bmatrix}$

Solution

(a) This is a compression by a factor of 1/2 in the y-direction.
(b) This is an expansion by a factor of 3 in the x-direction.
(c) This is an expansion by a factor of 3 in the x-direction followed by a reflection about the y-axis.
(d) This is a shear in the x-direction where $c = 1/2$.
(e) This is a reflection about the y-axis.
(f) This is a reflection about the line $y = x$ followed by a reflection about the x-axis.

PROBLEM 18-16 Determine whether each of the following matrices is a rotation, a reflection, a reflection followed by a rotation, or neither.

(a) $A = \begin{bmatrix} -\frac{\sqrt{3}}{2} & \frac{1}{2} \\ \frac{1}{2} & \frac{\sqrt{3}}{2} \end{bmatrix}$ **(b)** $B = \begin{bmatrix} \frac{\sqrt{3}}{2} & -\frac{1}{2} \\ \frac{1}{3} & \frac{\sqrt{3}}{2} \end{bmatrix}$ **(c)** $C = \begin{bmatrix} -\frac{1}{\sqrt{2}} & \frac{1}{\sqrt{2}} \\ \frac{1}{\sqrt{2}} & \frac{1}{\sqrt{2}} \end{bmatrix}$

Solution

(a) $(-\sqrt{3}/2)(1/2) + (1/2)(\sqrt{3}/2) = 0$ and $\det(A) = -1$. Therefore, matrix A represents a reflection followed by a rotation since A is orthogonal and $\det(A) = -1$.
(b) $(\sqrt{3}/2)(-1/2) + (1/2)(\sqrt{3}/2) = 0$ and $\det(B) = 1$. Therefore, matrix B represents a rotation: $\cos\theta = \sqrt{3}/2$, and $\theta = 30°$.
(c) $(-1/\sqrt{2})(1/\sqrt{2}) + (1/\sqrt{2})(1/\sqrt{2}) = 0$ and $\det(C) = -1$. Therefore, matrix C represents a reflection followed by a rotation.

PROBLEM 18-17 Express the rotation matrix $B = \begin{bmatrix} \sqrt{3}/2 & -1/2 \\ 1/2 & \sqrt{3}/2 \end{bmatrix}$ as a sequence of expansions, compressions, reflections and shears.

Solution Using the method of Example 4-11, we get

$$B = \begin{bmatrix} \frac{\sqrt{3}}{2} & -\frac{1}{2} \\ \frac{1}{2} & \frac{\sqrt{3}}{2} \end{bmatrix} = \begin{bmatrix} 0 & 1 \\ 1 & 0 \end{bmatrix} \begin{bmatrix} 1 & 0 \\ \sqrt{3} & 1 \end{bmatrix} \begin{bmatrix} 1 & 0 \\ 0 & -2 \end{bmatrix} \begin{bmatrix} \frac{1}{2} & 0 \\ 0 & 1 \end{bmatrix} \begin{bmatrix} 1 & \sqrt{3} \\ 0 & 1 \end{bmatrix}$$

$$= E_5 E_4 E_3 E_2 E_1$$

Now read from right to left. Therefore, B is equivalent to

(1) a shear in the x-direction with $c = \sqrt{3}$;
(2) a compression in the x-direction by a factor of 1/2;
(3) an expansion by a factor of 2 in the y-direction followed by a reflection about the x-axis;
(4) a shear in the y-direction where $c = \sqrt{3}$; and
(5) a reflection about the line $y = x$.

PROBLEM 18-18 Find the image of the square having vertices $A(1, 1)$, $B(-1, 1)$, $C(-1, -1)$, and $D(1, -1)$ under the transformation represented by the matrix $\begin{bmatrix} 1 & 3 \\ -2 & 1 \end{bmatrix}$.

Solution The image of A is

$$A' = \begin{bmatrix} 1 & 3 \\ -2 & 1 \end{bmatrix} \begin{bmatrix} 1 \\ 1 \end{bmatrix} = \begin{bmatrix} 4 \\ -1 \end{bmatrix}$$

The image of B is

$$B' = \begin{bmatrix} 1 & 3 \\ -2 & 1 \end{bmatrix} \begin{bmatrix} -1 \\ 1 \end{bmatrix} = \begin{bmatrix} 2 \\ 3 \end{bmatrix}$$

The image of C is

$$C' = \begin{bmatrix} 1 & 3 \\ -2 & 1 \end{bmatrix} \begin{bmatrix} -1 \\ -1 \end{bmatrix} = \begin{bmatrix} -4 \\ 1 \end{bmatrix}$$

The image of D is

$$D' = \begin{bmatrix} 1 & 3 \\ -2 & 1 \end{bmatrix} \begin{bmatrix} 1 \\ -1 \end{bmatrix} = \begin{bmatrix} -2 \\ -3 \end{bmatrix}$$

To find the total image of the square, we plot these four points and find the parallelogram having line segments $\overline{A'B'}$, $\overline{B'C'}$, $\overline{C'D'}$, and $\overline{D'A'}$ as sides (see Figure 18-16).

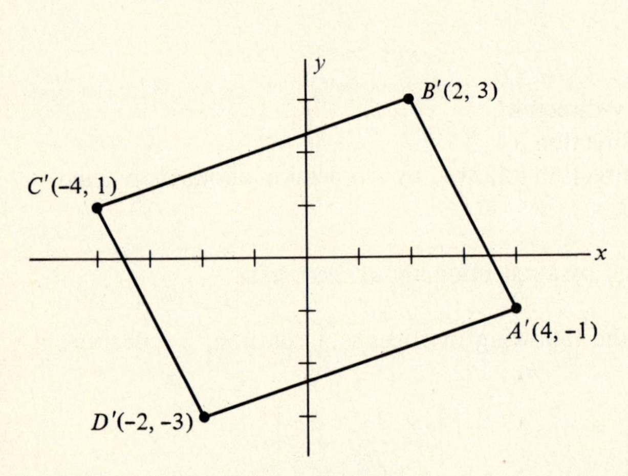

FIGURE 18-16

PROBLEM 18-19 Find the image of a triangle having vertices $A(3,0)$, $B(0,6)$, and $C(-3,0)$ under the transformation represented by the matrix $\begin{bmatrix} 3 & -2 \\ 1 & 2 \end{bmatrix}$.

Solution The image of A is

$$A' = \begin{bmatrix} 3 & -2 \\ 1 & 2 \end{bmatrix} \begin{bmatrix} 3 \\ 0 \end{bmatrix} = \begin{bmatrix} 9 \\ 3 \end{bmatrix}$$

The image of B is

$$B' = \begin{bmatrix} 3 & -2 \\ 1 & 2 \end{bmatrix} \begin{bmatrix} 0 \\ 6 \end{bmatrix} = \begin{bmatrix} -12 \\ 12 \end{bmatrix}$$

The image of C is

$$C' = \begin{bmatrix} 3 & -2 \\ 1 & 2 \end{bmatrix} \begin{bmatrix} -3 \\ 0 \end{bmatrix} = \begin{bmatrix} -9 \\ -3 \end{bmatrix}$$

Therefore, the image of the triangle ABC is triangle $A'B'C'$ which has vertices $A'(9, 3)$, $B'(-12, 12)$, and $C'(-9, -3)$.

PROBLEM 18-20 Find the matrix representation for each of the following transformations:

(a) expansion by a factor of 2 in the x-direction;
(b) expansion by a factor of 2 in the y-direction and a compression of 1/3 in the z-direction.

Solution

(a)
$$\begin{bmatrix} x' \\ y' \\ z' \end{bmatrix} = \begin{bmatrix} 2 & 0 & 0 \\ 0 & 1 & 0 \\ 0 & 0 & 1 \end{bmatrix} \begin{bmatrix} x \\ y \\ z \end{bmatrix}$$

(b)
$$\begin{bmatrix} x' \\ y' \\ z' \end{bmatrix} = \begin{bmatrix} 1 & 0 & 0 \\ 0 & 2 & 0 \\ 0 & 0 & \frac{1}{3} \end{bmatrix} \begin{bmatrix} x \\ y \\ z \end{bmatrix}$$

PROBLEM 18-21 Find the image of (3, 4, 5) if projected onto **(a)** the z-axis and **(b)** the yz-plane.

Solution

(a)
$$\begin{bmatrix} x' \\ y' \\ z' \end{bmatrix} = \begin{bmatrix} 0 & 0 & 0 \\ 0 & 0 & 0 \\ 0 & 0 & 1 \end{bmatrix} \begin{bmatrix} 3 \\ 4 \\ 5 \end{bmatrix} = \begin{bmatrix} 0 \\ 0 \\ 5 \end{bmatrix}$$

(b)
$$\begin{bmatrix} x' \\ y' \\ z' \end{bmatrix} = \begin{bmatrix} 0 & 0 & 0 \\ 0 & 1 & 0 \\ 0 & 0 & 1 \end{bmatrix} \begin{bmatrix} 3 \\ 4 \\ 5 \end{bmatrix} = \begin{bmatrix} 0 \\ 4 \\ 5 \end{bmatrix}$$

PROBLEM 18-22 Find the image of $(3, -4, 6)$ if **(a)** it is moved parallel to the y-axis by 2 times its distance from the yz-plane and **(b)** it is moved parallel to the y-axis by 3 times its distance from the xy-plane, then moved parallel to the x-axis by -2 its distance from the xz-plane.

Solution

(a)
$$\begin{bmatrix} x' \\ y' \\ z' \end{bmatrix} = \begin{bmatrix} 1 & 0 & 0 \\ 2 & 1 & 0 \\ 0 & 0 & 1 \end{bmatrix} \begin{bmatrix} 3 \\ -4 \\ 6 \end{bmatrix} = \begin{bmatrix} 3 \\ 2 \\ 6 \end{bmatrix}$$

(b)
$$\begin{bmatrix} x' \\ y' \\ z' \end{bmatrix} = \begin{bmatrix} 1 & 0 & 0 \\ 0 & 1 & 3 \\ 0 & 0 & 1 \end{bmatrix} \begin{bmatrix} 1 & -2 & 0 \\ 0 & 1 & 0 \\ 0 & 0 & 1 \end{bmatrix} \begin{bmatrix} 3 \\ -4 \\ 6 \end{bmatrix} = \begin{bmatrix} 1 & 0 & 0 \\ 0 & 1 & 3 \\ 0 & 0 & 1 \end{bmatrix} \begin{bmatrix} 11 \\ -4 \\ 6 \end{bmatrix} = \begin{bmatrix} 11 \\ 14 \\ 6 \end{bmatrix}$$

PROBLEM 18-23 Find the image of $(-5, 3, 4)$ if **(a)** it is reflected about the y-axis; **(b)** it is reflected about the xy-plane; and **(c)** it is reflected about the x-axis, then reflected about the yz-plane.

Solution

(a)
$$\begin{bmatrix} x' \\ y' \\ z' \end{bmatrix} = \begin{bmatrix} -1 & 0 & 0 \\ 0 & 1 & 0 \\ 0 & 0 & -1 \end{bmatrix} \begin{bmatrix} -5 \\ 3 \\ 4 \end{bmatrix} = \begin{bmatrix} 5 \\ 3 \\ -4 \end{bmatrix}$$

(b)
$$\begin{bmatrix} x' \\ y' \\ z' \end{bmatrix} = \begin{bmatrix} 1 & 0 & 0 \\ 0 & 1 & 0 \\ 0 & 0 & -1 \end{bmatrix} \begin{bmatrix} -5 \\ 3 \\ 4 \end{bmatrix} = \begin{bmatrix} -5 \\ 3 \\ -4 \end{bmatrix}$$

(c)
$$\begin{bmatrix} x' \\ y' \\ z' \end{bmatrix} = \begin{bmatrix} 1 & 0 & 0 \\ 0 & -1 & 0 \\ 0 & 0 & -1 \end{bmatrix} \begin{bmatrix} -1 & 0 & 0 \\ 0 & 1 & 0 \\ 0 & 0 & 1 \end{bmatrix} \begin{bmatrix} -5 \\ 3 \\ 4 \end{bmatrix} = \begin{bmatrix} 1 & 0 & 0 \\ 0 & -1 & 0 \\ 0 & 0 & -1 \end{bmatrix} \begin{bmatrix} 5 \\ 3 \\ 4 \end{bmatrix} = \begin{bmatrix} 5 \\ -3 \\ -4 \end{bmatrix}$$

PROBLEM 18-24 Find the image of $(2, \sqrt{3}, -5)$ under the following rotations: **(a)** a rotation of 30° about the x-axis and **(b)** a rotation of 30° about the x-axis followed by a rotation of 30° about the y-axis.

Solution

(a)
$$\begin{bmatrix} x' \\ y' \\ z' \end{bmatrix} = \begin{bmatrix} 1 & 0 & 0 \\ 0 & \cos 30^\circ & -\sin 30^\circ \\ 0 & \sin 30^\circ & \cos 30^\circ \end{bmatrix} \begin{bmatrix} 2 \\ \sqrt{3} \\ -5 \end{bmatrix}$$

$$= \begin{bmatrix} 1 & 0 & 0 \\ 0 & \frac{\sqrt{3}}{2} & -\frac{1}{2} \\ 0 & \frac{1}{2} & \frac{\sqrt{3}}{2} \end{bmatrix} \begin{bmatrix} 2 \\ \sqrt{3} \\ -5 \end{bmatrix} = \begin{bmatrix} 2 \\ 1 \\ -2\sqrt{3} \end{bmatrix}$$

(b)
$$\begin{bmatrix} x' \\ y' \\ z' \end{bmatrix} = \begin{bmatrix} \cos 30^\circ & 0 & -\sin 30^\circ \\ 0 & 1 & 0 \\ \sin 30^\circ & 0 & \cos 30^\circ \end{bmatrix} \begin{bmatrix} 1 & 0 & 0 \\ 0 & \cos 30^\circ & -\sin 30^\circ \\ 0 & \sin 30^\circ & \cos 30^\circ \end{bmatrix} \begin{bmatrix} 2 \\ \sqrt{3} \\ -5 \end{bmatrix}$$

$$= \begin{bmatrix} \frac{\sqrt{3}}{2} & 0 & -\frac{1}{2} \\ 0 & 1 & 0 \\ \frac{1}{2} & 0 & \frac{\sqrt{3}}{2} \end{bmatrix} \begin{bmatrix} 1 & 0 & 0 \\ 0 & \frac{\sqrt{3}}{2} & -\frac{1}{2} \\ 0 & \frac{1}{2} & \frac{\sqrt{3}}{2} \end{bmatrix} \begin{bmatrix} 2 \\ \sqrt{3} \\ -5 \end{bmatrix}$$

$$= \begin{bmatrix} \frac{\sqrt{3}}{2} & 0 & -\frac{1}{2} \\ 0 & 1 & 0 \\ \frac{1}{2} & 0 & \frac{\sqrt{3}}{2} \end{bmatrix} \begin{bmatrix} 2 \\ 4 \\ -2\sqrt{3} \end{bmatrix} = \begin{bmatrix} 2\sqrt{3} \\ 4 \\ -2 \end{bmatrix}$$

PROBLEM 18-25 Determine the geometric transformation accomplished in R_3 by using each of the following elementary matrices as a premultiplier.

(a) $\begin{bmatrix} \frac{1}{3} & 0 & 0 \\ 0 & 1 & 0 \\ 0 & 0 & 1 \end{bmatrix}$ **(b)** $\begin{bmatrix} 1 & 0 & 0 \\ 0 & -2 & 0 \\ 0 & 0 & 1 \end{bmatrix}$ **(c)** $\begin{bmatrix} 1 & 0 & 0 \\ 2 & 1 & 0 \\ 0 & 0 & 1 \end{bmatrix}$

Solution

(a) This matrix gives a compression by a factor of 1/3 in the x-direction.

(b) This matrix gives an expansion by a factor of 2 in the y-direction followed by a reflection about the xz-plane

(c) This gives a shear parallel to the y-axis by two times the distance of the point from the yz-plane.

PROBLEM 18-26 Classify each of the following matrices as a rotation, a combination of a rotation and a reflection, or neither.

$$\text{(a) } A = \begin{bmatrix} \frac{1}{\sqrt{3}} & \frac{1}{\sqrt{2}} & -\frac{1}{\sqrt{6}} \\ \frac{1}{\sqrt{3}} & -\frac{1}{\sqrt{2}} & -\frac{1}{\sqrt{6}} \\ \frac{1}{\sqrt{3}} & 0 & \frac{2}{\sqrt{6}} \end{bmatrix} \qquad \text{(b) } B = \begin{bmatrix} \frac{1}{\sqrt{3}} & -\frac{1}{\sqrt{6}} & \frac{1}{\sqrt{2}} \\ \frac{1}{\sqrt{3}} & -\frac{1}{\sqrt{6}} & -\frac{1}{\sqrt{2}} \\ \frac{1}{\sqrt{3}} & \frac{2}{\sqrt{6}} & 0 \end{bmatrix} \qquad \text{(c) } C = \begin{bmatrix} \frac{\sqrt{3}}{2} & 0 & \frac{1}{2} \\ \frac{1}{2} & \frac{\sqrt{3}}{2} & 0 \\ 0 & \frac{1}{2} & \frac{\sqrt{3}}{2} \end{bmatrix}$$

Solution

(a) The column vectors are orthonormal and $\det(A) = -1$. Therefore, A represents a combination of a rotation and a reflection.

(b) The column vectors are orthonormal and $\det(B) = 1$. Therefore, B represents a rotation.

(c) Since $\sqrt{3}/2(0) + 1/2(\sqrt{3}/2) + 0(1/2) = \sqrt{3}/4 \neq 0$, C is not orthogonal. Therefore, C represents neither a rotation nor a rotation with a reflection.

PROBLEM 18-27 Express the invertible matrix $A = \begin{bmatrix} 3 & 2 & 1 \\ 1 & 1 & 2 \\ 2 & 0 & 1 \end{bmatrix}$ as a sequence of expansions, compressions, reflections, and shears.

Solution Using the method of Example 4-11 we express A as a product of elementary matrices.

$$A = \begin{bmatrix} 3 & 2 & 1 \\ 1 & 1 & 2 \\ 2 & 0 & 1 \end{bmatrix} = E_9 E_8 E_7 E_6 E_5 E_4 E_3 E_2 E_1$$

where

$$E_9 = \begin{bmatrix} 0 & 1 & 0 \\ 1 & 0 & 0 \\ 0 & 0 & 1 \end{bmatrix} \qquad E_8 = \begin{bmatrix} 1 & 0 & 0 \\ 3 & 1 & 0 \\ 0 & 0 & 1 \end{bmatrix} \qquad E_7 = \begin{bmatrix} 1 & 0 & 0 \\ 0 & 1 & 0 \\ 2 & 0 & 1 \end{bmatrix}$$

$$E_6 = \begin{bmatrix} 1 & 0 & 0 \\ 0 & -1 & 0 \\ 0 & 0 & 1 \end{bmatrix} \qquad E_5 = \begin{bmatrix} 1 & 0 & 0 \\ 0 & 1 & 0 \\ 0 & -2 & 1 \end{bmatrix} \qquad E_4 = \begin{bmatrix} 1 & 0 & 0 \\ 0 & 1 & 0 \\ 0 & 0 & 7 \end{bmatrix}$$

$$E_3 = \begin{bmatrix} 1 & 1 & 0 \\ 0 & 1 & 0 \\ 0 & 0 & 1 \end{bmatrix} \qquad E_2 = \begin{bmatrix} 1 & 0 & -3 \\ 0 & 1 & 0 \\ 0 & 0 & 1 \end{bmatrix} \qquad E_1 = \begin{bmatrix} 1 & 0 & 0 \\ 0 & 1 & 5 \\ 0 & 0 & 1 \end{bmatrix}$$

Now read the geometric effect of $E_9 E_8 E_7 E_6 E_5 E_4 E_3 E_2 E_1$ from right to left. Therefore, A is equivalent to

(1) a shear parallel to y-axis by 5 times its distance from xz-plane;
(2) a shear parallel to the x-axis by -3 times its distance from the xy-plane;
(3) a shear parallel to the x-axis by 1 times its distance from the xz-plane;
(4) an expansion by a factor of 7 in the z-direction;
(5) a shear parallel to the z-axis by -2 times its distance from the xz-plane;
(6) a reflection about the xz-plane;
(7) a shear parallel to the z-axis by 2 times its distance from the yz-plane;
(8) a shear parallel to the y-axis by 3 times it distance from the yz-plane;
(9) a reflection about the line $y = x$.

PROBLEM 18-28 Find the image of the rectangular parallelepiped having vertices $A(0,0,0)$, $B(2,0,0)$, $C(2,1,0)$, $D(0,1,0)$, $E(0,1,2)$, $F(2,1,2)$, $G(2,0,2)$, and $H(0,0,2)$ under the transformation represented by the matrix $\begin{bmatrix} 1 & 0 & 2 \\ 2 & 1 & -1 \\ 3 & 2 & 3 \end{bmatrix}$.

Solution The image of A is

$$A' = \begin{bmatrix} 1 & 0 & 2 \\ 2 & 1 & -1 \\ 3 & 2 & 3 \end{bmatrix} \begin{bmatrix} 0 \\ 0 \\ 0 \end{bmatrix} = \begin{bmatrix} 0 \\ 0 \\ 0 \end{bmatrix}$$

The image of B is

$$B' = \begin{bmatrix} 1 & 0 & 2 \\ 2 & 1 & -1 \\ 3 & 2 & 3 \end{bmatrix} \begin{bmatrix} 2 \\ 0 \\ 0 \end{bmatrix} = \begin{bmatrix} 2 \\ 4 \\ 6 \end{bmatrix}$$

The image of C is

$$C' = \begin{bmatrix} 1 & 0 & 2 \\ 2 & 1 & -1 \\ 3 & 2 & 3 \end{bmatrix} \begin{bmatrix} 2 \\ 1 \\ 0 \end{bmatrix} = \begin{bmatrix} 2 \\ 5 \\ 8 \end{bmatrix}$$

The image of D is

$$D' = \begin{bmatrix} 1 & 0 & 2 \\ 2 & 1 & -1 \\ 3 & 2 & 3 \end{bmatrix} \begin{bmatrix} 0 \\ 1 \\ 0 \end{bmatrix} = \begin{bmatrix} 0 \\ 1 \\ 2 \end{bmatrix}$$

The image of E is

$$E' = \begin{bmatrix} 1 & 0 & 2 \\ 2 & 1 & -1 \\ 3 & 2 & 3 \end{bmatrix} \begin{bmatrix} 0 \\ 1 \\ 2 \end{bmatrix} = \begin{bmatrix} 4 \\ -1 \\ 8 \end{bmatrix}$$

The image of F is

$$F' = \begin{bmatrix} 1 & 0 & 2 \\ 2 & 1 & -1 \\ 3 & 2 & 3 \end{bmatrix} \begin{bmatrix} 2 \\ 1 \\ 2 \end{bmatrix} = \begin{bmatrix} 6 \\ 3 \\ 14 \end{bmatrix}$$

The image of G is

$$G' = \begin{bmatrix} 1 & 0 & 2 \\ 2 & 1 & -1 \\ 3 & 2 & 3 \end{bmatrix} \begin{bmatrix} 2 \\ 0 \\ 2 \end{bmatrix} = \begin{bmatrix} 6 \\ 2 \\ 12 \end{bmatrix}$$

The image of H is

$$H' = \begin{bmatrix} 1 & 0 & 2 \\ 2 & 1 & -1 \\ 3 & 2 & 3 \end{bmatrix} \begin{bmatrix} 0 \\ 0 \\ 2 \end{bmatrix} = \begin{bmatrix} 4 \\ -2 \\ 6 \end{bmatrix}$$

Therefore, the image of the given figure is the parallelepiped having vertices A', B', C', D', E', F', G', and H'.

PROBLEM 18-29 Determine which of the following matrices represent linear transformations which are isometries in R_2.

(a) $A = \begin{bmatrix} 1 & -2 \\ 2 & 1 \end{bmatrix}$ (b) $B = \begin{bmatrix} \frac{\sqrt{3}}{2} & \frac{1}{2} \\ -\frac{1}{2} & \frac{\sqrt{3}}{2} \end{bmatrix}$ (c) $C = \begin{bmatrix} \frac{3}{5} & \frac{4}{5} \\ \frac{4}{5} & -\frac{3}{5} \end{bmatrix}$

Solution

(a) The column vectors are orthogonal, but they are not orthonormal and $\det(A) \neq 1$ or -1. Therefore, A does not represent an isometry.
(b) The column vectors of B are orthonormal; thus B is orthogonal. Hence, B represents an isometry.
(c) The column vectors of C are orthonormal; thus C is orthogonal. Hence, C represents an isometry.

PROBLEM 18-30 Which of the following matrices represent linear transformations that are isometries in R_3?

(a) $A = \begin{bmatrix} \frac{1}{\sqrt{6}} & \frac{1}{\sqrt{2}} & \frac{1}{\sqrt{3}} \\ \frac{2}{\sqrt{6}} & 0 & -\frac{1}{\sqrt{3}} \\ \frac{1}{\sqrt{6}} & -\frac{1}{\sqrt{2}} & \frac{1}{\sqrt{3}} \end{bmatrix}$ (b) $B = \begin{bmatrix} 3 & 0 & -1 \\ 0 & 1 & 0 \\ 1 & 0 & 3 \end{bmatrix}$

Solution

(a) The column vectors of A are orthonormal; thus A is orthogonal and represents an isometry.
(b) The column vectors of B are orthogonal, but they are not orthonormal. Thus, matrix B is not orthogonal and does not represent an isometry.

Supplementary Problems

PROBLEM 18-31 Given the point $\begin{bmatrix} 2 \\ -3 \end{bmatrix}$, find its image under (a) a translation of -4 units in the x-direction and 3 units in the y-direction; (b) an expansion in the x-direction of a factor of 2; (c) a compression in the y-direction by a factor of 2/3; (d) a compression in the x-direction by a factor of 1/4; (e) a projection onto the x-axis; (f) a shear in the y-direction with $c = -1$.

PROBLEM 18-32 Given the point $\begin{bmatrix} 4 \\ 7 \end{bmatrix}$ find its image under (a) a translation by -4 units in the x-direction and 2 units in the y-direction; (b) an expansion in the x-direction by a factor of 3; (c) a compression in the y-direction by a factor of 4/7; (d) a projection onto the y-axis; (e) a shear in the x-direction with $c = 1/7$.

PROBLEM 18-33 Given the point $\begin{bmatrix} 4 \\ -3 \end{bmatrix}$, find its image under (a) a reflection about the x-axis; (b) a reflection about the y-axis; (c) a reflection about the line $y = x$.

PROBLEM 18-34 Given the point $\begin{bmatrix} -5 \\ 2 \end{bmatrix}$, find its image under (a) a reflection about the x-axis; (b) a reflection about the y-axis; (c) a reflection about the line $y = x$.

PROBLEM 18-35 Given the point $\begin{bmatrix} 4 \\ -3 \end{bmatrix}$, find its image under **(a)** a rotation counterclockwise through an angle of 45°; **(b)** a rotation counterclockwise through an angle of 60°; **(c)** a rotation clockwise through an angle of 30°.

PROBLEM 18-36 Given the point $\begin{bmatrix} 0 \\ -\sqrt{2} \end{bmatrix}$ find its image under **(a)** a rotation counterclockwise through an angle of 45°; **(b)** a rotation counterclockwise through an angle of 60°; **(c)** a rotation clockwise through an angle of 30°.

PROBLEM 18-37 Determine the geometric transformation accomplished in R_2 by using each of the following as a premultiplier.

(a) $\begin{bmatrix} 3 & 0 \\ 0 & 1 \end{bmatrix}$ **(b)** $\begin{bmatrix} 1 & 0 \\ -2 & 1 \end{bmatrix}$ **(c)** $\begin{bmatrix} 1 & 0 \\ 0 & -3 \end{bmatrix}$

PROBLEM 18-38 Determine the geometric transformation in R_3 by using each of the following as a premultiplier.

(a) $\begin{bmatrix} 1 & 0 & 0 \\ 0 & -1 & 0 \\ 0 & 0 & 1 \end{bmatrix}$ **(b)** $\begin{bmatrix} 1 & 0 & 2 \\ 0 & 1 & 0 \\ 0 & 0 & 1 \end{bmatrix}$ **(c)** $\begin{bmatrix} 1 & 0 & 0 \\ 0 & \frac{1}{2} & 0 \\ 0 & 0 & 1 \end{bmatrix}$

PROBLEM 18-39 Express $A = \begin{bmatrix} -1 & 4 & -3 \\ 1 & -3 & 3 \\ 0 & -2 & 1 \end{bmatrix}$ as a sequence of expansions, compressions, reflections, and shears.

PROBLEM 18-40 Find the image of $(2, \sqrt{3}, 5)$ under a **(a)** reflection about the y-axis; **(b)** reflection about the xz-plane; **(c)** reflection about the plane $y = x$.

PROBLEM 18-41 Find the image of $(2, \sqrt{3}, 5)$ under a **(a)** rotation of 30° about the y-axis; **(b)** rotation of 30° about the y-axis followed by a rotation of 30° about the x-axis, clockwise.

PROBLEM 18-42 Find the image of $(2, 1, 3)$ under a reflection about the plane $y = x$ followed by a rotation counterclockwise through an angle of 180° about the z-axis.

PROBLEM 18-43 Which of the following matrices represent linear transformations in R_2 or R_3 that are isometries?

(a) $A = \begin{bmatrix} \frac{1}{\sqrt{2}} & -\frac{1}{\sqrt{2}} \\ \frac{1}{\sqrt{2}} & \frac{1}{\sqrt{2}} \end{bmatrix}$ **(b)** $\begin{bmatrix} \frac{1}{\sqrt{2}} & 0 & -\frac{1}{\sqrt{2}} \\ -\frac{1}{\sqrt{2}} & \frac{1}{\sqrt{2}} & 0 \\ 0 & \frac{1}{\sqrt{2}} & -\frac{1}{\sqrt{2}} \end{bmatrix}$ **(c)** $C = \begin{bmatrix} \frac{\sqrt{3}}{2} & 0 & -\frac{1}{2} \\ 0 & 1 & 0 \\ \frac{1}{2} & 0 & \frac{\sqrt{3}}{2} \end{bmatrix}$

Answers to Supplementary Problems

18-31 **(a)** $\begin{bmatrix} -2 \\ 0 \end{bmatrix}$ **(d)** $\begin{bmatrix} \frac{1}{2} \\ -3 \end{bmatrix}$
(b) $\begin{bmatrix} 4 \\ -3 \end{bmatrix}$ **(e)** $\begin{bmatrix} 2 \\ 0 \end{bmatrix}$
(c) $\begin{bmatrix} 2 \\ -2 \end{bmatrix}$ **(f)** $\begin{bmatrix} 2 \\ -5 \end{bmatrix}$

18-32 **(a)** $\begin{bmatrix} 0 \\ 9 \end{bmatrix}$ **(d)** $\begin{bmatrix} 0 \\ 7 \end{bmatrix}$
(b) $\begin{bmatrix} 12 \\ 7 \end{bmatrix}$ **(e)** $\begin{bmatrix} 5 \\ 7 \end{bmatrix}$
(c) $\begin{bmatrix} 4 \\ 4 \end{bmatrix}$

18-33 (a) $\begin{bmatrix} 4 \\ 3 \end{bmatrix}$ (c) $\begin{bmatrix} -3 \\ 4 \end{bmatrix}$

(b) $\begin{bmatrix} -4 \\ -3 \end{bmatrix}$

18-34 (a) $\begin{bmatrix} -5 \\ -2 \end{bmatrix}$ (c) $\begin{bmatrix} 2 \\ -5 \end{bmatrix}$

(b) $\begin{bmatrix} 5 \\ 2 \end{bmatrix}$

18-35 (a) $\begin{bmatrix} 7/\sqrt{2} \\ 1/\sqrt{2} \end{bmatrix}$ (c) $\begin{bmatrix} 2\sqrt{3} - 3/2 \\ -2 - 3\sqrt{3}/2 \end{bmatrix}$

(b) $\begin{bmatrix} 2 + 3\sqrt{3}/2 \\ 2\sqrt{3} - 3/2 \end{bmatrix}$

18-36 (a) $\begin{bmatrix} 1 \\ -1 \end{bmatrix}$ (c) $\begin{bmatrix} -\sqrt{2}/2 \\ -\sqrt{6}/2 \end{bmatrix}$

(b) $\begin{bmatrix} \sqrt{6}/2 \\ -\sqrt{2}/2 \end{bmatrix}$

18-37 (a) Expansion in the x-direction by factor 3
(b) Shear in the y-direction with $c = -2$
(c) Reflection about the x-axis followed by an expansion in the y-direction by factor of 3

18-38 (a) Reflection about the xz-plane
(b) Shear parallel to the x-axis with $c = 2$
(c) Compression in the y-direction by factor of 1/2

18-39 One answer is

(a) shear parallel to the x-axis by 3 times its distance from the xy-plane
(b) reflection about the plane $y = x$
(c) shear parallel to the y-axis by -3 times its distance from the yz-plane
(d) shear parallel to the z-axis by -2 times its distance from the yz-plane
(e) shear parallel to the x-axis by -1 times its distance from the xz-plane

18-40 (a) $\begin{bmatrix} -2 \\ \sqrt{3} \\ -5 \end{bmatrix}$ (c) $\begin{bmatrix} \sqrt{3} \\ 2 \\ 5 \end{bmatrix}$

(b) $\begin{bmatrix} 2 \\ -\sqrt{3} \\ 5 \end{bmatrix}$

18-41 (a) $\begin{bmatrix} \sqrt{3} - 5/2 \\ \sqrt{3} \\ 1 + 5\sqrt{3}/2 \end{bmatrix}$ (b) $\begin{bmatrix} \sqrt{3} - 5/2 \\ 2 + 5\sqrt{3}/4 \\ 15/4 \end{bmatrix}$

18-42 $\begin{bmatrix} -1 \\ -2 \\ 3 \end{bmatrix}$

18-43 (a) Isometry in R_2 (c) Isometry in R_3
(b) Not an isometry

EXAM 5 (Chapters 15–18)

1. Find the characteristic equation of the matrix A where $A = \begin{bmatrix} 4 & -1 & 3 \\ 0 & 2 & 1 \\ 0 & 0 & 3 \end{bmatrix}$.

2. Find the eigenvalues of matrix A in question 1.

3. Find the eigenvectors associated with the eigenvalues in question 2.

4. Explain why matrix A in question 1 is diagonalizable.

5. Find a matrix P that diagonalizes matrix A in question 1.

6. Use P in question 5 to find a diagonalization of matrix A in question 1.

7. Find the standard form and name the conic represented by $16x^2 + 24xy + 9y^2 + 5x - 5y = 0$.

8. Find a single matrix that in R^2 compresses by a factor of 1/2 in the y-direction then expands by a factor of 3 in the x-direction.

9. Express the invertible matrix $A = \begin{bmatrix} 1 & 1 & 0 \\ 0 & 1 & 0 \\ 0 & 0 & 2 \end{bmatrix}$ as a sequence of expansions, compressions, reflections, and shears in R^3.

Solutions to Exam 5

1. $$\text{Det}(\lambda I - A) = \begin{vmatrix} \lambda - 4 & 1 & -3 \\ 0 & \lambda - 2 & -1 \\ 0 & 0 & \lambda - 3 \end{vmatrix} = 0$$

Hence, $(\lambda - 3)(\lambda - 2)(\lambda - 4) = 0$ is the characteristic equation.

2. $(\lambda - 3)(\lambda - 2)(\lambda - 4) = 0$ gives the eigenvalues of 3, 2, 4.

3. For $\lambda = 3$, $(3I - A)X = 0$. Therefore,

$$\begin{bmatrix} -1 & 1 & -3 \\ 0 & 1 & -1 \\ 0 & 0 & 0 \end{bmatrix} \begin{bmatrix} x \\ y \\ z \end{bmatrix} = \begin{bmatrix} 0 \\ 0 \\ 0 \end{bmatrix}$$

which has the solution

$$\begin{bmatrix} x \\ y \\ z \end{bmatrix} = \begin{bmatrix} -2t \\ t \\ t \end{bmatrix} = t \begin{bmatrix} -2 \\ 1 \\ 1 \end{bmatrix}, \quad t \neq 0$$

Therefore, $\begin{bmatrix} -2 \\ 1 \\ 1 \end{bmatrix}$ is a basis for the eigenvectors association with the eigenvalue 3. For $\lambda = 2$, $(2I - A)X = 0$. Therefore,

$$\begin{bmatrix} -2 & 1 & -3 \\ 0 & 0 & -1 \\ 0 & 0 & -1 \end{bmatrix} \begin{bmatrix} x \\ y \\ z \end{bmatrix} = \begin{bmatrix} 0 \\ 0 \\ 0 \end{bmatrix}$$

which has the solution

$$\begin{bmatrix} x \\ y \\ z \end{bmatrix} = \begin{bmatrix} t \\ 2t \\ 0 \end{bmatrix} = t\begin{bmatrix} 1 \\ 2 \\ 0 \end{bmatrix}, \qquad t \neq 0$$

Therefore, $\begin{bmatrix} 1 \\ 2 \\ 0 \end{bmatrix}$ is a basis for the eigenvectors assocation with the eigenvalue 2. For $\lambda = 4$, $(4I - A)X = 0$. Therefore,

$$\begin{bmatrix} 0 & 1 & -3 \\ 0 & 2 & -1 \\ 0 & 0 & 1 \end{bmatrix}\begin{bmatrix} x \\ y \\ z \end{bmatrix} = \begin{bmatrix} 0 \\ 0 \\ 0 \end{bmatrix}$$

which has the solution

$$\begin{bmatrix} x \\ y \\ z \end{bmatrix} = \begin{bmatrix} t \\ 0 \\ 0 \end{bmatrix} = t\begin{bmatrix} 1 \\ 0 \\ 0 \end{bmatrix}, \qquad t \neq 0$$

Therefore, $\begin{bmatrix} 1 \\ 0 \\ 0 \end{bmatrix}$ is a basis for the eigenvectors assocation with the eigenvalue 4.

4. Matrix A is diagonalizable because there are three distinct eigenvalues.

5. P is the matrix having basis eigenvectors associated with the eigenvalues 3, 2, 4 as corresponding columns.

Therefore, $P = \begin{bmatrix} -2 & 1 & 1 \\ 1 & 2 & 0 \\ 1 & 0 & 0 \end{bmatrix}$.

6. First find P^{-1}.

$$\left[\begin{array}{ccc|ccc} -2 & 1 & 1 & 1 & 0 & 0 \\ 1 & 2 & 0 & 0 & 1 & 0 \\ 1 & 0 & 0 & 0 & 0 & 1 \end{array}\right]$$

$$\xrightarrow{R_1 \leftrightarrow R_3} \left[\begin{array}{ccc|ccc} 1 & 0 & 0 & 0 & 0 & 1 \\ 1 & 2 & 0 & 0 & 1 & 0 \\ -2 & 1 & 1 & 1 & 0 & 0 \end{array}\right]$$

$$\xrightarrow[R_3 \to R_3 + 2R_1]{R_2 \to R_2 - R_1} \left[\begin{array}{ccc|ccc} 1 & 0 & 0 & 0 & 0 & 1 \\ 0 & 2 & 0 & 0 & 1 & -1 \\ 0 & 1 & 1 & 1 & 0 & 2 \end{array}\right]$$

$$\xrightarrow{R_2 \to 1/2 R_2} \left[\begin{array}{ccc|ccc} 1 & 0 & 0 & 0 & 0 & 1 \\ 0 & 1 & 0 & 0 & \frac{1}{2} & -\frac{1}{2} \\ 0 & 1 & 1 & 1 & 0 & 2 \end{array}\right]$$

$$\xrightarrow[R_3 \to R_3 - R_2]{} \left[\begin{array}{ccc|ccc} 1 & 0 & 0 & 0 & 0 & 1 \\ 0 & 1 & 0 & 0 & \frac{1}{2} & -\frac{1}{2} \\ 0 & 0 & 1 & 1 & -\frac{1}{2} & \frac{5}{2} \end{array}\right]$$

$$\therefore \quad P^{-1} = \begin{bmatrix} 0 & 0 & 1 \\ 0 & \frac{1}{2} & -\frac{1}{2} \\ 1 & -\frac{1}{2} & \frac{5}{2} \end{bmatrix}$$

$$\therefore \quad P^{-1}AP = \begin{bmatrix} 0 & 0 & 1 \\ 0 & \frac{1}{2} & -\frac{1}{2} \\ 1 & -\frac{1}{2} & \frac{5}{2} \end{bmatrix}\begin{bmatrix} 4 & -1 & 3 \\ 0 & 2 & 1 \\ 0 & 0 & 3 \end{bmatrix}\begin{bmatrix} -2 & 1 & 1 \\ 1 & 2 & 0 \\ 1 & 0 & 0 \end{bmatrix}$$

$$= \begin{bmatrix} 3 & 0 & 0 \\ 0 & 2 & 0 \\ 0 & 0 & 4 \end{bmatrix}$$

7. The matrix form of the quadratic equation is

$$[x \quad y]\begin{bmatrix} 16 & 12 \\ 12 & 9 \end{bmatrix}\begin{bmatrix} x \\ y \end{bmatrix} + [5 \quad -5]\begin{bmatrix} x \\ y \end{bmatrix} = 0$$

Thus, $A = \begin{bmatrix} 16 & 12 \\ 12 & 9 \end{bmatrix}$.

$$\det(\lambda I - A) = \begin{vmatrix} \lambda - 16 & -12 \\ -12 & \lambda - 9 \end{vmatrix} = 0$$

$$\lambda^2 - 25\lambda + 144 - 144 = 0$$

$$\lambda^2 - 25\lambda = 0$$

$$\lambda(\lambda - 25) = 0$$

Thus, the eigenvalues are 0 and 25. For $\lambda = 0$, we get

$$\begin{bmatrix} -16 & -12 \\ -12 & -9 \end{bmatrix}\begin{bmatrix} a_1 \\ a_2 \end{bmatrix} = \begin{bmatrix} 0 \\ 0 \end{bmatrix}$$

which has the solution

$$\begin{bmatrix} a_1 \\ a_2 \end{bmatrix} = \begin{bmatrix} 3t \\ -4t \end{bmatrix} = t\begin{bmatrix} 3 \\ -4 \end{bmatrix}$$

For $\lambda = 25$, we get

$$\begin{bmatrix} 9 & -12 \\ -12 & 16 \end{bmatrix}\begin{bmatrix} a_1 \\ a_2 \end{bmatrix} = \begin{bmatrix} 0 \\ 0 \end{bmatrix}$$

which has the solution

$$\begin{bmatrix} a_1 \\ a_2 \end{bmatrix} = \begin{bmatrix} 4t \\ 3t \end{bmatrix} = t\begin{bmatrix} 4 \\ 3 \end{bmatrix}$$

$\begin{bmatrix} 3 \\ -4 \end{bmatrix}$ and $\begin{bmatrix} 4 \\ 3 \end{bmatrix}$ are orthogonal, but we must change these to unit vectors. Therefore, the matrix P is $\begin{bmatrix} 3/5 & 4/5 \\ -4/5 & 3/5 \end{bmatrix}$. $\text{Det}(P) = 1$. Therefore,

$$[x' \quad y']\begin{bmatrix} 0 & 0 \\ 0 & 25 \end{bmatrix}\begin{bmatrix} x' \\ y' \end{bmatrix}$$

$$+ [5 \quad -5]\begin{bmatrix} \frac{3}{5} & \frac{4}{5} \\ -\frac{4}{5} & \frac{3}{5} \end{bmatrix}\begin{bmatrix} x' \\ y' \end{bmatrix} = 0$$

$$\therefore \quad [0 \quad 25y'] \begin{bmatrix} x' \\ y' \end{bmatrix} + [7 \quad 1] \begin{bmatrix} x' \\ y' \end{bmatrix} = 0$$

$25(y')^2 + 7x' + y' = 0$ or, completing the square, we get $(y' + 1/50)^2 = -7/25(x' - 1/700)$ which is the standard form of a parabola.

8. $\begin{bmatrix} 1 & 0 \\ 0 & 1/2 \end{bmatrix}$ compresses by a factor of 1/2 in the y-direction. $\begin{bmatrix} 3 & 0 \\ 0 & 1 \end{bmatrix}$ expands by a factor of 3 in the x-direction. Therefore,

$$\begin{bmatrix} 3 & 0 \\ 0 & 1 \end{bmatrix} \begin{bmatrix} 1 & 0 \\ 0 & \frac{1}{2} \end{bmatrix} = \begin{bmatrix} 3 & 0 \\ 0 & \frac{1}{2} \end{bmatrix}$$

9. Express A as a product of elementary matrices.

$$\begin{bmatrix} 1 & 1 & 0 \\ 0 & 1 & 0 \\ 0 & 0 & 2 \end{bmatrix} = \begin{bmatrix} 1 & 0 & 0 \\ 0 & 1 & 0 \\ 0 & 0 & 2 \end{bmatrix} \begin{bmatrix} 1 & 1 & 0 \\ 0 & 1 & 0 \\ 0 & 0 & 1 \end{bmatrix} = E_2 E_1$$

The matrix on the right represents a shear that moves each point parallel to the x-axis by an amount equal to its distance from the xz-plane. The matrix on the left represents an expansion by a factor of 2 in the z-direction. Either one may be performed first.

19 COMPUTER APPLICATIONS

THIS CHAPTER IS ABOUT

☑ Matrix Statements
☑ Selected Programs

This chapter is for those readers who use computers for computations in linear algebra. The Basic language is used. There are variations in the different versions of Basic, but the reader should be able to make the necessary changes after reference to the computer manual.

19-1. Matrix Statements

A. Matrix declaration

(1) DIM A(3, 4), B(2, 3) — This declares A to be a matrix with three rows and four columns and B to be a matrix with two rows and three columns.

(2) DIM A(M, N) — This declares A to be a matrix with M rows and N columns. When it is used, READ M, N and DATA with values for M and N should be included.

B. Matrix entry

(1) MAT READ A, B — If A is 3×4 and B is 2×3, then the first 12 data items are read into A on a row-to-row basis. The next six data items are read into B.

(2) MAT INPUT A — This allows the user to enter the desired matrix while the program is running.

C. Matrix output

(1) MAT PRINT A; — If A is 3×4, a semicolon indicates that A should be printed in 3 rows and 4 columns in packed format. If each entry in A contains the same number of digits, then this is very good. If the number of digits in entries varies, then the columns are not lined up.

(2) MAT PRINT A, — The comma assures that A is printed by rows and columns but only one number would appear in each print zone. They appear in columns with left adjustment.

(3) PRINT TAB (J * 6 − 5); C(I, J);

This permits the user to choose the width of each print zone. The 6 denotes the width of each zone and $J(6) - 5$ determines the beginning of each zone. Positive integers a and b may be used instead of 5 and 6 as long as $J(b) - a$ is positive for all values of J.

D. Matrix initialization

(1) MAT A = ZER — This sets all the entries in an existing $M \times N$ matrix A to zero.
(2) MAT A = IDN — This changes an existing square matrix A to the identity.
(3) MAT A = CON — This sets all the entries in an existing $M \times N$ matrix A to one.
(4) MAT B = ZER(3, 4) — This generates a 3×4 matrix B in which all entries are zeros.
(5) MAT B = IDN(4, 4) — This generates a 4×4 identity matrix B.
(6) MAT B = CON(5, 4) — This generates a 5×4 matrix B in which all entries are ones.

E. Matrix arithmetic

(1) MAT C = A	This replaces all the entries in matrix C by the entries in matrix A.
(2) MAT C = A + B	This adds matrices A and B and stores the result in C.
(3) MAT C = A − B	This subtracts matrix B from matrix A and stores the result in C.
(4) MAT C = A * B	This computes the product AB and stores the result in C.
(5) MAT C = (K) * A	This multiplies each entry in A by the real number K and stores the result in C.
(6) MAT C = INV(A)	This computes the inverse of matrix A and stores the result in C.
(7) MAT C = TRN(A)	This computes the transpose of matrix A and stores the result in C.

19-2. Selected Programs

EXAMPLE 19-1: Write a program without remark statements to find the sum of matrices A and B and print A, B and $A + B$.

$$A = \begin{bmatrix} 1 & 2 & 3 & 4 \\ 5 & 6 & 7 & 8 \\ 9 & 1 & 2 & 3 \end{bmatrix} \qquad B = \begin{bmatrix} 9 & 8 & 7 & 6 \\ 5 & 4 & 3 & 2 \\ 1 & -9 & -8 & -7 \end{bmatrix}$$

Solution

```
100  READ M, N
105  DIM A(M, N), B(M, N)
110  MAT READ A, B
115  MAT C = A + B
120  PRINT "MATRIX A IS"
125  PRINT
130  MAT PRINT A,
135  PRINT
140  PRINT "MATIX B IS"
145  PRINT
150  MAT PRINT B,
155  PRINT
160  PRINT "MATRIX A +B IS"
165  PRINT
170  MAT PRINT C,
175  DATA 3, 4
180  DATA 1, 2, 3, 4, 5, 6, 7, 8, 9, 1, 2, 3
185  DATA 9, 8, 7, 6, 5, 4, 3, 2, 1, −9, −8, −7
999  END
RUN

MATRIX A IS

1  2  3  4
5  6  7  8
9  1  2  3

MATRIX B IS

9   8   7   6
5   4   3   2
1  −9  −8  −7
```

```
MATRIX A + B IS

10   10   10   10
10   10   10   10
10   -8   -6   -4
```

EXAMPLE 19-2: Write a program without remark statements to find the product AB and print A, B and AB where matrices A and B are as follows.

$$A = \begin{bmatrix} 1 & 0 & 2 & 3 & 4 \\ 3 & 1 & -1 & 2 & 3 \\ 2 & 1 & 0 & 0 & 2 \\ 1 & 2 & 3 & 4 & 1 \end{bmatrix} \qquad B = \begin{bmatrix} 1 & 2 & 3 \\ 2 & 1 & -1 \\ 3 & 2 & 1 \\ -2 & 0 & 3 \\ 1 & -2 & -2 \end{bmatrix}$$

Solution

```
100  READ M, N, P
105  DIM A(M, N), B(N, P), C(M, P)
110  MAT READ A, B
115  MAT C = A * B
120  PRINT "MATRIX A IS"
125  PRINT
130  MAT PRINT A,
135  PRINT
140  PRINT "MATRIX B IS"
145  PRINT
150  MAT PRINT B,
155  PRINT
160  PRINT "MATRIX AB IS"
165  PRINT
170  MAT PRINT C,
175  DATA 4, 5, 3
180  DATA 1, 0, 2, 3, 4, 3, 1, -1, 2, 3, 2, 1, 0, 0, 2, 1, 2, 3, 4, 1
185  DATA 1, 2, 3, 2, 1, -1, 3, 2, 1, -2, 0, 3, 1, -2, -2
999  END
RUN

MATRIX A IS

1  0   2  3  4
3  1  -1  2  3
2  1   0  0  2
1  2   3  4  1

MATRIX B IS

  1   2   3
  2   1  -1
  3   2   1
 -2   0   3
  1  -2  -2
```

```
MATRIX AB IS

5 -2  6
1 -1  7
6  1  1
7  8 14
```

EXAMPLE 19-3: Write a program without remark statements to find A^{-1} and the product of A with A^{-1}, then print each of these where A is the invertible matrix.

$$A = \begin{bmatrix} 2 & 1 & 1 \\ 1 & 2 & 3 \\ 1 & 0 & 2 \end{bmatrix}$$

Solution

```
100  READ N
105  DIM A(N, N), B(N, N), C(N, N)
110  MAT READ A
115  MAT B = INV(A)
120  MAT C = A * B
125  PRINT "MATRIX A IS"
130  PRINT
135  MAT PRINT A,
140  PRINT
145  PRINT "THE INVERSE OF A IS"
150  PRINT
155  MAT PRINT B,
160  PRINT
165  PRINT "THE PRODUCT OF A AND A INVERSE IS"
170  PRINT
175  MAT PRINT C,
180  DATA 3
185  DATA 2, 1, 1, 1, 2, 3, 1, 0, 2
999  END
RUN

MATRIX A IS

2  1  1
1  2  3
1  0  2

THE INVERSE OF A IS

 .571429  -.285714   .142857
 .142857   .428571  -.714286
-.285714   .142857   .428571

THE PRODUCT OF A AND A INVERSE IS

1             .298023E-07  0
.596046E-07   1            0
.596046E-07   .596046E-07  1
```

note: Computer computations often result in decimal approximations. The exact values in this problem are

$$A^{-1} = \begin{bmatrix} \frac{4}{7} & -\frac{2}{7} & \frac{1}{7} \\ \frac{1}{7} & \frac{3}{7} & -\frac{5}{7} \\ -\frac{2}{7} & \frac{1}{7} & \frac{3}{7} \end{bmatrix} \qquad AA^{-1} = \begin{bmatrix} 1 & 0 & 0 \\ 0 & 1 & 0 \\ 0 & 0 & 1 \end{bmatrix}$$

EXAMPLE 19-4: Write a program to evaluate the determinant of the invertible matrix A where

$$A = \begin{bmatrix} 1 & 2 & 3 \\ 3 & -2 & 2 \\ 5 & 4 & 3 \end{bmatrix}$$

Solution

```
100  READ N
105  DIM A(N, N)
110  MAT READ A
115  MAT B = INV(A)
120  PRINT "THE DETERMINANT OF A IS", DET
125  DATA 3
130  DATA 1, 2, 3, 3, -2, 2, 5, 4, 3
999  END
RUN

THE DETERMINANT OF A IS   54
```

note: Some computer systems that recognize the DET function, as in this example, require that it be used after the MAT INV(A) statement. Furthermore, some systems require an argument such as DET(A). It would also be valuable to recall that if a square matrix is not invertible, then its determinant is zero.

EXAMPLE 19-5: Write a program to reduce matrix A to an equivalent upper triangular form and evaluate the determinant of A.

$$A = \begin{bmatrix} 4 & 5 & 6 \\ 2 & 0 & -1 \\ 3 & 1 & 4 \end{bmatrix}$$

Solution

```
100  READ N
105  DIM A(N, N)
110  MAT READ A
115  PRINT "MATRIX A IS"
120  PRINT
125  MAT PRINT A,
130  PRINT
135  REM IDENTIFY NONZERO LEADING ENTRIES
140  LET C = 0
145  LET S = 0
150  LET W = 1
155    LET C = C + 1
160    IF C > N OR C = N THEN 270
165    FOR R = C TO N
170      IF A(R, C) < 0 OR A(R, C) > 0 THEN 190
175      IF R = N THEN 155
180    NEXT R
185  REM INTERCHANGE ROWS IF NECESSARY
190  IF R = C THEN 230
195  LET S = S + 1
200  FOR J = C TO N
205    LET B = A(R, J)
210    LET A(R, J) = A(C, J)
```

```
215     LET A(C, J) = B
220   NEXT J
225   REM CHANGE ENTRIES BELOW LEADING ENTRIES TO ZEROS
230   FOR I = C + 1 TO N
235     IF A(I, C) = 0 THEN 260
240     LET X = A(I, C)/A(C, C)
245     FOR K = C TO N
250       LET A(I, K) = A(I, K) - (X) * A(R, K)
255     NEXT K
260   NEXT I
265   GOTO 155
270   FOR Q = 1 TO N
275     LET W = W * A(Q, Q)
280   NEXT Q
285   LET D = (-1) ^S * W
290   PRINT "AN UPPER TRIANGULAR FORM OF A IS"
295   PRINT
300   MAT PRINT A,
305   PRINT
310   PRINT "THE DETERMINANT OF A IS", D
315   DATA 3
320   DATA 4, 5, 6, 2, 0, -1, 3, 1, 4
999   END
RUN

MATRIX A IS

4  5   6
2  0  -1
3  1   4

AN UPPER TRIANGULAR FORM OF A IS

4    5     6
0  -2.5   -4
0    0     3.9

THE DETERMINANT OF A IS   -39
```

note: This method works on any square matrix whether singular or nonsingular.

EXAMPLE 19-6: Write a program to solve the following system of equations where the coefficient matrix A is invertible.

$$\begin{aligned} 2x + y - z &= -30 \\ x + 3y + 2z &= 40 \\ -x + 2y + z &= 10 \end{aligned}$$

Solution

```
100   REM A, B, X AND D ARE THE COEFFICIENT MATRIX, COLUMN OF
105   REM CONSTANTS, SOLUTION AND INVERSE OF A, RESPECTIVELY
110   READ N
115   DIM A(N, N), B(N, 1), X(N, 1), D(N, N)
120   MAT READ A
125   PRINT "THE COEFFICIENT MATRIX IS"
130   PRINT
135   MAT PRINT A,
140   PRINT
145   MAT READ B
```

```
150  PRINT "THE COLUMN OF CONSTANTS IS"
155  PRINT
160  MAT PRINT B,
165  PRINT
170  MAT D = INV(A)
175  MAT X = D * B
180  PRINT "THE SOLUTION IS"
185  PRINT
190  MAT PRINT X,
195  DATA 3
200  DATA 2, 1, -1, 1, 3, 2, -1, 2, 1
205  DATA -30, 40, 10
999  END
RUN

THE COEFFICIENT MATRIX IS

  2  1  -1
  1  3   2
 -1  2   1

THE COLUMN OF CONSTANTS IS

-30
 40
 10

THE SOLUTION IS

  4
 -8
 30
```

note: If the coefficient matrix is not invertible, then solve by Gauss–Jordan elimination by changing the augmented matrix to reduced row echelon form as in Example 19-7.

EXAMPLE 19-7: Write a program to change matrix A to reduced row echelon form, find its rank, and indicate whether it is invertible.

$$A = \begin{bmatrix} 1 & 3 & -6 & 7 \\ 2 & -1 & 2 & 0 \\ 3 & 2 & -4 & 7 \end{bmatrix}$$

Solution

```
100  READ M, N
105  DIM A(M, N)
110  MAT READ A
115  PRINT "MATRIX A IS"
120  PRINT
125  MAT PRINT A,
130  PRINT
135  REM IDENTIFY NONZERO LEADING ENTRIES
140  LET C = 0
145  LET S = 0
150    LET C = C + 1
155    IF C > N THEN 300
160    IF S = M THEN 300
165    FOR R = S + 1 TO M
170      IF A(R, C) < 0 OR A(R, C) > 0 THEN 195
```

```
175          IF R = M THEN 150
180       NEXT R
185    REM INTERCHANGE ROWS IF NECESSARY
190    REM DETERMINE RANK AND CHANGE LEADING ENTRIES TO ONES
195       LET S = S + 1
200       FOR J = C TO N
205          LET B = A(R, J)
210          LET A(R, J) = A(S, J)
215          LET A(S, J) = B
220       NEXT J
225       IF A(S, C) = 1 THEN 255
230       LET D = A(S, C)
235       FOR G = C TO N
240          LET A(S, Q) = A(S, Q)/D
245       NEXT Q
250    REM CHANGE ENTRIES ABOVE AND BELOW LEADING ENTRIES TO ZEROS
255       FOR I = 1 TO M
260          IF A(I, C) = 0 THEN 290
265          IF I = S THEN 290
270          LET X = A(I, C)
275          FOR K = C TO N
280             LET A(I, K) = A(I, K) - (X) * A(S, K)
285          NEXT K
290       NEXT I
295    GOTO 150
300    PRINT "THE REDUCED ROW ECHELON FORM OF A IS"
305    PRINT
310    MAT PRINT A,
315    PRINT
320    PRINT "THE RANK OF MATRIX A IS", S
325    PRINT
330    IF M - N < 0 OR M - N > 0 THEN 350
335    IF M - S > 0 THEN 360
340    PRINT "MATRIX A IS INVERTIBLE"
345    GOTO 999
350    PRINT "A IS NOT INVERTIBLE BECAUSE IT IS NOT SQUARE"
355    GOTO 999
360    PRINT "A IS NOT INVERTIBLE BECAUSE ITS RANKS IS LESS THAN M"
365    DATA 3, 4
370    DATA 1, 3, -6, 7, 2, -1, 2, 0, 3, 2, -4, 7
999    RUN
RUN

MATRIX A IS

1   3  -6   7
2  -1   2   0
3   2  -4   7

THE REDUCED ROW ECHELON FORM OF A IS

1   0   0   1
0   1  -2   2
0   0   0   0

THE RANK OF MATRIX A IS   2

A IS NOT INVERTIBLE BECAUSE IT IS NOT SQUARE
```

EXAMPLE 19-8: Write a program to find the coordinate vector of $[1 \quad 2 \quad 3]$ with respect to the ordered basis $\{[1 \quad 0 \quad 1], [0 \quad 1 \quad 1], [1 \quad 1 \quad 0]\}$.

Solution

```
100  REM N IS THE DIMENSION OF THE VECTOR SPACE
105  READ N
110  DIM A(N, N), B(N, 1), C(N, 1), D(N, N), T(N, N)
115  MAT READ A
120  PRINT "THE ORDERED BASIS IN COLUMN FORM IS"
125  PRINT
130  MAT T = TRN(A)
135  MAT PRINT T,
140  PRINT
145  MAT READ B
150  PRINT "THE GIVEN VECTOR IS"
155  PRINT
160  MAT PRINT B,
165  PRINT
170  MAT D = INV(T)
175  MAT C = D * B
180  PRINT "THE COORDINATE VECTOR IS"
185  PRINT
190  MAT PRINT C,
195  REM THE FIRST LINE OF DATA IS THE DIMENSION OF THE
200  REM VECTOR SPACE. THE NEXT N ROWS ARE THE ORDERED
205  REM BASIS VECTORS. THE LAST ROW IS THE GIVEN VECTOR.
210  DATA 3
215  DATA 1, 0, 1
220  DATA 0, 1, 1
225  DATA 1, 1, 0
230  DATA 1, 2, 3
999  END
RUN

THE ORDERED BASIS IN COLUMN FORM IS

1  0  1
0  1  1
1  1  0

THE GIVEN VECTOR IS

1
2
3

THE COORDINATE VECTOR IS

1
2
0
```

EXAMPLE 19-9: Write a program to find the transition matrix from basis B to basis S where

$$B = \{[2 \quad 1 \quad 1], [1 \quad 7 \quad 7], [4 \quad -1 \quad 0]\}$$

and

$$S = \{[0 \quad 1 \quad 1], [1 \quad 1 \quad 0], [1 \quad 0 \quad 1]\}.$$

Solution

```
100  READ N
105  DIM B(N, N), S(N, N), T(N, N), R(N, N), P(N, N), D(N, N), A(N, 1), C(N, 1)
110  MAT READ B
115  PRINT "THE ORDERED BASIS B IN COLUMN FORM IS"
120  PRINT
125  MAT T = TRN(B)
130  MAT PRINT T,
135  PRINT
140  MAT READ S
145  PRINT "THE ORDERED BASIS S IN COLUMN FORM IS"
150  PRINT
155  MAT R = TRN(S)
160  MAT PRINT R,
165  PRINT
170  PRINT "THE TRANSITION MATRIX FROM B TO S IS"
175  PRINT
180  MAT D = INV(R)
185  FOR J = 1 TO N
190     FOR I = 1 TO N
195        A(I, 1) = T(I, J)
200     NEXT I
205     MAT C = D * A
210     FOR I = 1 TO N
215     P(I, J) = C(I, 1)
220     NEXT I
225  NEXT J
230  MAT PRINT P,
235  REM THE FIRST LINE OF DATA IS THE DIMENSION OF THE VECTOR SPACE.
240  REM THE NEXT N ROWS ARE THE ORDERED BASIS VECTORS IN B.
245  REM THE LAST N ROWS ARE THE ORDERED BASIS VECTORS IN S.
250  DATA 3
255  DATA 2, 1, 1
260  DATA 1, 7, 7
265  DATA 4, -1, 0
270  DATA 0, 1, 1
275  DATA 1, 1, 0
280  DATA 1, 0, 1
999  END
RUN

THE ORDERED BASIS B IN COLUMN FORM IS

2  1   4
1  7  -1
1  7   0

THE ORDERED BASIS S IN COLUMN FORM IS

0  1  1
1  1  0
1  0  1

THE TRANSITION MATRIX FROM B TO S IS

0  6.5  -2.5
1   .5   1.5
1   .5   2.5
```

FINAL EXAM (Chapters 1–18)

1. Solve the following system by Gauss–Jordan elimination.

$$\begin{aligned} x + 2y - z &= 6 \\ 2x - y + 3z &= -13 \\ 3x - 2y + 3z &= -16 \end{aligned}$$

2. Find A^{-1} where $A = \begin{bmatrix} 1 & 3 & 1 \\ 0 & 1 & 2 \\ 1 & 0 & 1 \end{bmatrix}$.

3. Let A be a 5×5 matrix.

(a) If $A \xrightarrow{R_1 \leftrightarrow R_2} B$ and $\det(A) = 24$, then $\det(B) =$ ____________.
(b) If $A \xrightarrow{R_3 \to R_3 - 5R_2} B$ and $\det(A) = 29$, then $\det(B) =$ ____________.
(c) If $A \xrightarrow{R_2 \to 3R_2} B$ and $\det(A) = 24$, then $\det(B) =$ ____________.
(d) If A and B are similar and $\det(A) = 13$, then $\det(B) =$ ____________.
(e) If $a_{1j} = a_{2j}$ for $j = 1, 2, 3, 4, 5$ then $\det(A) =$ ____________.

4. Verify that the set of 2×2 matrices for form $\begin{bmatrix} a & 0 \\ 0 & b \end{bmatrix}$ is a subspace of the vector space of all 2×2 matrices.

5. If R_3 has the standard inner product, find $\langle \mathbf{a}, \mathbf{b} \rangle$ where $\mathbf{a} = [2 \quad 1 \quad 3]$ and $\mathbf{b} = [1 \quad 0 \quad 2]$.

6. If R_3 has the inner product $\langle \mathbf{a}, \mathbf{b} \rangle = a_1 b_1 + 2a_2 b_2 + 3a_3 b_3$, find $\|\mathbf{a}\|$, where $\mathbf{a} = [2 \quad 1 \quad 3]$.

7. If $\mathbf{a} = [2 \quad 1 \quad 3]$ and $\mathbf{b} = [1 \quad 0 \quad 2]$, find $\mathbf{a} \times \mathbf{b}$.

8. If $\mathbf{a} = [2 \quad 1 \quad 3]$, $\mathbf{b} = [1 \quad 0 \quad 2]$, and $\mathbf{c} = [1 \quad -2 \quad 0]$, find $\mathbf{a} \cdot (\mathbf{b} \times \mathbf{c})$.

9. Find a basis for V where V is the subspace of R_3 spanned by $\{[2 \quad 0 \quad 6], [0 \quad -1 \quad 2], [2 \quad 1 \quad 4], [1 \quad 1 \quad 1]\}$.

10. If $L: P_2 \to R_2$ is defined by $L(ax^2 + bx + c) = [a + b \quad 2c - a]$, show that L is linear.

11. If A is the matrix representation of the linear transformation L in question 10 with respect to the natural bases $S = \{x^2, x, 1\}$ and $T = \{[1 \quad 0], [0 \quad 1]\}$, find A.

12. Let $L: R^3 \to R^3$ be a linear transformation with ordered bases

$$S = \left\{ \begin{bmatrix} 2 \\ 0 \\ 1 \end{bmatrix}, \begin{bmatrix} 1 \\ 2 \\ 0 \end{bmatrix}, \begin{bmatrix} 1 \\ 1 \\ 1 \end{bmatrix} \right\} \quad \text{and} \quad T = \left\{ \begin{bmatrix} 6 \\ 3 \\ 3 \end{bmatrix}, \begin{bmatrix} 4 \\ -1 \\ 3 \end{bmatrix}, \begin{bmatrix} 5 \\ 5 \\ 2 \end{bmatrix} \right\}$$

Find the transition matrix from T to S.

13. If in question 12 the coordinate vector of $\mathbf{a}$ relative to the T basis is $\begin{bmatrix} 1 \\ 2 \\ -2 \end{bmatrix}$, find the coordinate vector of $\mathbf{a}$ with respect to S.

14. Let $L: R^5 \to R^4$ be defined by

$$L\left(\begin{bmatrix} a_1 \\ a_2 \\ a_3 \\ a_4 \\ a_5 \end{bmatrix}\right) = \begin{bmatrix} 0 & 0 & 2 & 0 & 2 \\ 0 & -1 & 0 & 0 & 1 \\ 0 & 1 & 0 & 1 & 1 \\ 0 & 5 & 0 & 1 & -3 \end{bmatrix} \begin{bmatrix} a_1 \\ a_2 \\ a_3 \\ a_4 \\ a_5 \end{bmatrix}$$

Find a basis for the kernel of L.

15. In question 14, find a basis for the range of L.

16. Let R^2 have the standard inner product. Use the Gram–Schmidt process to transform the basis $\left\{\begin{bmatrix} 1 \\ 2 \end{bmatrix}, \begin{bmatrix} -3 \\ 4 \end{bmatrix}\right\}$ into an orthonormal basis.

17. Find the eigenvalues of A if $A = \begin{bmatrix} 1 & 6 \\ 1 & 2 \end{bmatrix}$.

18. Find the bases for the eigenspaces associated with the eigenvalues in question 17.

19. Find a matrix P that diagonalizes A in question 17.

20. Find P^{-1} and then compute $P^{-1}AP$ to show that $P^{-1}AP$ is diagonal for A and P of questions 17 and 19.

Solutions to Final Exam

1. $$\begin{bmatrix} 1 & 2 & -1 & 6 \\ 2 & -1 & 3 & -13 \\ 3 & -2 & 3 & -16 \end{bmatrix}$$

$$\xrightarrow[R_3 \to R_3 - 3R_1]{R_2 \to R_2 - 2R_1} \begin{bmatrix} 1 & 2 & -1 & 6 \\ 0 & -5 & 5 & -25 \\ 0 & -8 & 6 & -34 \end{bmatrix}$$

$$\xrightarrow[R_3 \to 1/2R_3]{R_2 \to -1/5R_2} \begin{bmatrix} 1 & 2 & -1 & 6 \\ 0 & 1 & -1 & 5 \\ 0 & -4 & 3 & -17 \end{bmatrix}$$

$$\xrightarrow[R_3 \to R_3 + 4R_2]{} \begin{bmatrix} 1 & 2 & -1 & 6 \\ 0 & 1 & -1 & 5 \\ 0 & 0 & -1 & 3 \end{bmatrix}$$

$$\xrightarrow[R_3 \to -R_3]{} \begin{bmatrix} 1 & 2 & -1 & 6 \\ 0 & 1 & -1 & 5 \\ 0 & 0 & 1 & -3 \end{bmatrix}$$

$$\xrightarrow[R_2 \to R_2 + R_3]{R_1 \to R_1 + R_3} \begin{bmatrix} 1 & 2 & 0 & 3 \\ 0 & 1 & 0 & 2 \\ 0 & 0 & 1 & -3 \end{bmatrix}$$

$$\xrightarrow{R_1 \to R_1 - 2R_2} \begin{bmatrix} 1 & 0 & 0 & -1 \\ 0 & 1 & 0 & 2 \\ 0 & 0 & 1 & -3 \end{bmatrix}$$

Therefore, the solution is $\begin{bmatrix} x \\ y \\ z \end{bmatrix} = \begin{bmatrix} -1 \\ 2 \\ -3 \end{bmatrix}$.

2. $$\left[\begin{array}{ccc|ccc} 1 & 3 & 1 & 1 & 0 & 0 \\ 0 & 1 & 2 & 0 & 1 & 0 \\ 1 & 0 & 1 & 0 & 0 & 1 \end{array}\right]$$

$$\xrightarrow{R_3 \to R_3 - R_1} \left[\begin{array}{ccc|ccc} 1 & 3 & 1 & 1 & 0 & 0 \\ 0 & 1 & 2 & 0 & 1 & 0 \\ 0 & -3 & 0 & -1 & 0 & 1 \end{array}\right]$$

$$\xrightarrow{R_3 \to R_3 + 3R_2} \left[\begin{array}{ccc|ccc} 1 & 3 & 1 & 1 & 0 & 0 \\ 0 & 1 & 2 & 0 & 1 & 0 \\ 0 & 0 & 6 & -1 & 3 & 1 \end{array}\right]$$

$$\xrightarrow{R_3 \to 1/6 R_3} \left[\begin{array}{ccc|ccc} 1 & 3 & 1 & 1 & 0 & 0 \\ 0 & 1 & 2 & 0 & 1 & 0 \\ 0 & 0 & 1 & -\frac{1}{6} & \frac{1}{2} & \frac{1}{6} \end{array}\right]$$

$$\xrightarrow[R_2 \to R_2 - 2R_3]{R_1 \to R_1 - R_3} \left[\begin{array}{ccc|ccc} 1 & 3 & 0 & \frac{7}{6} & -\frac{1}{2} & -\frac{1}{6} \\ 0 & 1 & 0 & \frac{1}{3} & 0 & -\frac{1}{3} \\ 0 & 0 & 1 & -\frac{1}{6} & \frac{1}{2} & \frac{1}{6} \end{array}\right]$$

$$\xrightarrow{R_1 \to R_1 - 3R_2} \left[\begin{array}{ccc|ccc} 1 & 0 & 0 & \frac{1}{6} & -\frac{1}{2} & \frac{5}{6} \\ 0 & 1 & 0 & \frac{1}{3} & 0 & -\frac{1}{3} \\ 0 & 0 & 1 & -\frac{1}{6} & \frac{1}{2} & \frac{1}{6} \end{array}\right]$$

$$\therefore \quad A^{-1} = \begin{bmatrix} \frac{1}{6} & -\frac{1}{2} & \frac{5}{6} \\ \frac{1}{3} & 0 & -\frac{1}{3} \\ -\frac{1}{6} & \frac{1}{2} & \frac{1}{6} \end{bmatrix}$$

3. **(a)** -24 **(b)** 29 **(c)** 72 **(d)** 13 **(e)** 0

4. $\begin{bmatrix} a_1 & 0 \\ 0 & b_1 \end{bmatrix} + \begin{bmatrix} a_2 & 0 \\ 0 & b_2 \end{bmatrix} = \begin{bmatrix} a_1 + a_2 & 0 \\ 0 & b_1 + b_2 \end{bmatrix}$. Therefore, it is closed under vector addition. $k\begin{bmatrix} a & 0 \\ 0 & b \end{bmatrix} = \begin{bmatrix} ka & 0 \\ 0 & kb \end{bmatrix}$. Therefore, it is closed under scalar multiplication. Thus it is a subspace.

5. $\langle \mathbf{a}, \mathbf{b} \rangle = 2(1) + 1(0) + 3(2) = 8$

6. $\|\mathbf{a}\| = \sqrt{\langle \mathbf{a}, \mathbf{a} \rangle} = \sqrt{2(2) + 2(1)(1) + 3(3)(3)} = \sqrt{33}$

7.
$$\mathbf{a} \times \mathbf{b} = \begin{vmatrix} \mathbf{i} & \mathbf{j} & \mathbf{k} \\ 2 & 1 & 3 \\ 1 & 0 & 2 \end{vmatrix} = \begin{vmatrix} 1 & 3 \\ 0 & 2 \end{vmatrix}\mathbf{i} - \begin{vmatrix} 2 & 3 \\ 1 & 2 \end{vmatrix}\mathbf{j} + \begin{vmatrix} 2 & 1 \\ 1 & 0 \end{vmatrix}\mathbf{k}$$
$$= 2\mathbf{i} - \mathbf{j} - \mathbf{k} = [2 \;\; -1 \;\; -1]$$

8.
$$\mathbf{a} \cdot (\mathbf{b} \times \mathbf{c}) = \begin{vmatrix} 2 & 1 & 3 \\ 1 & 0 & 2 \\ 1 & -2 & 0 \end{vmatrix} = 1\begin{vmatrix} 1 & 3 \\ 0 & 2 \end{vmatrix} + 2\begin{vmatrix} 2 & 3 \\ 1 & 2 \end{vmatrix} = 4$$

9. Construct the matrix having the given vectors as rows and reduce to echelon form. The vectors forming the nonzero rows will constitute a basis for V.

$$\begin{bmatrix} 2 & 0 & 6 \\ 0 & -1 & 2 \\ 2 & 1 & 4 \\ 1 & 1 & 1 \end{bmatrix} \xrightarrow{R_1 \leftrightarrow R_4} \begin{bmatrix} 1 & 1 & 1 \\ 0 & -1 & 2 \\ 2 & 1 & 4 \\ 2 & 0 & 6 \end{bmatrix}$$

$$\xrightarrow[R_4 \to R_4 - 2R_1]{R_3 \to R_3 - 2R_1} \begin{bmatrix} 1 & 1 & 1 \\ 0 & -1 & 2 \\ 0 & -1 & 2 \\ 0 & -2 & 4 \end{bmatrix}$$

$$\xrightarrow[\substack{R_3 \to R_3 - R_2 \\ R_4 \to R_4 - 2R_2}]{R_2 \to -R_2} \begin{bmatrix} 1 & 1 & 1 \\ 0 & 1 & -2 \\ 0 & 0 & 0 \\ 0 & 0 & 0 \end{bmatrix}$$

Therefore, a basis for V is $\{[1 \;\; 1 \;\; 1], [0 \;\; 1 \;\; -2]\}$.

10.
$$L[(a_1x^2 + b_1x + c_1) + (a_2x^2 + b_2x + c_2)]$$
$$= L[(a_1 + a_2)x^2 + (b_1 + b_2)x + (c_1 + c_2)]$$
$$= [(a_1 + a_2) + (b_1 + b_2) \quad 2(c_1 + c_2) - (a_1 + a_2)]$$
$$= [a_1 + b_1 \quad 2c_1 - a_1] + [a_2 + b_2 \quad 2c_2 - a_2]$$
$$= L(a_1x^2 + b_1x + c_1) + L(a_2x^2 + b_2x + c_2)$$
$$L[k(ax^2 + bx + c)] = L[(kax^2 + kbx + kc)]$$
$$= [ka + kb \quad 2kc - ka]$$
$$= k[a + b \quad 2c - a]$$
$$= kL[(ax^2 + bx + c)]$$

Therefore, L is linear.

11.
$$L(x^2) = [1 \;\; -1] = 1[1 \;\; 0] + (-1)[0 \;\; 1]$$
$$L(x) = [1 \;\; 0] = 1[1 \;\; 0] + 0[0 \;\; 1]$$
$$L(1) = [0 \;\; 2] = 0[1 \;\; 0] + 2[0 \;\; 1]$$
$$\therefore \quad [L(x^2)]_T = \begin{bmatrix} 1 \\ -1 \end{bmatrix}$$
$$[L(x)]_T = \begin{bmatrix} 1 \\ 0 \end{bmatrix}$$
$$[L(1)]_T = \begin{bmatrix} 0 \\ 2 \end{bmatrix}$$
$$\therefore \quad A = \begin{bmatrix} 1 & 1 & 0 \\ -1 & 0 & 2 \end{bmatrix}$$

12.
$$\begin{bmatrix} 6 \\ 3 \\ 3 \end{bmatrix} = c_1\begin{bmatrix} 2 \\ 0 \\ 1 \end{bmatrix} + c_2\begin{bmatrix} 1 \\ 2 \\ 0 \end{bmatrix} + c_3\begin{bmatrix} 1 \\ 1 \\ 1 \end{bmatrix}$$

which has the solution $c_1 = 2,\ c_2 = 1,\ c_3 = 1.$

Therefore, $\begin{bmatrix} 6 \\ 3 \\ 3 \end{bmatrix}_S = \begin{bmatrix} 2 \\ 1 \\ 1 \end{bmatrix}.$

$$\begin{bmatrix} 4 \\ -1 \\ 3 \end{bmatrix} = c_1 \begin{bmatrix} 2 \\ 0 \\ 1 \end{bmatrix} + c_2 \begin{bmatrix} 1 \\ 2 \\ 0 \end{bmatrix} + c_3 \begin{bmatrix} 1 \\ 1 \\ 1 \end{bmatrix}$$

which has the solution $c_1 = 2,\ c_2 = -1,\ c_3 = 1.$

Therefore, $\begin{bmatrix} 4 \\ -1 \\ 3 \end{bmatrix}_S = \begin{bmatrix} 2 \\ -1 \\ 1 \end{bmatrix}.$

$$\begin{bmatrix} 5 \\ 5 \\ 2 \end{bmatrix} = c_1 \begin{bmatrix} 2 \\ 0 \\ 1 \end{bmatrix} + c_2 \begin{bmatrix} 1 \\ 2 \\ 0 \end{bmatrix} + c_3 \begin{bmatrix} 1 \\ 1 \\ 1 \end{bmatrix}$$

which has the solution $c_1 = 1,\ c_2 = 2,\ c_3 = 1.$ Therefore,

$$\begin{bmatrix} 5 \\ 5 \\ 2 \end{bmatrix}_S = \begin{bmatrix} 1 \\ 2 \\ 1 \end{bmatrix}$$

$$\therefore \quad P_{T \text{ to } S} = \begin{bmatrix} 2 & 2 & 1 \\ 1 & -1 & 2 \\ 1 & 1 & 1 \end{bmatrix}$$

13. $$[\mathbf{a}]_S = P_{T \text{ to } S}[\mathbf{a}]_T = \begin{bmatrix} 2 & 2 & 1 \\ 1 & -1 & 2 \\ 1 & 1 & 1 \end{bmatrix} \begin{bmatrix} 1 \\ 2 \\ -2 \end{bmatrix} = \begin{bmatrix} 4 \\ -5 \\ 1 \end{bmatrix}$$

14. Augmented matrix $= \begin{bmatrix} 0 & 0 & 2 & 0 & 2 & 0 \\ 0 & -1 & 0 & 0 & 1 & 0 \\ 0 & 1 & 0 & 1 & 1 & 0 \\ 0 & 5 & 0 & 1 & -3 & 0 \end{bmatrix}$

$$\xrightarrow{R_1 \leftrightarrow R_3} \begin{bmatrix} 0 & 1 & 0 & 1 & 1 & 0 \\ 0 & -1 & 0 & 0 & 1 & 0 \\ 0 & 0 & 2 & 0 & 2 & 0 \\ 0 & 5 & 0 & 1 & -3 & 0 \end{bmatrix}$$

$$\xrightarrow[\substack{R_3 \to 1/2R_3 \\ R_4 \to R_4 - 5R_1}]{R_2 \to R_2 + R_1} \begin{bmatrix} 0 & 1 & 0 & 1 & 1 & 0 \\ 0 & 0 & 0 & 1 & 2 & 0 \\ 0 & 0 & 1 & 0 & 1 & 0 \\ 0 & 0 & 0 & -4 & -8 & 0 \end{bmatrix}$$

$$\xrightarrow[R_4 \to -1/4R_4]{R_2 \leftrightarrow R_3} \begin{bmatrix} 0 & 1 & 0 & 1 & 1 & 0 \\ 0 & 0 & 1 & 0 & 1 & 0 \\ 0 & 0 & 0 & 1 & 2 & 0 \\ 0 & 0 & 0 & 1 & 2 & 0 \end{bmatrix}$$

$$\xrightarrow[R_4 \to R_4 - R_3]{R_1 \to R_1 - R_3} \begin{bmatrix} 0 & 1 & 0 & 0 & -1 & 0 \\ 0 & 0 & 1 & 0 & 1 & 0 \\ 0 & 0 & 0 & 1 & 2 & 0 \\ 0 & 0 & 0 & 0 & 0 & 0 \end{bmatrix}$$

$$\therefore \quad a_2 - a_5 = 0$$
$$a_3 + a_5 = 0$$
$$a_4 + 2a_5 = 0$$

Hence,

$$a_2 = a_5$$
$$a_3 = -a_5$$
$$a_4 = -2a_5$$

Therefore, if t and s are arbitrary, then the solution is

$$\begin{bmatrix} t \\ s \\ -s \\ -2s \\ s \end{bmatrix} = t \begin{bmatrix} 1 \\ 0 \\ 0 \\ 0 \\ 0 \end{bmatrix} + s \begin{bmatrix} 0 \\ 1 \\ -1 \\ -2 \\ 1 \end{bmatrix}$$

$\therefore$ a basis for the kernel is $\left\{ \begin{bmatrix} 1 \\ 0 \\ 0 \\ 0 \\ 0 \end{bmatrix}, \begin{bmatrix} 0 \\ 1 \\ 1 \\ -2 \\ 1 \end{bmatrix} \right\}$

15. Since only row operations were used in question 14 and the leading ones are in columns 2, 3, and 4, then the corresponding columns in the original matrix is the desired answer. Therefore, a basis for range L is $\left\{ \begin{bmatrix} 0 \\ -1 \\ 1 \\ 5 \end{bmatrix}, \begin{bmatrix} 2 \\ 0 \\ 0 \\ 0 \end{bmatrix}, \begin{bmatrix} 0 \\ 0 \\ 1 \\ 1 \end{bmatrix} \right\}.$

16. $$\mathbf{b}_1 = \begin{bmatrix} 1 \\ 2 \end{bmatrix}$$

$$\mathbf{b}_2 = \mathbf{a}_2 - \frac{\langle \mathbf{a}_2, \mathbf{b}_1 \rangle}{\langle \mathbf{b}_1, \mathbf{b}_1 \rangle} \mathbf{b}_1$$

$$\mathbf{b}_2 = \begin{bmatrix} -3 \\ 4 \end{bmatrix} - \frac{\left\langle \begin{bmatrix} -3 \\ 4 \end{bmatrix}, \begin{bmatrix} 1 \\ 2 \end{bmatrix} \right\rangle}{\left\langle \begin{bmatrix} 1 \\ 2 \end{bmatrix}, \begin{bmatrix} 1 \\ 2 \end{bmatrix} \right\rangle} \begin{bmatrix} 1 \\ 2 \end{bmatrix}$$

$$= \begin{bmatrix} -3 \\ 4 \end{bmatrix} - \frac{-3+8}{1+4} \begin{bmatrix} 1 \\ 2 \end{bmatrix} = \begin{bmatrix} -4 \\ 2 \end{bmatrix}$$

$\therefore \left\{ \begin{bmatrix} 1 \\ 2 \end{bmatrix}, \begin{bmatrix} -4 \\ 2 \end{bmatrix} \right\}$ is an orthogonal basis

$\therefore \left\{ \begin{bmatrix} \frac{1}{\sqrt{5}} \\ \frac{2}{\sqrt{5}} \end{bmatrix}, \begin{bmatrix} -\frac{4}{\sqrt{20}} \\ \frac{2}{\sqrt{20}} \end{bmatrix} \right\}$ or

$\left\{ \begin{bmatrix} \frac{1}{\sqrt{5}} \\ \frac{2}{\sqrt{5}} \end{bmatrix}, \begin{bmatrix} -\frac{2}{\sqrt{5}} \\ \frac{1}{\sqrt{5}} \end{bmatrix} \right\}$ is an orthonormal basis

17. $$\det(\lambda I - A) = \begin{vmatrix} \lambda - 1 & -6 \\ -1 & \lambda - 2 \end{vmatrix} = 0$$

$$\therefore \quad (\lambda - 1)(\lambda - 2) - 6 = 0$$

$$\lambda^2 - 3\lambda - 4 = 0$$

Hence, $(\lambda - 4)(\lambda + 1) = 0$. Thus, the eigenvalues are -1 and 4.

18. For $\lambda = -1$, $(-1I - A)X = 0$. Thus,

$$\begin{bmatrix} -2 & -6 \\ -1 & -3 \end{bmatrix} \begin{bmatrix} x \\ y \end{bmatrix} = \begin{bmatrix} 0 \\ 0 \end{bmatrix}$$

which has the solution

$$\begin{bmatrix} x \\ y \end{bmatrix} = \begin{bmatrix} -3t \\ t \end{bmatrix} = t \begin{bmatrix} -3 \\ 1 \end{bmatrix}, \qquad t \neq 0$$

Therefore, $\left\{ \begin{bmatrix} -3 \\ 1 \end{bmatrix} \right\}$ is a basis for eigenspace association with the eigenvalue -1. For $\lambda = 4$.

$$(4I - A) = \mathbf{0}$$

$$\begin{bmatrix} 3 & -6 \\ -1 & 2 \end{bmatrix} \begin{bmatrix} x \\ y \end{bmatrix} = \begin{bmatrix} 0 \\ 0 \end{bmatrix}$$

which has the solution

$$\begin{bmatrix} x \\ y \end{bmatrix} = \begin{bmatrix} 2t \\ t \end{bmatrix} = t \begin{bmatrix} 2 \\ 1 \end{bmatrix}, \qquad t \neq 0$$

Therefore, $\left\{ \begin{bmatrix} 2 \\ 1 \end{bmatrix} \right\}$ is the basis for the eigenspace associated with the eigenvalue 4.

19. P is the matrix having basis eigenvectors associated with the eigenvalues as columns. Therefore $P = \begin{bmatrix} -3 & 2 \\ 1 & 1 \end{bmatrix}$.

20. $$\left[\begin{array}{cc|cc} -3 & 2 & 1 & 0 \\ 1 & 1 & 0 & 1 \end{array}\right] \xrightarrow{R_1 \leftrightarrow R_2} \left[\begin{array}{cc|cc} 1 & 1 & 0 & 1 \\ -3 & 2 & 1 & 0 \end{array}\right]$$

$$\xrightarrow{R_2 \to R_2 + 3R_1} \left[\begin{array}{cc|cc} 1 & 1 & 0 & 1 \\ 0 & 5 & 1 & 3 \end{array}\right]$$

$$\xrightarrow{R_2 \to 1/5 R_2} \left[\begin{array}{cc|cc} 1 & 1 & 0 & 1 \\ 0 & 1 & \frac{1}{5} & \frac{3}{5} \end{array}\right]$$

$$\xrightarrow{R_1 \to R_1 - R_2} \left[\begin{array}{cc|cc} 1 & 0 & -\frac{1}{5} & \frac{2}{5} \\ 0 & 1 & \frac{1}{5} & \frac{3}{5} \end{array}\right]$$

$$\therefore \quad P^{-1} = \begin{bmatrix} -\frac{1}{5} & \frac{2}{5} \\ \frac{1}{5} & \frac{3}{5} \end{bmatrix}$$

Hence,

$$P^{-1}AP = \begin{bmatrix} -\frac{1}{5} & \frac{2}{5} \\ \frac{1}{5} & \frac{3}{5} \end{bmatrix} \begin{bmatrix} 1 & 6 \\ 1 & 2 \end{bmatrix} \begin{bmatrix} -3 & 2 \\ 1 & 1 \end{bmatrix}$$

$$= \begin{bmatrix} -\frac{1}{5} & \frac{2}{5} \\ \frac{1}{5} & \frac{3}{5} \end{bmatrix} \begin{bmatrix} 3 & 8 \\ -1 & 4 \end{bmatrix} = \begin{bmatrix} -1 & 0 \\ 0 & 4 \end{bmatrix}$$

APPENDIX A: GREEK ALPHABET

Α	α	Alpha	Ι	ι	Iota	Ρ	ρ	Rho
Β	β	Beta	Κ	κ	Kappa	Σ	σ	Sigma
Γ	γ	Gamma	Λ	λ	Lambda	Τ	τ	Tau
Δ	δ	Delta	Μ	μ	Mu	Υ	υ	Upsilon
Ε	ε	Epsilon	Ν	ν	Nu	Φ	ϕ	Phi
Ζ	ζ	Zeta	Ξ	ξ	Xi	Χ	χ	Chi
Η	η	Eta	Ο	o	Omicron	Ψ	ψ	Psi
Θ	θ	Theta	Π	π	Pi	Ω	ω	Omega

APPENDIX B: TRIGONOMETRY

1. Functions of Special Angles

θ	$0°$	$30°$	$45°$	$60°$	$90°$
$\sin\theta$	0	$1/2$	$\sqrt{2}/2$	$\sqrt{3}/2$	1
$\cos\theta$	1	$\sqrt{3}/2$	$\sqrt{2}/2$	$1/2$	0
$\tan\theta$	0	$\sqrt{3}/3$	1	$\sqrt{3}$	∞

2. Fundamental Identities

$$\sin\theta = \frac{1}{\csc\theta} \qquad \cos\theta = \frac{1}{\sec\theta} \qquad \tan\theta = \frac{1}{\cot\theta} \qquad \tan\theta = \frac{\sin\theta}{\cos\theta}$$

$$\sin^2\theta + \cos^2\theta = 1$$

$$1 + \tan^2\theta = \sec^2\theta$$

$$1 + \cot^2\theta = \csc^2\theta$$

APPENDIX C: ROW ECHELON FORM— THE LOWMAN METHOD

Proof: Let A be the $n \times m$ matrix defined as

$$A = \begin{bmatrix} a_{11} & a_{12} & a_{13} & \cdots & a_{1m} \\ a_{21} & a_{22} & a_{23} & \cdots & a_{2m} \\ a_{31} & a_{32} & a_{33} & \cdots & a_{3m} \\ \vdots & & & \cdots & \vdots \\ a_{n1} & a_{n2} & a_{n3} & \cdots & a_{nm} \end{bmatrix}$$

where $a_{11} \neq 0$. Applying the indicated elementary row operations on A, we get

$$\begin{bmatrix} a_{11} & a_{12} & a_{13} & \cdots & a_{1m} \\ a_{21} & a_{22} & a_{23} & \cdots & a_{2m} \\ a_{31} & a_{32} & a_{33} & \cdots & a_{3m} \\ \vdots & \cdots & & & \vdots \\ a_{n1} & a_{n2} & a_{n3} & \cdots & a_{nm} \end{bmatrix} \xrightarrow[R_n \to R_n - \frac{a_{n1}}{a_{11}} R_1]{R_2 \to R_2 - \frac{a_{21}}{a_{11}} R_1 \quad R_3 \to R_3 - \frac{a_{31}}{a_{11}} R_1} \begin{bmatrix} a_{11} & a_{12} & a_{13} & \cdots & a_{1m} \\ 0 & a_{22} - \dfrac{a_{12}a_{21}}{a_{11}} & a_{23} - \dfrac{a_{13}a_{21}}{a_{11}} & \cdots & a_{2m} - \dfrac{a_{1m}a_{21}}{a_{11}} \\ 0 & a_{32} - \dfrac{a_{12}a_{31}}{a_{11}} & a_{33} - \dfrac{a_{13}a_{31}}{a_{11}} & \cdots & a_{3m} - \dfrac{a_{1m}a_{31}}{a_{11}} \\ \vdots & \cdots & & & \vdots \\ 0 & a_{n2} - \dfrac{a_{12}a_{n1}}{a_{11}} & a_{n3} - \dfrac{a_{13}a_{n1}}{a_{11}} & \cdots & a_{nm} - \dfrac{a_{1m}a_{n1}}{a_{11}} \end{bmatrix}$$

$$\longrightarrow \begin{bmatrix} a_{11} & a_{12} & a_{13} & \cdots & a_{1m} \\ 0 & \dfrac{a_{11}a_{22} - a_{12}a_{21}}{a_{11}} & \dfrac{a_{11}a_{23} - a_{13}a_{21}}{a_{11}} & \cdots & \dfrac{a_{11}a_{2m} - a_{1m}a_{21}}{a_{11}} \\ 0 & \dfrac{a_{11}a_{32} - a_{12}a_{31}}{a_{11}} & \dfrac{a_{11}a_{33} - a_{13}a_{31}}{a_{11}} & \cdots & \dfrac{a_{11}a_{3m} - a_{1m}a_{31}}{a_{11}} \\ \vdots & \cdots & & & \\ 0 & \dfrac{a_{11}a_{n2} - a_{12}a_{n1}}{a_{11}} & \dfrac{a_{11}a_{n3} - a_{13}a_{n1}}{a_{11}} & \cdots & \dfrac{a_{11}a_{nm} - a_{1m}a_{n1}}{a_{11}} \end{bmatrix}$$

$$\longrightarrow \begin{bmatrix} a_{11} & a_{12} & a_{13} & \cdots & a_{1m} \\ 0 & a_{11}a_{22} - a_{12}a_{21} & a_{11}a_{23} - a_{13}a_{21} & \cdots & a_{11}a_{2n} - a_{1m}a_{21} \\ 0 & a_{11}a_{32} - a_{12}a_{31} & a_{11}a_{33} - a_{13}a_{31} & & a_{11}a_{3m} - a_{1m}a_{31} \\ \vdots & \cdots & & & \vdots \\ 0 & a_{11}a_{n2} - a_{12}a_{n1} & a_{11}a_{n3} - a_{13}a_{n1} & \cdots & a_{11}a_{nm} - a_{1m}a_{n1} \end{bmatrix}$$

Each of the nonzero entries in the second through the nth rows is the value of a second order determinant; we then choose

$$A_{22} = a_{11}a_{22} - a_{12}a_{21} = \begin{vmatrix} a_{11} & a_{12} \\ a_{21} & a_{22} \end{vmatrix}, A_{23} = a_{11}a_{23} - a_{13}a_{21} = \begin{vmatrix} a_{11} & a_{13} \\ a_{21} & a_{23} \end{vmatrix}, \ldots,$$

$$A_{2n} = a_{11}a_{2m} - a_{1m}a_{21} = \begin{vmatrix} a_{11} & a_{1m} \\ a_{21} & a_{2m} \end{vmatrix}$$

$$A_{32} = \begin{vmatrix} a_{11} & a_{12} \\ a_{31} & a_{32} \end{vmatrix}, A_{33} = \begin{vmatrix} a_{11} & a_{13} \\ a_{31} & a_{33} \end{vmatrix}, \ldots, A_{3m} = \begin{vmatrix} a_{11} & a_{1m} \\ a_{31} & a_{3m} \end{vmatrix}$$

We continue in this way until we get

$$A_{n2} = \begin{vmatrix} a_{11} & a_{12} \\ a_{n1} & a_{n2} \end{vmatrix}, A_{n3} = \begin{vmatrix} a_{11} & a_{13} \\ a_{n1} & a_{n3} \end{vmatrix}, \ldots, A_{nm} = \begin{vmatrix} a_{11} & a_{1m} \\ a_{n1} & a_{nm} \end{vmatrix}$$

We now write the reduced matrix as

$$\begin{bmatrix} a_{11} & a_{12} & a_{13} & \cdots & a_{1m} \\ 0 & A_{22} & A_{23} & \cdots & A_{2m} \\ 0 & A_{32} & A_{33} & \cdots & A_{3m} \\ \vdots & \cdots & & & \\ 0 & A_{n2} & A_{n3} & \cdots & A_{nm} \end{bmatrix}$$

If we continue the matrix reduction of this type, we will get finally a matrix of the form

$$\begin{bmatrix} a_{11} & a_{12} & a_{13} & \cdots & a_{1m} \\ 0 & A_{22} & A_{23} & \cdots & A_{2m} \\ 0 & 0 & B_{33} & \cdots & B_{3m} \\ \vdots & \vdots & & \cdots & \vdots \\ 0 & 0 & 0 & \vdots & K_{nm} \end{bmatrix} \xrightarrow[\substack{R_3 \to \frac{1}{B_{33}} R_3 \\ \vdots \\ R_n \to \frac{1}{K_{nn}} R_n}]{\substack{R \to \frac{1}{a_{11}} R_1 \\ R_2 \to \frac{1}{A_{22}} R_2}} \begin{bmatrix} 1 & \frac{a_{12}}{a_{11}} & \frac{a_{13}}{a_{11}} & \cdots & \frac{a_{1m}}{a_{11}} \\ 0 & 1 & \frac{A_{23}}{A_{22}} & \cdots & \frac{A_{2m}}{A_{22}} \\ 0 & 0 & 1 & \cdots & \frac{B_{3m}}{B_{33}} \\ \vdots & \vdots & & \cdots & \vdots \\ 0 & 0 & \cdots & 1 & \frac{K_{nm}}{K_{nn}} \end{bmatrix}$$

This is in row echelon form.

note: The proof looks complicated, but using the result is as easy as evaluating a second order determinant.

APPENDIX D: MODELS OF QUADRIC FORMS

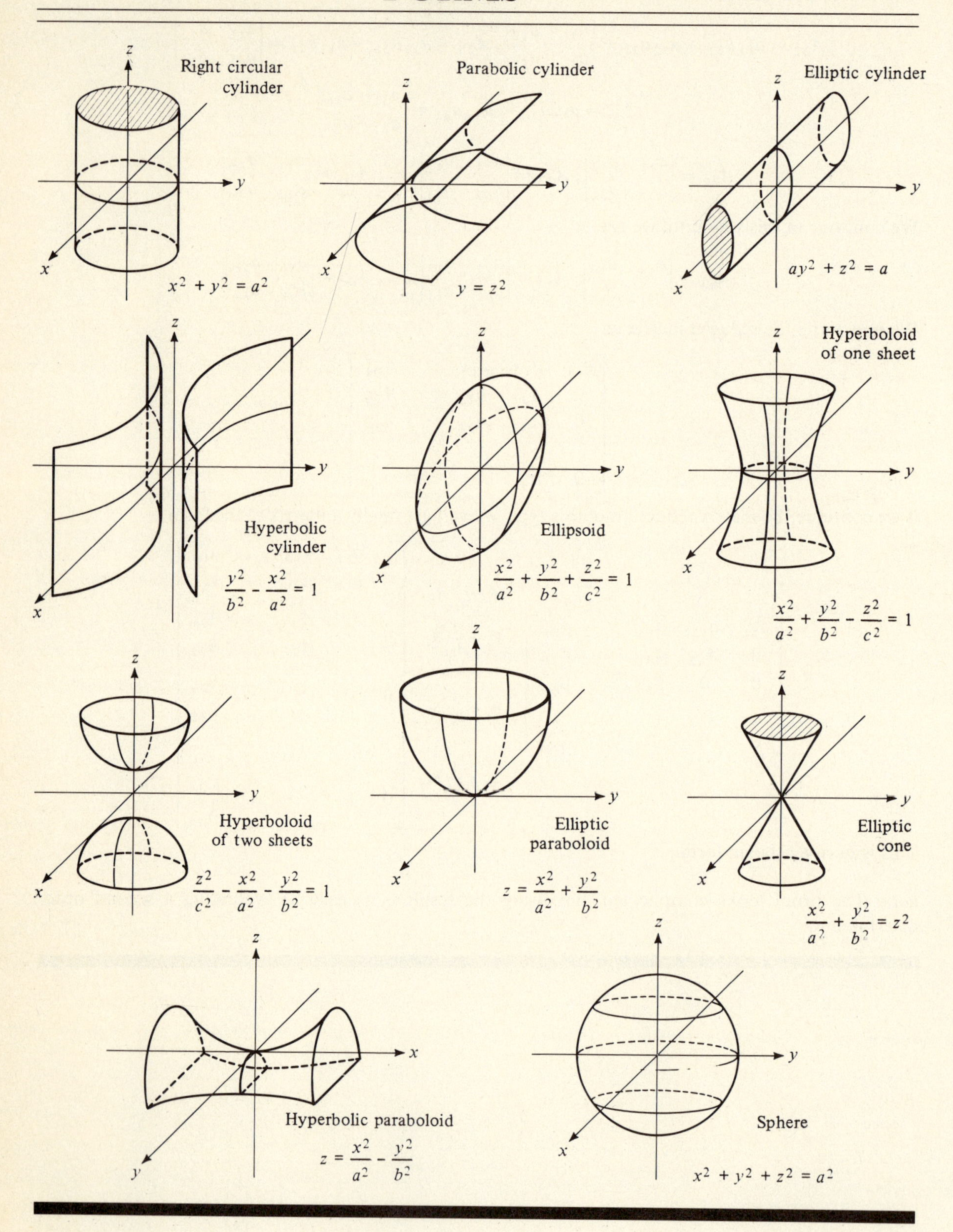

INDEX

Addition
of matrices, 7, 100
of vectors, 134, 214
Additive
identity, 134
inverse, 134
Adjacency matrix, 118
Adjoint matrix, 89
Angle between vectors, 215
Applications
airline routing, 118
business, 116–117
construction, 117
cryptography, 121
diet, 119
differential equations, 128, 380–381
engineering, 117, 120, 130
food services, 119, 123
graph theory, 118
maintenance, 130
manufacturing, 129
nutrition, 52, 119
probability, 111
production planning, 52, 119–122
sales, 117–118
scheduling, 133
tailoring, 120
technology, 35
See also Geometric: applications
Augmented matrix, 29

BASIC programs for finding
coordinate vector, 449
determinant of matrix, 445
inverse of matrix, 444
product of matrices, 443
rank of matrix, 447
reduced row echelon form, 447
solution for system of equations, 446
sum of matrices, 442
transition matrix, 449–450
upper triangular form, 445
Basis
change of, 277, 281
containing linearly independent subset, 165
definition of, 162
from spanning set, 167
natural, 162
ordered, 270
orthogonal, 233, 236–237
orthonormal, 233, 236–237
properties of, 164–167
standard. *See* Basis: natural

Cauchy–Schwarz inequality, 206
Cayley–Hamilton theorem, 354
Change of basis, 277, 281
Characteristic
determinant, 326
equation, 326
polynomial, 326
value. *See* Eigenvalue
vector. *See* Eigenvector
Circle, 386
Cofactor
definition, 70
expansion, 70–71
Column
matrix, 6
rank, 181
space, 181–185
Columns of a matrix, 6
Component, 136
Compression, 405, 414
Conics
degenerate, 386
quadratic forms of, 386
standard forms of, 386–387
Consistent system, 2, 5
Contraction, 405
Coordinate vector, 270
Cramer's Rule, 87
Cross product
definition, 219
determinant representation of, 219
properties of, 220

Definiteness
and eigenvalues, 392
of quadratic forms, 391–392
Dependent system, 3
Determinant(s)
cofactor expansion of, 70–71
Cramer's Rule, 87
definition of, 62
diagonal expansion of, 63
of elementary matrices, 95
evaluation of, 62, 70, 72
form of a circle, 124
form of a conic, 125, 128
form of a line, 123
form of a plane, 127
form of a sphere, 126
form of area of a triangle, 124
pivotal expansions of, 72
properties of, 64–69
row reduction expansion of, 68
Sylvester eliminant expansion of, 72
of 3×3 matrix, 63
of triangular matrices, 69
of 2×2 matrix, 62
Diagonal element, 6
Diagonal matrix, 12
Diagonalizable
linear transformation, 357
matrix, 357–358
Diagonalization procedure, 360
Differential equations, 128, 380–381
Dilation, 404
Dimension
of kernel, 258
of matrix, 181–182
of range, 258
of vector space, 163, 258
Directed line segment
initial point, 214
length of, 205
terminal point, 214
Direction
angles, 217
cosines, 217
numbers, 217
Distance between two vectors, 206
Dot product, 205
Double subscript notation, 9

Eigenproblem, 326
Eigenspace, 332
Eigenvalue(s)
of linear transformation
definition of, 329
equivalency with matrix eigenvalue, 330
of matrix
definition of, 326
procedure for finding, 327
properties of, 333–336
Eigenvector(s)
of linear transformation
definition of, 329
equivalency with matrix eigenvector, 330
of matrix
definition of, 326
procedure for finding, 327
properties of, 333–336
Elementary matrix
determinant of, 95
inverse of, 95–96
as postmultiplier, 95
as premultiplier, 94, 410, 418
Elementary row operations, 30
Elimination method, 3
Ellipse, 386
Ellipsoid, 390, 460
Elliptic cone, 390, 460

Elliptic paraboloid, 390, 460
Equation(s)
  linear, 1
    consistent, 2, 5
    Cramer's Rule, 87
    dependent, 3
    elimination method, 3
    equivalent systems, 3
    Gauss–Jordan elimination, 33
    Gaussian elimination, 33
    homogeneous, 43, 187
    inconsistent, 2, 5
    invertible matrix method, 45
    matrix form, 29
    nonhomogeneous, 188
    nontrivial solution, 43, 187
    solution set, 2
    trivial solution, 43
  nonlinear, 1
Equivalent
  matrices, 97
  statement, 46, 188–189
  systems, 3
Euclidean
  distance, 205
  inner product. *See* Dot product
Even permutations, 62
Expansion
  in x-direction, 404–414
  in y-direction, 404–414
  in z-direction, 414

Gauss–Jordan elimination, 33
Gaussian elimination, 33
Geometric
  applications, 35, 123–128, 214–231, 403–437
  vectors, 214
Gram–Schmidt process, 236

Homogeneous system
  nontrivial solution of, 43, 187
  properties of, 43
  solution space of, 163, 173
  trivial solution of, 43
Hyperbolic paraboloid, 390, 460
Hyperboloid
  of one sheet, 390, 460
  of two sheets, 390, 460

Identity matrix, 12
Inconsistent system, 2, 5
Indefinite quadratic form, 392
Inequality
  Cauchy–Schwarz, 206
  triangle, 207
Inner product
  definition, 203
  space
    definition, 203
    distance between vectors in, 206
    length, norm, or magnitude in, 205
    orthogonal basis for, 233, 236–237
    orthonormal basis for, 233, 236–237
  standard, 205
Inverse of a square matrix
  definition of, 36
  eigenvalues of, 334
  procedure for finding
    by adjoint method, 90
    by elementary row operations, 36
  properties of, 38–42
  and rank, 186
Invertible matrix
  definition, 36
  and rank, 186
  in solving equations, 45
Isometry, 421
Isomorphic vector spaces
  definition of, 255
  properties of, 255–256

Kernel, 256

Lagrange's identity, 220
Latent
  value. *See* Eigenvalue
  vector. *See* Eigenvector
Leading entry, 31
Length of a vector, 205
Line
  parametric equations of, 218
  symmetric equations of, 218
  vector equation of, 217
Linear combination, 143
Linear dependence
  definition of, 145
  properties of, 147
Linear equations, 1
Linear independence
  and basis, 162
  definition of, 145
  of an orthogonal set, 233
  properties of, 147
Linear span, 161
Linear transformation
  change of bases for, 277, 281
  contraction, 405
  definition of, 251
  dilation, 404
  dimension of, 258
  eigenvalue of, 329
  eigenvector of, 329
  injective, 254
  kernel of
    definition of, 256
    procedure for finding basis of, 305
  matrix of
    definition of, 275
    partitioned-matrix method for finding, 281
    procedure for finding, 276
  nullity, 304
  one-to-one, 254
  onto, 254
  opposition, 214
  projection, 406, 414–415
  properties of, 253–256
  range of
    definition of, 257
    dimension of, 304
    procedure for finding basis of, 305
  rank of, 304
  reflection, 408, 416–417
  representation of, 275–276
  rotation, 410, 415–416
  shear, 407, 415–416
  surjective, 254
Lower triangular matrix, 12
Lowman method for row reduction, 131–132

Magnitude of a vector, 205
Main diagonal element, 6
Matrices
  addition of, 7
  equality of, 7
  product of, 9
  rotation, 410, 417–418
  row equivalent, 30, 97
  similar
    definition of, 298
    determinants of, 299
    diagonalizable, 357
    procedure for finding, 298
    properties of, 299
Matrix
  adjacency, 118
  adjoint, 89
  augmented, 29
  characteristic
    equation of, 326
    polynomial of, 326
    value of, 326, 329
  coefficient, 28
  column(s)
    rank of, 181
    space of, 181
  definition of, 6
  determinant of, 62
  diagonal, 12
  diagonalizable, 357–258

echelon form of, 12, 31
eigenvalue of, 326, 329
eigenvector of, 326, 329
element of, 6
elementary
definition of, 92
as postmultiplier, 95
as premultiplier, 94, 410, 418
properties of, 94–96
of a graph, 118
identity, 12
inverse
definition of, 36
determinant value and, 87
procedure for finding, 36, 90
rank and, 186
invertible, 36, 87, 96, 186
of a linear transformation, 275–276
lower triangular, 12
main diagonal of, 6
multiplication of, 9
negative of, 8
noninvertible, 36
nonsingular, 36
order of, 12
orthogonal, 363
partitioned, 99
polynomial, 384
positive definite, 391–392
properties of matrix addition, 8
properties of matrix multiplication, 11
properties of scalar multiplication, 7
of a quadratic form, 384–385
rank of, 182
reduced row echelon form of, 32
representation of linear transformation, 275–276
row
definition of, 6
echelon form of, 31
rank of, 181
reduced echelon form of, 32
row space
definition of, 181
properties of, 181–185
scalar, 12
scalar multiplication, 7
single element, 6
singular, 36
skew-symmetric, 12
square, 6
stochastic, 111
symmetric, 12
transformation, 269
triangular, 12
tridiagonal, 12
transition
definition of, 271
nonsingularity of, 272
procedure for finding, 273
transpose of, 12
upper triangular, 12
zero, 12
Minor, 70
Multiplication
of matrices, 9, 100
scalar, 7
Multiplicity of eigenvalues, 343

Natural basis, 162
Negative definite, 391–392
Negative of a vector, 8, 134
Noninvertible matrix, 36
Nonsingular matrix, 36, 272
Nontrivial solution, 43, 187
Norm. *See* Vector: length of
Normalizing a vector. *See* Unit vector
Null space of a linear transformation, 304
Null space of a matrix
basis for, 305
dimension of, 305
Nullity, 304

Odd permutation, 62
One-to-One, 254
Onto, 254
Ordered
basis, 270
n-tuple, 2
Orthogonal
basis, 233, 236–237
diagonalization, 363–364
matrix, 363
set, 232
vectors, 215
Orthonormal
basis, 233
construction of, 236–237
coordinates with respect to, 235
of eigenvalues of symmetric matrix, 364
set, 232

Paraboloid, 390, 460
Parallel vectors, 214
Parallelogram law, 214
Parametric equations of a line, 218
Partitioned matrix, 99
Permutation
definition, 61
even, 62
inversion, 61
odd, 62
Perpendicular vectors, 215, 220
Pivotal expansion. *See* Sylvester eliminant
Plane
general equation of, 216
point-normal equation of, 216
vector equation of, 216
Polynomial
characteristic, 326
matrix of, 384
Position vector, 214
Positive definite matrix, 391–392
Positive definite quadratic form, 391–392
Postmultiplier elementary matrix, 95
Premultiplier elementary matrix, 94
Principal axis theorem, 385–386
Product
dot, 205
inner, 203
of matrices, 9
of scalar and matrix, 7
of scalar and vector, 134, 214
Programs in BASIC, 442–450
Projection
onto x-axis, 406, 414
onto y-axis, 406, 414
onto z-axis, 414
onto xy-plane, 415
onto xz-plane, 415
onto yz-plane, 415
orthogonal of any vector in V on W, 235
Proper value. *See* Eigenvalue
Proper vector. *See* Eigenvector

Quadratic equation
matrix form of, 385
standard form of, 386, 390

Rotation
of axes, 386, 410, 417–418
matrix, 411
about the origin, 417
Row
echelon form, 31
equivalent matrices, 30, 97
operations and determinants, 165–166
rank, 181
reduced echelon form, 32
reduction by Lowman's Method, 131–132
space
definition of, 181
properties of, 181–185
Rows of a matrix, 6

Scalar
  definition of, 7
  matrix, 12
  multiple of a matrix, 7
  multiple of a vector, 134, 214
Schmidt. *See* Gram–Schmidt process
Schwarz. *See* Cauchy–Schwarz inequality
Similar matrices
  definition of, 298
  determinants of, 299
  diagonalizable, 357
  procedure for finding, 298
  properties of, 299
Similarity, 298
Singular matrix, 36
Shear in
  x-direction, 407
  y-direction, 408
Shear parallel to
  x-axis by distance from
    xy-plane, 415
    xz-plane, 415
  y-axis by distance from
    xy-plane, 415
    yz-plane, 416
  z-axis by distance from
    xz-plane, 416
    yz-plane, 416
Skew-symmetric matrix, 12
Solution
  by Cramer's Rule, 87
  by Gauss–Jordan elimination, 33
  by Gaussian elimination, 33
  to homogeneous linear system, 43
  by invertible matrix method, 45
  set, 2
Space spanned by, 161
Span, 161
Spanning set, 159
Square matrix, 6
Standard basis. *See* Vector space: natural basis
Stochastic matrix, 111
Submatrix, 99
Subspace, 138
Subtraction, 215. *See also* Addition
Sum. *See* Addition
Sylvester eliminant, 72
Symmetric matrix
  definition of, 12
  diagonalization of, 363–364
  eigenspace of, 364
  eigenvalues of, 364
  eigenvectors of, 364
  procedure for othogonally diagonalizing, 366–367
  properties of, 364
System of
  differential equations, 128, 380–381
  homogeneous equations, 43, 187
  linear equations, 2

Trace, 299
Transformations
  compression, 405, 414
  contraction, 405
  dilation, 404
  expansion, 404, 414
  linear, 251, 403–421
  projection, 406, 414–415
  properties of in R_2, 410–412
  properties of in R_3, 418–420
  reflection, 408, 416–417
  rotation, 410, 417–418
  shear, 407, 415–416
  translation, 387, 403, 413
Transition matrix
  definition of, 271
  nonsingularity of, 272
  procedure for finding, 273
Transpose, 12
Triangle inequality, 207
Triangular matrix, 12
  lower, 12
  upper, 12
Trigonometric functions of angles, 457
Trivial solution, 43

Unit vector, 207
Upper triangular matrix, 12

Vector(s)
  addition, 134, 214
  angle between, 215
  basis, 162
  contraction of, 214
  coordinate, 270
  cross product of, 219
  definition of, 6, 134
  dilation of, 214
  dot product, 205
  eigenvector, 326, 329
  equality of, 7
  geometric representation of, 214
  initial point of, 214
  inner product of, 203
  length of, 205
  linear combination of, 143
  linear dependent set of, 145
  linear independent set of, 145, 233
  magnitude of, 205
  negative of, 134
  norm of, 205
  opposition of, 214
  orthogonal, 215, 363
    set of, 232
  orthonormal set of, 232
  parallel, 214
  parallelogram law, 214
  perpendicular, 215, 220
  position, 214
  scalar multiple of, 134, 214
  scalar multiplication of, 134, 214
  spanning set of, 159
  sum of, 134, 214
  terminal point of, 214
  unit, 207
  zero, 134
Vector space(s)
  $C(-\infty, \infty)$, 138
  ${}_mM_n$, 136
  P_n, 136–137
  R, 134–135
  R_n, 136
  R^n, 136
  $R(-\infty, \infty)$, 137–138
  additive inverse in, 134
  basis for
    containing linearly independent subset, 165
    definition of, 162
    from spanning set, 167
  definition of, 134
  dimension of, 163
  inverse in, 134
  isomorphic
    definition of, 255
    properties of, 255–256
  natural basis, 162
  negative vector in, 134
  ordered basis of, 270
  properties of, 142
  spanning set, 159
  standard basis. *See* Vector space: natural basis
  subspace, 138
  zero vector in, 134

Wronskian, 128

Zero
  matrix, 12
  vector, 134